HORTICULTURE

IN DIFFERENT AGRO-CLIMATIC CONDITIONS

HORTICULTURE

IN DIFFERENT AGRO-CLIMATIC CONDITIONS

— FOUR DECADES OF COORDINATED RESEARCH

(Volume I: Fruits, Plantation Crops, Spices, Medicinal and Aromatic Plants)

H.P. Singh
M.S. Palaniswami

WESTVILLE

WESTVILLE PUBLISHING HOUSE
NEW DELHI

H.P.Singh and M.S. Palaniswami (2010) Horticulture in Different Agro-Climatic Conditions
— Four Decades of Coordinated Research (Volume I) pp 278, Westville Publishing House, New Delhi

First Edition 2010

ISBN 978-81-85873-50-3 (Vol. I)

[redacted]

Published by Mrinal Goel, Westville Publishing House
47, B-5, Paschim Vihar, New Delhi – 110 063
Tel: 011-25284742, Fax: 011-25267469
Email: westville_2002@yahoo.co.in
Printed at Salasar Imaging Systems, New Delhi

Foreword

All India Coordinated Research Project (AICRP) under National Agricultural Research System is an unique mechanism for the development of location and situation specific technologies, considering that the needs of plants are determined by agro-climatic conditions. India being endowed with varied agro-climatic conditions, ranging from humid western Himalayan region to arid western plain to sub humid Sutlej Ganga alluvial plain to humid/semi arid Western Ghats and Karnataka plateau, needs region specific hybrids/varieties and production system management for enhancing profitability. With this view, AICRP programmes started in seventies for various crop commodities have played a pivotal role in providing region specific cultivars and technologies which has boosted the production of crops and horticultural crops. Indian Council of Agricultural Research is implementing 77 AICRPs programmes having centres in all the agro-climatic zones, across the country. Through the programmes of AICRP, there has been speedier development of technologies and many of the crops have been adopted even in the regions where it was not grown before.

Horticulture, which includes fruits, vegetables, spices, tubers, mushrooms, plantation crops, medicinal and aromatic plants and flowers, has 13 AICRPs, for various crops with 251 centres located across the country in different agroclimatic zones. These coordinating centres have played a significant role in the exchange of knowledge, germplasm and development of new cultivars and technologies for production and health management. Efforts made in the past have resulted in substantial increase in production and productivity of horticultural crops. Thus, horticulture has emerged as the best option for diversification. The achievements of all these AICRPs in the past four decades have been scattered, which has been compiled and edited in the form of book entitled "Horticulture in Different Agro-climatic Conditions". I compliment Dr H.P. Singh and Dr M.S. Palaniswami, for their concerted efforts in bringing out this publication having achievements of these AICRPs in last 4 decades in two volumes. I am sure, this book will be of immense value, to all those who are concerned with Horticulture Research and Development.

S. Ayyappan

Secretary, Department of Agricultural Research and Education and
Director General, ICAR, New Delhi

Preface

Horticulture, which includes fruits, vegetables, tubers, spices, medicinal and aromatic plants, plantation crops, mushroom and flowers has made rapid stride during the last two decades owing to growing demands and technological innovations. Currently, horticulture contributes 30.4% to GDP of agriculture from nine per cent crop area and has become most important option for diversification. Silent revolution in eighties, due to innovative technologies has become foundation for Golden Revolution (development of horticulture), which is now conspicuously experienced across the country. Production and productivity of horticultural crops have increased manifold. India has emerged as the second largest producer of fruits and vegetables, largest producer and consumer of spices and cashewnut besides highest productivity of banana, grapes, papaya and cassava. Productivity of many of horticultural commodities continue to be higher than world average. These changes in the scenario of horticulture have been possible due to the technological innovation and support of the Government for the development through plan investment in technology led development referred as 'Technology Mission'.

All India Coordinated Research Project (AICRP) for various crops commodity has played a significant role in Indian agriculture, since India is endowed with varying climatic conditions and needs location and situation specific technologies. AICRP, a mechanism to develop location and situation specific technology was adopted in late sixties and after its success in delivering results, it was extended to many other crops including horticultural crops during IV and V Plan periods which were strengthened subsequently in different Plan period. There are 13 AICRPs under horticulture, covering tropical fruits, subtropical fruits, arid fruits, cashew, palms, spices, medicinal and aromatic plants, vegetables, potato, tuber crops, mushroom, betelvine and floriculture. These AICRPs along with the main institutes are involved in generating region specific/value added varieties, production, technologies, mitigation strategies for pests and diseases and transfer of technologies in different horticultural crops. Experimentation and validation under AICRPs have helped in introduction of new horticultural crops where it was not grown earlier. AICRPs have developed 298 varieties of vegetables, 132 fruits, 112 varieties of tubers and 129 spices and plantation crops and large collections are conserved at various centres. Nutrient needs, water management, pest and disease management have been worked out which have impacted horticulture development. Since many of the information are scattered it was felt essential to bring the 4 decades work together. Thus efforts have been made to bring out this publication.

In this book, technological innovations, and salient achievements emancipated from the 13 AICRPs, since their inception, are embodied under different chapters covering its history, objectives, functioning, budget, centres, accomplishments including genetic resources, biodiversity, recommended varieties/hybrids, productivity enhancing technologies, planting material production and distribution, on farm demonstration, and technology transfer. Therefore, the book entitled "Horticulture in Different Agro-climatic Conditions – Four Decades of Coordinated Research" brought out in two volumes encompassing various crops and discipline is considered to be a most valuable publication and shall be useful to all those concerned with horticulture.

We are thankful to all the Project Coordinators for their support and cooperation in our endeavour to document the information. Our sincere thanks are due to Dr S Ayyappan, Director General, ICAR for his support and encouragement. This book is intended to serve as handy compendium and reference to scientists, policy makers, progressive farmers and those who are interested in horticulture in different agro-climatic conditions.

H.P. Singh
M.S. Palaniswami

About the Editors

Dr H.P. Singh has served horticulture in various capacities namely, Project Coordinator (Fruits); Director, NRC for Banana; Horticulture Commissioner, Govt. of India; Chairman, Coconut Development Board; Dean (Agriculture), GBPUA&T, Pantnagar; Vice Chancellor, Rajendra Agricultural University, Pusa and now Deputy Director General (Horticulture), ICAR.

He has also served international organisations like the Chairman of APCC, Jakarta, National Director for internationally-aided projects and Chairperson of Steering Committee of BAPNET of Bioversity International and implemented various international programmes. Besides, excellent academic record, Dr Singh has contributed to research and agricultural education which has earned him three international, twenty one national awards and five fellowships including the Fellow of NAAS. He has developed many cultivars and technologies which has shaped the horticulture. Recognising his immense contribution, OUAT, Bhubaneshwar, has conferred D.Sc. (*Honoris causa*) last year.

Dr Singh is the architect of Mission mode approach for horticultural development, starting from planning to execution which has been due to his insight. He has provided leadership to horticulture and is credited to have authored and edited 39 books and 30 bulletins, besides 175 research papers. He has widely travelled in India and abroad. His efforts have provided a strong foundation for horticultural development, leading to ***Golden Revolution*** in the country.

Dr M.S. Palaniswami commenced his academic carrier in Tamil Nadu Agricultural University in 1975 soon after his Post-graduation in Agricultural Entomology and in 1977 entered into ARS (ICAR). Last 33 years he has been in ICAR, working in research and development of Tropical Tuber and Root Crops. He is the Project Coordinator (Tuber Crops) since September, 2003. He has more than 150 publications in national and international journals and conferences. He has authored five books and 25 book chapters.

The Ph.D student under his guidance was given Jawaharlal Nehru Award for the outstanding research thesis of 2006. More than 22 varieties could be released under his coordination. The AICRP Tuber Crops under his leadership received the Chaudhary Devi Lal Outstanding AICRP Award 2007 from the ICAR. As the Head of division Crop Protection in CTCRI, he managed it effectively during 1997–2004. He has handled the projects funded by USDA/USIF, NATP and AP Cess Fund. The USDA, Washington has awarded special appreciation certificate for his contribution in the USIF/USDA project. His contribution to insect pheromone and kairomone of sweet potato weevil and biotypes of *Bemisia tabaci* whitefly is noteworthy.

He was the Editor/Chief Editor of Journal of Root Crops during 1989-95. He is the life member of Indian society for Root crops, Indian society for vegetable science, Association of Advancement for Pest Management in Horticultural Ecosystem and Association for Advancement of Entomology (Vice President). Visited Colombia, France, Malaysia, Singapore and USA for attending official and academic meetings.

Contents

VOLUME – I

(Fruits, Plantation Crops, Spices, Medicinal and Aromatic Plants)

VOLUME – II

(Vegetables, Potato, Tuber Crops, Mushroom and Floriculture)

Contributors

D. Pandey
CISH, Remankhera
P.O. Kakori
Lucknow – 227 107
Uttar Pradesh

D. Prasath
All India Coordinated Research Project on Spices
Indian Institute of Spices Research
Calicut
Kerala

H.P. Singh
KAB -II
ICAR
New Delhi – 110 012

Johny A. Kallupurakkal
All India Coordinated Research Project on Spices
Indian Institute of Spices Research
Calicut
Kerala

Khayal C. Thakur
AICRP, Central Potato Research Institute
Shimla – 171 001
Himachal Pradesh

Manjit Singh
AICMIP
DMR, Solan
Himachal Pradesh

Mathura Rai
AICRPVC, IIVR
Varanasi – 221 005
Uttar Pradesh

M.S. Palaniswami
AICRP on Tuber Crops
CTCRI
Thiruvananthapuram – 695 017
Keralka

M. Anandaraj
All India Coordinated Research Project on Spices
Indian Institute of Spices Research
Calicut
Kerala

M.G. Bhat
AICRPC
National Research Centre For Cashew
Puttur – 574 202, D.K.
Karnataka

Om Prakash
CISH, Remankhera
P.O. Kakori
Lucknow – 227 107
Uttar Pradesh

Prakash S. Naik
AICRP, Central Potato Research Institute
Shimla – 171 001
Himachal Pradesh

P. Naveen Kumar
Indian Agricultural Research Institute
New Delhi – 110 012

P. Patil
AICRPTF, IIHR
Hessaraghatta Lake Post
Bangalore – 560 089
Karnataka

P. Manivel
AICRMAP
National Resarch Center for Medicinal and Aromatic Plants
Boriavi, Anand
Gujarat

R.L. Mishra
Indian Agricultural Research Institute
New Delhi – 110 012

R.P. Tiwari
All India Coordinated Mushroom Improvement Project
ICAR
Solan
Himachal Pradesh

R.T. Reena
AICRPTF, IIHR
Hessaraghatta Lake Post
Bangalore – 560 089
Karnataka

Saravanan Raju
AINP on Betelvine
NRCMAP
Boriavi, Anand
Gujarat

Satyabrata Maiti
AINP
National Resarch Center for Medicinal and Aromatic Plants
Boriavi, Anand
Gujarat

Suchitra Pushkar
Indian Agricultural Research Institute
New Delhi – 110 012

S.D. Shikamany
AICRPTF, IIHR
Hessaraghatta Lake Post
Bangalore – 560 089
Karnataka

S. Arulraj
All India Coordinated Research Project On Palms
CPCRI
Kasaragod – 671 124
Kerala

T. A. More
AICRP AZF, Central Institute for Arid Horticulture
Bikaner – 334 006
Rajasthan

V. Ramesh
AICRP on Tuber Crops
CTCRI
Thiruvananthapuram – 695 017
Kerala

Acronyms

AAU	:	Assam Agricultural University
AC&RI	:	Agricultural College & Research Institute
AICFIP	:	All India Coordinated Fruit Improvement Project
AICRP Palms	:	All India Coordinated Research Project on Palms
AICRP Potato	:	All India Coordinated Research Project on Potato
AICRP	:	All India Coordinated Research Project
AICRPAF	:	All India Coordinated Research Project on Arid Fruits
AICRPF	:	All India Coordinated Research Project on Floriculture
AICRPM	:	All India Coordinated Research Project on Mushrooms
AICRPMAP	:	All India Coordinated Research Project on Medicinal and Aromatic Plants
AICRPS	:	All India Coordinated Research Project on Spices
AICRPSTF	:	All India Coordinated Research Project on Sub-tropical Fruits
AICRPTC	:	All India Coordinated Research Project on Tuber Crops
AICRPTF	:	All India Coordinated Research Project on Tropical Fruits
AICRPVC	:	All India Coordinated Research Project on Vegetable Crops
ANGRAU	:	Acharya NG Ranga Agricultural University
APHU	:	Andhra Pradesh Horticultural University
BAC	:	Bihar Agricultural College
BAU	:	Birsa Agricultural University
BCKV	:	Bidhan Chandra Krishi Viswavidyalaya
CARI	:	Central Agricultural Research Institute
CAU	:	Central Agricultural University
CAZRI	:	Central Arid Zone Research Institute
CIAH	:	Central Institute of Arid Horticulture
CISH	:	Central Institute for Sub-tropical Horticulture
CITH	:	Central Institute for Temperate Horticulture
CPCRI	:	Central Plantation Crops Research Institute
CTCRI	:	Central Tuber Crops Research Institute
DAT	:	Days After Treatment
Dr. BSKKV	:	Dr. Balasaheb Sawant Konkan Krishi Vidyapeeth
EFY	:	Elephant Foot Yam
g	:	Gram
GAU	:	Gujarat Agricultural University
GBPUAT	:	Govind Ballabh Pant University of Agriculture & Technology

HaNPV	:	*Helicoverpa armigera Nuclear Polyhedrosis virus*
HAU	:	Haryana Agricultural University
HC&RI	:	Horticultural College and Research Institute
HPKVV	:	Himachal Pradesh Krishi Viswa Vidyalaya
IARI	:	Indian Agricultural Research Institute
ICAR RC NEH	:	ICAR Research Complex for NEH Region
ICAR	:	Indian Council of Agricultural Research
IGKV	:	Indira Gandhi Krishi Viswavidyalaya
IIHR	:	Indian Institute of Horticultural Research
IISR	:	Indian Institute of Spices Research
JNKVV	:	Jawaharlal Nehru Krishi Viswa Vidyalaya
K_2O	:	Potassium
KAU	:	Kerala Agricultural University
kg	:	Kilogram
mg	:	Milligram
MPKV	:	Mahatma Phule Krishi Vidya Peeth
MPUAT	:	Maharana Pratap University of Agriculture and Technology
MSL	:	Mean Sea Level
N	:	Nitrogen
NAU	:	Navsari Agricultural University
NDUAT	:	Narendra Dev University of Agriculture & Technology
NPK	:	Nitrogen, Phosphorus and Potassium
NRC Litchi	:	National Research Centre on Litchi
NRC	:	Non-recurring Contingency
NRCC	:	National Research Centre on Cashew
NRCG	:	National Research Centre on Grapes
NRCM	:	National Research Centre on Mushroom
NRCO	:	National Research Centre on Orchids
NRCOG	:	National Research Centre on Onion and Garlic
NRCOP	:	National Research Centre on Oil Palm
NRCP	:	National Research Centre on Pomegranate
NRCSS	:	National Research Centre on Seed Spices
P_2O_5	:	Phosphorous
PAU	:	Punjab Agricultural University
ppm	:	Parts Per Million
q	:	Quintal
RAU	:	Rajendra Agricultural University
RC	:	Recurring Contingency
SKUAST	:	Sher-e-Kashmir University of Agricultural Sciences & Technology, Shalimar
t	:	Tonnes
TFP	:	Total Factor Productivity
TNAU	:	Tamil Nadu Agricultural University
TA	:	Travelling Allowance
TSS	:	Total Soluble Solids
UASB	:	University of Agricultural Sciences, Bangalore
UASD	:	University of Agricultural Sciences, Dharwad
UHS	:	University of Horticultural Sciences

1

Retrospective in Horticultural Research and Development

H.P. Singh and M.S. Palaniswami

India agriculture has made a rapid stride achieving self sufficiency in food recording 4.5 times increase in production from the base line of 1950-51 through Green Revolution. Our efforts have also resulted in achieving eight times increase in horticulture, six times in milk and nine times in fish from the base level of 1950-51. This has been possible due to technical interventions, which is also evident from the fact that area has remained static to 140 ± 2 million hectares for the last 40 years, but production has increased manifold. Pressure on the cultivable land for agriculture continues to be high as Indian agriculture supports 17% of world population and 11% of livestock only from 2.4% of global land and 4.5 % of water. India is the second largest country in the world with more than 1.16 billion population and with a population growth rate of 1.93%. Food production need in estimated to be 300 million tonnes by 2020 and to meet the ever increasing food demand, we have to go in for horizontal as well as vertical expansion. The vertical increase in production can be attempted with better value added varieties and farmer friendly biologically and environmentally sustainable management and production technologies. Looking into population growth, declining land and water resources coupled with challenges of climate change has created much greater concern to feed the growing population.

Horticulture sector covers 9% of the total crop area in the country and contribute 30.4% of the GDP in the agriculture sector. As a result of concerted horticulture research efforts since its inception, India has emerged as the second largest producer of fresh fruits and vegetables globally contributing 10 and 13% of the total world production of fruits and vegetables respectively. The total area under various horticulture crops during 2007-08 was 20.08 million ha with the production of 206.9 million tonnes (source: nhb.gov.in/ 2007-2008 accessed on 13 Nov 2009). The annual growth in area and production under horticulture covering fruits, vegetables, plantation crops, spices and flowers during the period 1990-91 to 2005 in India was 2.6 and 3.6% respectively. India ranks first in world production of banana, mango, sapota, guava, aonla and having the highest productivity in grapes and cassava. The availability of flowers has increased remarkably in all the cities especially of cut flowers.

ALL INDIA COORDINATED RESEARCH PROJECTS

The research needs of various agricultural crops are addressed through the All India Coordinated Research Projects (AICRPs) which was conceptualized during the reorganization of the ICAR in the year 1955-56. The experience gained through ***"All India Coordinated Wheat Rust Scheme in 1947"*** and ***"All India Coordinated Agronomic experiments Scheme in 1953"*** formed the basis of AICRP formulation. The AICRP operates on the national basis through inter disciplinary multilocation research approach to provide scope and opportunity for wider evaluation of the proven results of applied nature under different climatic zones. The main thrust of AICRPs is involvement of National agricultural Research System (NARS) including State Agricultural Universities, Central Agricultural Universities, ICAR Institutes, NGOs, KVKs etc., to develop region specific high yielding varieties, productivity enhancing technologies and mitigating strategies for locations specific major pests and diseases. The first AICRP was established on maize during 1957. Maize varieties and productivity enhancing technologies from the AICRP on Maize boosted the Maize production of nation significantly. This was the first success story of AICRP and it gains world wide appreciation. Subsequently AICRPs in rice and wheat in 1965, oil seeds and cotton in 1967 and on sorghum during 1969 came into being with support from Rockefeller foundation. AICRP has become an important land mark in the history of Indian agricultural research. During 1965-66, the performance of AICRP on maize was highly appreciated. The success of AICRP on maize brought laurels to the project due to its substantial contribution in enhanced food production in different parts of the country. In the annual report of ICAR, 1965-66, the AICRP was commended as 'a significant development in the field of agricultural research in India during the year under report was the formulation of All India Coordinated Research Projects for the improvement of agricultural crops and animals and also for investigation in the field of agronomy, soil science and agricultural engineering. The AICRP provided a major important forum and vehicle for placing agricultural research in the country on sound basis' (Randhawa, 1979). The contribution of AICRPs is signified from the focus given on organizing the scientific man power from various State Agricultural Universities (SAUs) and ICAR Institutions together with other institutional resources such as NGOs, KVKs etc. in a coordinated way in the verification and validation of technologies generated for different regions through coordinated research.

Agro-climatic Zone in the Distribution of AICRP

There is a wide variation in climate, soil and cropping systems in different agroclimatic zone of India. The ICAR has recognized 8 agroclimatic zones for effective land use planning (Table 1.1).

Table 1.1. Agro-climatic zones of India

Agro-climatic region	*States*
Humid Western Himalayan Region	Jammu and Kashmir, Himachal Pradesh, Kumaon and Garhwal in Uttarakhand
Humid Bengal-Assam Region	West Bengal and Assam
Humid Eastern Himalayan Region	Bay Island, Arunachal Pradesh, Nagaland, Manipur, Mizoram, Tripura, Sikkim, Meghalaya and Andaman and Nicobar Islands
Sub-humid Sutlez - Ganga Alluvial Plains	Punjab, Delhi, Uttar Pradesh plains and Bihar
Sub-humid to Humid Eastern and South Eastern Islands	Eastern Madhya Pradesh, Orissa and Andhra Pradesh
Arid Western Plains	Haryana, Rajasthan, Gujarat, Dadra and Nagar Haveli and Daman and Diu
Semi-arid Lava Plateaus and Central Islands	Maharashtra, Western Central Madhya Pradesh and Goa
Humid to semi-arid Western Ghats	Karnataka, Tamil Nadu, Kerala, Pondicherry and Lakshadweep Islands

The NBSS & LUP, Nagpur has delineated the country into 21 agro ecological regions as depicted below:

1. Western Himalaya, cold arid ecoregion with shallow skeletal soil and length of GP <90 days
2. Western Plain and Kutch Peninsula, hot arid and ecoregion with desert and saline soils and GP <90 days.
3. Deccan Plateau, hot arid ecoregion with mixed red and black soils and GP < 90 days
4. Northern Plain and Central Highlands, hot semi-arid ecoregion with alluvium-derived soils and GP 90–150 days
5. Central (Malwa) Highlands and Kathiawar Peninsular, hot semi-arid ecoregion with medium and deep black soils and GP 90-150 days
6. Deccan Plateau, hot semi-arid ecoregion with shallow and medium (inclusion of deep) black soils and GP 90–150 days
7. Deccan Plateau and Eastern Ghats, hot semi-arid ecoregion with red and black soils and GP 90–150 days
8. Eastern Ghats (TN uplands) and Deccan Plateau, hot semi-arid ecoregion with red loamy soils and GP 90–150 days
9. Northern Plains, hot sub-humid ecoregion with alluvium-derived soils and GP 150–180 days
10. Central Highlands (Malwa and Bundelkhand), hot sub humid ecoregion with medium and deep black soils and GP 90–150 days
11. Deccan Plateau and Central Highlands (Bundelkhand) hot sub humid ecoregion with mixed red and black soils and GP 150–180 days
12. Eastern Plateau (Chhattisgarh), hot sub-humid ecoregion with red and yellow soils and GP 150–180 days
13. Eastern (Chhota –Nagpur) Plateau and Eastern Ghats, hot sub-humid ecoregion with red loamy soils and GP 150–180 days
14. Eastern Plain, hot sub-humid ecoregion with alluvium-derived soils and G) 180–210 days
15. Western Himalayas, warm sub-humid (inclusion of humid) ecoregion with brown forest and podzolic soils and GP 80–210 (+) days
16. Assam and Bengal Plains, hot humid ecoregion with alluvium-derived soils and GP <210 days
17. Eastern Himalayas, warm perhumid ecoregion with brown and red hill soils and GP <210 days
18. North-eastern Hills (Purvachal), warm perhumid ecoregion with red and lateritic soils and GP <210 days
19. Eastern coastal Plains, hot sub-humid ecoregion with alluvium-derived sols and GP 150–210 days
20. Western Ghats and Coastal Plains, hot humid –perhumid ecoregion with red, lateritic and alluvium soil and >210 days
21. Islands of Andaman-Nicobar and Lakshadweep, hot perhumid ecoregion with red loamy sandy soils and GP <210 days

Growth pattern, flowering, fruiting, maturity and fruit quality are influenced by temperature. Shoots produced in cooler and warmer months differ in size of leaves and internodal length. Trees attain larger canopy in tropical climate than in subtropical climate. In subtropical regions, flowering reaches its peak in March, while in tropical region it is adjusted with soil moisture stress. Cool weather in subtropics and moisture stress in tropics are known to condition a major part of the shoots to flower at one time during the year. Since there is significant variability in different parts of the country, experimentation based on field performance would be required for gainful utilization agro-ecological zones. Wider adaptability, climatic acclimatization, response to production system and wide genetic base of horticultural and plantation crops are advantageous for extending the area under different zones for harnessing maximum potential.

Under varying climatic conditions and soils, AICRPs play a pivotal role in determining the suitability of crop as well as crop management strategies for the region towards higher productivity. Thus the AICRP provides scope and opportunity for wider evaluation of technologies for adoption; by virtue of multilocation and multidisciplinary research approach, AICRPs cover a wide range of agro-ecological conditions under a wider socio-economic background.

Distribution of AICRPs

In ICAR, there are 77 AICRPs under operation; 31 AICRPs crop based, 13 horticulture based, 12 under NRM, 14 under animal sciences, six under engineering and one in education. During IV Five Year Plan, the importance of horticulture was recognized and 11 AICRP projects were brought into effect during 1968-70 *viz.* potato, tuber crops, tropical fruits, cashew, spices, vegetables, sub-tropical fruits, palms, arid fruits, medicinal and aromatic plants and floriculture. During VI Five Year Plan, two more AICRPs added, *viz.,* mushroom and betel vine. In XI Plan the projects on Medicinal and aromatic plants as well as betel vine were combined together.

Review of Strengthening of AICRPs

During 1992, a workshop on management of change in AICRPs was conducted and suggestions made for future course of management were implemented. Subsequently Dr. H. K. Jain committee recommendations of 1996 and Dr. M. V. Rao committee recommendations of 2001 were implemented for effective functioning of AICRPs. Further, during May 16-18, 2005, a national review was conducted on management change in AICRPs for further improvement under the chairmanship of Secretary, DARE and DG, ICAR and the recommendations were adopted for optimum utilization of scientists and resources.

Functioning of AICRPs in Horticulture

The AICRPs in horticulture are functioning with the following main objectives:

- Collection and documentation of germplasm of different crops of horticultural importance from different parts of the region and maintaining them for evaluation and sharing with other centres for genetic enhancement / identifying their suitability in other regions and also to support as a 'National Active Germplasm Site'.
- To carry out multi location/ multi discipline evaluation of improved high yielding varieties/ selections suitable to different agro climatic conditions/ecosystem.
- To develop suitable technologies for enhanced production, suitable to the specific region/ conditions.

- To develop sustainable and environmental friendly pest/disease/weed management for each region suitable for each crop.
- To popularize and create awareness on the improved varieties and crop management technologies among the horticulture crops growers.
- To develop an organized production and supply of healthy planting material/seeds in liaison with State Horticulture Departments, NGOs, KVKs etc.
- To develop and strengthen effective linkages and partnership among SAUs, NGOs, department of agriculture/horticulture of each state etc.

AICRPs facilitate faster development of varieties /technologies suitable to each crop under each ecological / zones. It extends free and expeditious exchange of germplasm. AICRPs provide an effective platform for maximum utilization of varieties/technologies developed at one centre by others and also a platform for the scientists and extension workers, to share their experience and expertise for mutual benefits.

Management and Funding of AICRPs

Primarily for each AICRP, a Project Coordinator is appointed for coordinating the research project, with the following responsibilities.

- Formulation of location specific and need based research programmes in consultation with scientists of respective centers.
- Monitoring and coordinating research programmes identified for each region.
- Organize special group discussions between AICRP centres/SAU's to tackle common research problems in contiguous zones /centers.
- Conduct Annual / Biennial/ National Workshops / Group Meetings to review the progress of research in various AICRP centres and to finalize new technical programmes.
- Taking steps for recommendation and notification of varieties, production technologies and pest and disease management strategies for adoption to different regions/zones.

At ICAR level, the coordination work is being monitored and evaluated by the Deputy Director General (Horticulture) assisted by the Assistant Director Generals (Horticulture). The research programme in each project is periodically reviewed and monitored in the group meetings held annually/biennially. Budget utilization will be audited. Further there will be performance appraisal by QRT for each project.

Administratively, each AICRP centre is functioning under the respective universities on 75:25 (ICAR: SAU) budget sharing basis. From 1999, additional administrative and financial powers were delegated to the Project Coordinators by the ICAR, to disburse funds to the AICRP centers directly from the Coordinator's office. This helped in speedy implementation of the programmes. The revenue generated under the project will be shared 25% by the ICAR and 37.5% by the SAU of respective centre and 37.5% to be spent for R&D work of the concerned AICRP centres.

AICRP Centres Under Horticulture

The AICRPs under Horticulture division are on – Tropical Fruits, Sub Tropical Fruits, Arid Fruits, Cashew, Palms, Spices, Vegetables, Potato, Tuber Crops, Mushrooms, Medicinal and Aromatic Plants, Floriculture and Betle vine.

There are 251 centres under different AICRPs of horticulture. Tropical fruits has 26 centres, Subtropical fruits has 19, potato-25, tuber crops-17, cashew-15, palms-20, spices-25, vegetables -26, arid fruits-13, mushrooms-16, medicinal and aromatic plants-13, betel vine–11 and floriculture-21. these AICRP centres are inclusive of regular, voluntary and ICAR based centres. During XI plan 34 centres were newly added in different parts of the country under AICRPs of horticulture.

The 251 centres of AICRPs under horticulture division are located in eight different agro climatic zones of India *viz.,* (I) Humid western Himalayan region has 22 centres; (II) Humid Bengal and Assam basin – 21 centres; III- Humid eastern, Himalayan and bay islands – 14 centres; (IV) Sub-Humid Sutlej Ganga-Alluvial Plain – 38 centres; (V) Sub-Humid to Humid Eastern and South Eastern Uplands – 35 centres; (VI) Arid Western Plain – 34 centres; (VII) Semi Arid Lava Plateau and Central High Lands – 30 centres and (VIII) Humid to semi Arid Western Ghats and Karnataka Plateau – 57 centres (Fig. 1.1).

There are 251 centres of AICRPs under horticulture division spread over 26 states and one union territory of India (Fig. 1.2). There are 22 centres in Maharashtra, 24 in Karnataka, 20 in Tamil Nadu, 17 in Andhra Pradesh, 13 in Kerala, 12 each in West Bengal, Gujarat and Uttar Pradesh, 13 each in Bihar and Rajasthan, 10 in Orissa, 9 in Himachal Pradesh, 8 each in Assam and Chhattisgarh, 7 each in Punjab and Madhya Pradesh and Uttarakhand, 6 each in Haryana, Jharkhand, Meghalaya and Jammu & Kashmir, 4 in Arunachal Pradesh, 3 in Delhi, 2 in Manipur and remaining states have got one centre each. The distribution and location of centres under each project have been dealt under each corresponding chapter.

SALIENT ACHIEVEMENTS OF AICRPS IN HORTICULTURE

Germplasm Collection for Crop Improvement

During the period from 1970–2009 more than 30,000 accessions of different fruit crops, vegetables, spices and plantation crops, potatoes, tuber crops, medicinal and aromatic plants and flower crops were collected and maintained. AICRPs act as a national active germplasm site and extend their collection to other centres for crop improvement. Genetic resources include of 2194 accessions of fruits, 11,229 accessions of vegetables, 10,184 accessions in tuber crops, 6336 accessions of spices, 1304 accessions in plantation crops, 1327 accessions in medicinal and aromatic plants.

Significant Varieties

During the period, 1970–2009, there has been a lot of improvement in development of new varieties/hybrids, cultivation, production and protection strategies of different horticulture crops. A total of 672 high yielding varieties and hybrids of horticultural crops (fruits-17, vegetables-488, tuber crops-47, plantation crops-49, medicinal and aromatic plats-36, ornamentals-39 and mushrooms-5 were developed. As a result, the productivity of horticultural crops *viz.* banana, potato, cassava, cardamom, ginger, turmeric etc. has increased. The progress and growth of 13 AICRPs in horticulture are enunciated and elucidated in the respective chapters.

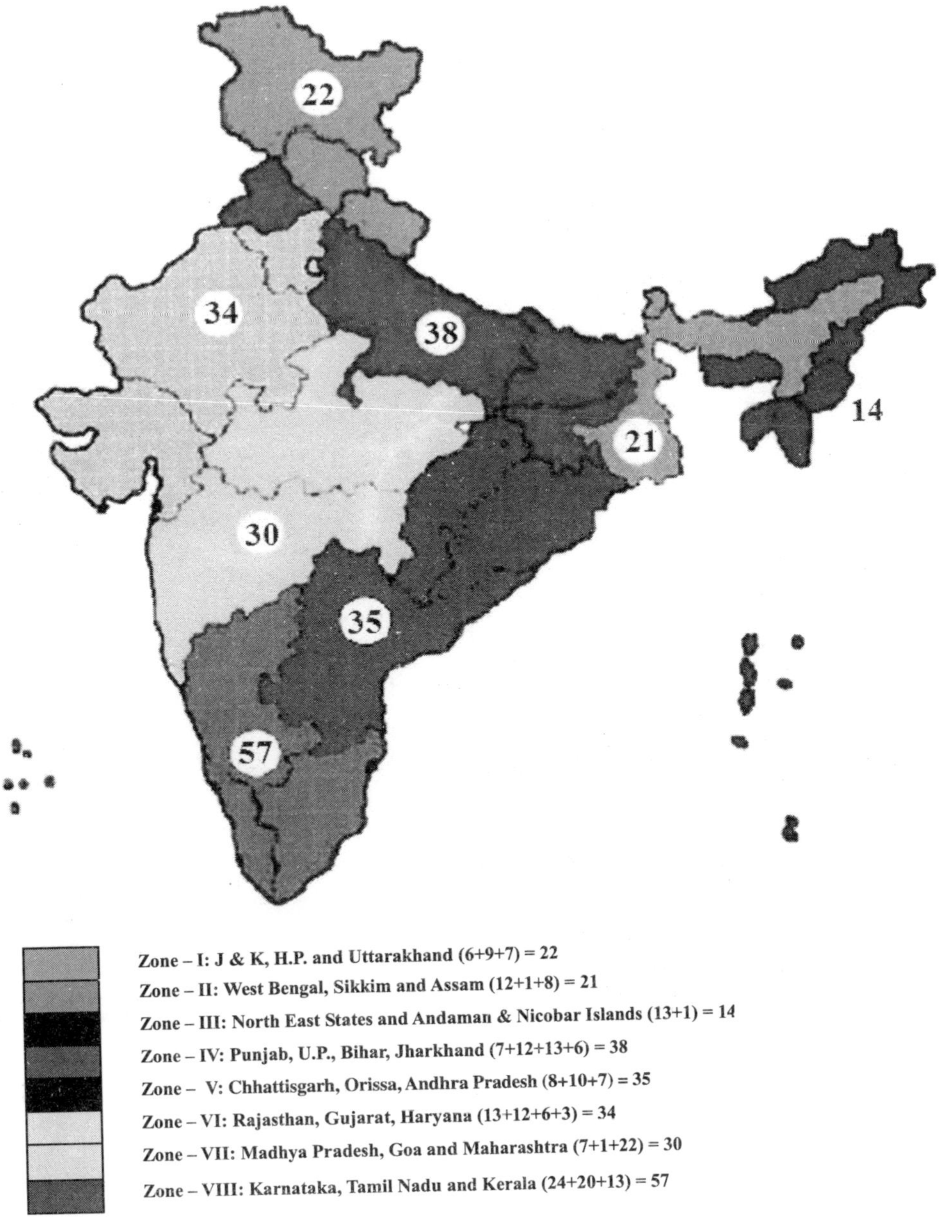

Fig. 1.1. Distribution of centres of AICRPs in horticulture under different agro-climatic zones of India

Fig. 1.2. Distribution of AICRPs centres under Horticulture Division in different states of India

Fruits

Mango cultivars Pusa Arunima (Export variety), and Pusa Surya (longer shelf life) and Mallika are recommended for commercial cultivation in south and east region India. In guava, varieties Shweta, export specific manjari naveen and Chittidar are recommended for commercial cultivation.

In banana ovue Bontha, BRS-1 and BRS 2 were identified for cultivation for Andhra Pradesh and Kerala. CO-6 and CO-7 papaya varieties were recommended for Tamil Nadu. In sapota PKM-4 and PKM 5 were found suitable for commercial cultivation in Tamil Nadu. Pineapple H7 and H8 were recommended for adopting in Kerala.

In pomegranate, variety Phule Arakta and Bhagawa have been reported as promising for Maharashtra and Andhra Pradesh and Karnataka, Jalore Seedless and Jodhpur Red for Rajasthan and Hybrid 61 (Mridula) for Maharashtra. Aonla varieties NA-7, Chakaiya, Kanchan and NA-6 have been found most promising uniformly all over the region. Date palm varieties Halawy; and Barhee suitable for desert and Medjool for making dry dates (*chhuhara*) are reported to be most suitable for date growing regions. 'Arka Sahan' custard apple hybrid. Variety Balanagar A.S.1 Island Gem, Bullocks Heart and Mammoth are recommended for Karnataka, Andhra Pradesh Tamil Nadu and Maharashtra. Fig varieties Condaria, Poona Fig and Dianna are reported as suitable for fig growing areas. In case of *ber,* var. Gola has been found to be suitable for very dry areas, Kaithali for slightly higher rainfall areas and Umran for moderate arid and semi-arid areas. Narendra Ber Selection-1 & 2 have been recommended for commercial cultivation in Uttar Pradesh.

Potato

JW-160 Kufri Pushkar – for plains and plateau regions, MS/92-2105 Kufri Arun - for Indo-Gangetic plains, SM/87-185 Kufri Shailja - for Hilly regions, HT/92-621 Kufri Surya - a heat tolerant hybrid suitable for French fries and best for Northern plains and warmer areas, MP/97-583 Kufri Chipsona-3 – excellent for chips and SM/91-1515 Kufri Himalini – suitable for North western and north-eastern hills are recommended for commercial cultivation.

Tuber Crops

Cassava varieties Sree Prakash, Sree Jaya and Sree Vijaya are suitable for AP and Assam. In taro (arvi) Ahina Kachu for Assam, Bhavpuri (KCS-2) and RNArvi-1 for AP, Narendra Arvi - 3 for UP, BKCol-2 (Indira arvi-1) for Assam and Chhattisgarh, Sonajuli and Jhankri for Orissa, Birsa Arvi-1 for Jharkhand and Sree Kiran for Kerala and in elephant foot yam — Narendra Asha for UP were recommended for commercial cultivation. In sweet potato, Indira Naveen(IGSP-4) (short duration) for Maharashtra, and Chhattisgarh; Indira Madhur and Indira Nandhini (IGSP-17) for Chhattisgarh, RNSakharkand-3 for AP, NDSP-10 for UP, Bidhan Jagannath for West Bengal and Birsa Sakharkand-1 for Jharkhand were released for cultivation.

Vegetable Crops

Tomato (Varieties/hybrids): BT-116-3-2 Det, NDT-3 Det., KS-118 Det. DVRT-2 Det., BT-20-2-1 Indet., NDT-9 Indet., NDTS-2001-3 Det, Mani Laima Det, IIVR Sel-l Det., IIVR Sel-2 Indet., BT-136 and VLT-32 & Hybrids – BSS-20 Indet., DTH-8 Det, CHTH-l Det., ARTH-128Indet., KTH-2 Indet., JKTH-3055Det., KTH-l, Nun-7730 Indet and TH01462 Det.

Brinjal : Punjab Sadabahar Long, NDB-28-2 Long, DBL-21 Long, KS-235 Round, ABSR-2 Small Long ,HABR-4 Round, IVBR-1 Round, HABL-1 Long, PB-66 Long and2Hybrids – ARBH-541 Long, PBH-6 Long, JBH-l Round, BH-l Round, BH-2 Round, VRBHR-I Round, IVBHL-54 Long, ARBH-786 Long,VNR-51 Small round, Navina Long and HABH-17 round.

Chilli and capsicum: AKC-86-39, BC-14-2 , RHRC Cluster Erect, PMR-57/88-K, LCA3, ASC-2000-02, KA-2, LCA-353, BC-25; Garden pea: VRP-2 Early Season, NDVP-12 Early Season, VRP-3 Mid Season, Organ Podded, VRP-5 Early, CHP-2 Midseason , VP-101 (Early) and PC-531 (Mid); Okra: 1DVR-3, DVR-4, HBH-142, SOH-152, SOH-10I6 and NBH-180; Onion: L-28 Rabi season, HOS-1 Rabi season and B-780-5-2-2; Garlic: VLG-7, DARL-52 and G-323; Cauliflower: KT-25 Snowball group and IVREC-2 Early; Cabbage: KGMR-l; Carrot: SKAUC-50; Muskmelon: NDM-15 and IVMM-3; Bitter gourd: PBIG-l, Pusa Hybrid-2 and NBGH-167; Pumpkin: CM-350 KAU and NDPK-24 and Cucumber: CHC-2, CH-20, PCUC-28, PCUCH-l, HybridNo1 Century Seeds and PCUCH-3.

Plantation Crops

In cashew a total of 27 high yielding cashew varieties have been developed and released for cultivation in different states with specific package.

In coconut, four cultivars and ten hybrids have been released for commercial cultivation in different states.

Spices

In spices improved varieties were recommended for commercial cultivation for different regions of India. In pepper seven varieties, in cardamom ten, coriander 22, turmeric 8, ginger 4, cumin 9, fennel 10 and fenugreek 15.

Medicinal and Aromatic plants In medicinal and aromatic plants location specific varieties for 22 medicinal and aromatic plants with their package and practices have been developed. Package and practices and four varieties of Isabgol for different zones have been developed.

Mushroom

In mushroom *Agaricus bisporus* – (White – Ooty 1) and in Oyster mushroom. *Pleurotus citrinopileatus* – (Pure white - CO1), *P. eous* – (Pink – A.P.K 1), *P. djamor* – White – (M.D.U 1), *P. flabellatus* – (White – M.D.U 2), *P. ostreatus* – (Ashy white – Ooty 1), P. *sajor caju* – (Ash – M 2), *P. florida* – (White – PF), *Hypsizygus ulmarius* – (Bluish white – CO 2), Calocybe *indica* – (Milky white – A.P.K 2) are commercially cultivated.

Floriculture

In floriculture Rose cv. Pusa Gaurav' is recommended for cut flower production. For loose flower production rose cvs. 'Neelambari' and 'Arunima' were found promising whereas 'Banjaran' proved to be the best for garden display. The exotic varieties of rose namely 'Skyline', 'Nobelesse' and 'Golden Gate' performed better. In gladiolus, Arka Kesar, 87-22-1, hybrid 84-7-11, 86-3-4, 84-4-8, Swarnima and Shagun have been found suitable for commercial cultivation. Carnation cvs Impala' 'Super Star', 'Veleta' and 'Fantasia' supplied with 4 hours extended lighting resulted in maximum flowering stem length and induced early flowering under poly house condition. In chrysanthemum 'Arka Ravi' 'Punjab Gold', Mother Teresa', 'CO.2' and Indira, 'Yellow Gold' (at Solan), 'Ratlam Selection' and 'Basanti Local', 'Sonali Tara', 'Sunil' 'Baggi' and 'IIHR Hybrid-11 (at Pune) are promising and recommended for cultivation. In orchids, Coconut husk was found as the best suitable medium for *Dendrobium* cv. 'Sonia' at Kalyani. In D*endrobium* orchid Sonia – 17, NPK 20:30:30, 10:30:30 or 10:20:20 at 0.2% spraying twice weekly were recommended for commercial cultivation.

GROWTH AND PROGRESS OF HORTICULTURE DURING LAST FOUR DECADES

India is the largest producer and exporter of spices and cashew and it ranks first in the production of coconut and arecanut. Andhra Pradesh, Bihar, Karnataka, Kerala, Maharashtra, Orissa, Tamil Nadu, Uttar Pradesh, West Bengal and Gujarat are important states in the forefront of horticultural production.

- The details of state-wise latest area, production and productivity (2007-08) of total horticulture crops are presented in Table 1.2.
- Major states of India contributing to the production of fruits, vegetables, plantation crops, spices, tubers, mushroom and flowers in horticulture sector are depicted in Table 1.2.

Table 1.2. Area and production of horticultural crops in major states during the year 2007-08

State	Total for Horticulture crops		
	Area '000 ha	Production '000 t	Productivity t ha^{-1}
Andhra Pradesh	1777.1	17888.3	10.1
Arunachal Pradesh	93.4	265.6	2.8
Assam	583.1	6089.5	10.4
Bihar	1121.4	17334.8	15.5
Chhattisgarh	426.1	3929.6	9.2
Goa	102.4	303.8	3.0
Gujarat	1048.5	13758.2	13.1
Haryana	320.4	3618.0	11.3
Himachal Pradesh	284.1	1899.1	6.7
Jammu & Kashmir	356.3	2832.9	8.0
Jharkhand	276.6	4021.9	14.5
Karnataka	1623.8	11541.3	7.1
Kerala	1876.2	10578.0	5.6
Madhya Pradesh	472.8	4525.7	9.6
Maharashtra	2208.8	18006.3	8.2
Manipur	59.9	395.1	6.6
Meghalaya	101.5	685.2	6.8
Mizoram	34.6	300.5	8.7
Nagaland	27.8	144.1	5.2
Orissa	1257.4	9975.9	7.9
Punjab	241.8	3950.3	16.3
Rajasthan	921.1	1840.8	2.0
Tamil Nadu	1220.2	19821.2	16.2
Tripura	78.8	970.4	12.3
Uttar Pradesh	1475.7	23744.6	16.1
Uttarakhand	274.6	1773.9	6.5
West Bengal	1658.0	25665.3	15.5
Total	20086.6	207012.1	10.3

Source: http://nhb.gov.in/2007-08.xls; accessed on 13/11/2009

- Maharashtra, Kerala, Andhra Pradesh, West Bengal, Karnataka and Uttar Pradesh rank in the order of larger area of cultivation under total horticultural crops. In terms of total production of Horticultural crops, Uttar Pradesh ranks first followed by West Bengal, Tamil Nadu, Andhra Pradesh, Bihar and Maharashtra; while the productivity was found maximum in Uttar Pradesh followed by Bihar (Table 1.2).

Fruits

The total area under fruit crops was 5.775 million ha with a production of 63.503 million tonnes. State wise area and production of fruits in India during 2007-2008 is encouraging. Maharashtra (16.88%), accounted for 17.39% of total fruit production of India followed by Andhra Pradesh, Tamil Nadu (11.85%), Gujarat (9.2%), Karnataka (7.22%) and Uttar Pradesh (5.91%). The national average productivity is 11.0 t ha^{-1}, Tamil Nadu recorded maximum (25.74 t ha^{-1}) followed by Gujarat (19.1 t ha^{-1}) and Karnataka (16.44) (Fig.1.2a; Table 1.3, 1.4). During 2004, among the area under total fruits, 39.52% are under mango and 10.67% under banana with a share in production of 23.54% and 32.91% respectively. The progress and growth in the fruits sector during 1970-2007 is commendable. The total fruit production increased three fold from the two fold increased area (Fig. 1.2a). The growth of some important fruit crops are given below.

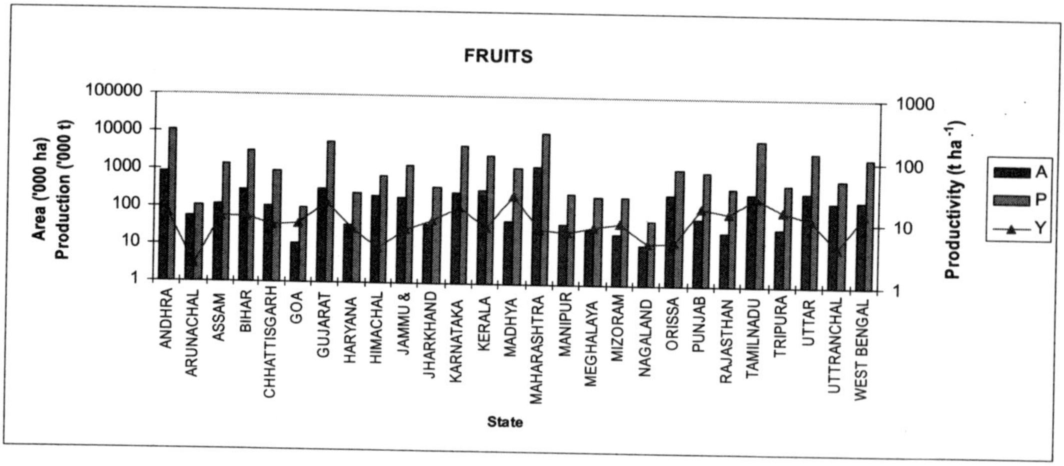

Source: http:/nhb.gov.in/2007-08.xl; accessed on 13/11/2009

Fig 1.2a. State wise area and production of fruits in India during 2007-08

Table 1.3. State wise area ('000 ha) and production ('000 t) of fruits in India for the year 2007-08

States	*Banana*		*Citrus*		*Grapes*		*Guava*		*Mango*		*Papaya*		*Pineapple*		*Sapota*		*Others*		*Total*	
	Area	*Prodn.*	*Area*	*Prodn.*	*Area*	*Prodn.*	*Area*	*Prodn.*	*Area*	*Prodn.*	*Area*	*Prodn.*	*Area*	*Prodn.*	*Area*	*Prodn.*	*Area*	*Prodn.*	*Area*	*Prodn.*
Andhra Pradesh	75.1	2254.3	236.5	2997.9	2.5	52.4	9.7	136.8	483.5	3865.2	14.4	1123.4	0.0	0.0	19.1	181.1	7.2	67.2	853.1	10722.3
Assam	44.1	610.9	19.0	163.4	0.0	0.0	4.4	79.2	4.5	39.6	7.0	107.1	12.7	195.7	0.0	0.0	26.2	180.8	122.7	1410.7
Bihar	30.5	1329.4	17.6	125.8	-	-	28.7	255.7	142.2	870.4	1.3	30.4	4.6	126.8	0.0	0.0	31.6	290.7	286.3	3252.4
Chhattisgarh	8.6	227.7	8.2	42.9	0.0	0.0	10.3	81.4	36.3	119.0	7.4	136.1	0.0	0.0	0.2	0.6	35.0	306.3	107.7	915.1
Gujarat	57.7	3157.7	34.6	360.2	0.0	0.0	8.6	131.1	109.6	930.1	11.2	489.0	0.0	0.0	26.5	258.9	53.1	466.1	306.9	5849.7
Himachal Pradesh	0.0	0.0	21.4	24.7	0.1	0.1	2.3	2.8	39.1	29.3	0.2	0.9	0.0	0.0	0.0	0.0	39.9	59.9	202.4	713.0
Jammu & Kashmir	0.0	0.0	11.7	19.1	0.2	0.5	0.0	0.0	9.8	17.8	0.0	0.0	0.0	0.0	0.0	0.0	46.8	129.9	194.9	1435.8
Karnataka	59.9	1513.3	11.7	250.3	11.1	209.0	7.7	166.7	132.0	1337.7	3.7	248.9	2.7	145.5	24.2	270.1	12.9	296.5	279.0	4587.9
Kerala	61.5	493.9	0.0	0.0	0.0	0.0	0.0	0.0	76.7	445.4	18.5	80.6	12.5	102.4	0.0	0.0	154.1	1457.5	323.3	2579.8
Maharashtra	80.0	4962.9	261.3	1627.7	45.6	1290.0	32.0	250.5	455.8	710.9	-	-	-	-	65.3	294.1	395.8	1315.3	1432.3	11047.6
Orissa	23.1	297.1	26.8	211.7	0.0	0.0	14.0	93.4	148.2	251.8	0.8	14.7	0.7	7.5	3.3	14.6	44.3	371.5	265.2	1275.1
Tamil Nadu	114.1	6116.5	11.4	38.6	2.8	83.5	8.5	93.3	136.6	753.6	0.4	66.4	0.7	22.8	6.8	168.8	10.8	177.4	292.5	7530.1
Uttar Pradesh	2.6	83.5	0.7	1.7	0.0	0.0	15.1	181.6	265.9	3256.6	0.2	15.9	0.0	0.0	0.0	0.0	31.2	217.4	316.0	3757.6
Uttarakhand	0.0	0.0	26.8	127.4	-	-	-	-	37.0	112.7	-	-	-	-	-	-	66.5	332.1	171.3	717.8
West Bengal	37.4	892.2	11.0	98.5	0.0	0.0	11.9	162.2	80.9	623.3	10.7	308.6	9.5	283.2	3.9	43.4	21.0	277.4	194.4	2766.6
Total	646.9	23204.8	843.2	7574.4	64.4	1677.1	178.2	1975.0	2205.3	13792.1	80.3	2685.9	80.4	1215.5	150.4	1238.1	1070.7	6862.0	5775.2	63502.9
Productivity ($t\ ha^{-1}$)	35.9		9.0		26.1		11.1		6.3		33.4		15.1		8.2				11.0	

Source: http://nhb.gov.in/2007-08.xls; accessed on 13/11/2009

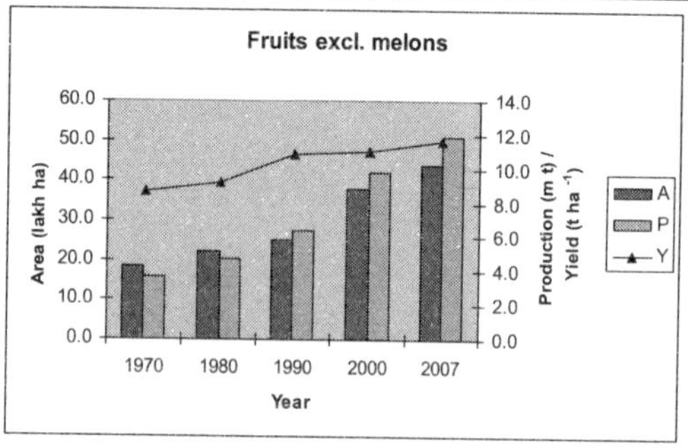

Fig. 3a

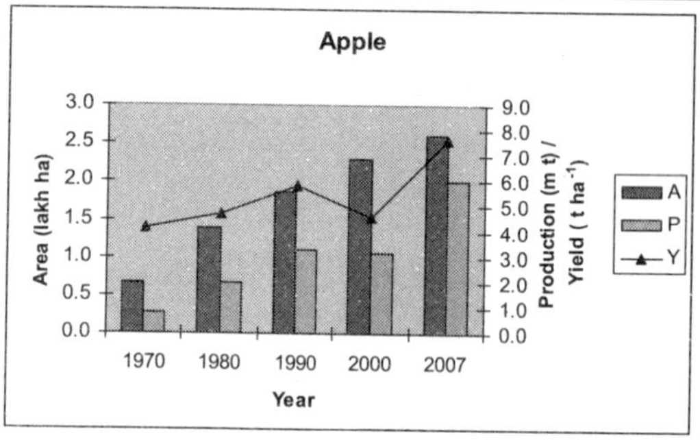

Fig. 3b

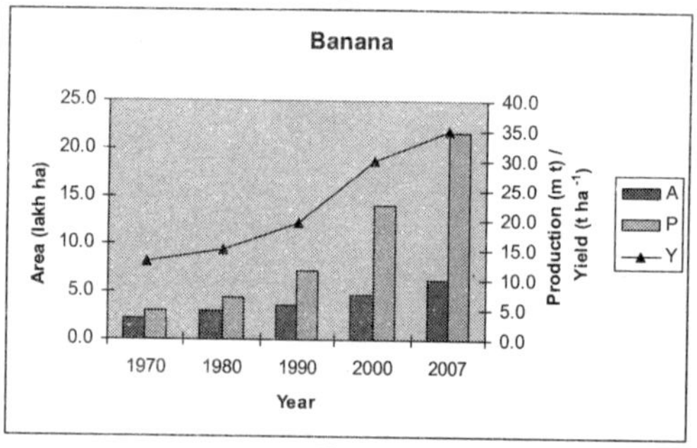

Fig. 3c

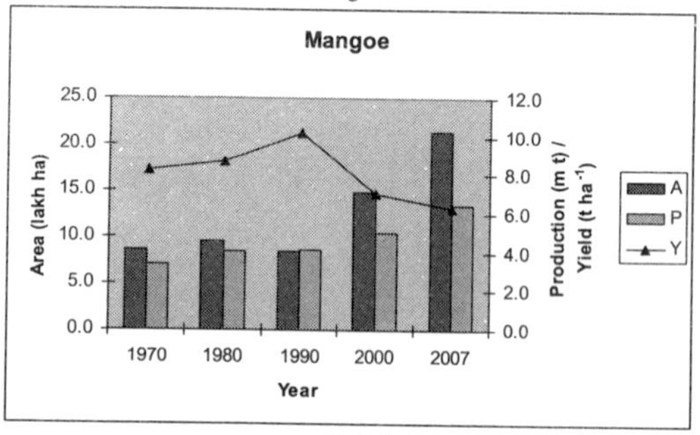

Fig. 3d

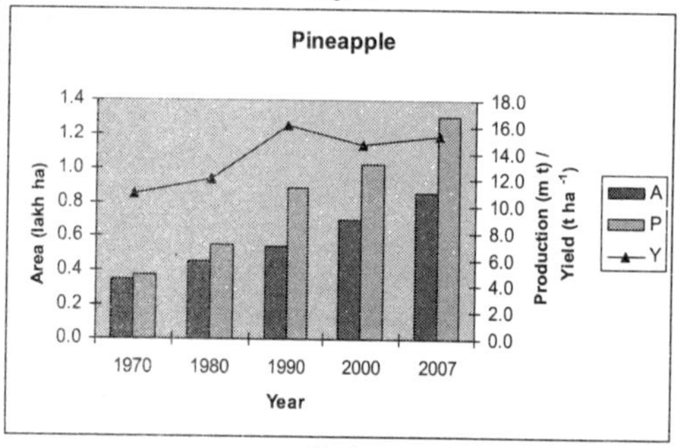

Fig. 3e

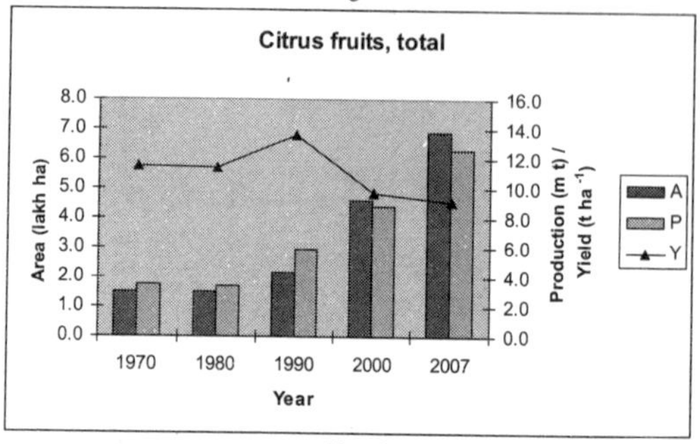

Fig. 3f

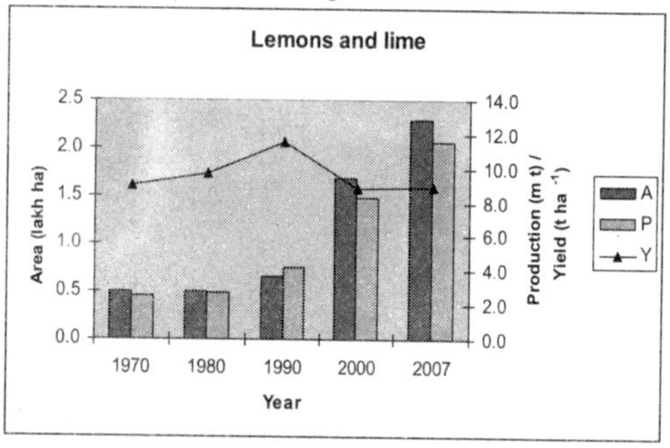

Fig. 3g

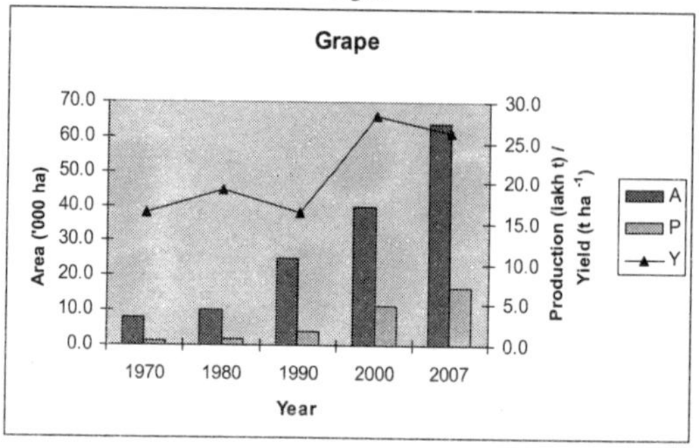

Fig. 3h

Source: http://faostat.fao.org; accessed on 27-01-2009

Fig. 1.3. Growth of different fruit crops in India during 1970-2007

Table 1.4. Crop wise area and production of major Fruits in India for the year 2007-08

Crops	*Area '000ha*	*Production '000t*	*Productivity t ha^{-1}*
Apple	263.9	2001.5	7.6
Banana	646.9	23204.8	35.9
Citrus	843.2	7574.4	9.0
Grape	64.4	1677.1	26.1
Guava	178.2	1975.0	11.1
Litchi	69.2	418.4	6.0
Mango	2205.3	13792.1	6.3
Papaya	80.3	2685.9	33.4
Pineapple	80.4	1215.5	15.1
Pome granate	122.2	858.1	7.0
Sapota	150.4	1238.1	8.2
Others	1070.7	6862.0	6.4
Total	5775.2	63502.9	11.0

Source: http://nhb.gov.in/2007-08.xls; accessed on 13/11/2009

Apple

During 2007-08, apple was grown in an area of 2.64 lakh ha with a production of 20.01 lakh tonnes. Jammu and Kashmir (1.19 lakh ha), Himachal Pradesh (0.92 lakh ha) and Uttarakhand (0.3 lakh ha) are major growing states. The total area under apple is maximum in Jammu and Kashmir with a production of 12.22 lakh tonnes followed by Himachal Pradesh (2.68 lakh tonnes) and Uttarakhand (1.23 lakh tonnes) (nhb.gov.in/2007-2008). The apple cultivation has shown four fold increase in area and eight fold increase in production from 1970 to 2007, especially in Jammu and Kashmir and Himachal Pradesh. The productivity of apple has been doubled, though there was a lull during 2000 (Fig. 3b).

Banana

Banana is one of the few crops where man made hybrids failed to excel the natural hybrids. Most of the varieties are natural hybrids and acclimatized to various agro-ecological condition. Cavendish clone, Robusta, Dwarf Cavendish, Shrimathi, Giant Governor and Mahalakshmi are commercially cultivated. Udhayam is promising variety for Tamil Nadu, Kerala, Andhra Pradesh, Karnataka, Bihar and Tripura. During 2007-2008 banana was cultivated in an area of 6.04 lakh ha covering 10.9% of the total area under horticultural crops. It ranks first in total production among fruits which is 209.97 lakh tonnes (35.25%). During 2007-2008 the area of cultivation increased to 6.47 lakh ha with a production of 232 lakh tonnes. India also occupies first position in banana production globally. Tamil Nadu contributes maximum area under banana cultivation and production (17.62 and 26.36% respectively) followed by Maharashtra, Andhra Pradesh, Kerala, Karnataka and Gujarat. The nation productivity of banana is commendable (35.9 t ha^{-1}). The Productivity of banana in Maharashtra (62.0 t ha^{-1}), Gujarat (54.7 t ha^{-1}) and Tamil Nadu (53.6 t ha^{-1}) surpasses national productivity (Table 1.3). The area under cultivation and production had increased three and seven times respectively and the productivity doubled from 1970 to 2007 due to adoption of technologies especially in major growing states like Tamil Nadu and Karnataka (Fig. 1.3c).

Mango

Mango, the king of fruits is the most important fruit accounting 38.2% of area with 21.7% share of total fruit production in the country. The major mango growing states are Andhra Pradesh, Maharashtra, Uttar Pradesh, Orissa, Bihar, Tamil Nadu, Karnataka and Gujarat. Andhra Pradesh ranks first in area and production followed by Uttar Pradesh and Karnataka. Andhra Pradesh and Uttar Pradesh constitute 50% of country's production (Table 1.3) and these two states produce a large number of export mango varieties. India's share in world production of mango is 54.2%. The area and production under mangoes increased to 2.5 and 2.0 times respectively from 1970 to 2007. the productivity showed a decline and it was mainly due to the new area of mango orchards which are now in prebearing stage (Fig. 1.3d).

Pineapple

Pineapple is grown over an area of 0.80 lakh ha with a production of 12.15 lakh tonnes. Kerala, North Eastern states and West Bengal are major states under pineapple. The national productivity is 15.1 t ha^{-1} and west Bengal ranks first in productivity (29.81 t ha^{-1}) (Table 1.3). The change of area of cultivation and progress in production during the period 1970 to 2007 in case Pineapple is significant. Area increased three times and production four times where as the productivity enhanced from 10 to 15 t ha^{-1} (Fig. 1.3e).

Citrus

Citrus fruit ranks second in total area with 8.43 lakh ha accounting for 14.6% with a production of 75.74 lakh tonnes (12%) of the total production. Maharashtra and Andhra Pradesh accounted for 31.0 and 28.0% of area respectively. Citrus fruits are mainly grown in Maharashtra, Andhra Pradesh and Gujarat (Table 1.3).There were conspicuous changes in the area and production trends of citrus fruits, lemon and lime during the period 1970 to 2007.The increased in area was about four times and that of production was more than six times. The productivity decline was mainly due to larger area under prebearing stage (Fig. 1.3 f and g).

Grapes

During 2007 – 2008, grapes occupied an area of 0.44 lakh ha and produced 16.77 lakh tonnes grapes. Maharashtra and Karnataka are the major important states under this crop with an area of 0.45 and 0.11 lakh ha respectively. Maharashtra alone contributes 77% of Indian production (Table 1.3). The area under grapes increased six fold and production more than eight times during 1970 to 2007. The productivity increased from 15 to 25 t ha^{-1} (Fig. 3h).

Guava

Guava accounts for 3.1 each of the total area and production respectively. The area under guava is 1.78 lakh ha with a production of 19.75 lakh tonnes. Maharashtra, Bihar, Orissa, Uttar Pradesh and West Bengal are the major areas under guava. Productivity in maximum in Karnataka (21.65 t ha^{-1}) followed by Gujarat (15.2 t ha^{-1}), Andhra Pradesh (14 t ha^{-1}) and Uttar Pradesh (12 t ha^{-1}) (Table 1.3).

Other Fruits

Maharashtra, Karnataka, Gujarat and Andhra Pradesh are major growing states of sapota. The total area under sapota was 1.5 lakh ha with a production of 12.38 lakh tonnes.

Pomegranate is grown over an area of 1.22 lakh ha with a production of 8.58 lakh tonnes. Maharashtra and Karnataka are major growing states.

Papaya was found grown over an area of 0.8 lakh ha with a production of 26.85 lakh tonnes. Kerala, Andhra Pradesh and Gujarat are major growing areas.

Litchi is gaining importance owing to its high economical returns and high export potential. It was grown over an area of 0.69 lakh ha with a production of 4.18 lakh tonnes. Bihar is the major growing state (http:// nhb.gov.in/2007-2008.xls) (Table 1.4).

Vegetables

There are 40 different vegetable crops belonging to solanaceous, cucurbitaceous, leguminous, cruciferous, root crops and leafy vegetables grown in tropical, sub tropical and temperate conditions of India. India is next only to China in area and production of vegetables. While among the vegetables, potato is grown in 22.89% area followed by onion (10.28%), tomato (7.33%) and brinjal (7.23%) with a share in production of 27.37%, 9.66%, 8.15% and 7.91% respectively. The total area of vegetables including root and tubers in India during 2007-2008 was estimated as 7.8 million ha and total production as 126 million tonnes (Table 1.5). The state wise area and production of major vegetables are presented in Table 1.6. West Bengal rank the list, followed by Uttar Pradesh, Bihar, Orissa, Maharashtra, Gujarat and Karnataka. Among vegetables, onion, brinjal, cabbage, cauliflower, tomato and peas are largely produced by Maharashtra, West Bengal, Orissa and Uttar Pradesh (Fig. 1.2b; Table 1.5, 1.6). The growth and progress of vegetables during 1970-2007 is noteworthy. The increase in area of vegetable was about two times where as production increased three times. The productivity increased from 7 to 13 t ha^{-1} (Fig. 1.4a). The area, production and productivity of tomatoes, egg plants, okra, cabbage and other brassicas, carrots, turnips, cucumbers and gherkins have shown significant growth and commendable contribution in the production. The productivity had almost doubled from 1970 to 2007 (Fig. 1.4a-4g). The progress and growth of some important vegetable crops are enunciated in the following paras.

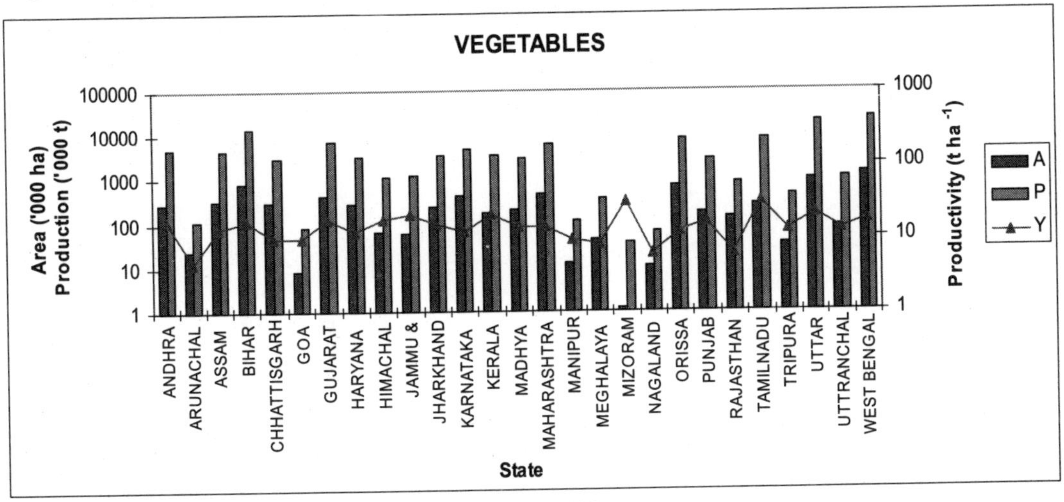

Source: http:/nhb.gov.in/2007-08.xl; accessed on 13/11/2009

Fig 1.2b. State wise area and production of vegetables in India during 2007-08

Table 1.5. State wise and crop wise area ('000 ha) and production ('000 t) of vegetables in India for the year 2007-08

State/UTs	***Brinjal***		***Cabbage***		***C.Flower***		***Okra***		***Peas***		***Tomato***		***Onion***		***S.Potato***		***Cassava***		***Others***		***Total***	
	Area	*Prodn.*	*Area*	*Prodn.*	*Area*	*Prodn.*	*Area*	*Prodn.*	*Area*	*Prodn.*	*Area*	*Prodn.*	*Area*	*Prodn.*	*Area*	*Prodn.*	*Area*	*Prodn.*	*Area*	*Prodn.*	*Area*	*Prodn.*
Andhra Pradesh	30.6	458.8	8.7	130.4	0.0	0.0	30.3	454.6	2.1	74.9	83.1	1579.0	35.8	608.7	2.7	54.3	17.8	355.3	66.5	924.4	**284.1**	**4769.5**
Assam	-	-	-	-	-	-	-	-	-	-	-	-	-	-	8.3	29.0	2.7	12.9	238.6	3917.7	**328.9**	**4474.2**
Bihar	54.6	1158.2	37.4	638.1	60.7	1023.9	57.2	707.3	9.0	57.6	46.2	921.9	51.3	1019.6	0.3	5.2			191.6	2516.3	**823.8**	**14067.8**
Chhattisgarh	22.3	327.5	11.3	177.4	16.6	251.9	23.0	202.4	9.9	44.8	37.7	404.1	8.8	136.7	3.7	32.6	0.0	0.0	128.3	1010.3	**292.6**	**2934.2**
Gujarat	55.8	987.7	23.0	396.2	17.4	314.2	41.5	365.9	0.0	0.0	30.8	739.6	84.3	2238.3	0.0	0.0	0.0	0.0	93.7	867.2	**411.7**	**7403.0**
Haryana	13.5	173.7	11.3	152.6	24.2	349.9	14.9	117.8	10.0	70.9	19.2	202.3	17.7	346.6	-	-	-	-	143.9	1511.1	**274.5**	**3277.1**
Jharkhand	19.0	379.8	11.4	183.0	20.7	331.9	24.5	343.2	-	-	17.5	350.2	12.1	242.1	-	-	-	-	93.6	1432.4	**238.8**	**3639.7**
Karnataka	15.6	352.6	6.8	145.2	2.8	48.5	8.8	68.4	1.1	14.8	47.2	1285.1	147.5	1169.7	3.9	45.5	1.2	13.0	110.2	1490.0	**406.1**	**5031.2**
Kerala	-	-	-	-	-	-	-	-	-	-	-	-	0.0	0.0	0.7	8.1	90.3	2492.1	75.9	978.8	**166.9**	**3479.0**
Madhya Pradesh	14.9	223.3	6.5	129.2	10.8	172.5	9.2	55.1	21.5	236.4	22.7	340.5	39.0	648.6	4.2	23.3	-	-	30.6	440.4	**209.4**	**2919.7**
Maharashtra	29.4	479.2	15.1	383.0	13.1	330.7	26.3	165.4	6.0	27.7	32.2	715.3	254.5	4003.1	0.0	0.0	-	-	59.6	152.3	**455.3**	**6454.9**
Orissa	129.0	1932.5	33.8	935.6	45.2	644.2	71.5	621.1	4.9	42.4	100.7	1344.2	28.8	262.4	47.2	404.1	0.0	0.0	186.8	1862.7	**660.8**	**8214.8**
Punjab	3.0	44.8	4.0	88.2	6.4	148.1	2.3	17.3	18.5	111.0	8.4	202.3	8.0	171.7	-	-	-	-	42.0	511.4	**171.6**	**2772.1**
Rajasthan	5.7	37.3	0.8	8.2	9.6	45.7	4.1	13.0	15.2	11.6	16.6	55.5	36.0	363.4	2.2	6.5	-	-	34.4	165.4	**135.7**	**818.9**
Tamil Nadu	6.6	73.4	2.5	141.2	0.7	13.9	3.9	29.1	-	-	24.3	307.0	32.1	280.3	1.2	20.5	151.5	6114.4	34.3	911.9	**262.7**	**7975.7**
Uttar Pradesh	3.1	96.5	2.2	68.7	7.7	158.4	9.2	102.1	158.1	1398.6	7.7	323.2	21.6	295.8	22.3	259.9	-	-	224.0	5992.2	**960.8**	**19790.3**
West Bengal	153.9	2734.9	73.2	2016.1	66.9	1682.1	71.5	815.3	20.9	123.8	51.1	956.7	18.7	248.8	23.3	235.9	-	-	432.8	3742.4	**1313.1**	**22456.8**
Total	**566.1**	**9595.8**	**265.4**	**5887.8**	**320.6**	**5796.6**	**409.0**	**4192.8**	**313.5**	**2560.0**	**571.7**	**10260.6**	**804.6**	**12157.2**	**125.8**	**1146.1**	**269.9**	**9053.9**	**2370.3**	**30773.7**	**7803.0**	**125887.1**
Productivity (tha^{-1})	**17.0**	**22.2**	**18.1**	**10.3**	**8.2**	**17.9**	**15.1**	**9.1**	**33.6**		**16.1**											

Source: **http://nhb.gov.in/2007-08.xls; accessed on 13/11/2009**

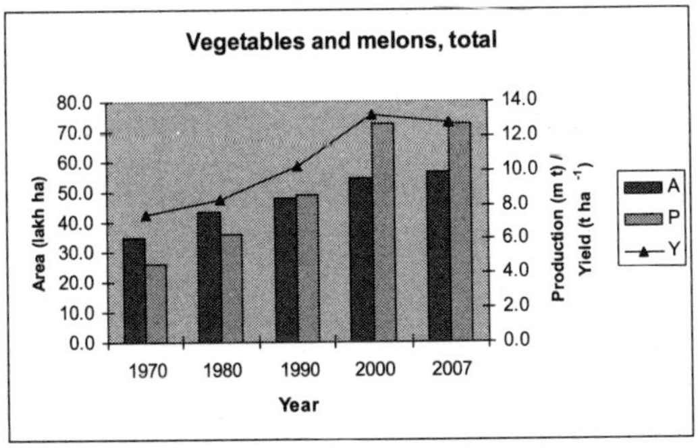

Fig. 4a

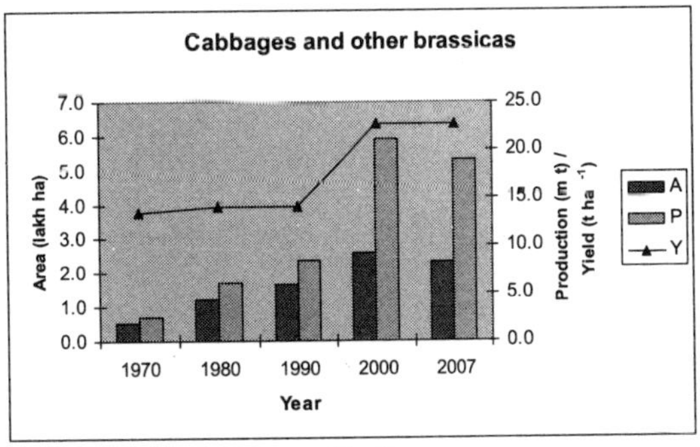

Fig. 4b

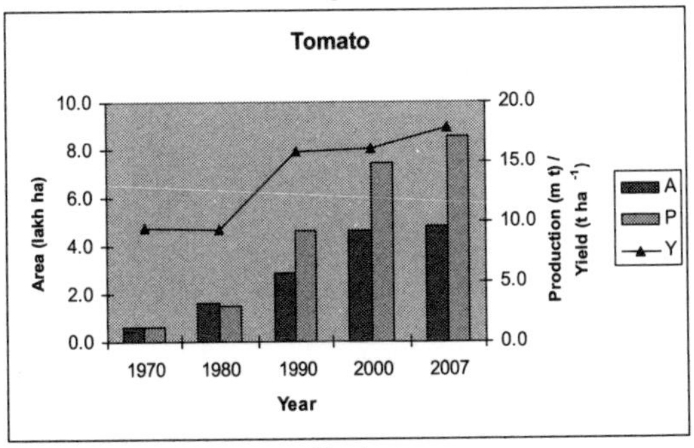

Fig. 4c

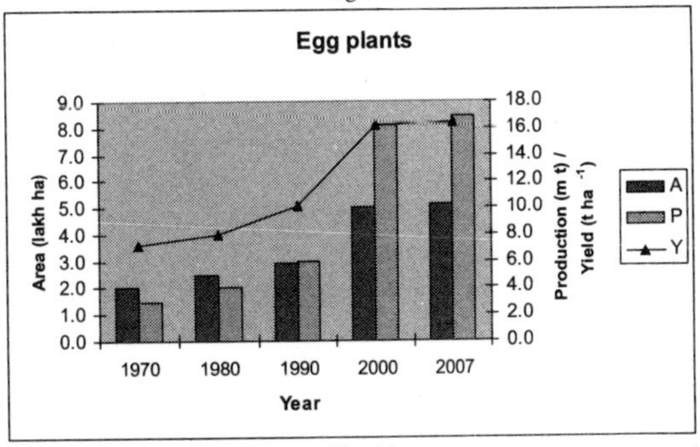

Fig. 4d

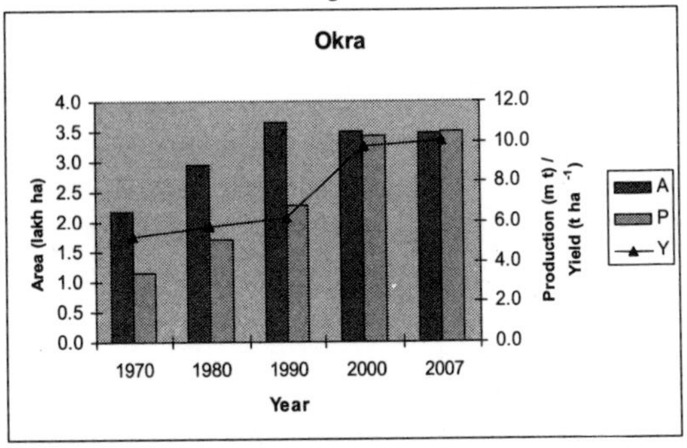

Fig. 4e

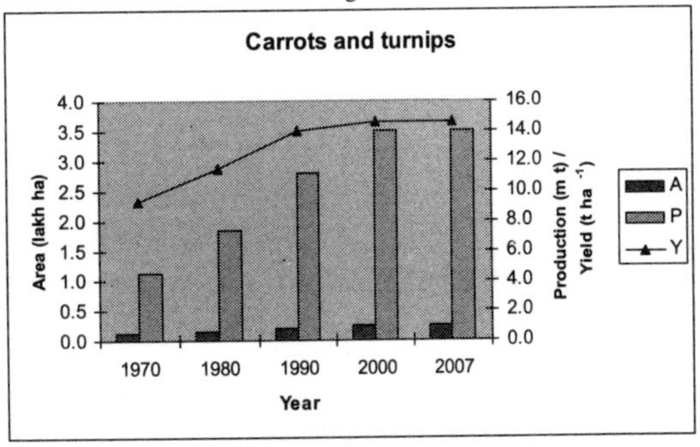

Fig. 4f

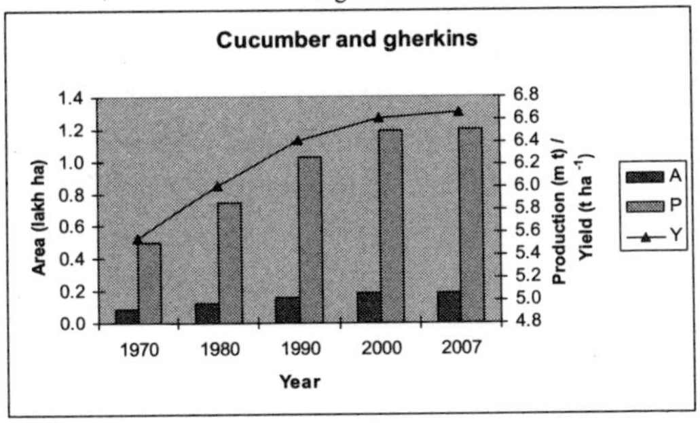

Fig. 4g

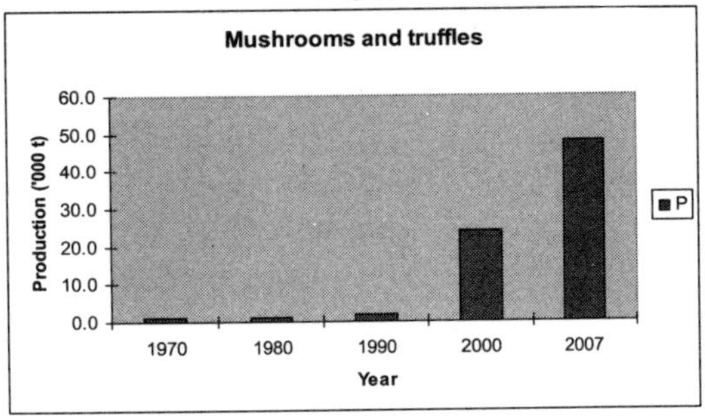

Fig. 4h

Source: http://faostat.fao.org; accessed on 27-01-2009

Fig.1.4. Growth of vegetable crops in India during 1970-2007

Table 1.6. Crop wise area and production of major vegetable and tuber crops in India for the year 2007-08

Crops	*Area '000 ha*	*Production '000 t*	*Productivity t ha*$^{-1}$
Brinjal	566.1	9595.8	17.0
Cabbage	265.4	5887.8	22.2
Cauli flower	320.6	5796.6	18.1
Okra	409.0	4192.8	10.3
Peas	313.5	2560.0	8.2
Tomato	571.7	10260.6	17.9
Onion	804.6	12157.2	15.1
Potato	1786.1	34462.6	19.3
Sweet potato	125.8	1146.1	9.1
Cassava	269.9	9053.9	33.6
Others	2370.3	30773.7	13.0
Total	7803.0	125887.1	16.1

Source: http://nhb.gov.in/2007-08.xls; accessed on 13/11/2009

Cauliflower

A primarily winter vegetable, cauliflower was grown over an area of 3.2 lakh ha with a production of 58.0 lakh tonnes. West Bengal, Bihar, Orissa Haryana and Jharkhand are major states contributing maximum to the area and production of the country (Table 1.5).

Cabbage

The total area under cabbage cultivation in India during 2007-08 was 2.65 lakh ha with a production of 58.88 lakh tonnes. West Bengal was the major state growing over an area of 0.73 lakh ha with a production of 20.16 lakh tonnes followed by Bihar (0.37 and 0.38 respectively), Orissa (0.33 and 9.35 respectively), Gujarat (0.23 and 3.96 respectively) and Maharashtra (0.15 and 3.83 respectively) (Table 1.5). The area under cabbage and other brassicas increased five times from 1970 to 2007 and the production ten fold (Fig. 1.4b).

Tomato

Tomato is the most important vegetable grown through out the country. The area under its cultivation during 2007-2008 was 5.72 lakh ha with a production of 102.6 lakh tonnes. The major states contributing to its cultivation in terms of area and production were Orissa (one lakh ha, 13.44 lakh tonnes respectively) followed by Andhra Pradesh (0.83 lakh ha, 15.79 lakh tonnes), West Bengal (0.51 lakh ha, 9.56 lakh tonnes), Bihar (0.46 lakh ha, 22 lakh tonnes), Karnataka (0.47, 12.85 Lakh tonnes), Maharashtra (0.3, 7.15), Gujarat (0.31, 7.39), Tamil Nadu (0.24, 3.07) lakh ha and Madhya Pradesh (0.23, 3.40 respectively) (Table 1.5). The growth of tomato cultivation was tremendous during the period 1970-2007. The increase in area and production was about eight times and the productivity doubled during the period (Fig. 4c).

Brinjal

Brinjal is the second major vegetable crop and during 2007-08, it was grown in an area of 5.66 lakh ha with a production of 95.96 lakh tonnes. Major Brinjal growing states were West

Bengal (1.54 lakh ha, 27.35 lakh tonnes respectively), Orissa (1.29, 19.32), Gujarat (0.56, 9.88), Bihar (0.54, 11.58), Maharashtra (0.29, 4.79) and Andhra Pradesh (0.30 lakh ha and 4.58 lakh tonnes) (Table 1.5). The area under cultivation of egg plant was enhanced two fold while the production was ten fold and the productivity was more than double in the last 2-3 decades (Fig. 4d).

Okra

Okra is an important vegetable crop cultivated through out the country. The area and production of okra were 4.09 lakh ha and 41.93 lakh tonnes respectively. The major growing states are Orissa, West Bengal, Bihar and Gujarat (Table 1.5). The growth of okra cultivation was progressive and significant (Fig. 4c).

Onion

Onion is an important commercial vegetable crop and was grown on an area of 8.04 lakh ha with a production of 121.57 lakh tonnes during 2007-08. The major contributing states in area and production was Maharashtra (2.54 lakh ha, 40.0 lakh tonnes respectively) followed by Karnataka (1.47, 11.70), Gujarat (0.84, 22.38), Bihar (0.51, 10.19), Madhya Pradesh (0.39, 6.48) and Andhra Pradesh (0.36 lakh ha and 6.08 lakh tonnes) (Table 1.5).

Peas

Peas a special vegetable crop known for its importance as a vegetable and source of vegetable protein. During 2007-08, it was grown over an area of 3.13 lakh ha with a production of 25.60 lakh tonnes. Uttar Pradesh contributed maximum in the area of cultivation and production (1.58 lakh ha and 13.98 lakh tonnes). Madhya Pradesh, West Bengal, Punjab and Rajasthan are the other peas growing states (Table 1.5).

Other Vegetables

Carrots and turnips are important temperate vegetables. Production of carrot and turnips increased three times from two fold enhanced area during 1970-2007, with almost doubling the productivity (Fig. 4f).

Cucumber and gherkins are important cucurbitaceous vegetables. During the period 1970 to 2007, the production enhanced from 5.4 to 6.6 million tonnes (Fig. 4g).

Mushroom

Production of all types of mushroom was estimated to about 0.4 lakh tonnes from Tamil Nadu, Maharashtra, Punjab, Haryana and Utter Pradesh during 2006-07. Tamil Nadu stands first in the total Indian mushroom production with an average annual production of more than 18,000 tonnes. With the sustained research, teaching and extension activities, mushroom cultivation has become one of the promising agricultural enterprises in Tamil Nadu since 1983. It has a high commercial value with in India and abroad. The growth and contribution of mushrooms during the period was meteoric and it was almost 50 fold from 1970 to 2007 (Fig. 4h).

Tuber Crops

Root and tubers including potato showed significant growth in enhancing the production and productivity during the period 1970 to 2007 (Fig. 5a).

Potato is grown in India in all states under very diverse conditions. Nearly 90% of potato was grown in vast Indo-Gangetic plains of north India where it is grown as winter crop (Oct-Mar) and above 6% area in the hills, where it is grown during summer. In India during 2007-08, an area of 17.86 lakh ha was found under potato with a production of 344.6 lakh tonnes (Table 1.6). Major potato growing states are Uttar Pradesh (4.78 lakh ha and 105.37 lakh tonnes respectively), West Bengal (4.07, 50.52), Bihar (3.22, 57.41), Assam (0.77, 5.04), Punjab (0.77, 13.13). We have reached self-sufficiency in potato seed. The area and production of potato has shown three and five fold respectively during the period 1970 to 2007 (Fig. 5b).

Cassava is a major industrial crop in Tamil Nadu and Andhra Pradesh and used as a secondary staple in Kerala and north – eastern states. An area of 2.70 lakh ha was observed under cassava during 2007-08 with a production of 90.53 lakh tonnes. Now the crop has been slowly spreading to Maharashtra and Gujarat. The productivity of cassava (33.6 t ha^{-1}) is the highest in the world. Kerala, Tamil Nadu and Andhra Pradesh contributed major share in the total area and production of cassava (Table 1.5). Though the area of cassava is decreasing, the productivity is increasing significantly due to the adoption of productivity enhancing technologies. The productivity doubled in 2007 compared to 1970. (Fig. 5c).

Sweet potato was grown in an area of 1.26 lakh ha with a production of 11.46 lakh tonnes during 2007-08. It is grown through out the country. Orissa, Uttar Pradesh and West Bengal contributed lion share in the area and production of sweet potato (Table 1.5). Though area and production under sweet potato have come down, the productivity of sweet potato has been increased to 9.1 t ha^{-1} (Table 1.6) (Fig. 5d).

Plantation Crops

Plantation crops are large group of crops like coconut, arecanut, cashew, cocoa grown over and area of 32.26 lakh ha (Table 1.7). Kerala ranks first among the states in area and production of plantation crops (Fig. 2c). Coconut is an important plantation crop. About one crore people are dependent on its cultivation, processing and related activities. Kerala, Tamil Nadu, Karnataka, Andhra Pradesh, Orissa, Maharashtra, Goa, Assam and Andaman and Nicobar islands are major growing states of India. The total area under coconut cultivation during 2007- 2008 was 19.40 lakh ha with a total nut production of 108.9 lakh tonnes. India leads in total production among 90 coconut producing countries. Kerala has the largest area under cultivation (8.70 lakh ha) and is followed by Karnataka (4.01 lakh ha), Tamil Nadu (3.74 lakh ha). Maximum productivity is noted in Tamil Nadu (10 t ha^{-1}) and least in Karnataka (2.8 t ha^{-1}) (Table 1.7). During the period 1970-2007 the area of coconut cultivation and production was doubled due to introduction of high yielding varieties/hybrids (Fig. 5e).

Arecanut is a commercial crop cultivated as sources of supari and was grown over an area of 3.86 lakh ha with a production of 4.76 lakh tonnes during 2007-08. Karnataka, Kerala, and Assam are major growing states of India. Meghalaya, Tamil Nadu, Tripura, Maharashtra, Andaman and Nicobar Islands, Maharashtra and Goa grow on a limited scale (Table 1.7).

Cashew has an important role in the Indian economy. It is grown over an area of 8.68 lakh ha with a production of 6.65 lakh t. Major cashew growing states are Andhra Pradesh, Maharashtra, Orissa, Tamil Nadu, Karnataka, Kerala and Goa (Table 1.7). The area of cultivation was increased by three fold and the production six fold and the productivity also steadily increased during the period 1970 to 2007 with intensification of superior varieties and productivity enhancing technologies (Fig. 5b).

Table 1.7. State wise area ('000 ha) and production ('000 t) of plantation crops in India for the year 2007-08

States/UT'S	*Cashew nut*		*Arecanut*		*Cocoa*		*Coconut*		*Total*	
	Area	*Prodn.*	*Area*	*Prodn.*	*Area*	*Prodn.*	*Area*	*Prodn.*	*Area*	*Prodn.*
Andaman & Nicobar	-	-	4.1	5.7	-	-	21.4	61.2	25.5	66.9
Andhra Pradesh	171.0	107.0	0.3	0.2	12.7	1.6	105.0	912.4	289.0	1021.2
Assam	15.0	12.0	70.3	68.8	-	-	19.0	105.3	104.3	186.1
Goa	55.0	31.0	1.7	2.6	-	-	25.5	87.2	82.2	120.8
Gujarat	4.0	4.0	-	-	-	-	16.4	95.2	20.4	99.2
Karnataka	103.0	56.0	168.0	224.0	7.2	2.8	401.0	1118.2	679.2	1401.0
Kerala	84.0	78.0	108.3	110.0	10.5	6.0	870.9	4165.9	1073.7	4359.9
Lakshadweep	-	-	-	-	-	-	2.7	36.5	2.7	36.5
Maharashtra	167.0	210.0	2.2	3.6	-	-	21.0	120.5	190.2	334.1
Meghalaya	-	-	12.1	16.5	-	-	-	-	12.1	16.5
Orissa	131.0	90.0	-	-	-	-	51.0	189.8	182.0	279.8
Pondicherry	5.0	2.0	0.1	0.1	-	-	2.1	13.2	7.2	15.3
Tamil Nadu	123.0	65.0	4.9	9.0	1.4	0.2	374.6	3736.4	503.9	3310.6
West Bengal	10.0	10.0	9.0	22.0	-	-	25.1	247.1	44.1	279.1
Total	868.0	665.0	386.6	476.0	31.8	10.6	1939.9	10893.9	3226.3	12045.5
Productivity (t ha^{-1})	0.8		1.2		0.3		5.6		3.7	

Source: http://nhb.gov.in/2007-08.xls; accessed on 13/11/2009

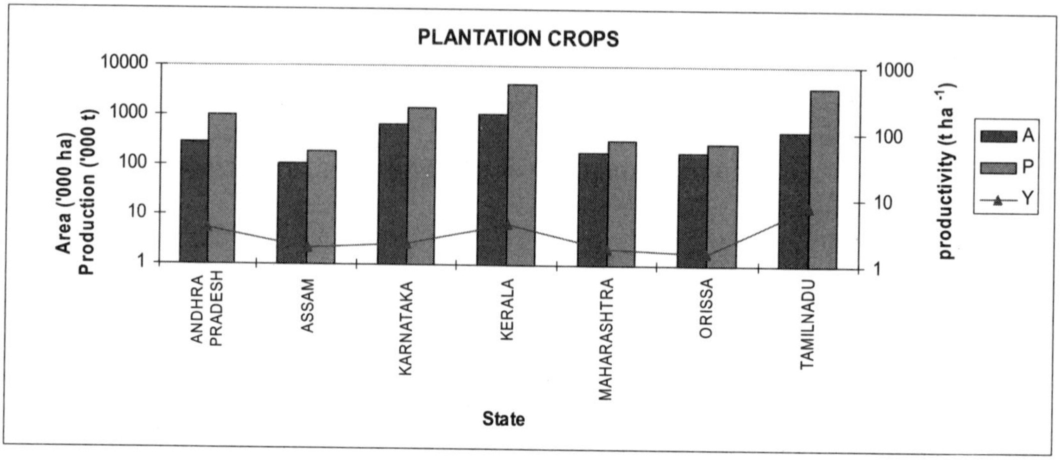

Source: http:/nhb.gov.in/2007-08.xl; accessed on 13/11/2009

Fig. 1.2c State wise area and production of plantation crops in India during 2007-08

Cocoa is grown over an area of 0.3 lakh ha with a production of 0.1 lakh tonnes and is largely grown in Andhra Pradesh (0.127 lakh ha), Kerala (0.105 lakh ha) and Karnataka (0.072 lakh ha) with a production of 0.016m, 0.060 and 0.028 lakh tonnes respectively. Cocoa is usually grown as a companion crop in coconut and arecanut gardens.

Spices and Aromatics

India is the largest producer, consumer and exports of spices and spice products in the world and produces more than 50 spices. Chilly is the major spice crop and account for one third production of total spices, followed by turmeric (14%), garlic (19%), seed spices (17%) and black pepper (27%). India is home of spices, producing a variety of spices like pepper, cardamom, ginger, turmeric and seed spices. Almost all states grow one or other spices. Major spices growing states are Andhra Pradesh, Kerala, Gujarat, Rajasthan, Maharashtra, Fig. 2d. Karnataka, Tamil Nadu, Orissa and Madhya Pradesh. In the area of cultivation, Rajasthan ranked first followed by Andhra Pradesh, Kerala, Gujarat and Karnataka. In terms of Production, Andhra Pradesh contributed maximum followed by Rajasthan, Gujarat, Karnataka and Tamil Nadu; while the productivity was higher in Haryana state followed by Arunachal Pradesh, Punjab and Meghalaya (Table 1.8).

Black pepper and small cardamom are largely confined to south India; cumin, coriander and fenugreek confined to northern states; large cardamom to Sikkim and turmeric and ginger is spread over many states. Spices are grown on area more than 26.0 lakh ha with an annual production of 41.0 lakh tonnes. The area and production of spices have increased during the period 1970-2007 especially ginger and pepper. The area under ginger was enhanced by five times where as the production increased 14 fold (Fig. 5g). In the case of pepper the production got doubled during the period (Fig. 5h).

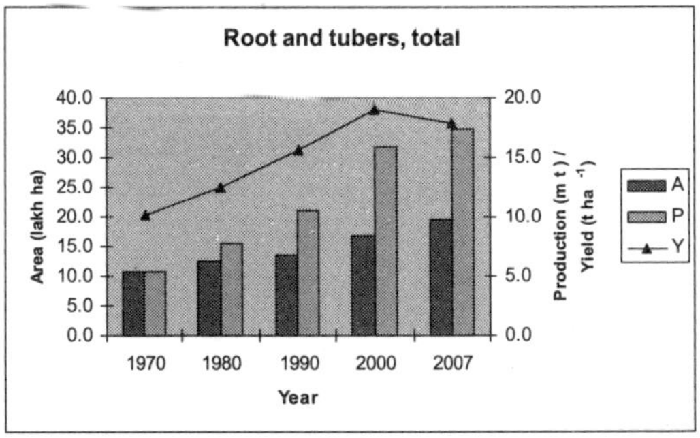

Fig. 5a

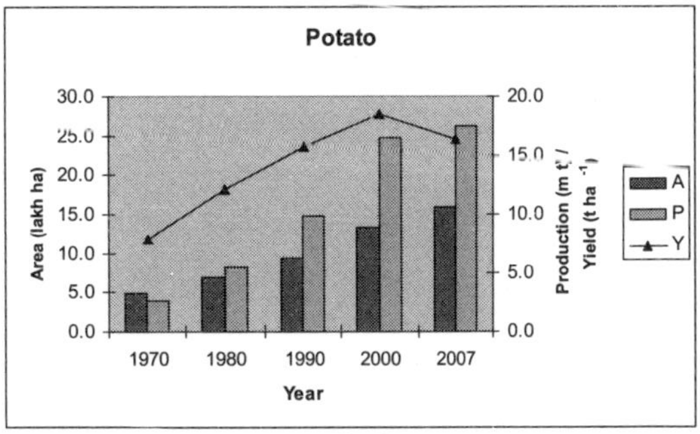

Fig. 5b

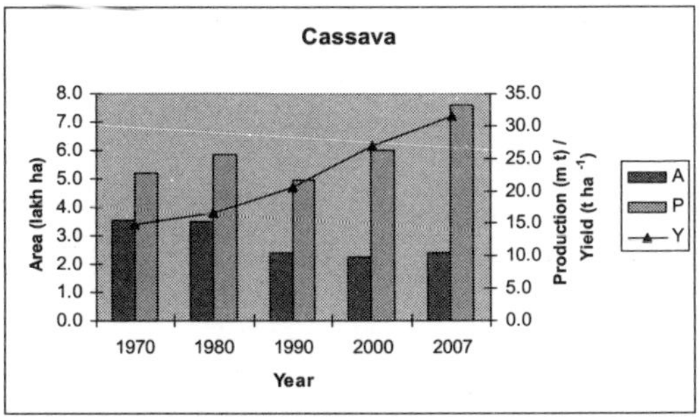

Fig. 5c

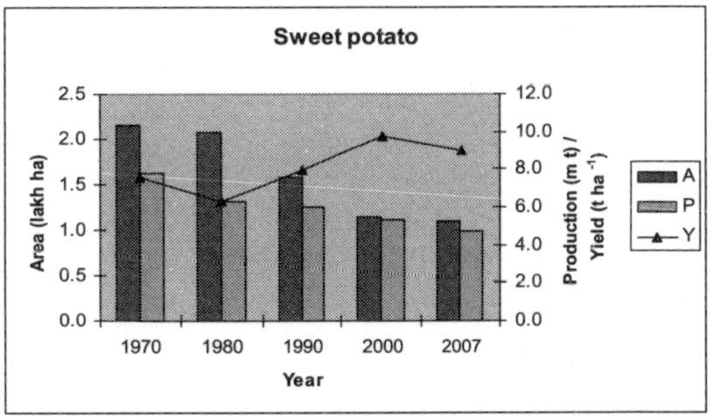

Fig. 5d

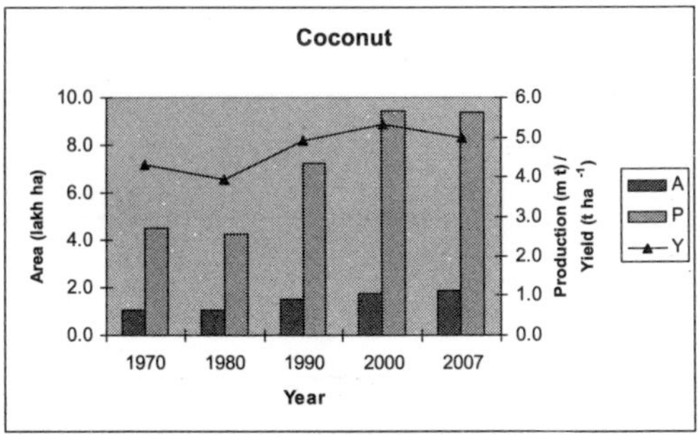

Fig. 5e

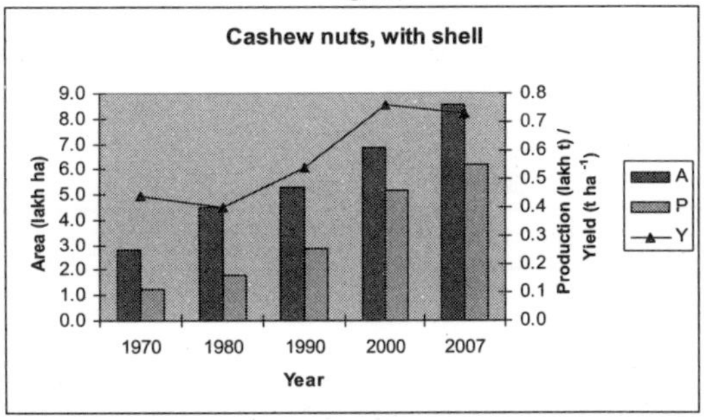

Fig. 5f

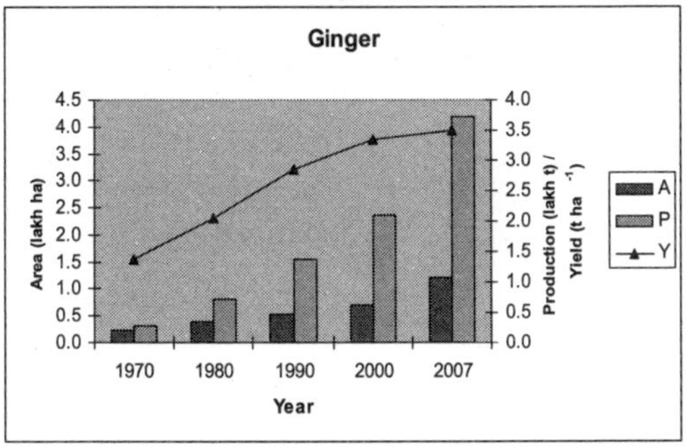

Fig. 5g

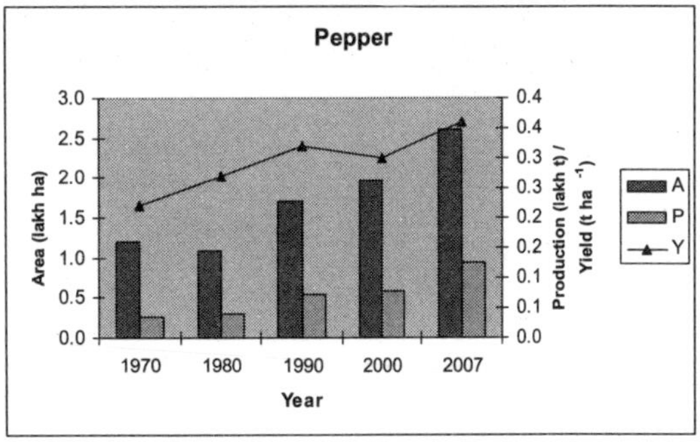

Fig. 5h

Source: http://faostat.fao.org; accessed on 27-01-2009mt = million tonnes

Fig. 1.5. Growth of root and tubers, plantation crops and spices in India during the period 1970-2007

Table 1.8. State wise area and production of spices in India for the year 2007-08

States	*Area'000 ha*	*Production'000 t*	*Productivity(t ha^{-1})*
Andhra Pradesh	317.8	1235.2	3.9
Arunachal Pradesh	8.2	47.5	5.8
Assam	27.2	18.5	0.7
Bihar	11.1	12.3	1.1
Chhattisgarh	11.9	7.8	0.7
Goa	0.7	0.2	0.3
Gujarat	299.8	356.8	1.2
Haryana	4.9	30.9	6.3
Himachal Pradesh	6.3	23.4	3.7
Jammu & Kashmir	3.6	0.6	0.2
Karnataka	235.2	344.9	1.5
Kerala	312.3	159.3	0.5
Madhya Pradesh	194.6	249.8	1.3
Maharashtra	114.3	100.2	0.9
Manipur	8.7	7.7	0.9
Meghalaya	18.4	80.9	4.4
Mizoram	9.0	38.3	4.3
Nagaland	4.5	26.2	5.8
Orissa	147.0	199.2	1.4
Pondicherry	0.1	0.5	5.0
Punjab	5.2	23.7	4.6
Rajasthan	556.4	520.6	0.9
Tamil Nadu	126.8	279.2	2.2
Tripura	4.5	9.4	2.1
Uttar Pradesh	56.8	166.9	2.9
Uttarakhand	3.3	2.9	0.9
West Bengal	79.0	114.4	1.4
Total	2603.2	4102.8	1.6

Source: http://nhb.gov.in/2007-08.xls; accessed on 13/11/2009

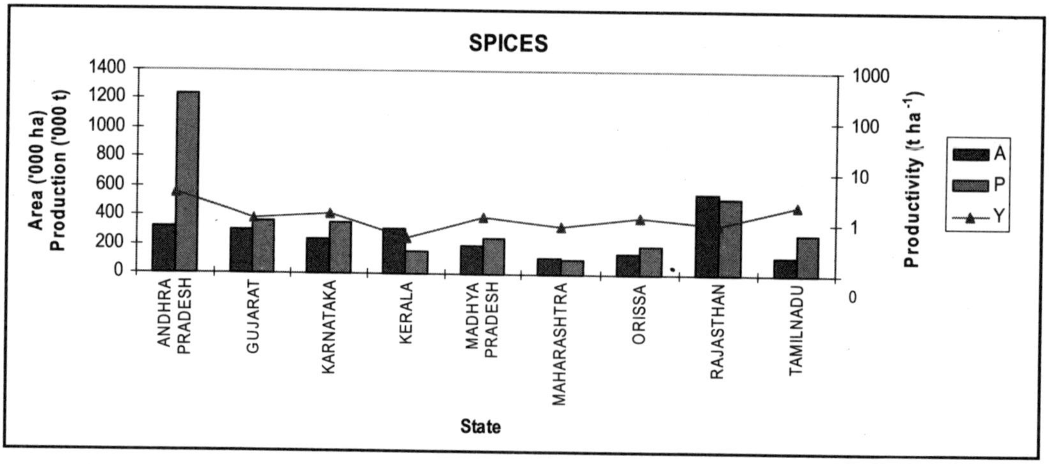

Fig. 1.2d State wise area and production of spices in India during 2007-08

Aromatic plants are grown largely in Rajasthan (1.98 lakh ha), Uttar Pradesh (1.337 lakh ha), Madhya Pradesh (0.196 lakh ha) and Chhattisgarh (0.115 lakh ha). Productivity is highest in Madhya Pradesh and Chhattisgarh (6 t ha^{-1}). The total area as per 2007-2008 statistics under aromatics is 3.86 lakh ha production is 3.29 lakh tonnes (Fig. 1.2e).

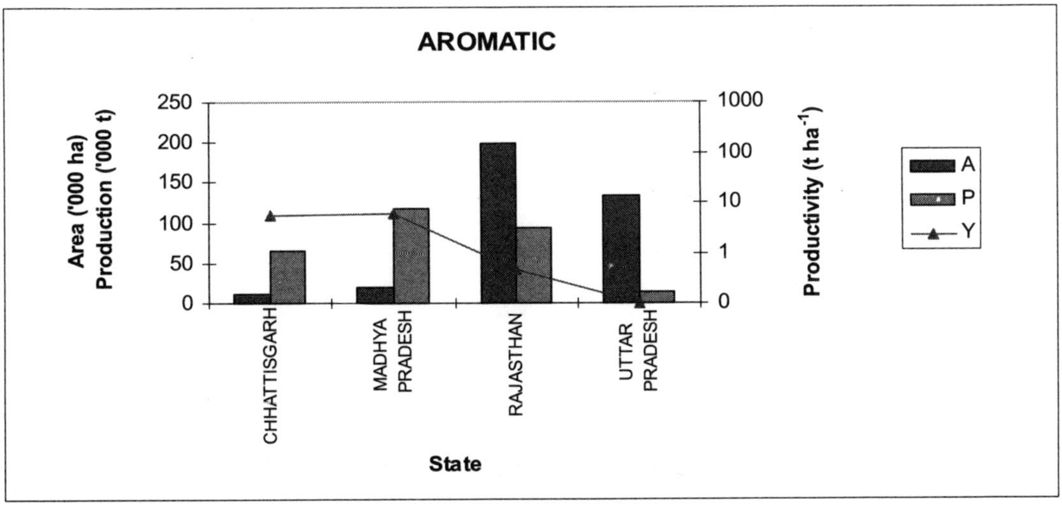

Source: http:/nhb.gov.in/2007-08.xl; accessed on 13/11/2009

Fig. 1.2e. State wise area and production of spices in India during 2007-08

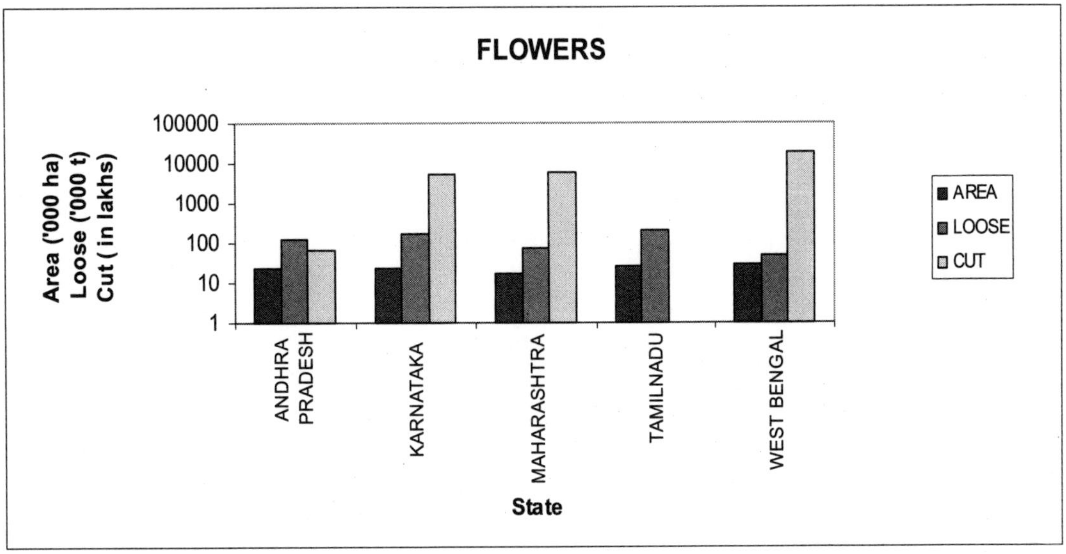

Source: http:/nhb.gov.in/2007-08.xl; accessed on 13/11/2009

Fig. 1.2f. State wise area and production of flowers in India during 2007-08

Floriculture

Flowers are important in socio-cultural and religious life of Indian people. India is known for traditional flowers such as jasmine, marigold, chrysanthemum, tuberose, crossandra and aster. Cut flowers like rose, gladiolus, carnation, anthurium, gerbosa, lilies and orchards have become popular. During 2007-08, the area under floriculture was 1.60 lakh ha with a production of 8.70 lakh tonnes of loose flowers and 434.21 million cut flowers. West Bengal, Tamil Nadu, Karnataka, Andhra Pradesh and Maharashtra are major states under floriculture Fig. 1.2f (Table 1.9). The transformation of Indian floriculture from push cart transportation to charted flight transportation is phenomenal. The export of flowers were at Rs 63 crores during 1996-97 and it is tripled to Rs 211 crores during 2004–05.

Table 1.9. State wise area and production of flowers in India for the year 2007-08

States	*Area'000 ha*	*Production '000 t*		*Productivity*
		Loose	*Cut*	*(t ha⁻¹)*
Andhra Pradesh	23.5	126.3	68.0	8.3
Bihar	0.2	2.3	11.0	66.5
Chhattisgarh	2.4	6.9	—	2.9
Gujarat	9.7	49.5	5063.0	527.1
Haryana	6.1	61.8	11.0	11.9
Himachal Pradesh	0.6	3.4	566.0	949.0
Jammu & Kashmir	0.3	1.3	218.0	731.0
Jharkhand	0.1	0.2	73.0	732.0
Karnataka	22.3	169.1	5550.0	256.5
Madhya Pradesh	2.6	1.5	—	0.6
Maharashtra	16.7	69.5	5728.0	347.2
Mizoram	0.1	0.0	37.0	370.0
Nagaland	0.0	0.0	17:0	0.0
Orissa	2.4	7.0	130.0	57.1
Punjab	1.0	77.9	—	77.9
Rajasthan	3.3	4.6	—	1.4
Tamil Nadu	26.7	214.4	—	8.0
Uttar Pradesh	8.4	12.4	3752.0	448.1
Uttarakhand	0.9	0.7	1456.0	1618.6
West Bengal	27.4	48.4	19680.0	720.0
Total	160.5	870.4	43421.0	276.0

Source: http://nhb.gov.in/2007-08.xls; accessed on 13/11/2009

Table 1.10. India's position in world production of important fruits and vegetables during 2007-208

Fruits	*Position*	*Vegetables*	*Position*
Banana	1	Peas	3
Mango	1	Brinjal	2
Guava	1	Cabbage	2
Lemon	1	Cauliflower	2
Papaya	4	Onion	2
Pineapple	7	Pumpkins and Gourds	2
Grapes	11	Potato	3
Orange	4	Lettuce	5
Citrus fruits/ Mosambi	8	Tomato	3
Apple	8		

Source: http://faostat.fao.org

CONCLUSION

Forty years of concerted and coordinated research activities have catapulted India into a formidable horticulture state to reckon with. India's position in the world production of major fruits and vegetables is encouraging (Table 1.10). We are much ahead in banana, mango, guava, lemon and to certain extent in papaya, pineapple and orange and citrus fruits. India is the second largest producer of fruits and contributing 10% in global fruit production after China. India occupies first place in production of banana, mango, guava and lemon. The productivity of grapes (26.1 t ha^{-1}) is highest in the world. India has higher national productivity in banana (35.9 t ha^{-1}) and papaya (33.4 t ha^{-1}) compound to world average productivity. Commercial horticulture is only a decade old, although traditionally it has been grown. Floriculture have been in fast pace of growth with a higher production. In vegetables also, India is in the forefront especially in the production of peas, brinjal, cabbage, cauliflower, onion, pumpkins and gourds, tomato and potato. All this could be achieved due to the indefatigable. Integrated and participatory approach of horticultural institutes of ICAR, AICRPs and state agricultural universities. AICRPs play a vital role in coordinating research at Central as well as at State level by involving ICAR research Institutes, SAUs, Non-Governmental Organizations, Krishi Vigyan Kendras, State Government Agri./ Hort./ Departments and other organizations located in different parts of the country. This partnership mode has resulted in the development of high yielding varieties/selections, improved production technologies and pest and disease management strategies suitable for each crop under different agro climatic regions of the country. AICRPs have contributed enormously over years in the field of food security, nutrition security and economic security and they have become models for national and international programmes. The contribution of AICRPs in the horticulture growth and development of India is commendable and praise worthy.

References

Singh, H.P. 2001. Horticulture Industry in India. *In: Hand book of Horticulture* (Ed.K. L. Chadha) ICAR, New Delhi. pp 1-20.

Randhawa, M.S. 1979. A history of the Indian Council of Agricultural Research ICAR, New Delhi. pp 94-315.

Web sites:

http://nhb.gov.in/2007-08.xls

http://faostat.fao.org

http://icar.org.in

2

Tropical Fruits

S.D. Shikamany, P. Patil and R.T. Reena

All India Coordinated Research Project (AICRP) was extended to tropical fruits during the Fourth Five-Year Plan (1969–74) on 31 July 1971. It was initially named as All India Coordinated Fruit Improvement Project (AICFIP). The project started to work on citrus, banana, papaya and pineapple as cell-I and on mango, guava and grape as cell-II. Later AICFIP was named as AICRP and Cell-I as AICRP on Tropical and Cell-II as AICRP on Subtropical Fruits.

The broad objectives of the project are

- Collection, conservation and evaluation of germplasm for its utilization as cultivars or as donor source of desirable traits in hybridization programme for improvement.
- Standardization of production technology, viz. rootstock, population density, nutrition, cropping system, weed control and water management under different ecological regions for identified fruits.
- Standardization of crop protection techniques through efficient management of insect pests and diseases.

During Fourth Five-Year Plan, the project was initiated at three centres on citrus [Gonikoppal (Karnataka), Shrirampua (Maharashtra) and Abohar (Punjab)] and banana [Yawal (Maharashtra), Coimbatore (Tamil Nadu) and Trichur (Kerala)]. Research on pineapple was started at three locations, viz. Jorhat (Assam), Trichur (Kerala) and IIHR, Bangalore (Karnataka) besides one on papaya at Coimbatore (Tamil Nadu). During Fifth Five-Year Plan, three new centres, viz. Akola (Maharashtra), Ambalavayal (Kerala) and Periyakulam (Tamil Nadu) were started to wok on citrus. In addition, the citrus bud wood certification schemes located at Tirupati (Andhra Pradesh), Kahikuchi now at Tinsukia (Assam), Ludhiana/Bhatinda (Punjab), IARI Sub-station, Pune (Maharashtra) were also merged with the coordinated project. In the Seventh Plan, four new centres, viz. Pusa, (RAU, Bihar) on banana and papaya, Gandevi (GAU, Gujarat) on banana and sapota, Kovvur (ANGRAU, Andhra Pradesh) on banana and Pedong (BCKV, West Bengal) on Citrus and pineapple were started from April 1987. The Ambalavayal centre in Kerala was closed with effect from April, 1986 and Periyakulam centre was allotted to take up work on sapota while, banana and papaya were allotted to Jorhat and Pune, respectively.

At Rahuri, ad-hoc schemes on acid lime and papaya were merged with AICRP on Tropical Fruits for continuation of the work. New crops were also allotted additionally to the existing centres to work (Periyakulam on sweet orange, Jalgaon on papaya and Pusa on acid lime). Since most of the work on agro-techniques of pineapple have been completed, except for the improvement, work on production technologies of pineapple was phased out. Work on improvement of pineapple is continued as there are promising hybrids and superior clones which would require further evaluation.

During Eight Plan period, a centre to work on banana and sapota was started at Dharwad (UAS, Karnataka) and work on sapota was additionally allotted to the existing Kovvur centre. Evaluations of promising lines of pineapple have been completed, hence pineapple was phased out during Ninth Plan. In lieu of this, jackfruit has been included at five centres, viz. Jorhat, Kannara, Kovvur, Mohanpur and Periyakulam.

At present, under this project, 17 centres are in operation (15 under State Agricultural Universities and 2 in ICAR Institutes). Of which, there are 8 centres working on citrus, 9 on banana, 4 each on papaya and sapota, and 5 on jackfruit (Table 2.1).

BUDGET AND STAFF

The combined budget for Cell 1 and Cell 2 during Fourth Plan was 56.4 lakh (ICAR & State share). The budget allocated to Cell 1 (present project) during the Fifth and Sixth Five-year plans was Rs. 89.8 lakh (ICAR State share) and Rs.169.61 lakh (ICAR & State share), respectively. During the Seventh Plan an outlay of 192.52 lakh (including ICAR and state shares) was provided. The centres were strengthened during Eigth and Ninth Plan with total allocation of Rs. 506.66 and Rs. 1061.43 lakh (ICAR and state share) has been allotted (Table 2.2a and b). During Tenth Plan, a total outlay of Rs.1833.90 lakh (ICAR and state share) has been allotted and it was Rs 5141.36 lakhs during XI plan. The present staff strength is 74 scientific, 79 technical, 18 administrative and 65 supporting staff with a total strength of 236 (Table 2.2c).

Table 2.1. Research centres of AICRP on tropical fruits

Centre	*Year of starting*	*Name of University/Institute*	*Centre address*	*Crops**	*Contact person/ E-mail id*
Akola	1976	Dr. Punjabrao Deshmukh Krishi Vidyapeeth, Akola, Maharashtra	AICRP (TF) Dr Punjabrao Deshmukh Krishi Vidyapeeth, Akola Maharashtra	Citrus (V plan)	Dr P.B. Ghode Sr. Horticulturist vc@pdkv.mah.nic.in
Arabhavi	1994	University of Agricultural Sciences, Dharwad, India.	AICRP (TF) Kithura Rani Channama College of Horticulture, Arabhavi Dist., Belgaum Karnataka	Banana Sapota (VIII plan)	Dr V.C. Kanamadi Horticulturist vckanamadi@yahoo.co.in
Chethalli	1970	IIHR Regional Station, Chethalli, Karnataka	Central Horticultural Experiment Station Chettalli, Karnataka	Citrus (IV plan)	Dr. H. Ravishankar Principal Scientist & Head, CHES chettalli@yahoo.co.in

(Contd.)

Centre	*Year of starting*	*Name of the University/Institute*	*Centre address*	*Crops **	*Contact person/ E-mail id*
Coimbatore	1971	Tamil Nadu Agricultural University, Coimbatore Tamil Nadu	Department of Fruit Crops Horticultural College & Research Institute, TNAU Coimbatore	Banana Papaya (IV plan)	Dr T.N. Balamohan Professor & Head tnb@tnau.ac.in
Gandevi	1987	Navasari Agricultural University, Gandevi Gujarat.	Fruit Research Station NAU, Gandevi Dist. Navasari, Gujarat	Banana, Sapota (VII plan)	Sh C.B. Patel Horticulturist frsgan@rediffmail.com
Jalgaon	1970	Mahatma Phule Krishi Vidyapeeth, Rahuri Maharashtra	Banana Research Station Jalgaon, Maharashtra	Banana (VIII plan)	Dr C.V. Pujari Horticulturist bhalerao@rediffmail.com
Jorhat	1972	Assam Agri. University Jorhat, Assam	Department of Horticulture Jorhat, Assam – 785 013	Banana (VII plan) Jackfruit (IX plan)	Dr K. Baruah Horticulturist kartik@aau.ac.in
Kannara	1971	Kerala Agrl. University Thrissur, Kerala	Banana Research Station KAU, Marakkal, Kannara P.O, Thrissur, Kerala	Banana (IV plan) Jackfruit (IX plan)	Dr Rema Menon Associate Professor brskannara@yahoo.com
Kovvur	1987	Acharya, N.G. Ranga Agricultural University Hyderabad Andhra Pradesh	Agril. Research Station, Kovvur, West Dist., Godavari, Andhra Pradesh	Banana (VII plan) Sapota, (VIII plan) Jackfruit, (IX plan)	Dr D. Madhava Rao Horticulturist arskovvur@rediffmail.com
Ludhiana	1975	Punjab Agricultural University Ludhiana, Punjab	Department of Horticulture PAU, Ludhiana, Punjab	Citrus (V plan)	Dr J.S. Randhawa Head sandeeppau@yahoo.com
Mohanpur	2000	Bidhan Chandra Krishi Vishwa Vidyalaya, Kalyani West Bengal, India.	Officer In-charge, AICRP (TF) Directorate of Research Build. BCKV, Kalyani, West Bengal	Banana Jackfruit (IX plan)	Dr B. Bandyopadhyay Officer In-charge biswanath_bckv@rediffmail.com
Periyakulam	1976	Tamil Nadu Agricultural University, Coimbatore Tamil Nadu, India.	Department of Fruit Crops & PHT Horticultural College and Research Institute Periyakulam, Tamil Nadu	Citrus (V plan) Sapota (VII plan) Jackfruit (IX plan)	Dr S. Balasubramanyan Prof. and Head deanhortpkm@tnau.ac.in
Pune	1968	IARI Regional Station Pune, Maharashtra	IARI Regional Station College of Agriculture, Pune Maharashtra – 411 005	Citrus Papaya (VII plan)	Dr S.K Sharma Senior Scientists unilksharma1959@yahoo.co.in
Pusa	1987	Rajendra Agricultural University, Samastipur Bihar, India	Horticultural Research Station RAU, Birauly, Dist: Samastipur Bihar – 848 113	Citrus (VIII plan) Banana Papaya (VII plan)	Dr P. K. Ray Senior Scientist pkrayraupusa@yahoo.com
Rahuri	1970	Mahatma Phule Krishi Vidyapeeth, Rahuri Maharashtra, India	AICRP on Citrus, Mahatma Phule Krishi Vidyapeeth Rahuri, Maharashtra – 413 722	Citrus (IV plan)	Dr S.N. Gohil Horticulturist satishbulbule2000@yahoo.co.in
Tinsukia	1975	Assam Agricultural University Jorhat, Assam	Citrus Research Station AAU, Gelapukhuri Road Tinsukia, Assam – 786 125	Citrus (IV plan)	Dr A. C. Barbora Chief Scientist amritbarbora@rediffmail.com
Tirupati	1975	Acharya N.G. Ranga Agricultural University Hyderabad Andhra Pradesh	AsICRP (TF), (Citrus) ANGRAUS.V Agricultural College Campus, Tirupati Andhra Pradesh – 517 502	Citrus (V plan)	Dr K. Purushotham Principal Scientist and Head kurubgopal@rediffmail.com

* Crops included in respective plan period

Table 2.2a. Financial allocation in various plans (Rs in lakh)

Head	*VI Plan*			*VII Plan*			*VIII Plan*		
	ICAR share	*State share*	*Total*	*ICAR share*	*State share*	*Total*	*ICAR share*	*State share*	*Total*
Salary	106.70	35.56	142.26	107.16	35.72	142.88	289.11	96.37	385.48
TA	2.45	0.82	3.27	3.42	1.14	4.56	10.19	3.40	13.58
RC	14.78	4.93	19.71	26.08	8.70	34.78	51.15	17.05	68.20
Works	2.70	0.90	3.60	1.20	0.40	1.60	6.63	1.87	7.50
Equipment	0.58	0.19	0.77	6.52	2.17	8.69	21.15	7.05	28.20
Vehicle	0.00	0.00	0.00	0.00	0.00	0.00	2.77	0.92	3.70
Total	127.21	42.40	169.61	144.38	48.13	192.52	380.0	126.66	506.66

Table 2.2b. Financial allocation in various plans (Rs in lakh)

Head	*IX Plan*			*X Plan*			
	ICAR share	*State share*	*Total*	*ICAR share*	*State share*	*Total*	*XI Plant Total*
Salary	660.13	220.04	880.17	1160.60	386.87	1547.47	4268.00
TA	19.54	6.52	26.06	25.00	8.33	33.33	66.67
RC	86.40	28.80	115.20	128.70	42.90	171.60	432.00
Works	16.42	5.47	21.89	16.00	5.33	21.33	143.34
Equipment	13.58	4.53	18.11	46.04	14.13	60.17	201.35
Vehicle	0.00	0.00	0.00	0.00	0.00	0.00	00.00
Total	796.07	265.36	1061.43	1376.34	450.16	1833.90	5141.36

Table 2.2c. Manpower in different plans

Category	*VI Plan*	*VII Plan*	*VIII Plan*	*IX Plan*	*X Plan*	*XI Plan*
Scientific	18	65	68	66	66	74
Technical	88	68	73	71	71	79
Administrative	18	15	16	16	16	18
Supporting	58	63	61	61	61	65
Total	182	211	218	214	214	236
Number of centres	13	16	17	17	17	26

During 11th plan, four new centres viz., Palghar (DBSKKV) for sapota and jackfruit, Bubaneshwar (QUA&T) for banana, Goriyakarma (BAU) for papaya and Sriganganagar (SKRAU) for Kinnow mandrin were approved in addition to one voluntary centre at Pasighat under CAU, Imphal for citrus. Besides the existing two co-opting centres, i.e. NRC for citrus and NRC for banana, two voluntary centres (Indian Institute of Horticultural Research, Bangalore and Central Institute for Sub-tropical Horticulture, Lucknow) were added. At present, there are 10 centres working on citrus, 11 on banana, 6 each on papaya and jackfruit and 5 on sapota.

GENETIC RESOURCES/BIODIVERSITY/GENE BANK MANAGEMENT

In citrus, a total of 12 accessions at Ludhiana, 108 at Rahuri, 45 at Tinsukia and 67 at Tirupati were maintained and evaluated. While in banana, 53 accessions at Arabhavi, 153 at Coimbatore, 63 at Gandevi, 11 at Jalgaon, 97 at Jorhat, 275 at Kannara, 105 at Kovvur, 82 at Mohanpur and 86 at Pusa were maintained and evaluated.

In sapota, a total of 26 collections at Arabhavi, 18 at Gandevi, 29 at Kovvur and 19 at Periyakulam are being maintained and evaluated. Survey for the elite types of jackfruit resulted in collection of 14 clones at Kovvur, 25 at Mohanpur and 19 at Periyakulam. Some regional-specific varieties are given in Table 2.2d.

Table 2.2d. Regional-specific varieties/value-added varieties with specification and important qualities

Crop	*Variety/hybrid*	*Pedigree*	*Centre*	*Year of release*
Acid Lime	Sai Sharbati	Kagzi Lime selection developed from local germplasm	Rahuri (Maharashtra)	1994
	Jai Devi	Kadayam type	Periyakulam (Tamil Nadu)	1990
	Balaji	Pureline selection	Tirupati (Andhra Pradesh)	2006
	PDKV Lime	Khagzi lime clone	Akola (Maharashtra)	2005
Lemon *Citrus limon Burm*	PAU Selection	Selection from two strains of baramasi lemon	Ludhiana (Punjab)	1990
Banana	Kovvur Bontha	Natural Selection	Kovvur (Andhra Pradesh)	1995
	BRS-1	Agniswar x Pisang lilin	Kannara (Kerala)	1998
	BRS-2	Vannan x Pisang lilin	Kannara (Kerala)	1998
	KBS-8 (AAA)	Selection	Kovvur (Andhra Pradesh	2007
Papaya	Coorg Honey Dew	Honey Dew	Chettalli (Karnataka)	1959
	CO-1	Selection from Ranchi	Coimbatore (Tamil Nadu)	1972
	CO-2	Selection from local type	Coimbatore (Tamil Nadu)	1979
	CO-3	CO-2 x Sun Rise Solo	Coimbatore (Tamil Nadu)	1983
	CO-4	CO-1 x Washington	Coimbatore (Tamil Nadu)	1983
	CO-5	Selection from Washington	Coimbatore (Tamil Nadu)	1985
	CO-6	Selection from Pusa Giant	Coimbatore (Tamil Nadu)	1986
	CO-7	Coorg Honey Dew x CP.85	Coimbatore (Tamil Nadu)	1997
Sapota	PKM-1	Selection from Guthi	Periyakulam (Tamil Nadu)	1980
	PKM-2	Guthi x Kirthibarthi	Periyakulam (Tamil Nadu)	1992
	PKM-3	Guthi x Cricket Ball	Periyakulam (Tamil Nadu)	1994
	PKM-4	Open pollinated seedling selection from the seedling progeny of PKM-1	Periyakulam (Tamil Nadu)	2002
	PKM-5	Selection from open pollinated seedlings of Virudhunagar local	Periyakulam (Tamil Nadu)	2007
Pineapple	H7	Kew x Ripley queen	Kannara (Kerala)	1998
	H8	Kew x Ripley queen	Kannara (Kerala)	1998

Acid lime Balaji

Acid lime Jai Devi

Sai Sharbati

PDKV lime

Kovvur Bontha

BRS-1

BRS-2

H-7

KBS-8 (AAA)

H-8

Varieties/hybrids of lime, banana and pine apple

Co-1

Co-2

Co-3

Co-4

Co-5

Co-6

Co-7

PKM-1

PKM-2

PKM-3

PKM-4

PKM-5

Varieties/hybrids of papaya and sapota

Lime

***Sai Sharbati*:** It is a kagzi lime selection developed from local germplasm. It has roundish smooth fruits with thin rind, uniform size and has good juice content (54.51%) with high TSS, ascorbic acid and acidity. It is tolerant to canker and tristeza.

***PDKV lime*:** Trees are medium and thorny, irregular branches, leaves green, pointed at tip, petioles narrowly winged, segment tightly adherent to each other. The fruit rind is thin, papery and smooth. Fruit shape is round with an average weight of 45-50 g and suitable for export.

Phule Mosambi (selection-4) is recommended for cultivation in western Maharashtra. It has recorded vigorous plant growth (27.03m^3 canopy volume), high yield potential (72.95 kg/plant) coupled with thinner rind (0.33 cm), medium-sized higher number of fruits (303 fruits/tree), better fruit quality (47.37% juice with 9.66 °B TSS and 62.31 mg/100g ascorbic acid content) and less pest and disease incidence (2.42% greening and 2.07% tristeza).

Phule Sharbati (RHR-L-124) is recommended for cultivation in western Maharashtra. It has recorded better plant growth (86.50m^3 canopy volume), less rind thickness (1.95mm), better fruit quality (52% juice content) and higher yield (52.19 t/ha, 11.28% increase over cv. Sai Sharbati) particularly during summer (cropping of 25-42%) having less disease and pest incidence (9.09 PDI for Citrus canker and 0.90 fleck/leaf for tristeza).

Banana

***BRS-1*:** Hybrid derivative by crossing the varieties Agniswar and Pisang Lilin and is medium-statured with medium green pseudostem characterized by reddish purple pigmentation, slight blotching and waxines. The plants grow to a height of 250cm and yield bunches weighing 12kg each. Bunches are semi-horizontal. The fruits are medium-sized with medium green colour when unripe and turn greenish-yellow on ripening. It is a dual-purpose variety suited for dessert and cooking. It is highly resistant to Sigatoka leaf spot, a major yield-reducing factor of banana in Kerala. It is also resistant to Panama wilt and insect pests such as rhizome weevil and pseudostem-borer.

BRS-2: Derivative of a cross between Vannan and Pisang Lilin, it is medium-statured and characterised by slight reddish purple pigmentation, moderate blotching and waxiness. The plant grows to a height of 315cm and yields good quality bunches with more than 8 hands and 118 fruits in 300 days. The hybrid can yield bunches with an average weight of 14kg each, though yield up to 16kg has been recorded. Bunches are pendulous with compactly arranged, short rounded fruits. Fully mature fruits are medium green, upon ripening it turns to golden-yellow and the fruit pulp is sweet. The hybrid is found to be ideal for ratooning. It can come up well under partial shade. It is highly resistant to Sigatoka leaf spot diseases, Panama wilt and insect pests such as rhizome weevil and pseudostem-borer. It has also built in resistance to nematodes.

KBS-8 (AAA): It is a selection from Kovvur Bontha (ABB) released for cultivation in Andhra Pradesh. It takes 439.70 days to harvest. Hands per bunch is 15.60, number of fingers per bunch is 312.70, average number of fingers per hand 22.75 and average bunch weight is 42kg.

Papaya

CO-1: It is a dioecious type with dwarf stature. Fruit is round with flattened base. The flesh is orange-yellow coloured with a TSS of 13-14° Brix. The average yield is 50-60 fruits per tree in a period of 20 months from planting.

CO-2: It is dioecious, dual purpose (for fruits and papain extraction), a pure line selection from a local type and is very popular commercial variety. Fruits are oblate and large in size, flesh orange coloured, soft with –TSS ranging from 13.5 to 14.5° Brix. The yield ranges from 80 to 100 fruits per tree each weighing 1.5-2.5kg. Latex yield per fruit is 25-30g.

CO-3: Gynodioecious hybrid, female plants and bisexual plants are found in the population, (CO-2 × Sun Rise Solo) suitable for table purpose. Fruits are pyriform and flesh is red. Tree yields 90-120 fruits each weighing 450-500g.

CO-4: Dioecious, medium-tall hybrid (CO-1 × Washington), it has purple petiole and stem. Fruit weighs 1.2-1.5kg and each tree yields 80-90 fruits.

CO-5: Dioecious type, selection from Washington, highly suitable for papain extraction. Petiole colour is pink. Each fruit weighs 1.5-2.0kg and yields 70-80 fruits per tree. It produces 50-60g of wet latex per fruit.

CO-6: Dioecious type, selection from Pusa Giant, it is suitable for consumption and latex extraction. Fruit is large, weighing 1.5-2.0kg. It yields 80-100 fruits/tree.

CO-7: Gynodioecious type (Coorg Honey Dew × CP.85) large fruited with deep red sweet flesh. Suitable for table purpose.

Sapota

PKM-1: Clonal selection from Guthi variety. Dwarf plant, high-yielding with medium-sized fruits having thick skin. The fruits are luscious with a TSS of 24° brix.

PKM-2: This hybrid combines the characters of high yield and good quality of PKM-1 (Guthi) and fruit size of Kirthibarthi. It yields 1,400 fruits weighing 80.0kg/tree with an estimated mean yield of 12,500 kg/ha. It yields 42.6 per cent more over PKM-1. The fruits are oval in shape and bigger in size having good quality with a high TSS (26.30° Brix).

PKM-3: Developed from the cross between Guthi and Cricket Ball, it yields 88.73kg/tree as compared to their parents namely Guthi (PKM-1) with 63.83kg/tree and Cricket Ball with 56.55kg/tree. The fruits are attractive with oval shape and bigger in size (91.3g) and bears mostly in clusters of more than three fruits. The quality of fruits is quite acceptable with a high TSS (26.8° Brix). The tree exhibits a characteristically cylindrical columnar growth habit, which makes it suitable for high-density planting. The fruits are harvested from November onwards. It is relatively tolerant to leaf spot (1–5%) as compared to PKM-1 and PKM-2 (26.50%).

PKM-4 (Spindle Sapota): It is open-pollinated seedling selection from the seedling progeny. It yields 1,260 fruits, weighing 108kg/tree.The fruits weigh 92 g having few seeds (2-3 seeds/

fruit) with 23° Brix TSS. Selected trees bears spindle-shaped fruits with smooth, light brown skin and good market appeal.

PKM (Sa)-5: It is a selection from open-pollinated seedlings of Virudhunagar local. Its features are compact tree canopy, narrow and clustering of leaves and suitable for high-density planting. Cluster bearing is 3.0/cluster. It has oval shaped fruits with coppery brown flesh. Has high total soluble solids (25.5% Brix) and suitable for preparation of dry flakes, milk shake powder and mixed fruit jam. It is a high-yielder (119.83 kg/tree/year, 18.70 tonnes/ha). As the leaves are waxy, narrow, small, upright and arranged in whorl form in cluster and rich in proline content, relative water content, epicuticular wax and chlorophyll stability index, it is suitable for vertisol and drought areas.

Pineapple

H7: Its (Kew × Ripley queen) fruit shape is cylindrical, tapering, slightly from near base. Fruit colour is dull yellow when ripe and flesh colour is pale yellow. It has mild external aroma, with good flesh taste, TSS (18.56°Brix) with less acidity (0.2%).

H8: This (Kew × Ripley queen) is similar to H-7 with smaller fruits (fruit weight with crown=1173g, crown weight 116.67g, fruit length 17.17cm, width 11.92cm, and pulp: peel ratio 1).

TECHNOLOGIES FOR ENHANCING PRODUCTION AND PRODUCTIVITY

Crop Improvement

Citrus

- Acid lime selections, viz. RHR-L-124 and RHR-L-122 at Rahuri and Tenali selection at Tirupati, continued to perform well with tolerance to canker. Khasi mandarin selection CRS - 4 at Tinsukia has selected for higher yield.
- Sweet orange Selection 2 and 4 at Rahuri and Himakuntala selection 1 at Tirupati continued to be promising in terms of growth and yield.
- Nagpur mandarin at Akola, Nucellar mosambi and Kodur Sathgudi sweet oranges at Rahuri and Tirupati, respectively, Vikram at Periyakulam and Kagzi lime local at Akola continued to be promising.

Banana

- Gandevi selection from Dwarf Cavendish is promising in Tamil Nadu, Gujarat and Maharashtra. KBS 8 clone selected from Dwarf Cavendish consistently recorded higher yield under Kovvur conditions.
- A high yielding clone, Manjeri Nendran-II, identified at Kannara, was suitable for annual cropping besides displaying field tolerance to Sigatoka leaf spot and pseudostem-borer. Among Kanthali clones at Mohanpur, clone I and clone II have been observed to be more potential.
- Promising hybrid H.212 (AB) [Karpooravalli × Pisang Lilin] showed superiority over Ney Poovan (AB) in yield, quality and tolerance to nematodes and another hybrid NPH-02-01 was found tolerant to *Fusarium* wilt and nematode (*Pretylenchus*). These are under multi locational trials.

- FHIA-01 (Gold Finger) is highly resistant against Sigatoka leaf spot, while, FHIA-01 and FHIA-17 is promising for yield and tolerance to nematode.

Papaya

- For the improvement of 9-1 (D) papaya (high-yielding red fleshed dioecious line) at Coimbatore, four superior single plant selections have been made in the BC_5F_1 generation for fruit yield and flesh colour. Crossed and sibmated fruits are harvested and seeds extracted for further raising of F_6 progenies.

Sapota

- Highest yield per tree recorded for PKM-4 and PKM-3. However, PKM-1 recorded significantly higher number of fruits besides showing resistance to bud-borer, leaf webber and leaf miner.

Jackfruit

- At Kannara, fruits from two clones were found good for all purposes (table purpose, chip making and culinary purpose), nine clones were found best for table purpose, four for making chips and two for culinary purpose. At Mohanpur, an extra early clone (Type – XVII) and high-yielding clone (Type-XV; Hazari and Type XXI) were identified for further studies.

Crop Production

Citrus

- Rangpur lime followed by Rough lemon were found superior for mandarin and sweet orange at most of the locations.
- Plant spacing of 6m × 6m (270 plants/ha) is ideal for superior quality fruit production having less pest and disease problem though the yield was less than 6 × 3m spacing for Nagpur mandarin. Plant spacing of 6m × 3m accommodating 550 plants/ha was found ideal for getting higher yields of Kinnow mandarin at Ludhiana and acid lime at Rahuri.
- Application of 1200g N, 400g P_2O_5 and 400g K_2O/plant/year for 17 years old Nagpur mandarin plants recorded highest yield at Akola.
- A fertilizer dose of 600g N, 225–300g P_2O_5, 450–600 g K_2O/with 7.5–15kg neemcake in Khasi mandarin at Tinsukia, sweet orange at Tirupati and acid lime at Periyakulam, Rahuri and Tirupati recorded higher growth and yield.
- At Rahuri, inclusion of biofertilizer *Azotobacter* and PSB each at 30g/plant to above treatment was found to be more productive with better fruit quality.
- Inclusion of VAM (500g/plant) + PSB (100g/plant) + *Azospirillum* (100g/plant) + *T. harzianum* (100g/plant) to 100% RDF/plant/year produced highest yield of Kinnow mandarin at Ludhiana, sweet orange and acid lime at Rahuri and Nagpur mandarin at Akola.
- Zinc is most critical micronutrient in citrus and preferred to apply in foliar form (0.5% $ZnSO_4$)
- For Kinnow mandarin, foliar application of 2% of 13:0:46 ($N:P_2O_5:K_2O$ respectively)

recorded higher fruit size and yield with high benefit:cost ratio at Ludhiana and Khasi mandarin at Tinsukia.

- Application of 75% recommended dose of nitrogen and potash through drip irrigation system was found ideal under Akola, Ludhiana, Rahuri and Tirupati conditions.
- Irrigation of Nagpur mandarin at 40% depletion recorded highest yield. Drip irrigation was found superior over other methods.
- Application of 75% recommended dose of nitrogen and potash through drip irrigation system was found ideal under Akola, Rahuri, Ludhiana and Tirupati conditions.
- Integrated method of weed control (hoeing followed by two sprays of glyphosate @ 1 kg/ha and hand weeding) recorded highest weed control efficiency (77.90%) with highest cost benefit ratio in Sathgudi sweet orange and Khasi mandarin under Tirupati and Tinsukia conditions respectively.
- Growing cowpea, blackgram and greengram in young Khasi mandarin orchard resulted in weed free condition besides giving additional income apart from favourable effects of these legumes on soil under Tinsukia conditions.
- Soil moisture stress from 15 September to 15 October followed by spraying of 50ppm ascorbic acid at the time of stress alleviation resulted in summer crop at Akola.
- Studies on regulation of fruit drop revealed that foliar application of copper oxychloride (0.3%) or carbendazim (0.1%) in combination with 2,4-D (10 ppm) or NAA (30 ppm) recorded significantly higher fruit retention (76.4–75.6%) as compared to the control (64.2%) in Kinnow mandarin. Similar results were also reported at Tinsukia for Khasi mandarin.
- Studies on regulation of flowering in acid lime revealed that application of GA_3 at 50 ppm during June and Cycocel at 1000 ppm during September followed by KNO_3 at 1 per cent (Akola) or KNO_3 at 2% (Periyakulam) during October recorded maximum number of flowers per shoot and yield.
- Medium pruning (removal of terminal shoots of 15-25 cm) in old Nagpur mandarin recorded significantly highest yield of 114.39 kg/tree/year as against only 79.25 kg/tree/year in control. Increased yield was due to more number of fruits and high average fruit weight.
- In acid lime medium pruning resulted in significantly higher number of fruits (920) with higher fruit yield (29.99 kg/tree).

Banana

- Plant spacing of 1.8m × 1.8m (3086) for Poovan and Kothia, 1.5x1.5m (4,444) for Rajapuri and Basrai, 1.2m × 1.5m × 2.4m (2808) for Gandevi selection, 1.2m × 1.2m × 2.0m (5,208) for Palayankodan, Dwarf Cavendish & Robusta, 2m × 3m (3 suckers/hill) (5,001) for Nendran and 1.8m × 3.6m (3 suckers/hill) (4,629) for Robusta with 300:90:450 g /$N:P_2O_5:K_2O$/hill per year.
- Plant spacing of 2m × 3m with 3 suckers/pit in Nendran and 1.8m × 3.6m with 3 suckers/pit in Robusta recorded higher yield and benefit cost ratio under Kannara conditions.
- Planting of banana from March to May, June and February are found ideal under Assam,

Kovvur and Mohanpur conditions, respectively. Planting of Nendran is remunerative during all the seasons under Kannara conditions.

- Combination of VAM (250 g/plant) + PSB (50 g/plant) + *Azospirillum* (50 g/plant) + *Trichoderma harzianum* (50 g/plant) with 75% recommended dose of fertilizer (RDF - 200:50:200g NPK/plant) for Rajapuri at Arabhavi, Basari at Gandevi, Karpurachekkarakeli at Kovvur are superior for yield.
- At Mohanpur, suckers weighing 1,500g is found to be better planting material (14.8 kg/bunch) for Martaman banana.
- Application of 150–200g N and 200g K_2O in vegetative phase and 50g N and 100g K_2O in reproductive phase were effective in Palayankondan, Rasthali, Grand Naine, Kothia, Rajapuri and Basrai banana under different agroclimatic conditions.
- Application of 25% N (50g) as organic source (FYM/green manure/neem cake) and 75% N (150g) as inorganic source enhanced the yield of banana at different agroclimatic regions.
- Application of 200g each N and K_2O (6 split doses) starting from 60 DAP at 30 days interval for Rajapuri banana, while 300g each N and K_2O/plant in three equal splits at 90, 120 and 150 days after planting for Gandevi selection recorded higher yield.
- Foliar spray of $ZnSO_4$ (0.5%) + $FeSO_4$ (0.2%) + $CuSO_4$ (0.2%) + H_3BO_3 (0.1%) at 3, 5 and 7 months after planting increased the yield of Robusta banana.
- Application of 200g nitrogen through different sources, viz. CAN, ammonium sulphate and urea in the ratio of 0:0:1, 0:1:0, 1:1:2, 1:2:1, 0:0:1, and 0:1:0, resulted in improved yield and benefit: cost ratio under Arabhavi, Jalgaon, Coimbatore, Kannara, Kovvur and Pusa conditions, respectively.
- Drip irrigation is found superior over other methods.
- Application of 75% recommended dose of fertilizer (RDF-200g each of N and K_2O/ plant/year) at Arabhavi (Robusta), Gandevi (Dwarf Cavendish), Jalgaon (Grand Naine) and Kannara (Robusta) and 50% RDF at Kovvur [Karpura Chakkerakeli (AAB, Mysore)] through drip is sufficient.
- In banana best intercrops are onion at Arabhavi, cowpea at Jalgaon, knol-khol and elephant foot yam at Jorhat, *Amorphophallus* and *Colocasia* at Kovvur and cabbage at Mohanpur.
- Foliar application of CCC (1000 ppm) at 4 and 6 months after planting recorded higher bunch weight and number of fingers for Robusta under Coimbatore conditions.
- Studies on chemical manipulation of banana for higher yield and quality revealed that application of 200g N/plant/crop with foliar spray of urea at 2% followed by bunch spray of 2,4-D (25 ppm) recorded higher bunch weight for Grand Naine at Jalgaon and Nendran at Kannara.
- Tissue culture plants proved superior in comparison to suckers in yield, reduction in crop duration and uniformity in harvest.
- To enhance the finger size and bunch weight besides its appearance, covering of bunches with white polythene sheets having 2-6% ventilation is recommended in sub-tropical regions during winter months.
- Potassium dihydrogen phosphate (0.5%) + urea (1%) applied at full bloom stage increases

bunch weight besides early maturation of fruits.

- Growing of cowpea significantly suppresses the weed growth. Integrated method of weed control (cross hoeing + weeding + double crop of cowpea incorporation or chemical spray of glyphosate @ 1-2 kg/ha) is more effective.

Papaya

- A spacing of 1.6m × 1.6m (3900 plants/ha) is economical for CO6 papaya to obtain high papain yield.
- Plants receiving inorganic nutrients alone (200g each of N, P_2O_5, K_2O/plant/year) give highest yield.
- In CO6, biofertilizer (*Azospirillum* and *Phosphobacteria* each at 20g/plant) and organic amendment improve the quality of fruits, latex yield and enzyme activity.
- Spraying of $ZnSO_4$ (0.5%) and boric acid (0.1%) once in two months improves the fruit and latex yield of 9 1(D) under Coimbatore conditions.
- Application of 75% recommended dose of fertilizer (RDF-150g each of N and K_2O/plant/year) at Coimbatore for CO2 give higher yield (33.6 fruits/tree with 1.09 kg of mean fruit weight).

Sapota

- The germination in khirnee is significantly enhanced by soaking the seeds in cow dung slurry for 24 hours or seed treatment with KNO_3 (1.0%) for 8 hours. June to October is best time for grafting.
- The growth of khirnee seedlings is enhanced by the application of GA_3 (200ppm).
- Inarching method of grafting gives good success. June to October is the best time for grafting.
- Significantly highest graft success is recorded when bud sticks are wrapped with newspaper and sealed in perforated polythene bags and stored under refrigerated conditions for 4–8 days besides extending the shelf-life of scion to 11 days for soft wood grafting for Kalapatti.
- Plant spacing of 8m × 4m for 16 year old PKM-1 sapota at Periyakulam give highest yield.
- Application of 600g N, 200g P_2O_5 and 400g K_2O/plant in PKM 1 and Kalipatti give the highest yield.
- Application of 75% recommended dose of fertilizer through drip gives higher yield for Kalipatti sapota.
- At Gandevi, application of 25kg FYM/5kg vermicompost with 300g N and 200g K_2O (75% RDF) with 60g P_2O_5/plant/year recorded higher yield of fourteen year old Kalipatti sapota. However, application of 5kg vermicompost with 150g N and 150g K_2O (75% RDF) with 40g P_2O_5/plant/year for young orchards (less than 10 years) give significantly higher growth and yield for Kalipatti at Arabhavi and PKM-1 at Arabhavi and Periyakulam conditions.

Jackfruit

- Grafting studies revealed that inarching recorded maximum graft take at Kannara (90%), Kovvur (86%) during September and at Mohanpur (85%) using seedling of 2-4 months, during August. While at Periyakulam, graft take was more (50%) in softwood grafting using 4 months old rootstock under 50% shade during October.

Pineapple

- Plant spacing of 30cm × 60cm × 90 cm or 22.5cm × 60cm × 75cm, accommodating 53-63 thousand plants/ha increases the fruit yield to 85–105 tonnes/ha.
- Application of 25 ppm ethephon along with urea (2%) and calcium or sodium carbonate (0.04%) induces > 90% flowering.
- Strategies for pest and disease management have been developed.

Citrus

- Blackfly in citrus is effectively controlled by spraying of Acephate (0.1125%) or Imidacloprid (0.005%) or releases of 4–6 eggs per shoot of *Mallada boninensis*, while, white fly is controlled by spraing of Ethion (0.1%).
- Leaf miner in citrus is effectively controlled by spraying of Imidacloprid (0.005%), followed by NSKE (5%) / *Bacillus thuringiensis* (B.t) at 0.1%.
- Lemon butterfly is effectively controlled by spraying of Endosulfan (0.07%) or *Bacillus thuringiensis* (0.05-0.1%)/Carbaryl (0.15%)/Thiodicarb (0.075%).
- Chlorpyriphos (0.05%) is effective to control mealybug.
- For the effective control of leaf folder, spray of acephate (0.1125%) followed by acephate (0.1125%) or neem oil (1%) and for mealy bug, Chlorpyriphos (0.05%), while for fruit sucking moth, fish oil rosin soap (1%) or neem oil (1%)/ or Karanj oil (1%) is recommended.
- Ethofenprox (0.075%) is found to be consistently effective in the management of citrus psylla.
- Acetamiprid (0.01%) or actara (0.01%) is highly effective in controlling the aphids, while Aldicarb (2 kg/ha) or neem cake (7.5 kg/plant) controls the nematode population in acid lime.
- Among natural enemies for insect pests, predatory beetles like *Chilocorus* spp. and *Brumes* sp. are predominant. *Chrysopa* is highly effective in controlling the aphids. *Verticillium lacanii* effectively controls the green scales during monsoon and winter seasons (80%) under Chethalli conditions.
- For the integrated management of foot rot and gummosis of sweet orange at Rahuri, two sprays of Al Phosetyl (0.2%) was effective.
- Pruning followed by Chlorothalonil (0.2%) spray at monthly interval from May onwards is found effective towards twig blight and Dinocap (0.2%) for felt disease.
- For management of dry root rot of citrus, Chlorothalonil [(0.2%) or Hexconazole (0.4%)] + Carbendazim (0.2%) is effective. Four sprays of Carbendazim (0.1%) at monthly interval is effective for pre-harvest stem end rot.
- Pruning followed by spraying of copper oxychloride (COC) @ 0.3% followed by four

sprays of streptocycline (100 ppm) + COC (0.3%) is recommended for the control of canker in acid lime. Growth of *Xanthomonas citri* is effectively controlled by *Aspergillus niger* cultural filterate followed by turmeric or tagetes leaf extract (5%) under *in-vitro* conditions, while at Rahuri and Periyakulam, tagetes leaf extracts (5%) is more effective than *Aspergillus niger*.

- Studies on the management of *Phytophthora* revealed that treatment with Bordeaux paste to the trunk followed by soil application *Trichoderma viride* followed by Al-phosetyl (0.2%) spray is effective.
- The treatments, trunk injection of tetracycline (100 ppm) during summer with good nutrition (10 kg FYM, 10 kg neem cake, 600g N, 300g P_2O_5, 400g K_2O/plant/year) was found effective for greening, alternatively, hot water treatment (45°C), followed by bud dip in tetracycline HCl (2000 ppm) for 1hour or injection of tetracycline HCl (100 ppm) with foliar application of $ZnSO_4$ (0.5%) is also effective.

Banana

- Effectiveness of chopped pseudostem pieces spread at ground level for the management of pseudostem borer is high only when the field population of the pest is medium to high. Otherwise, use of neemazal at 1% EC (5 ml/litre) or Carbaryl (0.2%) is ideal.
- Scarring beetle was effectively controlled by acephate (0.1125%) either spray or application of 20 ml in the heart of the plant.
- *Helicotylenchus multicinctus* is most predominant and frequently occurring nematode noticed. Paring followed by dipping in acephate (0.1125%) for 45–60 minutes and incorporation of sunhemp at 10g/m^2 (seed rate) at 90 days after sowing give significant reduction in nematode population in soil and root. For disinfection of suckers, paring and pralinage with carbofuran at 40g/sucker is effective.
- The combined treatment of *Paecilomyces lilacinus* along with the organic amendment, neem cake @ 250g/ plant or FYM @ 500g/plant, give maximum reduction of nematodes in banana cv. Robusta under Coimbatore conditions.
- Effective control of Sigatoka leaf spot has been achieved by three sprays of Propiconazole (0.1%) at 30 days interval or alternate spray of Tebuconazole (0.1%)/ Carbendazim (0.1%) and Mancozeb (0.2%) or Propiconazole (0.1%) at 30 days interval or with *Bacillus subtilis* or *Pseudomonas fluorescens* (0.5%) also found effective.
- Studies on induction of Systemic Acquired Resistance (SAR) for the management of leaf spot disease in banana at Kovvur reveals that although SAR elicitors could reduce disease severity they could not sustain the elicitation required for matching the efficacy of the fungicide.
- Application of neemcake (0.5kg/plant) or Carbofuran (40g/plant) and drenching with copper oxychloride (0.5%) gave good control of rhizome rot. Alternatively, dipping the suckers in copper oxychloride (0.5%) and streptomycin sulphate (300 ppm) for 30 minutes is also effective.

Papaya

- Combined application of neem cake (250g) with Carbofuran (1g ai) and *Pseudomonas fluorescens* (4g) give the highest fruit yield besides maximum reduction of nematode population in papaya cv. CO2 under Coimbatore conditions.

Sapota

- Application of Carbosulfan (0.05%) or Fenobucarb (0.1%) or Polytrin-C (0.044%) or profenophos (0.075%) or Acephate (0.1125%) or Endosulfan (0.07%) when sprayed three times at 15 days interval contribute to maximum reduction in incidence of leaf webber, bud worm and fruit flies and recorded higher yield.
- Two sprays of Carbendazim (0.1%) or Propineb (0.2%) or Zineb (0.2%) or Mancozeb (0.2%) or copper oxychloride (0.2%) or Thiophanate methyl (0.1%) at 60 days interval is effective in controlling leaf spot disease.

Pseudostem pieces for the management of pseudostem-borer at Kannara

Jackfruit

- Survey studies for the incidence of pests and diseases indicate that leaf-eating caterpillar, stem borer, spittlebug and its frothy secretion are noticed in jackfruit-growing regions of Tamil Nadu and Kerala. The incidence of fruit rot, leaf spot and anthracnose disease are noticed. Among these, *Rhizopus* fruit rot and anthracnose are major one.
- At Mohanpur, inflorescence rot, leaf spot, pink disease, die-back, fruit rot and fungal wilt are noticed.

TECHNOLOGIES TRANSFERRED, EXTENSION SERVICE RENDERED, MITIGATION TECHNIQUES TO ADDRESS THE CHALLENGES OF FARMING COMMUNITY

Banana

Paired row system of planting (1.2cm × 1.2cm × 2.0 m) and planting three suckers hill at a spacing of 1.8m × 3.6m; application of 150 × 200g N (25% organic and 75% inorganic source) and 200g K_2O in vegetative phase and 50g N and 100 g K_2O in reproductive phase along with foliar spray of $ZnSO_4$(0.5%) + $FeSO_4$(0.2%) + $CuSO_4$(0.2%) + H_3BO_3(0.1%) at 3,5 and 7 months after planting is optimum. Drip irrigation is found superior and fertigation helps in 25% nutrient saving.

Citrus

Plant spacing of 6m × 3m with application of 800-1200g N, 300-400g P_2O_5, 400-600g K_2O/plant/year with 7.5–15kg neem cake is found cost effective. Foliar application of $ZnSO_4$(0.5%) is also found critical. Soil moisture stress (30 days) followed by spray of Ascorbate (50 ppm) at the time of stress alleviation results in summer cropping.

Papaya

Spacing of 1.6m × 1.6m with 200g each of N:P_2O_5:K_2O/plant/year is ideal besides reducing nematode population with use of neem cake. Spraying of $ZnSO_4$(0.5%) and H_3BO_3(0.1%) in two months improves fruit and latex yield.

Paired row system of planting in banana

Sapota

Plant spacing of 8m × 4m is ideal. Application of 600:200:400g N: P_2O_5 :K_2O /plant/ year is optimum. Fertigation with 75% RDF has been found ideal.

Extension service rendered: The staffs working under All India Coordinated Research Project on Tropical Fruits are actively involved in interacting with farmers/developmental officers (agri-horticulture officers). Staff members of AICRP are regularly attending Zonal Research Advisory Committee meeting/ exhibitions/melas and departmental meeting and involved in disseminating the technology through these agencies to farmers. In addition, for better linkage with their clients, the staffs of AICRP are also involved in delivering the radio talks and demonstration of their technology through doordarshan. Many popular articles/ bulletins have also been published by the staff both in english and regional languages.

BENEFIT REAPED BY FARMERS/CONSUMERS AND IN TURN TO NATION AS A WHOLE

Crop Improvement

Output: Promising genotypes, viz. RHR-L-122 and RHR-L-124 in acid lime, CRS – 4 in Khasi mandarin, Selection-2 and selection 4 in mosambi, Himakuntla selection-1 in sweet orange, Manjeri Nendran-II, KBS-4 and KBS 8, Kanthali clone-I and clone II, NPH-02-01 (H 201 × Anaikomban) and H-212 in banana have been identified based on yield potential, quality parameters and resistance to insect pests and diseases.

At Tirupati, acid lime selection Tenali has been released as Balaji and at Akola, Kagzi lime clone has been released as PDKV lime for its yield potential. The spindle shaped sapota identified at Periyakulam from an open pollinated progeny of PKM-1 has been released as PKM-4 and is suitable for flakes preparation.

Outcome: Coordinated approach has helped in identifying the promising genotypes and also release of the regionally preferred varieties. This helped to increase the yield level in tropical fruits, thus achieving nutritional security (see Box).

Crop	Technology	Area covered ('000'ha)	Area covered % of area	Addition to national income (Rs in lakh)
Banana	Development of varieties/hybrids (Kovvur Bontha, BRS-1 and BRS-2).	238.365	45	2287.5
Citrus	Development of varieties (Khagzi lime clone-PDKV lime, Sai-Sharbati, Balaji and Jaidevi).	320.58	45	5583.5
Sapota	Development of variety/hybrids (PKM-1, 2, 3 and 4).	13.31	10	1655
Papaya	Development of varieties/hybrids (CO-1,2,3,4,5,6 and 7)	32.85	45	8.5

Crop Production Technologies

Output: The AICRP centres have developed region-specific technologies on tropical fruits. These include HDP in banana and sapota, nutrient management in citrus, banana and sapota, fertigation in citrus, banana, papaya and sapota and suitable intercrops in banana. In high-density planting of banana, increased density is through more number (2-3) of suckers are planted/pit than just decreasing the spacing between the plants. The technologies developed in production aspects helped in reducing the input costs besides increasing the productivity per unit area. The impact of some of the technologies on national income is estimated as below.

Crop	Technology	Area covered ('000'ha)	Area covered Total Area (%)	Addition to national income (Rs lakh)
Banana	Ideal planting period	52.97	10	60156
	Tissue culture plants as planting material	105.94	20	1290
	HDP	105.94	20	117695
	Nutrient dose	317.82	60	70734
	Fertigation	132.425	25	87718
	Intercropping	79.455	15	42391
	Weed management	211.88	40	3931
Citrus	Planting density	106.86	15	2394.5
	Root stock trial	178.1	25	2740
	Nutrient dose	284.96	40	6489.5
	Fertigation	249.34	35	468
	Intercropping	142.48	20	793
	Weed management	71.24	10	1365
Sapota	Planting density	26.62	20	58
	Nutrient dose	26.62	20	20
	Fertigation	13.31	10	5
Papaya	Planting density	7.3	10	7
	Nutrient dose	14.6	20	5
	Fertigation	7.3	10	2

Crop Protection Technologies

Output: The AICRP Centres have engaged in surveying insect pests and disease problems at different agroclimatic regions. Based on the information, effective management strategies for the control of same were developed. Non-chemical methods have also been developed using bioagents [*Bacillus thuringiensis* (B.t) for leafminer and lemon butterfly in citrus, *Mallada boninensis* (4-6 eggs per shoot) for blackfly in citrus, *Paecilomyces lilacinus* and *Trichoderma viride* for *Phytophthora* in citrus and nematodes in banana and *Pseudomonas fluorescens* for nematodes/Sigatoka leaf spot in banana and papaya] and botanicals [neem oil for leaf folder in citrus, NSKE for citrus leaf miner, Neemazal for pseudostem-borer of banana, neem cake for nematodes of banana and papaya].

Intercropping in citrus-ragi in sweet orange

Intercropping in citrus-sunnhemp in sweet orange

Outcome: The survey results helped in understanding incidence pattern of different insect pests and disease problems at different agro-climatic regions of the mandate crops and helped to formulate the effective strategies in managing the biotic stress. The technologies of crop protection enabled in curtailing losses of yield due to the biotic stress besides producing the pesticide-free produces in tropical fruits (see Box).

Crop	*Technology*	*Area covered*		*Addition to national income (Rs lakh)*
		(000' ha)	*Area (%)*	
Banana	Management of insect pests with chemicals/ botanicals	314.82	60	45870
	Management of diseases with chemicals/botanicals	344.305	65	67908.5
Citrus	Management of insect pests with chemicals/botanicals	498.68	70	15500
	Management of diseases with chemicals/botanicals	284.96	40	4385
Sapota	Management of insect pests and diseases with chemicals/botanicals	46.585	35	30
Papaya	Management of nematodes and diseases with chemicals/botanicals	14.6	20	3

3

Subtropical Fruits

Om Prakash and D. Pandey

The main objective of AICRP (Subtropical Fruits, STF) is to carry out systematic research on national and regional problems in growing mango, guava and litchi, adopting multidisciplinary and multilocation coordinated approach for improving the productivity and quality of fruits and increasing production in the country. In fact, AICRP (STF) is to serve as a national grid of integrated cooperative research by bringing together the national institutes, the state agricultural universities (SAUs) and other concerned organizations for identifying the gaps in research and planning and execution of relevant experiments. The AICRP (STF) envisages following technical programmes:

- Collection, evaluation, documentation and utilization of genetic resources
- Improvement in plant type by breeding
- Standardization of propagation techniques and rootstocks
- Standardization of agro-techniques
- Standardization of use of bioregulators
- Management of pests, diseases and other disorders.

CENTRES OF AICRP ON SUBTROPICAL FRUITS

Centres, crop dealt with and man power under the project are presented in the following tables.

Table 3.1. Centres and crops

Centre	*Establishment year*	*Plan*	*Crops assigned*	*Remarks*
CISH, Lucknow	1972*	IV	Mango, guava	Continuing
IARI, New Delhi	1970*	IV	Mango, grape	Continuing
IIHR, Bangalore	1970*	IV	Mango, grape	Continuing
HETC, Basti	1970	IV	Mango, guava	Closed since1990
HETC, Chaubatia	1970	IV	Temperate fruits	Closed since 1991
MACS(ARI), Pune	1970	IV	Grape	Continuing

(Contd.)

Centre	*Establishment year*	*Plan*	*Crops assigned*	*Remarks*
FRS, Sangareddy	1971*	IV	Mango, guava	Continuing
BAC, Sabour/Pusa	1971*	IV	Mango, guava and litchi	Continuing
RFRS, Mashobra	1971	IV	Temperate fruits	Closed since1991
TNAU, Coimbatore	1971	IV	Grape	Closed since1988
PAU, Ludhiana	1972	IV	Grape, low-chilling temperate fruits	Continuing
GRS, Hyderabad	1975	V	Grape	Continuing
RGRS, Sharbo	1975	V	Grape	Closed since1988
RFRS, Vengurla	1975*	V	Mango	Continuing
SKUA&T, Shalimar	1976	V	Temperate fruits	Closed since1991
FRS, Rewa	1976*	VI	Mango, guava	Continuing
GBPUA&T, Pantnagar	1988*	VII	Litchi, low-chilling temperate fruits	Continuing
AES, Paria	1988*	VII	Mango	Continuing
BCKV, Mohanpur	1988*	VII	Mango, litchi, guava	Continuing
RCA, Udaipur	1988*	VII	Mango, guava	Continuing
MPKV, Rahuri	1989	VII	Grape	Continuing
HC&RI, Periyakulam	1994*	VIII	Mango	Continuing
HARP, Ranchi	1998**	VI	Guava, litchi	Continuing
NRCL, Muzaffarpur	2001**	IX	Litchi	Continuing

*Centres presently working under AICRP (STF)

** Co-optive centres

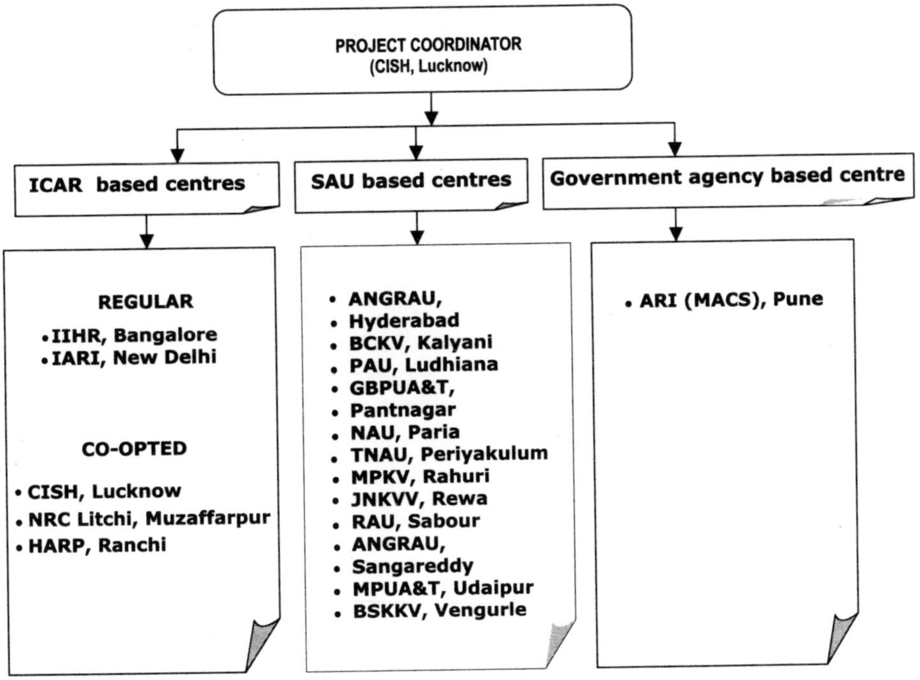

Fig.3.1. Organogram of AICRP (STF)

Man power (Plan wise)

Table 3.2. Approved staff under AICRP (STF)

Category	*IV Plan*	*V Plan*	*VI Plan*	*VII Plan*	*VIII Plan*	*IX Plan*	*X Plan*
Scientific	25	32	50	51	45	45	45
Technical	22	54	40	50	43	44	44
Administrative	-	07	07	08	08	8	8
Supporting	-	32	33	42	32	31	31
Total	47	125	130	151	128	128	128

RESEARCH ACHIEVEMENTS

GENETIC RESOURCES

The genetic resources comprising the cultivars, seedlings, species and related wild types were acquired from indigenous as well as by introduction from abroad (Table 3.3).

The management of genetic resources is essential in order to identify and release suitable varieties for commercial cultivation and to make available the parental material for breeding at present as well as in future.

Fig. 3.2. Centres of AICRP (STF)

Table 3.3. Total collection of genetic resources at different centres

Centre	Number of collections							
	Mango	Grape	Guava	Litchi	Peach	Plum	Pear	Almond
FRS, Sangareddy	430		27					
BAC, Sabour	74			15				
IARI, New Delhi	93	62						
AES, Paria	130							
IIHR, Bangalore	256	103	60					
FRS, Rewa	145		50					
RFRS, Vengurle	233							
MPKV, Rahuri		28						
ARI(MACS), Pune		1284						
PAU, Ludhiana		31			10		50	
RCA, Udaipur	14		19					
HC&RI, Periyakulam	19							
CISH, Lucknow	715	14	96	35				
GBPUA&T, Pantnagar	250		15	21	18	46	44	17
GRS, Rajendranagar			44					
HARP, Ranchi			52	10				
NRCL, Muzaffarpur				40				
BCKV, Mohanpur	18			10				
Total	2361	1600	231	46	28	46	94	17

Important Collections

Mango

In addition to the above, the clones of important varieties of mango, viz. Alphonso, Chausa, Dashehari, Kesar and Langra were collected at Vengurle, Paria, Rewa, Lucknow and Sabour centres respectively. A high-yielding clone of Dashehari, i.e., 'Dashehari 51' having regular-bearing habit with good quality fruits and higher yield has been identified at Lucknow for commercial cultivation. At FRS, Sangareddy, three clones of Benishan, viz. Pedda Benishan, Allahabad Benishan, Early Benishan and one clone of Banganpalli namely Ratti Banganpalli are promising. At FRS, Rewa, clone no. 2 of Chausa was promising with respect to yield and fruit quality. At BAC, Sabour, two chance seedling selections, viz. Subhash from Zardalu and Maneka from Gulabkhas were selected and released for commercial cultivation. Two clones of Himsagar, viz. T19 and T20 were found promising at BCKV, Mohanpur. These selections were made for good fruit quality, high yield and longer shelf-life of fruits. A high-yielding (1,800 fruits/tree) seedling of good quality fruits was identified at Periyakulam. A dwarf type (Malviya Bhog) was collected from Varanasi at IARI, New Delhi. Two local mango seedlings were also identified for pickle and juice making at HC&RI, Periyakulam.

Grape

Important collection of grape included the varieties of world fame, a number of *Vitis* and related species, F_1 hybrids and mutants. The assessment of hybrids and mutants has provided high-yielding material resistant to biotic and abiotic stresses.

Varieties

The genetic resources have been evaluated for both vegetative and fruiting characters. On this basis, some of the varieties have been released for cultivation. The released and promising genetic resources are being described.

Mango

'Lal Sundari' was observed to be distinctly dwarf, regular in bearing, resistant to powdery mildew and free from malformation under field conditions. 'Khajuria', 'Totapuri', and 'Vashibadami' at Paria, 'Zardalu' at Sabour, 'Sunderja', 'Dahiyar', 'Gajaria', 'Madhu Kupi' Rewa and 'Saheb Pasand' at CISH, Lucknow, were found promising. 'Kesar' has been recommended for commercial planting in Maharashtra from Vengurle and 'Sunderja' in eastern Madhya Pradesh from Rewa. "Bangalora", a south Indian variety, performed excellent in yield at BAC, Sabour and FRS, Rewa. At IARI, New Delhi, an exotic cultivar 'Eldon' showed excellent performance with medium-sized fruits (250g), good TSS content (17° Brix), attractive apricot peel with redish blush and long shelf-life (10–18 days). Similarly, 'Mallika' and 'Bangalora' produced maximum yield at most of the centres. Varietal trials indicate that 'Dashehari ripen about a month earlier in Andhra Pradesh than northern states which gives distinct advantage of capturing early market. Apart from that seedling germplasm, i.e., Peddapur-16 and Parigi-16 were recorded maximum yield at Sangareddy, Seedling No. 20/80 at BAC, Sabour whereas, Latif Aliwala and Surkhuroo-1 and Surkhuroo-2 at GBPUA&T, Pantnagar. At HC&RI, Periyakulam, Apple mango (KMI-5), Vellaikaachi (KMI-2), Kalayan Kaachi (KMI-3), Chengavarrukai (KMI-4) and Gundu (KMI-5) were found promising. Genetic donors identified on the basis of assessment are given below:

Dwarfing	*Amrapali, Creeping, Rumani, Kerla Dwarf and Ladavio*
High-yielding	Vashi Badami, Neelum, Banganpalli, Bangalora, Suvarnrekha, Mulgoa, Totapuri, Amrapali and Dashehari
High fruit weight	Neelgoa, Langra, Vashi Badami, Dashehari, S.B. Chausa, Latcampoo, Mallika and Fazali
High TSS	Saheb Pasand and Dashehari
Attractive peel colour	Eldon, Suvarnrekha, Sensation, Vanaraj, Janardan Pasand, Tomy Atkins, and Keitt, Kant.
Processing	Padri, Saheb Pasand and Ram Kela
Resistant to powdery mildew	Totapari Small Red, Totapuri, Tomy Atkins, Vashi Badmi and Hur
Resistant to malformation	Elachi, Dehiyar and Meghlantan

Grape

Flame Seedless at Ludhiana and CISH, Lucknow, Pusa Urvasi, Beauty Seedless and Gold at IARI and Kata Kurgan, Kishmish Red, Kishmish Charni, Kishmish Beli at Rajendranagar, Hyderabad, were found promising among the seedless in collection. At PAU, Ludhiana, early ripening mutant of Perlette was selected from a grower's field. In seeded group, Anab-e-Shahi and Madhu Angoor at GRS, Rajendranagar, and E-9-3 (white seeded) and E-21-28 (coloured seeded) at MPKV, Rahuri and Pusa Navrang at IARI, New Delhi and CISH, Lucknow, has given good yield. At Pune, screening of about 1083 F_1 hybrids of different cross combinations and 93 mutants have given resistant material to downy and powdery mildews. Some outstanding mutants are 'Seedless Anab-e-Shahi', even maturing 'Kali Sahebi' and disease resistant 'Gulabi'. The identified important donors are as follows:

For North India	
• Early ripening	'Beauty Seedless' 'Flame Seedless' 'Perlette' and Early Perlette
• High TSS	'Beauty Seedless'
• High yield	'Rubi Red' 'Beauty Seedless'
For South India	
• High T.S.S.	'E-29-7' 'Golden Queen' 'Athens', 'Bangalore Blue', 'Concord', 'Catawaba' and 'Karandu'
• High yield	'E-26-8', 'E-30-14', 'E-29-5', 'Dilkush' and 'Angur Kalan'
• Suitable for juice	'E-16/20', 'G-27/5', 'E-4/15'
• Suitable for raisin	'E-29/5', 'E-12/7', 'E-31/5', 'Arkavati', 'Kishmish Charni', Tas-e- Ganesh'
• Suitable for wine	'E-22/10', 'E-28/2', (White hybrids), 'E-18/10', 'E-16/20', 'E-18/8', (Red hybrids)
• Pests and diseases resistant	'Bangalore Blue', 'Catawba', 'Concord', 'Khalili', 'Cambell'
• Drought and salt tolerant	'Digraset'-Dogridge

Guava

Allahabad Safeda' and L-49 (Sardar) have been assessed as high-yielding cultivars at most of the centres. However, Lalit and Sweta released from CISH, Lucknow, and CISH-G-1 a promising seedling selection, have shown promise at CISH, Lucknow. At FRS, Rewa the cultivar Chittidar, Dhareedar and Rewa-72 are being recommended for commercial cultivation. Some newly released cultivars are as under:

Arka Amulya: This variety was evolved at the IIHR, Bangalore. Tree is semi-vigorous and heavy-yielder. Fruit is medium (180–200g) with white and sweet (TSS 12° Brix) pulp, a few soft seeds (100 seeds weighing 1.80g) and keeping quality good.

Arka Mridula: It is a seedling selection of 'Allahabad Safeda' released from the IIHR, Bangalore. Fruit is smooth and medium (180–200g) with soft seeds (weight of 100 seed is 1.60g), white and sweet pulp (TSS 12.5 °Brix) and good pectin content (1.04%). Keeping quality good.

Dhareedar: It is selected from old seedling orchard of Fruit Research Station, Kuthulia, Rewa. Tree is vigorous, medium tall (4.5-5.5m) with erect and upright branching and flat crown. Leaf medium in size (10.2cm × 4.7cm), elliptic ovate to oblong. Fruit medium to large (7.78cm × 8.23cm) with 159g mean fruit weight, roundish ovate, 5–7 raised lines on the surface of mature fruit, peel greenish yellow, pulp soft, sweet (TSS 11.7%, reducing sugar 4.31%, non-reducing sugar 5.42%, acidity 0.32%, ascorbic acid 200.08mg/100g fruits).

Kohir Safeda: This variety was evolved at FRS, Sangareddy. Tree is large and bears profusely (90-100kg/tree). Fruit is large with hard seed.

Lalit: This is a chance seedling selection from the population of a red-pulped guava collected from Allahabad area and raised at the CISH, Lucknow. Tree is heavy yielder with 24% higher yield than Allahabad Safeda. Fruit is attractive saffron yellow with red blush and medium in size (185g). Pulp is firm and pink with good blend of sugar and acid. It is suitable for both table use and processing for making beverage and jelly. Pink colour in the beverage made from its pulp remains stable for more than a year during storage.

Pant Prabhat: This variety was selected at the GBPUA&T, Pantnagar (Uttarakhand). Plant growth is upright with broad leaves, tree highly productive (100–125kg), fruit round, peel smooth and light yellow in colour, fruit medium (150-175g), pulp white, seeds small and soft as compared to Sardar, taste sweet with pleasant flavour, ascorbic acid content varies from 125mg (rainy season) to 300mg/100g fruit weight (winter season). TSS varies from 10.5 to 13.5 °Brix.

Safed Jam: This variety was evolved at FRS, Sangareddy. Tree is medium in height and good yielder (80.90kg). Fruit is round with thin peel and good taste. Seed is soft and located at the core.

CISH-G-1: This is a seedling selection of made by progeny grown from seeds of coloured guava. Tree medium height with sparse foliage. Fruit medium-sized; attractive red coloured with soft seed, TSS 15° Brix and having good shelf-life.

Sweta: This is a chance seedling selection of Allahabad Safeda released during 2005 at CISH, Lucknow. High-yielding attractive guava variety with globose fruits, medium sizes, creamy white flesh, high TSS (12.5-13.2° Brix) and vitamin 'C' (300mg/100g pulp) with good keeping quality.

Temperate fruits: The varieties and selections recommended for cultivation are given Table 3.4.

Table 3.4. Varieties and selections recommended for cultivation in temperate fruits

S. No.	*Fruit*	*Variety/Selection*	*Characteristics*
1.	Apple	'Tydeman's Early Worcestor',	Early-season
		'Starking Delicious',	Mid-season
		'Starkrimson'	Mid-season
		Lord Lambourne'	Mid-season
2.	Pear	'Baghu Ghosa, Punjab Beauty	Mid-Season
		'Gola', Pant Pear-17, Pant Pear-3	Low-chilling
3.	Almond	"Non Pariel'	High-yielding
		'California Papershell'	Thin-shell
		'HS-8'	Low-chilling
4.	Apricot	'Nugget'	Early-season
		'New Castle'	Low-chilling
5.	Plum	'Black Chamba"	Good-quality
		Victoria'	Good-quality
		'Sweet Early',	Early-season
		'Titron', Satluj Purple	Low-chilling
6.	Peach	'Sure Crop'	High-yielding
		'Flordasun', Early Grande, Florda Prince	Low-chilling

Documentation: The list of descriptors has been prepared on the pattern of International Plant Genetic Resources Institute (IPGRI) and supplied to various centres for recording of data. A catalogue on 152 cultivar of mango using 54 characteristics have been prepared by IIHR, Bangalore and 256 cultivars using 56 characters at CISH, Lucknow.

Utilization

Mango

Evaluation at different centres has enabled the identification of suitable cultivars for commercial cultivation in different region (Table 3.5).

Table 3.5. Cultivars recommended for commercial cultivation in different regions

Region	*Mango*	*Grape*
Northern India	Dashehari, Dashehari-51, Chausa, Langra, Lucknow Safeda, Bombay Green	Perlette, Beauty Seedless, Flame Seedless
Western India	Kesar, Alphonso, Pairi, Zardalu, Vashi Badami.	Thompson Seedlesss, Beauty Seedless
Southern India	Neelum, Bangalora, Banganpalli, Mulgoa, Suvarnrekha	Thompson Seedless, Anab-e-Shahi
Eastern India	Himsagar, Langra, Fazali, Bombay Green, Kishan Bhog, Suvarnrekha, Zardalu	

Multilocational trials of commercial varieties of different regions have shown that some of the north Indian varieties are poor performer under south Indian conditions and they put forth excessive vegetative growth, e.g., 'Zardalu' at Sangareddy. However, the performance of 'Neelum' and 'Bangalora' under north Indian conditions is desirable in yield and regularity of bearing. These remain dwarfest among the varieties being tested in the trial. Mango cv. Amrapali has been found suitable for high-density planting in northern India at IARI, New Delhi.

Mango

The mango genetic resources have been further utilized as Kerala Dwarf was found dwarfest with agreeable fruit quality. Hence, this has been used in breeding programme to impart dwarfing characters to the progeny. Various coloured varieties, viz. Eldon, Janardan Pasand, Lal Sundari Suvarnarekha, Sensation, and Vanraj have been used in hybridization for developing varieties having red peel colour. The cultivars Totapuri, Amrapali and Neelum have been used to impart regular bearing in hybrids. In addition to this, Elaichi and Bhadauran have been used in mango breeding programme for imparting resistance to mango malformation.

Guava

'Allahabad Safeda', Chittidar and 'Sardar' (L-49) cultivars of guava are suitable for cultivation throughout the country.

Litchi

The evaluation of germplasm revealed that Shahi outyielded at HARP, Ranchi, Rose Scented at GBPUA&T, Pantnagar, Bombai at BCKV, Mohanpur, Dehradun and Purbi at CISH, Lucknow, China and Shahi at RAU, Pusa.

Grape

(i)The grape cv. Digraset has been utilized as a rootstock for salt and drought tolerant. The two *Vitis* spp., viz. *Vitis longi*, *Vitis tielifolia* and two rootstocks, viz., berlanderii × riparia and riparia × rupestris have shown promise for drought tolerance. Apart from that four cultivars, viz. Bianshirai, Bhokari, Black Monukka and Motia were found to be promising for their drought tolerance capacity (ii) three hybrids, viz. H-533 (Tiliefolia × Gulabi), H-8 (Khalili × Champbell) and H-137 (James × Champbell) have been found resistant to mildews at ARI (MACS) Pune. These hybrids will be further utilized in backcrossing programme to improve the fruit quality.

IMPROVEMENT/BREEDING

Mango

Improvement of mango through breeding has been continuing at IIHR, Bangalore; IARI, New Delhi; RFRS, Vengurle and AES, Paria. At IIHR, Bangalore, to impart dwarfing character to the progeny variety, Kerala Dwarf has been used as a parent. Besides, polyembryonic cvs. Chandrakaran, Kensington, Peach and Vellaikolumban have been crossed with the cultivars Kesar, Banganpalli and Alphonso as female parents in the crossing programme. As a result of breeding, 18 hybrids have been released as varieties (Table 3.6).

Table 3.6. Mango hybrids as varieties

Variety	*Parentage*	*Releasing Centre & Important Characteristics*
		BAC, Sabour
Al Fazali	Alphonso x Fazali	Good quality and late maturing, free from malformation and fruit fly.
Sundar Langra	Langra x Sunder Pasand	Moderate and regular bearing.
Safari	Gulabkhas x Bombai	Medium fruit size with red blush on the base.
Jawahar	Gulabkhas x Mahmud Bahar	Semidwarf, precocious and regular bearer.
		AES, Paria
Neelphonso	Neelum x Alphonso	Regular bearer, dwarfer, late maturing, suitable for table and juice.
Neeleshan	Neelum x Baneshan	Dwarf, late maturing with good keeping quality.
Neeleshwari	Neelum x Dashehari	Dwarf, late maturing with non fibrous, firm texture and very good fruit quality.
		FRS, Sangareddy
Manjeera	Rumani x Neelum	Regular, heavy bearing in bunches and suitable
AU Rumani	Rumani x Neelum	for high density planting.Fruit large, heavy yielder and fibreless.
		IARI, New Delhi
Mallika	Neelum x Dashehari	Heavy bearer, big fruit size and good quality fruits.
Amrapali	Dashehari x Neelum	Regular bearer, dwarf suitable for high density planting.
Pusa Arunima	Amrapali x Sensation	Regular bearer, high quality, good yield, attractive colour.
		IIHR, Bangalore
Arka Aruna	Banganpalli x Alphonso	Dwarf, regular, suitable for high density planting and free from spongy tissue.
Arka Anmol	Alphonos x Janardhan Pasand	High quality and free from spongy tissue.
Arka Puneet	Alphonso x Banganpalli	Red blush, high keeping quality and free from spongy tissue.
Arka Neelkiran	Alphonso x Neelum	Regular bearer, late harvesting, free from spongy tissue.
		RFRS, Vengurle
Ratna	Neelum x Alphonso	Regular bearer and free from spongy tissue.
Sindhu	Ratna x Alphonso	Regular bearer and free from spongy tissue, early maturity, seedless.
		HC&RI, Periyakulam
PKM-1	Chinnasuvarnarekha x Neelum	Regular bearer, high yielder and bearing in clusters.
PKM-2	Neelum x Mulgoa	Regular bearer with excellent fruit quality.
		CISH, Lucknow
Ambika	Amrapali x Janardan Pasand	Fruit is bright yellow with dark red blush, flesh firm with scanty fibre, regular bearer and suitable for export.

Two cultivars, Amrapali and Sindhu have been highly acclaimed because of their commercial outstanding characteristics. 'Amrapali' produces high quality fruits and is regular bearer of dwarf plant type being suitable for high-density planting. 'Sindhu' bred at RFRS, Vengurle, is a fancy variety as it is considered the first seedless variety of mango. Its flesh is firm, deep orange and fiberless. It has high pulp to stone ratio (26:1) and tends to bear regularly. The stone is very small and papery with an average weight of only 6g. The variety has good qualities of 'Alphonso' and freedom from spongy tissue disorder. At GBPUA&T, Pantnagar, two seedling selections, viz. Pant Sinduri and Pant Chandra, were made recently.

Pant Sinduri: It is a seedling selection from unknown origin. Tree semi-dwarf, spreading type, regular bearer, leaf yellowish green, inflorescence pink coloured, average fruit weight (200g), fruit shape elliptical with light pink colour while rest of the fruit is yellowish-green colour, TSS (18° Brix). The fruit matures from last week of may to first week of June and having yield potential of 1.5 q/tree.

Pant Chandra: It is a seedling selection of unknown origin. Tree vigorous with upright growth habit, irregular bearer, panicle colour light yellow, leaf colour dark green, average fruit weight (160g), fruit shape elliptical oval and colour of fruit at maturity is light green, TSS (19° Brix), fruit mature in the mid of June with yield potential of 1.5 q/tree. Number of promising hybrids are under evaluation (Table 3.7) which will give even better varieties than the past.

Table 3.7. Evaluation of promising mango hybrids

Name of the Institute	*Hybrid Number*	*Parentage*	*Characteristics*
IARI, New Delhi.	H-13-7	Amrapali x Sensation	High quality, good yield, attractive colour
	H-8-11	Amrapali x Lal Sundari	High quality, good yield, attractive colour
IIHR, Bangalore	H-11-2	Alphonso x Neelum	High fruit quality
	H-20-8	Alphonso x Neelum	High fruit quality
	R1P 8	Alphonso x Kensington	Good quality and high pulp recovery
CISH, Lucknow	CISH-M-2	Dashehari x Chausa	Fruit dark yellow in colour with firm flesh and scanty fibre, late season variety with good commercial value
	H-39	Amrapali x Vanraj	Regular bearer, fruits attractive red colours, good fruit quality and export potential
	1084	Amrapali x Janardan Pasand	Fruits having deep red colour regular bearing, dwarf type high good shelf life with export potential
	949	Amrapali x Vanraj	Fruit having dark red colour on peel, regular bearing, dwarf and late in maturity having export potential

Grape

Breeding is being carried out at IIHR, Bangalore; IARI, New Delhi and MACS, Pune. The emphasis is given on development of early ripening and seedless varieties at IARI, New Delhi, whereas at IIHR, Bangalore, the objectives are to evolve seedless table and wine varieties resistant to mildews. At Pune, resistance to both biotic and abiotic stresses is taken into consideration. The promising hybrids/mutants under evaluation have been given in Tables 3.8 and 3.9.

Table 3.8. Evaluation of promising grape hybrids

Hybrid	*Parentage*	*Promising Characteristics*
H-233	Anab-e-Shahi x Catawba	Fruit quality and disease resistance.
H-245	Anab-e-Shahi x Catawba	Fruit quality and disease resistance.
H-246	Anab-e-Shahi x Catawba	Fruit quality and disease resistance.
H-146	Cheema Sahebi x Catawba	Fruit quality and disease resistance.
H-335	Gulabi x Catawaba	Fruit quality and disease resistance.
H-87	Gulabi x V. lanta	Fruit quality and disease resistance.
H-321	Gulabi x James	Fruit quality and disease resistance.
H-554	Khalili x Bhokari	Good fruit quality and yield.
H-580	Khalili x Bhokari	Good fruit quality and yield.
H-1342	Concord x Kismish Beli	Good fruit quality and yield.
H-1301	James x Kismish Beli	Good fruit quality. and yield
H-541	Khalili x Bhokari	Salt tolerance
G-68/2	Black Champa x Rubired	Wine
G-68/4	Black Champa x Rubired	Wine
G-102/1	Queen of Vineyard x Thompson Seedless	Seedless, crispy and flavored pulp
Arka Shewta (E-11/29)	Anab-e-Shahi x Thompson Seedless	Table grape and seedless
Arka Majestic (E-14/20)	Angur Kalan x Black Champa	Table grape
Arka Chitra (E-5/4)	Angur Kalan x Anab-e-Shahi	Table grape
Arka Soma (E-9/3)	Anab-e-Shahi x Queen of Vineyards	Wine grape
Arka Trishna (E-21/128)	Bangalore Blue x Convent Large Black	Wine grape
Arka Krishna (E-30/16)	Black Champa x Thompson Seedless	Juice grape
(17-50)	Banqui Abyad x Beauty Seedless	Table and raisin
(76-1)	Hur x Cardinal	Table and rasin
Pusa Urvashi (76-64)	Hur x Beauty Seedless	Table and rasin
Pusa Navrang (76-2)	Madelein Angevine x Rubired	Juice and wine
(76-3)	Madelein Angevine x Rubi Red	Juice and wine
(75-32)	Banaui Abayad x Perlette	Juice and wine

Table 3.9. Promising mutants of grapes

Mutant	*Promising Characteristics*
AM-552,610	Berry weight and resistant to downy mildew
AM-527, 586	High TSS
AM-302	Seedless, high TSS
GM-487, 491, 502	Even maturity
GM-511,534, 534, 538, 484	High TSS
GM-466,502, 511, 540	Resistant to fungal diseases
KM-383, 388, 437	Even maturity
KM-403, 410, 383	Large berries
KM-383, 432, 450	High TSS
ARI-302 (Anab-e-Shahi Seedless mutant)	Seedless

Interspecific hybridization has resulted in desirable hybrids for maturity, fruit quality and disease resistance. The donor parents like, 'Catawba', 'James', and *V. lanata* have been used for introducing fruit quality and disease resistance in otherwise promising and well adopted cultivars. Mutants of 'Anab-e-Shahi', 'Kali Sahebi' and 'Gulabi', have been selected for even maturity, berry characteristics and disease resistance.

In order to extend the crossing season, 'off' season crossing of 'Hur' × 'Beauty Seedless' was attempted at IARI, New Delhi, which gave 313 F_1 seeds. To study the possibility for parthenocarpically development of berries with sub seedless hybrids, viz., BA × Victory-(73-23), BA × Per. (72-160) and Katta Kurghan × Gold-(75-9) and functionally female seedless hybrid B.A. × BS-(72-30), BA × BS-(70-45) and BA × Per. (75-4) were selfed for two years. There was no berry set in functionally female hybrid while in subseedless hybrid, there were normal seeds/seedless berries. Subseedless hybrids when emasculated and bagged, the berries failed to develop parthenocarpically.

Guava

Breeding work in guava carried out at IIHR, Bangalore has resulted in a good selection and in identifying few hybrids. Arka Mridula is an outstanding, quality, high-yielding selection from the open-pollinated progeny of 'Allahabad Safeda'. Its flesh has high TSS (12.5 Brix) and good keeping quality. The details about the promising hybrids under evaluations at BAC, Sabour are given in Table 3.10.

Table 3.10. Details of promising guava hybrids

Hybrid Number	*Parentage*	*Characteristics*
Hybrid 1	Allahabad Safeda x Seedless	Heavy yielder, soft seeds
Hybrid 45	Allahabad Safeda x Sardar	Seeds soft, white flesh and good size fruits
Hybrid 53	Allahabad Safeda x Sardar	Seeds soft, white flesh and good size fruits
Hybrid 58	Allahabad Safeda x Sardar	Seeds soft, white flesh and good size fruits
Hybrid 84	Allahabad Safeda x Sardar	Seeds soft, white flesh and good size fruits
Hybrid 105	Allahabad Safeda x Sardar	Seeds soft, white flesh and good size fruits
Hybrid 161	Apple Colour x Allahabad Safeda	Red skin colour, excellent quality soft seeds

Interspecific hybridization have been started in order to develop a wilt resistant dwarfing rootstocks. *P. chiennsis* have been crossed to *P. molle* and *P. guajava* cv. Beaumont. Some of the hybrids of these crosses are under observation at IIHR, Bangalore. It was observed that one F_1 hybrid seedling of *P. chinensis* × *P. guajava* was dwarf (40.1cm) with short internode (0.5cm). This hybrid produced suckers and may used as dwarfing rootstock for guava. Recently at CISH, Lucknow, one inter specific hybrid between *Psidium guajava* × *Psidium molle* has been developed which have tolerance against guava wilt and also graft compatible with different cultivars of guava.

Litchi

As a planned hybridization programme at BAC, Sabour, a cultivar Sabour Madhu (Purbi × Bedana) has been developed. This is a very late maturing variety. Fruit is medium with 68.6% pulp, 21.0% TSS with light red colour. Apart from that another cultivar, i.e., Swarna Roopa has been developed through clonal selection at HARP, Ranchi. It is an early maturing cultivar. It bears 18-20 attractive fruits in cluster, protuberances completely flattened at ripening. Fruit is round, medium (3.0 × 2.7cm, 18.95g), highly sweet (TSS 19.0%) and contains high pulp content (76.62%) with medium acidity (0.39%). The pulp contains total sugar (12.5%), reducing sugar (8.5%) and ascorbic acid (38.30mg/100g). It has small and flat seed (3.1g) with dark brown colour. Fruit is ready to harvest in during third week of May under Ranchi conditions.

Temperate Fruits

Apple is the most important temperate fruit. Therefore, attempts have been made for its improvement through breeding at three centres, viz. HETC, Chaubattia; RFRS, Mashobra and RRS, Shalimar. At Chaubattia, Early Shunberry was crossed with Red Delicious and Chaubattia Princess and Chaubattia Anupam varieties have been released. At Mashobra, the famous variety Ambri was crossed with Richared, Starking Delicious, Red Delicious and Royal Delicious which resulted into the release of four varieties, i.e., Ambrich, Ambstarking, Ambred and Ambroyal. Likewise, at Shalimar, hybridization was done with an objective to improve the variety Ambri, which was crossed with Red Delicious and Golden Delicious varieties. Lal Ambri (Red Delicious × Ambri) and Sunheri (Ambri × Golden Delicious) varieties were evolved and released for cultivation. In other temperate fruits, Pant Peach 1 and Pant Plum 1 have been identified as promising lines at GBPUA&T, Pantnagar. 'Florda Prince' and 'Early Grande' peach have been released by PAU, Ludhiana, for general cultivation in Punjab.

PRODUCTION TECHNOLOGY

Mango

Propagation methods: From the trials carried out on propagation methods in different agro-climatic regions, veneer grafting has been found suitable for Uttar Pradesh and Hyderabad region of Andhra Pradesh, Veneer grafting and patch budding for Madhya Pradesh and epicotyl grafting for Konkan region of Maharashtra. Patch budding and epicotyl grafting for mango in eastern Madhya Pradesh and Konkan region respectively, have been adopted on commercial scales. At BAC, Sabour; FRS, Sangareddy and FRS, Rewa, veneer grafted trees gave the maximum yield, whereas trees raised by air layering produced more yield at GBPUA&T, Pantnagar.

The studies on the performance of plants propagated by different methods have shown that in the initial stage veneer grafted plants grew faster than that of other methods. But in the long run the differences among the methods with respect to growth were non-significant except that the air layered plants make slightly less growth. But yield differences attained these years (1987–2001) showed non-significant differences at all locations.

Rootstocks and inter-stocks: The need of a dwarfing clonal rootstock has been long felt in mango, particularly for high-density planting. Long-term rootstock trials conducted at various centres have given varying results. 'Vellaikolamban' recorded the lowest values of growth as well as fruit yield of 'Alphonso' scion, whereas 'Olour' gave the maximum yield and canopy height, at IIHR, Bangalore. 'Latra' proved to be a dwarfing rootstock, while 'Kalapadi' gave the maximum yield in its combination with 'Bombai' at BAC, Sabour. Olour closely followed by random rootstock and Kalepad (Dwarf) recorded minimum growth on Langra scion at FRS, Rewa. 'Rumani' rootstock gave low values for various growth characters with 'Dashehari' scion at CISH, Lucknow.

At IARI, New Delhi, minimum growth was obtained in cv. Amrapali by using 'Anupam' as an interstock. Interestingly, this interstock gave the tallest plants with 'Mallika' scion. 'Chausa' as interstock, gave the highest yield of 'Amrapali' scion. 'Mahmooda Vikarabad' as an interstock

Pant Sinduri

Ambika

Hybrid-1084

Hybrid-39

Hybrid-949

Neelshan (GUJ)

Neelshwari

Neelphonso

Popular varieties of mango

Lalit

Pant Prabhat

Shweta

Hybrid 161

Popular varieties of guava

Rose scented litchi

Late seedless litchi

Popular varieties of litchi

for 'Benishan' and 'Dashehari' scions gave shortest plant at Sangareddy. 'Royal Special' also proved to be dwarfing rootstock at Sangareddy. At CISH, Lucknow, among various combinations tried with different interstock on 'ST-9' and 'Chandrakaran', rootstocks with 'Dashehari' scion. ST-9 produced dwarfer plant while 'Kurkkan' on 'Chandrakaran' gave taller plants. Interstock imparted significant differences in yield on same rootstock. On 'Chandrakaran', rootstock 'Movandan' interstock gave maximum yield while for 'ST-9' rootstock 'Pahutan' interstock gave more yield.

High density–planting: High-density trials in mango particularly with 'Amrapali' have been planted at IARI, New Delhi; FRS, Rewa and BAC, Sabour. High-density trial laid out at IARI, New Delhi, with 'Amrapali' (1600 plants/ha.) has continued to out yield (22.96 ha.) the other varieties, i.e., 'Mallika' (833 plants/ha) and Neelum (2000 plants/ha.) 'Amrapali' appeared to be the most suitable variety for high-density orcharding in northern India. In planting system cum high-density trial, double hedge row system has given significantly higher yield than other systems at most of the centres.

Pruning: At IARI, New Delhi, 'Amrapali' responded to pruning very well. Pruning increased the number of fruits by 56.6%. At Pantnagar, high-density trial pruning after harvest gave the maximum yield, followed by the application of Paclobutrazol (3.2-4.0ml./m. canopy diameter) in Dashehari.

Nutrition: Leaf sampling technique for monitoring nutrient requirements has been standardized. The fertilizer trials have not given consistent results. No response of nutrient application was observed at CISH, Lucknow, in 'Dashehari' and 'Langra'. However, N is required to be applied on the lateritic soils or poor soils deficient in N.

Intercropping: Intercropping studies conducted over a number of years have shown the highest monetary return from black gram-wheat in mango system in cereal and pulses and brinjal-onion-mango in vegetables. At FRS, Sangareddy, intercropping with brinjal and *bhindi* (okra) in *kharif* and onion and tomato in rabi season in young mango plantation was found to be an optimum combination in terms of production. Onion-radish-lentil was found the optimum inter crop combination at BAC, Sabour.

Use of plant regulators: The experiments on regulation of shoot growth and flowering habits with the help of paclobutrazol, i.e., cultar which have been conducted at different centres on different mango cultivars. Results over number of years have shown that this chemical, as in case of many other fruits, can help in successfully overcoming the problems of irregular bearing. The soil application of the chemical has proved superior to foliar application and the soil application has also shown residual effect over a number of years. The application of cultar through soil @ 5 or 10g during July-August has reduced the vegetative growth, induced 3-4 weeks early, profuse flowering, regular and higher yield than the control trees. In case of 'Alphonso' at Vengurle, a regular yield which was 2.8 times more than the control was obtained with the treatment. Similarly, application of Paclobutrazol (3.2–4.0 ml/m canopy diameter) through soil has given the maximum yield at FRS, Sangareddy; IIHR, Bangalore; CISH, Lucknow; BCKV, Mohanpur; HC&RI, Periyakulam and BAC, Sabour.

Based on the results the soil application of Cultar @ 5g/tree on 'Alphonso' cultivar of mango for Maharashtra is recommended for regular bearing, early flowering and higher yield.

Studies on the effect of photosynthesis improving chemicals in mango have shown that

20ml cytozyme gave significantly higher yield when compared with control. Whereas, at Sabour 1ml/10 litres, Vipul gave maximum yield in 'Langra'.

Control of insect pests: The survey and surveillance of insect pests attacking mango and the crop losses resulting thereby have shown that the hopper, mealy bug, fruit fly, shoot-borer, shoot gall psylla and leaf-webber are serious pests of mango. The extent of crop losses varies from region to region and year to year. The insecticides control measures have been standardized for all the pests. The efficacy is affected because of untimely and non-spray of surrounding orchards of stray mango trees allowing the increase in population of pests for subsequent attack.

Hopper (*Amaritodus atkinsoni*). This is a most serious pest of mango, which directly affects the yield by reducing fruit set. It is widely prevalent all over the country. The adults and nymphs suck the sap from the tender shoots, panicles and secret honey dew which causes the development of sooty mould on leaves and panicles. The appearance of black foliage is a clear cut indication of its attack.

The hopper can be controlled by spraying of Imidacloprid (0.005%), methyl parathion (0.025%), Monocrotophos (0.025%), Fenitrothion (0.025%) or Carbaryl (0.2%) before the opening of flowers and after fruit set.

In IPM, lowest hopper population was recorded in module-I, i.e., first spray at panicle emergence stage with Imidacloprid (0.005%), second spray of NSKE (5%) after 21 days of first spray and third spray with Endosulphan (0.07%) when fruits are at pea. The gave good results against mango hopper at RFRS, Vengurle, BAC, Sabour, AES, Paria and FRS, Sangareddy.

At CISH, Lucknow, ETL study of mango hopper, revealed that less than 10 hoppers per infloresence is below the economic injury level.

Mealy bug (*Drosicha mangiferae*) is another major pest which causes severe damage to mango crop throughout the country. The nymphs and adults suck the plant sap, reducing the vigour of plant and also infest the crop during flowering season and if, the control measures are not taken timely the entire crop may be damaged. The mealy bug can be controlled through cultural practices such as summer ploughing of orchards, which exposes depositing egg masses to sun resulting in their destruction. Soil treatment of methyl parathion dust 2% or aldrin dust 10% @ 250gms/tree around the trunk and spraying of 0.05% Monocrotophos or 0.2 % Carbaryl or 0.05% methyl parathion has been found very useful in controlling early instar nymphs of the mealy bug.

Fruit fly (*Bactrocera dorsalis*) The oriental fruit fly is one of the most serious pest of mango in the country which has created problems in the export of fresh mangoes. The female fly punctures the outer wall of the mature fruit with the help of its pointed ovipositor and inserts eggs in small clusters inside the mesocarp of mature fruits. The larva hatching out of the eggs feeds upon the pulp of the fruits and the fruit start rotting before ripening. Such fruits finally drop down or become unmarketable after harvesting.

At IIHR, Bangalore, it has been found that mango fruit fly can be controlled effectively by integrating three components, viz. spraying with Deltamethrin (0.002%) during pre-ovipositor stage (first week of April), collection of fallen infested fruits and plugging of basins to destroy pupae wooden block methyl eugenol traps soaked in ethanol, methylo eugenol, Malathion

(6:4:1) for 72 hours and hung on trees @ 10 traps/ha from April to July reduced the infestation of fruit flies to great edict at CISH, Lucknow.

Shoot-borer (*Chlumetia transversa*): This is a pest of wide occurrence which causes damage through its larva which after hatching enters the midrib of leaves and later into young shoots or sometimes into the inflorescence stalk. The affected plant parts wilt and dry up adversely affecting the growth and fruiting of trees.

Based on three years data generated at RFRS, Vengurle, two sprays of quinalphos (0.5%) at 15 days interval starting from the emergence of new growth are recommended for the management of the shoot borer.

Shoot gall psylla (Apsylla cistellata): The pest is very serious in many parts of northern India, particularly in tarai region of Uttar Pradesh, North Bihar and West Bengal. As a result of feeding of nymphs during August-September, the buds develop into hard conical green galls inhibiting the development of floral buds and thereby fruit set. Three sprays of Quinalphos or Monocrotophos (0.05%) at ten days interval during the month of August have been found to be the most effective in controlling this pest.

Leaf webber (Orthaga euadrsalis): The pest has become serious only in the recent past. The caterpillars on hatching feed on leaf surface by scrapping and later web tender shoots and leaves together and feed within. Infestation begins in the month of April and continues up till December.

For the control of leaf webber, Carbaryl (0.2%) or Monocrotophos (0.04%) or Decamethrin (0.015%) may be sprayed thrice, first by the end February, second in the first week of March and third in the third week of March.

Diseases

Powdery mildew, anthracnose and bacterial canker are serious diseases of mango. Screening of mango varieties against these diseases has shown that some of them are resistant under field conditions. The resistance is however to be ascertained by artificial inoculation.

Powdery mildew: The disease is found in all mango-growing regions and can destroy the crop completely in severe infection. It attacks all parts of the panicle and because of tenderness, floral bud, flowers and fruit set are more affected. Sometimes young leaves and shoots are also attacked. High humidity, cloudy weather and low night temperatures(17-23.86 °C) favours the incidence of mildew.

For the control of powdery mildew, three sprays of Bayleton (0.1%) are recommended at fortnightly interval beginning from the panicle emergence at CISH, Lucknow. Four varieties, viz. Totapuri, Totapuri Red Small, Vashi Badami and Ladavio have shown field resistance against powdery mildew at BAC, Sabour.

At CISH, Lucknow, minimum, optimum and maximum temperatures for the germination of spores were 10, 21 and 32 °C respectively with more luxuriant growth in shade and spores dislodged with the increase in light. At BAC, Sabour, Mahmood Bahar, Kesar, Neelum, Samarbehist Chausa, Sensation and Bangalora showed tolerance to powdery mildew having 1–15% disease intensity. Similarly RFRS, Vengurle, incidence of mildew was observed on vegetative flush and the time required for the maximum disease manifestation ranged from 21–45 days. Fungal spores remained viable without any colateral host up to 3 months.

Anthracnose: The various symptoms of anthracnose include leaf spots blossom blight, twig blight and fruit rot which occur under favourable climatic conditions of high humidity, high temperature (24–32 °C) and frequent rains at the time of flowering and vegetative growth. The fungus perpetuates both on living and dead leaves and twigs. Anthracnose can be controlled by spraying of Bavistin (0.1%), or Dithocarbamate (0.2%). Before storage, the fruits can be dipped in benomyl solution (1000ppm) to avoid infection.

Bacterial canker: The infection takes place on leaves, leaf stalk, twigs, and branches and on fruits. Initially water soaked lesions appear which later turn dark brown developing into canker. The affected fruit become unmarketable. Streptocycline (300ppm) alone gave the best control of bacterial canker in mango at CISH, Lucknow.

Mango malformation: This malady is very serious in northern India, particularly around Delhi and Punjab. It has defied the solution so far because no remedial measures have been evolved. The incidence varies greatly from variety to variety, region to region and year to year. The symptoms of malformation have been produced by the application of morphactin and malformin. The fungus *Fusarium moniliforme* var. *subglutinans* has been isolated from the malformed shoots panicles but the fungus could not reproduce the malady. The varieties Bhadauran, Elachi Dahiyar and Meghlantan have been found free from malformation. The malformation is not a problem in warmer growing regions such as South and West India.

Antimalformin substances did not give any control of malformation at Rewa and Lucknow. At CISH, Lucknow, de-blossoming alone initiated the healthy panicles to the extent of 80%. At IARI, New Delhi, and CISH, Lucknow, application of NAA (200ppm) in the first week of October followed by de-blossoming in the late December or January continued to be the best treatment for reducing the incidence of malformation.

Spongy tissue: The break down of flesh tissues has been frequently observed in Alphonso variety on ripening. The tissues get de-generated, become tasteless, lose firmness and turn into spongy nature, hence, called spongy tissue. The disorder has adversely affected the export of the famous 'Alphonso' mango. It has been established that the post-harvest exposure of fruit to light of 620–780 nm gave maximum incidence of spongy tissue likewise, advancement at maturity increase the incidence of sponginess. Therefore, mulching, sod culture and optimum time of harvest (50–75% maturity) reduce the incidence of spongy tissue in 'Alphonso.' Growth regulators and other chemicals have not been successful in checking the disorder. Newly developed hybrids, viz. Ratna, Sindhu, Arka Aruna, Arka Anmol, Arka Puneet and Arka Neelkiran have been reported to be almost free from this malady.

Grape

Propagation and rootstock: Normally, grapes are propagated by hard wood stem cuttings doing away the necessity of grafting methods and rootstocks. But the problems of salinity, soil moisture stress and insects affecting the root system have created the need for use of suitable rootstock.

A rootstock trial was therefore, conducted at IIHR, Bangalore, and GRS, Rajendra Nagar, Hyderabad, using different rootstocks for the commercial varieties Anab-e-Shahi, Gulabi and Thompson Seedless. The following conclusion were drawn:

- Chip budding is the best grafting method wherever, the rootstock plants are to be used.
- The commercial variety Gulabi can be used as rootstock to increase productivity and

rootstock 1613 for combating the nematode problem.

- Generally, Dogridge rootstock increased the scion vigour in both Anab-e-Shahi and Thompson Seedless.
- There was no effect of rootstocks on quality of berries.
- Rootstocks, Dogridge, Digraset and 1613 are recommended for sodic soils and where irrigation water has high chloride content.
- St. George identified as a drought tolerant rootstock.
- Thompson Seedless raised on Dogridge rootstock produced the maximum number of canes per vine and proved superior in terms of number of fruiting canes and bunches per vine at GRS, Rajendranagar.

Training: Training of vines is very important component of grape growing. Among various training systems, bower system was found to be the best for exploiting the productivity fully in all varieties of grapes, particularly in Thompson Seedless in south Indian conditions 'Black Champa' is better suited for head and Kniffin system of training than 'Gulabi' and 'Bangalore Blue'. The cultivars trained on head system at IARI, New Delhi, showed that Beauty Seedless, Cardinal and Perlette are more productive than others. The highest yield per hectare for 'Perlette' trained on Head system can be obtained at spacing of 2 × 2m.

Pruning: At IARI, New Delhi, vines bearing three shoots for each cluster produced uniformly coloured sweet berries which ripened three days earlier. Thus, shoot cluster ratio of 3:1 has been standardized for improving the quality of 'Beauty Seedless'.

At PAU, Ludhiana, pruning to 4 buds in Perlette, 7-8 buds in Thompson Seedless and 5 buds in Anab-e-Shahi were found for good yield and regular fruits.

The best time of pruning *Vinefera* varieties Anab-e-Shahi and Thompson Seedless under Bangalore condition was 20 October. Pruning twice for a single crop is advantageous over skipping the back pruning. Double cropping is not recommended in 'Thompson Seedless'.

Nutrition: On the basis of nutritional studies conducted at IIHR, Bangalore, a dose of 600g each of N, P_2O_5, K_2O along with 120kg FYM per vine has been recommended for Anab-e-Shahi in Karnataka. Likewise, a dose of 500g each of N P_2O_5 and K_2O/ha was found to be optimum and is recommended for 'Thompson Seedless. Split applications of N were not found beneficial. However, 40% application of the annual dose of K_2O in April pruning, 30% at October pruning and the rest 30% after 45 days of October pruning were found to be favourable for fruit yield.

Nutritional survey conducted by MPKV, Rahuri, for Thompson Seedless, Manik Chaman and Sonaka in different part of Maharashtra revealed that N, P and K contents varied between 1.12 to 2.80%, 0.50 to 1.10% and 1.52 to 2.68%, respectively, and none of the vineyard was found deficient in these elements.

Root studies are very important for the placement of fertilizers in soil. All the cultivars studied in general, had greater concentration of feeding roots nearer the trunk at 50 cm radial distance and at a shallow depth of 0–15cm in soils. Highest root density was observed in vines trained on Bower system. In Thompson Seedless, however, it was more in case of Head system under Bangalore conditions. In general, irrespective of variety both root density and activity were found to decrease with increasing depth and radial distance.

At GRS, Rajendranagar, application of higher levels of NPK increased the concentration of these elements in grape leaves. The higher levels of N (75g/plant/year) and P (50g/plant/year) had deterioty effect on yield while K had a positive effect.

At PAU, Ludhiana 75g N, 96g P_2O_5 and 60g K_2O/vine/year of age with basal dose of 15kg FYM/vine/year recorded the highest fruit yield at 5 years age in Perlette.

Root studies on 'Perlette' grape using P^{32} showed maximum feeding roots at 25cm depth followed by 50cm and 80cm depth. The roots were laterally distributed up to 60cm while in case of 'Anab-e-Shahi' at Ludhiana, the most feeding roots were concentrated at 50 cm distance both horizontally and vertically.

Water management: Water management studies were started during the seventh plan are being conducted at Rahuri and Rajendranagar, Ludhiana and Bangalore. However, at Bangalore the effect of water stress at different timings on grape production has indicated that the stress imposed during 30–60 days as well as 110–140 days after October pruning considerably reduces the yield. Stress during the ripening period, i.e., 100–140 days after October pruning seems to be favourable for higher TSS content. The stress imposed prior to April pruning seems to have increased number of cluster slightly.

At IARI, New Delhi, Dormex (1.5% a.i.) applied on dormant buds in January followed by GA_3 (30-45ppm) application at 50% bloom and CPPU (0.1%, a.i.) at fruit set resulted in early ripening in 'Perlette' by 8 days, 'Beauty Seedless' by 10 days and 'Pusa Seedless' by 11 days and increased berry weight and yield over control.

Use of plant growth regulators: Early ripening of berries is the key to successful grape growing particularly in northern India. The experimental trials conducted at New Delhi and Ludhiana, have proved that Dormex (1.5%) application in early January induces early bud burst as well as early ripening of berries. 'Pusa Seedless', 'Beauty Seedless' and 'Perlette' ripened 16 days, 12 days and 9 days earlier than control, respectively, under Delhi conditions.

At PAU, Ludhiana, fruit quality of 'Perlette' grape was considerably improved with pre-bloom flower bud thinning with plastic brush retaining only 100–125 flower buds per panicle, followed by dipping cluster in GA_3 (40ppm) and stem, girdling at 4 mm berry size stage and repeating with GA_3 application six days later. Similarly the 'Flame Seedless' grape showed uniform ripening and developed deep pink peel with 400ppm ethephon application at colour break stage.

Cycocel (200ppm) and IBA (100ppm) applied as spray checked effectively the drop of floral buds and flowers in Thompson Seedless. However, Cycocel (1000 ppm) spray at five leaf stage in the same variety checks apical dominance.

Gibberelic acid (GA_3), 15 ppm during flowering and 40-60 ppm at berry set stage increases berry length and reduces cluster compactness.

Control of pests: Flea beetle, mealy bug and stem-borer are important insect pests found damaging the grapes.

Flea beetle: It feeds on tender growth causing considerable damage to the plant. The experiments have proved that aldicarb (Temik) 10g followed by phorate (Thimet) 10g and phosalone (Zolone) 40% D were very effective in controlling the grubs and pupae. While Fenvalerate (20C) at 0.02% followed by carbaryl (50WP) at 0.15%, Cypermethrin (10EC) at 0.02% capable in controlling the adults of flea beetle.

Mealy bug: The pest causes damage by sucking the plant sap. Within 24 hours of dichlorovos

(0.20%) spray, 75.81% mortality was recorded. The effectiveness of systematic insecticide viz., rogor and monocrotophos was noticed only after ten days of application. The mealy bug population was effectively chccked by releasing the predator beetle *Cyptolaemus* at 15–20 days interval starting from 15 December.

Stem-borer: The borers can be destroyed by injecting 1ml petrol or by inserting half tablet of aluminum phosphide in each hole.

Root-knot nematodes population in 'Anab-e-Shahi' was reduced up to 60 days after the application of 65g Carbofuran per vine just after pruning in basins.

Control of diseases: The control of fungal diseases is very important in grapes because they cause serious damage to the plant.

Downy mildew: This is the most serious disease infecting all aerial parts of the vine. Initially light yellow translucent spots appear on upper leaf surface which gradually spread to all other plant parts. Ridomil (0.04%) and Galben (0.04%) are very effective in controlling the downy mildew. A large number of hybrids and mutants have been found resistant to this mildew which are being utilized as parents in developing the resistant varieties. At GRS, Rajendranagar, and MPKV, Rahuri, CME 151 (0.01%), Ridomil MZ (0.2%) and metalaxyl-mancozeb (0.2%) were found superior among various fungicides for the control of downy mildew.

Powdery mildew: Whitish gray patches appear on leaves, tender shoots and berries. The Fungicides Bayleton (0.1%), Calixin (0.07%), Anvil (0.05%), Systhane (0.04%) and Topas (0.025%) have been found effective to control powdery mildew.

At MPKV, Rahuri, metalaxyl mancozeb (0.2%) alone or in combination with bioagent was found to be the most effective for controlling downy mildew. Tridimefon (0.1%) and Penconazole (0.05%) were found effective against powdery mildew at MPKV, Rahuri and microsul (0.2%) at GRS, Rajendranagar.

Anthracnose: The anthracnose can be controlled by spraying of Carbendazim (0.1%) or Thiophenate-M (0.1%) at 15 days interval.

Guava

Propagation: At Faizabad, *Psidium* species were propagated by stooling. Maximum rooting was observed in stool layers with 7500ppm IBA + NAA. Cent per cent rooting and maximum establishment were obtained in *P. chinensis* while rooting was poor in *P. friedreschthalianum*.

Planting system: Various planting systems, viz., square, hedge row, double hedge row, paired and cluster planting, are being compared for plant growth and yield characteristics. Double hedge row planting (high-density planting) has given the highest yield at Rewa, Pantnagar, Sabour and Udaipur. The trial is continuing. At Ranchi, highest yield (143.87 q/ha) was obtained with maximum plant density (1,060 plants/ha).

Nutrition: Nutritional experiments conducted at Basti have shown that the application of 50g N, 30g P_2O_5 and 30g K_2O/plant/year of age in cv. Allahabad Safeda has resulted in maximum plant growth whereas, 75g N, 20g P_2O_5 and 20g K_2O/plant/year of age have given maximum yield in 'Sardar' guava. At GBPUA&T, Pantnagar, and FRS, Rewa, maximum yield was found in N_2P_2K (75g N, 50g P_2O_5 75g K_2O/tree/year) in cv. Pant Prabhat, Allahabad Safeda, while at RCA, Udaipur, maximum yield was recorded with the fertilizer dose of 40g N, 20g P_2O_5 and 20g K_2O/tree/year in cv. Sardar. The spray of urea combined with zinc sulphate has caused early ripening of fruits by 10 days.

Crop regulation: The aim of crop regulation in guava is to get winter season crop instead of rainy season crop or both crops partly. A number of treatments have been tried during the past two decades with varying results but all of them have succeeded in increasing the winter crop. At Pantnagar, three-fourths shoot pruning and half shoot pruning increased winter season crop as well as total yield. At Faizabad, urea (10%) spray gave the highest yield in winter season but flower drop was only 20–30%, whereas total flower drop was obtained with 1500ppm ethephon.

At BAC, Sabour and RCA, Udaipur, foliar spray of 15% fertilizer grade urea at 50% bloom stage followed by second spray of the same 10 days later produced the highest yield.

On the basis of crop regulation studies, pruning of three-fourths upper position of current season shoots in the first week of May has been recommended for commercial adaptation for the *Tarai* region of Uttar Pradesh and Uttarakhand.

Wilt disease: Undoubtedly, the most serious limitation of guava cultivation is wilt disease. Fully grown up trees bearing full crop start wilting and drying suddenly in a period of few years, the orchard is wiped out. The first step is to establish pathogenicity of the wilt disease. At number of centres, i.e. Lucknow, Basti, Kalyani, Ranchi, a number of pathogens have been isolated from the roots and twigs of wilt affected plants. The most common pathogens belonged to *Fusarium* spp. At CISH, Lucknow, different potent fungi, viz. *Gliocladium roseum, G. penicilloides, G. virens, Fusarium oxysporum* and *Acremonium* sp. were artificially inoculated produced wilt symptom. All these pathogen produced wilt symptom but *Gliocladium roseum* was found most potent pathogen in causing wilt on artificial inoculation. It cause around 90% wilting after 180 days of inoculation.

The technique for inoculation for successful reproduction of disease was standardized. Stem hole inoculation technique proved quite successful. Maximum of 31–34 °C, minimum of 23–25 °C and relative humidity of 87–98% with heavy rainfall was found conducive for wilt development. For control of wilt, Pusa Mrida (*Aspergillus niger* Strain AN 17) were tried in seedling as well as on old plant. The treated plants attained more vigour, high number of foliage development, more plant height and more stem girth as compared to untreated guava seedlings. In old plants, treated plants are perfectly healthy till now at BCKV, Mohanpur; CISH, Lucknow and HARP, Ranchi.

Litchi

Many trials on production technology were started at Pusa in 1988 and thereafter. It will therefore, take a few more years to give any tangible information.

Propagation: At Pusa, three propagation methods, viz., air layering, cutting and stooling were compared. The success in layering was 85.30 and 90.38% during July and August, respectively. The success in other months was either very low or nil. No success was obtained in propagation by stem cuttings. The rooting was also poor in stooling. At Ranchi, in layering the highest number of roots and growth were recorded in June and July and it took less time for development of roots in these months.

Planting system: Five planting systems, i.e., square, hedge row, double hedge row, paired planting, and cluster planting, were compared.

Double hedge row system of planting resulted in maximum growth of litchi plants, followed by cluster planting and the least in paired system of planting at RAU, Pusa. Double hedge-row system of planting also gave the highest yield at BAC, Sabour and GBPUA&T, Pantnagar.

Training system: In a comparison of three training systems, i.e., central leader, modified central leader and open vase with the control, the height and spread of the plants, girth, of trunk, primary and secondary branches were maximum in control plants at most of the centres.

The central leader system of training gave the maximum plant growth in litchi at BCKV, Mohanpur. Modified leader system of training gave significantly higher yield than control.

Pruning trial: A pruning trial comprising six treatments has been laid out at Pusa, Pantnagar and BCKV, Mohanpur.

At Mohanpur, harvesting of fruits with 7.5 cm terminal shoots acted as shoot pruning and increased fruit size, weight, TSS content and yield.

Fruit cracking: It is a serious problem in litchi. Various treatments, viz. NAA (20ppm), ethrel, (500ppm), GA_3 (10ppm), $Ca(NO_3)_2$ (10%) or $CaCl_2$ (10%) were found to reduce cracking of fruits at GBPUA&T, Pantnagar. At RAU, Pusa, fruit cracking in litchi cv. Shahi was greatly reduced when the orchard soil was kept moist with regular irrigation at field capacity level and atmospheric humidity was maintained around 41–46% (RH) level by sprinkling water.

Mite: The infestation and spread of litchi mite has been minimized by removal of infested twigs after harvesting in June and spraying of Dicofol (0.05%) twice at seven days interval on flushing in October.

Leaf roller and fruit borer: At BCKV, Mohanpur, the incidence of leaf roller and fruit borer (*Platyplea aprobola*) insect, was very high (86.56%, 27.0-65%). These insect pests could, however, be controlled effectively with sprays of Carbaryl (0.15%).

Temperate Fruits

Propagation: Generally, temperate fruits are easily propagated except nut fruits, particularly walnut, which is main temperate nut fruit of the country. Annular budding has given the best result (41.34% success) in mid-August, while the success in patch and forkert budding were lower in walnut.

Rootstock: RFRS, Mashobra, the effect of six rootstocks, viz. MM 109, M 13, MM 108, MM 111, M 25 and seedling was found significant on growth and yield of scion cultivars Royal Delicious, Golden Delicious and Ruspippin.

An interstock trial was laid out at HETC, Chaubattia, with intermediate pieces of M 9, M7, M 26, M 13, M 16, MM 104, MM 106 and control (no intermediate piece). All the growth attributes were found superior in M 16 in comparison with other interstocks. The maximum yield was also recorded on M 16 interstock. M 9 and M 25 showed dwarfing effect. The scion variety was Red Delicious.

In another rootstock experiment using Red Delicious as a scion variety, M 9 rootstock was observed to be brittle, having poor anchorage and uneven swelling at the union.

Training: At HETC, Chaubattia, a training trial was laid out using Red Delicious plants on M 7 rootstocks in 1971. It has been found that the plants trained by central leader and modified central leader system produced significantly more vegetative growth in comparison with open centre system and control. All the training systems gave better performance than control in respect of yield and qualitative characters of fruits. Amongst various training systems, the plants trained by modified central leader systems gave quality fruits.

Nutrition: The trial on standardization of N, P, K level in apple at Mashobra showed no significant difference in fruit maturity, TSS and acidity due to treatments. However, significant differences were observed in extension growth, yield and foliar N and K status. A consistent trend in yield increase was not obtained by increasing phosphorus levels.

Nitrogen deficiency can be overcome in apple by urea sprays at the rate of 0.5% from bud burst to harvesting at weekly intervals. Fertilizer placement either in pockets or bands has not been found to be superior to conventional method of broadcasting.

The pot experiment has revealed that different sources of nitrogen produced non-significant effect on growth and leaf nutrient status of apple except Zn. However, different levels of nitrogen has significant effect on growth and plant nutrient status. Extension growth significantly increased with the increase in the levels of nitrogen and maximum growth was recorded at 300 mg N/kg dry soil. Leaf N content also significantly increased with increase in levels of N, although the leaf N content represented level above the normal range.

The pot experiment on effect of different levels of P_2O_5 on growth and availability of other nutrients, has revealed that different treatments produced significant differences on extension growth, leaf P, Zn and Mn content but N, K, Fe and Cu content were not affected. Extension growth was maximum at 100g P_2O_5 treatment and control. Leaf Zn content was decreased by high level of phosphorous but the differences were not significant. Addition of lime had a synergistic effect on the uptake of Zinc. Leaf Mn contents increased with the increase in the levels of P_2O_5 but decreased with the addition of lime.

Experiment on the effect of some micronutrient on fruit quality, nutrient status and storage behaviour of apricot cv. New Castle was carried out and analysis of micro and macro nutrient elements was done. It was observed that none of the treatments had any significant effect on fruit N, P, Ca, Mg and Fe contents. However, P, Mn, Zn and Cu content underwent a significant change due to various treatments. Micronutrient mixture containing boric acid (0.1%), zinc sulfate (0.25%), and manganese sulphate (0.1%) gave the best results.

In an experiment to see the effect of different rates and carrier of nitrogen on growth and availability of other nutrient elements in young apple plants, it was observed that different sources of nitrogen produced non-significant effect on growth and foliar nutrient status except Fe and Mn content. However, different levels of nitrogen had significant effect on foliar nutrient status. Increasing levels of N has antagonistic effect on the uptake of Mn content in the levels of young apple plants and the differences were significant.

Mulching and weed control: In mulching trial at HETC, Chaubattia, data on apple variety Royal Delicious showed that all mulches increased height, thickness of stock and scion as compared to control remarkably. The soil temperature and soil moisture records indicated that all the mulches had proved significantly better. However, black polythene and oak leaves mulches gave good results than other mulches in this experiment.

Studies on post-emergence chemical weed control in apricot orchards showed that all the treatments were significantly effective in controlling weeds as compared with control. Among all these treatments gramoxone spray gave the best performance.

Use of plant regulators: The plant regulators were sprayed on peach variety Crawford Early at Chaubattia to study their effect on yield, ripening, storage and other parameters. Spraying with alar (1000ppm), ethrel (500 and 1500ppm), CCC (1500ppm) hastened fruit ripening by about 6 days but decreased the fruit firmness. Alar and ethrel did not cause the percentage loss in size and weight of fruits during storage but there were losses in these attributes due to CCC treatments.

In general, ethrel application on standing crop increased the dehulling percentage after 144 hours of the spray in walnut. However, leaves of tree turned yellowish and dropped earlier as compared to control. Ethrel at 2000ppm and 1500ppm gave the maximum percentage of dehulled walnuts as compared to other concentrations.

Control of insect pests: For the control of San Jose scale, tree spray oil (3%) is considered quite effective treatment. In orchard, the pest has been controlled by Chlorpyriphos (0.05%).

In nursery, 1g of Aldicarb or Carbofuran per plant has provided hundred percent control of the scale in one year old heavily infested plant. Both these chemicals are potent nematicides and have also been found effective in suppressing wooly aphids.

Peach trees sprayed with Chlorpyriphos (0.05%), Fenvalerate (0.025%), Formothon (0.04%), Methyl dematan (0.025%) Decamethrin (1.25), BPMC (0.05%) and Monocrotophos (0.04%) had no incidence of leaf curl.

Control of diseases: The powdery mildew of apple is a stubborn disease that can be controlled only by very early sprays in the season. Subsequent sprays are ineffective and the infected shoots are required to be removed by pruning to check the disease spread. Karathane (0.05%) and saprol (0.05%) have been found equally effective in checking the mildew on Janathan variety.

The collar and root rots are serious diseases of apple, which can only be controlled by employing resistant rootstocks. Among 13 rootstocks, MM 102, MM 793, MM 104, MM 110, M 13, and M 2 have been found to be moderately resistant and rootstocks MM 115, MM 106 and carb C are moderately susceptible. Bavistin (0.02%) and dithane Z-78 (0.4%) were found effective in controlling the incidence of the disease.

Gummosis is a serious problem in almond. Foliar spraying with borax (0.3%) + blitox (0.3%) and $CuSO_4$ (0.3%) checked effectively the shoot and fruit surface gummosis as compared with other treatments and control. Therefore, spraying with borax in combination with blitox is recommended for the control of gummosis in almond.

All the seven fungicides, viz., Bavistin (0.03%) Hexacap (0.025%), Difolatan (0.2%), blitox (0.3%), Wettable sulfur (0.2%) and Dithane-M-45 (0.25%) checked leaf blotch disease of almond significantly as compared with control and were at par with each other.

TECHNOLOGIES DEVELOPED AND THEIR IMPACT

Mango

- Mango cultivars Pusa Arunima and Pusa Surya released by IARI, New Delhi are recommended for commercial cultivation.
- Mallika is recommended for commercial cultivation in south and east Indian region.
- 'Dashehari-51', a regular bearing superior clone of Dashehari with higher yield by about 38% and ripening 15 days later to Dashehari has been found suitable for cultivation under Lucknow conditions.
- 'Vashi Badami' has been found free from mango leaf hopper, gall midge and shoot borer and recommended for release for cultivation in Valsad district of Gujarat.
- Bangalora, Mallika, Neelum, Suvarnarekha, Van Raj and Benishan have been reported to perform very well under south Telengana conditions. These cultivars can therefore, be grown commercially in this region.
- Veneer grafting has been found suitable for commercial propagation of mango in Uttar Pradesh, Uttarakhand, Madhya Pradesh, Bihar and Andhra Pradesh. Patch budding has

also been found very successful in Madhya Pradesh.

- Regular bearing could be obtained in mango cv. Dashehari grown in high density planting (2.5m × 2.5m) with the help of paclobutrazol @ 2.0 ml/tree of 10 year age as soil drench and regular pruning after harvesting under Tarai region of Uttarakhand.
- Placement of an annual dose of 400g of nitrogen, 200g of P_2O_5 and 400g of K_2O/tree in 40cm deep trench dug 2 metres away from the trunk of Dashehari gives high yield under north Indian conditions.
- Application of Paclobutrazol @ 5.0–10.0g a.i. (20 to 40ml)/tree/year through soil drench about three months before expected flowering time has been found effective in inducing regular bearing and increasing productivity in mango cv. Langra. The dose of 5g should be applied to trees less than 20 years in age and 10g for those more than 20 years.
- Soil application of paclobutrazol @ 3–5 ml/m canopy diameter through soil drench 90 to 120 days before bud-burst in Alphonso mango has been recommended for getting early and regular flowering and increased yield in Maharashtra.
- Heading back of crowded branches and centre opening with use of Paclobutrazol in over crowded mango orchard is recommended for commercial adoption.
- One spray of Imidacloprid (0.005%) is recommended for spraying of panicle emergence stage to control mango hopper and other pests.
- Two sprays of Quinalphos (0.05%) at 15 days interval starting from the emergence of new flush are recommended for the management of mango shoot/borer.
- Three sprays of Quinalphos (0.05%) or Monocrotophos (0.05%) at 10 days interval starting from the last week of July to the first week of August is recommended for the management of mango shoot gall psylla in the Jharkhand region.
- Cultivar Dahiyar and Meghlantan were found resistant to malformation for several years and can be used in breeding programme.
- Cultivar Ellaichi is recommended as resistant donor for mango malformation while cultivars Neelgoa reported susceptible and its propagation may be discouraged in hot spot areas.
- Bayleton (0.05%), systhane (0.1%), punch (0.04%) and karathane (0.1%) proved effective in controlling powdery mildew in mango.
- Carbendazim (0.1%) can be recommended for the control of anthracnose of mango.
- Application of bavistin (0.01%) without sticker is equally effective against anthracnose disease of mango.

Grape

- Perlette, Pusa Seedless, Beauty Seedless, Muscat Oliver and Flame Seedless have been listed as promising varieties with respect to yield and fruit quality under North Indian conditions.
- In grape, chip budding gives the maximum success under Bangalore and Hyderabad conditions.
- Perlette trained on head system, the cheapest method of training, gave the highest yield at a spacing of 2.0m × 2.0m under Ludhiana conditions. Black Champa is better suited

for head and kniffin systems of training than Gulabi and Bangalore Blue under Bangalore conditions. Bower system of training was, however, found to be the best in long run out of head, telephone and Kniffin systems of training tried for exploiting full productivity in all varieties of grape, particularly Thompson Seedless, Muscat Hamburg, Perlette and Anab-e-Shahi in production of quality fruits in Karnataka, Tamil Nadu and Punjab. However, Geneva double curtain system proved more beneficial than bower system for Thompson Seedless in North interior of Karnataka.

- Pruning twice for a single crop is advantageous. The best time of pruning for fruiting of *Vinifera* varieties under Bangalore conditions is around 20 October and double cropping is not recommended in Thompson Seedless.
- Perlette vine pruned at 3-4 buds gave more yield than longer or shorter pruned vines under Ludhiana conditions. Rejuvenation of old and unproductive Perlette vines by looping the main stem at head level has given the best results. Vines rejuvenated by this method have given the highest yield of best quality grapes.
- Regulating bud load by retaining 350 buds for the first kilogram weight of pruned material and 40 buds for every additional kilogram is recommended for maintaining the vigour and yield in Anab-e-Shahi at Coimbatore.
- Application of 900g N, 300g P_2O_5 and 900g K_2O together with 8 kg FYM is recommended for a vine of Anab-e-Shahi more than five years in age and spread to 3.0m × 4.5m area under Bangalore conditions.
- Application of 600g of N, 200g of P_2O_5, 600g of K_2O and 120kg FYM per vine is recommended for Anab-e-Shahi in Karnataka.

Nutrients Application

(750 vines/ha planted at 4.5m x 3.0 m spacing)

Age (years)	*Variety*	*Amount of nutrition (kg/ha/year)*			*Centre*
		N	P_2O_5	K_2O	
3-5 years	Anab-e-Shahi	500	125	750	IIHR, Bangalore
	Thompson Seedless	500	125	750	
Above 5 years	Anab-e-Shahi	500	500	1000	
	Thompson Seedless	500	500	1000	

The fertilizers to be applied in split doses as below

	N	P_2O_5	K_2O
After April pruning	60%	50%	-
After October pruning	40%	50%	50%
At fruit setting	–	–	50%

- Replacement of 50% of pan evaporation through 4 drippers placed around the vine area at the distance of 50 cm is recommended for optimum irrigation in Thompson Seedless grape under Bangalore conditions.
- Imposition of moisture stress 40–70 days after back pruning is recommended for increasing fruit yield without affecting quality adversely in grape Anab-e-Shahi and

Thompson Seedless under Bangalore conditions.

- In grape, application of Dormex (49% aqueous solution of Hydrogen Cyanamide) @ 1-1.5% a.i has been recommended on commercial scale for hastening the bud-burst and ripening of grapes under north Indian conditions to escape ripening grapes from rains.
- Ringing and use of GA_3 at pre-bloom stage increased fruit set and improved the fruit quality in Perlette grape at Abohar conditions.
- Dogridge imparted maximum vigour to the scion 'Thompson Seedless' but with much reduced yields under normal conditions of Bangalore.
- For the control of downy mildew of grape, four sprays of rodomil MZ or Aliete (0.2%) at 15 days interval starting from third leaf stage after pruning proved to be effective under Bangalore conditions.
- Topas (0.05%), Bayleton (0.1%) and Systhane (0.05%) control powdery mildew and carbendazin (0.1%) for the control of anthracnose in grape.
- In grape, powdery mildew could be controlled with three sprays of Bayleton (0.1%) applied at 15 days interval. Bavistin (0.1%) or Topsin (0.1%) controlled anthracnose in grape under Bangalore and Hyderabad conditions.

Guava

- Cultivar Chittidar is being recommended for cultivation in Madhya Pradesh.
- In the initial years, double hedge row system was found superior for high-density planting of guava in Andhra Pradesh, Bihar, Madhya Pradesh and Rajasthan.
- Pruning of three-fourths portion of current season growth in guava in the first week of May has been recommended for crop regulation under tarai region of Uttar Pradesh (now Uttarakhand) to avoid rainy season crop for getting good quality and increased winter season crop.
- The *Psidium cujavillus* rootstock has been reported to increase guava yield by 2-2.5 times.
- *Aspergillus niger* mixed in FYM and applied in root zone is recommended for control of diseases.

Litchi

- Double hedge row system of planting in initial years is recommended for raising high-density planting in litchi under Pusa (Bihar), Ranchi (Jharkhand) and Pantnagar (Uttarakhand) conditions.
- Irrigation at every 14 day interval starting from pea-stage of fruit development in April coupled with 4-5 sprayings of water on whole tree at 7–10 day intervals beginning 15 April during fruit growth and development is recommended as general practice in litchi under Pantnagar conditions. Three sprayings of 1.0 % Borax on tree at 15 days intervals beginning 10 April reduces the extent of cracking of fruits in litchi significantly and improves colour development of fruit peel and TSS and sugar content in fruits.
- For control of fruit borer in litchi first spray should be given with endosulphon (0.07%)

at red colour initiation stage, followed by Azadiractin 1500ppm (0.2%) spray after 10 days of first spray.

Low Chilling Temperate Fruits

- Gola (Pant Pear 18, Pant Pear 17 and Pant Pear 3) can be undertaken for cultivation in Tarai area of Uttar Pradesh and Uttarakhand.
- Peach cvs. Early Grande and Florda Prince, Plum cv. Satlaj Purple and pear cv. Punjab Beauty have been recommended for cultivation under Punjab conditions.
- 250g N together with 200g each of P_2O_5 and K_2O/tree of the age of 5 year age and above is recommended for peach cv. Sharbati under Pantnagar conditions of Uttarakhand.
- In almond, leaf curl aphid was controlled with the application of Thiometon (0.025%), Phosphamidon (0.02%), methyl demeton (0.025%) or Dimethoate (0.03%) at pre bloom stage. Almond mealy bug was controlled by soil drenching of Phosalone (0.07%), Farmothion (0.025%) or Quinalphos (0.02375%) in May
- Virus free bud wood material of apple cultivar was available to be supplied to growers.

Publications from the Project

- World Bibliography on Mango-1970
- List of Germplasm Collection Mango, Grape, Guava, Pome, Stone fruits and Walnut at various Horticultural Research Centres in the Country-1973
- Germplasm Collection of Pome, Stone and Nut Fruits in India-1976
- Germplasm Collection of Mango, Grape and Guava in India-1976
- Bibliography of Work Done on Temperate Fruits in India-1976
- Germplasm Collection of Pome, Stone and Nut Fruits in India-1976
- World Bibliography on Mango (Supplementary list)-1979
- List of Mango Workers of India-1979
- World Bibliography on Mango (IInd Supplementary list)-1986
- World Bibliography on Litchi-1986
- World Bibliography on Guava-1988
- Germplasm Collection of Pome, Stone and Nut Fruits in India-1988
- Descriptors of Jack Fruit Germplasm Collection-1988
- Descriptors for Seedling Mango Germplasm Collection-1988
- Horticulture Development of District Raigarh (MP.). A Report submitted to MOS for DARE. — November, 1988
- Germplasm Collection of Mango, Grape, Guava and Litchi-1990
- A Decade of Research on Pests of Horticultural Crops 1980-90. Report submitted in the Group Discussion of Entomologist Working in Horticultural Coordinated Project, held at IISR, Lucknow, during August 12-13, 1991
- Genetic Resources of Mango (*Mangifera indica* L.), Guava (*Psidium guajava* L.) and Litchi (*Litchi chinensis* Somn.) at different Centres of AICRP — 2002
- Fruit Varieties Released by Centres of All India Coordinated Research Project on

Subtropical Fruits – 2002

- Recommended Technologies on Fruits (Mango, Grape, Guava, Litchi, Pear, Plum, Peach and Almond) in AICRP (STF) – 2002
- The Litchi: Accomplishments of Research (AICRP–STF) – 2006.

WORKSHOP/GROUP WORKER'S MEETINGS

Venues	*Duration*
PAU, Ludhiana (Punjab)	April, 1969
PAU, Ludhiana(Punjab)	May, 1971
College of Agriculture, Solan (Himachal Pradesh)	June 6–10, 1973
Telhan Bhawan, Hyderabad (Andhra Pradesh)	May, 24–28,1976
Ranikhet, Chaubatia (Uttarakhand)	November, 17–20, 1976
Panaji, Goa	May, 2–5, 1979
Punjab Rao Krishi Vidyapeeth, Nagpur (Maharashtra)	29th August to 1st September, 1984
KKV, Dapoli (Maharashtra)	December, 18–20, 1986
NBPGR, Pusa (New Delhi)	April, 12-13, 1988
RAU, Pusa, Bihar (Bihar)	May, 23st25, 1988
APAU, Hyderabad (Andhra Pradesh)	September, 18–20,1991
RCA, Udaipur (|Rajasthan)	30th November to 3rd December, 1994
GBPUA&T, Pantnagar (Uttarakhand)	March, 21–23, 1996
BCKV, Kalyani (West Bengal)	January, 15–17, 1998
GAU, Navsari (Gujarat)	February, 21–23, 2000
NRC, Pune (Transfer of grape experiments) (Maharashtra)	24th February, 2001
AC&RI (TNAU), Madurai (Tamil Nadu)	January, 10–12, 2002
ANGRAU, Hyderabad (Andhra Pradesh)	May, 13–15, 2004
MPUA&T, Udaipur, (Rajasthan)	June, 2–4, 2006

4

Arid Zone Fruits

T.A. More

Organised research on arid zone fruits was initiated by the ICAR in 1976 on ad-hoc basis under an AP Cess Fund Scheme, while research on some selected fruits in the semi-arid areas was started at 10 centres, viz. Abohar (PAU), Jodhpur (CAZRI), Chandan Well near Jaisalmer (Government of Rajasthan), Hisar (HAU), Khedoi (GAU), Rahuri (MPKV), Yercaud (TNAU), Lahaul (HPKVV), Bangalore (IIHR) and Udaipur (RAU) in the country. During sixth-Five Year Plan, this ad-hoc scheme was merged as such and named as Cell-III of the AICFIP. During seventh Five Year Plan, Cell-III of the AICFIP was restructured to form an independent project entitled "All India Coordinated Research Project on Arid Zone Fruits (AICRP on AZF)". The centre from Chandan Well was shifted at Bikaner under Rajasthan Agricultural University. The Yercaud and Udaipur centres were shifted to Arupukottai (TNAU) w.e.f. 1st April, 1983 and Jobner (RAU) w.e.f 24.09.82, respectively. Three centres (Lahaul – HPKVV and Bhojka – RAU, Jodhpur – CAZRI) were closed down. Anantapur (APAU, 1983), Dantiwada (GAU, 1983), Faizabad (NDUAT, 1987), Lucknow (CISH, 2004), Bikaner (CIAH, 2004) were established as new centres. At present, 13 centres, 10 under SAUs in 9 States and 3 under ICAR Institutes in three states are functioning. The Project Coordination Unit, initially established at Hisar was shifted in 1993 to Bikaner (Rajasthan) and is presently stationed at CIAH, Bikaner.

CENTRES

The list of AICRP – AZF centres, their locations, year of establishment and state-wise distribution are given in Tables 4.1 and 4.2.

Table 4.1. List of AICRP AZF centres, their locations, affiliation and year of establishment

Centre	*Univ./inst./centre*	*Location*	*Year of Starting*
Project Coordination Unit	Central Institute for Arid Horticulture, Bikaner	Bikaner (Rajasthan)	1993
Regional Fruit Research Station	Punjab Agriculture University, Ludhiana	Abohar (Punjab)	1976
Dryland Agril. Research Station	ANG Ranga Agriculture University, Hyderabad	Anantapur (AP)	1983
Regional Research Station	Tamil Nadu Agriculture University, Coimbatore	Aruppukottai (TN)	1983
Regional Research Station	CCS Haryana Agriculture University, Hisar	Bawal(Haryana)	1989
Deptt. of Horticulture	N.D.University of Agriculture and Technology Kumarganj, Faizabad	Kumarganj, Faizabad (UP)	1987
Dept. of Horticulture	Mahatma Phule Krishi Vidyapeeth, Rahuri	Rahuri, (Maharashtra)	1976
Date Palm Research Station	Rajasthan Agric. University, Bikaner	Bikaner (Rajasthan)	1983
Dept. of Horticulture, SKN COA	Rajasthan Agric. University, Bikaner	Jobner(Rajasthan)	1982
Dryland Agril. Research Station	SDAU, Sardar Krushinagar	Palanpur, (Gujarat)	1983
Date Palm Research Station	SDAU, Sardar Krushinagar	Mundra (Gujarat)	1978
Division of Fruit Crops	Indian Institute of Horticultural Research, Bangalore	Bangalore (Karnataka)	1991
Division of Perennial Crops	Central Arid Zone Research Institute, Jodhpur	Jodhpur(Rajasthan)	1976
Central Institute for Arid Horticulture	Central Institute for Arid Horticulture, Bikaner	Bikaner (Rajasthan)	2004
Central Institute for Sub-Tropical Horticulture	Central Institute for Sub-Tropical Horticulture, Lucknow	Lucknow (UP)	2004

Table 4.2. State-wise distribution of AICRP on arid zone fruit centres and their mandate crops

Centre	*State*	*Mandarte crops*
Regional Fruit Research Station, Abohar	Punjab	Date palm
Dryland Agril. Research Station, Anantapur	Andhra Pradesh	Ber, Custard Apple, Pomegranate, Aonla, Tamarind and Phalsa
Regional Research Station, Aruppukottai	Tamil Nadu	Ber, Custard Apple, Aonla, Tamarind and Bael
Regional Research Station, Bawal	Haryana	Ber, Aonla, Date Palm, Phalsa, Lasora, Mulberry
Department of Horticulture, NDUAT, Faizabad	Uttar Pradesh	Ber, Aonla, Bael, Jamun
Department of Horticulture, MPKV, Rahuri	Maharashtra	Ber, Pomegranate, Custard Apple, Fig, Jamun, Tamarind, Aonla
Date Palm Research Station (RAU), Bikaner	Rajasthan	Date palm
Department of Horticulture, SKN College of Agriculture, Jobner	Rajasthan	Ber, Pomegranate, Aonla
Regional Agricultural Research Station, Sardarkrushinagar	Gujarat	Ber, Aonla, Custard Apple, Pomegranate
Date Palm Research Station, SDAU, Mundra	Gujarat	Date palm
Division of Fruit Crops, IIHR, Bangalore	Karnataka	Pomegranate, Custard Apple and Fig
Central institute for Arid Horticulture, Bikaner	Rajasthan	Aonla, Ber and Bael
Central Institute for Sub tropical Horticulture, Lucknow	Uttar Pradesh	Aonla and bael

GENETIC RESOURCES/BIODIVERSITY/GENE BANK MANAGEMENT

Systematic work on introduction, collection, conservation and evaluation of germplasms got an impetus with the starting of the All India Coordinated Research Project on Arid Zone Fruits. The present status of germplasm collections being maintained at the centres of this project including those at other places in the country are given here which will serve as an important reference for plant breeders.

Ber (*Ziziphus* spp.)

Regional Research Station, TNAU, Aruppukottai, Tamil Nadu

Accession No.	Spp/cultivars	Source/places	Year
Zm-1	Banarsi	Banglaore	1982
Zm-2	Umran	Kaithal	1982
Zm-3	Kaithli	Kaithal	1982
Zm-4	Gola	NA	1982
Zm-5	Kathaphal	NA	1982
Zm-6	Safeda Rootstock	NA	1982
Zm-7	Aruppukottai local	Aruppukottai	1982
Zm-8	Kakadia	NA	1982
Zm-9	Kakadia Rootstock	NA	1982
Zm-10	Periyakulam local	Periyakulam	1982
Zm-11	Hachi Rootstock	NA	1985
Zm-12	Hachi Rootstock	NA	1985
Zm-13	Hachi Rootstock	NA	1985
Zm-14	Hachi Rootstock	NA	1985
Zm-15	Sendurai Rootstock	Sendurai	1985
Zm-16	Sendurai Rootstock	Sendurai	1985
Zm-17	Sendurai Rootstock	Sendurai	1985
Zm-18	Sendurai Rootstock	Sendurai	1985
Zm-19	Sendurai Rootstock	Sendurai	1985
Zm-20	Sullakarai local	Virudhunagar	1988
Zm-21	Seb	NA	1989
Zm-22	Mundia	NA	1989
Zm-23	Guli	NA	1989
Zm-24	Sivakasi local	Sengamalanatchiyarpuram	1989
Zm-25	Kalayanputhur	Palani	1993
Zm-26	Periyakulam Local (II)	Periyakulam Local (II)	1993

CCSHAU, Regional Research Station, Bawal, Rewari, Haryana

Accession No.	Spp/cultivars	Source/places	Year
YSKB-1	*Zizyphus mauritiana*	Rasgan	2003
YSKB-2	*Zizyphus rotundifolia*	Jagmal Her	2003
YSKB-3	*Zizyphus rotundifolia*	Baliwas	2003
YSKB-4	*Zizyphus mauritiana*	Tigara	2003
YSKB-5	*Zizyphus mauritiana*	Ramgarh	2003
YSKB-6	*Zizyphus mauritiana*	Ramgarh	2003
YSKB-7	*Zizyphus mauritiana*	Bhusawar	2003
YSKB-8	*Zizyphus mauritiana*	Bhusawar	2003
YSKB-9	*Zizyphus mauritiana*	Bhusawar	2003
YSKB-10	*Zizyphus mauritiana*	Bhusawar	2003
YSKB-11	*Zizyphus mauritiana*	Weir	2003

(Contd.)

Accession No.	Spp/cultivars	Source/places	Year
YSKB-12	*Zizyphus mauritiana*	Deeg	2003
YSKB-13	*Zizyphus mauritiana*	Beechpuri	2003
YSKB-14	*Zizyphus mauritiana*	Beechpuri	2003
YSKB –15	*Zizyphus mauritiana*	Beechpuri	2003
YSKB-16	*Zizyphus mauritiana*	Beechpuri	2003
YSKB-17	*Zizyphus mauritiana*	Aazi Kalan	2003
YSKB-18	*Zizyphus mauritiana*	Aazi Kalan	2003
YSKB-19	*Zizyphus mauritiana*	Aazi Kalan	2003
YSKB-20	*Zizyphus mauritiana*	Hodal	2003

Central Arid Zone Research Institute, Jodhpur, Rajasthan

Spp/cultivars	Source/places	No. of collection
Z. mauritiana	Rajasthan, Punjab, Uttar Pradesh, Haryana,	-
Z.nummularia	Rajasthan	-
Z. mauritiana var. rotundifolia	Rajasthan	-
Z.spinachristi	Exotic	01

Narendra Deva University of Agriculture & Technology, Faizabad, U.P.

Spp/cultivars	Source/places	Year
Ponda	Aligarh	1993-94
Sanaur-2	Aligarh	1990-91
Agra Gola	Aligarh	1990-91
Mundia Murahra	Aligarh	1990-91
Dandan	Aligarh	1990-91
Chhuhara	Aligarh	1990-91
Darakhi	MPKV, Rahuri	1993-94
Seedless	MPKV, Rahuri	1993-94
Peondi	MPKV, Rahuri	1993-94
Vilayati	MPKV, Rahuri	1993-94
Banarasi Karaka	Varanasi	1982-83
Banarasi Peondi	Varanasi	1982-83
Gola	HAU, Hisar	1982-83
Godhan	HAU, Hisar	1982-83
Bagwadi	HAU, Hisar	1982-83
Jalandar	HAU, Hisar	1982-83
Kaithali	HAU, Hisar	1982-83
Kali	HAU, Hisar	1982-83
Kakrola Gola	HAU, Hisar	1982-83
Kala Gola	HAU, Hisar	1982-83
Katha	HAU, Hisar	1982-83
Narma	HAU, Hisar	1982-83
Mirchia	HAU, Hisar	1982-83
Pathan	HAU, Hisar	1982-83

(Contd.)

Spp/cultivars	Source/places	Year
Popular Gola	HAU, Hisar	1982-83
Reshmi	HAU, Hisar	1982-83
Sanaur-5	HAU, Hisar	1982-83
Safeda Rohtak	HAU, Hisar	1982-83
Safeda Select	HAU, Hisar	1982-83
Tikadi	HAU, Hisar	1982-83
Tikadri	HAU, Hisar	1982-83
Umran	HAU, Hisar	1982-83
ZG-3	HAU, Hisar	1982-83
Sandhaura Narnaul	HAU, Hisar	1982-83
Sanaur-3	HAU, Hisar	1982-83
Seb	CAZRI, Jodhpur	1982-83
Katha Rajasthan	CAZRI, Jodhpur	1982-83
Narendra Ber Sel.-1	Aligarh	1993-94
Narendra Ber Sel-2	Aligarh	1993-94
Narendra Ber Sel-3	Aligarh	1993-94
Narendra Ber Sel-4	Aligarh	1993-94
Narendra Ber Sel-5	Ghazipur	1993-94
Narendra Ber Sel-6	Ghazipur	1993-94
Narendra Ber Sel-7	Ghazipur	1993-94
Narendra Ber Sel-8	Ghazipur	1993-94

S.K.N.College of Agriculture, R.A.U, Jobner, Rajasthan

Spp./cultivars	Source/places	Year
Umran	-	-
Gola	-	1985
Kaithali	-	-
Chomulocal	-	-
Mundia	-	-
Jogia	-	-
Hussain Rejo (Chinese)	CIAH,Bikaner	1999
Chandni Supari	CIAH,Bikaner	1999
Saphar Chandni	CIAH,Bikaner	1999
Ponda	CIAH,Bikaner	1999
Jogia	CIAH,Bikaner	1999
Akharota	CIAH,Bikaner	1999
Gola	CIAH,Bikaner	1999
Tikadi	CIAH,Bikaner	1999
Gulkhasi	CIAH,Bikaner	1999
Phalisa Alwari	CIAH,Bikaner	1999
Nilgiri	CIAH,Bikaner	1999
Thornless	CIAH,Bikaner	1999
Katha	CIAH,Bikaner	1999
Katha Bombay	CIAH,Bikaner	1999
Tabes Taso	CIAH,Bikaner	1999

(Contd.)

Spp./cultivars	Source/places	Year
Meharoon	CIAH,Bikaner	1999
Dharkhi	CIAH,Bikaner	1999
Lakhan	CIAH,Bikaner	1999
Ilaichi	CIAH,Bikaner	1999
Noki	CIAH,Bikaner	1999
Mirchia	CIAH,Bikaner	1999
Pathani	CIAH,Bikaner	1999
Chhuara	CIAH,Bikaner	1999
Najuk	CIAH,Bikaner	1999
Kheera	CIAH,Bikaner	1999
Z.G.-3	CIAH,Bikaner	1999
Kathaphal	CIAH,Bikaner	1999
Kishmis	CIAH,Bikaner	1999
Chanchal	CIAH,Bikaner	1999
Tasinga	CIAH,Bikaner	1999
Kharki-1	CIAH,Bikaner	1999
Ali Raj	CIAH,Bikaner	1999
Banarasi Peondi	CIAH,Bikaner	1999
Gutka	CIAH,Bikaner	1999
CAZRI Gola	Godhra	1997
Sanur-5	Godhra	1997
Dandan	Godhra	1997
Ashapuri-1	Nadbai, Raj.	2003
Sev	Jobner	2003
Sukhawani	Jobner	2003
Pusa Prolafic	Nadbai, Raj	2003
Ashapuri-2	Asalpur	2003
Pemadi	Bhusawar	2003
Supari	Weir	2003
Chencho	Bhusawar	2003
Khirni	Bhusawar	2003
Kukari	Bhusawar	2003

Mahatma Phule Krishi Vidyapeeth, Rahuri, Maharashtra

Spp./Cultivar	Spp./cultivar	Spp./cultivar
Bagwadi	Guli	Noki
Bahadurgarhia	Gutka	Pathani
Banarasi	Illaichi	Ponda
Banarasi Karaka	Illaichi	Popular Gola
Banarasi pewandi	Jhajjar selection	Safarchandi hathed
Betawadi	Jhajjar special	Safeda Rohitki
Betwadi Hathed	Jogia	Sakhari hathed
Bhawnagari	Jallundhary	Safarchandi
B.S. 75-1	Kadaka	Sanaur-1
B.S. 75-2	Kaithali	Sonaur-2
B.S. 75-3	Bahula chameli	Sanaur-3

(Contd.)

Spp./cultivars	Spp./cultivar	Spp./cultivar
Chinese	Kalagola	Sanaur-4
Chhuhara	Katha	Sanaur-5
Chanchal	Kathaphal	Sanaur-6
Chameli	Khari No.1	Sandhura narnaw
Gola	Ladu	Seedless
Gobindgarh special	Chameli	Seo Bahadurgarhia
Glori	Mehrun	Shamber(Hatched)
Foliso alwari	Mehrun(khedi)	Seo
Darakhi No.2	Muria Maharana	Surti
Dandan	Nalgarhi	Tasbasto
Kharki No.1	Narikeli	Thornless
Gola Gurgaon No.3	Narma	Z.G.No.4
Golar	Nasuk	Z.G.No.3
Gorfa	Nehru Mandal	Villaiti
		Umran

S.D. Agricultural University, Sardarkrushinagar, Gujarat

Spp./cultivars	Source/place	Year
Mundia Mahera	HAU, Hissar	1985
Kaithli	HAU, Hissar	1985
Umran	PAU, Ludhiana	1984
Gola	CAZRI, Jodhpur	1986
Seb	CAZRI, Jodhpur	1986
Seo	HAU, Hissar	1985
Mundia	CAZRI, Jodhpur	1986
Banarasi Karka	HAU, Hissar	1985
Tasbtaso	MPKV, Rahuri	1986
Laddu	MPKV, Rahuri	1986
Jhajjar selection	MPKV, Rahuri	1986
Chinese	MPKV, Rahuri	1986
Manuki	MPKV, Rahuri	1986
Sukavani	Valbhipur	1985
Dandan	MPKV, Rahuri	1986
Safeda-1	MPKV, Rahuri	1986
Surti Kantha	GAU, Pariya	1985
Kharek	GAU, Pariya	1985
Jogiya	GAU, Pariya	1985
Narma	GAU, Pariya	1985
Deshi	NBPGR, New Delhi	1985
Vikas	GAU,Talod	1985
Rama	GAU,Talod	1985
Mehrun	GAU,Talod	1985
Ajmeri	MPKV, Rahuri	1986
Sanour-1	MPKV, Rahuri	1986
Banarsi	MPKV, Rahuri	1986

(Contd.)

Spp./cultivars	*Source/place*	*Year*
Chanchal	MPKV, Rahuri	1986
Shamber	MPKV, Rahuri	1986
Sandhura Narma	MPKV, Rahuri	1986
Bhavnagari	MPKV, Rahuri	1986
Badami	MPKV, Rahuri	1986
Selection-1	Avada	1985
Selection-2	Vishnagar	1984
Selection-3	Vishnagar	1984
Selection-4	Dangarva	1985
Selection-5	Jagudan	1984
Selection-6	Akhaz	1984
Selection-7	Prantij	1984
Selection-8	Prantij	1984
Selection-9	Langhanj	1984
Selection-10	Langhanj	1984
Selection-11	Langhanj	1984
Selection-12	Langhanj	1984
Noki	MPKV, Rahuri	1986
Surti	MPKV, Rahuri	1986
Darakhi-1	MPKV, Rahuri	1986
Banarasi Pevandi	MPKV, Rahuri	1986
Safeda Rohatki	MPKV, Rahuri	1986
Jhajjar special	MPKV, Rahuri	1986
Glori	MPKV, Rahuri	1986
Deshi Alwar	MPKV, Rahuri	1986
BS-75-3-1	MPKV, Rahuri	1986
Narikeli	MPKV, Rahuri	1986
Kakrola	MPKV, Rahuri	1986
Mirchia	MPKV, Rahuri	1986
Jollundh	MPKV, Rahuri	1986
Kala Gola	MPKV, Rahuri	1986
Chhuhara	MPKV, Rahuri	1986
Illaichi	MPKV, Rahuri	1986
Kadaka	MPKV, Rahuri	1986
Reshmi	MPKV, Rahuri	1986
Akrota	MPKV, Rahuri	1986
Gola Gurgoan	MPKV, Rahuri	1986

Central Institute for Arid Horticulture, Bikaner, Rajasthan

Spp./cultivars	*Source/place*	*No. of collection*	*Year*
Z.mauritiana	Rajasthan, Gujarat	01	1995
Z.spinacristi	Jodhpur	01	1996-97
Z.jujuba	Phagli, HImachal Pradesh	01	1997-98
*Z.mauritiana var.*Rotundofolia	Gujarat, Rajasthan, Uttarakhand	22	1995-96
Z.nummularia	Rajasthan	01	1995-96
Cultivars/genotype	Rajasthan, Gujarat, Uttar Pradesh Delhi, Punjab,Haryana, Uttarakhand	338	1995-96 to 2003

Aonla/Indian Gooseberry (*Emblica officinalis*)

Agricultural Research Station (ANG. Ranga AU), Anantapur, Andhra Pradesh

Spp./cultivars	*Spp./cultivars*	*Spp./cultivars*
NA-6	Kanchan	Pune
PD	NA 10	NEO- 4
NEO-4	Chakiya	NKT
NA-7	Lucknow	ATPS -3
ATPS-2	Local	ATPS- 4

Regional Research Station, TNAU, Aruppukottai, Tamil Nadu

Acc No.	*Spp./cultivars*	*Source/place*	*Year*
EO-1	Maudurai selection	Madurai	1984
EO-2	Panpozhi	Panpozhi	1983
EO-3	Srivilliputhur	Srivilliputhur	1981
EO-4	Coimbatore	Coimbatore	1981
EO-5	Banarsi	Bangalore	1982
EO-6	Palur	Palur	1988
EO-7	Virudhunagar	Virudhunagar	1988
EO-8	Mettupalayam	Mettupalayam	1988
EO-9	Chakaiya	Faizabad	1988
EO-10	Banarsi	Kallar	1990

Central Arid Zone Research Institute, Jodhpur, Rajasthan

Spp./cultivars	*Source/place*
Chakaiya	Faizabad, Anand
Kanchan	Faizabad, Anand
NA 7	Faizabad, Anand
Krishna	Faizabad, Anand

Narendra Deva University of Agriculture & Technology, Faizabad, UP

Spp./cultivars	*Source/place*	*Year*
Banarasi	Pratapgarh	1978-79
Chakaiya	Pratapgarh	1978-79
Francis	Pratapgarh	1978-79
Narendra Aonla-4	Pratapgarh	1982-83
Narendra Aonla-5	Pratapgarh	1982-83
Narendra Aonla-6	Pratapgarh	1982-83
Narendra Aonla-7	Pratapgarh	1982-83
Narendra Aonla-8	Pratapgarh	1982-83
Narendra Aonla-9	Pratapgarh	1982-83

(Contd.)

Spp./cultivars	Source/place	Year
Narendra Aonla-10	Agra	1988
Narendra Aonla-12	Manda	1992-93
Narendra Aonla-13	Dawrahra	1992-93
Narendra Aonla-14	Sindhora Mirzapur	1992-93
Narendra Aonla-15	Pratapgarh	1993-94
Narendra Aonla-16	Pratapgarh	1993-94
Narendra Aonla-17	Pratapgarh	1993-94
Narendra Aonla-18	Gondey, Pratapgarh	1994
Narendra Aonla-19	Newada	1995
Anand-1	G.A.U., Anand	1994-95
Anand-2	G.A.U., Anand	1994-95
Narendra Aonla-19	Pratapgarh	2000
Narendra Aonla-20	Bhadausi Pratapgarh	2000
Narendra Aonla-21	Atarea	2001
Narendra Aonla-22	Bhadausi	2001
Narendra Aonla-23	Sindhora, Mirzapur	2001
Narendra Aonla-24	Dighia, Allahbad	2001
Narendra Aonla-25	Dighia, Allahabad	2001

SKN College of Agriculture, Jobner, Rajasthan

Spp./cultivars	Source/place	Year
Banarasi	–	–
Chakaiya	Faizabad	1989
Krishna	Faizabad	1989
Kanchan	Faizabad	1989
NA-7	Faizabad	1995
NA-6	Faizabad	1995
NA-10	Faizabad	1995
Anand-1	GAU	2002
Anand-2	–	–

Mahatma Phule Krishi Vidyapeeth, Rahuri, Maharashtra

Spp./cultivars	Source/place	Spp./cultivars	Source/place
Kanchan	Faizabad	N.A. 6	Faizabad
Krishna	Faizabad	N.A. 10	Faizabad
Chakaiya	Faizabad	Bansi Red	Faizabad
Francis	Faizabad	Anand-1	Gujarat
Banarasi	Faizabad	Anand-2	–
Neelam	Faizabad		

S D Agricultural University, Sardarkrushinagar, Gujarat

Spp./cultivars	*Year*	*Spp./cultivar*	*Year*
Selection-1	1990	Selection-7	1990
Selection-2	1990	Selection-8	1990
Selection-3	1990	Selection-9	1990
Selection-4	1990	Selection-10	1990
Selection-5	1990	Selection-11	1990
Selection-6	1990	Selection-12	1990

Central Institute for Arid Horticulture, Bikaner, Rajasthan

Spp./cultivar	*Source/place*	*No. of collection*	*Year*
Cultivars/Genotypes	UP, Gujarat, Rajasthan	22	1995-98
Genotypes	H.P.	12	2003
BSR-1	Selam	01	2003
Collections	Rajasthan	09	2003

Tamarind (*Tamarindus indica*)

Agricultural Research Station, Anantapur, Andhra Pradesh

Spp./cultivar	*Source/place*	*Spp./cultivars*	*Source/place*
PKM 1	TN	S 102	Salem
JK 1	–	S 163	Salem
N 1	–	B 63(Bommidi)	–
Pratistan	MS	V 2 (Vellore)	Vellore
ATPS 1 (Selection)	Anantapur	P 11 (Pallochi)	–
ATPS –2	–	V 1(Vellore)	–
V-29 (Vellore)	–	H 77(Hosur)	–
V 59	–	SSS 414	–
Urigam 112	–	SSS 456	–

Regional Research Station, TNAU, Aruppukottai, Tamil Nadu

Accession No.	*Spp./cultivar*	*Source/place*	*Year*
TI-1	Aruppukottai collection	Aruppukottai	1987
TI-2	Periyakulam collection	Periyakulam HRS	1987
TI-3	Anchity collection	Anchity	1988
TI-4	Madurai collection	Madurai	1987
TI-5	Ettayapuram collection	Ettayapuram	1987
TI-6	Kovalangulam collection	Kovilangulam	1987
TI-7	Palayamapatti collection	Palayampatti	1987
TI-8	Mettupalayam	Mettupalayam	1987

(Contd.)

Accession No.	Spp./cultivar	Source/place	Year
TI-9	Poovani collection	Poovani farm	1987
TI-10	Kallar collection	Kallar	1987
TI-11	Faizabad collection	Faizabad	1987
TI-12	Sendurai Kottapatti(S.K.) collection	Sendurai	1988
TI-13	Sendurai(S) collection	Sendurai	1988
TI-14	Ayyalur(A) collection	Ayyalur	1988
TI-15	Rajapalayam collection	Rajapalayam	1988
TI-16	Uluppakudi-1 collection	Uluppakudi	1992
TI-17	Kokilapuram collection	Kokilapuram	1992
TI-18	Uluppakudi-2 collection	Uluppakudi	1992
TI-19	Karikundu-1 collection	Karikundu	1992
TI-20	Punnapatti collection	Punnapatti	1992
TI-21	Vemparpatti collection	Vemparpatti	1992
TI-22	Karantham-alai collection	Karanthamalai	1992
TI-23	Karikundu-2 collection	Karikundu	1992
TI- 24	Uluppakudi-3	Uluppakudi	1992
TI-25	Oddukkam collection	Oddukkam	1992
TI-26	Kovilangulam collection	Kovilangulam	1993

Date Palm *(Phoenix dactylifera* L.)

Date Palm Research Centre, RAU, Bikaner, Rajasthan

Spp./cultivars	Source/place	Year	Spp./cultivars	Source/place	Year
Halawy	USA	1978	Shamran	Abohar	1979
Barhee	USA	1980	Hayami	Abohar	1979
Medjool	USA	1970	Bint-a-isha	Abohar	1979
Khalas	USA	1979	Sedami	Abohar	1979
Sayar	USA	1982	Nagal Hilali	Abohar	1979
Zahidi	USA	1979	Zagloul	Abohar	1979
Khadrawy	USA	1979	Sewi	Abohar	1979
Barshi	USA	1987	Bikaner	Abohar	1979
Khuneizi	Oman	1982	Medini	Abohar	1979
Nagal	Oman	1982	Migraf	Abohar	1979
Khusab	Oman	1987	Abdul Rehman	Abohar	1979
Hatemi	Saudi Arabia	1987	Gizaz	Abohar	1979
Tayar	Saudi Arabia	1987	Suria	Abohar	1979
Ruziz	Saudi Arabia	1987	Sri Ganaganagar	Abohar	1979
Amiri	Egypt	1988	Muscat	Abohar	1979
Sakloti	Egypt	1988	Hamara	Abohar	1979
Agaloni	Egypt	1988	Umshok	Abohar	1979

Central Arid Zone Research Institute, Jodhpur, Rajasthan

Spp./cultivars	Source/place	Spp./cultivars	Source/place
Halawy	USA	Migraf	USA
Shamran	USA	Umshok	USA
Medjool	USA	Muscat-2	USA
Khadrawy	USA	Bikaner	-
Abdul Rehman	Abohar	Nagal Hilali	USA
Barhee	USA	Sayer	USA
Zahidi	USA	Zagloul	Abohar
Hayani	USA	Saddami	Abohar
Medni	USA	Pakistan	USA

Datepalm Research Station, GAU, Mundra, Kutch, Gujarat

Spp./culticars	Source/place	Year	Spp./culticars	Source/place	Year
Zahidi	USA	1980	Selection-13	Dhrab	1983
Halawy	USA	1980	Javantri	Tera vill.	1984
Barhee	USA	1980	Kotho	Tera	1984
Sayer	USA	1980	Gulchati	Tera	1983
Medjool	USA	1980	Trofo	Tera	1984
Khadrawy	USA	1980	Sopari	Dhrab	1983
Khalas	USA	1980	F.R.S. khedoi-7	Khedoi	1984
Selection-3	Dhrab local	1983	F.R.S. khedoi-10	Khedoi	1984
Selection-6	Dhrab Vill.	1983	Yakub Ishak Turk	Dhrab	1984

Regional Fruit Research Station PAU, Abohar, Punjab

Spp./cultivars	Source/place	Year
Hillawi(Local)	Bahadurgarh	1956
Khadrawi	Bahadurgarh	1956
Shamran	Bahadurgarh	1956
Barhee	USA	1957
Dayri	USA	1957
Deglet Noor	USA	1957
Hayany	USA	1957
Hilali	USA	1957
Hillawi	USA	1957
Iteema	USA	1957
Khadrawy	USA	1957
Khalasa	USA	1957
Mektoom	USA	1957
Medjool	USA	1957
Saidy	USA	1957
Sayer	USA	1957
Thoory	USA	1957

(Contd.)

Spp./cultivars	Source/place	Year
Zahidi	USA	1957
Bint Aisha	Egypt	1959-60
Hayani	Egypt	1959-60
Samani	Egypt	1959-60
Sewi	Egypt	1959-60
Zaglool	Egypt	1959-60
Aali	Khairpur(Pak)	1959-60
Halwain	Khairpur(Pak)	1959-60
Khudain	Khairpur(Pak)	1959-60
Abdul Rehman	Adan	1964
Gizaz	Adan	1964
Hamara	Adan	1964
Sabiah	Adan	1964
Suria	Adan	1964
Medini	Adan	1964
Migraf	Adan	1964
Umsho	Adan	1964

Central Institute for Arid Horticulture, Bikaner, Rajasthan

Accession No.	Spp./cultivar	Source/place	No. of collection	Year
EC402388-90	*P.dactylifera*	Iraq	03	1997
	P.reclinata	Argintina	01	-
	P.dactylifera	Cochin	06	-
AHDP 001-006	Collection	Gujarat	06	2003
	Cultivars	Punjab, Gujarat,Rajasthan	50	1997-98–2003

Pomegranate (*Punica granatun*)

Agricultural Research Station, ANGRAU, Anantapur, Andhra Pradesh

Spp./cultivars	No. of collection	Spp./cultivars	No. of collection
Ganesh	1/1-9	Tabest	12/1-10
G-137	3/1-9	GKVK-1	13/1-5
Muscat	4/1-9	Lupania	13/6-10
Jodhpur Red	5/1-6	Dholka	14/1-5
Jalor seedless	5/7-9	Kalisherin	14/6
P-23	7/1-9	Utkal	14/7-9
Agan	8/1-15	Yerkad	14/10
Basin seedless	8/6-9	Bedana Sadana	15/1-3
Dorasut	9/1-9	Suren Anar	15/4-6
Alan	10/1-10	Virupakshi	15/7-8
Speen Danedar	11/1-8	Madakasira	15/9-10
Kali Sharin	11/9-10	Rubi	-

Indian Institute of Horticultural Research, Bangalore, Karnataka

Spp./cultivar	Spp./cultivar
Ganesh	Muskat
Kabul Yellow	G-137
Double Flower	Dholka
Nana	P-23
Keshar	P-26
Ruby	Nayana
Amlidana	Nimali
Daru	Lopkapur
Lokapur	Daya
Ganesh Mutant	Kalpitya
Jodhpur Red	Gulesha Rose Pink
Jalore Seedless	Yercaud

Mahatma Phule Krishi Vidyapeeth, Rahuri, Maharashtra

Spp./cultivar	Spp./cultivar	Spp./cultivar	Spp./cultivar
Alandi	Yerchad local	Jyoti (GKVK)	Gulsha red
Ganesh	Siah shirin	Jalore seedless	Achik Dana
G-137	Boseka linski	Alah	Kazakaki Anar
Muskat	Bedana sedana	G.R.Pink	Kazaki Anar
P-13	Damini(Pune)	Shrin Anar	Kali shirin
P-16	Kabul cenoor	Jodhpur Red	Khog
P-23	Kabul IIHR	Kabul Yellow	Lupania
P-26	Bedanathin skin	Patna-5	Uthukul
Agah	Kandhari	Sural Anar	Sur Sukkar
A.K.Anar	Tabesto	Cranedo-de-Elcho	Phule Arakta
Speen Danedar	Malta	Dorsata Malas	Ruby red
Concentration white	Spin sakarin	Gulsha	Greece pomegranate
Dholka	Basein seedless	Surk Anar	Bhagawa
Kabul	Yercand	Mridula	K.R.S.
Yercaud ARS	Bedana suri		

Central Institute for Arid Horticulture, Bikaner, Rajasthan

Spp./cultivar	Source/place	No. of collection	Year
Punica granatum	Raj.,M.S.,Guj., H.P.	145	1995-2003
Cultivars	Iran	06	2000

Fig (*Ficus carica)*

Indian Institute of Horticultural Research, Bangalore, Karnataka

Spp./cultivar	Source/place	Spp./cultivar	Source/place
Poona	–	Pasquel	–
Deanna	–	Brown Turkey	–
Excel	–	Ex.China	China
Conadria	MS	Ex.Greece	Grace
Daulatabad	–	Ex.California	California
Zardirniz	–	Ficus hispida	–
Dinkar	MS		

Mahatma Phule Krishi Vidyapeeth, Rahuri, Maharashtra

Spp./cultivar	Spp./Cultivar
Poona Fig	Deanna
Dinkar	Maisaram
Excel	Timla Fig
Conadria	

Custrd Apple (*Annona sp.*)

Indian Institute of Horticultural Research, Bangalore, Karnataka

Spp./cultivar	Spp./cultivar	Spp./cultivar
British Guinea	Red Sitaphal	*A.glabra*
Balanagar	Washington-97	*A.muricata*
Barbadose Seedling	Washington-05	Islandgem (*A.atemoya*)
Mammoth	A.cherimola	Pinksmammoth(*A.atemoya*)
Local Sitaphal	A.reticulata	Bullcoks heart (*A.atemoya*)

Agricultural Research Station, ANGRAU, Anantapur, Andhra Pradesh

Spp./cultivar	Spp./cultivar	Spp/cultivar	Spp./cultivar
Atemoya	ATPS-6	Red Sitaphal x Pond apple	Pythpta-5
Balanagar	Red Seetaphal	Red Sitaphal x Pond apple	Pythpta-6
ATPS-1(Anantapur selection)	ATPS-7	Balanagar(Sangareddi)	Molkalmur 1
Bullocks heart	Tirupati-1	Alemoya x Washington	Molkalmur 2
Chance seedling	Tirupati-2	Bangalore	Molkalmur 3
ATPS-2	Madanapalli-1	Bellary	Molkalmur 4
ATPS-3	Madanapalli-2	Ramaphal	CRIDA, Hyderabad
Island gem	Rayadurg	Hyderabad	Molkalmur 7
ATPS-4	Gudipadu-1	Pythotal-1	Molkalmur 8
Pink's mammoth	Gudipadu	Pythpta-2	Molkalmur 9
ATPS-5	Atemoya x Balanagar	Pythpta-3	CRIDA,Hyderabad
Washington	Atemoya x Balanagar	Pythpta-4	Arka Sahan(Hybrid)

Regional Research Station, TNAU, Aruppukottai, Tamil Nadu

Accession No	*Spp./cultivar*	*Source/place*	*Year*
AS-1	Courtallam	State Hort.Farm,Courtallam	1989
AS-2	Pond apple	IIHR, Bangalore	1984
AS-2A	Pond apple red	IIHR, Bangalore	1984
AS-3	Mammoth	IIHR, Bangalore	1984
AS-4	British Guinea	IIHR, Bangalore	1984
AS-5	Washington 10700	IIHR, Bangalore	1984
AS-6	Aruppukottai local	Aruppukottai	1984
AS-7	Washington 98797	IIHR, Bangalore	1984
AS-8	Balanagar	IIHR, Bangalore	1984
AS-9	Bangloare (Local sita)	IIHR, Bangalore	1984
AS-10	Balanagar graft	IIHR, Bangalore	1984
AS-11	Mettupalayam	Mettupalayam	1990
AS-12	Yercaud	Yercaud	1990
AS-13	Virudhunagar	Virudhunagar	1990
AS-14	Atemoya graft	Periyakulam	1990
AS-15	Ram sita	Periyakulam	1990

Mahatma Phule Krishi Vidyapeeth, Rahuri, Maharashtra

Spp./cultivar	*Spp./Cultivar*	*Spp./cultivar*
Balanagar	Madhu	Mammoth
Washington 98787	Chittorgarh	Pink mammoth
Washington 107005	British guinea	Bullock heart
Yellow sitaphal	Barbadose	Salem selection
Red sitaphal	Island Gem	Arka Sahan

S.D. Agricultural University, Sardardrushinagar, Gujarat

Spp./cultivar	*Source/place*	*Year*	*Spp./cultivar*	*Source/place*	*Year*
Selection-1	-	1988	Selection-12	Bhavnagar	1988
Selection-2	-	1988	Selection -13	Junagadh	1988
Selection-3	-	1988	Selection-14	Vanthali	1988
Selection-4	-	1988	Selection-15	Vanthali	1988
Selection-5	-	1988	Selection-16	Bhavnagar	1988
Selection-6	-	1988	Selection-17	Boriavi	1988
Selection-7	-	1988	Selection-18	Boriavi	1988
Selection-8	Dhoraji	1988	Selection-19	Padia	1988
Selection-9	Junagadh	1988	Selection-20	Boriavi	1988
Selection-10	Junagadh	1988	Selection-21	Ratanpur	1988
Selection-11	Junagadh	1988	Selection-22	-	1988

Phalsa (*Grewia subinaequalis*)

Central Arid Zone Research Institute, Jodhpur, Rajasthan

Spp./cultivar	Source/place	No.of collection
Phalsa genotypes	Agra, Udaipur Ajmer	3

SKN. College of Agriculture, Jobner

Spp./cultivars	*Source/place*	*Year*
Phalsa	Bhusawar	2003

Central Institute for Arid Horticulture, Bikaner, Rajasthan

Spp./cultivars	*Source/place*	*No. of collection*	*Year*
Phalsa	Rajasthan, Uttar Pradesh	04	2001
Phalsa	Goa	01	2003

Bael (*Aegle marmelos*)

Regional Research Station TNAU, Aruppukottai, Tamil Nadu

Acc. No.	*Spp./cultivar*	*Source/place*	*Year*
NB-4	*A. marmelos*	NDUAT, Faizabad	1988
NB-5	*A. marmelos*	Faizabad	1988
NB-9	*A. marmelos*	NDUAT, Faizabad	1988
AM-1	*A. marmelos*	–	1989
AM-2	*A. marmelos*	–	1989

Central Arid Zone Research Institute, Jodhpur, Rajasthan

Spp./cultivars	*Source/place*	*No.of collection*
A.marmelos	Faizabad	05

Central Institute for Arid Horticulture, Bikaner, Rajasthan

Spp./cultivars	*Source/place*	*No. of collections*	*Year*
Bael cvs.	Faizabad	06	1996-97
Bael cvs.	Pant nagar	04	1998
Bael cvs.	Jodhpur	01	2001
Collection	Udaipur(Rajasthan)	01	2003
Collection	S.K.Nagar, Uttar Pradesh	04	2003
Collection	Ajmer	01	2002

Narendra Dev University of Agriculture & Technology, Faizabad, U.P.

Spp./cultivar	*Source/place*	*Year*
Narendra Bael-1	Faizabad	1982-83
Narendra Bael-2	Faizabad	1982-83
Narendra Bael-3	Basti	1982-83
Narendra Bael-4	Barabanki	1982-83
Narendra Bael-5	Gonda	1982-83
Narendra Bael-6	Deoria	1982-83
Narendra Bael-7	Basti	1982-83
Narendra Bael-8	Deoria	1982-83
Narendra Bael-9	Basti	1982-83
Narendra Bael-10	Barabanki	1982-83
Narendra Bael-11	Lucknow	1982-83
Narendra Bael-12	Mirzapur	1982-83
Narendra Bael-13	Itawan	1983-84
Narendra Bael-14	Jaunpur	1993-94
Narendra Bael-15	Jaunpur	1993-94
Narendra Bael-16	Newada, Pratapgarh	1993-94
Narendra Bael-17	Sultanpur	1993-94
Narendra Bael-18	Bahraich	1993-94
Narendra Bael-19	Moredabad	1994-95
Narendra Bael-20	Moradabad	1994-95
Narendra Bael-21	Vijaipur Allahabad	1994-95
Narendra Bael-22	Basti	1994-95
Narendra Bael-23	Basti	1994-95
Narendra Bael-24	Basti	1994-95
Narendra Bael-25	–	–

Lasora or Goonda (*Cordia myxa*)

CCSHAU, Regional Research Station, Bawal Rewari, Haryana

Accession No.	*Spp./cultivar*	*Source/place*
2019	Lasora *Cordia myxa*	Khear Murar
2020	Lasora *Cordia myxa*	Khear Murar
2021	Lasora *Cordia myxa*	Bawal
2022	Lasora *Cordia myxa*	Bawal
2023	Lasora *Cordia myxa*	Bawal
2024	Lasora *Cordia myxa*	Bawal
2025	Lasora *Cordia myxa*	Nagal Sahbazpur
2026	Lasora *Cordia myxa*	Bawal
2027	Lasora *Cordia myxa*	Tihara
2028	Lasora *Cordia myxa*	Rewari
2029	Lasora *Cordia myxa*	Rewari
2030	Lasora *Cordia myxa*	Jatuwas
2031	Lasora *Cordia myxa*	Bharawas

(Contd.)

Accession No.	Spp./cultivar	Source/place
2032	Lasora *Cordia myxa*	Bawal
2033	Lasora *Cordia myxa*	Dharchana
2034	Lasora *Cordia myxa*	Paranpura
2035	Lasora *Cordia myxa*	Dabari
2036	Lasora *Cordia myxa*	Giglana
2037	Lasora *Cordia myxa*	Giglana
2038	Lasora *Cordia myxa*	Manheti
2039	Lasora *Cordia myxa*	Khol
2040	Lasora *Cordia myxa*	Khol
2041	Lasora *Cordia myxa*	Khol
2042	Lasora *Cordia myxa*	Atali Mandi
2043	Lasora *Cordia myxa*	Khod
2044	Lasora *Cordia myxa*	Narnaul
2045	Lasora *Cordia myxa*	Narnaul
2046	Lasora *Cordia myxa*	Nasibpur
2047	Lasora *Cordia myxa*	Bucholi
2048	Lasora *Cordia myxa*	Rajawas
2049	Lasora *Cordia myxa*	Rajawas
2050	Lasora *Cordia myxa*	Jawahar Nagar
2051	Lasora *Cordia myxa*	Loharu
2052	Lasora *Cordia myxa*	Loharu
2053	Lasora *Cordia myxa*	Badwa
2054	Lasora *Cordia myxa*	Arya Nagar
2055	Lasora *Cordia myxa*	Dhani Kutubpur
2056	Lasora *Cordia myxa*	Dhani Kutubpur
2057	Lasora *Cordia myxa*	Hansi
2058	Lasora *Cordia myxa*	Jatu luhari
2059	Lasora *Cordia myxa*	Chang
2060	Lasora *Cordia myxa*	Charkhi Dadari
2061	Lasora *Cordia myxa*	Gothara
2062	Lasora *Cordia myxa*	Motala Kalan
2063	Lasora *Cordia myxa*	Lala
MKD-15	Indian Cherry *Coridia rothi*	MamdiaAsampur
MKD-19	Indian Cherry *Cordia myxa*	Menethi
MKD-21	Indian Cherry *Cordia myxa*	Bhojpur
MKD-25	Indian Cherry *Cordia myxa*	Mahendergarh
MKD-26	Indian Cherry *Cordia myxa*	Mahendergarh
MKD-51	Indian Cherry *Cordia myxa*	Azamgarh
MKD-64	Indian Cherry *Cordia myxa*	Hansi(Katla Bagh)

Central Institute for Arid Horticulture, Bikaner, Rajasthan

Accession No.	Spp./cultivar	Source/place	No.of collection	Year
-	Lasora (*C. myxa*)	Raj. Guj.	33	1999-2000
-	Lasora	Raj. Guj.	32	2000-2001

Central Arid Zone Research Institute, Jodhpur, Rajasthan

Spp./cultivar	*Source/place*	*No. of collection*
Gonda (*Cordia myxa*)	Jodhpur, Pali, Ajmer, Chittorgarh	04

S.K.N. College of Agriculture (RAU), Jobner, Rajasthan

Spp./cultivar	*Source/place*	*No. of collection*	*Year*
Lasora	Bhusawar	02	2003
Lasora	Asalpur	02	2003

Miscellaneous Fruits

Karonda *(Carisa carandas)*

Central Arid Zone Research Institute, Jodhpur. Rajasthan

Spp./cultivar	*Source/place*
Karonda	Rajasthan

Central Institute for Arid Horticulture, Bikaner, Rajasthan

Spp./cultivar	*Source/place*	*No.of collection*	*Year*
Collections	Rajasthan, Gujarat, Goa, Madhya Pradesh	07	2001-03

Khirni

S.K.N.College of Agriculture, Jobner, Rajasthan

Spp./cultivar	*Source/place*	*No. of collection*	*Year*
Khirni	Bhusawar Gujgarh	02	2003

Central Institute for Arid Horticulture, Bikaner, Rajasthan

Spp./cultivar	*Source/place*	*No. of collection*	*Year*
Khirni	Gujarat	01	1997-98

Jamun

Mahatma Phule Krishi Vidyapeeth, Rahuri, Maharashtra

Spp./cultivar
Gokak-1
Gokak-2
Gokak-3

CCSHAU, Regional Research Station, Bawal, Rewari, Haryana

Accession No.	*Spp./cultivar*	*Source/place*	*Year*
MKD-1	Ker *(Capparis decudua)*	Rewari	2002
MKD-2	Shivlingi (Bryonopsis laciniosa)	Nangal Shapur	2002
MKD-3	Ker *(Capparis decudua)*	Mohanpur	2002
MKD-4	Ker *(Capparis decudua)*	Mohanpur	2002
MKD-6	Ker *(Capparis decudua)*	Kutina	2002
MKD-9	Ker *(Capparis decudua)*	Govind Puri	2002
MKD-16	Ker *(Capparis decudua)*	Ghothra	2002
MKD-18	Ker *(Capparis decudua)*	Mazara Bhalki	2002
MKD-22	Ker *(Capparis decudua)*	Israna	2002
MKD-30	Ker *(Capparis decudua)*	Dulana	2002
MKD-31	Ker *(Capparis decudua)*	Jhanjhariwas	2002
MKD-33	Ker *(Capparis decudua)*	Kharolli	2002
MKD-34	Ker *(Capparis decudua)*	Kharolli	2002
MKD-37	Ker *(Capparis decudua)*	Reva	2002
MKD-38	Ker *(Capparis decudua)*	Kopper	2002
MKD-39	Ker *(Capparis decudua)*	Thali	2002
MKD-45	Ker *(Capparis decudua)*	Rathiwas	2002
MKD-46	Ker *(Capparis decudua)*	Rathiwas	2002
MKD-47	Ker *(Capparis decudua)*	Madhogarh	2002
MKD-48	Ker *(Capparis decudua)*	Azamgarh	2002
MKD-49	Ker *(Capparis decudua)*	Azamgarh	2002
MKD-50	Ker *(Capparis decudua)*	Azamgarh	2002
MKD-52	Ker *(Capparis decudua)*	Loharu	2002
MKD-53	Ker *(Capparis decudua)*	Loharu	2002
MKD-54	Ker *(Capparis decudua)*	Loharu-Gigan	2002
MKD-55	Ker *(Capparis decudua)*	Obra	2002
MKD-56	Ker *(Capparis decudua)*	Badva	2002
MKD-57	Ker *(Capparis decudua)*	Badva	2002
MKD-58	Ker *(Capparis decudua)*	Bhanria	2002
MKD-59	Ker *(Capparis decudua)*	Gangua	2002
MKD-60	Ker *(Capparis decudua)*	Arya Nagar	2002
MKD-61	Ker *(Capparis decudua)*	Arya Nagar	2002
MKD-62	Ker *(Capparis decudua)*	Arya Nagar	2002
MKD-63	Ker *(Capparis decudua)*	Kharad	2002
MKD-65	Ker *(Capparis decudua)*	Jita Khedi	2002
MKD-66	Ker *(Capparis decudua)*	Bhivani khera	2002
MKD-67	Ker *(Capparis decudua)*	Bhivani khera	2002
MKD-68	Ker *(Capparis decudua)*	Bhivani khera	2002
MKD-69	Ker *(Capparis decudua)*	Kalanaur	2002
MKD-70	Ker *(Capparis decudua)*	Kherdi	2002
MKD-71	Ker *(Capparis decudua)*	Kherdi	2002
MKD-72	Ker *(Capparis decudua)*	Bond Khurd	2002
MKD-73	Ker *(Capparis decudua)*	Bond Khurd	2002
MKD-75	Ker *(Capparis decudua)*	Sank Road	2002
MKD-76	Ker *(Capparis decudua)*	Sank Road	2002

(Contd.)

Accession No.	*Spp./cultivar*	*Source/place*	*Year*
MKD-79	Ker *(Capparis decudua)*	Sohna	2002
MKD-80	Ker *(Capparis decudua)*	Alipur	2002
MKD-81	Ker *(Capparis decudua)*	Manesar	2002
MKD-82	Ker *(Capparis decudua)*	Manesar	2002

S.K.N. College of Agriculture, RAU, Jobner, Rajasthan

Spp./cultivar	*Source/place*	*No. of collection*	*Year*
Ker (*C.decidua*)	Asalpur Farm	03	2003

Peelu (*Salvedora spp.*)

CCSHAU, Regional Research Station, Bawal, Rewari, Haryana

Accession No.IC	*Spp./cultivar*	*Source/place*	*Year*
MKD-5	Peel *(Salvadora oleoides)*	Kutina	2002
MKD-7	Peel *(Salvadora oleoides)*	Kutina	2002
MKD-8	Peel *(Salvadora oleoides)*	Kutina	2002
MKD-10	Peel *(Salvadora oleoides)*	Mamdia Asampur	2002
MKD-11	Peel *(Salvadora oleoides)*	Mamdia Asampur	2002
MKD-12	Peel *(Salvadora oleoides)*	Mamdia Asampur	2002
MKD-13	Peel *(Salvadora oleoides)*	Mamdia Asampur	2002
MKD-14	Peel *(Salvadora oleoides)*	Mamdia Asampur	2002
MKD-17	Peel *(Salvadora oleoides)*	Ghothra	2002
MKD-20	Peel *(Salvadora oleoides)*	Rata	2002
MKD-23	Peel *(Salvadora oleoides)*	Chelsa	2002
MKD-24	Peel *(Salvadora oleoides)*	Chelsa	2002
MKD-27	Peel *(Salvadora oleoides)*	Jhanjhariwas	2002
MKD-28	Peel *(Salvadora oleoides)*	Jhanjhariwas	2002
MKD-29	Peel *(Salvadora oleoides)*	Jhanjhariwas	2002
MKD-32	Peel *(Salvadora oleoides)*	Nimbi	2002
MKD-35	Peel *(Salvadora oleoides)*	Reva	2002
MKD-36	Peel *(Salvadora oleoides)*	Reva	2002
MKD-40	Peel *(Salvadora oleoides)*	Manpura	2002
MKD-41	Peel *(Salvadora oleoides)*	Manpura	2002
MKD-42	Peel *(Salvadora oleoides)*	Rathiwas	2002
MKD-43	Peel *(Salvadora oleoides)*	Rathiwas	2002
MKD-44	Peel *(Salvadora oleoides)*	Rathiwas	2002
MKD-74	Peel *(Salvadora oleoides)*	Bond Khurd	2002
MKD-77	Peel *(Salvadora oleoides)*	Makrani	2002
MKD-78	Peel *(Salvadora oleoides)*	Makrani	2002

REGIONAL SPECIFIC VARIETIES

Ber

Thar Sevika: Developed by CIAH, Bikaner by hybridization from a cross Seb × Katha, Thar Sevika is an early-maturing variety. The fruits are juicy, sweet with TSS content of 22-24%. Fruits after consumption do not cause throat soaring, which is common in other cultivars. Average fruit yield is 30–32 kg/tree. The hybrid is also suitable for staggered picking which can be done up to third week of January.

Thar Bhubhraj: A selection from local material of Bhusavar area of Bharatpur district of Rajasthan, CIAH-Sel-1 is an early maturing cultivar having an average yield potential of 30–36 kg/tree. The fruits are ready for harvesting during last week of December-first week of January. The fruits are very juicy, sweet with TSS content of 22-23%.

Pomegranate

Phule Arakta: This variety has been developed by MPKV, Rahuri centre, by crossing ganesh and Gul-e-Shah Red. Tree is evergreen and spiny branches. The flowers are deep red in colour. Phule Arakta is heavy yielder and possess desirable fruit character. The fruits are bigger in size, sweet with soft seeds, bold red aril colurs and possess glossy dark red skin. It is susceptible to fruit spots and thrips.

Bhagawa: This variety has been developed by MPKV centre. This variety is also heavy yielder and possesses desirable horticultural characters. This variety matures in 180–190 days with average yield of 30.38 kg fruits/tree. Bigger fruit size, sweet, bold and attractive arils, glossy very attractive saffron colored thickh skin makes it suitable to distant market.

Mridula: This variety has been developed by MPKV centre. This variety is also heavy yielder and possesses desirable horticultural characters. Fruit size is dark red aril colour and matures in 150–160 days

Aonla

Narendra Aonla 6: It is developed by Kumarganj (Faizabad) centre. It is chance seedling of Chakaiya. Tree is tall, semi-spreading with sparse foliage. Fruit is small to medium, oval round, skin light green and late in maturity.

Narendra Aonla 7: It is also developed by Kumarganj (Faizabad) centre. It is seedling selection from open-pollinated strain of Francis. Tree is tall, semi spreading in growth habit with dense foliage. Fruits are medium to large, conical in shape with pappilate apex at pea stage. Skin is smooth, semi translucent, yellowish-green. Precocious and profuge-bearer.

Custard Apple

APK (Ca) 1: The custard apple accession AS1 has been released for commercial cultivation as **APK (Ca) 1** by Arrupukottai centre, during 2003. This variety is tolerant to drought. It has high yield potential of 14.9 kg/tree and yield 72 fruits/tree. The average fruit weight is 207.5 g. The fruits are medium in size, with TSS content 24.5 Brix.

Bael

Narendra Bael 5: This variety has been developed by Kumarganj (Faizabad) centre. It is early maturing cultivar, small in height (3–5 m), more spreading type. Fruits are sweet in taste

Thar Sevika Ber

Thar Bhubhraj Ber

Varieties of ber

Phule Arakta Pomegranate

Bhagwa Pomegranate

Varieties of pomegranate

Narendra Aonla 6

Narendra Aonla 7

APK (Ca)1 Custard Apple

Varieties of aonla and custard apple

(TSS: 35–38 °Brix) and less number of seeds (40–50). Average fruit weight is 900–1000 g and produce 50–60 kg/tree.

Narendra Bael-7: This variety has been developed by Kumarganj (Faizabad) centre. It is early-maturing cultivar, tall in height (5–7 m), less spreading type. Fruits are round and big in shape and average fruit weight is 3–4.5 kg.

Narendra Bael-9: This variety has been developed by Kumarganj (Faizabad) centre. It is early maturing cultivar, medium in height (4–6 m), more spreading type. Fruits are sweet in taste (TSS: 35–40 °Brix) and less number of seeds (30–35). Average fruit weight is 1–1.5 kg and produce 70–80 kg/tree.

TECHNOLOGIES DEVELOPED

Excellent publication on varietal descriptor of date palm, bael and ber were made at RFRS, Abohar; NDUAT (Faizabad) and CHES, Godhra, respectively. The varietal descriptor is valuable document for future researchers on crop improvement and germplasm conservation. Jobner Centre has prepared a document on entomological findings on Arid Zone Fruits.

Evaluation of available germplasm and trials conducted with selected varieties led to identification and recommendation of varieties of these fruits for different regions. In ber, var. Gola has been found to be suitable for very dry areas, Kaithali for slightly higher rainfall areas and Umran for moderate arid and semi-arid areas.

Pomegranate variety Phule Arakta and Bhagawa have been reported as the most promising for Maharashtra and Andhra Pradesh and Karnataka (MPKV, Rahuri and ANGAU, Anantapur), Jalore Seedless and Jodhpur Red for Rajasthan (RAU).

Aonla varieties NA-7, Chakaiya, Kanchan and NA-6 have been found most promising uniformly all over the region. Date palm varieties Halawy, Barhee and Medjool are reported to be most suitable for date-growing regions. Custard apple varieties Island Gem, Bullocks Heart and Mammoth are suitable in Karnataka, Andhra Pradesh and Maharashtra. Fig varieties Condaria, Poona Fig and Dianna are reported as suitable for fig-growing areas.

Some centres have made promising selections to suit specific situations and purposes. Notable among these are:

- **Aonla:** NDUAT has reported that selection NA-19, 20, 24 and 25 are found excellent with respect of fruit yield and quality characters.
- **Ber:** NDUAT reported Narendra Ber Selection-1 and 2 have been recommended for commercial cultivation in Utter Pradesh. HAU (Bawal) has observed field tolerance to powdery mildew and fruitfly under low inoculum pressure in variety B.S.1. 2.
- **Date palm:** Varieties Halawy and Barhee are reported as most suitable for desert, variety Medjool for making dry dates (*chhuhara*) at PAU - Abohar and RAU - Bikaner and varieties Zahidi and Hallawy for soft dates (*pind khajoor*) PAU - Abohar. SDAU - Mundra has located very high yielding, good quality clones from wild date palm groves of Kachchh area, which were supposedly developed by Middle East settlers a few centuries ago.
- **Custard apple:** Variety Balanagar is suitable under very low rainfall situation (ANGRAU - Anantapur) and Selection A.S.1 under moderate rainfall situation (TNAU - Aruppukottai).
- **Fig:** Varieties Condario and Dianna are best selections for situations of minimum irrigation availability (MPKV-Rahuri).

- **Pomegranate:** Two promising types Bhagwa and Phule Arakta have been recommended and released for cultivation from MPKV Rahuri and Ruby and Amlidana have been released for table purpose and anardana production, respectively.
- **Bael:** NDUAT has selected two promising lines, NB 5 and NB 9, which possess salinity/sodicity tolerance. HAU (Bawal) has identified some promising germplasms having good plant characters.
- Crop improvement studies have led to production of improved types of pomegranate and custard apple. The IIHR, Bangalore, has released Arka Sahan custard apple hybrid having smooth skin, attractive pulp colour and sweetness with less seeds. The MPKV, Rahuri has released Hybrid 61 (Mridula) of pomegranate having red aril colour, very soft seed and other desirable characters.

Standardization of Propagation Techniques

Ber: Most centres have reported *in situ* patch budding as the most suitable for high success rate and tree vigour. June - July is reported as the best season.

Custard apple: ANGAU, Anantapur centre, has standardized softwood-grafting technique, which gave maximum success when done during January – March under local conditions. Patch budding has been recommended with good success at TNAU, Aruppukottai centre.

Pruning and Training

Ber: Both time and severity of pruning for optimum yield varies at locations and with climatic conditions. February is best time for pruning in southern locations, April for central India and May in northern India.

Date Palm: The 1: 6 bunch leaf ratios recorded maximum number of strands per bunch, number of fruits per strands, average bunch weight and yield per palm at doka stage at Bikaner, while at Mundra 9 bunch per palm was found best in respect of fruit quality attributes and yield.

Standardization of Plant Density

Plant density for optimum yield under local conditions of the centres has been standardized.

The TNAU (Aruppukottai) centre reported planting at 8m × 3m and 8m × 4m spacing respectively for Banarasi and Kaithali varieties of *ber* for high-density plantation.

The IIHR, Bangalore, reported 5m × 2m as optimum spacing for high-density planting of pomegranate. The MPKV (Rahuri) reported 1,000 plants/ha as optimum in pomegranate.

Water Management

Use of black polythene mulch for moisture conservation in ber has been reported by most centres to be superior to all other mulches in promoting moisture status, shoot growth, flowering, fruit weight and yield. The local alternative materials found effective are paddy straw (NDUAT,-Faizabad), rice husk, groundnut husk (APAU, Anantapur), and coir pith (TNAU, Aruppukottai).

Mulching in date palm with black polythene has been found suitable, when laid after flowering in improving yield and fruit size (RAU, Bikaner) besides some local weeds like *bui* (*Aerva persica*), *shaniya* (*Crotolaria burhia*), and *ker* (*Capparis decidua*). Local grasses have also been used by some centres (HAU, Bawal, RAU, Jobner).

The highest fruit yield was obtained at 250% normal with 5% slope in ber (Arrupukottai).

Application of water through drip significantly increased the yield and quality parameters up to 60% daily evaporation replenishment at Bawal, while at Anantpur, plants irrigated with drip and mulch with irrigation schedule at IW/CPE ratio of 0.8 in pomegranate fruit crops. At Faizabad in aonla drip irrigation with 0.60 IW/CPE at 3rd day irrigation was found most effective.

Nutrition Management

At Abohar 4.4 kg urea/plant/year in two splits half before flower and rest at the emergence of spathe gave maximum fruit yield in fullgrown date palm.

The 1–3 leaves can be used for N, P and K estimation and third leaf from top can be used for micronutrient analysis for custard apple (Anantapur)

Under Bangalore condition, a fertilizer dose of 500 g N, 250 g P_2O_5 and 250 g K_2O per plant/year in two splits during May and another post-monsoon along with 3-4 basket of FYM and 1 kg neem cake for each plant of pomegranate is recommended.

In aonla, fertilizer dose with combination of 100 g N, 50 g P_2O_5, 50g K_2O/plant/year (1000 g N, 500g P_2O_5 and 500 g K_2O of 10 year old plant) is recommended for commercial cultivation in Uttar Pradesh.

In fertilizer dose with combination of 100 g N, 50 g P_2O_5, 50 g K_2O per plant per year (800 g N, 400 g P_2O_5 and 400 g K_2O of 8 year-old plant) is recommended for commercial cultivation of bael in Uttar Pradesh (Faizabad).

In bael, six months old leaves from middle portion of the shoot should be sampled to know the nutrient status of plants.

Maximum fruit yield in ber was obtained at Bawal when 50% of the nutrients were applied through chemical and rest through biofertilizers and FYM.

At Bikaner, fruit yield/palm at doka stage was highest under foliar spray of $FeSO_4$ plus thiourea treatment. The leaf sampling in date palm may be taken from middle portion of leaflets before flowering for nutrient estimation.

Flowering, Fruit Development and Ripening

The NDUAT (Faizabad) has reported the phenomenon of self-incompatibility in most of the commercial varieties of aonla and need of proportionate planting of pollinizer varieties. The NA-6 was found to be the best pollinizer for NA-7.

At Bangalore, hand-pollination with Squamosa (custard apple) pollen is recommended to get not only higher productivity, but also large sized symmetrical Arka Sahan fruits.

Pre-harvest application of ethrel @ 100 ppm at colour break stage was found to improve fruit weight and quality in date palm (RAU, Bikaner). Removal of one-third strands from the centre of the bunch at the time of pollination induced early maturity and improved quality of dates. Spray of ethrel (1500 ppm) at colour break stage gave uniform ripening and higher TSS in *doka* and *dang* fruits of date palm (PAU, Abohar). Application of 1000 ppm ethepon at colour turning stage induced early ripening by one week in Halawy variety of date palm.

Application of 750 ppm of ethephon at colour turning stage induced early ripening in ber (NDUAT, Faizabad). In pomegranate, 50 fruits/tree was found to be optimum for highest return per tree (MPKV, Rahuri). Defoliation of trees by two sprays of 0.5% potassium iodide at weekly intervals induced higher fruit yield in custard apple. Application of antitranspirants like power oil (1.5%), Kaolin (5.0–7.5%) and liquid paraffin (1%) have been found to improve yield and check fruit cracking in pomegranate and ber.

Pest and Disease Management

Major pests and diseases of arid zone fruits have been recorded and documented by the centres for their regions. Based on general distribution, severity and presumptive crop loss, the following pests and diseases are regarded as more serious.

Ber

Powdery mildew is very general in distribution; its severity is positively correlated with RH and negatively with maximum and minimum temperatures (CCSHAU, Bawal; RAU, Jobner), particularly during October — December (ANGAU, Anantapur). Areas receiving winter rains (coastal Tamil Nadu - Aruppukottai) escape the disease.

Isariopsis black leaf spot is very general in distribution and is more serious in hot and humid winter areas. *Alternaria* leaf spot has localized distribution.

Fruit fly (*Carpomyia vesuviana*) is general in distribution and fruit-borer (*Meridarchis scyrodes*) is predominant in southern and western parts. Bark-eating caterpillar, chaffer beetle and hairy caterpillar are other pests of localized importance.

Pomegranate

Pomegranate suffers from leaf and fruit spots caused by several pathogenic fungi. The more common fungi causing appreciable damage, particularly in fruits, are *Alternaria, Colletotrichum, Cercospora* and *Drechslera*. These are more serious in *mrig bahar* flush due to high humidity (MPKV, Rahuri).

Pomegranate wilt or quick decline caused by *Fusarium* is reported to be very serious in Maharashtra (MPKV, Rahuri). A similar disease has been reported in Karnataka and Andhra Pradesh (IIHR, Bangalore) but its causal agent has not been confirmed. Black necrosis of aril is reported to be very serious in Maharashtra. Oily leaf spot is also reported to be very serious in Maharashtra and Karnataka, but its causal organism has not been confirmed so far.

Pomegranate butterfly (*Virachola* sp.) is a more common pest causing damage to fruits in most areas. Bark-eating caterpillar (*Indarbela* sp.) and fruit-borer (*Deudorise* sp.) is also common in certain areas (MPKV, Rahuri and SDAU, SK Nagar).

Aonla

Rust is the most serious disease in Uttar Pradesh and Rajasthan. Rust affected fruits are prone to attack of several secondary fungal rots causing extensive damage (NDUAT, Faizabad). No serious pests are recorded in aonla.

Rust disease is found to cause damage in central and southern parts of the country with greater relative seriousness.

Date palm

Graphiola leaf spot is the only disease of concern having a general distribution. Rotting of sucker hearts in new plantations was reported. *Botryodiplodia* is a common pathogenic fungus (RAU, Bikaner). Scale insects (*Parlatoria blanchardi*) cause significant damage in some areas (GAU, Mundra; RAU, Bikaner; PAU, Abohar).

Three or four bird species cause considerable damage to date palm, pomegranate, ber, particularly in desertic areas of Rajasthan and Gujarat.

Custard Apple

No serious disease is reported. A case of dieback of twigs was seen at IIHR, Bangalore.

Host Resistance

Attempts have been made to locate host resistance/tolerance in germplasms to diseases and pests. The details are given below:

The MPKV, Rahuri, identified four powdery mildew tolerant varieties of *Z.mauritiana* lines. Attempts have been made to transfer the observed resistance to susceptible cultivated varieties by hybridization, rescue of abortive embryo and its multiplication through tissue culture. Common varieties showing powdery mildew tolerance in the field at MPKV Rahuri are Darakhi-1 and Darakhi 2, Guli, Vilaiti and Seedless. The CCSHAU, Bawal centre, has also reported a number of varieties as moderately resistant/tolerant to powdery mildew. Prominent among them are BS-1, Sanaur-3, Kathaphal, Kishmish, and Narma. These show tolerance against low disease pressure. Of these, BS-1 and Sanaur-3 showed some degree of tolerance to fruit fly also.

Screening of germplasms against leaf and fruit spots in pomegranate by MPKV, Rahuri centre revealed 5 varieties (Uthukul, Malta, Gul-e-Shah, Khog, Speen Danedar) as completely free from leaf spots and one variety (Yercaud HS) as resistant. Two varieties (Achik Dana and Alandi) showed some degree of resistance.

The MPKV, Rahuri centre has also reported field resistance in ber varieties against Isariopsis black leaf spot (9 varieties) and *Alternaria* leaf spot (6 varieties).

The CCSHAU, Bawal centre, has reported differences in the degree of fruitfly incidence in *ber* varieties. Some varieties (eg. Sanaur 3, Illaichi, BS 1, BS 2, etc.) were less susceptible.

The RAU, Bikaner centre has reported the least incidence of scale insect in Medjool date palm. Variations in intensity of Graphiola leaf spot in date palm have also been reported.

Epidemiological Studies

Epidemiological information on some pests and diseases has been gathered. The details are given below:

At Anantapur (Andhra Pradesh), incidence of powdery mildew in ber coincides with higher temperature and humidity during October. Progress of the disease has strong negative correlation with maximum and minimum temperature and positive correlations with rainfall during the vulnerable period (October — January) in Rajasthan (RAU, Jobner). Climatic factors promoting fruit fly incidence in ber have been studied. Minimum temperature of 5–8 °C and maximum temperature of 19–21 °C favour pest attack. Among the cultural practices promoting disease control, late pruning in May checked black leaf spot disease at TNAU, Aruppukottai centre. No correlation of disease incidence with pruning time was recorded by any of the centres in case of ber powdery mildew. *Graphiola* leaf spot of date palm is favoured by maximum temperature around 20 °C and RH of 70–80% (morning) and 50% (evening). Maximum incidence of scale insects was observed during October — March (RAU, Bikaner).

Management

- The Fig rust can be controlled by 3–4 sprays of 0.2% chlorothalonil or copper oxychloride (0.40%) during the crop susceptibility period.

- For the control of *Graphiola* leaf spot of date palm, four sprays of 0.4% copper oxychloride at 15 days interval during August to November has been found effective.
- For the control of ber fruit fly carbaryl 50 WP 0.1% has been found to be the best pesticide.
- Three to four sprays at 0.01% starting from peanut stage till maturity can check the fruit fly incidence effectively. Fenthion 100 EC 0.05% at an interval of 15 days checked the spread of disease. Other pesticides found effective are deltamethrin (0.002%) or fenvalerate (0.01%).
- Monocrotophos at 0.03% has been found to be effective for the control of ber fruit borer, which is relatively serious in southern and western India.
- Combined spray of wettable sulfur and carbaryl has been found highly effective for the control of both fruit fly and powdery mildew.
- For the control of date palm scales, Monocrotophos and Dimethoate at 0.05% concentration are reported to be the best pesticides.
- For the control of chaffer beetles one or two applications of carbaryl 50 WP at 0.2% or monocrotophos 36 WSC at 0.05% after the onset of monsoon has been recommended, by CCSHAU, Bawal Centre.
- For ber powdery mildew, 2–4 sprays of dinocap or Carbendazim or Triedomorph or Thiophenate methyl (0.1%) or wettable sulfur (0.2%) at 15–20 days interval were found as effective at during the most vulnerable period starting with the onset of the disease.
- Isariopsis black spot of *ber* can be checked by 2-3 sprays of Captafol or Mancozeb or copper oxychloride (0.2%) or Carbendazim (0.1%) at 15 days interval during the humid, vulnerable period.
- Three to four sprays of Chlorothalonil (0.2%) at 15 days interval starting from the initiation of symptoms were found to be effective for the control of aonla rust.
- For the control of leaf and fruit spots of pomegranate, spraying of Ziram (0.25%) or Bordeaux mixture (1%) or captan (0.2%) were found effective (MPKV, Rahuri). One prophylactic spray at flowering or fruit set followed by 4–9 sprays at 10–15 days interval, depending upon progress of the disease, especially during *mrig bahar,* were recommended. Other fungicides found effective for fruit spot include mancozeb (0.25%), captafol (0.2%) and carbendazim (0.1%).
- Pomegranate bark eating caterpillar can be effectively checked by spraying the infested bark with any of the pesticides like monocrotophos 36 WSC (0.08%), Fenvalerate (20 EC (0.04%), Quinalphos 25 EC (0.08%).
- Rotational use of Deltamelthrin 28 EC at 0.002% and carbaryl 50 WP at 21 days interval starting from fruit set has been found the most effective for the control of pomegranate fruit borer.

Attempts have also been made to control pests by non-pesticidal measures. Bagging of pomegranate fruits with butter paper gave good protection against pomegranate butterfly. Black beetle (*Chilocerus nigritus*) has been identified as an effective predator of date palm scale. Attempts have been made to control date palm weevil/scale with pheromone traps and bio control agents. For the control of bird damage in date palm, use of 600 gauge LDPE bags or 3 × 3 mm mesh reusable wire nets has been recommended.

Post-harvest Technology

Information on post-harvest pathology and preservation of arid zone fruits, particularly of date palm has been provided.

Ber:Fruits in storage in polypacks showed rotting due to occurrence of several fungi; mainly *Aspergillus* and *Penicillium* spp Ber varieties Kaithali and Sanduri were found less susceptible (TNAU, Aruppukottai).

Date Palm: Fruits were seen rotting in storage after harvest was mainly due to *Aspergillus* spp. Pre-harvest fungicidal sprays and margocide checked the rot to a great extent.

Suitable date palm varieties for dessert (Halawy, Barhee) or for processing as dry dates (Medjool) and soft dates (Khadrawy) have been identified.

Boiling of *doka* stage berries for 20 minutes followed by dehydration at 50° C for 70 hours is recommended for preparation of good quality dry dates (*chhuhara*).

Half *dang* stage berries can be processed to prepare soft dates (*pind khajoor*) by dehydrating at 50 °C for 70 hours.

Khalal stage of cv. Halawy has been found appropriate for making best quality dry dates in Kutch region of India.

Recipes have been standardized for pind khajoor, beverages, jam and date chutney.

TECHNOLOGY TRANSFER

IItem	*Particulars*
Ber	*Gola for high yield and early maturity, Umran for high yield*Narendra Ber Sel
(Ziziphus mauritiana var. rotundifolia)	1 and 2 have been recommended for commercial cultivation in Uttar Pradesh.
Propagation	**In-situ* budding: on *Z. nummularia* for dwarfing, on Z. rotundifolia for vigour*For rapid multiplication of *ber* under rainfed conditions, the seeds of rootstock are sown in 300 gauge polythene tubes of 25cm x 10 cm size, filled with a mixture of equal parts of FYM, clay and sand in the first week of April, 90 days old seedlings are budded during July so that budded plants become ready during August for field planting, taking only 120 days for raising an orchard.*In irrigated areas, *ber* seeds, after soaking in water for 48 hours or in concentrated sulphuric acid for 6 hours, are sown in the nursery bed during April, seedlings are budded during July — September and the budded plants transferred in polythene bags and sprayed with Waxol are planted in the field.
Pruning/training/plant architecture	*January — April in south, April in west and central and May in north India.
Water management	*Black polythene mulch was found superior in retaining soil moisture*Rain water harvesting at 5% surface slope have been recommended for obtaining higher fruit yield in ber.
Nutrition management	*In case of integrated nutrient management results were obtained when 50% of the nutrients were applied through chemical fertilizers and rest through bio-fertilizers and FYM. This treatment gave significantly higher plant height, spread, fruit weight and yield as compared to control*In ber, fertilizer dose with combination of 100 g N + 50 g P_2O_2 + 50 g K_2O/plant/year of age (800 g N+ 400 g P_2O_2+K_2O for 8 years old plants) is recommended for ber cultivation in Uttar Pradesh.

Contd.

Item	*Particulars*
Weed management	Weed management studies at Bawal in ber cv. Kaithali showed that application of Atrazine @ 25 kg/ha as pre-emergent spray and Paraquat 1.5 litre/ha as post emergence spray is the best for controlling weeds.
Pest/disease management	*One prophylactic spray at flowering with 7 subsequent sprays at 10 days interval with 0.2% wettable sulphur were effective for the control of powdery mildew in ber.*For the control of powdery mildew in *ber*, 2–4 sprays of 0.2% sulfex or 0.1% karathane or 0.2% Bavistin should be done at 20 days interval.*For the control of leaf spot, mancozeb (0.3%) found to be best fungicide followed by captafol (0.3%) and Topsin-M (0.1%).*Black leaf spot of ber can be controlled with Carbendazim (0.1%) or Captafol (0.2%) spray.*Control of *Isariopsis* leaf spot of ber can be done by 0.2% Difolatan or dithane Z-78 or Blitox-50 when sprayed 2-3 times at 15 days interval.*Two sprays with Monocrotophos 0.03% and third spray with 0.1% carbaryl at 15 days interval starting from pea size fruits controlled ber fruitfly effectively. *For the control of *ber* fruitfly, 2 sprays of 0.03% phosphomidon or dimethoate or monocrotophos should be done at monthly interval.*For the control of *ber* fruitfly, prophylactic sprays of 0.03% oxydemeton or dimethoate starting from the stage when 70-80% fruits have attained pea size may be done. The spray should repeat at one-month interval. To check the incidence of fruitfly at maturity or ripening stage, *ber* trees should be sprayed with 0.05% Malathion + 0.5% *gur* or sugar solution repeating the spray at weekly interval, if necessary. *Three sprays of Deltamethrin 0.002% or 0.05 % dichlorvos starting from pea size fruits controlled fruit borers. *Spray of 0.1% carbaryl 50 WDP or monocrotophos 0.05% proved effective to control chafer beetles.
Pomegrante **(*Punica granatum L.*)**	*Varieties of pomegranate viz. Bhagava (also known as Shendri, Ashtagandha, etc.) and Phule Arakta have been recommended and released for cultivation respectively.
Pruning/training/ plant architecture	It is recommended that planting of pomegranate be done on light soils i.e., upto 45 cm depth and at 4.5m x 3.0m distance (i.e. 740 plants/ha.) The planting distance of 5m ´ 5m is recommended under Bawal conditions.
Aonla (*Emblica officinalis*)	Varietel trial results reveal that in aonla cvs. NA-6 and NA-7 gave excellent performance with respect to plant structure, fruiting behaviour and fruit quality continuously throughout the experimentation.
Propagation	Patch budding during July-August proved most successful.
Training / plant architecture	Nil.
Water management	Drip irrigation at 0.60 IW/CPE ratio at 3rd day irrigation is recommended for aonla production under saline-sodic soil condition. This treatment reduced soil pH, ECe, ESP and available nitrogen along with better plant growth, fruit yield and quality of fruits.
Nutrition management	Foliar spray of Zn (0.5%) + boron (0.2%) + Cu (0.4%) during April and September months reduced fruit necrosis, improved plant growth and quality of aonla fruits cv. Francis.Fertilizer dose with combination of 100 g N + 50 g P_2O_5 + 50 g K_2O per plant per year (1000 g N + 500 g P_2O_5 + 500 g K_2O for 10 years old plants) is recommended for commercial cultivation in Uttar

(Contd.)

Item	Particulars
	Pradesh.Combined spray of $ZnSO_4$ (0.5%) + Thiourea (0.25%) reduced fruit drop, increased fruit yield and fruit quality in aonla.Aonla cvs. NA-6, NA-7 and NA-10 can be growth successfully in sodic soil up to 35.0 ESP and saline soil upto 10.0 dSm^{-1} ECe.
Pest/disease management	Pre-harvest spray of Mancozeb (0.3%) thrice at 15 days interval from mid of September gave excellent response to check aonla rust disease.
Date palm	Halawy and Barhee cultivars of date palm can be recommended for commercial cultivation in the marginal arid regions.
Propagation	Rooted suckers weighing more than 10 kg should be planted after treating them with 10000 ppm IBA at the bottom for better survival in field.
Pruning/training /plant architecture	*Thinning of date palm bunches by removal of 1/3 strands from their centres, gave best quality dates and hastened maturity of fruits.
Pest/Disease management	Cover the date palm bunches with transparent alkathene (600µ gauge low-density polythene) at colour break stage of dates). Transparent alkathene covering also protect the dates against rain in late maturity varieties.
Post-harvest technology	It is recommended that in order to produce best quality of chhuhara from fruits of Halawy variety harvested at khalas stage, the fruits should be first dipped in boiling water for 20 minutes and then dried in air circulation at 45 °C for 60-65 hours. It is recommended that, in order to produce pind khajoor from fruits of Barhee variety, the fruits should be dipped in boiling water for 60 seconds and then dried in air circulating tray at 45 °C for 60–65 hours (year-2002).It is recommended that, low to medium quality of date varieties are useful for good quality beverage with 13.0% TSS by adding ginger powder @ 5 g per liter juice. Local red variety is superior for making jam. TSS of jam should be 70% with 0.3% acidity. Low to medium quality of date variety is useful for preparing good quality date chutney. Ingredients of chutney should be 1 kg. Date pulp + 600g.sugar + 50 g. Tamarind pulp + 20 ml limejuice.
Custard apple	This accession was released as APK (Ca)1 custard apple and reccomended for commercial cultivation in Tamil Nadu. It is tolerant to drought and hence suitable for cultivation under dry land vertisol. Its yield potential is 14–15 kg/tree. It bears 72 fruits per tree weighing 207.5 g each. The fruits are medium in size, heart shaped and sweet in taste with a TSS of 24.5° Brix.
Propagation	Softwood grafting durig March-April in south India.
Nutrition management	Under nutritional trials in Annona cv. Arka Sahan, preliminary data of one year revealed that a dose of 150 g N, 100 g P_2O_5 and 100 g K_2O (nitrogen as Ammonium Sulphate) gave maximum fruit yield of 23.52 kg/plant and also fruit number (83). This was closely followed by treatment of 75 g nitrogen, 50 g P_2O_5, 50 g K_2O, 5 baskets of FYM and 3 kg vermicompost, where total number of fruits were 60 and fruit yield was 16.52 kg/plant. For obtaining highest yield, 30 g each of Azotobacter + Azospirillum + P solubilization culture and standard dose of NPK (250:125:250 g/plant) can be applied while

(Contd.)

Item	*Particulars*
	considering the reduction of cost of cultivation due to reduction in 50 per cent application of NPK, the treatment with Azotobacter, Azospirillum, Phosphobacteria and VAM + P solubilization culture + 50% NPK is found to be beneficial with high yield.
Fig	In germplasm evaluation, the varieties of Fig, viz., Dinkar, Excel, Diana, Conadria, Timla and Chalisgaon Selection have been added.
Propagation	Chip budding and wedge grafting can be used as commercial method of propagation.
Nutrition management	Application of 900 g, 250 g and 250 g NPK per plant to Poona Fig spaced at 5 x 5 m was found optimum in Maharashtra.
Pest/disease management	*Fungicides chlorothalonil (0.2%) and copper oxychloride (0.4%) were the most effective for the control of fig rust*Fig rust can be effectively controlled by 4 sprays of 0.1% Dithane M-45.
Bael	NB-17, NB 5 and NB 9 are the promising cultivars. Bael cvs. NB-5 and NB-9 can be grown successfully in sodic soil up to 30.0 ESP and saline soil up to 5.0 dSm^{-1} ECe.Flowering behaviour studies in bael reveal that flower bud emergence started in 3rd week of May and maximum in NB–4 (29 days) and minimum in NB-7 (15 days). Anthesis was maximum between 6-7 AM. The stigma was maximum on the day of anthesis. The fruit set and retention was maximum with cv. NB-9 followed by NB-5.
Propagation	Patch or modified ring budding during July — September.
Nutrition management	In bael, six months old leaves from middle portion of the shoot should be sampled to know the nutrient status of plants.
Phalsa	
Pruning/training/plant architecture	Pruning trial in bush phalsa, the plants pruned at 20 cm. height from ground level gave highest fruit yield and quality.

5

Cashew

M.G. Bhat

The All India Coordinated Spices and Cashewnut Improvement Project (AICS & CIP) was started during the IV Five-Year Plan in 1971 with its headquarters located at the Central Plantation Crops Research Institute, Kasaragod. During the VII Plan, the ongoing project (AICS & CIP) was bifurcated into two separate projects, one on cashew and another on spices. The headquarters of the independent All India Coordinated Research Project (AICRP) on cashew was shifted to the newly established National Research Centre for Cashew, Puttur, in 1986.

The AICRP on Cashew has presently eight centres and one sub-centre; of which four were started at the inception of AICS & CIP in 1971 [Bapatla (ANGRAU the then APAU); Madakkathara (KAU, shifted from Anakkayam); Vengurla (BSKKV, the then KKV) and Vridhachalam (TNAU)]. Besides these four centres one centre was located in CPCRI, Regional Station, Vittal, DK (Karnataka) but later it was discontinued from AICRP-Cashew. During the V Plan, one centre at Bhubaneswar (OUAT) and in the VI Plan, two centres, one at Jhargram (BCKVV) and another at Chintamani (UAS) were added. During VIII Plan, one centre at Jagdalpur (IGAU) and a sub-centre at Pilicode (KAU) were also started. These centres of AICRP on Cashew are located in ten cashew-growing states of the country and are under the administrative control of different SAUs.

i) Cashew Research Station
Acharya NG Ranga Agricultural University
Bapatla – 522 101, District, Guntur
Andhra Pradesh

ii) Cashew Research Station
Department of Horticulture
Orissa University of Agriculture and Technology
Bhubneswar – 751 003, Orissa

iii) Agricultural Research Station
AICRP on Cashew
University of Agricultural Sciences
Chintamani – 563 125, Kolar District
Karnataka

iv) SG College of Agricultural and Research Station
Indira Gandhi Agricultural University
Kumharwand Jagdalpur, District, Bastar
Chhatisgarh – 494 005

v) Regional Research Station
Bidhan Chandra Krishi Vishwa Vidyalaya
Jhargram Farm, Jhargram – 721 507
Midnapore (West) District, West Bengal

vi) Cashew Research Station
Kerala Agricultural University
Madakkathara – 680 651
District, Thrissur
Kerala

vii) Regional Agricultural Research Station
Kerala Agricultural University
Pilicode – 671 353, District, Kasaragod
Kerala

viii) Regional Fruit Research Station
Dr. Balasaheb Sawant Konkan Krishi Vidyapeeth
Vengurla – 416 516
District, Sindhudurg
Maharashtra

ix) Regional Research Station
Tamil Nadu Agricultural University
Vridhachalam – 606 001
District, Cuddalore, Tamil Nadu

x) Birsa Agricultural University
Ranchi – 834 006
Jharkhand

xi) Agrl. Experiment Station
Navsari Agricultural University
Paria
Valsad District Gujarat

xii) Kither Rani Chennamma College of Horticulture
CIHS, Arabhavi – 591 310
Karnataka

xiii) ICAR RC for Goa

xiv) ICARC for NEH
Barapani, Shilong
Meghalaya

The objectives of the AICRP on Cashew are:

- Evolving high-yielding varieties with good kernel quality and tolerant to biotic and abiotic stresses. (the earlier objective before 2007 was : evolving high-yielding varieties with export grade kernels, tolerant/resistant to pests and diseases)
- Standardising agro-techniques for the crop under different agro-climatic conditions, and
- Evolving cost-effective and efficient pest and disease management practices.

GENETIC RESOURCES

All the centres of AICRP on Cashew have Regional Cashew Field Gene Banks (RCFGBs), wherein germplasm accessions are maintained and their performance is being evaluated. Most of the centres are involved in survey and collection of cashew germplasm having desirable yield and other attributes which are conserved in the RCFGBs and are being characterized as per IPGRI descriptors. Now the total germplasm collections come to 1,261. The centre-wise collections are given (Table 5.1).

Table 5.1. Cashew germplasm holdings in different centres

Region	*Centres*	*No. of Accessions*
East Coast	Bapatla	132
	Bhubaneswar	95
	Jhargram	106
	Vridhachalam	269
West Coast	Madakkathara	148
	Pilicode	45
	Vengurla	277
Maidan tract/others (Plains Region)	Chintamani	128
	Jagdalpur	61
	Total	1261

- Cashew germplasm collections at Bapatla were enriched by adding collections from plantations of East Godavari, Vishakapatnam and Srikakulam districts and also from other states. Two accessions, viz., 10/19 and 3/33, identified as high yielders were included in the MLT-92 trial at this centre. A total of 132 genotypes have been presently conserved in the Regional Cashew Field Gene Bank (RCFGB), of which T 268 and T 15/4 gave nut yield of more than 5 kg/tree in the fifth harvesting.
- At Bhubaneswar, among the 47 germplasms planted during 1975, the best yielders for cumulative average yield (5 harvesting) were M 17-4 (5.48kg), M 26/2 (5.42kg), M 16/3 (4.95kg) and T.No.135 (4.19kg) during 1980's and 1990's.
- A total of 106 germplasm accessions are being evaluated at Jhargra. Of which, NRC 137, NRC 138 and NRC 193 were superior in terms of yield characteristics. The overall best performance was in JGM 38 which gave an annual yield of 14.70 kg/tree and had a shelling percentage of 28.30.
- A total of 269 cashew germplasm accessions are conserved at RCFGB, Vridhachalam, which were collected from forest plantations of Tamil Nadu, viz., Narimanam, Andimadam, TAFCORN, Pudukottai, Andhra Pradesh, Karnataka, Kerala and Maharashtra. The highest cumulative yield (3 harvesting) of 8.27kg/tree was recorded in M 33/3 followed by 6.83kg/tree in M 44/3 (VRI-2) (4 harvesting). Out of the germplasm accessions evaluated, TK-1 recorded the highest annual nut yield of 3.63 kg /tree, followed by SL–1 (3.12 kg/tree). M26/2, NF40 were found promising having 28.73 and 25.44kg cumulative nut yield per tree. Eighteen accessions were found to be promising, having consistency in nut weight (>6.0g) and shelling (>26.0%). Survey was conducted in the Tsunami affected cashew plantations of Cuddalore and Nagapattinam districts by team of scientists from NRCC and RRS, Vridhachalam. Four cashew trees having possibility of salt tolerance were identified and grafts were prepared from these trees and planted for evaluation.
- A total of 148 accessions are being maintained in RCFGB at Madakkathara. Out of the 93 accessions evaluated, M-1-1, A-6-1 and A-26-2 were identified as promising types in respect of yield, nut and kernel characters. These accessions recorded the high nut yield of more than 17.0kg /tree and had a nut size ranging from 6.35-7.75g, with a kernel weight of 1.75-1.83g and an export grade of W 280. Out of the clonal progenies evaluated during 1992–96, H-3-9, Anakkayam-1, H-1596 and A-26-2 for yield; K-30-1, H-8-15 for nut weight and Brazil-239, Brazil-241 for apple weight were identified as promising. Exotic cashew seeds were procured from Malavi and Brazil during 1990's and were evaluated. A precocious dwarf (KGN-1) suitable for high-density planting was identified at Madakkathara during 1997–2001. However, on further evaluation it was found that the dwarf nature was not consistent in different centres.
- Presently, 45 germplasm accessions have been conserved at the RCFGB, Pilicode, which represents the major cashew growing tracts of Kerala. Of these, PCKC-9, TBP -1, and BLM-1 yielded more than 5.0 kg of nuts in the fourth harvesting. Till date, cashew accessions from Northern Kerala having higher percentage of bisexual flowers have been collected and are being evaluated for utilization in hybridization programme.

- At Vengurla, 277 varied cashew accessions have been collected and conserved. Out of the accessions evaluated for yield performance, 80 accessions have been planted in conservation block at closer spacing of 4m × 4m. Of these, 22 cashew accessions have been identified as promising in terms of nut yield and apple characters during 1990s. Three cashew types having more than 10.0g nut weight (Jumbo nuts) have been collected from different locations in Sindhudurg dist. The germplasm accessions from Goa (Balli-2) gave highest cumulative yield of 10.18kg in five years. The promising accessions are being utilized in hybridization.
- The hybrid, H 303, gave the highest yield of 1.90 kg /tree and the number of panicles/ m^2 was maximum (13.16) in T 44/3 at Vengurla centre.
- Presently, 128 germplasm accessions are available at Chintamani centre. About 25 accessions of locally identified grafts from Siddlaghatta and Srinivaspur taluks having desirable yield traits are ready for planting and evaluation. Accessions 35/3 ARSC and 41/3 ARSC recorded the maximum annual nut yield of more than 25 kg in the 13th harvesting at Chintamani centre.
- The AICRP-Cashew centre located at Jagdalpur has got a collection of 61 germplasm accessions. Of these, CARS-10 was found to be tolerant to low temperatures of about 2–2.5 °C and did not have any leaf shedding as in other accessions.

REGIONAL SPECIFIC VARIETIES

Based on the evaluation of diverse germplasm collections in different centres of AICRP on Cashew, so far 27 cashew varieties have been released under AICRP-Cashew. The centres at Madakkathara, Vengurla, Bapatla and Vridhachalam have several varieties with good yield performance. The varieties which are having lower shelling percentage or lower nut weight have been withdrawn from recommendation for cultivation. Some of the varieties have been proved to perform equally well in adjoining states or in a different location. For instance, BPP-8 developed and released from Bapatla (Andhra Pradesh), is performing well in Andhra Pradesh and also in Orissa and West Bengal. Dhana variety of Madakkathara (Kerala in West Coast) is doing well not only in Kerala but also in Orissa (in East Coast) and plains region of Karnataka. VRI-3, the variety of Vridhachalam (Tamil Nadu) has performed very well in Kerala. Vengurla varieties are recommended to neighbouring state of Goa and also Uttar Kannada district of Karnataka as they are performing well there.

The names and characteristics of 27 released varieties are given in Table 5.2:

At Bapatla, cluster bearing and bold nut types were identified and collected from Guntur and Vishakapatnam districts (Simhachalam). Out of high-yielding types, T.No.30/1, 3/33 and 10/19 were identified to be promising during 1992–96.

Two high-yielding hybrids which were semi-tall, had cluster bearing habit and recorded shelling percentage of more than 25.0 were released by APAU as BPP-1 and BPP-2 from Bapatla centre during 1982–91. The hybrid H-25 recorded a nut yield of 4.50 kg /tree followed by H-14 which gave a nut yield of 3.70 kg in the second harvest at Bapatla during 2002–06.

Under multi-location trials at Bhubaneswar, H 320 and H 303 obtained from AICRP-Cashew Vengurla gave highest yield of more than 9.0 kg/tree/year in the eight harvest. Kernel recovery in these hybrids was maximum (>30.0%) during 2002–06.

Table 5.2. Parentage, place of release, year of release, yield, nut weight, salient features/characteristics and area for which recommended in respect of released cashew varieties

Variety	*Hybrid/ selection*	*Parentage*	*Place of release*	*Year of release*	*Yield (kg/tree)*	*Nut wt (g)*	*Salient features*	*Recommended for*
BPP-1*	Hybrid	T.No.1 x T.No.273	Bapatla	1980	10.0	5.0	Semi-tall, cluster bearing, shelling percentage 27.5, W 400 grade	Andhra Pradesh (withdrawn)
BPP-2*	Hybrid	T.No.1 x T.No.273	Bapatla	1980	11.0	4.0	Semi-tall, cluster bearing, shelling percentage 25.7, W 450 grade	Andhra Pradesh (withdrawn)
BPP-3*	Selection	3/3 Simhachalam	Bapatla	1980	11.0	4.8	Shelling percentage 28.1, W 400 grade	Andhra Pradesh (withdrawn)
BPP-4	Selection	9/8 Epurupalam	Bapatla	1980	10.5	6.0	Poor shelling percentage (23%), W 400 grade	Andhra Pradesh
BPP-5*	Selection	T.No.1	Bapatla	1980	11.0	5.2	Poor shelling percentage (24%), W 400 grade	Andhra Pradesh (withdrawn)
BPP-6	Selection	T.No.56	Bapatla	1980	10.5	5.2	Poor shelling percentage (24%), W 400 grade	Andhra Pradesh
BPP-8	Hybrid	T.No.1 x T.No.39	Bapatla	1993	14.5	8.2	Shelling percentage 29%, W 210 grade	Andhra Pradesh, Orissa, West Bengal
Bhubaneswar-1	Selection	WBDC-V (Vengurla 36/3)	Bhubaneswar	1989	10.5	4.6	Cluster bearing, high shelling percentage (32%), W 320 grade	Orissa
Chintamani-1	Selection	8/46 Taliparamba	Chintamani	1993	7.2	6.9	Shelling percentage 31%, W 210 grade	Karnataka – Maidan parts
Jhargram-1	Selection	T.No.16 of Bapatla	Jhargram	1989	8.5	5.0	Shelling percentage 30%, W 320 grade	West Bengal
Anakkayam-1*	Selection	T.No.139 of Bapatla	Anakkayam	1985	12.0	6.0	Vigorous type, short duration of flowering, shelling percentage 28%, W 280 grade	Kerala (withdrawn)
Madakkathara-1	Selection	T.No.39 of Bapatla	Madakkathara	1987	13.8	6.2	Shelling percentage 26.8%, Kernel weight 1.6g, W 280 grade	Kerala
Madakkathara-2	Selection	Neduvellur material	Madakkathara	1987	17.0	7.3	Shelling percentage 26.2%, Kernel weight 2g, W 240 grade	Kerala
K-22-1	Selection	Kottarakkara-22 (Layer-23)	Kottarakkara	1987	13.2	6.2	Shelling percentage 26.5%, Kernel weight 1.6g, W 280 grade	Kerala
Dhana	Hybrid	ALGD-1 x K-30-1	Madakkathara	1993	17.5	9.5	Shelling percentage 28.0%, cluster bearing, W 210 grade	Kerala, Orissa, Plains region of Karnataka
Kanaka	Hybrid	BLA-139-1 x H-3-13	Madakkathara	1993	19.0	6.8	Early variety, shelling percentage 31.0%, W 210 grade	Kerala
Priyanka	Hybrid	BLA-139-1 x K-30-1	Madakkathara	1995	16.9	10.8	Shelling percentage 26.5%, kernel weight 2.2g, W 180 grade	Kerala

(Contd.)

Variety	*Hybrid/ selection*	*Parentage*	*Place of release*	*Year of release*	*Yield (kg/tree)*	*Nut wt (g)*	*Salient features*	*Recommended for*
Vengurla-1	Selection	Ansur-1	Vengurla	1974	19.0	6.2	Shelling percentage 31%, Early flowering type, W 240 grade	Maharashtra, Goa and Karnataka
Vengurla-2*	Selection	WBDC-VI (Vengurla 37/3)	Vengurla	1979	24.0	4.3	Short flowering and fruiting phase, shelling percentage 32.0, W 320 grade	Maharashtra and Goa (withdrawn)
Vengurla-3*	Hybrid	Ansur-1 x Vetore-56	Vengurla	1981	14.4	9.1	High sex ratio, poor shelling percentage 27%, W 210 grade	Maharashtra and Goa (withdrawn)
Vengurla-4	Hybrid	Midnapore Red x Vetore-56	Vengurla	1981	17.2	7.7	Shelling percentage 31%, cluster bearing, high sex ratio, W 210 grade	Maharashtra, Goa & Karnataka
Vengurla-5*	Hybrid	Ansur Early x Mysore Kotekar 1/61	Vengurla	1984	16.6	4.5	Shelling percentage 30.0, very high sex ratio, W 400 grade	Maharashtra and Goa (withdrawn)
Vengurla-6	Hybrid	Vetore-56 x Ansur-1	Vengurla	1991	13.8	8.0	Shelling percentage 28%, W 210 grade	Maharashtra and Goa
Vengurla-7	Hybrid	Vengurla-3 x M 10/4 (VRI-1)	Vengurla	1997	18.5	10.0	Shelling percentage 30.5%, kernel weight 2.9 g, W 180 grade	Maharashtra, Goa & Karnataka
VRI-1 *	Selection	Vazhisodanaipalayam material	Vridhachalam	1981	7.2	5.0	Shelling percentage 28%, medium sized tree, W 320 grade	Tamil Nadu (withdrawn)
VRI-2 *	Selection	T.No.1668 of Kattupalli	Vridhachalam	1985	7.4	5.1	Wide adaptability only variety released at national level, shelling 28.3%, W 320 grade percentage 28.3% W 320 grade	Tamil Nadu (withdrawn)
VRI-3	Selection	M 26/2 – Edayanchavadi material	Vridhachalam	1991	10.0	7.2	Shelling percentage 29.1%, kernel grade W 210 grade	Tamil Nadu and Kerala

* Withdrawn from recommendation due to poor performance/lower nut size/lesser shelling percentage

The highest cumulative yield of more than 30.0 kg / tree / year up to the 13th harvest was recorded at Chintamani, in H 320, H 303 and NRCC Sel 2. Kernel recovery was maximum (29.3%) in T No. 3/28 during 2002–06. A promising hybrid H-1608 (Dhana) from Cashew Research Station, Madakkathara, gave a cumulative yield of 29.26kg/tree (5 harvesting) and had a nut weight of 7.42g at Chintamani, indicating its suitability for cultivation in the plains region of Karnataka.

At Jagdalpur, H-303 had a consistent yield performance during 2002–06 with an annual nut yield of 1.5 kg/tree, while shelling percentage was maximum (32.5%) for both Vengurla-4 and VRI-1. In the multil-ocation trials at Jhargram during 1990's, the highest yield was observed in VTH 59/2 (8.2kg), H-2/16 (6.3kg) and VTH-30/4 (6.2kg), which were also suitable for the plains region. BPP-2 gave highest nut yield of 3.07 kg /tree, while cumulative nut yield was higher in Vridhachalam-2 (8.41 kg.) and highest nut weight was recorded in H-303 (9.37 g). The multilocational trials hinted that certain varieties from coastal tracts could perform equally well in plains region also.

Trials conducted at Madakkathara indicated a significant positive correlation between nut characters, viz. nut weight, nut volume, length: breadth ratio and seedling vigour. Four hybrids viz., H-1597, H-1598, H-1608 and H-1610, were identified as high-yielders, of which H-1598 and H-1608 were released as "Kanaka" and "Dhana", respectively during 1990's from Madakkathara centre.

Various cashew varieties or accessions were evaluated under multilocational trials at Vengurla. Of these, M 44/3 (VRI-2), followed by M-10/4 (VRI-1) were found to be superior in growth and yield characters during 1980's and 1990's while, Vengurla-5 gave significantly higher yield upto 23.7 kg/tree, followed by Vengurla-2, H-1598 and H-1600. Out of the 23 hybrids developed, 3 hybrids had nut weight exceeding 10.0g and shelling percentage of more than 28.00. These hybrids in the first harvest could yield up to 2.0kg of nuts per tree.

Hybridization trials of Vridhachalam have been done using VRI-2 and VRI-3 as a common parent, both of which have more than 28% shelling and till date, 17 hybrids have been produced and are being evaluated. The hybrid H-10 recorded the maximum nut yield of 6.42 kg/tree at Vridhachalam and this also had maximum cumulative yield of 57.52 kg for 9 harvestings. The hybrids H-14 and H-15 had a shelling percentage of more than 26.0 at Vridhachalam during 2002–06.

Efforts are on at most of the centres of AICRP-Cashew to develop region specific varieties which have medium plant height, medium to bold nut size, higher number of flowering laterals and higher ratio of bisexual flowers.

TECHNOLOGIES FOR ENHANCING PRODUCTION AND PRODUCTIVITY

Since its inception, a total of 27 high-yielding cashew varieties have been developed and released for cultivation by different centres of AICRP Cashew.

At Madakkathara, it was established that cashew is amenable for air-layering, grafting and budding, in order to obtain true-to-type planting material. All the centres are presently adopting the softwood grafting technique for vegetative method of propagation of cashew which was standardized simultaneously by several centres of AICRP Cashew and NRCC. An innovative "Trench Method/zero energy sand humid chamber" for propagation of soft wood grafts was standardized at Bapatla, during 1982–96. The percentage of graft success was found to be above 80% even during February-March.

Trials on evaluation of fertilizer doses, fertilizer application in high-density plantation, evaluation of efficacy of drip irrigation and organic farming approaches have been laid out at several centres of AICRP on Cashew.

Spacing trials were conducted to arrive at an optimum planting density for maximizing the productivity. The planting density of 156 trees/ha was found to be profitable and hence was recommended for cultivation. Highest nut yield of 4.19kg/ tree was obtained with 200 tree/ha and a fertilizer dose of 225 : 75 : 75 Kg NPK/ha in Pilicode under spacing and fertilizer trials during 2002-06.

Application of 500g N; 125g P_2O_5 and K_2O each/tree/year was found to be suitable in most of the centres. During 1998–91, NPK fertilizer experiments conducted at Bapatla revealed that application of nitrogen increased cashew yields significantly over control. However, effect of phosphorous and potash were not significant.

At Jhargram, nut yield increased significantly with spraying of 2% urea along with Endosulfan (0.05%) thrice at flushing, panicle emergence and fruit setting stages during 1990's. Foliar application of urea at 4% and 3% along with normal soil application of fertilizer lead to higher yields of 90% and 80% respectively, at Vridhachalam during 1990's than the conventional soil application of fertilizers without urea spray. Foliar application of urea (2%) along with insecticidal application of Endosulfan (0.05%), increased the number of laterals/m^2, number of nuts/plant and also yield/plant at Bhubaneswar during 1990's.

In order to obtain higher returns from the cashew plantations during the initial cropping period, perennial trees like acacia, casuarinas, eucalyptus, cassia and Melia were planted at a spacing of 2m × 2m, whereas mango and cashew were planted at a spacing of 4m × 4m as intercrops at Bhubaneswar. The growth of eucalyptus was found to be better as compared to other perennial trees at Bhubaneswar during 1982–91. During the later part of crop growth, the perennial trees led to shading effect on cashew leading to reduction in nut yield.

Alternatively annual crops were evaluated at most of the centres to obtain higher returns during the initial cropping period. Intercropping with ginger, turmeric, cluster-bean, black-gram, horsegram, groundnut, vegetables and medicinal plants with cashew as main crop during the initial stage of orchard development were evaluated and found to be profitable. These crops were recommended for the economic upliftment of farmers at different locations.

Trials were conducted for management of tea mosquito bug (TMB) and cashew stem and root borer (CSRB) as well as other foliar pests which cause severe crop losses in different regions, during different years. Various insecticides were evaluated during different years for their efficacy. Effective spray schedule for the management of tea mosquito bug and other minor pests of cashew was standardized using monocrotophos (0.05%) at flushing, Endosulfan (0.05%) at flowering (now Endosulfan is withdrawn from recommendation) and Carbaryl (0.1%) at fruiting. During 1990's, it was found that skipping the first spray was as effective as adopting the recommended three spray schedule. Monocrotophos (0.05%) at flushing and Carbaryl (0.1%) at flowering and fruiting were found effective in controlling the TMB, shoot tip caterpillar, leaf and blossom webber as well as other foliar pests in majority of the AICRP Centres. Due to the controversy of aerial application of Endosulfan, this insecticide was withdrawn from usage in cashew cultivation.

For the control of Cashew Stem and Root Borer (CSRB) infestation, swabbing of neem oil (5%) up to one meter height of trunk twice in a year along with soil application of Sevidol (4G)

75g/tree was reported to be effective in several centres. However, due to lack of consistency, prophylactic trials were discontinued and post extraction prophylaxis (PEP) was evaluated which involved application of pesticide after removal of pest stages from infested trees. Chlorpyriphos (0.2%) and monocrotophos (0.2%) were found to be effective in preventing re-infestation by the pest. Up to 78.0% of treated trees did not have reinfestation by the pest when chlorpyriphos (0.2%) was applied as PEP treatment.

CROPPING SYSTEM APPROACH

In order to enhance the returns from unit cultivated area, different crops such as vegetables, tuber crops and pulses as well as medicinal plants have been evaluated over the years in different centres as intercrops in cashew orchard in the early stage of orchard life. A net profit of Rs. 13,650/ha in blackgram and Rs. 13,000/ha in green-gram was recorded at Bapatla during 2002–06 when additional fertilizer was applied to intercrops. In intercropping trial at Bhubaneswar, sesamum and horsegram were found to be suitable intercrops during initial cropping period of cashew. Highest net profit was recorded in pigeonpea Rs. 21383/ha which also had the highest cost benefit ratio of 1 : 1.63 at Jhargram during 2002–06.

The highest returns from intercrops evaluated at Madakkathara was from Tapioca (Rs 44,310/ha), which also had the highest cost benefit ratio 1 : 2.16 during 2002–06.

Results of raising forest tree species as intercrops in cashewnut indicated an adverse effect on growth and yield of main crop due to shading effect at Vengurla during 1990's. Hence, alternatively vegetable crops were evaluated Vengurla Centre. Of these, net returns from cucumber as intercrop was the highest (Rs. 33659/ha) which resulted in C: B ratio of 1 : 3.17 at Vengurla during 2002–06. At Vridhachalam intercropping with groundnut, consistently led to highest returns during 2003-04 and 2004-05 cropping seasons.

STRATEGIES FOR PEST AND DISEASE MANAGEMENT

Trials on management of tea mosquito bug (TMB) and other foliar pests using newer insecticides, evaluation of curative treatments for management of cashew stem and root borers (CSRB) as well as evaluation of cashew germplasm for tolerance to pest attack are being conducted at most of the AICRP Centres.

Effective spray schedule for the management of tea mosquito bug and other minor pests of cashew has been standardized using monocrotophos (0.05%) at flushing, Endosulfan (0.05%) at flowering (now Endosulfan is withdrawn from recommendation) and Carbaryl (0.1%) at fruiting, During 1990's it was found that skipping the first spray was as effective as adopting the recommended three spray schedule. The new insecticides Profenophos, Triazophos, L- Cyhalothrin and Chlorpyriphos were found to be on par with the recommended spray schedule in managing TMB and other foliar pests at most of the centres. At Chintamani, application of Chlorpyriphos (0.05%), Triazophos (0.1%), Ethofenprox (0.015%) and Profenophos (0.05%) were at par with the standard spray in controlling TMB, leaf miner, leaf and blossom webber and inflorescence thrips during 1990's.

Chemical control trials at Bapatla indicated effective management of foliage pests by Monocrotophos (0.05%) and Endosulfan (0.05%) followed by Carbaryl (0.15%) at foliage, panicle emergence and fruiting stages, respectively during 1990's.

To optimize pesticide usage, skipping of first spray for control of TMB and giving only two sprays at flowering and fruiting periods were found to be as effective as recommended three spray schedule at Vengurle during 1992–96. At Bhubaneswar, L-cyhalothrin was the most effective insecticide in managing leaf thrips, apple and nut borer and also shoot tip caterpillar, which is a regionally important pest in east coast region. Most of the AICRP-Cashew Centres reported L-cyhalothrin and Profenophos as the most effective pesticides for managing the foliar pests.

Under prophylactic trials on management of Cashew Stem and Root Borer (CSRB), swabbing of tree trunk up to a height of 1m. twice during April-May and November-December with neem oil 5% or kerosene and coal tar (2:1) or swabbing neem oil 5%, followed by application of Sevidol granules @ 75g/tree as pre-treatment prophylaxis (PTP) reduced percentage of fresh infestation to 5.6% at Bapatla during 1997–2001.

At Madakkathara under prophylactic control of stem and root borer, swabbing neem oil (5%) on collar portion up to one meter height thrice during a year and exposed stem and roots was found to be effective during 1992–96 at Madakkathara.

Lindane (0.2%) led to maximum number of tree without reinfestation (66.67%) followed by carbaryl (1.0%) (58.33%) while under untreated check 41.67 per cent of trees had no reinfestation by CSRB at Bapatla during 2002–06.

Treating CSRB infested trees with chlorpyriphos (0.2%) and carbaryl (1.0%) after grub removal was found to have the least percentage of treated trees with reinfestation (40.0%) by the pest at Jagdalpur during 2002–06.

Cent per cent of infested trees treated with chlorpyriphos (0.2%) and lindane (0.2%) as PEP treatment were without reinfestation of CSRB at Vengurla during 2002–06.

Maximum control of CSRB without reinfestation of treated trees was obtained by treatment with chlorpyriphos (0.2%), lindane (0.2%) and carbaryl (1.0%) at Vridhachalam during 2002–06.

TECHNOLOGY TRANSFER

Softwood grafting of cashew was standardized simultaneously at NRC for Cashew Nut and at several AICRP-Cashew centres. Grafting with 90 days old scion on 30–45 days old rootstocks during January, February, July to December is recommended for commercial multiplication with highest graft success depending on location.

Several cashew varieties have been released by AICRP-Cashew of which some of the salient released varieties are mentioned here. BPP-8, a bold nut cashew variety with a good shelling percentage was released from Cashew Research Station, Bapatla, during 1993 which was found suitable for commercial cultivation in Andhra Pradesh, Orissa and West Bengal. Chintamani-1 (ARSC 5/10) a clonal selection from 8/46 Taliparamba, having a yield potential of 14 kg/tree, medium nut size and is a late-bearing variety was recommended during 1993.

In comparative yield trials (CYT), on the basis of yield and nut characters, Vengurla-3 was recommended for Maidan region of Karnataka during 1992–96 from Chintamani centre. Under hybridization and selection programme, hybrid H-1591 was released as "Priyanka" for Kerala state. Hybrid H-1598 is also found to be promising and released by state varieties release committee as Kanaka. Hybrid H-1608 was released as Dhana from Madakkathara centre during 1992–96.

Control of major foliar pests of the region, viz. tea mosquito bug, leaf miner and leaf folder was achieved by spraying Monocrotophos (0.05%), Endosulfan (0.05%) and Carbaryl (0.1%) at new flush stage, panicle stage and inflorescence stage respectively. This has been recommended for the respective region by AICRP Centres during 1990's. Regarding synthetic pyrethroids tested against tea mosquito 0.01% Permethrin, 0.0075% Cypermethrin, 0.025% Decamethrin and 0.01% Fenvalerate were found equally effective, and hence suggested for pest management.

Monocrotophos (0.05%), Endosulfan (0.05%) and Carbaryl (0.1%) are the most effective both in reducing the intensity of tea mosquito and other foliage pest infestation and was recommended during late 1980's. However, Endosulfan was withdrawn from recommendation. Other newer pesticides were evaluated of which, L-cyhalothrin and Profenophos were found effective in managing TMB and other foliar pests.

Swabbing the tree trunk with HCH (0.2%) or aldrin (0.1%) was found to give better prophylactic control against stem and root borer during 1982–91 and was suggested as a plant protection measure at Bhubaneswar. For the control of stem and root borer, swabbing the trunk with 0.1% BHC (50 WP) as a prophylactic treatment was recommended by Vengurle centre during 1982–91.

The ban on HCH and Aldrin led to withdrawal of these two pesticides from recommendation. Prophylactic treatment was found to be unsuitable due to the high cost involved and inconsistency of the results. Hence, post-extraction prophylaxis (PEP) was evaluated in which swabbing chlorpyriphos (0.2) or monocrotophos (0.2%) could lead to maximum number of trees without re-infestation. Chlorpyriphos (0.2%) was found to consistently perform better than other pesticides in majority of the AICRP Centres.

Cashew grafts of recommended elite varieties were supplied regularly to different Government agencies and farmers by different AICRP-Cashew centres. This has helped in covering large potential areas with quality planting material to ensure better productivity. Higher productivity would help in meeting the raw nut requirement of the processing industries for export and indigenous consumption of kernels.

BENEFITS REAPED BY FARMERS CONSUMERS

Large number of high-yielding cashew varieties have been developed and released by AICRP-Cashew Centres. Several released cashew varieties were found suitable to other areas for which they were not initially recommended. The BPP-8 performed well in Orissa and West Bengal, Dhana performed well in Orissa and plains region of Karnataka. Vengurla-6 and Vengurla-7 have performed well in Goa and Vri-3 has performed well in Kerala. This indicated that research benefits have percolated to different cashew growing tracts of the country.

The Centres of AICRP on Cashew have been producing the cashew grafts of high-yielding varieties suitable for the respective region and these cashew grafts are being supplied to the cultivators and developmental agencies. Application of 1000g N and 250g K_2O/plant increased the nut yield significantly over control. At Madakkathara, on the basis of performance in the CYT, the cashew types BLA-139-1, BLA-39-4, H-3-17, NDR-2-1, K-22-1 and M-44-3 were recommended as improved selections for cultivation in Kerala. As a result of breeding, four hybrids, viz., H-1598, H-1600, H-1608 and H-1610 were identified as most promising from among 14 hybrids and these have been selected for multi-location trial from 1982–1991.

The variety released during the IX Plan in 1997 is Vengurla-7 which is a hybrid developed by the Regional Fruit Research Station, Vengurle (KKV). This hybrid is a cross between Vengurla-3 and M 10/4 (VRI-1). This variety is recommended for the states of Maharashtra, Goa and Karnataka (1997–2001). Two varieties, viz., VRI-2 (M 44/3) and VRI-3 (M 26/2), were released during 1982–91 from the Vridhachalam Centre. At Vengurla for minimizing fruit drop and increased fruit set and yield, three rounds of 3 per cent urea along with insecticides was recommended during 1982–91.

For effective control of tea mosquito and flower thrips, three rounds of spraying each at vegetative flushing, panicle emergence and fruit set with 0.05% monocrotophos, 0.05% endosulfan (now withdrawn from recommendation) and 0.05% diamethoate respectively, was recommended during 1990's. By the technologies developed by the centres of AICRP-Cashew and transferred to the farmers, the farmers have been able to get higher yield of cashew. Thus the production of raw cashewnuts in India has increased over years. This has helped cashew processing industries in reducing the quantity of import of raw nuts. Finally, the consumers and the nation get the benefit.

MILESTONES AND LAMP POSTS OF THE PROJECT

Release of a large number of high-yielding cashew varieties (27) suitable to certain regions and also some adaptable to different regions across states is an important milestone. Other milestone is the standardization of softwood grafting technique to multiply true-to-type planting material. Now over 70 lakh cashew grafts are produced in India by public and private sector using this technique. A large number of agro-techniques such dose of fertilizers, suitable spacing, pruning technique, control measure for serious insect pests have been developed.

6

Palms

S. Arulraj

Coconut, oil palm and palmyrah occupy a predominant place in the Indian rural economy. In order to strengthen the research in this vital sector of palms, the All India Coordinated Coconut and Arecanut Improvement Project was started in 1972. During the Seventh and Eighth Plan periods, oil palm and palmyrah were also included as mandate crops and the project was renamed as All India Coordinated Research Project on Palms. The project has, at present, 16 centres including its head quarters at Kasaragod (Kerala); Arsikere, Gangavathi (Karnataka); Veppankulam, Killikulam, Aliyarnagar, Aduthurai (Tamil Nadu); Ambajipeta, Vijayarai, Pandirimamidi (Andhra Pradesh); Ratnagiri, Mulde (Maharashtra); Bhubaneswar (Orissa); Jagdalpur (Chhattisgarh); Mondouri (West Bengal) and Kahikuchi (Assam). Out of the 16 centres, 10 are conducting research on coconut, four on oil palm and two on palmyrah.

SCENARIO ANALYSIS

Coconut, oil palm and palmyrah play an important role in the sustainability of fragile ecosystem in coastal and island regions. More than 20 million people in rural areas are engaged in production, processing and marketing of these crops. The long-term nature of research on these crops, the prospects of higher returns from research investment and the likely distribution of research benefits to the small holders, make it all the more imperative to develop a long-term perspective.

The average coconut productivity in India is still at a low level (6,776 nuts/ha/year) as compared to the production potential of 27,300 nuts/ha/year. This indicates the scope for increasing the productivity by bridging the yield gap through the use of quality planting material and adoption of better management practices. Though the coconut-growing regions have substantially increased their total nut production, there a need to develop location-specific agro-techniques for sustaining and improving the yield levels.

Oil palm is one of the highest oil yielding crops that could help the country to minimize the edible oil imports with the cultivation of location-specific oil palm hybrids along with the required management practices. Under optimum management conditions, farmers would be able to realize an yield of more than 20 tonnes of FFB/ha, yielding about 4 tonnes of edible oil/ha.

Palmyrah palm, adorns the dry landscape of the semi-arid regions of Tamil Nadu, Andhra

Pradesh, Orissa, West Bengal, Bihar, Karnataka and Maharashtra. India has nearly 102 million palms and half of the trees are in Tamil Nadu. Out of 51.9 million trees in Tamil Nadu, more than 50% of palms are concentrated in southern districts of Thoothukudi, Tirunelveli, Virudhunagar and Ramnad districts, while Thoothukudi district alone has a major share of 10 million trees. The palm offers vast scope for increasing the employment potential and thus the standard of living among the poor rural community.

Opportunities in Sector

Coconut offers wide scope in terms of improvement in coconut productivity and also vast employment potential through the adoption of appropriate coconut-based farming systems. Large-scale adoption of value-addition technologies in coconut sector is still a less exploited field in India as compared to other countries.

Palmyrah palm ranks first (among the available palm species) in yielding sugar as well as other edible and non-edible products. Many value-added products can be prepared from fruit and other parts of the palm. The palmyrah palm also serves as a windbreak and arrests sand dunes. Palmyrah palm could be grown in soil with sub optimal levels of physico-chemical characteristics. Tamil Nadu is a potential centre for the growth and development of "Palm Products Industry" to a greater extent so as to attract foreign exchange by way of export of palm products.

Objectives

- To identify, conserve and utilize elite gene sources for useful traits in coconut, oil palm, palmyrah and underutilized palms from different agro-climatic regions
- To evolve crop management technologies for location-specific coconut-based farming systems
- To evolve location-specific crop management technologies for oil palm cultivation
- To standardize location-specific biocontrol based integrated pest and disease management schedules for coconut.

History

The All India Co-ordinated Coconut and Arecanut Improvement Project was sanctioned in 1970 by the ICAR and the project started functioning in 1972. The time line of the Project over the years is presented here under:

1970	All India Coordinated Coconut and Arecanut Improvement Project was sanctioned by the ICAR
1972	The Project started functioning and 12 centres located in eight states to the project
1975	Five more centres added
1977	Konark Centre (OUAT) added
1980	Experiments in Konark shifted to Patha Farm
1980	Mondouri (BCKV) Centre in West Bengal added
1982	Kahikuchi Centre (AAU) in Assam sanctioned under VI Plan
1985	Kahikuchi centre started functioning

1987	Jalalgarh Centre in Bihar was recommended for closure in the VII Biennial Workshop
1988	Razole centre merged with Ambajipeta centre
1990	Four oil palm centres at Vijayarai (Andhra Pradesh), Mulde (Maharashtra), Gangavathy (Karnataka) and Aduthurai (Tamil Nadu) added
1990	Pilicode, Mahuva and Dapoli centres were closed during VII Plan. Coimbatore- coconut programmes shifted to Aliyarnagar
1990	Arecanut centre at Coimbatore closed
1992	Research work on arecanut phased out
1995	Andamans Centre closed during VIII Plan
1995	Two centres for Palmyrah research sanctioned during VIII Plan – Pandirimamidi in Andhra Pradesh and Killikulam in Tamil Nadu started
2000	Programmes of Konark centre shifted to Bhubaneshwar due to super cyclone

FUNCTIONAL PROFILE OF AICRP ON PALMS

Details of the 16 coordinating centres functioning under the AICRP on Palms are:

State	*Centre/Location*	*University*	*Areas of Research*
Andhra Pradesh	Agricultural Research Station Ambajipeta – 533 214 East Godavari District	Acharya N.G. Ranga Agricultural University, Rajendranagar Hyderabad – 500 030	Coconut Crop Improvement Agronomy, Pathology and Entomology experiments
	Agricultural Research Station Vijayarai – 534 475 West Godavari District	Acharya N.G. Ranga Agricultural University, Rajendranagar Hyderabad – 500 030	Oil Palm Crop Improvement and Agronomy experiments Palmyrah Crop Improvement Pathology experiments
	Horticultural Research Station Pandirimamidi Ramapachodaram PO – 533 288 East Godavari District	Acharya N.G. Ranga Agricultural University, Rajendranagar Hyderabad – 500 030	
Assam	Horticultural Research Station Kahikuchi, Guwahati – 781 017 Kamrup District	Assam Agricultural University Jorhat – 785 013, Assam	Coconut Crop Improvement and Agronomy experiments
Chhatisgarh	Saheed Gundadhoor College of Agriculture & Research Station Kumharawand Farm Jagadalpur – 494 005, Chhatisgarh	Indira Gandhi Krishi Vishwavidyalaya Raipur – 492 012	Coconut Crop Improvement experiments
Karnataka	Agricultural Research Station Arsikere – 573 103, Hassan District	University of Agricultural Sciences GKVK, Bangalore – 560 065	Coconut Crop Improvement Agronomy and Pathology experiments
	Agricultural Research Station Gangavathy – 584 227 Raichur District	University of Agricultural Sciences, Krishi Nagar Dharwad – 580 005	Oil Palm Crop Improvement and Agronomy experiments
Kerala	Central Plantation Crops Research Institute Kasaragod – 671 124	Indian Council of Agricultural Research	Coconut Crop Production

(Contd.)

State	Centre/Location	University	Areas of Research
Maharashtra	Regional Coconut Research Station, Bhatye – 421 612 Ratnagiri District	Dr Balasaheb Sawant Konkan Krishi Vidyapeeth, Dapoli – 415 712 Ratnagiri District	Coconut Crop Improvement Agronoy and Entomology Experiments
	Agricultural Research Station Mulde – 416 520, Kudal Taluk Sindhudurg District	Dr Balasaheb Sawant Konkan Krishi Vidyapeeth, Dapoli – 415 712 Ratnagiri District	Oil palm Crop Improvement and Agronomy experiments
Orissa	Department of Horticulture Orissa University of Agriculture & Technology Bhubaneswar – 751 003	Orissa University of Agriculture & Technology, Bhubaneshwar – 751 003	Coconut Crop Improvement and Agronomy experiments
Tamil Nadu	Coconut Research Station Aliyarnagar – 642 101 Coimbatore District	Tamil Nadu Agricultural University Coimbatore – 641 003	Coconut Crop Improvement Agronomy, pathology and Entomology experiments
	Coconut Research Station Veppankulam – 614 906 Thanjavur District	Tamil Nadu Agricultural University Coimbatore – 641 003	Coconut Crop Improvements Agronomy and Pathology experiments
	Tamil Nadu Rice Research Institute Aduthurai – 612 101 Thanjavur District.	Tamil Nadu Agricultural University Coimbatore – 641 003	Oil palm Crop Improvement and Pathology experiments
	Agricultural College & Research Institute, Killikulam – 628 252 Vallanad, Tuticorin District	Tamil Nadu Agricultural University Coimbatore – 641 003	Palmyrah Crop Improvement and Pathology experiments
West Bengal	Department of Plantation Crops Faculty of Horticulture, BCKV Mondouri Kalyani – 741 235 Nadia District	Bidhan Chandra Krishi Viswavidyalaya Mohanpur, Nadia – 741 252 West Bengal	Coconut Crop Improvement and Agronomy experiments

BUDGET ALLOTMENT DURING TENTH PLAN

Considering the relative importance of these crops, 73.84% of budget was allotted to coconut, while oil palm crop received 14.65% of budget and palmyrah centres were allotted 11.51% .

Crop	Budget allotted (Rs in lakh)	(%)
Coconut	498.42	73.84
Oil palm	98.88	14.65
Palmyrah	77.70	11.51
Total	675.00	100.00

MAJOR ACHIEVEMENTS

- In coconut, twelve cultivars and twelve hybrids have been released based on their performance at different locations for commercial cultivation in the respective states.
- In addition to the released cultivars and hybrids, a few types are in the final stages of evaluation at different locations. In the East Coast of Tamil Nadu, the Andaman Ordinary is performing well. In the Maharashtra coast at Ratnagiri, Laccadive Ordinary, Pratap

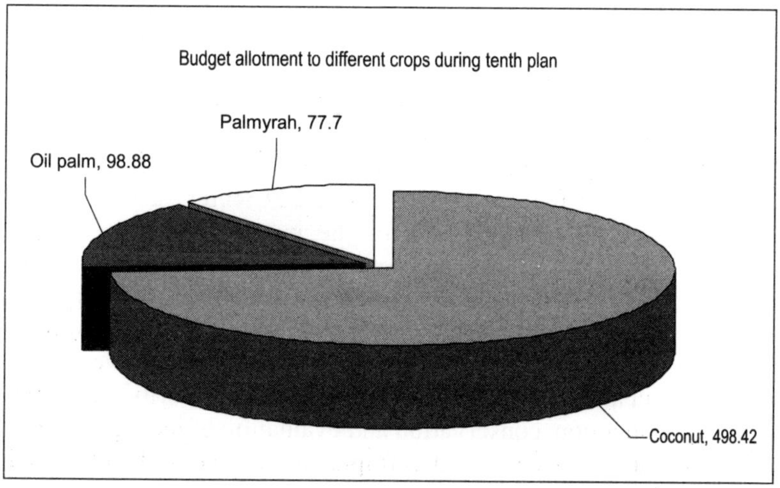

and Kerasankara have shown stable performance as suitable varieties of the tract. At Ambajipeta, Philippines Ordinary is performing well.

- In West Bengal, Jamaica Tall and in Assam, Assam Green Tall (Kamrupa) are suggested for commercial cultivation.
- The hybrids showing promise in different centres include WCT × GBGD, COD × WCT and WCT × COD.
- Three arecanut cultivars were released for cultivation based on multilocation trials.
- Nutrient management techniques for littoral sandy soils of West Coast region have been standardized.
- Nutritional requirements for D × T hybrids for each agroclimatic zone have been worked out.
- Recommendation on biomass recycling within the system/use of low-cost technologies on lignin degrading fungi/use of epigeic earthworms was offered for commercial exploitation.
- Irrigation schedule based on evaporation demand has been standardized for interior and coastal regions of Tamil Nadu and Konkan coast of Maharashtra.
- Successful and economically-viable coconut-based cropping systems have been identified for different states. In general, nut yield increased in mixed farming compared with monocrop of coconut. The net profit under mixed farming Rs. 50,000–65,000/ha.
- Plant-protection packages for the management of basal stem rot, leaf-eating caterpillar, red palm weevil and rhinoceros beetle have been developed.
- Fertilizer and irrigation requirements for oil palm in major states have been worked out.
- Promising germplasm types in palmyrah available in Andhra Pradesh and Tamil Nadu were collected and a genebank has been established.

Genetic Resources

Coconut

Sixty-four coconut accessions are maintained in nine coordinating centres for further evaluation. The germplasm maintained at different coordinating centres are Aliyarnagar 42; Ambajipeta 12; Jagdalpur 10; Kahikuchi 12; Mondouri 10; Ratnagiri 27 and Veppankulam 40. These materials include primarily the types received from CPCRI for trials at Centres. In addition, locally popular types are also maintained in the gene bank. Based on these trials, 12 varieties have so far been released for commercial cultivation in the respective regions.

Considerable variability exists in the local germplasm available in different coconut-growing regions of the country. This has to be collected and conserved in the wake of their destruction due to crop diversification and urbanization. Hence, the project on survey and collection of local germplasm in coconut and palmyrah was implemented as a priority area under the AICRP. Emphasis was given on collection, conservation and evaluation of local germplasm in coconut at seven centres. A total number of 81 local germplasm types could be collected at different regions for further evaluation.

Palmyrah

Under the AICRP on Palms, palmyrah germplasm collection programme, representing the variability available in the crop in different regions of the country, is the most important thrust area in palmyrah research. Survey to collect elite germplasm with emphasis on leaf production, dwarfness and yield of inflorescence sap is in progress. At present, 187 types are maintained at Killikulam (Tamil Nadu), while 176 accessions are maintained at Pandirimamidi (Andhra Pradesh). Observations on biometrical parameters indicated wide variability available among germplasm materials.

At Killikulam centre, a total number of 187 germplasm types are maintained in the germplasm bank in different blocks. Among 13 accessions under evaluation, the mean weight of fruit ranged from 210–1510g with a mean of 653.38g, length of fruit ranged from 14.7–23.20cm with a mean of 16.13cm. The circumference ranged from 25.8–44.0cm, weight of seed was from 57.0–283.30g with a mean of 135.88g. The accession AP 10 collected from Bynagundam of West Godavari district showed superiority over other accessions in expressing higher weight of fruit (1510g) and seed (283.30g). The size of the fruit was also good.

A total number of 176 accessions are being maintained at Pandirimamidi centre and the evaluation of these accessions is in progress. The data recorded on the accessions planted during 1991 revealed that maximum plant height was observed in accession 1(6.03m) and stem girth of 1.76m in accession 4. Number of leaves produced ranged from 7.4–12 among the accessions. Further, lamina and petiole lengths were found maximum in accession 2 (1.004m and 1.38m respectively). However, lamina breadth was found higher in accession 8.

The palm BF34 yielded 186 litres of inflorescence sap (neera) in a tapping duration of 75 days, followed by BF23, which recorded 138 litres in a tapping duration of 63 days. The palm BF34 was found to record more yield consistently for a period of seven years. Studies were conducted on growth and development of palmyrah fruit. The mean data revealed that the inflorescence length was 90cm. From flowering to maturity as tender fruit, it took 62 days. The total bunch weight was 5.6 kg. The individual tender fruit weight was 677.3g and fruit weight 1.361 kg. The flesh weight was 225.1g. Weight of individual tuber was 162.6g, length 22.7cm and circumference 9.5cm.

The inflorescence sap characteristics like reducing sugars (286–856 mg/100 gm), protein (16–25 mg/100 ml) and vitamin C (7.87–15.74 mg/100g) were analyzed.

Growth and development studies in palmyrah were initiated at both the centres during the current year. At Killikulam Centre, it was observed that the rate of leaf growth is comparatively more in juvenile stage, followed by early juvenile and seedling stages. Generally, the growth stage of plant advances, the growth rate seems to increase rapidly. Similarly, time taken for unfurling of leaf is also shorter in juvenile, followed by early juvenile stage. The growth of leaf ceases (temporarily) as the stalk appears. It takes about 98,125 and 141 days for unfurling of leaf in juvenile, early juvenile and seedling stages, respectively. Similarly, 104 and 131 days are taken for the appearance of stalk in juvenile and early juvenile stages respectively.

Indiscriminate removal of palmyrah leaves results in significant reduction in fruit yield/ inflorescence sap yield of palmyrah. However, leaves form the major component of the raw materials of palmyrah and hence a study was initiated during 2006 to fix the appropriate level of defoliation without affecting inflorescence sap yield or fruit yield.

Region Specific Varieties

Coconut

As a result of research efforts and multi-locational testing of coconut germplasm and hybrids, 12 each of hybrids and cultivars in coconut which can yield up to 2.3–4.4 tonnes of copra/ha/ year, were developed and recommended for release. Details of released varieties that are cultivated in large scale are given in the table.

Hybrid/Varieties Recommended	*Nut Yield/ Palm/Year*	*Copra Yield*		*Copra Yield*	*Oil Content*
		Mean nut (g)	*Mean/Palm (kg)*	*(tonnes/ha)*	*(%)*
Chandra Sankara Kerala, (COD x WCT) Karnataka, Tamil Nadu	116	215	25	4.4	68
Kera Sankara, Kerala, (WCT x COD), Karnataka	108	187	21	3.5	68
Chandra Laksha, Kerala, (LCOT x COD), Karnataka	109	195	21	3.7	69
Laksha Ganga Kerala (LCOT x GBGD)	108	195	21	3.7	70
Ananda Ganga Kerala (ADOT x GBGD)	05	216	21	3.6	68
Kera Ganga, Kerala (WCT x GBGD)	100	201	21	3.5	69
Kera Sree 112, Kerala (WCT x MYD)	112	216	24	4.2	66
Kera Sowbagya Kerala, (WCT x SS Apricot)	130	195	25	4.3	65
VHC-1, Tamil Nadu (ECT x CGD)	98	135	13	2.3	70
VHC-2, Tamil Nadu (ECT x MYD)	107	152	16	2.9	69
Godavari Ganga, Andhra Pradesh	140	150	21	3.7	68
Assam Green Tall, Assam	101	162	16	2.8	64
Arasampatti Tall, Tamil Nadu	126	131	16	2.8	67
Pratap, Maharashtra	150	152	23	4.0	67

Based on the multi-locational trials for evaluation of germplasm and coconut hybrids, eight varieties and two hybrids were approved and released for the respective regions:

Coconut Variety/hybrid	*Accession number*	*Regions for which recommended*
Kalyani Coconut-1	IND 031 S	Rainfed and Irrigated areas of West Bengal
Gauthami Ganga	IND 003 S	Coastal zone of Andhra Pradesh
Konkan Bhatye Coconut Hybrid-1	IND 003 S x IND127 S	Irrigated area of Konkan coastal zone of Maharashtra
Kahikuchi Coconut Hybrid-1	IND 058 S x IND 069 S	Irrigated areas of Assam
CPCRI Coconut Selection-6 CCS-6	IND 006 S	West coast region of Kerala and east coast region of Tamil Nadu and Andhra Pradesh as well as Andaman and Nicobar Islands.
CPCRI Coconut Selection-9 CCS-9	IND 069 S	Rainfed and irrigated areas of Kerala, Tamil Nadu and West Bengal. Comes up well in different types of soil including sandy, sandy loam and red sandy loam with pH ranging from 6.0 to 8.0.
CPCRI Coconut Selection-8 CCS-8	IND 004 S	Coastal zone of Andhra Pradesh and Tamil Nadu, Konkan region of Maharashtra and Bastar region of Chhattisgarh.
CPCRI Coconut Selection-4 CCS-4	IND 016 S	West coast region of Kerala, Karnataka and Maharashtra and interior and coastal zone of Tamil Nadu and Andhra Pradesh
CPCRI Coconut Selection-5 CCS-5	IND 002 S	West coast region of the country and of West Bengal.
CPCRI Coconut Selection-7 CCS-7	IND 049 S	Root (wilt) disease affected areas of Kerala

Characteristic Features of Released Varieties and Hybrids

- ***Laccadive Ordinary(Chandrakalpa):*** The cultivar has been introduced from Lakshadweep Islands. The palm is similar to WCT in nut characters.The average nut yield is 100 with a copra content of 176 g/nut . The oil content is 72%. This cultivar was released in 1985 under the name "Chandrakalpa".
- ***Benawali Green Round (Pratap):*** This is a popular cultivar in Goa, Konkan and Coastal Maharashtra. The bunches are heavy with closely packed nuts. The average annual yield ranges from 139 to 160 nuts with a copra content of 150g/nut, oil content being 59%.
- ***Chowghat Orange Dwarf(COD):*** This indigenous dwarf cultivar is found sparsely cultivated through the West Coast of India, particularly in Chavakkad area of Thrissur in Kerala. The palms have a characteristic orange colour on leaf petioles, spadices and nuts. This cultivar commences flowering in about 4 years after planting with an average yield of 63 nuts/palm per year. The tender nut water of this cultivar is sweet with a total sugar content of 7g/100ml and this cultivar has been recommended and released for tender nut purpose.
- ***Chandra Sankara(COD × WCT):*** This hybrid is between COD × WCT and was released by CPCRI, Kasaragod in 1985. It is a early bearing and high yielding hybrid with an average annual yield of 116 nuts per palm. The copra content is 215g/nut.

- ***Chandra Laksha(LO × COD):*** This is a tall × dwarf hybrid with an annual yield of 109 nuts per palm. This hybrid comes to bearing in about 6 years.
- ***Kera Sankara(WCT × COD):*** This hybrid comes to bearing in 4-5 years and attains steady bearing by the 6th or 7th year after planting. The mean annual yield is 108 nuts/palm with a copra content of 187g/nut.
- ***Laksha Ganga (LO × GB):*** This hybrid was released by Kerala Agricultural University. It comes to bearing in about 5 years. The mean yield is 108 nuts/palm/year and copra content is 195g/nut. The oil content is 70%.
- ***Ananda Ganga(AO × GB):*** This is a hybrid between Andaman Ordinary and Gangabondam with an annual average yield of 95 nuts. The copra content is 216 g/nut and oil content is 68%.
- ***Kera Ganga(WCT × GB):*** This is yet another hybrid released by KAU. The average annual yield is 100 nuts/palm. The copra content is 201g/nut and oil content is 69%.
- ***Kera Sree(WCT × MYD):*** This is a recently released hybrid from KAU. The annual mean yield is 112nuts/palm with a copra content of 216g/nut.
- ***Kera Soubhagya (WCT × SSG):*** This is a cross between West Coast Tall and Straight Settlement Green. Comes to bearing in about 5-6 years with an annual yield of 116 nuts/palm. Copra content per nut is 196g and oil content is 65%.
- ***VHC-1 (ECT × GB):*** This is a hybrid between East Coast Tall and Green Dwarf released by Tamil Nadu Agriculture University. This hybrid comes to bearing in about four years, with an average of 98 nuts/palm/year. The copra content is 135g/nut.
- ***VHC-2 (ECT × MYD):*** This hybrid was also released by TNAU. It gives around 107 nuts/palm/year with a copra content of 152 g/nut.

Salient Characteristics of Coconut Varieties and Hybrids

- ***Kalyani Coconut-1:*** In addition to better nut yield and oil yield, the tender nut of the coconut accession IND 031 S higher quantity of 360 ml of nut water as compared to 260 ml in ECT and also higher levels of total sugar, potassium and sodium content in nut water. West Bengal, being a tender coconut production area, IND 031 S suits for dual purpose as a tendernut coconut variety and a commercial variety for oil. Hence, IND 031 S was approved for release as a dual purpose variety with the name "Kalyani Coconut-1" for West Bengal.
- ***Gauthami Ganga:*** Gangabondam (IND 003 S) dwarf cultivar is from East Godavari district in Andhra Pradesh. The nuts are medium-sized and papaya shaped. This cultivar starts bearing by 3rd or 4th year after planting. The IND 003 S is one of the best known dwarf types contributing to the development of a number of hybrids like Keraganga, Anandaganga, Lakshaganga and Godavari Ganga. IND 003 S is also a high yielding coconut variety (12,813 nuts/ha/annum) with sweet taste for nut water (sugar content 6.40g/100ml) with a very high quantity of nut water (446 ml) and high potassium content (2035 ppm). Performance of the variety at farmers' fields is also good. Hence, this variety is recommended for release. It could also be noted that the variety is performing well in Andhra Pradesh, Tamil Nadu and Maharashtra. Considering the wide adaptability,

Gangabondam was released with the name "Ganga Gauthami".

- ***Konkan Bhatye Coconut Hybrid-1:*** The IND 003 S × ECT hybrid seedlings, supplied from by Ambajipeta and tested at Ratnagiri Centre has yielded 20,300nuts/ha/annum. The overall mean yield was 116 nuts/palm/year (for the last 10 years). This hybrid is yielding 24.5% more nut than Pratap, 61% higher than IND 069 S and 73% higher than ECT. The hybrid gives a copra yield of 3.43 tonnes/ha and oil yield of 2.30 tonnes/ha. As this hybrid is performing better than the earlier released Pratap variety as well as the 11 other hybrids tested under this project, this hybrid was approved for release as "Konkan Bhatye Coconut Hybrid-1" for the Konkan coastal region.
- ***Kalpa Dhenu (CCS 6):*** IND 006 S is giving 26.07% higher copra yield and 21.44% higher oil yield as compared to IND 069 S. The tender nut quality is also good. The oil extracted from the copra of this variety has higher lauric acid content. It is also expected that this variety would perform well in both East Coast region and Andaman area. Hence, the variety was approved for release as a "national variety" with the name CCS-6 (CPCRI Coconut Selection-6). The popular name (to be given by CPCRI) would include the place of origin of this accession, i.e. Andaman.
- ***Kera Keralam (CCS 9):*** West Coast Tall (IND 069 S) is the common tall cultivar extensively cultivated along the west coast of India. The WCT palms normally come to bearing in about 5–7 years. The average annual yield ranges from 50–100 nuts per palm with a mean of 80 nuts under rainfed conditions, while the variety yields 150–200 nuts/palm per annum under irrigated conditions. The copra content is 176 g and oil content 68%. Coconut accession IND 069 S is performing well in Kerala, Tamil Nadu and West Bengal. The accession responds well to management and any incremental level of management in terms of water and nutrients results in better productivity. Since WCT is performing well at some centres of AICRP on Palms in addition to CPCRI centres, it is approved for release as a "national variety" by the Central Varietal Release Committee.
- ***Kera Bastar (CCS 8):*** The accession IND 004 S has performed well in Ambajipeta (Andhra Pradesh), Ratnagiri (Maharashtra), Veppankulam (Tamil Nadu) as well as in the non-traditional area of Jagdalpur (Chattisgarh). The accession could give a nut yield of 110 nuts/palm/year with a copra yield of 2.97 tonnes/ha and oil yield of 2.04 tonnes/ha. Under these circumstances, this accession is recommended for Kerala, Andhra Pradesh, Tamil Nadu and Chattisgarh. Since the accession IND 004 S is performing well in a few centres of AICRP on Palms, it is approved for release as a "national variety" by the Central Varietal Release Committee.
- ***CCS 4:*** Based on the good performance of coconut accession IND 016 S at Kasaragod (Kerala) with higher copra and oil yield 40.11% and 38.05% respectively than popular accession IND 069 S and better performance at Ambajipeta (Andhra Pradesh) Veppankulam (Tamil Nadu) and Aliyarnagar (Tamil Nadu) and Ratnagiri (Maharashtra), IND 069 S is approved for release as a "national variety" by the Central Varietal Release Committee.
- ***CCS 5:*** The accession IND 002 S performs well (copra and oil yield was 16.01 and 13.45% higher than IND 069 S) at Kasaragod (Kerala) as well as at Mondouri Centre of

AICRP on Palms (West Bengal). The tender nut quality is also good. Hence, IND 002 S is approved for release as a "national variety" by the Central Varietal Release Committee.

- ***CCS 7:*** Considering the high volume of nut water (290 ml), oil content (65.5 %) with an yield potential of 2.4 tonnes of copra/ha, IND 049 S is approved for release for cultivation in the root (wilt) disease affected region of Kerala. It is to be noted that the variety is also amenable for high density planting @ 204 palms/ha.
- ***Coconut Hybrid - IND 058 S × IND 069 S:*** In view of the good copra yield as well as oil yield at Kahikuchi Centre in Assam and availability of limited number of released coconut varieties for North East Region, the hybrid IND 058 S × IND 069 S is approved for release as a State variety in Assam.

Arecanut

Two arecanut selections from germplasm collection, viz., VTL-11 and VTL-17 were released as Sumangala and Sreemangala. CAL-17 and SAS-1 were recommended for release and cultivation in Andaman Islands and Sirsi area, respectively.

Mohitnagar cultivar of arecanut was recommended for release for cultivation in northern part of West Bengal, coastal Karnataka and Kerala. Performance of the arecanut cultivars released in the AICRP on Palms in presented in Table.

Cultivar	*Fresh Nut Yield*	*Chali (dry) Nut Yield (kg/palm/year)*
Sumangala	12.93	3.28
Sreemangala	12.82	3.10
Mohitnagar	15.08	3.67

Oil Palm

Oil palm, as a potential edible oil generating crop, is being promoted to bridge the edible oil deficit in the country. Its cultivation is gaining momentum in Andhra Pradesh, Karnataka and Tamil Nadu. In order to improve the productivity of oil palm crop on par with other competing countries, it is essential that location-specific oil palm hybrids with a production potential of more than 20 tonnes FFB/ha are identified. Evaluation on performance of tenera hybrid combinations of oil palm in different agroclimatic regions, since 1991, could result in identification of an high-yielding hybrid type 124D × 266P yielding 24.06 tonnes FFB/ha at Mulde centre.

Similarly, hybrid 115D × 291P is performing well both at Vijayarai and Mulde centres. The hybrid is yielding 13.80 tonnes FFB/ha at Vijayarai centre and 20.32 tonnes FFB/ha at Mulde centre.

Evaluation of dura type of oil palm germplasm for drought tolerance was undertaken at Gangavathi and Mulde centres 1998 onwards. At Gangavathi, the lipid peroxidation was estimated for the stress period (April 2007) and significant differences were observed between the genotypes for lipid peroxidation. The genotype ZS-1 recorded lower OD value of 0.099 which was significantly lower than all the other genotypes except ZS-8 (0.128) and TS-4 (0.131) where in it remained on par. On the other hand, the genotype ZS-5 recorded highest OD value of 0.347. Based on the results, it could be inferred that the genotype ZS-1 could be more stress tolerant than other genotypes.

At Vijayarai centre (Andhra Pradesh), the trials on comparative performance of oil palm planting material obtained from four sources. The performance of oil palm planting materials obtained from different sources, viz., Papua New Guinea, IRHO, Costa Rica and Palode was found to be good. However, maximum FFB yield was recorded in Palode (59.76 kg/palm and 8.5 tonnes/ha) and the lowest of 36.82 kg/palm and 5.3 tonnes/ha was recorded in Costa Rica.

Technologies for Enhancing Production and Productivity

Coconut

Fertilizer schedule for Konkan coast of Maharashtra, black soils of maidan tract of Karnataka, sandy loam soils of Tamil Nadu and east coast of Andhra Pradesh were developed.

Recommended dose of NPK g/palm/year	
Maidan tract of Karnataka	0:450:900
Konkan coast of Maharashtra	750:225:900
Tamil Nadu	560:320:1200
Andhra Pradesh	650:450:1200

Nutrient management techniques for littoral sandy soils of West Coast region and Orissa coast region have been standardized. A fertilizer dose of 500g N, 250g P_2O_5 and 750g K_2O is recommended for tall variety of coconut in West Bengal conditions. Nutritional requirements for D × T hybrids for each agroclimatic zone have been worked out.

The Integrated Nutrient Management trials at Aliyarnagar centre proved the utility of use of organic sources of nutrients as compared to chemical fertilizers. It was seen that the treatments '100% composted coir pith – CCP - '(186nuts/palm, 'Farmers practice' (185nuts/palm)and '50% CCP+50% chemical fertilizer' (184nuts/palm) recorded significantly more number of nuts/palm than the treatments 'neem cake + bone meal + Ash' (164nuts/palm) and 'recommended dose of chemical fertilizer' (163 nuts/palm). Application of composted coir pith on 100% N basis recorded maximum net income of Rs. 58,752/ha and benefit cost ratio of 2.31. Application of 50% CCP + 50% chemical fertilizers recorded a net income of Rs 55,298/ with a benefit: cost ratio of 2.20 and this is followed by farmers with Rs.54,930 and 2.14 respectively. Application of chemical fertilizers and the treatment neem cake + bone meal + ash recorded the lowest net income of Rs.37755/- and Rs.28552/- respectively. These treatments recorded the lowest benefit cost ratio of 1:80 and 1:49 respectively. However at Ambajipeta centre, the treatment on the use of recommended dose of chemical fertilizers gave the highest yield, whereas in Veppankulam centre, it was found that FYM +100% NPK gave significantly higher yield (123.4/palm) than all other treatments. Similar result was obtained with respect to cumulative mean nut yield for the last seven years (1996–2003), recording maximum of yield of 129.8nuts/palm. Recommended dose of NPK alone recorded 109 nuts/palm. It is thus evident that application of organic (or) inorganic source alone is not sufficient to meet the requirement of palm for higher productivity. The following Integrated nutrient management packages are recommended for coconut varieties in different regions:

- In interior Tamil Nadu region, 100 % N requirement for coconut crop could be supplied

Pratap

Aliyar Nagar Tall

Jamica Tall

Assam Green Tall

ECT x MYD

WCT x COD

Kalyani Coconut 1

Gauthami Ganga

Konkan Bhatye Coconut Hybrid 1

Kalpa Dhenu

Kera Keralan

Kera Bastar

Kalpa Prathiba

CPCRI Coconut Selection 7

Kahikuchi Coconut Hybrid 1

Varieties hybrids of coconut

Kera Keralam

Kera Bastar

Kalpa Mitra

Kalpa Raksha

Coconut+ pepper+ banana + turmeric

Coconut + pepper + pineapple

Coconut + pepper + perennial-brinjal

Andrographis paniculata in coconut garden

through composted coir pith and the balance requirement of P and K could be applied as chemical fertilzers.

- In coastal Andhra Pradesh region, 50% N requirement for coconut crop could be supplied through composted coir pith and the balance requirement of N, P and K could be applied as chemical fertilzers.
- In *Maidan* tract of Karnataka, 100% N requirement for coconut crop could be supplied through composted coir pith and the balance requirement of P and K could be applied as chemical fertilzers.
- In irrigated coconut gardens in Orissa, 50% N requirement for coconut crop could be supplied through composted coir pith and the balance requirement of N, P and K could be applied as chemical fertilzers.
- For organic coconut production in Konkan region of Maharashtra, application of vermicompost (prepared from coconut leaves and other available biomass) @ 50kg per bearing coconut palm during the last week of May or first week of June by basin method is recommended.
- In coastal Tamil Nadu region, 50% N requirement for coconut crop could be supplied through composted coir pith and the balance requirement of N, P and K could be applied as chemical fertilzers.

The following Integrated nutrient management packages are recommended for cultivation of hybrid coconut in different regions:

- A fertilizer dose of 1000:250:1000gram NPK/palm/year is recommended for the cultivation of Dwarf × Tall coconut hybrids in interior Tamil Nadu region.The 50% of recommended nitrogen dose could be applied as organic manures and the balance NPK could be applied as chemical fertilizers.
- A fertilizer dose of 500:500:2000g NPK/palm/year is recommended for the cultivation of Dwarf × Tall coconut hybrids in irrigated coconut gardens in Assam. The 50% of recommended nitrogen dose could be applied as organic manures and the balance NPK could be applied as chemical fertilizers.
- A fertilizer dose of 1000:500:1000gram NPK/palm/year is recommended for the cultivation of Dwarf × Tall coconut hybrids in irrigated coconut gardens in the new Alluvial Zone of West Bengal. The 50% of recommended nitrogen dose could be applied as organic manures and the balance NPK could be applied as chemical fertilizers.
- A fertilizer dose of 1000:500:2000g NPK/palm/year is recommended for cultivation of Dwarf × Tall coconut hybrids in Konkan coastal region of Maharashtra. The 50% of recommended nitrogen dose could be applied as vermicompost and the balance NPK could be applied as chemical fertilizers.

At Veppankulam, irrigating the palms helped in reducing button shedding and increasing the yield. Irrigation once in 10 days gave 52% higher yield in Tamil Nadu.

Irrigation schedules based on evaporative demand (Eo) has been standardised for the judicious use of water for irrigation to coconut palms in Pollachi and Thanjavur regions of Tamil Nadu given as follows:

- Pollachi region (Coimbatore district, Tamil Nadu)
  The quantity of water to be applied per palm per day through drip irrigation will be 65 litres during January—May, 30 litres during June–July, 55 litres during August–September and 35 litres during October—December for palms growing on red sandy loam soil in Pollachi region.

- Thanjavur region(Thanjavur district, Tamil Nadu).
  The quantity of water to be applied per palm per day through drip irrigation is 80 litres per day during March—September and 50 litres per day during October—February for Thanjavur region for sandy loam soil.

In the trial on drip irrigation-cum-fertilizer on young coconut palms, at Aliyarnagar centre, drip irrigation at 100% pan evaporation and basin irrigation have recorded maximum nut yield of 162 and 158 nuts/palm/year. Drip irrigation at 66% Eo has recorded an yield of 127 nuts/palm/year. No irrigation has recorded a yield of 83 nuts/palm/year. Economic analysis revealed that drip irrigation at 100% Eo recorded a maximum net income of Rs. 44,469/ha and a benefit: cost ratio of 2.08 against the control with Rs. 15,598 and 1.54 respectively. Basin irrigation, which was on par with drip 100% Eo, recorded Rs. 42,294 and 2.01 respectively.

Similar encouraging results were obtained at Ratnagiri centre also wherein the mean cumulative yield/palm for last six years (1998 – 2003) showed that palms under basin irrigation treatment recorded 94 nuts/palm which is significantly higher than the yield recorded by all other treatments except treatments having irrigation gives significantly lower cumulative yield than other treatments.

At Veppankulam centre, drip irrigation at 100% Eo and basin irrigation had a similar effect on growth, yield attributes and nut yield. These two treatments registered significantly higher number of functional leaves, bunches harvested, female flower production and nut yield and it is superior to all other treatments. The cumulative mean nut yield for five years (1998 – 2003) showed that maximum number of nuts (126.1/palm) was recorded in drip irrigation at 100% Eo and it was followed by basin irrigation (116.5 nuts/palm). Though the nut yield in drip system (100% Eo) was only 8% higher than basin method, water savings in drip system was 44.6% more than basin method.

Technology for soil moisture conservation was developed.

- Mulching the basin with coir dust @ 50 kg/palm/basin
- Burying coconut husk in 2 or 3 layers in 0.5m wide trenches opened at 1.8m away from the bole, keeping the inner portion of the husk facing upwards
- Growing green manure crops like sunnhemp, Daincha, Pureria, Calapagonium, Mucuna sp., Desmanthes, Desmodium, *Sesbania rostrata,* etc., in the basin and incorporation of the same along with fertilizers
- Addition of tank silt to light textured soil whenever it is easily available for increasing water-holding capacity
- Incorporation of cattle manure/compost/organic matter @ 50 kg/palm/year to increase water retaining capacity of the soil
- Spreading common salt in the basin at 2 kg/palm to increase leaf turgidity

- Deep planting of new seedlings can be adopted so that the root system will be at deeper layer, which can tap moisture from lower layers
- Before planting coconut seedlings, pits may be filled with 2 or 3 layers of coconut husk at the bottom, which can help in conserving moisture
- Microirrigation like drip irrigation or surface watering through hollow bamboo pipes burried under pots may be done to utilize limited water most effectively.
- In heavy soils of Andhra Pradesh, deep ploughing has been found to conserve more water. However in light textured soils, inter-culture may be avoided. Zero tillage has been found to conserve soil as well as moisture in sloppy terrain.

Oil Palm

Field trials on assessing the irrigation and nutrient requirements for oil palm were conducted at all the four oil palm centres under AICRP on Palms. At Aduthurai centre in coastal Tamil Nadu region, a combination of basin irrigation along with application of fertilizers @ 1200:600:2700g NPK/palm/year was found to be the best and registered higher FFB yield of 15.68 tonnes/ha/year than the rest of the combinations. The above combination also recorded highest net returns (Rs. 31,082/ha) and benefit: cost ratio (1:56). Substantial improvement in nutrient availability was observed due to addition of fertilizers and irrigation. Application of fertilizer @ 1200:600:2700g NPK/palm/year influenced soil N (244.2 kg/ha), soil P (39.44 kg/ha) and soil K (268.89 kg/ha) phenomenally over rest of the treatments due to addition of sufficient quantity of nutrients. Basin irrigation increased soil N (233.5 kg/ha), soil P (36.57 kg/ha) and soil K (248.98 kg/ha) significantly over the control.

At Gangavathy centre (Karnataka), water-use efficiency differed significantly among the irrigation treatments. Drip irrigation recorded significantly higher water-use efficiency of 7.99 kg FFB/ha mm as compared to 5.73 kg FFB/ha mm in basin irrigation. Similarly, the WUE was significantly higher in all fertilizer treatments as compared to the control. The WUE ranged from 7.44 kg FFB/ha mm in case of 400:200:900g NPK to 8.8 kg FFB/ha.mm in 1200:600:2700g NPK as against 4.43 kg FFB/ha mm in the control.

Encouraging results were obtained at Mulde centre (Maharashtra) also wherein drip irrigation increased the number of bunches (7.02) significantly over the control and conventional method of irrigation (6.33). The FFB yield (139.51 kg/palm/year) was maximum in drip irrigation, followed by basin irrigation (I_1) (127.35 kg/palm/year). Application of fertilizer also showed significant effect on average weight of bunch and FFB yield. Yield of FFB, average number of FFB and bunch weight were maximum (146.64 kg/palm/year, 6.99 and 20.81kg/bunch, respectively). The yield increase was significant and showed a graded trend.

At Vijayarai centre (Andhra Pradesh), cumulative mean yield recorded for the last 11 years from 1992–2003 revealed that the different treatments responded well. There was an incremental increase to both irrigation and fertilizer levels. Basin irrigation recorded maximum FFB yield/palm (85.8 kg) and 194% increase in yield over no irrigation treatment. Different fertilizer treatments have recorded 128.9, 155.3% and 192.3% increase in yield respectively over no fertilizer level.

Similar results were obtained at all the four centres. Based on these trials, following fertilizer schedules are recommended for oil palm cultivation in different regions:

- Coastal Tamil Nadu region : 1200: 600: 2700 gram NPK/palm/year
- Thungabadhra Command area in Karnataka: 1200: 600: 1200g NPK/palm/year
- Konkan coastal region in Maharashtra State: 1200: 600: 2700 gram NPK/palm/year
- Coastal Andhra Pradesh region : 1200: 600: 2700g NPK/palm/year.

Performance of oil palm under drip irrigation with a NPK dose of 1200:600:2700 g/palm

Cropping System Approach

Basically coconut is a small holders' crop and 90% of the total area is under small holder with less 0.2 ha. Under the present situation (where the price fluctuation is too high) growing coconut alone without any component crop/animals cannot bring farmers additional income for their satisfactory living. Thus, coconut-based cropping/farming systems involving cultivation of compatible crops in the interspaces of coconut and its integration with other enterprises like dairy, poultry, etc., leads to considerable increase in production and productivity per unit area, by more efficient utilization of sunlight, soil, water and labour.

Scope for Multiple Cropping

A spacing of 7.5 m × 7.5 m in the square system is recommended for coconut (175 palms/ha), but the spacings adopted by farmers vary widely and in the *malnad* region of Karnataka much wider spacing are adopted with densities as low as 120 palms / ha. Experimental evidences have shown that a sole crop of coconut, at the recommended spacing of 7.5 m × 7.5 m does not fully utilize the available resources like soil, air space and sunlight.

Horizontal Space (soil space)

Coconut palm like all monocots has a typical adventitious root system. Under favourable conditions, as many 4,000–7,000 roots are found in middle aged palms. About 74% of the roots produced by a palm under good management did not go beyond 2m lateral distance and 82% of the roots were confined to the 31–120cm depth of soil. Recent studies have confirmed that more than 80% of the root activity was confined to a lateral distance of 2m from the trunk. Thus, active root zone of coconut is confined to about 22% of the available land area and the remaining area could be profitably exploited for raising subsidiary crops.

Air Space/Solar Radiation

The venetian structure of coconut crown and orientation of leaves allow part of the incident solar radiation to pass through the canopy and fall on the ground. The leaves in palm crown are not randomly distributed, but clumped around few widely spaced growing points. This non-random distribution also leads to low extinction coefficient of around 0.65 for PAR. About 56% of the sunlight is transmitted through the canopy during the peak hours (10–16 hours) in palms aged around 25 years. The diffused sunlight facilitates growing a number of shade tolerant crops in the interspaces. Age, spacing, soil fertility status, varietal characteristics, leaf area and time of the day influence the light penetration through the canopy. Based on growth habit of palm and the amount of light transmitted through its canopy, life span of coconut palm could be divided into three distinct phases, from the point of view intercropping. Planting till full development of canopy (below 8 years): good light transmission initially, but decreasing with age - suitable for growing annuals/biennials.

Young palms (8–25 years): maximum ground coverage and low canopy - poor light availability - not suitable for multiple cropping. Mature trees (more 25 years): Increase in trunk height; reduction in crown size - light transmission increasing with age - ideal for raising annual and/or perennial crops.

Based on the extensive trials conducted at all the 10 coconut research centres in AICRP on Palms, successful and economically-viable coconut-based cropping systems have been identified for different states. Inter planting nutmeg, cinnamon and clove in coconut gardens recommended to increase the net economic returns in the Konkan Region of Maharashtra.

Coconut and pineapple

Tree spices as mixed crops under Ratnagiri (Maharashtra) conditions, especially coconut, nutmeg crop combination yielding a net profit of Rs. 69,593/ha. In coconut-based high-density multispecies cropping system, trials conducted at Ratnagiri centre, cinnamon, nutmeg, clove, black pepper, garcinia and allspice have been planted in separate blocks of 0.20 ha each of coconut plantation. The yield of coconut gradually increased over a period after planting the mixed crops irrespective of the crop block. The maximum increase in yield was recorded in clove plot (93.88%), followed by allspice (86.08%), cinnamon (72.30%) and nutmeg (70.18%). While nutmeg intercropped plot could give a net profit of Rs. 1.23 lakh, cinnamon intercropped plot gave a net profit of Rs. 84,529 and in the control plot (coconut only), a sum of Rs. 31,349 was obtained as the net profit.

Coconut and turmeric

At Kahikuchi (Assam), coconut in combination with black pepper, banana, Assam lemon and ginger yielded a net profit of Rs.64,948/ha. At Kahikuchi centre, highest net return/ha (Rs.1,30,468/ha) was recorded under coconut-black pepper-turmeric cropping system, followed by coconut-black pepper-ginger intercropping, while the lowest net return of Rs.23,462 only was obtained under control plot (coconut-black pepper). At Veppankulam (Tamil Nadu), companion crops like banana, okra, seedless lime and tubers performed well with a cost: benefit ratio of 1:3.48. An average of 161 nuts/palm were recorded in mixed farming plot compared to monocrop of coconut (132.5 nuts/palm).

At Veppankulam centre, in general, the coconut yield increased due to intercropping due to its companion effect such as microclimate, less soil temperature, litter fall and frequent irrigation. Based on economics, coconut + banana (Poovan) + *C. parviflorus* (Sinukizhangu) +

okra were found to be the best multiple cropping system for adult coconut groves of the east coast region. At Mondouri (West Bengal), intercropping in coconut garden with turmeric, ginger, yam and colocasia and vegetables like okra and amaranthus were started for the first time. Very encouraging results are obtained on their performance.

Coconut-nutmeg mixed cropping coupled with rearing silkworm, growing chillies, potato, grams as intercrops and double cropping of potato-wheat, Frenchbean-wheat and chilli wheat were profitable crops in coconut gardens in the maidan tract of Karnataka.

In Tamil Nadu, cropping system models with banana and turmeric were more profitable.

In Assam, black pepper, mandarin orange, lemon, coffee and pineapple were found to be most remunerative crop combination in coconut garden.

At Jagdalpur centre in Chattisgarh, among the seven models tried, coconut + chrysanthemum + carnation could give the highest returns, indicating the scope for the introduction of floricultural crops as intercrops in coconut gardens to improve the income of farmers.

Coconut and Nutmag

There exists excellent scope for diversification in traditional intercrops cultivated with coconut. New initiatives for introducing medicinal and aromatic plants will transform the economy of coconut farmers whose income sustainability is threatened due to uncertain prices of copra and nuts. Hence, feasibility trials for such ventures were taken up in 9 Cooperating Centres. Citronella (*Cymbopogon winterianus*), pipali (*Piper longum*), Vedailota (*Paederia foetida*), patchouli (*Pogostemon cablin*) and Madhu shaleng (*Deeringia amranthoides*) were introduced for commercial cultivation. Based on similar results obtained from the nine centres, full-fledged trials in randomized block design were initiated at the centres to study the quality and yield parameters of selected species and the allelopathic effect.

Coconut based mixed farming and multi-species cropping have not only increased coconut yield but also provided continuous income and employment generation and land appreciation.

Coconut and Banana

Recommendation on biomass recycling within the system/use of low-cost technologies on lignin degrading fungi/use of spigeic earthworms was offered for commercial exploitation.

Strategies for Pest and Disease Management

Coconut

The etiological studies on Tatipaka disease of coconut in Andhra Pradesh confirmed the presence of mycoplasma like organisms. Survey was conducted in East Godavari district of Andhra Pradesh to identify Tatipaka disease affected palms and the diseased palms were removed to reduce the inoculum potential and also to prevent further spread of the disease.

Root regeneration in an infected palm by making rings of 3cm width above the bole and heaping a mixture of river sand and farmyard manure and/or river silt around the rings produced new roots and checked further deterioration of the diseased palm. Application of zinc sulphate at the rate of 2 kg/palm twice in a year and subsequently at the rate of 1 kg/palm/year increased zinc, iron, calcium and potash contents in leaves and reduced disease intensity of diseased palms.

The coconut basal stem rot commonly referred as *Ganoderma* wilt in Andhra Pradesh and Thanjavur wilt in Tamil Nadu is confirmed to be caused by *Ganoderma lucidum/ Ganoderma applanatum* at Veppankulam and Ambajipeta centres. An early diagnosis method was developed for this disease. The coconut basal stem rot commonly referred as Ganoderma wilt in Andhra Pradesh and Thanjavur wilt in Tamil Nadu is confirmed to be caused by *Ganoderma lucidum/Ganoderma applanatum* at Veppankulam and Ambajipeta centres. An early diagnosis method was developed for this disease

- Root feeding of 6 ml Calixin was recommended for controlling basal stem rot in Andhra Pradesh.
- Removal of palms affected severely by basal stem rot by taking isolation trenches to separate healthy and disease affected palms, field sanitation, irrigating the palms during summer helped to reduce the intensity of the disease.
- For the management of the disease, root feeding of 2ml Tridemorph in 100ml of water at quarterly intervals for one-year along with 5kg of neemcake application per year was recommended. If aureofungin sol is used for root feeding, 2g of aureofungin sol + 2g copper sulphate in 100ml of water at quarterly intervals for one year is to be given and the treatment has to be repeated once in two years.
- Tapping for inflorescence sap helped to reduce the intensity of the disease and increased the nut yield.
- For the management of basal stem rot disease of coconut, applying 50kg FYM mixed with 500g *Trichoderma harzianum* and 200g phophobacteria once in a year to the base of palms in 2 m radius basin and root feeding 100ml of 2% Calixin at quarterly intervals have been recommended for Thanjavur region.

In sero diagnosis studies on basal stem rot of coconut conducted at Ambajipeta centre, in simple serological tests, i.e., slide agglutination test and glass capillary tube test, the produced antiserum positively reacted (precipitation formation) with cultures of *G. applanatum* and *G. lucidum* and with sap from basal stem rot affected coconut palm root tissues. However, simple serological tests failed to detect basal stem rot disease infection in stem tissues of diseased coconut palm. The indirect form of Elisa was found to be more sensitive in detecting the *Ganoderma* spp. in tissues from root, basal portion of stem and up to 1m height from basal portion of stem of coconut palm.

Studies on the early detection of basal stem rot through the use of indicator plants, at Ambajipeta centre gave highly useful results. Red gram has been identified as an indicator plant for early detection on the incidence of Ganoderma wilt disease in coconut at Ambajipeta. Among the 13 plant species screened against basal stem rot disease in sick soil, red gram plants showed splitting of the bark at the crown region. In the confirmation trial conducted at Veppankulam centre, the appearance of external splitting symptom on basal stem of red gram plants was observed one month after sowing in 20% of the red gram plant population near the diseased palms. In red gram plants near healthy palms, this symptom was not noticed.

The pathogenicity studies conducted at Veppankulam centre revealed that the seedlings inoculated with basal stem rot isolate TKT 1 showed wilting symptoms 45 days after inoculation. Seedlings inoculated with the isolate TKT 2 showed sporophore production at collar region of seedlings 45 days after inoculation. The sporophore production was also observed in seedlings inoculated with isolates CRS 1, CRS 4 and PV 2 two months after inoculation.

The basal stem rot management studies being conducted at Aliyarnagar, Ambajipeta, Arsikere and Veppankulam centres indicated the possibility of using biocontrol agents in the management of basal stem rot disease. At Ambajipeta centre, among the three native biocontrol agents, *T. hamatum* (50g) + neem cake (5kg) was found highly effective in arresting the spread of basal stem rot disease, the vertical spread of disease being 11.3cm followed by *T. viride* (50g) + neem cake (5kg) and *T. harzianum* (50g) + neem cake (5kg) with disease spread of 16.7 and 15.3cm respectively as against 68.7cm in untreated control. It was also noted that the biocontrol agents performed better when used in combination with neem cake than either alone or in combination with FYM. At Arsikere centre, it was observed that the disease progress was less in palms receiving Hexaconazole (1%-100ml) root feeding at quarterly intervals along with soil application of neem cake (@ 5kg/palm/year) + *Trichoderma viride* (@ 100g/palm/year in two doses). However, progress of disease in control treatments was of higher proportion and a few palms died especially in rainfed coconut gardens.

Under the varietal screening trials at Veppankulam centre, 10 coconut genotypes were tested for their reaction to basal stem rot disease of coconut by planting them in BSR hot spot area of Thambikkottai village in Thanjavur district. Among the 10 coconut genotypes, ECT × BSR tolerant ECT registered a higher rate of survival (66.7%) as compared to other genotypes in BSR sick soil. ECT recorded 40% survival. In ECT × BSR tolerant ECT hybrid, out of 12 surviving palms, only two palms were found to be infected with BSR (16.7%). The mean disease index was also low in ECT × BSR tolerant ECT (36.71) as compared to ECT (51.34). The hybrid ECT × BSR tolerant ECT recorded a higher nut yield of 141 nuts/palm during the year 2002–03, which also recorded a higher cumulative mean nut yield of 132.13 nuts/palm when compared to other genotypes.

Effective management of crown chocking disease of coconut with application of borax has been demonstrated in West Bengal and Assam. Application of 50g borax in soil along with adequate fertilizers and green manures/FYM could control the crown chocking disease.

Studies were undertaken at Ambajipeta centre on biocontrol of bud rot disease of coconut. Dual culture technique was employed to test the efficiency of native fungal bioagents, viz., *Trichoderma* spp. and bacterial bioagent *P. fluorescens* on *P. palmivora* under *in vitro* conditions. All the *Trichoderma* spp, viz., *T. viride*, *T. harzianum*, *T. hamatum* and *P. fluorescens* were found to inhibit the mycelial growth of *P.palmivora* with percent inhibition ranging from 46–88%. Among the three species of *Trichoderma* tested, maximum inhibition of pathogen was recorded with *T. viride* with an inhibition of 82.03%. This was followed by *T. hamatum* and *T. harzianum* with an inhibition of 75.39 and 71.30% respectively. The *P. fluorisescens* inhibited the mycelial growth of *P.palmivora* up to 50%.

Field experiments on the control of stem bleeding disease in coconut indicated that

Trichoderma spp. was effective against stem bleeding disease. Of different treatments, the maximum decrease in bleeding (exudation) patch (perimeter) was obtained when *T. harzianum* was applied as a paste on the bleeding patch coupled with basal application of the same bioagent (50g) in combination with 5kg neem cake (10.6cm decrease), followed by *T. viride* (9.8cm decrease) and *T. hamatum* (8.3cm decrease) when applied as a smearing paste in combination with soil treatment (50g) by mixing in 5 kg neem cake.

Root feeding with carbendazim 2% @ 100 ml/palm at quarterly intervals significantly reduced the leaf blight disease (14.6% disease reduction). This was found to be on par with foliar spraying with Bordeaux mixture 1% (13.1% disease reduction). In the control, maximum disease intensity of 41.60% was observed.

Severe incidence of leaf-eating caterpillar *Opisina arenosella* can be effectively controlled by the integrated pest management strategy including release of parasitoids and root feeding and spraying of Monocrotophos. Spraying of 0.05% Monocrotophos or 0.05% Endosulphan or 0.05% Phosalone or root feeding of 10ml of Monocrotophos with 10ml water were recommended. The *Brachymeria nosatoi* is an effective host specific pupal parasitoid of black-headed caterpillar with highest parasitization. This newly introduced parasitoid is being successfully multiplied and supplied to farmers from Ambajipeta centre.

Based on the trials conducted at Ratnagiri centre on IPM for black-headed caterpillar, field release of larval parasite *Goniozus nephantidis* at the rate of 20% of the estimated pest population or 3,500 parasites per hectare is recommended. Though parasitoids effected up to 60% biological suppression, which is sufficient to control pest under check, if the pest population goes on increasing, 2-3 releases of parasitoids at fortnightly interval are necessary.

In Integrated Pest Management trials for leaf-eating caterpillar at Aliyarnagar, it was observed that from an initial pre-release mean population of 785.60, the mean population was reduced drastically to 210.50 21 days after giving the first phase root feeding of Azadirachtin F 5%. Further reduction in population to 48.20 was obtained 21 days after the release of two larval parasitoids and one pupal parasitoid. It was observed that during the second phase of Azadirachtin F 5% and biocontrol release treatments, there was a significant reduction in population level which reached a very low level of 4.35 per palm. There was significant decline in populations of larvae, pre-pupae and pupae after imposing the treatment for two times successively. Studies on the effect of release of baculovirus inoculated rhinoceros beetles on leaf and spindle damage at Ambajipeta centre revealed that the leaf and spindle damage decreased to an extent from 40to 15% in leaf and 73–47% in spindle damage in Ambajipeta village. In the studies conducted with rhinolure traps, it was observed that each rhinolure sachet exhausted in about 60 days. Number of beetles trapped/trap ranged from 1–9.3 with an average of 12.6/trap/month. Peak catches of beetles were obtained during April, May, June, September and October. It was also found that, the trapped beetles were over numbered with females.

The trial on the management of eriophyid mite in coconut gardens conducted at Ratnagiri centre in Maharashtra revealed that adoption of recommended package for eriophyid mite management could reduce the eriophyid mite infestation from 66.58–6.96%.

Similarly, the IPM package for the management of leaf-eating caterpillar carried out at farmer's garden could reduce the mean population level from 459.91 to 205.63 per palm at 21 days after root feeding Monocrotophos. Drastic reduction in mean pest population per palm was achieved from 205.63–4.35 palms after the release of bio-agents for six times consecutively at 21 days interval. Consequently, there was an increase in the level of parasitisation of all the parasitoids, viz. Braconids, Bethylids and Chalcids from 4.16, 2.03 and 0.75% to 26.05, 14.85 and 6.40% respectively.

From the data obtained in the integrated pest management demonstration plots being maintained by Ambajipeta centre, it was evident that, a progressive decrease of leaf and spindle damage due to rhinoceros beetle over a period after implementation of IPM could be achieved. On an average, within a period of 18 months, an average reduction of 56.54 and 62.83% in leaf and spindle damage respectively could be achieved in all the study areas, whereas in control plots an increase in leaf and spindle damage (63.00% and 13.00% spindle) was recorded.

Recommendations for Disease Management

A few of the disease management recommendations approved for commercial adoption are listed below:

Integrated Disease Management for Basal Stem rot

- The diseased part of the garden should be isolated from healthy area by digging isolation trench (1m deep and 0.5m width). Injury or damage to roots and pruning or cutting of roots should be avoided to prevent infection through injured roots.
- Removal and burning of diseased and dead palms along with roots is recommended. The pit for replanting should be filled up with a mixture of soil and farmyard manure in equal quantities along with 50g of *Trichoderma viride* talc powder + 1 kg neem cake.
- Talc formulation of *Trichoderma viride* (50g) in combination with 5 kg neem cake/palm/year should be applied to all palms in a garden where diseased palms are noticed. Even if one diseased palm is noticed in a garden, talc formulation of *T.viride* (50g) in combination with neem cake (5kg) per palm/year should be imposed to all the palms in the garden.
- Frequent watering/irrigation should be done during summer months. While irrigating, care should be taken to avoid flow of water from diseased trees to others. Basin system of irrigation to individual palms should be adopted.

Management of Stem Bleeding

Smearing of talc powder paste formulation of *Trichoderma viride* on bleeding patches on the stem is recommended. Paste formulation can be prepared by adding 25 ml of water to 50 g of talc powder formulation of *Trichoderma viride*.

Management of Leaf Blight

Root feeding with carbendazim 2% @ 100 ml/palm at quarterly interval is recommended to manage leaf blight disease in coconut in interior Tamil Nadu.

Red Gram as Indicator for Basal Stem Rot

Red gram plant is a good indicator plant for early diagnosis of basal stem rot disease prone coconut gardens. In *Ganoderma* prone areas, red gram shows bark splitting symptom within a period of 3 months. Red gram could be raised in basins of coconut in *Ganoderma* prone/ suspected soils and immediately after observing the symptoms on red gram, the recommended integrated disease management package against basal stem rot disease in coconut is to be adopted.

Infectivity Nature of *Phytophthora palmivora*

The *Phytophthora palmivora*, bud rot pathogen in coconut, is found to infect coconut, oil palm and palmyrah. Hence, care should be taken that while taking management practices against bud rot in coconut, palmyrah palms existing in coconut gardens may also be treated as the primary source of inoculum of *Phytophthora palmivora*.

Palmyrah

Survey on bud rot, leaf blight disease caused by *Pestalotia palmarum,* leaf spot caused by *Stigmina palmivora* and bud rot caused by *Phytophthora palmivora* was conducted at Killikulam centre. Maximum incidence of leaf blight disease was noticed in Nanguneri village of Tirunelveli district (33.33%) Disease Index (PDI) and minimum incidence (18.00 PDI) was observed in Keelavallanadu village of Thothukoodi district. Leaf spot disease was recorded at four places with PDI ranging from 9–20. Bud rot disease incidence was noticed at five places and the incidence ranged from 8–20%. Maximum incidence was noticed in Keelavallnadu village where the disease incidence was 20%.

Bud rot affected palmyrah palm

An experiment was laid out at Killikulam centre on the eco-friendly management of nursery diseases of palmyrah with nine treatments. Soil application of *Trichoderma viride* showed maximum reduction in tuber rot incidence. It was followed by soil application of neem cake where disease incidence was 13.33 as against 43.33% in control.

Survey carried out for tuber rot disease in Rajanagaram Mandal in Andhra Pradesh indicated that the disease incidence ranged from 2–17%. Survey at Nidadavolu and Chagallu Mandals of West Godavari district in Andhra Pradesh revealed that the incidence of tuber rot ranged from 0.5–3.0%.

Leaf blight disease was observed in South Tamil Nadu region with the PDI ranging from 3–37. Leaf spot disease was observed with PDI ranging from 6.00–25.30.

Overall Impact

The All India Coordinated Research Project on Palms is coordinating research on perennial crops like coconut, oil palm and palmyrah (including arecanut in earlier phase) and has made impressive impact on production, productivity, profitability and sustainability of the respective cropping systems.

Coconut

India is one of the largest producers of coconut in the world. At present, the country is producing 12,597 million nuts from an area of 18.4 lakh ha and the average productivity of 6847 nuts/ha, highest in the world. The country is self-sufficient for its domestic demand for raw nuts and its derived products such as copra and coconut oil. The demand forecast for 2025 AD is about 20,572 million nuts which could be achieved through concerted research and developmental programmes. Decade-wise statistics on area, production and productivity of coconut in India (1950-51–2000-01) given as follows.

Decade-wise data on area, production and productivity of coconut in India (1950-51 to 2000-01)

Year	*Area ('000 ha)*	*Production (million nuts)*	*Productivity (nuts/ha)*
1950-51	626.5	3281.7	5238
1960-61	714.4	4639.1	6466
	(14.51)	(41.36)	(23.44)
1970-71	1045.45	6075.0	5811
	(45.73)	(30.95)	(-10.13)
1980-81	1083.3	5942.0	5485
	(3.62)	(-2.19)	(-5.61)
1990-91	1513.9	9700.2	6407
	(39.75)	(63.25)	(16.81)
2000-01	1839.8	12597.3	6847
	(21.53)	(29.87)	(6.87)

(Figures in parenthesis are the % increase/decrease over previous decade)

The coconut productivity has increased by 30.71% as compared to 1950-51. However there were fluctuations in the inter decennial data due to incidence of drought, cyclones, pests and diseases. The AICRP on Palms centres conduct research in crop improvement, crop production and Crop Protection disciplines, which resulted in appropriate technologies relevant to different agro-climatic regions to increase the productivity level of coconut. These technologies have been effectively transferred to the field by the transfer of technology network of SAUs and State Departments of Horticulture/Agriculture. The salient achievements made by the AICRP-Palms centres are highlighted below.

Evaluation of coconut germplasm for their performance in different agro-climatic regions is one of the priority areas of research under AICRP on Palms. The germplasm maintained at these centres are being used for location-specific breeding programmes. Based on these trials, 24 high-yielding varieties/hybrids were released. The performance details of these varieties/ hybrids are presented in the earlier Chapter. The increase in the net returns from the high yielding varieties/hybrids varied from Rs. 11,000 to Rs. 22,000/ha as compared to the West Coast Tall. At present, the rate of adoption of these varieties is only 20%, thus this technology could produce an additional of 17,220 lakh nuts worth of Rs 80 lakh/annum.

Research studies conducted by AICRP on Palms centres have proved that a well balanced combination of organics and inorganics could sustain the yield level of coconut palms in long run. A blanket recommendation of chemical fertilizers for coconut has been standardized as 500:320:1200g of N: P: K given in two split doses during May-June and September-October. This combined with 25 kg of organic manures will give a sustainable yield of 80 nuts/palm/

year as compared to 40 nuts/year in unfertilized gardens. The additional net returns is Rs. 20,000/ha and this technology is adopted by 40% of the coconut farmers in the country.

During late 1990's, the price of coconut and its products had experienced a steep fall causing serious economic loss to farmers. Forecasting these problems, AICRP-Palms had viewed its research in crop management with farming system perspective. Long-term experiments conducted at different locations had proved that adoption of suitable coconut-based farming/cropping system could reduce the degree of production and price risks in coconut farming. The increased net returns over coconut monocrop was up to Rs. 1.3 lakh are recorded at Kahikuchi centre in Assam with a crop combination of coconut, black pepper and turmeric.

This technology is being adopted by 20% of the coconut farmers in the country to the desirable level and by another 40% in a less scientific manner. During the recent price crash for coconut and its products, those farmers who have adopted these models could realize better stability to the gross farm income.

Agronomists of AICRP on Palms have standardized the irrigation practices for Tamil Nadu, Karnataka and Maharashtra primarily, through the use of micro-irrigation system. This technology has gained importance in water scarce areas of northern Kerala, parts of Tamil Nadu, Karnataka and Maharashtra and the overall increase in net returns by adoption of this technology varies from Rs. 35,00 to Rs. 55,00/ha/annum.

Research efforts have resulted in the standardisation the control/managerial measures for major coconut pests like rhinoceros beetle, red weevil, eriophyid mite, leaf-eating caterpillar and root grub. Suitable prophylactic/control measures if taken could save from the death the coconut palm and economic advantages in terms of benefit: cost ratio ranges from 1.30 to 1.83 depending on severity of pest incidence and frequency of control measures adopted. This technology is widely being adopted by coconut farmers in Kerala, Tamil Nadu, Karnataka and Andhra Pradesh.

The extent adoption of other technologies such as management of Ganoderma wilt, stem bleeding, bud rot, Tatipaka disease, etc., ranged from 20-50% and the benefit: cost ratio of adoption of these technologies is above 1.5.

Arecanut

India is the largest producer of arecanut in the world as compared to other major arecanut producing countries such as Bangladesh, Sri Lanka and Indonesia. The present level of production is to the tune of 3.35 lakh tonnes from an area of 2.90 lakh ha and the average productivity of 1155 kg/ha is the highest in the world.

Area, production and productivity of arecanut in India

Year	*Area (ha)*	*Production (tonnes)*	*Productivity (kg/ha)*
1960-61	113,008	95,468	845
1970-71	167,300	141,000	843
1980-81	184,500	191,400	1,037
1990-91	217,008	230,500	1,099
1997-98	268,700	333,900	1,243
2000-01	290,000	335,000	1,152

Source: Directorate of Economics and Statistics, New Delhi

Based on studies conducted under All India Coordinated Coconut and Arecanut Improvement Project, three cultivars have been released for large-scale cultivation. They are Sumangala, Sreemangala and Mohitnagar which give 64, 59 and 84% higher yield respectively than the local varieties. The realized additional net return varied from Rs. 25,000/ha to Rs. 25,000 to Rs. 55,000/ha under varied price situation of arecanut.

Oil palm

The potential of high yield per hectare and the special characteristics of the palm oil enable it to compete with other dozen major vegetable oils in the market. Globally palm oil is the leading oil in the export trade accounting for 35% of the total oils and fats bought and sold the world over. Oil palm is recognized as a crop of prime importance in wiping out the edible oil deficit in India.

India's present edible oil consumption is over nine million tonnes. Since the production always fall short of the requirement, we are forced to resort to import resulting in huge drain on foreign exchequer. India's population increase is 1.5 million each year. At present, the per capita consumption rate of oils and fats is only 8 kg as against the world average of 16 kg. The growth in consumption of oils and fats in India was 2.1% against a population growth of 1.9% indicating the growing affluence in India which was backed by a GDP increase between 5 per cent and 9%. With the rise in standard of living and burgeoning population, the consumption is likely to tremendously increase. At a consumption rate of 10 kg, almost 10 million tonnes of edible oil will be required to cater to the needs of our population.

A crop like oil palm, which gives steady income, is the best answer to circumvent this problem. Moreover, this palm can produce ten times the yield of any other annual crop obviously reducing the area required for production of required quantity of oil. Therefore, the future scope for oil palm development in India is very bright and its contribution to the edible oil pool of the country will be substantial in view of the crop's many advantages like highest oil yield, eco-friendliness, sustainability of income, freedom from theft and buy back guarantee.

Realizing the importance of oil palm research, ICAR is strengthening its research efforts through the four AICRP on Palms centres located in Tamil Nadu, Andhra Pradesh, Karnataka and Maharashtra in collaboration with NRC for Oil Palm. The multi-locational trials conducted at these centres helped us in identifying promising tenera hybrids and also to evolve an optimum irrigation-cum-fertilizer schedule for oil palm crop.

Palmyrah

Palmyrah palm, adorns the dry landscape of semi arid regions of Tamil Nadu, Andhra Pradesh, Orissa, West Bengal, Bihar, Karnataka and Maharashtra. India has nearly 102 million palms and half of the trees are in Tamil Nadu. Out of 51.9 million trees in Tamil Nadu, more than 50% of palms are concentrated in southern districts of Thoothukudi, Tirunelveli, Virudhunagar and Ramnad, while Thoothukudi district alone has a major share of 10 million trees.

Among the four sugar yielding plants, viz. *Borassus flabellifer* (palmyrah), *Phondeix sylvestris* (date), *Caryota urens* (sago) and

Cocos nucifera (coconut), palmyrah palm ranks first in yielding sugar as well as other edible and non-edible produces. The inflorescence sap or neera collected from the inflorescences is a major source of sweetening agent and it replaces cane sugar in palmyrah growing tracts of Tamil Nadu and other states. The endosperm of immature seednuts from young fruit is a delicacy during summer like that of tender coconut. Matured fruits are roasted and the mesocarp is consumed. Many value-added products can also be prepared from the fruit. From the non-edible seed nut, one can have an edible sweet spongy hastorium or apocole. Palmyrah is a fodder crop as well, since the mesocarp of tender fruit is relished by cattle.

Palmyrah is also a source of fibre. Three types of fibres from petiole, a tough fibre from leaf base and leaf are obtained. Mesocarp of the fruit also yields fibre. Several industries based on palmyrah fibre are functioning in Tamil Nadu, Andhra Pradesh and West Bengal. Varieties of fancy and utility articles are made from palm leaves, veins and roots. Mainly women are employed in such cottage industries. Palm leaves have been used to write scripts from time immemorial. Mature leaves are cured and used for thatching while senesed leaves are utilized as fuel.

Several products of palmyrah have medicinal value. Besides sugar, neera contains minerals such as calcium, phosphorus and iron. It has vitamins like ascorbic acid, thiamine, riboflavin and niacin. Neera is a laxative promoting digestion and diuretic serving to keep the body cool. Palm candy is a medicine to get relief from sour throat and dry cough. It is added to milk and taken as a beverage. The spadix as well as the root has medicinal value.

Palmyrah palms also serves as a wind-break and arrest sand dunes in their lands. The mature outer wood of the stem is used in construction as pillars, beams and rafters. Realizing the importance of palmyrah in the economy of the backward regions of rural economy, palmyrah research is being strengthened under the All India Coordinated Research Project on Palms. Palmyrah germplasm collection programme was the most important thrust area in palmyrah research. Survey and collection of palmyrah germplasm is in progress in Tamil Nadu (Killikulam) and Andhra Pradesh (Pandirimamidi) besides studies on pest and diseases associated with the crop. Survey to collect elite germplasm with emphasis on leaf production, dwarfness and yield of neera is in progress. So far, 187 accessions were collected at Killikulam (Tamil Nadu) and 176 at Pandirimamidi (Andhra Pradesh). A palmyrah accession (BF-34) yielding 287.8 litres of neera in a tapping duration of 85 days has been identified. The neera characteristics like reducing sugar (286–856 mg/100 g) protein (16–25mg/100 ml) and vitamin C (7.87–15.74 mg/100 g) were analyzed.

Crop protection studies were undertaken at both Killikulam and Pandirimamidi centres. Survey and surveillance programmes on the incidence of pests and diseases in palmyrah palms in major palmyrah-growing regions is in progress. Control measures for tuber rot (apocolon) in the nursery caused by *Rhizoctonis solani* has been standardized.

Research programmes being implemented under the AICRP on Palms are expected to improve the palmyrah production and productivity in the near future.

MANAGEMENT EFFICIENCY INDICATORS

Resource Generation

The All India Coordinated Research Project on Palms Centres are maintaining experimental farms for conducting the assigned technical programme. Apart from the experiments, the centres are also producing seedlings and planting material for their clientele depending on the local demands. A few centres are also producing large quantities of planting material of spices which could be planted as mixed crop in coconut gardens. Income generated through the sale of farm products during the last five years is presented as follows:

Year-wise resource generation from AICRP on Palms Centres

Year	*Resource generated (Rs)*
1999-2000	9,91,612
2000-2001	6,60,958
2001-2002	10,71,625
2002-2003	10,83,554
2003-2004	10,03,554
2004-2005	
2005-2006	
2006-2007	

Manpower Management

The present staff strength of AICRP on Palms is 82, of which 35 are scientists. Centre-wise details of different categories of staff are furnished below.

Staf strength of AICRP (Palms) as on 31-12-2007

Centre	*Scientific*	*Technical*	*Administrative*	*Supporting*	*Total*
Aliyarnagar	4	1	-	1	6
Ambajipeta	5	2	1	2	10
Arsikere	2	2	–	–	5
Bhubaneshwar	2	1	–	–	3
Jagadalpur	2	1	–	–	4
Kahikuchi	2	1	–	–	3
Mondouri	2	1	–	–	3
Ratnagiri	3	2	1	15	21
Veppankulam	5	2	1	1	9
Aduthurai	1	2	–	–	3
Gangavathi	1	2	–	–	3
Mulde	1	2	–	–	3
Vijayarai	1	2	–	–	3
Killikulam	2	1	–	1	4
Pandirimamidi	2	1	–	1	4
Total	35	23	3	21	82

Budget Allocation

Relative quantum of funds sanctioned for salaries, infrastructure, equipment and operations are givenbelow (Rs lakh - ICAR share)

Staff strength of AICRP (Palms) as on 31-12-2007

	Tenth Plan			Eleventh Plan Proposal		
	Non-Plan	*Plan*	*Total*	*Non-plan*	*Plan*	*Total*
A. Recurring						
Pay and Allowances	–	500.00	500.00	–	1166.95	1166.95
TA		20.00	20.00	–	35.00	35.00
Contingency	–	104.00	104.00	–	211.50	211.50
Total A	–	**624.00**	**624.00**	–	**1413.45**	**1413.45**
B. Equipment	–	36.00	36.00	–	89.31	89.31
Works	–	15.00	15.00	–	70.81	70.81
Furniture	–	–	–	–	3.60	3.60
Total B	–	**51.00**	**51.00**	–	**163.72**	**163.72**
Grand Total	–	**675.00**	**675.00**	–	**1577.17**	**1577.17**

INTERACTION AND LINKAGES

Taking into cognizance, the problems faced by farmers and the need to improve the agricultural production scenario in the country, various programmes are being implemented by different agencies for assisting coconut cultivators. At the national level, the frontline extension programmes of All India Coordinated Research Project on Palms and CPCRI, programmes of Ministry of Agriculture through Coconut Development Board and Directorate of Extension and at the state level different Agriculture/Horticulture departments, private agencies and NGOs play a critical role in the implementation of various transfer of technology programmes for agricultural development.

Keeping the above in view, AICRP on Palms is interacting with different categories of clientele like scientists, extension personnel, farmers, students, officials, etc., and maintain effective and active collaboration, coordination and cooperation with several ICAR institutions and also development organizations both at national and international levels and implementing various programmes on a regular basis to disseminate the research results among the cultivators. The following are the major interaction programmes implemented by the AICRP centres.

Frontline Demonstrations

To popularize the production and protection technologies evolved at AICRP centres and also to work out the economic viability of technologies under farmers' conditions, frontline demonstrations were organized at the AICRP centres. Details of the frontline demonstrations organized during the Tenth Five Year Plan are given as under.

Centre	Topic	No. of Frontline Demonstrations
Aliyarnagar	Management of rhinoceros beetle	1
	Management of opisina	1
	composted coir pith	3
Ambajipeta	Management of rhinoceros beetle	1
	Management of opisina	1
	composted coir pith	3
Ratnagiri	Management of rhinoceros beetle	1
	Management of opisina	1
	composted coir pith	3
Aduthurai	Nutrient and irrigation management in oil palm	4
Vijayarai	Nutrient and irrigation management in oil palm	4
Mulde	Nutrient and irrigation management in oil palm	2
Veppankulam	Basal stem rot management	3
Arsikere	Basal stem rot management	3

Training Programmes

All the AICRP (Palms) centres are serving as the knowledge source for the respective states with reference to the crops handled by them. They offer a series of institutional and off-campus training programmes for the benefit of field level extension personnel, farmers, farm women and rural youth.

The pre- and post-knowledge tests indicated that the training programmes conducted by AICRP on Palms could enhance the knowledge levels of participants at significant levels. Further, feed-back received from the participants helped us to improve the training courses. A list of topics of training programmes offered at different centres during the last five years are given here.

Centre	Title of the Training Programme
AICRP Palms Centre, Mondouri (Kalyani)	Palm climbing and production technologies on coconut
	Coconut production technologies in West Bengal
	Advanced coconut production technologies in West Bengal
	Resource person training on coconut based cropping system in West Bengal
	Resource person training on seedling production technologies of coconut in West Bengal
	Resource person training on pests and diseases management on coconut
	Resource person training on vermicompost production technologies on coconut waste
AICRP-Palms Centre, Veppankulam	Three day training programme on DxT hybridization technique to agricultural Officers, tree climbers, etc.
	Five days training programme on root feeding technologies in coconut to coconut farmers, rural youths, etc.
	One-day training programmes on coconut cultivation to farmers.
	One-day training programmes on IPM in coconut
AICRP-Palms Centre, Ambajipeta	Training to officers of the Department of Agriculture on root feeding technique
	Skill demonstration on basal application of biocontrol agent *Trichoderma viride* 50 g + neem cake 5 kg/palm against basal stem rot disease of coconut
	Interaction programmes with the farmers on coconut production and protection technologies through farmers meet and field visits

(Contd.)

Centre	*Title of the Training Programme*
	Training to officers of the Department of Agriculture through T&V meetings on coconut production and protection technologies
	AMC level training programmes to farmers on coconut production and protection technologies
AICRP-Palms Centre, Aduthurai	Training on oil palm and palmyrah
AICRP-Palms Centre, Ratnagiri	Vermicomposting and mushroom production
	Training programmes on coconut organized in collaboration with Coconut Development Board, Department of Agriculture and Nationalized Banks to farmers.
	Cultivation and post-harvest technology of spice crops to farmers
AICRP-Palms Centre, Mulde	Preservation of fruits and vegetables.
	Six months training on nursery management
AICRP-Palms Centre, Pandirimamidi	Importance of soil testing in horticultural crops
	Production technology of horticultural crops
AICRP-Palms Centre, Killikulam	Cultivation of fruit crops and medicinal plants under waterlands
	Cultivation of fruit crops and medicinal plants under waterlands
	Technologies for coconut cultivation
	Training on watershed development

Publications

To popularize the technologies evolved at the AICRP centres based on long period of research, publication were regularly brought out to provide the same as reference materials for extension workers and progressive farmers. Some of the publications recently brought out by the AICRP Centres are listed below:

Title	*Centre*	*Year*
Oil Palm Tree (Tamil)	AICRP-Palms Centre, Aduthurai	1999
A Treatise on Palmyrah	AICRP-Palms Centre, Killikulam	1999
Tengina besaya (Kannada) (Coconut Cultivation)	AICRP-Palms Centre, Arsikere	2000
Survey and assessment of coconut pest damage (A training manual for IPM on coconut pests)	AICRP-Palms Centre, Aliyarnagar	2001
Oil Palm Research in Konkan	AICRP-Palms Centre, Mulde	2002
Prospects of Coconut based Cropping System in West Bengal	AICRP-Palms Centre, Mondouri	2001
Coconut black headed caterpillar, Opisina arenosella Walker - A Study in the East coast of Andhra Pradesh	AICRP-Palms Centre, Ambajipeta	2003
'Kamrupa' - A high yielding coconut cultivar (Folder)	AICRP-Palms Centre, Kahikuchi	2003
Aliyar Nagar - 1 - A promising coconut cultivar (Folder)	AICRP-Palms Centre, Aliyarnagar	2003
Basal stem rot disease of coconut - symptoms and control measures (Tamil)	AICRP-Palms Centre, Veppankulam	2003
Biocontrol of Basal stem rot (Ganoderma wilt) disease of coconut	AICRP-Palms Centre, Ambajipeta	2004
Coconut based cropping system for Konkan region	AICRP-Palms Centre, Ratnagiri	2005
Integrated nutrient management in coconut	AICRP-Palms Centre, Ratnagiri	2006
Recent technologies in coconut disease management	AICRP-Palms Centre, Ambajipeta	2007
Eriophyid mite management	AICRP-Palms Centre, Ambajipeta	2007
Integrated nutrient management in coconut	AICRP-Palms Centre, Ambajipeta	2007

THRUST AREAS

Considerable variability exists in the local germplasm of coconut and palmyrah in different states. This has to be collected and conserved in the wake of their destruction by crop diversification and urbanization. Hence, survey and collection of local germplasm of coconut and palmyrah for evaluation needs to be strengthened.

Maximization of Coconut Productivity

Considering the fact that the cost of production of coconut in India is higher than the production cost in other coconut-growing countries and prevailing price for coconut/coconut oil in India is almost three times higher than that of international price, considerable thrust will be given during the Eleventh Five-Year Plan to improve the productivity and reduce the cost of inputs for profitable coconut cultivation. In addition to the continuation of the long-term experiments initiated during Ninth and Tenth Five-Year Plan periods, a few projects aiming at improving the productivity and reducing the cost of inputs for profitable horticulture would be introduced.

Genetic improvement of coconut has not received the attention that it deserved mainly due to some unique problems encountered with the crop. The perennial habit, long juvenile period, extensive out crossing and heterogeneous nature, difficulties in clonal propagation, large area required for systematic experimentation and long gestation period for obtaining confirmed results are some of the barriers for limited progress in coconut breeding for improvement. Among several breeding methods, exploitation of heterosis had maximum impact in coconut improvement. Systematic work on hybridization in India was started in 1930s and the discovery of hybrid vigour (Patel, 1937) received considerable attention in production of coconut hybrids. After that, a large number of hybrid combinations of Tall × Dwarf (T × D) and Dwarf × Tall (D × T) have been evaluated over the years and the programme resulted in the release of 12 coconut hybrids in India. At present, hybrids are being commercially cultivated in about 6% of coconut area in country.

Based on the trials conducted in different countries, it was generalized by coconut breeders that "whenever higher profitability in long-term is to be preferred from coconut cultivation with the consequence of needing at least 3-4 additional years to pay off investment cost, the T × T hybrids should be preferred". There is another argument in favour of T × T hybrids. The fruits collected from the D × T hybrids cannot be used as seed nuts (due to segregation of dwarfism genes). However, in the case of T × T hybrids, the nuts collected in the plantations can still be used as seed nuts, although their yield capacity will be lower than that of hybrids. A recent point of view is to recommend T × T hybrids for ensuring profitable cultivation of intercrops, as their rapid growth would permit better development of the lower vegetation layers. However, the studies on the performance of Tall × Tall hybrids has not received much attention. Trial conducted at CPCRI, Kasaragod, as well as Coconut Research Station, Veppankulam, also indicated that Tall × Tall hybrids could improve the productivity of coconut in different agroclimatic regions, especially with reference to higher nut yield, more copra productivity and higher coconut oil production per unit area. Hence, there is a need to initiate a programme for testing the performance of T × T hybrids in different coconut growing-regions. Care would be taken to include locally better performing tall type as one of the parents in each cross combination. Crossing programme would be undertaken at CPCRI, Kasaragod, and seedlings would be provided to nine coconut centres for further evaluation.

Improvement of Tender Coconut Yields

Multiple products of coconut palms are being recognized in most of the coconut-growing countries. One potential product that is presently gaining popularity is tender coconut. It is reasonably forecasted that out of the 12,800 million nuts being produced in India, about 3,000 million nuts could be used for tender nut purposes for domestic consumption as well as for export. Although all the existing varieties could be made use for tender nut production, short stature, palm varieties are more appropriate for the purpose because of easiness in harvesting and possibility of accommodating higher palm population in an unit area. In India, a few dwarf variety palms known for producing tender coconut with large volume of sweet water are available. Cultivation of such dwarf variety palms has to be promoted for ensuring a regular supply of tender coconuts in domestic market. Research conducted in India so far could result in identification of Chowghat Dwarf Orange and Gangabondam as ideal tendernut varieties. However, yield levels of these two cultivars are medium to low in different regions. Hence, there is a need to identify high-yielding dwarf types preferably D × D hybrids specific to different agro-climatic regions.

The greater the share of dwarfism in a coconut hybrid, the better is its precocity and the more the smaller canopy will permit high-density planting. However, one school of thought is that, in general, dwarfs are not very hardy and have little drought resistance. Possibly, these shortcomings are transmitted partly to the hybrid. In addition, on the scale of world coconut, dwarfism remains a marginal phenomenon. Variability of dwarfs is inferior to that of talls. The long-term potentialities of dwarf hybrids are therefore probably limited. The precocity and rapid emission of bunches of most dwarf ecotypes are of major importance. Hence, D × D hybrids could also play a greater role in crop improvement needs in coconut. Location specific D × D cross combinations would be identified for nine coconut centres. Crossing programme would be undertaken at CPCRI, Kasaragod, and the hybrid seedlings would be provided to the AICRP centres for further evaluation.

Improvement in Water and Nutrients Productivity

Fertilizer and water are important critical inputs in coconut and oil palm production and play a greater role in deciding the yield performance. Among various components of cost of production of coconut and oil palm, in addition to harvesting charges, expenditure incurred towards fertilizer and irrigation are major components. Fertigation is an efficient method of fertilizer application through drip system of irrigation. Drip fertigation helps to increase the fertilizer-use efficiency, make savings in fertilizer costs, reduce labour requirement, enables uniform, precise and direct application to root zone and correction of micronutrient deficiencies etc. Soluble fertilizers, viz. urea (for nitrogen), DAP (for phosphorus and nitrogen) and muriate of potash (for potassium) could be combined and supplied through drip irrigation, after filtering the solutions before adding to fertilizer tank. The fertilizer should be applied in active absorption zone for which the emitter should be placed 1 m away from the bole of palms. Studies conducted at CPCRI, Kasaragod indicated that considerable saving of fertilizers and irrigation water could be achieved through fertigation. Compared to conventional method, we can save fertilizers up to 50% because of higher fertilizer use efficiency. Similar results are expected from oil palm cultivation also. Hence, fertigation trials are proposed to be initiated in all the 10 coconut centres and four oil palm centres during Eleventh Five Year Plan.

Location-specific Coconut Based Farming Systems

Utility of an integrated farming system model in coconut has been effectively demonstrated at Kasaragod. The system involves adoption of selected soil and moisture conservation measures to utilize the available rainfall in the region, to economise the irrigation water through drip irrigation system and an integrated nutrient management system with major emphasis on farm waste utilization. Successful crops chosen based on coconut based cropping system trials laid out in coordinating centres would be included as a component in proposed farming system, to be laid out in all the 12 coconut centres. The study would enable us to evolve an integrated farming system model for location which would be further replicated among the coconut community in the region.

Region-specific Cocoa Types

Cocoa is one of the most ideal mixed crops that could be grown profitably in coconut gardens. At present, cocoa is cultivated in about 18000 ha with an annual production is about 12,000 tonnes and productivity of 560kg/ha. The Indian chocolate market is about 20,000 tonnes at present and is growing at about 15%/year. The production of cocoa beans meets only 30% of the demand projected by the processing industry in India. Hence, there is an urgent need to promote cocoa cultivation in the country.

The clonal trials conducted at CPCRI, Kasaragod, resulted in identification of a few promising lines that could improve the cocoa productivity. These lines are to be tested in different agro-climatic regions for cultivation as a mixed crop in coconut gardens. Hence, it is proposed to initiate a new trial on the "evaluation of new cocoa accessions suitable for cultivation as a mixed crop in coconut gardens".

Location-specific Pepper Types

Cocoa is one of the most ideal mixed crops that could be grown profitably in coconut gardens. The clonal trials conducted at IISR, Kozhikode, resulted in the identification of a few promising lines that could improve the pepper productivity. These lines are to be tested in different agro-climatic regions for cultivation as a mixed crop in coconut gardens. Hence, it is proposed to initiate a new trial on the "evaluation of new pepper accessions suitable for cultivation as a mixed crop in coconut gardens".

Identification of Location-specific Medicinal and Aromatic Plants

There exists excellent scope for diversification in traditional intercrops/mixed crops cultivated with coconut. New initiative on introducing medicinal and aromatic plants, flowering and ornamental plants will transform the economy of coconut farmer whose income sustainability is threatened due to uncertain prices of copra and nuts.

Biocontrol of Diseases and Pests

Adequate thrust is to be given for strengthening our research efforts for evolving more effective biological control measures for the management of coconut pests and diseases. Certain leads have been obtained on evolving an integrated disease management package for the management of basal stem rot disease. The package needs further refinement to enhance its acceptability to farmers.

"National Live Palmyrah Germplasm Repository" could be established at Killikulam and molecular characterization programme is to be initiated. The possibility of support from IPGRI under the scheme on "Under utilized country specific crops" is proposed to be explored for this purpose.

CONCLUSION

We must strengthen the process of imparting dynamism and optimism in the farm sector, as was done in the. Accelerated agricultural progress based on the enhancement of productivity, sustainability and profitability through farming systems diversification, sustainable intensification, quality upgradation and value-addition, is vital not only for food security and poverty eradication, but also for national sovereignty. Growth of coconut, oil palm and palmyrah sectors both in terms of GDP and livelihood security with social equity has been ever challenging and more so in the present context of inadequate public sector investment and services and pressure of globalization of agriculture leading to a greater demand for highly knowledge-intensive services. Investments in coconut, oil palm and palmyrah sectors in the country need to be rationalized and appropriate knowledge driven institutional reforms are to be brought in. Appropriate framework for technology development and dissemination would be very much needed for transforming the sector from the present approach of its sustenance as a way of life to a vibrant economic activity with a sense of pride for future generations. As coconut, oil palm and palmyrah are playing a major role in the rural economy, especially among the millions of small holders in coastal regions, any improvement in productivity to be brought about through the strengthening of the AICRP on Palms would directly help in increasing the income of small holders and thus would enable them to raise their standard of living.

7

Spices

M. Anandraj, Johny A. Kallupurakkal and D. Prasath

The All India Coordinated Research Project on Spices (AICRPS) is the largest research network in the country through which a nation wide collaborative and interdisciplinary research is being carried, linking ICAR system with the SAUs. The AICRPS initiated in 1971 is vested with the mandate to conduct research on 12 spice crops, viz. black pepper, cardamom, large cardamom, ginger, turmeric, cumin, coriander, fennel, fenugreek, clove, nutmeg and cinnamon. Scientific research on these crops are carried out under the AICRPS in 19 coordinating centers located in 15 states with respective SAUs, representing major agro-climatic regions located in 15 states of India. The headquarters of the project is at Indian Institute of Spices Research, Calicut.

The ICAR launched a major programme for the first time to organize spices research in the country by establishing the All India Spices and Cashew nut Improvement project in 1971. This project was started with four centres exclusively for spices, grew steadily to 12 centres in 1986 (VII Plan) and subsequently to 20 centres by the end of VIII Plan. In order to give adequate and undivided attention to spices combined project was bifurcated. The AICRP on Spices came into existence in 1986.

Objectives

- Evolving high-yielding varieties, resistant/tolerant to pests and diseases
- Multi-locational testing of new varieties under varied agroecological situations
- Developing location-specific varieties and standardizing agro-technologies through integrated nutrient management for increasing sustainable productivity
- Evolving suitable pests and disease management technologies through integrated pest and disease management
- To collect, conserve and study the biodiversity of spices in the country
- To act as an interface between ICAR/SAUs/IISR, DASD systems.

ORGANIZATIONAL SET UP

This project is now spread over in 15 states in the country with 19 research centres and 6 participating voluntary centres (see the Map and Table 7.1). They are located in 15 SAUs and one at ICRI (Spices Board), Gangtok. The centres of AICRPS represents 13 major agroclimatic regions of India. The project has staff strength of 51 including scientists and 32 technical/ auxillary ones.

At present there are110 programmes in 12 crops on which research is being conducted. The utility and efficiency of the programmes are constantly being reviewed critically in different workshops/group meetings.

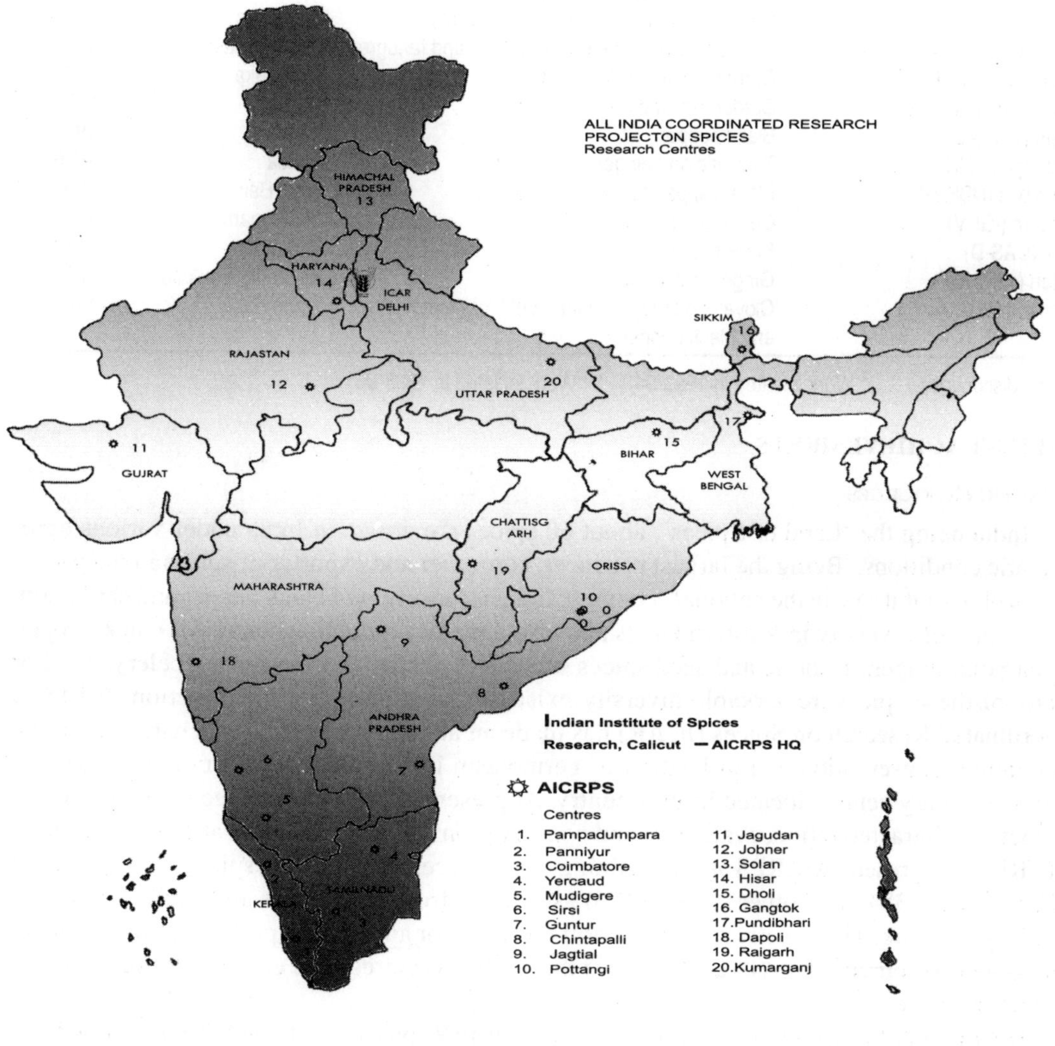

The AICRPS in the country

Table 7.1. AICRPs centres and mandate crops

Centre	Crops handled	States	Year of starting
Chinthapalli (ANGRAU)	Black pepper and ginger ***	Andhra Pradesh	1981
Coimbatore (TNAU)	Coriander, fenugreek and turmeric ***	Tamil Nadu	1975
Dapoli (KKV)	Black pepper, Nutmeg, clove and cinnamon	Maharashtra	1995
Dholi (RAU)	Turmeric, coriander and fenugreek	Bihar	1993
Guntur (ANGRAU)	Coriander and fenugreek	Andhra Pradesh	1975
Hisar (CCSHAU)	Coriander, fennel and fenugreek	Haryana	1993
Jagtial (ANGRAU)	Turmeric	Andhra Pradesh	1986
Jagudan (GAU)	Cumin, coriander, fennel, and fenugreek	Gujarat	1975
Jobner (RAU)	Cumin, coriander, fennel, and fenugreek	Rajasthan	1975
Kumarganj (NDUAT)	Turmeric, ginger, fennel coriander, and fenugreek	Uttar Pradesh	1995
Mudigere (UAS-B)	Cardamom and black pepper**	Karnataka	1971
Pampadumpara (KAU)	Cardamom and black pepper **	Kerala	1971
Panniyur (KAU)	Black pepper	Kerala	1971
Pottangi (OUAT)	Turmeric and ginger	Orissa	1975
Pundibari (NBKVV)	Black pepper, turmeric and ginger	West Bengal	1996
Raigarh (IGKV)	Coriander, turmeric and ginger	Chattisgarh	1996
Sirsi (UAS-D)	Black pepper	Karnataka	1981
Solan (YSPHUF)	Ginger and turmeric *	Himachal Pradesh	1971
Yercaud (TNAU)	Clove**, nutmeg**, cinnamon**, cardamom * and black pepper **	Tamil Nadu	1981

* Crop discontinued ** Work initiated in 1993-94 *** Work initiated in 1995-96.

SALIENT ACHIEVEMENTS

Genetic Resources

India being the 'Land of Spices', about 60 spices are grown in India under various agro-climatic conditions. Being the largest producer, consumer and exporter of spices in the world, spices play a vital role in the national economy. India is the origin of black pepper and cardamom with centre of diversity in Western Ghats and also a major production centre for black pepper, cardamom, ginger, turmeric and seed spices such as coriander, cumin, fennel, celery, etc. For many of these spices, remarkable diversity exists in the country. In this direction, All India Coordinated Research on Spices (ICAR) has made an attempt for collection, characterization, evaluation, conservation and utilization of germplasm in 12 spice crops, through 19 centres and 8 voluntary centres located in the country. At present, 11 research projects on germplasm collection, characterization, evaluation and conservation are being operated at 19 centres under AICRPS. At present AICRPS is maintaining 7,001 accessions (major spices - 2984; minor (seed) spices –3841 and other spices - 176) assembled from indigenous and exotic sources in *ex-situ* gene bank. The genetic resources form the basis for any crop improvement programme. The genetic resources of spices assembled at the AICRPS centres were enriched through further collection surveys.

By utilizing these germplasm, about 75 varieties in 9 spice crops have been developed and released for cultivation in various agroclimatic regions of the country through AICRPS with high yield, pest/disease resistance and quality attributes. Besides, some of the lines are under

A high yielding pepper from Sirsi

Unique germplasm collections. a) Turmeric with unusually large size of mother rhizome b) A farmer's collection 'Ademane pepper'

Effect of NAA on growth of coriander

pipeline for release and several lines have been identified for yield, quality and pest/disease and drought resistance or tolerance from the germplasm of spices Table 7.2).

Table 7.2. Genetic resources of spices at AICRPS centres (as on 31-03-2006)

Crop/Centre	*Indigenous*	*Exotic*		*Total*
		Cultivated	*Wild and Related Species*	
		BLACK PEPPER		
Panniyur	144	22	3	169
Sirsi	110	6	1	117
Chintapalle	56	-	-	56
Yercaud	116	8	-	124
Pundibari	11	-	-	11
Dapoli	66	-	-	66
Dholi	7	-	-	7
Total	**510**	**36**	**4**	**550**
		CARDAMOM		
Pampadumpara	121	-	-	121
Mudigere	263	2	-	265
Total	**384**	**2**		**386**
		GINGER		
Pottangi	167	2	3	172
Solan	271	-	2	273
Dholi	42	-	-	42
Kumarganj	29	-	-	29
Pundibari	34	-	-	34
Raigarh	15	-	-	15
Dapoli	3	-	-	3
Total	**561**	**2**	**5**	**568**
		TURMERIC		
Pottangi	171	22	-	193
Jagtial	273	-	-	273
Dholi	78	2	-	80
Bhavanisagar	-	-	-	-
Raigarh	42	-	-	42
Kumarganj	108	-	-	108
Pundibari	116	14	-	130
Solan	171	-	-	171
Coimbatore	255	-	-	255
Total	**1214**	**38**	**-**	**1252**
		CLOVE		
Pechiparai	22	-	-	22
Thadiyankudisai	-	-	-	-
Dapoli	3	-	-	3
Total	**25**	**-**	**-**	**25**

(Contd.)

Crop/Centre	*Indigenous*	*Exotic*		*Total*
		Cultivated	*Wild and Related Spcies*	
		NUTMEG		
Pechiparai	20	-	-	20
Thadiyankudisai	-	-	-	-
Dapoli	92	-	-	92
Total	112	-	-	112
Cinnamon				
Pechiparai	12	-	-	12
Thadiyankudisai	-	-	-	-
Dapoli	11	-	-	11
Total	**23**	**-**	**-**	**23**
		CASSIA		
Pechiparai	4	-	-	4
Dapoli	6	-	-	6
Total	**10**	**-**	**-**	**10**
		CORIANDER		
Jobner	683	-	112	795
Jagudan	71	-	20	91
Coimbatore	224	-	-	224
Guntur	243	-	5	248
Hisar	179	-	-	179
Dholi	85	-	-	85
Raigarh	20	-	-	20
Kumarganj	65	-	-	65
Total	1570	-	137	1707
Cumin				
Jobner	366	-	10	376
Jagudan	210	-	7	217
Kumarganj	19	-	-	19
Total	**595**	**-**	**17**	**612**
		FENNEL		
Coimbatore	3	-	-	3
Jobner	191	-	12	203
Jagudan	93	-	4	97
Hisar	74	-	-	74
Dholi	44	-	-	44
Kumarganj	36	-	-	36
Total	**441**	**-**	**16**	**457**
		FENUGREEK		
Jobner	357	-	8	365
Jagudan	63	-	-	63
Coimbatore	255	-	-	255
Guntur	125	1	-	126
Hisar	86	-	-	86

(Contd.)

Crop/Centre	Indigenous	Exotic		Total
		Cultivated	Wild and Related Species	
Dholi	97	-	-	97
Raigarh	13	-	-	13
Kumarganj	60	-	-	60
Total	**1056**	**1**	**8**	**1065**
Paprika				
Coimbatore	27	-	-	27
IISR, Calicut	26	-	21	47
Total	53	-	21	74
Vanilla				
Coimbatore	2	-	-	2
IISR, Calicut	53	12	-	65
ICRI, Myladumpara	35	-	-	35
Total	**90**	**12**	**-**	**102**
Grand total		7,001		

Major spices	-	2756
Tree spices	-	228
Seed spices	-	3841
Other spices (Paprika and vanilla)	-	176
Grand Total	-	7001

Crop Improvement

The germplasm were subjected to evaluation under initial evaluation trial (IET) and coordinated varietal trial (CVT) and resulted in evolving 50 improved varieties in major seed spices. Apart from this, promising lines were identified in each major seed spices for yield and quality attributes and also resistant/tolerant sources against major pest and diseases.

TECHNOLOGIES FOR ENHANCING PRODUCTION AND PRODUCTIVITY

Black Pepper

- Grafting of *P.nigrum* (laterals, runners, topshoots) on *P.colubrinum* rootstock was standardized and pepper grafts are being evaluated (Table 7.3).
- For vegetative propagation, the middle one-third portion of runner shoot of the vine which has higher rate of growth and production of spikes. Dipping cut end of the vine in IBA (1000 ppm) for 45 seconds produced more number of roots and better shoot growth.
- For trailing pepper, silver oak is identified as a suitable standard which also provided shade and support to the vine hence recommended for Andhra Pradesh.
- Elephant foot yam, Colocasia, ginger and turmeric are as intercrops in black pepper garden.
- Under planting is recommended after 18–22 years old plantation.
- Drip irrigation from December to April @ 2 litres/day, increased number of spikes, length and green berry yield.

Table 7.3. High-yielding spices varieties

Variety/Year	*Pedigree/Parentage of Release*	*Institution/University*	*Yield kg/ha (dry)*	*Salient Features*	*Recommended State/Region*
			BLACK PEPPER		
Panniyur 1 1967	Inter-cultivar hybrid of Uthirankotta x Cheriyakaniyakadan	Pepper Research Station (KAU), Panniyur, Karimbam	1,242	Vigorous growing vine. Do not tolerate shade. Medium maturity group. Long spikes, close setting of berries, bold berries, oleoresin 11.8%, piperine 5.3%, essential Kerala oil 3.5%, dry recovery 35.3%	Kerala, Karnataka and Tamil Nadu
Panniyur 2 1991	Clonal selection from open pollinated progeny of Balankotta	-do-	2,570	Shade tolerant, suitable for intercropping, medium maturity group, medium quality oleoresin 10.9%, high piperine (6.6%), essential oil 3.4%, dry recovery 35.7%	Kerala and Karnataka
Panniyur 3 1991	Inter-cultivar hybrid of Uthirankotta x Cheriyankaniyakadan	-do-	1,953	Suitable for all pepper growing regions, performs well under open situation. Late maturity group. Long spikes and bold berries, piperine 5.2%, oleoresin 12.7%, essential oil 3.1%, dry recovery 27.8%	Kerala, Karnataka and Tamil Nadu
Panniyur 4 1991	Clonal selection from Kuthiravally type II	-do-	1,277	Stable yield er, performs well under adverse condition also, tolerant to shade, late maturity, 4.4% piperine, 9.2% oleoresin, 2.1% essential oil and 34.7% dry recovery	Kerala, Karnataka and Tamil Nadu
Panniyur 5 1996	Clonal selection from open pollinated progeny of Perumkodi	-do-	1,110	Suitable for both mono and mixed cropping in coconut/arecanut gardens, shade tolerant, medium maturity, tolerant to nursery disease. Long spikes, piperine 5.3%, oleoresin 12.33%, essential oil 3.8% and dry recovery 35.7%	Kerala
Panniyur 6 2000	Clonal selection from Karimunda type III	-do-	2,127	A vigorous vine. Tolerant to drought and adverse climatic conditions, stable and regular bearer, medium maturity group. Suitable for open condition as well as partial shade, spike 6-8cm, more number of spikes/unit area, close setting and attractive bold berries, piperine 4.9%, oleoresin 8.27% and essential oil 1.33% and 33.0% dry recovery.	Kerala and Karnataka
Panniyur 7 2000	Open pollinated progeny of Kalluvally	-do-	1,410	Vigorous vine and a regular bearer, long spike, a hardy type vine, tolerates adverse climatic condition, suitable for open and shaded conditions, very long spike (16–24 cm) high piperine content (5.6%), oleoresin 10.6%, essential oil 1.5%, and 34.0% dry recovery.	Kerala, Karnataka and Tamil Nadu
PLD –2 1996	Clonal selection from Kottanadan	NRC for Oil Palm, Regional Station, ICAR, Palode, Pacha (Post), Kerala	2,475	Late maturity high quality cultivar, contains piperine 3.0%, oleoresin 15.45%, essential oil 4.8%. Suitable for plains and higher elevations.	Kerala

(Contd.)

Variety/Year	Pedigree/Parentage of Release	Institution/University	Yield kg/ha (dry)	Salient Features	Recommended State/ Region
			GINGER		
Suprabha 1988	Clonal selection from Kunduli local	High Altitude Research Station, OUA & T Pottangi, Koraput (Dist), Orissa	16.6	Plumpy rhizome, less fibre, wide adaptability, suitable for both early and late sowing, duration 229 days. 8.9% oleoresin, 4.4% crude fibre, 1.9% essential oil and 20.5% dry recovery.	Orissa and adjoining states
Suruchi 1990	Clonal selection from Kunduli local	-do-	11.6	Profuse tillering, bold rhizome, suitable for rainfed/irrigated conditions duration 218 days. 10.9% oleoresin, 2.0% essential oil 3.8% crude fibre, 23.5% dry recovery	Orissa, Central and South India.
Surabhi 1991	Induced mutant of Rudrapur local	-do-	17.5	Plumpy rhizome, dark skinned yellow fleshed, suitable for both irrigated/rainfed, duration 225 days. 10.2% Oleoresin, 2.1% essential oil, 4.0% crude fibre, 23.6% dry recovery.	Orissa
Himgiri 1996	Clonal selection from Himachal collection	Department of Vegetable Crops, YSPUH &F, Nauni, Solan Himachal Pradesh	13.5	Best for green ginger, less susceptible to rhizome rot disease, suitable for rainfed condition. 4.29% oleoresin, 1.6% essential oil, 6.05% crude fibre, 20.2% dryrecovery, 230 days duration.	Himachal Pradesh
			TURMERIC		
CO.1 1983	Vegetative mutant by x-ray irradiation of Erode local	Dept. of Spices and Plantation Crops, HC & RI, TNAU Coimbatore – 641 003 Tamil Nadu	30.5	Bold and bright orange yellow rhizomes, curcumin 3.2%, oleoresin 6.7%, essential oil 3.7%, dry recovery 19.5%, suitable fordrought prone, hilly areas saline and alkaline areas. Crop duration 270 days. Plants are robust, vigorous and taller (40-60cm) with more leaves (25-32 Nos) and tiller production 3.7 to 5.0	Tamil Nadu
BSR.1 1986	Clonal selection from Erode local irradiated with x rays	-do-	30.7	Bright yellow rhizome, curcumin 4.2%, oleoresin 4.0%, essential oil 3.7%, dry recovery 20.5%, crop duration 285 days, suitable for drought prone areas of Tamil Nadu.	Tamil Nadu
BSR.2 1994	Induced mutant from Erode localinsects	-do-	32.7	A high yielding short duration variety (245 days) with bigger rhizomes, resistant to scale	Tamil Nadu
Krishna 1983	Clonal selection from Tekurpeta collection	Maharashtra Agrl. University Kasba, Digraj Maharashtra	9.2	Plumpy rhizomes, curcumin 2.8%, oleoresin 3.8%, essential oil 2.0%, dry recovery 16.4%, duration 240 days. Moderately resistant to pests and diseases.	Maharashtra
Sugandham 1984	Clonal selection from germplasm	Spices Research Station GAU Jagudan Mehsana, Gujarat	15.0	Thick, round rhizomes with short internodes. Curcumin 3.1%, oleoresin 11.0%, essential oil 2.7%, dry recovery 23.3%, duration 210 days. Moderately tolerant to pest & diseases.	Gujarat
Roma 1988	Clonal selection from T.Sunder Korapurt (Dist), Orissa	High Altitude Research Station OUA&T, Pottangi	20.7	Suitable for both rainfed and irrigated condition. suitable for hilly areas and late season planting. Curcumin 6.1%, oleoresin 13.2%, essential oil 4.2% and dry recovery 31.0%, duration 250 days.	Orissa, Tamil Nadu, Himachal Pradesh, Andhra Pradesh and Kerala

(Contd.)

Variety/year	*Pedigree/Parentage of Release*	*Institution/University*	*Yield kg/ha (dry)*	*Salient Features*	*Recommended State/Region*
Suroma 1989	Clonal selection from T. Sunder by x- ray irradiation	-do-	20.0	Round and plumpy rhizome, field tolerance to leaf blotch, leaf spot and rhizome scale, curcumin 6.1%, oleoresin 13.1%, essential oil 4.4% and dry recovery 26.0%, duration 253 days	Orissa, Tamil Nadu, Himachal Pradesh, Andhra Pradesh, and Kerala
Ranga 1992	Clonal selection from Rajpuri local	-do-	29.0	Bold and spindle shaped mother rhizome, suitable for late planting and low lying areas. Moderately resistant to leaf blotch and scales, curcumin 6.3%, oleoresin 13.5%, essential oil 4.4% and dry recovery 24.8%, duration 250 days.	Orissa, Tamil Nadu, Himachal Pradesh, Andhra Pradesh and Kerala
Rasmi 1992	Clonal selection from Rajpuri local	-do-	31.3	Bold rhizomes, suitable for both rainfed and irrigated condition, early and late sown season, curcumin 6.4%, oleoresin 13.4%, essential oil 4.4% and dry recovery 23.0%, duration 240 days.	Orissa, Tamil Nadu, Himachal Pradesh, Andhra Pradesh and Kerala
Rajendra Sonia1989	Selection from local germplasm	Dept. of Hort., Tirhut College of Agriculture, RAU Dholi – 843121, Bihar		Bold and plumpy rhizome, grows widely under all north Indian conditions. Curcumin 8.4%, essential oil 5.0% and dry recovery 18.0%, duration 225 days.	Bihar
Megha turmeric 1996	Selection form Lakadong type	ICAR R.C. NEH Region Shillong, Meghalaya	23.0	Suitable for the North East hill and North West Bengal. Bold rhizomes, high curcumin content 6.8% and dry recovery 16.37%, duration 300-315 days.	Meghalaya Uttar Bengal Krishi
Suranjana (TCP-2) 2000	Clonal selection from local types of West Bengal	Viswa Vidyalaya, North Bengal Pundibari (PO) – 736 165, Dist. Cooch Behar, West Bengal	(Pot. yield 29.0)	Suitable for open and shaded conditions, sole or intercrop, suitable for rainfed as well as high rain fall areas. Curcumin 5.7%, oleoresin 10.9%, essential oil 4.1%, dry recovery 21.2%, duration 235 days, tolerant to leaf blotch and rhizome rot. Resistant to	
Suguna 1991	Selection from germplasm collected from Assam	Indian Institute of Spices Research, ICAR, Calicut – 673 012, Kerala	29.3	Short duration type (190 days), curcumin 4.9%, oleoresin 13.5%, essential oil 6.0% and dry recovery 20.4%, field tolerance to rhizome rot.	West Bengal Kerala and Andhra Pradesh
Suvarna 1991	Selection from germplasm collected from Assam	-do-	17.4	Bright orange coloured rhizome with slender fingers. Maturity 200 days, field tolerant to pest and diseases Curcumin 4.3%, oleoresin 13.5%, essential oil 7.0% and dry recovery 20.0%.	Kerala, Karnataka and Andhra Pradesh
Sudharsana 1991	Selection from germplasm collect from Singhat, Manipur	-do-	28.8	High yielding variety, short duration type (190 days). Field tolerant to rhizome rot. Curcumin 5.3%, oleoresin 15.0%, essential oil 7.0% and dry recovery 20.6%.	Kerala and Andhra Pradesh
IISR Prabha 1996	Open pollinated progeny selection	-do-	37.47	High yielding variety, curcumin content 6.5%, oleoresin 15.0%, essential oil 6.5% and dry recovery 19.5%, crop duration 205 days.	Kerala and Tamil Nadu

Contd.

Variety/year	Pedigree/Parentage of Release	Institution/University	Yield kg/ha (dry)	Salient Features	Recommended State/Region
IISR Prathiba 1996	Open pollinated progeny selection	-do-	39.12	High quality line, 6.2% curcumin content with high yield, 16.2% oleoresin, 6.2% essential oil, 18.5% dry recovery, crop duration 225 days.	Kerala, Tamil Nadu and other states
IISR Alleppy Supreme 2004 (identified)	A clonal selection from Alleppy turmeric	-do-	35.45.58 (dry)	Shows tolerance to leaf blotch disease. Rhizomes contain 5.55% curcumin, 16.0% oleoresin, 19.0% dry recovery, crop duration 210 days	Kerala (rainfed) Maharashtra, Karnataka and N. Bengal (irrigated)
IISR Kedaram 2004 (identified)	Clonal selection from germplasm	-do-	34.5 5.28 (dry)	Tolerant to leaf blotch disease, Rhizomes contain 5.5% curcumin, 13.6% oleoresin, maturity 210 days and 18.9% driage.	Kerala (rainfed) Maharashtra, Karnataka and N. Bengal (irrigated)
Kanthi 1996	Clonal selection from Mydukur variety of Andhra Pradesh	Dept. of Plantation Crops and Spices, College of Hort., KAU, Vellanikkara – 680 656, Trichur, Kerala	37.65	Erect leaf with broad lamina, big mother rhizomes with medium bold fingers and closer internodes. Medium duration. Curcumin content (7.18%), oleoresin 8.25%, essential oil 5.15%, dry recovery 20.15%, duration 240–270 days.	Kerala
Sobha 1996	Clonal selection from local type germplasm	-do-	35.88	Mother rhizome big with medium bold and closer internodes. Inner core of rhizomes is dark orange like Alleppey. More territory rhizomes. Dryage 19.38%, curcumin content (7.39%), oleoresin (9.65%), essential oil (4.24%), medium duration 240–270 days.	Kerala
Sona 2002	Clonal selection from local germplasm	-do-	4.02 (dry)	Orange yellow rhizome, medium bold with no territory fingers. Best suited for central zone of Kerala. Rhizome medium bold, field tolerant to leaf blotch. Curcumin 7.12%, essential oil 4.4%, oleoresin 10.25%. 18.88% dry recovery, medium duration. 240–270 days.	Kerala
Varna 2002	Clonal selection from local germplasm	-do-	4.16 (dry)	Bright orange yellow rhizome, medium bold with closer inter nodes, territory fingers present. Suited to central zone of Kerala. Field tolerant to leaf blotch, curcumin 7.87%, essential oil 4.56%, oleoresin 10.8%. 19.05% dry recovery, medium duration 240–270 days.	Kerala

(Contd.)

Variety/year	Pedigree/Parentage of Release	Institution/University	Yield kg/ha (dry)	Salient Features	Recommended State/Region
			CINNAMON		
YCD.1 1996	Clonal selection from OP seedlings progenies of Sri Lankan type.	Hort. Research Station, (TNAU), Yercaud, Dist. Salem, Tamil Nadu	Dry bark yield of 360kg/ha (143.9g dry bark/plant)	Good bark recovery, adapted to wide range of soil and rainfed conditions. Bark oil 2.8%, leaf oil 3.0%, bark recovery 35.3%.	High ranges at 500–1000 m above MSL, Tamil Nadu
PPI (C)-1 2003	Selection from OP seedlings progeny introduced from Sri Lanka seed supplied from IISR.	HRS, (TNAU) Pechiparai Kanyakumari Dist., Tamil Nadu	Fresh bark yieldof 980 kg/ha.	Bark, higher oil recovery from the bark (2.9%) and leaf oil recovery of 3.3%, bark oil 2.9%, leaf oil 3.3%, bark recovery 34.22%. Suitable for cultivation in high rain fall zones and hill regions of Tamil Nadu at an altitude range of 100-500m MSL.	High rainfall zones and hill ranges of Tamil Nadu.
Konkan Tej 1993	Seedling selection from progenies of Sri Lankan accessions.	Regional Coconut Research Station, Dr. BSKKV, Vengurle, Bhatye (Dist) Ratnagiri, Maharashtra.	Fresh bark yield of 334g/ plant (102g dry bark/plant)	Superior qualities with 3.2% bark oil with bark recovery 29.16%, cuminaldehyde in bark oil 70.23, eugenol in bark oil 6.93%, leaf oil 2.28%, eugenol in leaf oil 75.5%, yields 4.10kg fresh bark. Bark recovery 51.78%.	Konkan region of Maharashtra
Sugandhini (ODC-130) 2000	Single tree selection from Wynad loca collection. A Sri Lankan type	Aromatic and Medicinal Plants Research Station (KAU), Odakkali, Asamanor, P.O., Ernakulam	640 kg (1.2 kg fresh bark/tree, 18.34 kgfresh leaves/tree/ year)	Recommended for cultivation for leaf oil production, bark oil 0.94%, leaf oil 1.6%, bark recovery 51.0%, cinnamaldehyde in bark oil 45.0%,eugenol in leaf oil 93.7%; released mainly for leaf oil purpose. Densely foliage.	Midlands and high lands of Kerala and South India
			NUTMEG		
Konkan Sugandha 1998	Single plant selection from local seedling population.	Regional Fruit Research Station, Vengurla, Maharashtra.	200-526 fruits/ tree	Adaptable in Konkan region. Tree canopy is conical and compact.	Maharashtra
Konkan Swad 2003	Selection from nutmeg seedlings from Ratnagiri dist.	Regional Coconut Research Station, Bhatye (Dr. BVSKU), Ratnagiri.	761.38 fruits/tree	Adapted to Konkan region with warm, humid conditions as well as shade provision. Canopy erect, conical shape. Contain 39.8% essential oil in seed and 10.9% in mace. No incidence of pest & disease are noticed.	Konkan region of Maharashtra.
Guj. Cor.1 1974	Selection from germ plasm	Main Spices Research Station, GAU, Jagudan,	1100	Suitable for early sowing, erect plant, moderately tolerant to wilt and powdery mildew, round bold grains, essential oil 0.35%, duration 112 days.	Gujarat

(Contd.)

Variety/year	Pedigree/Parentage of Release	Institution/University	Yield kg/ha (dry)	Salient Features	Recommended State/Region
		Mehsana Dist., Gujarat			
Guj.Cor.2 1985	Reselection from Co.2	-do-	1450	Semi spreading type, suitable for early sowing, moderately tolerant to powdery mildew, oblong, essential oil 0.40%, lodging and shattering resistant.	Gujarat
Co.1 1977	Selection from germplasm Koilpatti local	Dept. of Spices and Pl. Crops, HC and RI, TNAU, Coimbatore, Tamil Nadu	440	A variety with small statuted plant, duration 110 days, suitable for rainfed areas and for greens and grains, small grain, essential oil 0.27%.	Tamil Nadu
Co.2 1982	Reselection from culture P2 of Gujarat	-do-	520	A dual purpose variety, suitable for saline, alkaline and drought prone areas, maturity 90-100 days, seeds oblong, medium 0.40% essential oil.	Tamil Nadu
Co.3 1991	Reselection from Acc.695 of IARI, New Delhi	-do-	650	A dual purpose variety, good yielder, medium sized grains, suitable for rainfed & irrigated condition, rabi as well as kharif season. Field tolerant to powdery mildew, wilt & grain mould. Seed oil content ranges from 0.38 to 0.41%, 85-95 days duration.	Tamil Nadu, Gujarat and Andhra Pradesh
Co(CR)4 2002	Reselection from germplasm ATP 77 Guntur collection	-do-	600	Early maturing variety (65-70 days), suitable for both rainfed and irrigated condition; grains oblong and medium; essential oil content is 0.4%; field tolerant to wilt & grain mould.	Tamil Nadu
Rajendra Swati1988	Pure line Selection from Muzaffarpur collection	Dept. of Hort., Tirhut College of Agriculture, RAU Dholi Muzaffarpur, Bihar	1300	Medium sized plant with fine, aromatic round grains, essential oil 0.65%. Suitable for intercropping, field tolerance to aphids.	Bihar
RCr.41 1988	Recurrent half sib selection from local type from 'Kota'	SKN College of Agrl., R AU, Jobner Jaipur, Rajasthan	1270	A tall erect plant type with thick stem. Grows well under irrigated condition, resistant to stem gall, wilt and moderately resistant to powdery mildew; small seeds (9.3 g/1000 seed), essential oil 0.25%. Long duration variety (130-140 days).	Rajasthan
RCr 20 1997	Recurrent half sib selection from Jaipur local	-do-	900	Medium sized bush plant, suitable for rainfed crop or limited moisture conditions and heavy soils of south Rajasthan. Moderately resistant to stem gall, bold grains, essential oil 0.25%. Early maturity, duration 100-110 days	Rajasthan
RCr 435 2003 (Identified)	Half sib selection from local germplasm from Jalore	-do-	1000	Plants are bushy, erect, bold seeds, medium size, contain essential oil 0.33%; medium maturity; adapted for irrigated condition, moderately resistant to root knot and powdery mildew. Medium maturity, duration 110-130 days.	Rajasthan

(Contd.)

Variety/year	Pedigree/Parentage of Release	Institution/University	Yield kg/ha (dry)	Salient Features	Recommended State/Region
RCr 436 2001	Half sib selection from local germplasm from 'Kota'	-do-	1100	Plants semi dwarf, bushy type; with quick early growth and bold big seeds, resistant to root rot & root knot nematodes; most suitable for limited moisture condition and heavy soils of south Rajasthan, essential oil 0.33, early maturing (90-100 days).	Rajasthan
RCr 684 1999 (Identified for national release)	Mutation breeding by gamma rays. Induced mutant of Rcr-20	-do-	990	A variety, resistant to stem gall and less susceptible to powdery mildew. Adapted to medium heavy textured soil, essential oil 0.32% and sandy loam soil under irrigations. Seeds of the variety are bold. Plants are tall and erect with higher number of seeds/umbel. Medium maturity (110-120 days).	Rajasthan
RCr-4462001 (Identified for national release)	Half sib selection from local type from Jaipur local.	-do-	1200	Plants tall, are leafy erect with higher number of seeds per umbel. Seeds are medium in size and 0.33% volatile oil content. Moderately Resistant to stem gall and wilt under rainfed, limited moisture or irrigated conditions, medium maturity (110-130 days).	Rainfed areas in Rajasthan
Sadhana 1989	Mass selection from local Alur collection	Regional Agrl. Research Station, (ANGRAU), Lam, Guntur, Andhra Pradesh	1025	Dual purpose, semi-erects variety; suitable for rainfed condition, field tolerance to diseases and white fly, mites & aphids, essential oil 0.2%. A mid late variety 95-110 days duration withstands moisture stress, responded well to input management under optimum moisture level.	Andhra Pradesh
Swathi 1989	Mass selection from Nandyal germplasm	-do-	855	Plants medium size semi erect type. Early maturing variety (80-85 days), suitable for rainfed condition, and late sown season, essential oil 0.30%. Medium size oval grain. Field tolerant to white fly, moderately tolerant to disease. It suits well to the areas where the soil moisture retentiveness in comparably less, being early maturity. It escapes powdery mildew disease.	Rainfed areas in Andhra Pradesh
Sindhu 1991	Mass selection from germplasm Warangal local	-do-	1000	Oval medium breakable grains, essential oil (0.4%); suitable for rainfed areas, tolerant to wilt, powdery mildew as well as drought condition, medium duration (100-110 days).	Andhra Pradesh
Hisar Anand 1994	Mass selection from Haryana collection	Dept. of vegetable Crops, CCSHAU, Hisar, Haryana.	1400	A medium tall, dual purpose variety, oval medium size seeds; essential oil 0.35%, wider adaptability to different soil conditions. Resistant to lodging due to spreading habit	Haryana
Hisar Sugandh 2001	Mass selection from indigenous germplasm	-do-	1400	Suitable for irrigated conditions. Resistant to stem gall disease	Haryana & Rajasthan
Hisar Surabhi 2004 (Identified)	Mass selection from local germplasm	-do-	1800	Bushy erect plant type, seed medium, oblong; oil content 0.425%, tolerant to frost, less susceptible to aphids, medium duration (130–140 days)	Haryana & other states

Contd.

Variety/year	*Pedigree/Parentage of Release*	*Institution/University*	*Yield kg/ha (dry)*	*Salient Features*	*Recommended State/Region*
Azad Dhania-1 1996	Mass selection from Kalyanpur germplasm collection	C.S.Azad Univ. of Agric. and Technology, Kanpur, Uttar Pradesh	1000	Erect, early branching, number of umbellates per umbel 5, essential oil 0.29%, tolerant to moisture stress, tolerant to powdery mildew and aphids, maturity 120-125 days.	Uttar Pradesh
Pant Haritima, 1993	Selection from local type Pant Dhania	Dept. of Veg. Sciences, Govind Ballabh Pant Univ. of Agri. and Techn, Pantnagar Udham Singh Nagar (Dist) Uttaranchal	1200	A tall erect plant, a dual purpose type, good yielder of leaf, smaller seeds with high oil (0.4%), resistant to stem gall, maturity 155-160 days.	UP, Bihar, Haryana, Bengal, Assam.
DWA-3 1999	Pure line selection from Karnataka collection.	University of Agricultural Sciences, Dharwad, Karnataka	400	A dual purpose variety and for seed production in rabi crop, essential oil 0.27%. Moderately tolerant to powdery mildew, black clay, soils are best suited.	Karnataka
CIMPOS-33	Selection from germplasm introduced from Bulgaria	Central Institute of Medicinal Plants, Regional Centre, Bangalore,	2100	Tall erect, compact, profusely branching and flowering, grains small and bold, essential oil 1.3%. Mainly recommended for oil production.	Uttar Pradesh
			CUMIN		
5-404 1952	Slection from local germplasm	Spices Research Station GAU, Jagudan, Mehsana (Dist) Gujarat	350 kg	An erect plant, medium size fruit, contain 2.2% essential oil, 7.7% crude fibre. Moderately tolerant to powdery mildew.	
Mc.43 1970	Selection from germplasm	-do-	580	Plants semi spreading, grains bold lustering withstand lodging and shattering moderately tolerant to *Fusarium* wilt, *Alternaria* blight and powdery mildew. Essential oil content 2.7%, crude fibre 15.5%.	Gujarat
Guj. Cumin 1 1983	Selection from local germplasm (Vijaypur-5)	-do-	550	Plants bushy and spreading, grains bold, linear oblong; withstand shattering and lodging, moderately tolerant to wilt powdery mildew and blight, 3.6% essential oil, 14.25% crude fibre.	Gujarat and Rajasthan
Guj Cumin 2 1992	Pure line selection from M2 irradiated seeds of MC 43	-do-	620	Bushy plant, good branching habit, grains bold, lustrous medium sized, tolerant to wilt and blight and powdery mildew, 4.0% essential oil, 22.1% crude fibre.	North Gujarat & Sourashtra region of Gujarat
Guj. Cumin 3 2000	Recurrent selection derived from W.German entry EC-232689	-do-	620	Bushy dwarf plant, fruit medium sized, frost wilt resistant variety suitable for winter season limited irrigation. Higher essential oil (4.4%) content.	Gujarat and Rajasthan
Guj. Cumin-4 2003	Natural crossing followed by selection	-do-	1250	The first wilt resis tant variety, normal, seed appearance and no splitting seed habbit	
RZ-19 1988	Recurrent single plant progeny selection from collection of Ajmeer.	SKN College of Agriculture RAJAU, Jobner Dist. Jaipur, Rajasthan	560	Erect plant, pink flowers, bold, lustrous grain, gray pubescent, tolerant to wilt and blight suitable for late sowing season. Duration 140-150 days.	Rajasthan and Gujarat

(Contd.)

Variety/Year	Pedigree/Parentage of Release	Institution/University	Yield kg/ha (dry)	Salient Features	Recommended State/Region
RZ-209 1995 (Identified)	Recurrent single plant progeny selection from Jalore.	-do-	650	A variety shown some resistance with and blight discuss. Maturity 120–130 days.	
RZ-223 (Identified)	Mutation breeding in UC-216	-do-	600	Wider adaptability, resistant to wilt, superior in yield and seed quality over RZ-19. Plants bushy, semi erect, long bold attractive seeds, with 3.0 to 3.5% volatile oil content, medium duration (120–130 days)	
			FENNEL		
S-7-9 1956	Selection	Spices Research Station, GAU, Jagudan, Mehsana (Dist), Gujarat	1100 kg	A bushy plant with big umbel, moderately tolerant to blight, contain 1.2% essential oil. Duration 210 days, 24.0% crude fibre.	Gujarat
PF – 35 1973	Selection from local germplasm	-do-	1280	Plant tall and spreading, moderately tolerant to leaf spot, leaf blight and sugary diseases.	Gujarat
Guj. Fennel 1 1984	Pure line selection from Vijaypur local	-do-	1720	Plant tall and bushy, shattering and lodging, suitable for early sowing and as *rabi* crop, reasonably tolerant to drought, moderately tolerant to sugary disease, oblong, medium bold and dark green seeds.	Gujarat
Guj. Fennel 2 1997	Pedigree selection from local germplasm	-do-	1940	Plants bushy, bold grains, rich in volatile oil (2.4%) suitable for both rainfed and irrigated condition.	North Gujarat and Saurashtra region of Gujarat.
Guj. Fennel -II 2003	Selection based on individual plant progeny performance from local germplasm	-do-	2489	A medium maturity type (148 days) adopted to rabi season under irrigation; seeds medium bold, contains 1.8% volatile oil.	Gujarat under irrigated condition
Co.1 1985	Reselection from PF 35	Dept. of Spices and Plant. Crops, HC and RI, TNAU Coimbatore, Tamil Nadu	570	Medium statured, diffuce branching, suitable for intercropping and border cropping with chilli and turmeric. Suitable for drought, prone, water logged saline and alkaline conditions. Duration 220 days.	Tamil Nadu
RF 101 2001	Recurrent half sib selection from local germplasm collection from Jobner	SKN College of Agriculture, RAJAU, Jobner, Jaipur, Rajasthan	1550	Erect, medium tall nature, medium maturity type (150-160 days) with long bold grains, most suitable for loamy and black cotton soil	Rajasthan

(Contd.)

Variety/year	Pedigree/Parentage of Release	Institution/University	Yield kg/ha (dry)	Salient Features	Recommended State/ Region
RF 125 2003 (Identified)	Recurrent half sib selection is an exotic collection EC 243380 from Italy.	-do-	1700	Plants are short statured with compact umbels and long bold seeds when green presence of denser view of plants. Duration 110-130 days, tolerant to sugary disease.	Rajasthan
.Hisar Sawrup 2004	Mass selection from ind igenous germplasm of Haryana	Dept.of Vegetable Crops, CCS HAU, Hisar, Haryana	1600	Plants grow upright, spreding, gives a bushy appearance. A late maturity type (175-185 days) grain long and bold with 1.6% volatile oil content, resistant to lodging, no shattering of grains	Haryana under irrigation condition
			FENUGREEK		
Co.1 1982	Reselection from TG-2336 introduced for North India	Dept. of Spices and Plant. Crops, HC & RI, TNAU Coimbatore Tamil Nadu	600	A quick growing short, dual purpose, early maturing variety (80-95 days) tolerant to root rot disease. Seed contain 21.7% protein.	Tamil Nadu
Co 2 1999	Selection from CF 390	-do-	480	Short duration dual purpose variety, field tolerant to *Rhizoctonia* root rot disease, suitable for both *kharif* and *rabi* season. Early maturity, short duration (85-90 days).	Tamil Nadu
Rajendra Kanti 1988	Pure line selection from Reghunathpur collection	Dept. of Hort., Tirhut College of Agrl., Dholi Muzaffarpur, Bihar	1250	Medium sized bushy plant; early maturity (65-90 days), suitable for intercropping in both kharif and rabi season; field tolerant to *cercospora* leaf spot, powdery mildew and aphids. Seed contains 9.5% protein.	Bihar
Rajendra Abha (Kasuri methi)	NA	Dept. of Hort., Tirhut College of Agrl., Dholi Muzaffarpur, Bihar	NA	*NA*	Bihar
Hisar Sonali 1994	Pure line selection from local germplasm	Dept.of Vegetable Crops (CCS HAU), Hisar Haryana	1700	Tall and bushy vigorous growing variety, dual purpose variety, late maturity (140-145 days), suitable for cultivation under irrigated condition. Moderately resistant to root rot and aphids.	Haryana and Rajasthan
Hisar Suvarna 2001	Pure line selection from local germplasm	-do-	1600	A quick growing, erect and tall, dual purpose, medium maturity (130-140 days), moderately resistant to percospora and powdery mildew. Wider adaptability suitable for cultivation through out the country.	Haryana and Rajasthan
Hisar Madhavi 2001	Pure line selection from local germplasm of UP	-do-	1900	A quick growing, erect and tall dual purpose, medium maturity (130-140 days) variety, moderately resistant to powdery mildew and to downy mildew. A variety with under adaptability suitable for both irrigated and rainfed conditions.	Haryana and Rajasthan

(Contd.)

Variety/year	Pedigree/Parentage of Release	Institution/University	Yield kg/ha (dry)	Salient Features	Recommended State/Region
Hisar Mukta 2001	Pure line selection natural green seed coated mutant line from UP	-do-	2000	A quick growing seed type variety medium maturity (135-140 days), moderately resistant to downy mildew and powdery mildew. Erect and tall plants. Wide adaptability, suitable for both irrigated and rainfed conditions.	Haryana
Guj Methi 1 1999	Recurrent selection based on pure line selection from J.Fenu 102	Spices Research Station GAU, Jagudan Mehsana Dist, Gujarat	1860	The first variety from Gujarat released for the state. Plant dwarf.	Gujarat
RMt.1 1990	Pure line selection from Nagaur local	SKN College of Agriculture, Jobner, Rajasthan	1400	Vigorous semi erect medium size moderately branched growth habit, maturity 140-150 days, medium sized bold and attractive typically yellow coloured grains, moderately resistant to root, knot nematode and powdery mildew and aphids. Contain 0.2% diogenin, 21.0% seed protein.	Rajasthan
RMt-143 1997 (Identified)	Pure line selection of local collection of Jodhpur.	-do-	1600	Moderately resistant to powdery mildew, seeds bold yellow colour 140-150 days maturity, suitable for heavier soils.	Rajasthan
RMt 303 1999 (Identified for national release)	Mutation breeding from variety RMt 1	-do-	1900	Medium maturity (145-150 days) variety, seeds bold less susceptible to powdery mildew	Rajasthan
RMt-305 (Identified)	Mutation breeding from variety RMt-1	-do-	1300	First determinant type, multipoded, early maturing wider adaptability resistant to powdery mildew and root knot nematodes, seeds bold attractive and yellow, contain 21.7% protein, duration 120-125 days.	Rajasthan
Lam sel. 1992	Selection from germplasm collection of Uttar Pradesh	Regional Agrl. Res. Station ANGRAU, Guntur, Andhra Pradesh	740	Dual purpose early maturing variety (85-90 days), bushy type with medium height (40 cm), more number of branches and green matter. When cultivated for green leaves purpose it gives an average green yield of 12 tonnes/ha, field tolerant to major pests and diseases, seed contain 53 % protein.	Andhra Pradesh
Pant Ragini 2001	Selection from local germplasm	G.B Pant University of Agriculture and Technology Pantnagar, Dist. U S Nagar, Uttarakhand	1200	A dual purpose tall bushy type resistant to downy mildew and root rots, medium maturity (170-175 days), seed contain 2.0 to 2.5% essential oil.	Uttar Pradesh and Uttaranchal.

* Yield kg/ha (dry)

- A low-cost technology for mass multiplication of *Trichoderma* sp. as biocontrol agent for the management of *Phytophthora* foot rot has been developed.
- Fertilizer schedule for different states/cropping systems/regions were standardized.

Fertilizer recommendations for black pepper

Region /state/cropping system	*NPK (g/vine)*
Panniyur and similar areas	50:50:200
Medium fertile soil	100:40:140
Arecanut and pepper mixed cropping system for heavy rainfall, Sirsi region of Karnataka	200:80:280
Karnataka (under irrigated conditions)	150:60:210
Andhra Pradesh (under rainfed conditions)	50:50:150
Chintapalle area of Andhra Pradesh (high altitude areas)	100:60:160
Laterite soils of Kerala, where soil N and K are low	140:55:270

Application of one-third of the above dose is recommended for the first year and two-thirds during the second year and full dose from third year onwards (the P dose may be either increased to 100g/vine/year or decreased to 25g/vine/year depending on P status of soil). Fertilizers are best applied in two splits, first half in May-June with the receipt of rains and second half in August-September after cessation of heavy rainfall. Application of micronutrients, zinc, boron and molybdenum @ 5, 2 and 1 kg/ha in soil during September increased the yield and quality of pepper. Addition of 1 kg lime is also recommended.

- Mulching the basins of pepper vines with saw dust, coconut husk or dry leaves or polythene sheets reduced the causality of vines, reduced spike shedding and increased the yield by 30.70%.
- Irrigating the vines at IW/CPE ratio of 0.25 (@ 100 litres) from December to March at 8–10 days interval increased the yield.
- Irrigation and fertilizer schedules for pure plantation and pepper + arecanut mixed cropping system as well as HPT were developed.
- *Ailathus malbaricum* was found to be the best standard for pepper in drought prone areas.

Cardamom

- Technology was developed for rapid clonal multiplication of planting material, optimum time for seed capsule collection, seed germination, raising nursery, scarification of seed for better germination, enhancing seed germination in inter, optimum time of planting, spacing, seed treatment for producing disease-free seedlings, nursery disease management, high-density planting, cultivation under artificial shade, and enhancing productivity by maintaining bee colonies were developed.
- Scarification of seeds for ten minutes with 25% nitric acid followed by Bavistin treatment.
- Covering the nursery-beds with polythene sheets soon after sowing was found to enhance seed germination by nearly 30% in the event of sowing seeds very late in the season (November-December).

- Paddy straw as mulch for seed beds was found to be the best and most economical.
- A spacing of 1.8 m × 1.8 m (3025 plants/ha) was found to be optimum for cardamom.
- High density planting with a spacing of 1.8 m × 0.9 m (6170 plants/ha) indicated early and high yields in the initial 3-4 years, but removal of alternate rows after 5-6 years was found inevitable for facilitating easy cultural operations.
- The quick method of clonal multiplication through proliferation of suckers by planting at closer spacing of 0.9 × 0.9 m was developed which gives a multiplication ratio of 4-5 planting clonal units per tiller could be obtained per year, which works out to be about 40,000 planting units/hectare.
- Trench system of planting resulted in a higher yield (1294.0 kg dry capsules/ha) as compared to pit system (256.7 kg/ha).
- Spacing of 2m × 2m is recommended for Kerala for the Malabar and Vazhuka types.
- For Karnataka, a general spacing of 1.8 m × 1.8 m is suggested. A spacing of 1.8m × 1.8 m (3025 plants/ha) and 2.7 m × 2.7 m (1371 plants/ha) are recommended for Malabar and Mysore types, respectively for Tamil Nadu.
- Planting of cardamom (Malabar type) with a high density of 5,000 plants/ha under rainfed condition in trench system with a fertilizer dose of 120:120:240 kg NPK/ha is recommended for Karnataka.
- Studies on the integrated nutrient management using organic and inorganic manures on the yield of cardamom at Mudigere during the last four years confirmed that 100% inorganic manure alone recorded maximum green capsule yield (707 kg/ha).
- Variety Mudigere 1 (Malabar type) is best suited for high density planting (1m × 1m spacing). A closer spacing of Malabar types (1m × 1m) also gave higher yield.
- Coffee – Cardamom and black pepper combination with yield potential of 850, 450 and 750 kg/ha, respectively was found to be feasible for Kodagu district in Karnataka.
- A closer spacing of 1m × 1 m for Malabar types gave higher yields, closely followed by 1.5m × 1.5m spacing under rainfall conditions of Shevroys hills in Tamil Nadu.
- Mudigere and Pampadumpara centres have come out with independent recommendation on the integrated nutrient management using organic and inorganic manures, viz. 75:75:150 kg NPK/ha and 100:100:175 kg NPK/ha, respectively. Increase in the dosage of fertilizer (150:150:225 kg NPK/ha) resulted in obtaining 644 kg/ha (against 547 kg/ha). However, the recommended dosage gave the highest benefit : cost ratio of 4:66 (Karnataka).
- Application of boron in the form of disodium tetraborate @ 20 kg/ha or molybdenum in the form of sodium molybdate @ 0.25 kg/ha mixed with appropriate quantity of FYM @ 2 kg/plant at the onset of monsoon (1 May or first week of June) increased the cardamom yield by 20%. The economic analysis revealed that highest marginal returns (Rs. 4,980) were realized by the application of borax and molybdenum followed by the application of molybdenum or boron alone.
- The combined effect of irrigation at 75% available soil moisture and higher light intensity (65–70%) resulted in substantial increase in yield.

- Application of ethrel at 250 ppm under higher light intensity (15 K lux) gave highest number of suckers (50) as against 25 suckers (without growth regulator) at normal light intensity (5 k.lux).
- Spraying 25 ppm NAA or 2.5 ppm 2, 4 –D twice at monthly interval soon after fruit setting (June-July) minimized fruit drop.

Ginger

- Cultivation practices like size of seed rhizome, planting season optimum sowing time, bed size, spacing, raised bed method of planting, advantages of mulching, type of mulch, optimum time of harvesting for quality produce, weed control, compatible crop combination, after care, cultural operations etc. were standardized for maximizing the yield.
- Seed rhizome size, rate as well as planting and spacing were standardized for different states.

Seed size/rate and spacing recommendations for ginger

State	*Recommendations*
Kerala	15g bit with atleast one viable bud, seed rate 1,500 kg/ha and 20cm x 20 cm to 25cm x 25cm spacing at a depth of 4-5cm.
Orissa	15-20 g piece having 2-3 healthy buds, seed rate 1800 kg/ha, 25cm x 20cm spacing at 5cm depth.
Himachal Pradesh	25-30g having 2-3 buds, seed rate 1,800–2,250 kg/ha at 30cm x 20cm spacing.
Bihar	Seed rhizomes 18-20 g size with atleast 2 buds at 30 cm x 20 cm spacing.
Andhra Pradesh	30 x 20 cm spacing and seed rate of 1,700 kg/ha.

- Depending upon the method of cultivation, seed rate standardized as 1,500 kg/ha for Kerala and Karnataka, 1,800 kg/ha for Orissa, 1,800-2,250 kg/ha for Himachal Pradesh and 1700 kg/ha for Andhra Pradesh. Seed rhizome should be 2.5–5.0cm in length, weighing 20-25 g with two viable healthy buds. Size of seed rhizome recommended as 15 g piece for Karnataka, 20g for Orissa and Bihar and 25–30g for Himachal Pradesh.
- Seed treatment with biocontrol agents is recommended for control of seed-borne disease.
- Seed treatment of rhizomes with insecticides, fungicides and bactericides for 30 minutes before planting to prevent rhizome rot.
- Burning soil in ginger beds prior to planting combined with optimum use of pesticide ensures over 95% germination and low incidence of rhizome rot disease has been recorded.
- Optimum sowing time has been fixed to be the first fortnight of April (Orissa).
- Soybean and red gram are the best comparison crop with ginger as the main crop in Orissa.
- Ginger is a high nutrient requiring crop and application of both organic manure and fertilizers are essential. Soil organic matter has positive correlation with the yield.

Manures and fertilizers recommended for ginger

Region	*Fertilizer/manure*
Kerala	FYM 30 tons/ha, NPK 70:50:50 kg/ha. Full dose of phosphorous and 50% K may be applied as basal dose. Half the quantity of N applied at 60 DAS. The remaining quantity of N and K applied at 60 DAS.
Karnataka	FYM / compost 25 tons/ha, NPK 100:50:50 kg/ha. Apply the entire dose of P and K at planting. Half quantity of N to be applied at 30-40 DAS and other half at 60–70 DAS.
Orissa	FYM 25 tons/ha, NPK 125:100:100 kg/ha. Full P and half K applied as basal in furrows before planting and N and K in 2 splits at 45th and 90th day.
Himachal Pradesh	FYM 20–30 tonnes/ha. CAN @ 400 kg/ha, NPK 100:50:60 kg/ha, superphosphate 15 kg/ha. Apply superphosphate and potash at the time of planting and N in 3 equal doses first at the time of planting and subsequent 2 doses at 1 month interval, K_2O also applied in splits half at sowing and other half at rhizome initiation.
Bihar	FYM 20–30 tonnes/ha, NPK @ 60:60:120 kg/ha.
Andhra Pradesh	FYM 20–30 tonnes/ha, 75:50:50 N, P_2O_5, K_2O kg/ha
Chattisgarh	FYM 20–30 tonnes/ha, NPK @ 150:120:120 kg/ha
Sikkim	manure 40–60 tonnes/ha
Meghalaya	FYM 10 tonnes/ha + 60 : 90:60 N, P_2O_5, K_2O

- Application of neem cake 2 tonnes/ha together with inorganic fertilizers increased the availability of nutrients in soil, increased yield and reduced incidence of rhizome rot.
- Application of Zn (20 kg/ha) Bo (10 kg/ha), Mg (10 kg/ha) and Jagromin (chelated form of micronutrients) along with recommended dose of NPK (100:50:50 kg/ha) resulted in increased yield.
- Soil application of borax @ 5 kg/ha recommended for increased yield in West Bengal.
- Technologies for seed rhizome storage were standardized and demonstrated.

Turmeric

- Package of practices for profitable turmeric cultivation has been evolved and popularised in turmeric-growing states.
- Mother rhizomes followed by bold primary rhizomes have been identified as the best planting material.
- Mulching with paddy straw or green leaves increased the rhizome yield.
- Mixed cropping with chilli, colocasia, brinjal, cereals, (maize and ragi) for Andhra Pradesh and K2 chilli as border crops in Tamil Nadu are recorded. Crop rotation: turmeric-maize (Tamil Nadu), turmeric rotated with maize or paddy (Andhra Pradesh and Orissa), turmeric and radish (Maharashtra). Sequential cropping – turmeric – banana – sugarcane – turmeric, turmeric – banana – paddy – turmeric (Tamil Nadu).
- Cultivation practices-size of seed rhizome (20–25g), planting season, bed size, raised bed method of planting, spacing, advantages of mulching, type of mulch, optimum time of harvest for quality produce, weed control, compatible crop combination, after care, cultural operations etc are standardized.
- Recommendation given for minimizing rhizome rot in field by pre-sowing seed treatment

with insecticides, fungicides and bactericides for 30 minutes before planting perfected.

- Processing and production of quality dry turmeric (8–10 % moisture) was standardized and popularised.
- Long-term storage effect on quality deterioration evaluated.
- Increasing levels of nitrogen increased the rhizome yield. Nitrogen @ 150 kg/ha provided highest rhizome yield (41,916 kg fresh rhizome/ha) at 30 × 20 cm spacing at Uttar Pradesh.
- Daughter rhizome of 25 g size sown at 30 × 20 cm spacing and sowing on third week of May and mulched with sesame leaves has been recommended.

Spacing recommended for turmeric

State	*Recommended spacing*
Karnataka, Maharashtra	30 x 30 cm or 45 cm x 30 cm
Orissa	25 x 25 cm or 20 x 25 cm
Tamil Nadu	25 x 25 cm in ridge, furrow and flat bed system
Uttar Pradesh, Bihar and Andhra Pradesh	30 x 20 cm 30 x 20 cm/45 x 15 cm

Fertilizer recommendations for turmeric in different states

States	*Recommendations*
Andhra Pradesh	200-300kg N, 125-150 kg P_2O_5, 100-150 kg K_2O in 3 splits *viz.* 60 N, 60 P and 60 K kg/ha as basal dose, 60 N, 65 K kg/ha at 60 DAS, 60 kg N /ha at 120 DAP.
Tamil Nadu	200:200:200 kg NPK/ha as basal and top dressing with 25 kg N /ha + 18 kg K /ha at 30, 60, 90th and 120th DAP.
Orissa	60:30:90 kg NPK/ha in 2 split dose as basal, whole FYM and half K on furrows before planting (7.2 kg/m2 and green mulch 2.5 kg/m2), half N at 45 DAS, half N and half K at 90 DAP.
Kerala	30:30:60 kg NPK/ha, full and half K applied as basal dose, 2/3rd after 30 days and rest N at 60 DAP
Madhya Pradesh and Bihar	175:150:150 kg NPK100:50:150 kg NPK/ha. Application of 15 kg/ha iron sulphate and 20 kg/ha of zinc sulphate is also recommended.
Chattisgarh and	150:125:125 kg NPK/ha, in 3 split doses 30, 60 and 90 days after planting.
West Bengal	60:60:20 NPK kg/ha and 20-25 q FYM/ha

Coriander

- Twenty varieties were released for cultivation. Recently LCC-128 and a leafy variety EC-232666 identified as promising ONES at Guntur.
- In North India, coriander is grown as rabi crop, sown in October-November. In Southern part, summer crop is planted in May where as rabi crops in October. A seed rate of 14–16 kg/ha for RCr-41, RCr-20 and 12 kg/ha for RCr-436 for Rajasthan and a seed rate of 12–15 kg/ha for Andhra Pradesh were recommended.
- The spacing, sowing time, seed rate recorded and conducted for different states.

Recommendations of spacing for coriander

State	Spacing
Gujarat	- 40cm x 15 cm or broadcasting
Tamil Nadu	- 15cm x 15 cm
Andhra Pradesh	- 25cm x 15 cm
Uttar Pradesh	- 30cm x 10 cm
Rajasthan	- 30cm x 10 cm

State-wise sowing time recommended for coriander

State		Time Recommended
Rajasthan	:	First fortnight of October or first week of November/mid of October to mid of November
Gujarat	:	Last week of October to first fortnight of November
Tamil Nadu	:	June–July (*kharif*) September – October (*rabi*)
Andhra Pradesh	:	First fortnight of October to first fortnight of November, Rainfed)
Bihar	:	First week of October
Uttar Pradesh,Haryana,Chattisgarh	:	Middle of October to first week of November, first week of November, middle of October

- For raising a successful seed crop of coriander, varieties Narnaul selection or Pant Haritma should be planted at 30 × 20 cm along with 60 kg N/ha 37.5 kg P_2O_5/ha.
- Seed treatment with Azotobacter + 50% of the recommended dose of nitrogen gives higher seed yield.
- For raising a successful seed crop of coriander in Haryana, three irrigations at branching, flowering and seed filling stage along with a fertilizer dose of 60 kg N/ha and 40 kg P_2O_5/ha is recommended.
- Application of CCC @ 100 ppm as foliar spray at 40 DAS significantly improved the yield and yield attributing characters in coriander (Hisar). First November sowing + 60 kg N/ha + one green leaf cutting recorded significantly higher seed yield in addition to green leaf yield in the variety Hisar Anand.
- Application of pendimethalin at 0.5–0.75 kg/ha as pre-emergence in combination with one bunch weeding at 45 DAS control the weed and increases the yield. Oxyfluorefen was found toxic to coriander crop. The highest seed yield of coriander was obtained with four irrigations at 30, 60, 90 and 120 DAS and pendimethalin @ 1.0 kg/ha as pre-emergence. Azotobacter strain Msx-9 was found most effective for improving growth and seed yield in coriander.
- The highest seed yield of coriander was obtained with four irrigations at 30, 60, 90, 120 DAS and pendimethalin @ 1.0 kg/ha as pre-emergence.
- Soil application of Mn SO_4 @ 25 kg/ha and $CuSO_4$ @ 0.50% and soil + foliar application of $FeSO_4$ @ 5 kg/ha + 0.125%, Mn SO_4 @ 12.5 kg/ha and $CuSO_4$ @ 12.5 kg/ha gave significantly higher yield.

State-wise recommendation of fertilizer & manure for coriander

State	Recommendations
Rajasthan	10-20 t of well decomposed FYM + 60: 30: 40 Kg NPK/ha + neem cake. The FYM is applied at the time of field preparation, out of which 20 kg N, 30 kg P_2O_5 and 20 kg K_2O/ha is applied at the time of sowing and the rest in two equal splits one at time of irrigation and second at flowering stage.
Gujarat	20 t FYM + 60: 60: 30 kg NPK/ha. N applied in two split doses as basal and at flower initiation stage and P & K as single basal dose in last ploughing.
Tamil Nadu	10-15 t FYM + 20: 40: 40 NPK/ha.
Andhra Pradesh	10-15 t FYM + 30 : 40 : 20 kg NPK/ha in single dose in last ploughing
Bihar	60: 40: 20 kg NPK/ha under irrigation and apply N in two equal splits at sowing and 60 DAS.
Uttar Pradesh	60: 30 : 30 NPK /ha.
Haryana	60 kg N & 37.5 kg P_2O_5/ha.

The highest seed yield of coriander was obtained with 4 irrigation at 30, 60, 90 & 120 DAS and pre emergence application of pendimethalin @ 1.0 kg/ha.

Cumin

Sowing time, seed rate & spacing recommendations for cumin

State	*Sowing Time**	*Seed Rate***	*Spacing Line Sowing - Row Spacing*
Gujarat	First week of November	16 – 20 kg/ha	22.5 to 30 cm or 15 cm x 30 cm
Rajasthan	15th – 30th November	12 – 15 kg/ha	30 x 30 cm or 20 x 30 cm
Tamil Nadu	First fortnight of October	8 – 10 kg/ha	30 x 15 cm/15 x 10 cm

* Sowing around 15th October is the most appropriate time fir reducing disease incidence.

** Before sowing soak the seeds in water for 24 hours and pretreat with fungicide *viz.* captaf/agrosan/difoltan/bavistin/ thiram @ 3 g/kg of seed to reduce the wilt.

- Production technology aspects such as seed rate, sowing time/season, planting, method of sowing, spray pre sowing seed treatment, irrigation regeneration, etc., standardised for major growing States.
- For weed control, pre-emergent application of herbicide terbutryn 0.5 kg a.i/ha or oxidiazone 0.5 kg/ha or pendimethalin 1.0 kg/ha has been recommended.
- Plants must be thinned out to obtain a spacing of 12–15cm in broadcast crop and 5cm in line sown crop during first weeding.
- Light irrigation to be given after sowing and second irrigation at 8–10 days after germination and thereafter 3–5 times at 15–25 days interval increase the yield.
- Three years crop rotation ie. cluster bean-cumin-cluster bean–wheat-cluster bean mustard is recommended for cumin cultivation to obtain higher yield and to reduce the wilt incidence.

Fertilizers and manures recommended for cumin

State	*General*	*Recommendations*
Gujarat	30:30:20 kg NPK/ha i.e. ½ N and 30 kg P_2O_5 and 20 kg K may be added at the time of land preparation and 50 kg N broadcasted with irrigation.	30kg N and 15kg P/ha. Half N and full P+10 tonnes com post/ FYM as basal dose and the balance N given month after sowing as top dressing. 25:20:20 kg/ha is also recommended.
Rajasthan		30 kg N and 20 kg P/ha. The N is given in two splits at 30 and 60 DAS and full P as basal +10 t compost/ FYM at sowing.
Tamil Nadu		70:60:40kg NPK/ha 30kg N and P and K are applied as basal and the remaining N in two equal splits at 30 and 60 DAS

Fennel

Nursery sowing during June-July for transplanting in August-September and September-October for transplanting in November are recommended.

State-wise recommendations of sowing time

State	*Recommendations*
Rajasthan	Depending upon the prevailing climatic conditions from first fortnight of October to mid-November for direct sowing. In transplanted crop nursery in June, transplanting in August/September
Gujarat	In transplanted crop first week of September in nursery and transplanting in second half of October and first fortnight of October for direct sowing.
Bihar	*Rabi*: First week of September in nursery and transplanting in first week of October.
Haryana	First fortnight of October

State-wise recommendations of spacing and seed rate for fennel

State	*Spacing*	*Seed Rate**	
		Direct Sowing (kg/ha)	*Transplanting (kg/ha)*
Rajasthan	60cm between the rows and 20cm within row	10-12	3-4
Gujarat	*Rabi* : 45cm x 10cm and *kharif* : 60cm x 40cm	9-12	3-5
Haryana	30cm x 20 cm/90cm x 60cm /100cm x 60cm is recommended for seedling transplanting	9-12	3-5

* For better germination, soak the seeds in water and treat with bavistin/ thiram/agrosan GN @ 2 g/kg of seeds.

State-wise fertilizer recommendations for fennel

State	*Recommendations*
Gujarat	N 45 kg/ha, P_2O_5 30 kg/ha. For *rabi* crop 90 kg N, 45 kg P_2O_5, 20 kg/ha K_2O and 25 t FYM (as basal dressing). N is given in 3 splits ie. 50% N & P as basal, 25% N at 30 DAS and balance 25% at 60 DAS. For *Rabi* crop, 90:30:00 NPK/ha. 45 kg N and 30 kg P/ha as basal dose and 22.5 kg N/ha as second split at 30 DAS. After completion of weeding, thinning and operation 22.5 kg N/ha as second split at 60 DAS.
Rajasthan	For soils of poor fertility status: 90 kg N, 40 kg P_2O_5/ha as basal and 30 kg N after 30 DAS and 30 kg N 60 DA transplanting in the irrigation.
Haryana	50 kg N and 30 kg/P_2O_5. N is applied in 3 splits, 50% as basal and at 30 and 60 DAS with irrigation.
Bihar	60 kg N, 40 kg P_2O_5 and 20 kg K_2O.
Uttar Pradesh	60 : 50 : 10 kg NPK/ha.
	15 kg N, 50 kg P and 10 kg K/ha as basal dose followed by 15 and 30 kg N/ha as top dressing in splt dosed.

- Integrated weeding can be done with pre-emergent weedicide pendimethalin @ 1.0 kg/ha supplemented with one hand weeding in 50 DAS (for Rajasthan and Haryana).
- For producing chewing type of fennel, umbels are harvested 30–40 days after pollination when seed size is just half of the fully developed seeds.
- Fennel variety, HF-33, should be sown/transplanted during first fortnight of October as delay reduce yield. In fennel, seed quality decreases with increasing umbel order. Seeds from main and primary umbels are vigorous and better in performance than secondary and tertiary order umbels (Hisar).
- Intercropping of fennel and coriander with potato, tomato and cauliflower has been recommended.
- Among biofertilizer Azotobacter (strain Ms x 9) was found most effective for improving growth and seed yield.
- The response of fennel to Zn, Fe, Mn and Cu revealed significantly higher seed yield over control. Application of Zn SO_4 (0.50%), Fe SO_4 (0.25%) + 0.125% and Mn SO_4 (12.5 kg/ha + 0.25%) registered higher seed yield.
- For successful seed crop of fennel variety PF-35, sowing should be done during third week of October with irrigation at crown stage and main umbel development stage at Hisar.
- For Bihar, fennel sown in nursery during first week of September and transplanted during first week of October is superior as compared to other dates of sowing.
- Intercropping trial of fennel in Bihar, fennel + raddish was found to be more remunerative per unit area.
- Sowing time/ sowing season: *rabi* May-June (sowing directly), *kharif* October-November (transplanting) rainfed (June-August-Nursery).

Fenugreek

- Soak the seeds in water up to 6–8 hours and treat with Bavistin or Agrosan GN or Thiram @ 3 g/kg of seed before sowing to prevent root rot. Seeds are drilled or dibbled in line at a depth less than 50 cm or broadcasted.

- Seed treatment with *Rhizoctonia meliloti* culture is recommended.
- Technology for increasing prc luction, *viz.* seed rate, sowing period, spacing and fertilizer package were recommended for major fenugreek growing state (see Box).

State-wise recommendations for fenugreek

State	Seed Rate(kg/ha)	Sowing Period	Spacing (cm x cm)	Fertilizer/ha
Rajasthan	20-25	First week of October to last week of November	25 x 10 row	10 FYM at the time of field prepara tion. 40 kg N, 40 kg P_2O_5 (20 kg N & 40 kg P_2O_5/ha at the time of sowing)
Gujarat	20-25	Last week of September to First week of October	15 x 10 and 30 x 10	40 kg P_2O_5/ha. 40 kg N, 25 kg S and in two equal splits as basal and top
Bihar	20-25	Middle of October	20 x 10	40 kg N, 60 kg P_2O, 20 kg K_2O
Uttar Pradesh	15-20	October – November	30 x 10	60 : 50 : 10 kg NPK/ha
Tamil Nadu	-	First week of October	15 x 10 or 22.5 x 10	N 50 kg/ha, P 25 kg/ha and 40 kg ha is recommended
Haryana	-	Middle of October up to November (First week of November is ideal)	30 x 10	40 kg/ha P
Andhra Pradesh	30	First week of October	—	10 t FYM/ha, 30 kg N, 40 kg P_2O_5 and 20 kg K_2O as basal dose

* 10-15 tonnes FYM is to be applied at the time of field preparation

- Irrigate the field immediately after sowing. The crop requires 6-7 irrigations in light soils and 4-5 irrigation in heavy soil. The crop should be irrigated at IW/CPE ratio of 1.0 and irrigation should be imposed on 30,70, 90 and 110 DAS.
- Two weedings and hoeings, first at 10-20 DAS, together with thinning, second weeding at 50 DAS. Integrated weed management using pre-sowing application of pendimethalin @ 1 kg/ha (Haryana), pre-plant application of Fluchlorin @ 0.75 kg/ha in 500-600 litres of water supplemented with hand weeding once in 50 DAS.
- Seeds are harvested in the 45 day after anthesis, when the seeds attain physiological maturity (Tamil Nadu) and the moisture level of storing seed is standardized at 7.73%.

Tree Spices (Nutmeg, Cinnamon and Clove)

- Grafting of scions is recommended on root stocks at leaf-stage. In nutmeg, application of 100 kg FYM, 400 g N, 300 P_2O_5 and 1200 g K_2O/tree/year and 50 g in each of *Azospirillum* and phosphobacteria recorded highest yield.
- In clove, application of 100 kg FYM, 400 g N, 350 g P_2O_5 and 1200 g K_2O/tree/year and 50 g each of *Azospirillum* and phosphobacteria resulted in the highest green bud yield/ buds per cluster. Dripping of 8 litres of water per day per plant during dry month is beneficial for clove.

Pests and Disease Management

Black Pepper

Nursery diseases/pest and management in black pepper

Disease/pest and causal organism	*Management/controlmeasures*
i) Leaf rot and blight (*Rhizoctonia solani)*	Removal of affected cuttings/ spraying and drenching the cuttings with 0.2% copper oxychloride or Bavistin (0.2%) or Bordeaux mixture (1%) at biomonthly intervals.
ii) Basal wilt (*Sclerotium rolfsii, Phythium* sp.)	Removal of the affected cuttings and spraying Bordeaux mixture (1%) and drenching the cuttings with 0.2% copper oxychloride at monthly intervals.
iii) Phytophthora foot rot (*Phytophthora capsici* F.*piperi)*	Dip the cuttings in culture suspension of *T.harzianum* and *T.viride.* Fortnightly application of 1.0% Bordeaux mixture, or Captafol (1.0 %) solarization of nursery soil mixture (3.3 k lux) fortified with VAM and *Trichoderma.*
iv) Root knot (M*eloidogyne incognita*) Root necrosis and rot (*Radopholus similis*)	Add Phorate @ 1.0 g per bag. VAM incorporation in nursery Root necrosis and rot (*Radopholus similis*) ture/fumigation with methyl bromide @ 500 g/tonne of nursery soil mixture under polythene cover for 24-48 hours –5-15 days before planting of the cuttings.
v) Scale insects *Lepidosaphes piperis*)	Spray Monocrotophos 0.05% or Dimethoate 0.1%
vi) Leaf gall thirps *(Liothrips karnyi)*	Apply Carbofuran 3 G or Phorate 10 G @ 3 g a.i/vine during June-July and September-October spraying 0.05% Monocrotophos or dimethoate on tender flushes.
vii) Top shoot borer (C*ydia hemidoxa*)	Spray Monocrotophos 0.05% on tender terminal shoots

- Package for management of *Phytophthora* foot rot and other diseases and pests were formulated and popularised (sea Box).

Recommendations for management of major diseases of black pepper

Disease	*Management schedules*
Phytophthora foot rot	Provide adequate drainage, grow grass cover, avoid trialing of runner shoots on the ground, careful farm operations without injuring the underground parts of the vine. Improve the microclimate by shade regulation, lopping off the branches of live standards to ensure better light penetration and to reduce humidity.Remove the affected vines along with the root system and burn to check the inoculum build up. Drench the spot with 0.2% copper oxychloride or 1% Bordeaux Mixture @ 5-10 litre/vine. Spraying Bordeaux mixture (1.0%) and drenching copper oxychloride (0.2%) twice during June and September. Apply *T. viride* (50 g/vine) mixed with 1kg neem cake followed by Akomin 0.2% (potassium phosphonate) spray in August or spray 0.3% potassium phosphonate @ 3 l/vine and drench @ 5litre/vine twice during the season (before onset of monsoon and during second week of August) in addition to cultural practices.Spraying Bordeaux mixture 1% at the time of flowering did not affect the fruit setting.

(Contd.)

Disease	*Management Schedules*
Slow decline (slow wilt)	Application of Carbofuran 3G or Phorate 10 G (30 g/vine) plus Bordeaux mixture spraying and copper oxychloride (0.2%) drenching plus soil application of neem cake @ 2 kg/vine. Drench the basin with 1% copper oxychloride @ 5-10 litres/vine 2-3 times during monsoon season.
Black berry (pollu) disease	Spray Bordeaux mixture (1.0%) or captofol (1%) once before flowering starts (July) and at berries formation stage (August). A combination spray of quinalphos 0.1% and zyneb 0.2% once in June and again in September to control pollu fungus and pollu beetl
Root knot and burrowing nematodes	Apply phorate 10G, Carbofuran 3G, Aldicarb @ 3 g a.i./ vine twice a year, May -June and September-October and cover with a thin layer of soil.
Anthracnose	Apply combination of Fungicides Carbendazium (0.1%) and Mancozeb 0.1%. Carbendazim 2 g/l followed by SAAF 2g/l (Carbendazim + Mancozeb)
Pests	Management schedules
Pollu beetle	Spray Endosulfan (0.05%) or Quinalphos (0.025%) or Dimethoate or Monocrotophos (0.05%) twice in June-July (21- 30 days after setting of berries) and September-October.
Top shoot borer	Spray endosulfan (0.05%) when new shoots emerge
Leaf gall thrips	Spray Monocrotophos or Endosulfan or Dimethoate (0.05%) or Phosphamidon (0.03%)
Scale insects	Two sprays of monocrotophos (0.05%) or dimethoate (0.05%) or quinalphos (0.05%) at 15 days interval after harvest of berries or fish oil and neem oil.
Mealy bug	Drench Monocrotophos (0.1%) or Dimethoate (0.05%) or Quinalphos (0.05%) at 15 days interval or fish oil and neem oil

- Novel method of raising nursery using polythene sheet covering instead of normal practice of using leaf mulch soon after planting for enhanced sprouting of the cutting in the nursery.
- Drying of pepper on clean cement floor or in HDPE film is recommended for obtaining clean and quality produce.

Cardamom

- Seed treatment with Captan (@ 4 g/kg of seed) was found to increase germination and protect seedlings from 'seed rot' and 'damping off' diseases.
- Thrashing the old leaves and suckers during May-June, spraying with Nuvarar (June-July) 35% EC, 1.25 ml /litre. up to 1.5 l plant followed by another spray with zolone 25% EC, 2 ml/litre, 30-40 days later and removal of borer affected suckers from the ground were effective in suppressing both thrips and borer in cardamom.
- In a year-six rounds of insecticides application are recommended for the management of cardamom thrips, shoot and capsule-borer. Bee safe, Endosulfan (0.05%) and quinalphos (0.05%) are advised during peak flowering period as well as skipping of in insecticidal application coinciding monsoon.

Recommendations for management of major diseases of cardamom

Disease	*Management Schedules*
Katte	Eradicate and rogue infected plants; adopt phytosanitation to control aphid vector.
Rhizome rot / Azhukal disease/capsule rot	Adopt phytosanitation, regulate excess shade and thrashing to remove senile parts. Spray and drench with Bordeaux mixture 1% or copper oxychloride 0.2% before onset of monsoon (May/June) and after the end of monsoon (August).
Seedling rot, damping off and leaf spot	Soil solarization; incorporate VAM in the nursery stage, treat seed with carbendazim + captan @ 2 g/kg of seed for 60 min. Spray mancozeb 0.05%. Provide good drainage, drench with Bordeaux mixture against damping off. Two sprays of captofol (0.3%) or copper oxychloride or carbendazim (0.1%) at 10 days interval against leaf spot.
Root-knot nematode	Apply phorate 10 G @ 2.5 g a.i. / clump during May / June and November / December or aldicarb, carbofuran @ 5-10 kg a.i/ha.

Recommendations for management of major insect pests of cardamom

Insect pest	*Management Schedules*
Thrips	Spray Monocrotophos 0.05% and Phosalone 0.05% alternatively at bimonthly intervals.
Shoot/panicle and capsule borer	Spray Monocrotophos 0.05% or Quinalphos/endosulfan/fenthion (0.075%) twice at January/ February and September/October.
Shoot fly	Spray Quinalphos 0.03% or Phosalone 0.05% or Monocrotophos 0.05% or ekalux (0.03%) or apply Carbofuran 0.3 kg a.i/ha.
Root grub	Apply Carbofuran 3 G @ 8-10 kg/ha during June/July and November/ December or phorate 2 to 4 g a.i/plant in soil or Chloropynfos @ 0.07%.

Ginger

Disease management in ginger

Disease	*Management Schedules*
Rhizome rot	Seed treatment with Mancozeb (indofil M-45 - 0.25%) + Carbendazim (0.1%) for 60 minutes; dry under shade for 24 hours before sowing/storage. Dip seed rhizome in copper oxychloride (Blitox-50 - 0.3%) before storage.
Leaf spot	Spraying Mancozeb 0.2% or Bordeaux mixture 1.0%.
Bacterial wilt	Treatment of seed rhizomes with streptocyclin 200 ppm for 30 minutes, and shade dried before planting. Beds should be drenched with Bordeaux mixture 1% or copper oxychloride 0.2%.

Pest management measures in ginger

Pest	*Management Schedules*
Shoot borer	Spray dimethoate (0.1%) or quinalphos (0.1%) at bimonthly intervals or malathion (0.1%) or monocrotophos (0.075%) during July — October at monthly intervals.
Rhizome scalein storage/field	Seed dipping in Quinalphos (0.025%) or Fenthion (0.05%) before storage + foliar spraying of Phosphomidon (0.05%) in the field at 6th and 7th month after planting.
Rhizome fly	Spray Quinalphos (0.02%) at 60 DAS followed by soil application of furadan 10kg/ha.
Nematode	pests. Use Thimet @ 10-15 kg/ha or phorate 10 kg/ha at the time of field preparation, treat infected rhizomes with hot water (50°C) for 10 minutes, use nematode free seed rhizome and solarizing ginger beds for 40 days.

Turmeric

Recommendations for management of major diseases and pests of turmeric

Disease/Pest	Management Schedules
Leaf blotch	Spray 1% Bordeaux mixture or copper oxychloride or Carbendazim (0.1%) or Mancozeb (2.5 g/l) at 500 litre/ha or Captan (0.2 ml) at monthly interval. Treat seed rhizome with Bavistin, Mancozeb and carbendazim (0.2%) for 60 minutes and shade dry before planting.
Leaf spot	Spray 1% Dithane M-45 (2.5 g/litre of water) at 15-20 days interval or copper oxychloride (0.25%) or Bavistin (0.1%) or Chlorothalonil (0.2%) or Zineb 0.3% or Bordeaux mixture 1%.
Rhizome rot and wilt	Treat seed rhizome with Dithane M-45 (0.25%) or bavistin (1.0%) for 30 minutes prior to storage and at the time planting. Drenching of infected plants/ beds with blitox-50 (0.3%) or Mancozeb (0.3%) or 1% Bordeaux mixture or copper oxychloride (0.25%).
Nematodes	Use healthy nematode free planting material, soil application of aldicarb at one kg/ha twice (3rd and 5th month after planting) followed by irrigation. Application @ 10-15 kg/ha at the time of field preparation is also advocated.
Storage rot	Seed treatment with dithane M–45 (0.25%) or bavistin (1.0%)
Rhizome scale	Seed dressing with quinalphos (0.075%) for 30 minutes before storage and also before sowing + spraying of phosphomidon (0.05%) in field.
Shoot borer	Spray Malathion (0.1%) or Monocrotophos (0.075%) or Rogor or Ekalux (0.05%) or dimethoate or Quinalphos during July to October at monthly interval.
Thrips	Spray dimethoate (0.05%) or methyl dimethion (750 ml/ha).
Caterpillar	Spray malathion (20 ml/litre water).
Rhizome fly and maggot	Spray Quinalphos (0.02%) at 60 DAP.

Coriander

Management of major diseases and insect pests of coriander

Disease/Pest	*Management Schedules*
Powdery mildew	(1) Two sprays of 250 litres with Karathane 48EC/Dinocap/ Bavistin (0.01%) at disease initiation and 10 days after with Carbendazim (0.1%) or (2) Two spray of wettable sulphur (@ 2.5 g/litre), one at the time of flower initiation and seed 15 days later or (3) dusting of sulphur @ 25 kg/ha thrice. (4) Ecofriendly techniques of using 5% onion leaf extract as foliar spray (Coimbatore) three times can also protect the plants. (5) A cost effective ecofriendly biocontrol method by spraying of neem seed kernel extract (NSKE) @ 5% thrice, first spray immediately after the appearance of disease and the subsequent two sprays at 15 days interval.
Stem gall	Selection of disease-free seeds, collection and destruction of affected plants, growing coriander variety RCr-41, seed treatment with Agrosan GN/Thiram (2.5 g/kg) or Bavistin (1 g/kg) of seed. Spraying of Bayletenten or Systhane (0.4 ml/litre of water) repeat after 20 days.
Grain mould	Spray carbendazim (0.2%) 20 days after flowering (grain formation stage and again at grain hardening stage or captan (0.2%) twice at grain set and at 15 days interval.
Wilt complex	Sow in the first and second week of November for reducing disease incidence. Seed treatment with *Trichoderma viride* formulation with CFU 10^6 @ 4 g/kg seed and neem cake application @ 150 kg/ha following foliar spray of hexaconazone (0.05%). Biocontrol agent *T. viride* in combination with foliar spray of either Hexacomazole or Thiophenate methyl on 25th, 40th and 55th DAS and soil application of neem cake @ 150 kg/ha. Use resistant varieties - Co-3, Sadhana,

(Contd.)

Disease/Pest	*Management Schedules*
	Surabhi, Guj. Cumin 1 and 2. Adopt crop rotation, give deep summer ploughing, use disease-free seeds. Seed dressing as well as soil drenching with Carbendazim 0.2%. Seed dressing with 1.5 g Bavistin + 1.5 g Thiram / kg seed and spraying twice with carbendazim 0.1%. Another eco-friendly biocontrol method by seed treatment with *Pseudomonas fluorescens* @ 10 kg/ha plus soil application of neem cake @ 150 kg/ha.
Root rot	Drenching Brassicol or Bavistin 0.1% twice at 20 days intervals.
Aphids and white fly Heliothis caterpillar	Endosulphan (0.05%) twice just at the time of floral initiation and 15 days there after with monocrotophos or Quinalphos (0.1%) or Dimethoate (1.5%) or methyl demethon (2.01%) or phosphomidon (0.5%) thrice @ 400-500 litres/ha at 10 days interval.Foliar application of 0.03% phosphomidon or Dimethoate or methyldematon (0.05%)
Red mites	Foliar application of sulfex 3 gm or dicofol 3 ml/litre of water, 2-3 times at 10 days interval or spraying endosulfan 0.07% twice just at the time of floral initiation and 15 days after or two sprays with monocrotophos/quinalphos (0.1%) at 15 days interval.
Frost	One light irrigation can save the crop in irrigated areas. In rainfed areas, spray 0.1% H_2SO_4 can save the crop and smoking around the border of the coriander field.

Cumin

Recommendations for management of major diseases and pest of cumin

Disease/Pest	*Management Schedules*
Wilt	Use healthy and disease free seeds. Seed treatment with a mixture of Carbendazim + Captan (1:1) @ 4g/kg of seed, soil solarisation during summer months and dry summer ploughing. Organic amendments with neem cake @ 1 tonne/ha. Integrated management by *T. harzianum* application as seed treatment as well as soil application along with spraying of topsin (0.07%) or indofil (0.3%) and neem oil. Three year crop rotation sequence to accommodate cumin once in three years such as cumin-clusterbean-wheat - clusterbean-mustard is recommended to reduce wilt menace of cumin in Rajasthan.
Blight	The disease appears at flowering and spreads rapidly in cloudy ,cool weather. As a prophylatic measure seed dressing with carbendazim or captan (0.1%) as well as foliar application of any fungicides such as Indofil M-45 (0.2%)/ blitox-50, cupramar or fytolan or blue copper or bavistin (0.1% each) or copper oxychloride (0.1%) at the time of flowering followed by fortnightly intervals if there is rain or heavy dew. Seed dressing with Captan or Thiram @ 2.5 g /kg seed accompanied by spraying the crop with Mancozeb @ 0.2% thrice at an interval of 15 days from 45 DAS can also be adopted.
Powdery mildew	Adopt control measures at the initial stage of appearance of disease symptoms. Dusting of sulphur @ 25 kg/ha or spray 400-500 Lits/ha of wettable sulphur or dinocap (0.25%) or karathane LC or calaxin (0.05 to 0.1%) and repeat after 15-20 days commencing from 45 days after germination.
Aphids	Endosulfan has been considered relatively safe which can be sprayed @ 0.07 % concentration. However careful spray of 0.03% Phosphamidon (100 EC) or dimethoate (0.03%) or monocrotophos (0.036%) @ 400-500 litres/ha also is recommended. Repeat the spray after 10 days if required.

Fennel

Recommendations for management of major diseases and pests of fennel

Diseases/Pests	Management Schedule
Powdery mildew	Spray Karathane (0.1%) or wettable sulphur (0.25%) once at flower initiation stage and another after 15 days (500-800 Litres). Dusting with sulphur 300 mesh (20-25 kg/ha) also controls the disease.
Sugary disease	Vector control by spraying with Phosphomidon (0.2%) or Dimethoate (Rogar) (0.33%) or Quinalphos/ DDVP @ 0.05% at 15 days interval (3 sprays) commencing from 50% emergence of flowering to check aphids that spread the disease. Stop application of fertilizer as well as irrigation.
Ramularia & Alternaria	Spray Bavistin (0.1%) or Indofil M-45 (0.2%) or any copper fungicides (500-800 litres./ha) twice at 60 & 90 DAS (500-800 litres./ha.)
Stem & root rot	Treat the seeds with bavistin @ 2g/ kg of seeds.seed.
Gall midge	Five sprays of the insecticide (Methylodemeton 0.025%, Monocrotophos –0.036%, Thiometon 0.03% and DDVP 0.05%) at 15 days interval commencing from emergence of flowers.
Weevil andCaterpillar	Spray 500-700 litres of Dimethoate 30EC (0.03%) or 0.02% Phosphamidon 85 WSC.
Aphids	Spray endosulfan (0.07 %) in the evening hours or dimethoate (0.03%) or Phosphamidon (0.1%) malathion (0.1%) or 0.03% methyl demeton @ 500-700 litres/ha.
Frost damage	In frost prone areas spray with 0.1% solution of sulphuric acid .Irrigation of the crop prior to incidence of frost, creating smoke cover in early morning before sunshine.

Fenugreek

The plant-protection strategies in fenugreek against major diseases and pests were standardized.

Recommendations for management of major diseases and pests of fenugreek

Disease/Pest	*Management Schedules*
Root rot	Seed treatment with thiram or captan @ 2 to 3 g/kg. Application of neem cake @ 150 kg/ha combined with seed pelleting with antagonistics like *T. viride*, *T. harzianum* (take formulation 4 g/kg of seed) followed by soil application of neem cake 150 kg/ha. Drenching carbendazim or brassicol (0.1%) once at initial appearance and after one month. Deep summer ploughing of field and adoption of crop notation.
Powdery mildew	Dusting with 300 mesh sulphur (25 kg/ha) or by two spraying of wettable sulphur @ 3 g/lit. or Dinocap (0.25%) or Karathane L.C. (0.1%) twice at flowering stage and 15 days interval commencing from 40 DAS or spraying dinocap or Karathane (0.1%) @ 400-500 Lits. /ha.
Cercospora leaf spot	Two sprays of carbendazim (0.05%)/ mancozeb (0.02%)/Zineb (0.3%)
Downey mildew	Spray 400-500 Lits. solution of 0.2%, Blitox/Difolatan/Phytolan. Repeat the spray after 10 days if needed. Spray of zineb (3g/litre) at 15 days interval is also recommended.
Damping off	Spray carbendazim (1g/Lit.) Cultural operation like crop rotation and removal of diseased plants. To minimise the source of infection seed treatment with benomyl (3g/kg seed) or carbendazim (3 g/kg seed) is recommended for reducing the seedling mortality.
Rust	Adopt good cultural operation and field sanitation. Spray of fungicide Mancozeb (0.2%) and bayleton (0.05%).
Aphids	Spray Endosulfan (0.07%) or Dimethoate (0.3%) or 0.33% Phosphomidon C or 1.0% Malathion or monocrotophos 400-500 litres/ha.
Leaf eating caterpillar & pod borers	Spray 0.05% Quinalphos or Endosulfan 0.075% or Monocrotophos (0.04%)

TECHNOLOGIES TRANSFERRED

Black Pepper

Pepper Research Station, (Kerala Agricultural University) Panniyur, Kerala

Released varieties

Seven high yielding varieties of black pepper (Panniyur-1, Panniyur-2, Panniyur-3, Panniyur-4, Panniyur-5, Panniyur-6, Panniyur-7) were released for cultivation.

Crop production and management

- Rate of growth and production of spikes were maximum in plants raised from cuttings of the middle one-third portion of runner shoots.
- Dipping the cuttings in 1000 ppm IBA before planting in polybags produced good rooting resulting in better shoot growth.
- A technique for rapid multiplication of pepper was developed.
- Application of NPK @ 50:50:150 g vine^{-1} is recommended. Application of lime (1 kg/vine) increases the yield by 40%.
- Pongalyam (*Ailanthus malabarica;* now A. *triphysa*) a suitable standard for pepper in drought-prone areas.
- Mulching the basins of vines with saw dust, coconut husk, dry leaves or polythene sheets reduces the casualty of vines and spike shedding, and increases yield by 30–70%.
- Irrigating the vines at IW/CP ratio of 0.25 (100 litres/vine) from December to March at an interval of 8–10 days increases berry yield substantially. Irrigating the vines using drip irrigation with 2 l of water/vine during summer results in better yield.

Disease management

- Nursery diseases can be managed by spraying and drenching with Potassium phosphonate 0.2% (Akomin) followed by drenching with Bordeaux (1%) mixture. Before planting, the cuttings could be dipped in a spore suspension of *Trichoderma harzianum,* and Bordeaux mixture (1%) could be sprayed after the emergence of leaves.
- *Phytophthora* foot rot disease can be managed by following phytosanitation, drenching with copper oxychloride (0.2%) followed by two sprayings of Bordeaux mixture (1%) during second week of July and September, or potassium phosphonate (0.3%). Potassium phosphonate (0.3%) may be applied as foliar spray and drenching in second round. Application of neem cake 1kg vine^{-1}during June and Phorate 3G @ 3 g a.i. vine/may also be done. The disease could also be controlled by spraying Bordeaux mixture (1%) as first round in June and Potassium phosphonate (0.2%) spray as second round in August or Potassium phosphonate (3 ml/litre) spraying and drenching, followed by soil application of neem based *Trichoderma*.

Pest management

- Pollu (*Longitarsus nigripennis*) infection can be effectively managed by two spraying of Quinalphos (0.05 %), during July-August and September-October.

Agricultural Research Station (Pepper), (Univ. of Agril. Sciences, Dharwad), SIRSI Karnataka

Crop production and management

- A simple technique for large scale propagation of black pepper cuttings and a low cost propagating structure using polyethylene covers were developed.

Disease management

- *Phytophthora* foot rot disease of black pepper grown as mixed crop in arecanut plantation can be managed by spraying the vines with Potassium Phosphonate (0.3%) @ 3 litres/ vine and drenching the soil @ 5 llitres/vine twice during the season (i.e., before onset of monsoon and again during second week of August) in addition to following all the cultural practices or spraying potassium phosphonate (0.03%) @ 3 litres/vine twice in a season and application of *Trichoderma viride* to the basin of the vine @ 50g/vine along with neem cake 1 kg and FYM 5 kg during the first week of June. First round of treatment is to be given just before the onset of monsoon, i.e., June first week and second round during the month of August. Wherever nematode incidence is high, Phorate 3 G @ 3 g/ vine can be given during September-October or following all cultural practices, application of neem cake 1kg/vine, Phorate 3 G @ 30g/vine, spraying with Bordeaux mixture (1%) @ 3 litre/vine and drenching the soil @ 5 litre/vine before onset of monsoon as first round (June), or spraying Akomin (0.4%) or Ridomil MZ-72 WP (100 ppm) @ 3 l $vine^{-1}$ and drenching the soil @ 5 l $vine^{-1}$ as second (August) and third (September) rounds respectively.
- Nursery diseases in black pepper can be controlled by spraying and drenching with Bordeaux mixture (1%) at 15 days interval.

Cardamom Research Station, (Kerala Agricultural University), Pampadumpara, Kerala

Disease management

- Carbendazim 2g/litre followed by SAAF 2 g^{-1} (Carbendazim + Mancozeb) are effective for managing anthracnose disease.
- Quick wilt disease of black pepper can be managed by adopting phytosanitary measures and applying of copper oxychloride (0.2%) as soil drench and spraying Bordeaux mixture (1%). Spraying Bordeaux mixture (1%) at the time of flowering did not affect the fruit setting.
- Solarization of potting mixture and fortifying with *Trichoderma* and VAM are highly effective in reducing foot rot incidence in the black pepper nursery. It also helped in early and better sprouting of the cuttings. There is a marked increase in the number of roots and length of roots of the cuttings planted in solarized soil compared to non-solarized soil.

Management of insect pests

- Two sprayings with either Dimethoate (0.05%) or Monocrotophos (0.05%) at fortnightly interval after the harvest of berries were effective for the management of black pepper mussel scale, *Lepidosaphes piperis* at high ranges of Idukki district.

Zonal Agricultural Research Station (University of Agricultural Sciences, Bangalore) Musifwew, Karnataka

- Spraying and drenching with Ridomil MZ 72 WP (1.25g/litre) and application of biocontrol agent, *Trichoderma harzianum* 50 g with 1 kg of neem oil cake during the pre and post monsoon effectively controlled *Phytophthora* foot rot disease.

Department of Horticulture, Dr Balasaheb Sawant Konkan Krishi Vidyapeeth, Dapoli, Maharashtra.

A rapid multiplication technique for black pepper on soil mound was developed and standardized.

Cardamom

Zonal Agricultural Research Station, (University of Agricultural Sciences, Bangalore) Mudigere, Karnataka

High yielding varieties

Two high-yielding variety of cardamom (Mudigere-1, Mudigere-2) were released for cultivation.

Crop production and management

- A spacing of 1.8m × 1.8 m (3020 plants/ha) was found optimum under Mudigere condition.
- A quick method of clonal multiplication through proliferation of suckers was standardized. This involves planting suckers in trenches at a closer spacing of 0.9m × 0.9 m, fertilizer application of NPK @ 75:75:150 kg/3,000 plants in two equal splits, following regular plant protection measures, weeding, irrigation and mulching. Multiplication ratio of 4-5 planting clonal units per tiller could be obtained per year, which will work out to be about 40,000 planting units/ha.
- A fertilizer dose of NPK 37.5:37.5:50.75 kg/ha for one year old crop during September and NPK 75:75:150 kg/ha from second year onwards in two split doses (during May and September) is optimum.
- Application of boron in the form of disodium tetraborate @ 20 kg/ha or molybdenum in the form of sodium molybdate @ 0.25 kg/ha mixed with FYM @ 2 kg/plant at the onset of monsoon increased cardamom yield by 20%.

Pest management

- Spraying monocrotophos (0.05%) in March, followed by two sprays of phosalone (0.05%) during May and August is recommended for effective control of cardamom thrips and borer.
- Root grub (*Basilepta* sp.) infestation in the cardamom nursery could be checked by incorporating Furadan 3G @ 6 to 8 kg/acre.
- Dithane M-45 (0.25%) + Ridomil (0.1%) are most effective fungicides for controlling damping off and leaf spot disease in cardamom nursery. Three sprays of carbendazim (0.1%) at an interval of 20 to 25 days from June onwards effectively checks the leaf spot in cardamom nursery.

Cardamom Research Station (Kerala Agricultural University), Pampadumpara – 685 553, District Idukki, Kerala

High yielding varieties

- Two high-yielding varieties, PV-1, PV-2 suitable for high ranges of Idukki was developed and released.

Crop production and management

- Application of inorganic fertilizer NPK @ 100 : 100 : 175 kg/ha in two splits during two monsoon periods recorded maximum cardamom yield under rain fed conditions.

Pest management

- Six rounds of insecticidal application (February, March, April, May, August, October) is recommended for the management of cardamom thrips (*Sciothrips cardamomi*) and shoot and capsule-borer (*conogethes punctiferalis*). Insecticides safe for bees such as endosulfan (0.05%) and Quinalphos (0.05%) are to be used during peak flowering period (June–September) or avoid insecticidal application during monsoon.

Ginger

Department of Vegetable Crops (Dr YS Parmar University of Horticulture and Forestry), Solan, Himachal Pradesh

High yielding varieties

A high-yielding variety Himgiri, highly suitable for vegetable purpose was released for cultivation.

Crop production and management

The rhizomes of ginger can be stored in pits (1m × 1m × 1m) constructed underground with bricks in a spacious room. A thick layer of dry sand (8 cm) is put at the bottom of the pit on which the rhizomes of ginger are loosely packed. A thick layer of dry grass (2 cm) is placed on the rhizomes. A perforated plastic pipe (3.5 cm diameter) is kept in the centre of the pit to provide free aeration during storage. The pits are covered on the top with wooden planks plastered with a mixture of mud and cowdung.

Disease management

The following recommendations were made for the integrated management of rhizome rot of ginger:

- Rhizome selection
- Rhizome treatment with Dithane M-45 (0.25%) + Bavistin (0.1%) for 60 minutes and soil application of Thimet 10 G (12 kg/ha) or rhizome treatment with dithane M-45 (0.25%) + Bavistin (0.1%) + Durmet (0.2%) followed by durmet spray (0.1%) after 2 months of planting
- Rhizome treatment with *Trichoderma harzianum* in cowdung slurry (0.5%) or soil application of *Glomus mosseae*
- Storage of seed rhizomes in pits with sand layer

- Rhizome treatment with Unilax 72 WP (0.25%) at planting time
- Rhizome treatment with dithane M-45 (0.25%) + bavistin (0.1%) + durmet (0.2%) is recommended for the control of storage rot.

Tirhut College of Agriculture, (Rajendra Agricultural University), Dholi, Bihar

- Rhizome rot disease can be reduced by treating the seed rhizomes with Ridomil MZ @ 3.5g/litre for one hour.

Department of Vegetable Science, Narendra Dev University of Agricultural and Technology, Kumarganj, Faizabad, Uttar Pradesh

- Hot water treatment of seed rhizomes at 51 °C for 10 minutes and application of *T. harzianum* mixed with neem cake checks rhizome rot of ginger.

High Altitude Research Station, (Orissa University of Agriculture and Technology), Pottangi, Orissa

High yielding varieties

Ginger varieties 'Suprabha' and 'Suruchi' were released by Pottangi centre. Another variety Suravi is also recommended. Suprabha and Suravi are released at national level.

Crop production and management

- Optimum time for planting ginger is first fortnight of April, with a fertilizer dose of NPK 125:100:100 kg/ha. and 25t FYM/ha.
- Technology for storage of seed rhizomes was developed. Ginger can be stored in pits (1×1×1) lined with stones and inner walls and bottom filled with dry sand (10 cm thick). Disease free rhizomes are selected immediately after harvest and soaked in mancozeb or bavistin (0.3%) for 30 minutes and also treated with malathion (0.05%) and then dried for 48 hours under shade. These treated rhizomes are placed in pits leaving 10–15 cm space on the top, covered with wooden planks and plastered with cowdung, leaving a small space for aeration.
- Zero energy cool chamber for storing fresh ginger was recommended. It is a double walled brick structure, filled with sand between the walls. The sand is kept moist by frequent watering. This structure is found ideal for storing fresh ginger. The loss in weight of rhizomes is only 23% after storing for 4 months while ginger stored in open condition is shrunken in 4 months.

Turmeric

High Altitude Research Station (Orissa University of Agriculture and Technology), Pottangi, Orissa

High yielding varieties

Four high yielding varieties of turmeric, Roma, Suroma, Ranga and Rasmi, have been released, out of which Roma is a national release.

Crop production and management

Manure and fertilizer requirement of turmeric was worked out to be FYM 7.5 kg/m^2 to supplement green mulch and NPK @ 60 : 30: 90 kg/ha.

Regional Agricultural Research Station, (Acharya N.G. Ranga Agricultural University), Jagtial, Andhra Pradesh

The biocontrol agents, *Trichoderma viride* and *Pseudomonas fluroscence* applied to seed and soil along with recommended NPK and FYM was found effective in the management of rhizome rot of turmeric.

Department of Horticulture, Uttara Banga Krishi Vishwavidyalaya, Pundibari, West Bengal

A turmeric variety, TCP - 2 was released, which yields 29.2 tonnes/ha. The variety is tolerant to leaf blotch, moderately resistant to rhizome rot, resistant to rhizome scale and moderately resistant to shoot borer.

Regional Agricultural Research Station (Indira Gandhi Agricultural University), Raigarh Chattisgarh

The optimum fertilizer dose recommended is 150:125:125 kg NPK/ha.

Nutmeg

Department of Horticulture, Dr Balasaheb Sawant Konkan Krishi Vidyapeeth, Dapoli, Maharashtra

Nutmeg variety "Konkan Sugandha" was released.

Horticultural Research Station (Tamil Nadu Agricultural University), Yercaud, District Salem, Tamil Nadu

Application of FYM 100 kg, N 400 g, P_2O_5 300 g and K 1200 g /tree/year and 50 g each of *Azospirillum* and phosphobacteria 15 days before the application of inorganic fertilizer increased fruit yield, fruit weight and tree girth.

Clove

Horticultural Research Station (Tamil Nadu Agricultural University), Yercaud, District Salem, Tamil Nadu

Application of 50 g each of *Azospirillum* (10^{-6} cfu) and phosphobacteria (10^{-5} cfu) with FYM 100 kg, N 400 g, P_2O_5 350g, and K_2O 1200 g/tree/year in two equal splits in May-June and October-November increased the yield.

Cinnamon

Department of Horticulture (Dr Balasaheb Sawant Konkan Krishi Vidyapeeth), Dapoli, District Ratnagiri, Maharashtra

Cinnamon variety "Konkan Tej" was released. This variety yields up to 334 kg bark, superior in quality with high oil (3.2%), cinnamaldehyde (70.23%) and eugenol (6.93%) in leaf oil. Bark yield is 300 g/plant/year.

Horticultural Research Station (Tamil Nadu Agricultural University), Yercaud, District Salem, Tamil Nadu

Cinnamon variety YCD-1 and PP1 (C)-1 were released for cultivation from Yercaud and Pachiparai centers respectively.

Coriander

SKN College of Agriculture, (Rajasthan Agricultural University), JOBNER, District Jaipur, Rajasthan

High yielding varieties

Six high-yielding varieties of coriander, RCr-41, RCr-20, RCr-435, RCr-436, RCr- 684 and UD-4 46, were released.

Crop production and management

- Sowing during 25 October to 4 November at 30 cm row spacing is recommended. The optimum seed rate is 15-20 kg, but a seed rate of 14-16 kg/ha for coriander varieties RCr - 41 and RCr-20 and 12 kg/ha for RCr - 436 is recommended.
- Foliar application of MnSO4, ZnSO4 and CuSO4 each @ 0.50% and soil application of FeSO4 @ 5 kg/ha or foliar application @ 0.125% was recommended to get higher seed yield under micronutrient deficient sandy loam soils. Foliar application should be done at pre-flowering stage.
- The quality of coriander depends on the stage of harvesting. The proper stage of harvesting was found to be when all the seeds are fully grown, green coloured and when 50% seeds turn yellow.

Disease management

- Minimum wilt incidence with the maximum seed yield was observed with *Trichoderma harzianum*_application through seed treatment as well as soil application.
- Coriander varieties RCr - 441, RCr - 435, RCr - 436, UD - 446 and UD - 684 were found resistant to root knot nematodes.

Regional Agricultural Research Station (Acharya N.G. Ranga Agriricultural University), Guntur, Andhra Pradesh

High yielding varieties

Two medium late-maturing high-yielding coriander varieties, Sindhu and Sadhana and one early-maturing, high-yielding variety, Swati, have been released.

Disease management

Technologies for management of powdery mildew, grain moulds in coriander were developed and recommended.

Department of Vegetable Crops, (Chaudharay Charan Singh Haryana Agril. University), Hisar Haryana

High yielding varieties

Three high-yielding varieties of coriander Hisar Anand, Hisar Sugand and Hisar Surabhi were released.

Crop production and management

- The highest seed yield of coriander was obtained with four irrigations at 30, 60, 90 and 120 days after sowing and application of pendimethalin 1.0 kg/ha at pre-emergence.
- The optimum fertilizer dose worked out for seed yield of coriander was 60 kg N/ha in two equal doses, i.e., half at sowing and half at 60 DAS or in three equal splits, one-third at sowing time, one-third at 30 DAS and remaining one-third at 75 DAS.

- For raising a successful seed crop of coriander varieties Narnaul selection or plant Harithma should be planted at 30 × 20 cm along with 60 kg N/ha and 37.5 kg P_2O_5/ha.
- Application of CCC @ 100 ppm as foliar spray 40 DAS significantly improved the yield and yield attributary characters.
- The treatment combination of 1 November date of sowing + 60 kg N/ha + one green leaf cutting recorded significantly higher seed yield in addition to green leaf yield in coriander variety Hisar Anand.
- Inter cropping of coriander with potato, tomato and cauliflower has been adapted.
- For weed control in coriander, pendimethalin @ 0.5-0.75 kg/ha as pre Emergence in combination with one hand weeding at 45 days after sowing was found most effective for controlling weed and increasing seed yield. Oxyfluorfen was found toxic to the crop.
- Azotobacter strain MSX-9 was found most effective for improving growth and seed yield.

Horticulture College and Research Institute, (Tamil Nadu Agricultural University), Coimbatore, Tamil Nadu

Crop production and management

Planting in the middle of October is recommended.

Disease management

Seed treatment with *T. viride* 4 g/kg of seed and foliar spray of Hexaconazole (0.05%) at 25, 40 and 55 days after sowing effectively controlled the wilt disease.

Tirhut College of Agriculture (Rajendra Agricultural University), Dholi, Bihar

Seed treatment with bavistin @ 1 g/kg seed effectively checked the stemgall disease.

Department of Vegetable Science , (Narendra Dev University of Agricultural and Technology), Kumarganj, Uttar Pradesh

Crop production and management

- Sowing coriander during the middle of October and a spacing of 30cm × 10cm is recommended.
- Application of NPK 60 : 30 : 30 kg/ha gave maximum seed yield of 18.20 q/ha with medium spacing of 30cm × 10 cm.

Disease management

Cor - 2, Cor - 6, Cor - 19, Cor - 20, and Cor - 21 were found moderately tolerant to stem gall disease in coriander.

Cumin

SKN College of Agriculture, (Rajasthan Agricultural University), Jobner, District Jaipur, Rajasthan

High yielding varieties

Cumin variety RZ-19, an erect plant with bold and lustrous grain with an yield of 500 kg/

ha was released for cultivation. The variety is tolerant to wilt and blight diseases and is suitable for late sowing.

Disease management

- Integrated management schedule for the wilt management recommended basal application of *T. harzianum* with and without carbendazim and neem cake as soil application reduced the incidence of wilt. Minimum incidence of wilt (23.80%) with seed yield of 131.8 g/plot was recorded in plots where *T. harzianum* was applied through seed treatment and as soil application spraying with Mancozeb 0.3%.
- Epidemiological study of alternaria blight in cumin revealed that minimum incidence of wilt was observed when sowing is done on 10 November followed by sowing on 20 November, whereas sowing on 20 December sowing showed maximum wilt incidence (32.8%).
- Three year crop rotation ie., cluster bean – cumin – cluster bean – wheat – cluster bean – mustard recorded minimum wilt incidence.

Main Spices Research Station, Sardarkrushinagar Dandiwada Agricultural University Jagudan – Gujarat

High yielding varieties

Cumin varieties, viz. Guj. Cum-1, Guj. Cum-2, Guj.Cum-3 and Guj. Cum-4 were released.

Crop production and management

- Five irrigations and interculturing next day after irrigation are recommended for higher yield.
- Highest yield of cumin (Guj. Cumin-1) can be ensured by 5 irrigation schedules (at sowing time, after 8 days and remaining three irrigation at 30, 50 , 70 DAS).

Disease management

- Three wilt resistant entries were identified and one was released (Guj.Cumin-3, Acc.1136, Acc.1145, Acc.1165).
- Spraying mancozeb (25 g/10 litres) was effective for the control of cumin blight.

Fennel

SKN College of Agriculture, (Rajasthan Agricultural University), Jobner, Rajasthan

High yielding varieties

- UF-125 of fennel was released.
- RF-101 erect, medium tall, medium maturity variety with long bold grains, most suitable for loamy and black cotton soil was released.

Crop production and management

- Sowing fennel, PF-35 during third week of October at a spacing of 30 × 20 cm is recommended.
- Application of micronutrients had good impact on fennel production. Soil application of $ZnSO_4$ @ 20 kg/ha (10.11 q/ha), foliar application $ZnSO_4$ @ 0.50% (9.27 q/ha), $FeSO_4$

@ 0.25% (9.92 q/ha), $MnSO_4$ @ 0.50% (9.42 q/ha) and $CuSO_4$ @ 0.50% (9.45 q/ha) and soil + foliar application of ZnSO4 @ 10 kg/ha + 0.25% (8.89 q/ha), FeSO4 @ 5.0 kg/ha + 0.125% (9.12 q/ha) and $MnSSO_4$ @ 12.5 kg/ha + 0.25% (9.49 q/ha) is recommended.

- UF-125 (released as RF-125) can be used for oil extraction. UF-134 is also promising line with respect to grain yield and volatile oil content. UF-143 had 40% higher yield as compared to local variety.

Main Spices Research Station, Sardarkrushinagar Dandiwada Agricultural University, Jagudan, Gujarat

High yielding varieties

Guj. Fennel -1, Guj. Fennel – 2, Guj. and Fennel -11 were released.

Crop production and management

For *rabi* fennel, eight irrigations and application of NPK 90:30:20 kg/ha are recommended for attaining maximum yield.

Department of Vegetable Crops (Chaudharay Charan Singh Haryana Agricultural University), Hisar, Harayana

Crop production and management

- For successful seed crop of fennel variety PF-35, sowing during third week of October, while for PF-33 first fortnight of October is optimum with irrigation at crown stage and main umbel development stage. To obtain higher and quality seed of fennel, the crop should be spaced at 30 × 20 cm and pruned to retain umbel up to second order.
- Sowing the crop in October resulted in minimum seed infestation of chalcid wasp.
- Inter cropping of fennel with potato, tomato and cauliflower has been adopted.
- Azatobacter strain MSX-9 was found most effective for improving growth and seed yield.
- In fennel seed quality decreases with the increasing umbel order. Seeds from main and primary umbels are vigorous and better in performance then the secondary and territory order umbels.
- Application of Pendimethalin @ 1.0 kg/ha as pre-emergence treatment + one hand-weeding at 50 DAS recommended for the control weed in fennel.

Fenugreek

SKN College of Agriculture, (Rajasthan Agricultural University), Jobner, Rajasthan

High yielding varieties

- Four high yielding varieties, viz. RMt – 303, RMt-1, RMt-143 and RMt-305 were released.
- The first determinant type of fenugreek UM - 305 was also released. This variety has a potential of high yield, early maturing, dwarf and resistant to diseases.

Crop production and management

- Sowing fenugreek variety RMt - 1 on 31 October is optimum.

- Sowing fenugreek variety RMt - 1 by the last week of October and UM-305 up to 15 November with a row spacing of 25-30 cm was recommended for semi-arid conditions.
- Application of N 40 kg and P_2O_5 40 kg/ha and seed treatment with *Rhizobium meliloti* was recommended.

Main Spices Research Station, Sardarkrushinagar Dandiwada Agricultural University, Jagudan, Gujarat

Gujarat Methi-1 was released having a yield potential of 1864 kg/ha.

Department of Vegetable Crops, (Chaudharay Charan Singh Haryana Agricultural University), Hisar, Haryana

High yielding varieties

Four varieties Hisar Suvarna, Hisar Madhavi, Hisar Mukta and Hisar Sonali were identified and recommended for cultivation.

Crop production and management

- Sowing time of fenugreek extends from mid September to mid January with yield reduced to half at last sowing. Sowing fenugreek Hisar Sonali during first week of November at a spacing of 30cm × 10 cm was recommended. This variety gave highest return when sown on 16 November with three times leaf cutting.
- Application of pendimethalin at 0.50 kg/ha at pre-emergence in combination with one hand weeding at 40 days after sowing was found effective for controlling weeds and increasing seed yield.
- Treating the seeds and seedlings with *Azotobacter* increased the yield.
- In fenugreek green cutting results in reduction in yield. The crop should be grown exclusively either for green or for seed. However, in variety Hisar Sonali, there is increase in seed yield with one cutting when inoculated with *Rhizobium.*

Disease management

Sulfex was superior to karathane in reducing the incidence of powdery mildew disease.

Horticulture College and Research Institute, (Tamil Nadu Agricultural University), Coimbatore, Tamil Nadu

Co-1 and Co-2 varieties were released.

Crop production and management

A closer spacing (15cm × 10 cm) gave maximum grain yield when the crop is same on 5 October, followed by the crop sown on 20 October.

Disease management

Seed treatment with *Trichoderma viride* (talc formulation with cfu 10^6) at the rate of 4 g/kg of seed, followed by soil application of neem cake 150 kg/ha was found effective against root rot disease in fenugreek.

Department of Vegetable Science, (Narendra Deva University of Agricultural and Technology), Kumarganj, Uttar Pradesh

Crop production and management

NPK @ 60: 50: 50 kg/ha produced maximum seed yield of 23.27 q/ha with medium spacing of 30cm × 10 cm.

Disease management

NDM - 7, NDM - 8, NDM - 11, UM - 304, UM - 321 of fenugreek were moderately resistant to powdery mildew at Kumarganj.

List of Coordinating Centres

AICRPS Centre	Telephone	Fax/Email
Andhra Pradesh		
Regional Agricultural Research Station Acharya N.G. Ranga Agricultural University Chintapalle 531 111, District Visakha	(08937) 238244 (O)	Fax: (08937) 238244 Gram: AGRIVARSIT Yadr_hatzone@yahoo.co.in
Andhra Pradesh		
Regional Agricultural Research Station Acharya N.G. Ranga Agricultural University Guntur 522 034	(0863) 2524053/ 2524017 (O)	Fax: (0863) 2524073/2524017 gkalidasu@yahoo.com
Andhra Pradesh		
Regional Agricultural Research Station Acharya N.G. Ranga Agricultural University Jagtial 505 327 District Karimnagar, Andhra Pradesh	(08724) 277281, 277283 (O)	Fax: (08274) 277283 Gram: AGRIVARSIT Yadr_rarsjgl@yahoo.com
Bihar		
Department of Horticulture Tirhut College of Agriculture Rajendra Agricultural University Dholi 843 121, Musaffarpur Bihar	(06274) 240632 (O)	Fax: 0621-2293-227 Gram: DHULI COLLAGR ltirhutag@sify.com
Chattisgarh		
Regional Agricultural Research Station Indira Gandhi Agricultural University Boirdadar Farm, Raigarh 496 001 District Chattisgarh	(07762) 222402 (O)	Fax: (07762) 224886/222402 crgrars@rediffmail.com
Gujarat		
Main Spices Research Station Sardarkurshinagar Dantiwada Agricultural University Jagudan 382 710 District Mehsana, Gujarat	(02762) 285337/285342 (O)	Fax: (02762-285337 aicrps_jagudan@yahoo.com

(Contd.)

AICRPS Centres	Telephone	Fax/Email
Haryana		
Department of Vegetable Crops CCS, Haryana Agricultural University Hisar 125 004, Haryana	(01662) 237720-26 (Ext. 4207) (O)	Fax: 01662-234613 Gram: Agrivarcity Hisaraicrpspices@hau.ernet.in
Himachal Pradesh		
Department of Vegetable Crops Dr YSP University of Horticulture and Forestry Nauni, Solan 173 230 Himachal Pradesh	(01792) 252329 (O)	Fax: (01792) 252030/252242 Gram: Vanudyan solanspices@yahoo.com
Karnataka		
Regional Research Station University of Agrlcultural Sciences, Bangalore Mudigere 577 132 District Chickmangalur, Karnataka	(08263) 228135 (O)	Fax: 08263-228135 kmdevaraju@gmail.com
Agricultural Research Station (Pepper) University of Agricultural Sciences Dharwad Sirsi 581 401 District Uttara Kannada, Karnataka	(08384) 226797 (O)	Fax: 08384-226146/228411 sirsipepper@sancharnet.in ars_sirsipepper@rediffmail.com
Kerala		
Cardamom Research Station Kerala Agricultural University PAMPADUMPARA 685 553 District Idukki, Kerala	(04868) 236263 (O)	entojoe2003@yahoo.co.in crsp1956@yahoo.co.in
Pepper Research Station Kerala Agricultural University Panniyur, PB No.113 Kanjirangadu (P.O) Karimbam (Via), Taliparamba 670 142 District Cannanore, Kerala	(0460) 2227287 (O)	Fax: 0460-2227287 prskan@sancharnet.in
Maharashtra		
Department of Horticulture Konkan Krishi Vidyapeeth Dapoli 415 712 District Ratnagiri, Maharashtra	(02358) 282563282064/282026 282415/282130Ext.221/225)	Fax: 02358-282074/282414 Gram: PRINAGRI, DAPOLI uday_gadre@sancharnet.in
Orissa		
High Altitude Research Station Orissa University of Agricultural and Technology Pottangi 764 039 District Koraput Orissa	(06853) 252565 (O)	Fax: 06853-223348/220226 dillipidash@yahoo.com
Rajasthan		
Dept. of Genetics & Plant Breeding SKN College of Agriculture Rajasthan Agricultural University Jobner 303 329 District Jaipur Rajasthan	(01425) 254041 (O)	Fax: (01425) 254022 Gram: AGRICOL JOBNER evd_sastry@yahoo.com

(Contd.)

AICRPS Centres	Telephone	Fax/E. mail
Tamil Nadu		
Horticultural Research Station YERCAUD 636 602 District Salem Tamil Nadu	(04281) 222456 (O)	Fax : (04281) 222387 laksh567@yahoo.co.in
Department of Spices and Plantation Crops Horticultural College and Research Institute Tamil Nadu Agril. University Coimbatore 641 003 Tamil Nadu	(0422) 2431222/ 6611284 (O)	Fax: (0422) 2445414 Gram: FARMVA Rspices@tnau.ac.in
Uttar Pradesh		
Department of Vegetable Science Narendra Dev University of Agricultural and Tech. Narendra Nagar Post, Kumarganj Faizabad 224 229, Uttar Pradesh	(05270) 262066, 262076,	Fax: (05270) 262097/262331 262164, 262067 (O) nduat@up.nic.in raviprakashsaxena@rediffmail.com
West Bengal		
Department of Horticulture Uttar Banga Krishi Viswa Vidyalaya North Bengal Campus, Pundibari P.O, District Cooch Behar West Bengal 736 165	(03582) 222545/270246/ 270249/270157 (O)	Fax: 03582-270246/270249/ 225471 Gram: PUNDIBARI janajc@rediffmail.com pundibari@yahoo.co.in

Voluntary Centres

The Associate Director Regional Agricultural Research Station Kerala Agricultural University Ambalavayal 673 593 District Wynad, Kerala	08173-260421(0)	
The Professor and Head Horticultural Research Station Tamil Nadu Agricultural University Pechiparai 629 161 Kanyakumari District Tamil Nadu	04651-281191/281192 (O)	swarnapiria2002@yahoo.co.in
The Scientist-in-charge Regional Research Station Spices Board (Govt. of India), Donigal Post, Sakleshpur Karnataka 573134	08173-344281 244007 / 244124 (O)	
The Director Indian Cardamom Research Institute Myladumpara, Kailasanadu Idukki District, Kerala 685553	04868-237268/ 237206 (O)	

8

Medicinal and Aromatic Plants

Satyabrata Maiti and P. Manivel

Medicinal and aromatic plants have contributed over centuries for alleviating human ailments. The tradition of using indigenous knowledge to treat human diseases gradually shifted to obscurity with the development of modern-day science, especially in parts of world witnessing rapid technological developments. In areas, where impact of modern science was unnoticeable the traditional system of medicine continued to play dominant role in health maintenance. The last three decades, however, produced some reversals. The limitations of modern medicines, especially to treat chronic human ailments became more obvious. The continued onslaught of killer diseases like cancer, AIDS and nibbling diseases like diabetes and arthritis drifted interest of people to natural ways of maintaining health. With information technology developing more than any other science, the failure of promises of modern medicines became more and more apparent. People in developed countries started exploring age-old traditions. In this context, the plant-based medicines, ones lost in the oblivion, came to occupy important place in the health management of modern man.

Medicinal and aromatic plants provide new window of opportunities to congruently strengthen health, food, nutritional, economic and agro-ecological securities. Business opportunities are extending due to diversified usages of plants in pharma, neutraceuticals, cosmetic and agro-chemical industries. Endowed with a wide spectrum of phyto-geographic and agroclimatic zones and a rich knowledge of folklore heritage, India happens to harbour a rich reservoir of valuable genetic resources of medicinal and aromatic plants. Traditional knowledge of their use in healthcare has resulted in the application of a large number of plant species in a variety of formulations of Ayurveda, Unani and Sidha systems of medicine. Besides, out of an estimated 100 medicinal and aromatic plant species, whose products occupy a prominent place in the world market of modern medicines, flavour and fragrances, nearly half of them are endemic to India. Thus, the country is uniquely positioned to capture the new opportunities available both locally and globally in providing increasingly sought-after naturally buffered holistic healthcare alternatives.

Today, nearly 95% of the industries' demand for over 700 medicinal and aromatic plant species is being met through harvests from wild sources. In majority of the cases the harvest from wild exceeds the sustainable quantity available from the natural resources and involves destructive collections causing irretrievable loss of biodiversity, thus closing options for future generations. The critical situation of over-collection of a large number of medicinal and aromatic plant species from wild has triggered worldwide concern to initiate appropriate steps. Moreover, the limitations of wild plant material to give consistent and uniform quality has made it relevant to produce raw material through planned and suitably organized cultivation using modern agricultural practices, or at least to ensure collection from wild through good gathering or good harvesting practices. As demand for medicinal and aromatic plants materials and their quality standards are moving higher, it is important to develop strategies, which not only ensure steady supply of quality materials but also lessen the pressure on natural resources.

Globally, trade of medicinal and aromatic plants is currently valued at about 150 billions. India accounts for less than 1% of the world's trade, whereas China accounts for about 33%. Given the agroclimatic suitability, variability, biodiversity richness, treasure of traditional knowledge and historic pattern of medicinal and aromatic plants in India, the prospects for intensifying and diversifying the country's medicinal and aromatic plants industry are unmatched. Therefore, it is extremely important to not only conserve the rich medicinal and aromatic plants biodiversity but also to cultivate it for utilization by the industries to sustainably meet their domestic and export demands and also to maintain desired quality standards. Shift from collection mode to cultivation for utilization will ensure purity, authenticity and availability of medicinal and aromatic plants for pharma and aroma industries including polyherbals for nutraceuticals. In this direction, ICAR has implemented the programme on "All India Coordinated Project on Medicinal and Aromatic Plants".

ESTABLISHMENT OF AICRP-MAP

The ICAR sanctioned a small scheme to work on over two dozen important medicinal and aromatic plants during Fourth Five Year Plan Period (1969–1974). In the Fourth Plan period, two centres were identified to be located at ICAR Institutes namely at Indian Agricultural Research Institute, New Delhi, and Indian Institute of Horticultural Research, Bangalore, where a small base on medicinal plants was created through a Cess Fund Scheme (at Division of Plant Introduction, IARI) and a Ford Foundation supported scheme (on *Discorea* spp.) at IIHR, Bangalore. To begin with, the project had five coordinating centres located in Delhi and one each in Karnataka, Himachal Pradesh, Gujarat and Madhya Pradesh. During Fifth, Sixth and Seventh Plans, four new Centres were started one each in Madhya Pradesh, Rajasthan, Uttar Pradesh and Kerala. However, during the Eighth Plan period, two new centres were started one each in Haryana (Hisar) and Maharashtra (Akola), making to 11 coordinating centres. Out of the 11 centres of AICRP, nine are based in SAUs and the remaining two in ICAR institutes (IIHR, Bangalore and NBPGR, Delhi). During IX Plan Jain Committee recommended conversion of AICRP on M&AP to a networking project as 'All India Networking Project on Medicinal and Aromatic Plants' (AINRPMAP) from the Tenth-Plan and that has been approved by the ICAR.

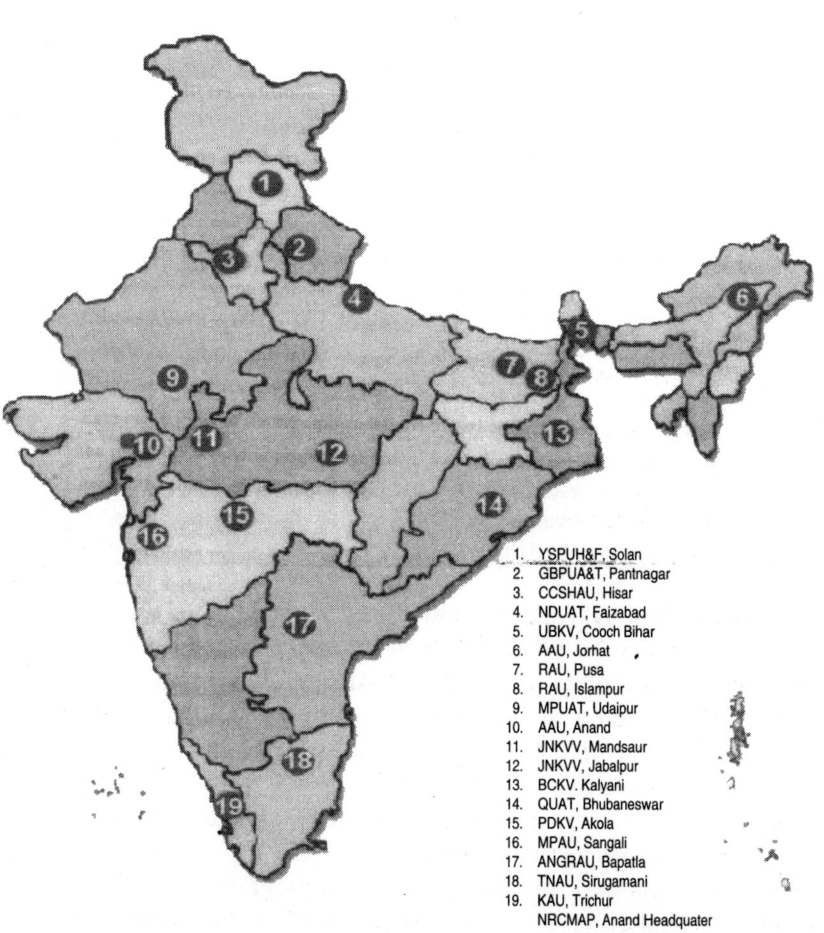

Fig. 8.1 AINRP on MAP centres after merging with the AINRP on Betelvine centres (Nos. 1, 2 (Bharsar), 3, 4, 5 (Kalimpong), 9, 10, 11, 15, and 19 are MAP centres)

Objectives

- Conserving and plant introduction studies with a view to develop agro-technologies and making available, in a decentralized manner, quality raw materials of many medicinal plants as per the therapeutic requirements of the community based PHCs located in various agroclimatic zones of the country.
- Developing high-yielding, superior quality strains of selected medicinal and aromatic plants which have market potential–both local and export.
- Establishment of medicinal plants gardens.
- Development of appropriate agrotechnologies, cropping pattern and post-harvest technology for economically important medicinal and aromatic plants for use in their commercial cultivation under different agroclimatic conditions.
- Coordination and organization of nucleus and breeder seed/planting material multiplication programme.

Centres Under the Project

Table 8.1. Research centres of AICRP on Medicinal and Aromatic Plants

AICRP Centre	*Institutions/SAUs*	*State*	*Year of* Starting
Akola	Dr. Panjabrao Deshmukh Krishi Vidyalaya	Maharashtra	1997
Anand	Anand Agricultural University	Gujarat	1975
Bharsar	G. B. Pant University of Agriculture & Technology	Uttarakhand	2004
Faizabad	N. D. University of Agriculture & Technology	Uttar Pradesh	1980
Hisar	CCS Haryana Agricultural University	Haryana	1992
Kalimpong	Uttar Banga Krishi Viswavidyalaya	West Bengal	2005
Mandsaur	Jawaharlal Nehru Krishi Vishwa Vidyalaya	Madhya Pradesh	1973
Solan	Dr. Y. S. Parmar University of Horticulture & Forestry	Himachal Pradesh	1974
Trichur	Kerala Agricultural University	Kerala	1987
Udaipur	Maharana Pratap University of Agriculture & Technology	Rajasthan	1980
Ranchi	Birsa Agricultural University	Jharkhand	2008
Raipur	Indra Gandhi Agricultural University	Chhattisgarh	2009

The AICRP on Medicinal and Aromatic Plants started in the Fourth Five Year Plan with its Headquarters at the IARI, New Delhi. The Headquarters was shifted to NBPGR from IARI in 1976 and again it was shifted to NRCMAP, Anand, Gujarat, in 1995. The Director of the National Research Centre for Medicinal & Aromatic Plants is having the additional charge of the Project Coordinator of the AINRP on Medicinal and Aromatic Plants. The Project Coordinator is responsible for:

- Formulation of location-specific and need-based research programmes in consultation with scientists of respective centres.
- Assist, advice, identify and monitor research programmes.
- Organize special group discussions between AICRP centres/SAUs to tackle common research problems in contiguous zones/centres.
- Conduct Annual/Biennial National Workshops/Group Meetings to discuss progress of research of AICRP in various states.
- Review and advice on Cess Fund *Ad-hoc* / NATP / NAIP Research Schemes on medicinal and aromatic plants and clearance of proposals for funding schemes by ICAR.

At ICAR level, the coordinating work is being monitored and evaluated by the Deputy Director-General (Horticulture) assisted by the Assistant Director-General (Horticulture II). Dr. Rajendra Gupta at IARI, New Delhi first took the charge of the Project Coordinator. Thereafter, Dr K.C. Dalal and Dr B.R. Tyagi took the charge of the Project Coordinator, respectively. At present, Dr Satyabrata Maiti is having the additional charge of the Project Coordinator since February 4, 1999.

Administratively, each AICRP centre is functioning under the respective universities on 75:25 (ICAR: SAU) budget sharing basis. From 1999, additional administrative and financial powers were delegated to the Project Coordinator by the ICAR, to disburse funds to the AINRP centres directly from the Coordinator's Cell. This helped in speedy implementation of the programmes.

Research programmes for each AINRP centre are formulated during Workshops/ Group Meetings by the scientists from respective centres facilitated by the Project Coordinator and other experts. Two types of programmes are designed: multilocational trials and station trials. Multilocational trials are those which have uniform treatments that are tested across locations. The station trials are generally location-specific and cannot be repeated in other centres. The formulated programmes are then presented in the plenary session for discussion and approval. The approved programmes are then allotted to different AINRP centres for execution.

SIGNIFICANT ACHIEVEMENTS

Genetic Resources Management

Biodiversity is the vital and major factor for improvement of biological productivity. Increasing preference and demand for plant-based medicines, flavours and fragrances, and human activities such as urbanisation, intensive agriculture and industrialisation of biodiversity rich areas have posed an unprecedented pressure on wild populations of medicinal and aromatic plants of the country. Such a situation obviously demands that apart from conservation action based on *in situ* strategy the available genetic diversity of medicinal and aromatic plant species must be collected and used for their *ex situ* conservation, authentication, biological and chemical characterization, evaluation, documentation, enhancement, prospection for the target specific bioactive molecules and developing improved cultivars with increased yield, pest/disease resistance and meeting international quality standards concerning content and composition of secondary products of the selected medicinal and aromatic plants employing classical as well as biotechnological tools.

The germplasm collection of different medicinal and aromatic plants was initiated in 1969 since inception of the AICRP on MAP by different participating centres. Considerable progress has been made in collection of germplasm of these medicinal plant species. After the establishment of AICRPMAP the collection, maintenance and evaluation of MAPs germplasm have been done very systematically and the present germplasm status at different AICRPMAP centres is presented in Table 8.2.

Table 8.2. Germplasm maintained at different AINRP on MAP centres (2007-08)

Crop	*Centre*	*No. of Accessions*
Aloe sp.	Anand	20
	Trichur	14
	Udaipur	5
	Faizabad	2
	Hisar	25
Asalio (*Lepidium sativum*)	Anand	10
	Mandsaur	10
Asoka (*Saraca asoca*)	Trichur	42
Asparagus (*Asparagus* sp.)	Hisar	24
Ashwagandha (*Withania somnifera*)	Mandsaur	118
	Udaipur	52

(Contd.)

Crop	*Centre*	*No. of Accessions*
	Hisar	50
	Anand	6
Babchi (*Psoralia corylifolia*)	Faizabad	22
Brahmi (*Bacopa monniera*)	Trichur	29
Barmasi (*Catharanthus* sp.)	Anand	6
Chitrak (*Plumpago rosea*)	Trichur	25
Chlorophytum tuberosum	Anand	1
Citronella (*Cymbopogon winterianus*)	Hisar	6
Gentiana kurroo	Solan	9
Geranium (*Pelargonium graveolense*)	Udaipur	6
Giloe (*Tinospora cordifolia*)	Trichur	12
Guggal (*Commiphora wightii*)	Anand	32
Gurmar (*Gymnema sylvestre*)	Hisar	9
Henbane	Solan	14
Heracleum	Solan	10
Hypericum perforatum	Solan	5
Isabgol (*Plantago ovata*)	Anand	39
	Mandsaur	80
	Hisar	83
Kacholam (*Kaempferia galangal*)	Trichur	12
Kalmegh (*Andrographis paniculata*)	Anand	5
	Hisar	13
	Trichur	6
	Anand	6
Lemon grass (*Cymbopogon flexuosus*)	Hisar	40
Liquorice (*Glycyrrhiza glabra*)	Hisar	3
	Anand	2
Long pepper (*Piper longum*)	Trichur	67
Lotus (*Nelumbo nucifera*)	Trichur	24
Matricaria	Solan	4
Mint (*Mentha* sps)	Hisar	7
	Solan	9
	Faizabad	01
Mucuna species	Hisar	4
	Solan	7
Neelamari (*Intigofera tinctoria*)	Trichur	25
Opium poppy (*Papaver somniferum*)	Faizabad	40
	Mandsaur	235
	Udaipur	90
Palmarosa (*Cymbopogon martini* cv. *motia*)	Hisar	64
	Faizabad	01
Plantago sp.	Anand	5
	Udaipur	16
	Anand	12
	Hisar	30
	Trichur	20
	Faizabad	16

Contd.

Crop	*Centre*	*No. of Accessions*
Picrorhiza kurroa	Solan	25
Plantago sp.	Anand	5
	Udaipur	16
	Anand	12
	Hisar	30
	Trichur	20
	Faizabad	16
Podophyllum hexandrum	Solan	12
Safed musli (*Chlorophytum borivilianum*)	Anand	18
	Hisar	12
	Mandsaur	24
	Udaipur	
Senna (*Cassia sp.)*	Anand	3
Sylibum *(Silybum marianum*)	Anand	10
Tulsi (*Ocimum* spp.)	Hisar	18
Valeriana jatamansi	Solan	30
Vetiver (*Vetiveria zizanioides*)	Trichur	37

Table 8.3. List of herbal garden (medicinal and aromatic plants garden) established at different AICRP-MAP centres.

AICRP Centres	*No. of Species Maintained*
PDKV, Akola	The herbal garden possesses more than 340 medicinal plant species comprising 84 trees, 79 perennial/shrubs, 38 creepers/climbers, 31 bulbs/rhizomes and 76 annual/seasonal. These species have been registered under IC numbers allotted by NBPGR, New Delhi.
AAU, Anand	About 275 species of medicinal and essential oil importance are also maintained in field gene banks. Perennial species, numbering about 50, are maintained in "long-term field gene banks".
GBPUA& T, Bharsar	Presently around 60 species of different high-value medicinal and aromatic plants have been conserved in herbal garden.
NDUAT, Faizabad	About 103 germplasm are being maintained
CCSHAU, Hisar	The herbal garden at Hisar has 156 species of medicinal plants comprising 41 tree species, 75 herb and shrub species, 15 creeper species, 15 aromatic plants and 10 under utilized plant species.
UBKV, Kalimpong	It is a new centre and the establishing herbal garden has already been intiated
JNKVV, Mandsaur	About 100 medicinal plant species
YSPUHF, Solan	About 59 medicinal plant species
KAU, Trichur	About 250 species are being maintained
MPU A & T, Udaipur	Herbal garden at Udaipur has 450 species out of which 178 are herbs, 182 shrubs, 52 climbers and 38 trees.
NRCMAP, Anand (HQ)	Tree species: 110; Shrubs: 40; herbs: 240 including 40 climbers

Regional Specific Varieties

Earlier as most of the medicinal plants were collected from its natural habitat, there was no systematic cultivation of these crop species. Due to increased demand and shrinking supply from natural habitats, commercial cultivation of some of the high demand medicinal plant species was started. Even though, some species were cultivated from earlier times, from the inception of AICRP-MAP improvement of some of important species has been initiated and progressed. In the beginning all the varieties developed and released for commercial cultivations were mostly from introduction from its natural habitats or foreign countries or selection from local varieties. The total list of varieties so for developed through this project are listed in the Table 8.4. As every species needs some specific environment for its growth and development varieties were developed for particular areas. Still they can be cultivated in other areas also if suitable growing conditions are available.

Table 8.4. List of medicinal and aromatic plant varieties developed / released by different AINRP centres

Crop	*Variety*	*Developed by*	*Year of Release*
		Medicinal Plants	
Chlorophytum borivilianum (safed musli)	Jawahar Safed musli	405 AINRP on M &AP, Jawaharlal Nehru Krish Vishwa Vidyalaya, Mandsaur	2004
Catharanthus roseus (periwinkle)	Prabhat	AINRP on M&AP, CCS Haryana Agricultural University, Hisar	2002
Cassia angustifolia Vahl. (senna)	ALFT-2	AINRP on M &AP, Gujarat Agricultural University, Anand	1989
Diaoscorea floribunda Mart. & Gal. (yam)	FB(C)-1	Indian Institute of Horticulture Bangalore	1974
Arka Upakar		Indian Institute of Horticulture, Bangalore	1980
Digitalis lanata Ehrh. (foxglove)	D 76	AINRP on M&AP, Y.S.Parmar University of Horticulture and Forestry, Nauni Solan	1991
Glaucium flavum (yellow horned poppy)	H 47-3	AINRP on M&AP,Y.S.Parmar University of Horticulture and Forestry, Nauni Solan	1991
Glycyrrhiza glabra L. (liquorice)	Haryana Mulhatti-1	AINRP On M&AP, CCS Haryana Agricultural University, Hisar	1989
Hyoscyamus muticus L.) (Egyptian Henbane	HMI-80-1	AINRP on M&AP, College of Agriculture, Indore	–
Lepidium sativum L. (cress)	GA 1	AINRP M &AP, Center, Gujarat Agricultural University, Anand	1998
Rauvolfia serpentine Benth. Ex Kurz. (Sarpagandha)	RI-1	AINRP on M&AP, College of Agriculture, Indore	–
Papaver somniferum L. (Opium Poppy)	Jawahar Aphim 16	AINRP on M&AP, College of Agriculture, Mandsaur	1984
	Kirtiman	AINRP on M&AP,ND University of Agriculture and Technology, Faizabad	1990
	Jawahar Opium 539	AINRP on M&AP, College of Agriculture, Mandsaur	1997
	Jawahar Opium 540	AINRP on M&AP, College of Agriculture, Mandsaur	1998
	Chetak Aphim	AINRP on M&AP, Rajasthan Agricultural University, Udaipur	1994
	Trishna	NBPGR, New Delhi	
	NOP-1	AINRP on M&AP, ND University of Agriculture and Technology, Faizabad	
	Ghazipur Local	AINRP on M&AP,ND University of Agriculture and Technology, Faizabad	
Piper longum L. (long Pepper)	Viswam	AINRP on M& AP, College of Horticulture, Trichur	1996

(Contd.)

Crop	Variety	Developed by	Year of Release
Plantago ovata Forsk. (isabgol)	Gujarat Isabgol- 1	AINRP on M &AP, Center, Gujarat Agricultural University, Anand	1976
	Gujarat Isabgol-2	AINRP on M &AP, Center, Gujarat Agricultural University, Anand	1983
	Haryana Isabgol-5	AINRP On M&AP, CCS Haryana Agricultural University, Hisar	1989
	Jawahar Isabgol-4	AINRP on M&AP, Mandsaur	1996
	Gujarat Isabgol- 3	S.K. Nagar Krishi Viswavidyalaya, Gujarat	2006
Solanum laciniatum Act.	NH 88-12	AINRP on M&AP,Y.S.Parmar University of Horticulture and Forestry, Nauni Solan	1991
Solanum viarum Dunal Syn. *S. khasianum* Clarke. (khasi kateri)	Arka Sanjeevani	Indian Institute of Horticulture, Bangalore	1989
	Arka Mahima	Indian Institute of Horticulture, Bangalore	1992
Withania somnifera Dunal (Aswagandha)	Jawahar Asgand-20	AINRP on M&AP, Mandsaur	1989
	Jawahar Asgand-134	AINRP on M&AP, Mandsaur 458 001	1998
	Aromatic Plants		
Cymbopogon flexuosus Nees ex. Steud Wats. (Lemon Grass)	NLG-84	AINRP on M&AP,ND University of Agriculture and Technology, Faizabad	1994
C. martinii Stapf.Var. motia (Palmarosa)	Rosha Grass-49	AINRP On M&AP, CCS Haryana Agricultural University, Hisar	1989
	CI-80-68	AINRP on M&AP, Indore 452 001	–
Jasminum grandiflorum L. (Jasmine)	Arka Surabhi	Indian Institute of Horticulture, Bangalore	1993
Mentha spicata L. (Spearmint)	Punjab Spearmint-1	AINRP on M&AP, Y.S.Parmar University of Horticulture and Forestry, Nauni Solan	1991
Valeriana jatamansi DC. (Mushakbala)	Dalhousi Clone	AINRP on M&AP, Y.S.Parmar University of Horticulture and Forestry, Nauni, Solan	1994
Vetiveria zizanioides (L) Nash. (Vetiver)	Hyb-8	NBPGR, New Delhi	–

Apart from varietal wealth development some salient progress made in crop improvement, crop production, crop protection, phyto-chemistry and post-harvest management of medicinal plant species are presented here.

Aloe *(Aloe vera)*

The *Aloe barbadensis* is an important medicinal plant of India, belonging to the family Liliaceae. Frequently growing along roadsides and waste places, it prefers dry areas. It is native of North Africa, Canery Islands and Spain and is growing wild on the coasts of Maharashtra, Gujarat and Southern India. In addition to the medicinal value, the mucilaginous gel from the leaf is extracted and processed for use in several cosmetics. Demand for such products has increased tremendously during the last decade and hence its large-scale cultivation has been initiated in some states. However, main bottleneck faced by farmers is lack of quality planting material, which is due to low multiplication rate.

The salient achievements made in this crop are:

- Highest aloin and gel yield was obtained from five year-old plants at Udaipur.
- The method for estimation of barbaloin was standardized at Trichur and Udaipur.

- Karyotype study of germplasm showed that all the accessions were of diploids with somatic chromosome number 2n = 2x = 14. Total chromosome length (TCL) of diploid chromosome complement ranged from 84.58 μ to 197.50 μ and average chromosome length (ACL) ranged from 14.11 *μm* to 16.04 μ.
- At Hisar, out of 99 collections evaluated, germplasm KC/OP-49, KC/OP-15, KC/OP-25 and KC/OP-40 had significantly higher leaf yield per plant than the local check HAV-1. Whereas, genotypes KC/OP-25, KC/OP-36, KC/OP-15, Morni Hill-1, KC/OP-44 and KC/OP-53 had significantly higher mucilage than HAV-1.
- Comparison of pathological and biochemical basis of four bacterial pathogens (which cause soft rot of aloe) isolated from varied geographical locations of our country reveled that even though all were gram negative rods and usually appeared single in smear and produced circular, creamy-white, translucent colonies with smooth surface and convex elevation, the pathogenicity of cultures showed wide variability in culture in terms of pathogenic reaction with 10 different hosts. Biochemically also they were different.

Asalio *(Lepidum sativum)*

Asalio, belonging to family Cruciferae, is an erect herbaceous glabrous annual and suitable for cultivation in winter season. The demand of this crop is increasing as low water requiring medicinal herb. The crop is traditionally cultivated in small areas in the drier tract of Gujarat, Rajasthan and Madhya Pradesh. It is cultivated as culinary vegetable. The roots, leaves and seeds are used for different medicinal preparations; the roots are bitter, acrid and useful in treatment of secondary syphilis and tenesmus. The leaves are stimulant, diuretic and antibacterial and are useful in scorbutic diseases and hepatopathy. The seeds are bitter, depurative, tonic, aphrodisiac, ophthalmic and diuretic and are useful in poultices for sprains and in leprosy, skin disease, dysentery, diarrhoea, dyspepsia, etc.

The salient research achievements are:

- About 20 germplasm are being maintained at Anand and Mandsaur.
- A high-yielding variety, GA-1, was released in 1998 from AINRP on M &AP centre, Gujarat Agricultural University, Anand, for general cultivation. It yields about 1,352 kg/ha which is about 14-15% higher yield than the local check variety at Anand.
- Its germplasm is being maintained at Anand and Mandsaur and each centre has 10 germplasm.
- At Anand centre, the agro-techniques for cultivation have also been developed.
- Nine germplasm collected from farmers' fields at Mandsaur, Neemuch, Ratlam and Jabalpur districts were evaluated for seed yield and lines MLS-7, MLS –1, MLS-3, MLS-5 and MLS-6 were high seed-yielding accessions.
- At Mandsaur, sowing by broadcasting method gave higher seed yield of 2,437 kg ha^{-1} than line sowing (2,417 kg ha^{-1}). A seed rate of 6 kg ha^{-1} gave highest seed yield of 2,521 kg ha^{1} followed by 8 and 10 kg ha^{-1}, both giving same yield of 2,479 kg ha^{-1}.
- Based on two years study, cropping system comprising Safed Musli in Kharif season and Asalio (*Lepidium sativum*) in Rabi season gave the highest net profit (Rs. 5 lakhs) with CBR 1:2.49 at Anand.

Ashwagandha (*Withania somnifera* Dunal)

Ashwagandha is one of the important medicinal plants commercially cultivated in north western region of Madhya Pradesh for a long period as a dry land crop in late *kharif* season. It belongs to family Solanaceae. The withaniols, somniferin and several other alkaloids present in roots and to some extent leaves and seeds are used in ayurvedic and unani medicines, particularly for hiccups, bronchitis, rheumatism, dropsy, several female disorders, stomach and lung inflammation and skin diseases. The development of high-yielding varieties with high alkaloid content is the future thrust area for research.

The salient research achievements made in this crop are:

- The available germplasm (118) of Ashwagandha are being maintained at Mandsaur, Udaipur (52), Hisar (50) and Anand (6).
- Two varieties Jawahar Asgand-20 (JA-20) and JA-134 were released from Mandsaur. These two varieties are elite selections of local varieties from farmers' field. JA-20 is a high seed-yielding and JA-134 is a high quality root-yielding varieties.
- The available germplasm were evaluated for yield and qualitative characters at different centres. At Mandsaur, out of 67 germplasm lines tested entries MWS-100, MWS-130, 137, 205, 223, 331 and 332 were found superior. Whereas, at Udaipur, six genotypes, viz. RAs-23, RAs-34, WS-90-141, WS-90-111, WS-90-124 and WS-222 had performed better over the best check JA-134.
- At Akola, yield of total alkaloids was found significantly highest at 100% flowering stage (6.58 kg/ha). At Udaipur, total alkaloid content was found to be maximum (0.66%) when harvested at maturity stage.
- At Faizabad, variety JA-134 showed delayed maturity (180 days) than JA-20 (173 days) but recorded lower fresh root yield (865 kg/ha) than JA-20 (1174 kg/ha). The dry root yield was 422 kg/ha in the variety JA-20 and JA-134.
- Production and protection technologies were standardized for this crop at Mandsaur. Overnight seed soaking with MH 30 at 250 ppm increased dry root and seed yield. Plant density of 20 lakh/ha (spacing 20cm × 2.5cm) is optimum for better quality root yield. Fertilizer dose of 20 kg N, 40 Kg P_2O_5 and 20 kg K_2O/ha is suitable for better quality root yield. Vermicompost at 5 tonnes/ha responded well to harvest ideal fasciculated roots, though under inferior soil, 50% recommended dose of and rest 50% through organic manure can also be given.
- At Indore, Ashwagandha responded very favourably to application of 2.5 tonnes/ha of FYM along with 12.5 kg N + 25 kg P/ha. Seed treatment of the crop with Bavistin at 0.1% yield maximum plant population leading to higher dry root yield. Two sprays of Dithane for leaf blight recorded higher yield.
- For obtaining higher root yield, alkaloids yield and monetary return from Ashwagandha, the application of 5 tonnes FYM/ha is recommended at Akola region. Similarly, seed rate of 9 kg/ha is recommended for obtaining higher root yield and monetary return.
- At Akola, the root yield was found significantly influenced by the harvesting time and significantly the highest root yield was recorded with the harvesting at 100% flowering

stage. The fibre content was found to increase with the harvesting time and the highest fibre content was recorded at maturity stage. The total alkaloids content and it was significantly higher at 50% flowering stage followed by 100% flowering stage.

- For Udaipur regions, sowing and harvesting of Ashwagandha during 37th and 7th meteorological weeks, respectively are appropriate dates. Application of 30 kg N, 30 kg P_2O_5 and 20 kg K_2O/ha are optimum doses. Application of 0.5 kg/ha Isoproturon as pre-emergence herbicide followed by hand weeding and hoeing at 45 DAS is best preposition to control weeds. Two sprays of Mepiquate Chloride 1000 ppm or Chlormequate Chloride 250 ppm at flower initiation stage and 15 days after first spray improves yield and quality of roots.
- Mancozeb was found to be effective in controlling *Alternaria* leaf blight which increased root yield also at Indore and Mandsaur.
- *Fusarium* sp. was found responsible for seedling mortality at Anand.

Asparagus (*Asparagus racemous*)

The plant is a member of Liliaceae. The fasciculated roots are medicinally important. Saponins are active ingredient. Dry powder of root is used for the preparation of many Ayurvedic vital tonics. The plant is mainly propagated by seeds and it is a perennial crop. The cultivation of the crop is spread throughout India, however, in limited scale.

- Germplasm were collected from wild and a total of 104 germplasm of Asparagus are being maintained at Anand (6), Faizabad (28), Hisar (24) and NRCMAP (46). The preliminary evaluation of these germplasm is in progress.
- Agro-technologies for its commercial cultivation are being developed at Anand centre.
- At Akola yield of fresh and dry tubers was significantly influenced with the spacing and the highest tuber yields (fresh and dry) were obtained with the plant spacing of 60 × 60 cm. The durations of crop significant influenced number of shoots, length of tubers, fresh and dry tuber weight/plant and yield of fresh and dry tubers. At Akola, saponin content decreased gradually with the increase in storage period and significantly lowest content was noticed at 12 months after harvesting.

Asoka (*Saraca asoca*)

Asoka is a small evergreen tree belonging to the family Caesalpiniaceae and is a sacred tree of Hindus and Buddhists. It occurs almost throughout India up to an altitude of 750 m. Bark containing tannin and catechol is useful in menorrhagia (due to fibroids), leucorrhoea and haemorrhagic dysentery. Leaves are paripinnate and used in impure blood. Flowers are orange to orange-yellow, very fragrant in dense axillary corymbs. The dried flowers are used against diabetes and haemorrhagic dysentery and seeds are used for treating bone fractures.

- The systematic collection of germplasm was initiated at Trichur centre and 42 germplams are available at present. Out of the 42 accessions two are identified as early flowering types. Two accessions one from Trichur district and other from Trivandrum district are identified as higher bark producers represented by increased height, number of leaves and girth of stem.

Babchi (*Psoralea corylifolia*)

The plant belongs to the subfamily Papilionaceae of Leguminosae family. Seeds are used in indigenous system of medicine for the treatment of leucoderma, leprosy, psoriasis, etc. Psoralen present in pericarp of seed is therapeutically important. The crop thrives well in loamy soil having a pH range of 6.5-7.5 with good organic matter.

- At Faizabad, 22 germplasm of babchi have been collected from various parts of Uttat Pradesh. They were evaluated in multilocational testing and IC-111226 has been found the best at national level.
- Ten improved lines were evaluated at various locations and IC-111239 at Udaipur and IC-111256 at Hisar performed well for seed yield.
- From Udaipur centre a variety "Prapap Babchi-1" has been recommended for release for Rajasthan. The variety is suitable for marginal land and required minimum inputs for cultivation. The plant height is 164 $\pm$ 70 cm depends upon soil type and availability of water. Flower colour is bluish purple, leaf colour green, seed colour black, seed size bold and days to maturity (240-280 days). The seed yield is 28.0 q/ha with psorelin content 1.37%.
- At Udaipur sowing of 40cm × 30cm spacing with 40 kg N and 20 kg P_2O_5/ha gave significantly higher number of inflorescence per plant, seed yield, psoralen content and psoralen yield compared to its lower levels of fertility. Spacing of 40cm × 30cm produced significantly highest seed yield, psoralen content and psoralen yield. One spray of malic hydrazide (500 ppm) at 30 DAS produces 13.91q seeds/ha which is higher by 102.4% over water spray.
- At Faizabad, stem-rot disease is controlled by bioagent (*Trichoderma harzianum*).
- The IPM package tested consisted of soil application of neem cake (4q ha^{-1}) + seed treatment with *T. viride* (4g/kg) + spraying with *Trichoderma* suspension (50 g/litre) + spraying of carbendazim (0.05%) in December-January significantly reduced the collar rot incidence compared to the control plot at Faizabad.

Brahmi (*Bacopa monnieri*)

Brahmi is a creeping succulent herb with ascending branches grows wild on damp places and marshy lands and distributed in major parts of the plains of India. It belongs to the family Scrophulariaceae. The whole plant is useful for nervous disorders such as insanity, epilepsy, nervous breakdown etc. The fresh juice is considered as brain tonic.

An extensive exploration was carried out all over Kerala and 29 accessions were collected and maintained at Trichur. These accessions were evaluated and observed considerable variation between them. They were grouped into five clusters and the accession No. 14 has been identified as superior as it yielded higher biomass and bacoside content. The accessions with higher biomass yields with higher bacoside content were identified from Aluva (Ernakulam district) and Kottakkal (Malappuram district). The systematic evaluation and its domestication are in progress.

Chirayita (*Swertia* sp.)

Chirayita is a member of the family Genetianaceae. Among different Swerita species,

Swertia chirayita is commonly used for drug. The drug (chiretta) is obtained from the derived plants and is used against chronic fever, anemia, bronchial asthma, and as a bitter tonic. *Swertia* grows well in moist temperate Himalayas from Kashmir to Bhutan and in Khasi hill in Meghalaya. It is a critically endangered species and highly prized in the Indian system of medicine, is valued as a bitter tonic, blood purifier, hepatic stimulant etc.

- At Bharsar, three type of variability were observed among the collected materials of Chiryita, such as plants of short leaved type, plants of narrow leaved type and plants of broad and long leaved type.
- Extensive survey was conducted for collection of chirayita population in Darjeeling district at UBKV, Kalingpong, and high population density of *S. chirayita* and *S. bimaculata* were noticed at Sukhiapokhri (6,400ft) and Sonada (6,800ft).
- Comparative study of species – *Swertia chirayita* and *S. bimaculata* — revealed that there was a distinct difference between them for plant height, leaf (arrangement, size, length, breadth and purple pigmentation), seed (colour and size) and root length.
- Seed germination studied in chirayita at Kalimpong indicated that seeds of *S. chirayita* germinated only in treatment with sterilized soil: sand: leaf manure (8:4:1 v/v) + GA_3 500 ppm and with sterilized soil: sand: leaf manure (8:4:1 v/v) + GA_3 700 ppm. The degree of seedling mortality was extremely higher in *Swertia chirayita* than in *Swertia bimaculata.*
- Chirayita can be propagated by seeds. Seedling growth is slow hence 4-6 leaved seedlings are pricked into polybags for further growth. After about 8–10 months in polybags, seedlings are field transplanted at a spacing of 30cm × 45cm. In second or third year, plants develop aerial shoots which bear flowers. After seed setting the entire plant dies and the whole plant is harvested. Periodically some radicle leaves should also be harvested.
- Seed germination studied in chirayita at Kalimpong indicated that seeds of *S. chirayita* germinated only in treatment with sterilized soil: sand: leaf manure (8:4:1 v/v) + GA_3 500 ppm and with sterilized soil: sand: leaf manure (8:4:1 v/v) + GA_3 700 ppm. The degree of seedling mortality was extremely higher in *Swertia chirayita* than in *Swertia bimaculata.*

Chitrak (*Plumbago rosea*)

Chitrak is a perennial shrubby plant found throughout India but indigenus to sikkim or Khasi Hills. It belongs to the family Plumbaginaceae. It is known as *lal chitrak* as it bears red flowers in terminal and auxiliary racemes. It is grown for roots ad root barks. The active principle of root is plumbagin. It has stimulant action on central nervous system, plain muscles and on secretion of sweat, urine and bile. The leaves and roots are also used to treat infections and digestive problems such as dysentry.

- Twenty-five genotypes of the species were studied at Trichur. Evaluation of clustering pattern of 17 accessions indicated that clustering pattern did not follow the geographical distribution. Maximum contribution of character towards the divergence was through the number of flowering branches. Tall plants with ovate type of leaves with dark green colour and maximum number of roots are the plant architecture for higher root yield.

- At Trichur, application of FYM at 10 tonnes/ha along with azospirillum and PSB at 0.5 g/plant is the best combination for higher root yield and plant nutrient status in *Plumbago rosea.* A combined application of organic manures with biofertilizers gave higher root yield and maintained relatively higher plant nutrient status compared to their application alone in *Plumbago rosea.*
- At Trichur, a study on the influence of combination of organic manures and biofertilisers on yield and quality of *Plumbago rosea* revealed that combined application of FYM at 10 tonnes/ha with Azospirillum and PSB at 0.5 g/plant was better combination for maximum root yield and plumbagin content in *Plumbago rosea.* Irrigation at IW/CPE of 1.0 (approximately 7 days interval) during December—May is optimum for higher root yield and plumbagin content in plumbago rosea and also maintained relatively higher moisture content in root zone. The effect of spacing on yield and quality of *Plumbago rosea* revealed that spacing of 50cm × 30 cm maintained optimum plant population and higher plumbagin content as well as yield.
- Chitrak is a perennial shrubby plant found throughout India grown for roots and rootbarks. The active principle of root is plumbagin. At Trichur, application of 15 tonnes/ha of FYM along with *azospirillum* at 8 months after planting, recorded tallest plant (13.76 cm) followed by10 tonnes/ha of FYM with *azospirillum* (13.03 cm).The N, P and K content of the plant were highest when the plants were applied with 10 tonnes/ha of FYM with azospirillum and PSB. Biomass yield (193.33 g/plant) was highest when applied with 10 tonnes/ha of FYM with *azospirillum,* followed by 64.33 g/plant and 116.66 g/plant, respectively with 10 t ha^{-1} FYM with *azospirillum* and PSB. A maximum root yield of 3.7 t ha^{-1} was recorded when applied with 10 tonnes/ha FYM with *azospirillum* which was at par with 15 tonnes/ha FYM with azospirillum and PSB (3.2 tonnes/ha).

Yam *(Dioscorea deltoidea)*

It is an endangered species and important source of diosgenin (steroidal sapogenin), which is used for the production of corticosteroids and antifertility compounds. From solan centre, a variety 'Chamba clones' has been identified and agrotechnique for cultivation of this species has also been developed.

- Rhizomes of this species should be planted in first fortnight of March under sub-temperate conditions and in first fortnight of April under temperate conditions. The planting distance between rows and plant should be 60 cm × 45 cm. Harvesting after three years of planting during December gives approximately 200-225 q/ha of fresh rhizomes at Solan.

Gentiana (*Gentiana kurroo*)

Gentiana kurroo (Indian Gentiana) is a small perennial herb belonging to family Gentianaceae. The plant is naturally distributed in Kashmir and North Western Himalayas. It is commonly known as *karu* or *kutiki.* The plant grows well in cooler climate. The rhizome and roots are used for the therapeutic purposes. It is used for preparation of bitter tonic for improving appetite and stimulating gastric secretion. The drug is also reputed for its use against fever, urinary tract complaints and hepatic disorders. Gentiopurin and gentianic acid are active ingredients of the plant. Propagation of plant is mainly by seed as well as by cuttings or by rhizome. The plant is not under cultivation and raw drug is collected the natural plant population is facing stress and its conservation status is highly vulnerable. Indian Gentian is reported endangered by different authours. Sustaninable harvesting methods are required for its conservation. Attempts are initiated at Solan for bringing plants under cultivation.

- Gerrmplasm of this species is being maintained at Solan (Table 8.2). Eleven collections were evaluated at Solan and the collection from Kanjiiar Dhar was identified on the basis of its highest root and leaf yield and bitter content.
- Reproductive biology of Gentiana in detail has been worked out. Nature of pollination, duration of bud initiation to fruit dehiscence, peak flowering time, time taken for pollen dehiscence, stigma receptivity and pollen viability duration were studied. Controlled cross resulted in maximum fruit set, seed yield per fruit, seed yield per plant and number of seeds per fruit. Higher fruit set has been observed in open-pollination (80.57%) compared to the controlled selfing (7.25%).
- Domestication of the species was conducted. Initial transplantation in polybags containing a mixture of sand + soil + FYM, when seedlings were 4-6 leaved, followed by transplantation in the main field was proved to be successful as it ensured least transplantation shock and seedling establishment was also quite good (approximately 80–90%) at Solan.
- Solan centre has successfully domesticated and standardized propagation techniques for large-scale cultivation. It is propagated by both seeds as well as rhizome cuttings, about 50g of seed is sufficient for one hectare and sowing time is February-March. Seedlings should be pricked into polythene bags. Seedlings in polybags are ready for field transplantation after 5-6 months, maintaining 30cm × 30 cm spacing. Rootstock should be harvested after three years and approximately 100 kg/ha of dry yield is obtained at Solan.
- Application of NPK at 80:40:30 kg/ha produced maximum leaf and root yields at Solan.

Glory Lily (*Gloriosa superba*)

The plant belongs to the family Liliaceae and listed in the threatened category and locally known as Kalihari. Seeds and fleshy roots are medicinally important. Colchicine is the chemical ingredient, which is responsible for therapeutic action. Due to its action on spindle fibre during cell division, the plant has been identified as an important anticancer therapy agent. It is a perennial herb with weak stems climbing on supports with the help of leaf tip tendrils. The colchicine present both in tubers and seeds is used for treatment of gout and rheumatism and also for producing polyploids in plant breeding. It was recently introduced to commercial cultivation in limited areas of Tamil Nadu and Himachal Pradesh. Colchicines are used in modern and Ayurvedic medicines and also in biological studies. It is grown as a *kharif* crop in well-drained and humus rich soil.

- Different methods of pollination studied at Solan showed that though all the methods tried resulted in seed setting, controlled selfing between different flowers on the same plant gave the best result in terms of percentage fruit set, number of seeds/fruit, weight of seeds/fruit, higher 100 seed weight (2.66 gm), colchicine content (0.8%) and maximum seed germination (46.25%). Increasing the number of hand pollinations (from one to three) resulted in increased seed yield.

- It can be propagated both by sexual and asexual means. Seed sowing can be done during February-March, which can be enhanced to 50% by treating the seeds with 4000-ppm thiourea. One-year-old seed raised tubers, or tuber splits (cut at the neck region) should be field transplanted during March-April at a spacing of 45cm × 60cm. Application of NPK (40:50:75) at the time of planting and additional dose of 80 kg/ha after 8 weeks of planting is required for good harvest. An yield of 150 kg/ha during first year and nearly 250-300 kg/ha of seeds after second year can be obtained.

Guggal *(Commiphora wightii)*

Guggal is a small tree, endemic to drier tract of Rajasthan and Gujarat belongs to the family *Burseraceae*. As the plant generally dies after tapping of gum the natural habital of this plant is shrinking and hence it is entered in the endangered category of the Red Data Book. Now efforts are being made by different agencies including Forest Department to preserve guggal in its natural form. The oleo-gum-resin obtained by tapping the plant is highly priced for its two major sterols, viz. gugglusterone-Z and gugglusterone-E and is used as in indigenous drug against rheumatism and arthritis.

- Air-layering for quick vegetative propagation, gum tapping during January-February is the right time for tapping. Root feeding of ethepon (40% aqueous solution) enhances gum production at Anand.
- At Anand, *Botryodiplodia theobromae* (fungus) and *Pseudomonas* sp. (Bacteria) were found pathogenic to guggal plant.

Basant *(Hypericum perforatum)*

The plant is a rhizomatous perennial herb cultivated for its essential oil and medicinal purpose. It belongs to the family Hypericaceae. It is grown in comparatively cooler limate. Hypericum possesses astringent, expectorant and diuretic properties and is used in pulamonary and urinary troubles, diarrhoea, hysteria and nervous depression, haemoptysis and other haemorrhages, and jaundice.

- Solan centre is maintaing about 5 germplasm in this species. It is mostly used for treating depression and healing wounds, burns, etc. It is propagated both by seeds as well as slips, propagation by seed is easy. Seeds are sown during November-December or February-March under polythouse/glasshouse conditions and 100g of seeds are sufficient for a hectare of area. Transplanting should be done during February-March at a spacing of 30cm × 40cm. Application of 90 kg CAN/ha gives favourable effect on herb yield. During first year, branches are mostly straggling which become erect in next year. Harvesting one-year-old plantation provides dry herbage yield of 1-1.5 tonnes/ha.

Isabgol (*Plantago ovata*)

Isabgol is an important medicinal crop traditionally grown as *Rabi* crop in Gujarat, Madhya Pradesh and Rajasthan as the required cool and dry climate usually prevails during growth in this region. This species is distributed in Canary Islands, Mediterranean regions of Sothern Europe and North Africa to Pakistan. It belongs to the family Plantaginaceae. Husk is medicinally useful against constipation and irritation in the digestive tract. India is the sole exporter of isabgol husk in the international market. Creation of genetic variability, development of high-yielding varieties with downy mildew disease resistance and also increased yield through production and protection management, are important research areas in this crop. Apart from traditional isabgol (blond psyllium), black psyllium (*Plantago indica*) is also has market demand and work has been initiated in this species also.

- The germplasm of this species are maintained at Anand, Mandsaur, Hisar and Udaipur centrers besides in NRCMAP, Anand (Table 8.2).
- Two varieties (GI-1 and GI-2) from Anand centre, one each from Hisar (HI-5), Mandsaur (JI-4) and S.K. Nagar (GI-3) has been released for commercial cultivation (Table 8.4).
- At Udaipur, germplasm accessions PB-31, RI-153, PC-6 for seed yield, PB-3-1 and RI-149 for high swelling factor and PS-19 and HI-1 for downy mildew resistance were identified. Besides, MIB-124, Palampur-2 and MIB-122 were tolerant to cold injury. Two lines, PS-19 and HI-1, were found immune/highly resistant to downy mildew. At Anand, DM-1 was found highly resistant to downy mildew. Entry, J.I.80 and J.I.56, had significantly higher seed yield than the GI-2 as best control.
- Downy mildew disease is a serious problem in this crop. The yield reduction is heavy which is inversely related to the disease incidence. Three subsequent sprays of Metalaxyl 0.2%, Mancozeb 0.3% and again Mancozeb 0.3% were found to be significantly superior and having less disease (32.5%) and maximum seed yield (1,584 kg) at Anand. Fungicidal seed treatment with Apron at 4,6 and 8 g/kg were found non-significant for downy mildew disease and seed yield.
- At Anand, isabgol can be economically grown with minimum of three irrigations (first at sowing then 30 days and 70 days after sowing). Split application of nitrogen was not advantageous. Whole of the nitrogen should be applied as basal at the rate of 30 kg/ha. Spraying of Isoproturon either two days before or two days after sowing at of 0.5 kg.a.i./ha effectively controlled weeds.
- At Anand, maximum yield was recovered from the treatment of *Tephrosia purpurea* (900 kg/ha) followed by sun hemp (670 kg/ha) and fallow (620 kg/ha).
- At Akola, Isabgol variety G1 could be grown after harvest of short duration *kharif* crops of groundnut, soybean and pigeonpea to get maximum net return. It is recommended to sow isabgol crop during 10–20 November and seed rate 4 kg/ha for getting higher seed yield and monetary returns in Akola region. For obtaining maximum seed yield and monetary returns, isabgol G-1 should be applied with 50 kg N and 30 Kg P_2O_5 per hectare.
- At Udaipur, Isabgol yield on clay loam soil is higher by 74.4 % compared to sandy loam soil. Sowing during 10–20 November with 4kg seeds/ha at 30cm × 5cm spacing and application of 10 tonnes FYM, 30 kg N and 15 kg P_2O_5 ha are appropriate. Single super phosphate, when applied as source of P produces 8.5% higher seed compared to DAP. Application of 3 irrigations at 25, 50 and 75 days after sowing (DAS) are sufficient. Application of Isoproturon at 0.375 kg/ha as pre-emergence followed by one hand weeding at 45 DAS controls weeds efficiently. GI-2 is the most suitable varieties under

Rajasthan conditions. Delay in sowing from 10 November to 20 November resulted in significant decreases of various yield attributes, seed yield and husk yield of isabgol.

- At Udaipur, germplasm PS-19 and HI-1 were found to be immune, whereas, DM-11, P-79, Palampur-3 and PB-62 were resistasnt to downy mildew. The incidence of downy mildew remains low if sowing is done during 25 November–10 December Downy mildew initiated when the average maximum and minimum temperatures goes between 29.6 and 10.6 °C, respectively. For downy mildew control seed treatment with Metalaxyl at 6g/kg seed + spray of 0.25% Metalaxyl at 25 DAS + spray of 0.3% Mancozeb at 40 and 50 DAS is an efficient method.
- Application of either 25 or 50 kg N/ha and 2 irrigations at 25 and 75 days after sowing is the best for isabgol crops at Hissar.
- At Faizabad, highest yield (3463 kg/ha) was recorded at 20 cm spacing with 75 kg ha^{-1} of N followed by 15 cm spacing (3249 kg/ha) also with 75 kg/ha of N.
- At Mandsaur, highest grain yield of 856 kg/ha was recorded when the crop was sown on 21 November. Highest seed yield was recorded with three irrigations (tillering + before ear head emergence + after earhead emergence) which produced the yield level of 762 kg/ha. Whereas at Udaipur, seed and straw yields of isabgol were significantly higher due to application of three irrigations (at tiller initiation + full tiller + 75% flowering).
- At Udaipur, initiation and spread of downy mildew infection were higher when average temperature was 20 °C. This observation may be used for disease forecasting and decision making in its management.
- During the field survey four new diseases were recorded and their pathogenicity was proved. They were identified as *Fusarium solani* (finger rot), *Rhizoctonia solani* (crown rot), *Phyllosticta* sp. (leaf spot), *Curvularia* sp. (leaf spot).

Kacholam (*Kaempferia galanga*)

The *Kaempferia galanga*, commonly known as aromatic ginger, sand ginger or resurrection lily, is a monocotyledonous plant in ginger family. It is found primarily in open areas in southern China, Taiwan, Cambodia and India, but is also widely cultivated throughout Southeast Asia. Kaempferia galanga has a peppery camphorous taste. It is one of four plants known as galangal, and is differentiated from the others by the absence of stem and dark brown rounded rhizomes, while the other varieties all have stems and pale rosebrown rhizomes. The rhizomes of aromatic ginger have been reported to include cineol, borneol, 3-carene, camphene, kaempferol, kaempferide, cinnamaldehyde, p-methoxycinnaamic acid, ethyl cinnamate and ethyl p-methoxycinnamate. Extracts of the plant using methanol have shown larvicidal activity against the second stage larva of dog roundworm (*Toxocara canis*). It was also found to be effective as an amebicide in vitro against three species of Acanthamoeba which cause granulomatous amebic encephalitis and amebic keratitis. The rhizomes of the plant, which contains essential oils, have been used in Chinese medicine as a decoction or powder for treating indigestion, cold, pectoral and abdominal pains, headache and toothache. Its alcoholic maceration has also been applied as liniment for rheumatism.

- Germplasm evaluation in Kacholam identified a better variety which was quantitatively and qualitatively superior among other genotypes at Trichur. Twelve germplasm lines are being maintained and evaluation is in progress. The standardization of production physiology of *Kaempferia* has been done.

- Combined application of FYM at 10 tonnes/ha and NP at 37.5:20 kg/ha Azospirillum or Azotobacter was beneficial for economizing oil production. FYM at 20 tonnes/ha and NPK at 40:20:20 Kg/ha was beneficial for higher yield in *Kaempferia galanga*
- Production physiology has revealed that improvement of growth at 2nd month through improvement in LAI, RGR, CGR etc improves the final yield of the crop. No significant effect was noticed on yield and yield attributes with the application of Ca, Fe, Mn and S. Significant difference was noticed between treatments on rhizome yield in kacholam with the application of organic and inorganic source combinations.

Kalmegh (*Andrographis paniculata*)

Kalmegh is an annual herb belonging to the family Acanthaceae. This species is native to Tropical Asia and occurring in the plains of Himachal Pradesh to Assam, Mizorum, southward through peninsular India. Commonly found growing in moist, shady places, uncultivated fields and boundaries. Leaves and stems are used as bitter tonic possess antityphoid and antibiotic properties for treatment of dysentery, cholera, diabetes, consumption, influenza, bronchitis, swelling and itches, plies and gonorrhoea. Kalmegh is cultivated in limited areas of several parts of India as a *Kharif* crop. Collection and evaluation of germplasm, understanding the reproductive biology, development of good agricultural practices including stress physiology, post-harvest behaviour, etc., are some of the important areas for research.

- Six accessions of *Andrographis* collected from different parts of India were evaluated at Anand center for quality and yield. Significantly higher yield was obtained in accession number 3 (2844 kg/ha.) However, non-significant difference was found in the case of andrographoloide content among the accessions studied.
- Twenty-nine selections of kalmegh were evaluated for earliness based on days taken to 50% flowering at AAU, Anand and were classified as late and early flowering types. Selections 16, 17, 18, 20 and 23 were early (60 days) compared to control (87 days) and selections 3, 4 and 10 were late maturing.
- Cultivation of kalmegh has been successfully grown as rainfed crop in medium black soil of Indore region.
- The research conducted at Anand centre revealed that kalmegh can be profitably cultivated during *kharif* season. Transplanting the 45 days old seedling in the first fortnight of July and harvesting at the second fortnight of November gave highest biomass production (4,017 kg/ha) of good quality.
- In Kalmegh, it has been established that application of castor cake to provide 40 kg/ha nitrogen could produce highest herbage yield (3,740 kg/ha) and 96.8 kg/ha of Andrographoloid yield.
- Two collections from Uttar Pradesh were evaluated at national level and significant differences were observed for herb yield at Faizabad.
- At Akola planting time on 1st July and harvesting during 1–16 November were suitable

for obtaining higher herbage yield and total yield of Andrographolide and iron. The Periodic evaluation of Kalmegh at Akola centre indicated that the dry foliage yield was maximum at 135 days after transplanting (DAT), however, the content of Andrographolide and iron was maximum during 60–90 DAT. The total yield of Andrographolide and iron was maximum during 90–135 DAT.

- Transplanting of 40 days old seedlings on July 1st and harvesting on November 1st is appropriate when judged in terms of dry herbage yield at Udaipur region.
- Effect of date of planting and harvesting on yield and quality revealed that at Anand and Udaipur, 1st July transplanting and 16th November harvesting significantly increased the herb yield, but quality parameters were non significant. At Hissar, planting during third week of July and harvesting at third week of November was found superior. At Faizabad, maximum herbage yield (12,737 kg/ha) was obtained from 1st July plating and 16 November harvesting. At Hisar, plant height and primary branches significantly increased when crop was sown on 1st July with 20 × 20 cm Spacing.
- At Anand, Kalmegh crop when harvested at 60 DAP, i.e., at flower initiation stage, reasonable content of Andrographoloid could be obtained with higher number of leaves and branches.
- Changes in active principle of dried herbs due to storage were monitored at Akola under different storage conditions. Though moisture content of the material varied significantly, difference in andrographoloide content was found to be non-significant. However, andrographolide content was found to decrease with the storage period and significantly lowest content was recorded after 12 months period.
- At Akola, the spacing of 30cm × 15cm had significantly higher plant height, fresh and dry foliage yield. Significantly higher Andrographolide content was recorded at 15 days harvest after 50% flowering. The iron content was significantly higher with harvest at 50% flowering and the content successively decreased with the increase in harvesting time. The interaction effect of spacing and harvesting time was found significant for seed yield. Significantly higher seed yield than control was observed with the treatment combination of 30cm × 30cm spacing and harvest at 45 days after 50% flowering (209 kg/ha).
- At Faizabad, application of FYM at 7.5 tonnes/ha and harvesting at 150 DAS gave maximum economic yield in kalmegh.
- Method for estimation of Sennasoide content through calorimetry HPLC has been stadardised in Anand.
- Maximum andrographolide yield was obtained in 15cm × 15cm spacing and harvesting at 60 OAP, i.e., at flower initiation stage produced highest number of leaves and andrographolide at Anand.
- At Faizabad, on the basis of 3 years of experimentation data, it was concluded that application of FYM at 7.5 tonnes/ha and harvesting at 150 DAS gave maximum economic yield in kalmegh.
- At Akola, the spacing of 30cm × 15cm had significantly higher plant height, fresh and

dry foliage yield. Significantly higher Andrographolide content was recorded at 15 days harvest after 50% flowering. The iron content was significantly higher with harvest at 50% flowering and the content successively decreased with the increase in harvesting time. The interaction effect of spacing and harvesting time was found significant for seed yield. Significantly higher seed yield than control was observed with the treatment combination of 30cm × 30 cm spacing and harvest at 45 days after 50% flowering (209 kg/ha).

Kuth (*Picrorhiza kurro*)

An endangered temperate herbaceous perennial herb, it grows naturally throughout, alpine Himalayas between altitudinal range of 3,000 and 5,000m elevation. Traditionally, the roots are used to treat anemia, jaundice, abdominal pains etc. Propagation is done by seeds as well as stolon cuttings.The plant belongs to family Scrophulariaceae. It is a prostrate perennial herb with bitter rhizome and leaves. Root and stem are used for medicine purpose. Rhizome is used against fever and gastralgia. The powdered rootstock is used in abdominal pains, liver ailments and also as laxative.

- Solan centre maintains about 25 germplams of *Picrorhiza kurroa.*
- At Solan, detailed reproductive biology study has been conducted. Cytological study showed that this species is diploid with (2n) = 34 chromosomes and the species as genomic allotetraploid with 2n = 34 chromosomes. The flower structure and response to selfing and open-pollination studies revealed the species to be a strongly cross-pollinated one.
- Transplanting should be done during May-June with planting density of 1,10,000 plants/ ha. Harvesting after two years, during September-October, yields 450-600 kg/ha (seed raised) and 850-1050 kg/ha (vegetatively propagated) of dry rootstock.

Lemongrass (*Cymbopogon flexuosus*)

The perennial grass belongs to the family Poaceae. It is an aromatic plant and owes its name to the strong lemon like aroma of essential oil derived from the leaf. It is cultivated for its essential oil, and is extensively used in perfumery, flavouring and medicine. It also has antibacterial properties. The species is considered to have originated in India. Warm humid climate favours better plant growth.

- Evaluation of improved genotypes carried out Faizabad and Hissar showed that NLG-84 was found the best in terms of herbage yield, oil content and oil yield at Faizabad. Whereas at Hissar, Hl-7 showed highest herb yield; HL-11, highest oil content; HL-2, highest per centage of citral.
- Twenty-one genotypes of lemongrass genotypes were tested at Hissar and highest herb yield per plant was found in HL-10, followed by HL-9 against the best check OD-58. The genotype CKP-25 recorded the highest oil content and oil yield.

Liquorice (*Glycyrrhiza glabra*)

Liquorice is a hardy herb or undershrub with thick rootstock having many branched stems or roots and belongs to the family Papilionaceae. The crop is grown well under Haryana condition. The dried, peeled and unpeeled underground stems and roots constitute the drug which contains glycyrrhizic acid is responsible for therapeutic action and used as expectorant.

Liquorice extracts are also used as a condiment or as flavouring for tobacco, confectionery, beverages, jams and marmalades. Ammoniated glycyrrhizin, a product of liquorice extract, is used as flavour in sweet sauces, reconstituted vegetable proteins, chocolates, chewing gum and diet colas.

- Haryana Mulhatti-I (EC 114303) has been released from Hisar for general cultivation.
- Two accessions, E.C. 21950 and E.C. 25371, at Anand and three accessions, HM1, EC120170 and EC 19741 at Hissar are being maintained.
- At Hisar, fresh stolon yield was higher in genotype HMK-6-2 followed by HMK-1-3, HMK-7-1, HMK-7-4, HMK-1-2 and HMK-7-5 against check HM-1 (94.44 q/ha).
- It has been established that thin (0.1-0.5 cm dia) roots of two years old crop contained more than 9% glycyrrhizic acid compared to 5%, in such roots of one-year-old crop. It was further revealed that the thinner roots contained more glycyrrhizic acid compared to the thick ones of same age group. Glycyrrhizic acid content increased with increasing age and maximum content was recorded after 2 years of plant growth.
- Variation was prominently observed in germination with respect to sowing time. The germination was maximum in the last week of June sowing (57.5-91%) followed by July (52.8–68%) and January sowing (55.2–64.2%). Maximum germination of 23.8% was recorded with application of NPK (40:40:20 kg/ha) dose,s followed by 10 tonnes FYM + 10 kg Azatobactor per hectare (21.2%).
- Comparison of glycyrrhizin content in the thin (≤0.5 cm dia.), medium (0.5 to 1.0 cm dia.) and thick (1.0-2.0 cm diameter) roots collected from 1–3 years old crop indicated that active principle was reduced with root thickness during first two years crop. However, glycyrrhizin content increased with thickness in the three years old crop.
- At Hissar, its yield were 7662 kg/ha (January), 6853 kg/ha (Febenruary) and 5261 kg/ha (March) in different sowing dates, whereas plant spacing of 75cm × 45cm (6836 kg/ha), and 90cm × 30cm (7428 kg/ha) were performing better over 75cm × 30cm (5962 kg/ha) and 90cm × 45cm (61.40 kg/ha).
- Solar drying was found to be better method of drying liquorice roots compared to sun or shade drying as roots contained higher glycyrrhizin (9.8%) in this treatment.
- At Hissar, maximum fresh and dry stolon yield was obtained from the crop during June 20, 2003 planting followed by July and January, plantings when uprooted after two years. The interaction effect between date of sowing and plant spacing for both fresh and dry stolon yield indicated that on June 20 planting with 90cm x 30cm produced maximum yield.
- Intercropping studies in liquorice indicated that during *kharif* season, crops like green gram, clusterbean, cowpea, pearlmillet, ricebean, tulsi, kalmegh and ashwagandha could be cultivated at Hissar region.
- Fresh and dry stolon yield were higher in inorganic treatment of NPK at 40:40:20 kg/ha than treatment of FYM + PSB + Azotobactor at Hissar. Maximum glycyrrhizin content of 8.2% in NPK treatment was statistically at par with the FYM + PSB + Azotobactor (8.0%).

- Agro-technique has been developed at Solan. Meristem culture for rapid multiplication and harvesting procedure were developed at Anand. Viability of stolon cuttings can be preserved at least for 8 days if wrapped in moist gunny bag.
- Anaytical procedure and estimation method for estimation of monoammonium glycyrrhizinate and 18 B glycyrrhizinic acid, DGL crude fibre content and gllycyrrhizins has been developed.
- At Hisar, seeds treated with concentrated sulphuric acid for 150 seconds resulted in 60.6% germination. Variation was prominently observed in germination with respect to sowing times. The germination was maximum in June last week sowing (57.5–91 %).

Long Pepper (*Piper longum*)

Long pepper is shade-loving climber belonging to the family *Piperaceae*. The plant is naturally found growing in warmer parts of the country in deciduous to evergreen forest. The matured female spikes and roots are used as drug in Indian System of Medicines against diarrhoea, indigestion, jaundice, urticaria, abdominal disorders, hoarseness of voice, asthma, cough, piles, malarial fever, worms, sinusitis, cough, bronchial asthma and other respiratory disorders. The crop is in commercial cultivation in well-drained soil and humid climate in the states of Maharashtra, West Bengal, Kerala.

- At Trichur, 67 germplasm are being maintained. The evaluation of long pepper has resulted in an identification of bush type accession with an average height of 90cm with higher number of fruiting branches-medium sizes spikes of 2.4 cm mean length, 2.2cm circumference. Leaves are heart shaped purple tinge at nodes and an average yield of 651kg/ha dried spikes with a ratio of fresh to dry 3.96:1. A rigorous evaluation of germplasm accession of long pepper enabled us to release a high yielding variety adapted to irrigated coconut gardens in 1996 under the name "Viswam". *In-vitro* mutation breeding was also initiated in long pepper and promising results were obtained.
- Eleven selected genotypes were studied at Trichur and found that accession 8 is significantly superior among the selected accessions as well as hybrids for spike yield. For the piperin content, accession 4 was better. It was also found that tall plants with maximum number of fruiting branches will produce higher spike yield. Size of spike has positive association with piperin content. In an another evaluation of selected genotypes including hybrids showed that Accession No.2 and Accession No.7 recorded significantly higher dry spike yield compared to the check variety Viswam.
- The progeny of cross, Kanjoor × Nilambur cross showed significant variability for length of leaf, width of leaf, internodal length and vegetative branch, in comparison with parents.
- The crop production trails in long pepper have revealed that Azosdririllum, PSB and VAM were found to have a positive influence on most of the root characters. High-density planting at 30/90cm × 30cm combined with application of vermicompost and mulching resulted in maximum dry spike yield under irrigated conditions.

Madhunashini *(Gymnema sylvestre)*

The plant is a member of family Asclepinaceae. It is a stout woody climber distributed through India. The plant is considered antiperiodic, diuretic and stomachic. The leaves are believed to be hypoglycaemic and are an important ingredient of Ayurvedic formulations for diabetics. Gymnemic acid present in the plant is responsible for its therapeutic actions.

- Hisar centre is having about 9 germplams accession of *Gymnema sylvestre.*
- Studies revealed that in *Gymnema sylvestre* (Madhunashini), semi-hard wood laterals characterized by 50% browning of stem, was excellent for vegetative propagation.
- At Anand, three IBA treatments and three NAA treatments were tried. The length of longest root was recorded maximum (12.82 cm) with 600 ppm of IBA followed by 400 ppm of IBA (11.5 cm).
- Riboflavin, a cheap vitamine source, was found to be effective in increasing rooting in cuttings of Gymnema.

Matricaria (*Matricaria chamomilla* Linn.)

It is a member of Asteraceae family and a herb of great medicinal and aromatic importance. The crop has wide adaptability. It flowers abundantly in area experiencing mild cool climate during early stages of growth and warm dry weather at maturity. The dried flower on distillation yields essential oil that is used in perfumery as antiphlogestic, remedy for gastro-intestinal troubles and flavouring agent. It also has medicinal properties and used as expectorant, carminative. Suitable for growing in hills up to 1500m elevation. Flowers of this species are important source of chamazuline rich essential oil used in perfumery, flavouring and as a therapeutic agent in relieving spasmodic pains etc. Humus rich soil with an optimum pH of 7 is suitable for its cultivation.

- Four germplasm accessions of this species are being maintained at Solan. It is propagated by seeds. Seed sowing is done in November/December and transplantation should be done at 30cm × 30cm spacing in of February. Fully opened flowers should be harvested in April/May. Fresh flower yield is about 5-6q/ha with essential oil ranging from 0.3–1.3%.
- At Solan, Endosulfan (1.5 ml/litre) was found to be effective in reducing the population of bugs and increasing flower yield, followed by malathion and deltamethrin (1.0 ml/ litre)

Melissa *(Melissa officinalis)*

The plant is a temperate, aromatic herb belongs to the family Lamiaceae. The essential oil of this plant is used in perfumery. Leaves and stems are bitter and used against diseases of brain, heart and liver.

- Essential oil from leaves is used in flavoring and pharmaceutical products. Externally it is used in rheumatism and neuralgia. The species is propagated through stolon cuttings planted in February at a spacing of 30cm × 30cm. Fresh aerial herbage is harvested before flowering yielding about 110q/ha with essential oil content ranging from 0.12–0.29%.
- At Solan, application of NPK at 90:60:60 kg/ha and harvesting at 270 and 60 days thereafter was found optimum for getting higher herb yield, oil content and oil yield.

Mentha *(Mentha* spp.*)*

It is a member of family Lamiaceae and grows well in wetlands. The plant is an annual aromatic herb cultivated for its herbage. The herb yields essential oils containing d-sylvestrene, d-carvone and citronellol. It is used for stomach disorders, fever, cough and cold. It is also used in commercial preparations of mouthwash.

- In 1991, Solan centre released a variety, Punjab Spearmint-1 for commericial cultivation in sub-temperat and subtropical region of our country. It has a herbage yield potential of 96 q/ha and oil content ~ 0.50 %.
- Japanese mint (*Mentha arvensis*): varieties Gomti and Himalayan are the best varieties yielding maximum oil, followed by Shivalik at Hisar.
- The 10 genotypes of *Mentha* species were evaluated and EC-41911 have been identified for maximum herb and oil yield at Faizabad.
- Carvone rich essential oil derived from its herbage is used in confectionary, cough syrups and other therapeutic products like antiseptics, carminatives, stimulants, etc. It is propagated through stolon cuttings. Average fresh herbage yield is about 200-250q/ha, which gives 75-80kg/ha of essential oil.
- At Solan, five insecticides, viz. Endosulphan 0.05% (Thiodan 35 EC), Malathion 0.05% (Malathion 50 EC) Deltametharin 0.0025 % (Decis 2.8 EC), lantana seeds (8.0 %) and econeem (0.009%) were tested for the control of *Helicoverpa armigera* (Hub) and *Plusia orichalcea* Fab of *Mentha spicata*. Deltametharin (1.0 ml/litre was found to be effective in controlling both the species of the pest and increasing the yield of the crop.

Mucuna (*Mucuna pruriens*)

The plant belongs to the family Leguminaceae. Seeds are used in the Indian system of medicine against nerval disorders, spermatorrhoea, diabetis, etc. L-DOPA present in the seeds is used against Parkinson's disease. Well-drained clay loam soil is suited for its cultivation.

- Solan centre has identified a mucuna variety 'IC-127362' suitable for subtropical condition. It is white seeded with non stingy pod hairs with seed yield = ~ 697 g per plant and L-DOPA content = ~ 5.47 %.
- The white seeded accessions of *Mucuna* species performed far better as compared to *Mucuna pruriens* (black seeded) accessions in L-DOPA yield/plant (39.73g/plant in IC-127363 (white seeded) and 4.69g/plant in Punba collection (black seeded). However, the L-DOPA content was maximum in Srahan collection (black seeded) with 7.3% as compared to 5.76% in IC 127363 (white seeded accession).
- Propagation is through seed and seed rate is 40–45 kg/ha. Application of 15-20 tonnes/ ha FYM, 80-kg/ha phosphorus and 80-kg/ha potassium can be applied as basal dose. Nitrogen (100 kg/ha) is applied in three equal splits, first as basal, second after 30 days of germination and third after 60 days of crop age. The seeds are dibbled at a spacing of 90cm × 90 cm. Since it is twinning annual, so vines are provided support with the help of sticks. Seed yield is 15–40 q/ha. Under irrigated conditions when good support is provided yield of 50 q/ha can be obtained.

Opium Poppy (*Papaver somniferum*)

Opium poppy is an erect annual herb about 1 m high with long stalk and globular capsule and belongs to the family Papaveraceae. Opium poppy is commercially cultivated as *rabi* season crop in specific pockets of Madhya Pradesh, Uttar Pradesh and Rajasthan through licensing from Central Bureau of Narcotics, Government of India. The latex derived from the capsule contains many medicinally important alkaloids. Several analgesic, sedative, antispasmodic, hypnotic and anaesthetic drugs are manufactured form the opium latex, seeds also find its use for culinary purposes. Conservation and characterization of genetic diversity, development of improved cultivars and agro-technology for them, management of downy mildew disease, etc., are major research areas of this crop.

- Faizabad, Mandsaur and Udaipur centres are maintaining opium germplasm (Table 8.2).
- Constant breeding efforts were made AICRP on MAP, the fist variety Jawahar Aphim 16 was commercially released from Mandsaur centre during 1984. Subsequently, seven more varieties were released and these are: Kirtiman, Jawahar Opium 39, Jawahar Opium 540, Chetak Aphim, Trishna, NOP-1 and Ghazipur Local (Table 8.4).
- An opium poppy variety namely "Chetak Aphim" is developed by Udaipur centre is a short plant stature (75cm–82cm) a composite variety with complete white coloured non serrated petals, peduncle is non-hairy, capsule at the maturity turn cream coloured with brownish ting. The capsules become ready for lancing about 115–120 days and seeds are harvested 135–140 days after sowing. Its latex yield is about 55 kg ha^{-1} with morphine content of 12.5% and seed yield 9-10 q/ha. It is moderate resistant to downy mildew.
- From Faizabad based on the evaluation of germplasm, IC-42 has been identified and released at national level. The mass selection has been done in local population and two genotypes namely NOP-1 and NOP-4 has been identified. NOP-4 has been relesed as Kirtiman for Uttar Pradessh. The genotype NOP-1 has been identified as moderately resistant to downy mildew. The downy mildew resistant genotype Ghazipur local has been isolated and used as resistant parent to incarporate the resistancy. NOP-204 and NOP-449 with high latex yeilding have been idetified.
- At Faizabad centre, about 104% latex yield increase in hybrid opium poppy developed by crossing inbred lines (ND-20 × NOP-4) suggesting that the current yield barrier could be removed and making available hybrid varieties for cultivation. To develop hybrid variety in opium poppy new hybrids are being developed from the centres and are being evaluated.
- At Udaipur, initial studies on morphine content of flower petals of opium poppy variety chetak aphim showed that the morphine content was minimum in those petals which are likely to open next day (0.26%). It gradually increased in the petals collected on first day (0.46%), second day (0.55%), third day (0.64%) and fourth day (0.82%) of opening probably due to increased mobility of morphine due to certain metabolic activities. But it decreased in the petals collected on the day of falling from the flower either forth day or fifth day of opening (0.62%).
- Sowing of opium poppy from 30 October to 9 November 9 at 30cm × 10cm spacing using 5 kg seeds/ha and 10 tonnes FYM, 120 kg N and 40 kg P_2O_5 /ha is appropriate in Udaipur region. Splitting of nitrogen (¼ at sowing, ½ at rosette and ¼ at pre flowering

stages) gives higher yields. Growing of legume either as green manuring with sunhemp or groundnut for seed production during *kharif* followed by poppy in rabi proved economically beneficial. Irrigation at 5.0 cm depth gives 42.8% higher seed yield than irrigating at 7.5cm depth. Alternate day lancing produced 8.9 per cent higher latex compared to daily lancing. Late sown crop suffers more due to diseases and yield is substantially reduced.

- At Udaipur, irrigations were given to opium poppy at different stages, viz. stem elongation 30 days after sowing, rosette, bud, flower initiation, 50% flowering, late capsule, capsule maturity and at after lancing. The critical stages of irrigation were at rosette stage, stem elongation, bud and 50% flowering. Further, moisture stress at various physiological stages did not influence morphine, codeine and thebaine content of opium gum.
- The results of experiment on integrated nutrient management in opium poppy-Ashwagandha crop rotation involving organic sources at Udaipur indicated that increasing levels of FYM from 5 to 15 tonnes/ha significantly increased leaves per plant, gum, seed yield, capsule husk yield, morphine and thebaine content of gum. However, response of applied FYM was noted upto 10 tonnes/ha for plant height and capsules/plant. Codein content of gum was not influenced by FYM application.
- Integrated approach of sulphur nutrition (40 kg/ha), seed treatment with Apron 35 SD at 8g/kg and application of systemic (0.2% Ridomil MZ) fungicide at 35, 55 and 75 DAS for management of downy mildew at three locations, viz. Faizabad, Mandsaur and Udaipur produced lowest disease severity and highest latex yield. At Udaipur, the best treatment was also proved the most profitable.
- At Faizabad, the maximum latex yield (24.37 kg/ha) was recorded with 75% vermi compost + P and K. The maximum seed yield (6.76 kg/ha) was recorded in 75% FYM (N) + doses of P & K. At Mandsaur, the highest latex yield of 55 kg/ha was recorded under RD ($N_{150} P_{75} K_{40}$ kg/ha).
- Intergrated downy mildew management with a combined treatment of seed treatment with Metalaxyl at 6 g/kg seeds and 3 spray of Metalaxyle + Mancozeb at 0.2% at 35, 55 and 75 days after sowing significantly reduced primary and secondary infection of downy mildew at Mandsaur. Treatment also increased the latex, seed as well as husk yields.
- At Udaipur, Downy mildew foliar (localized) infection in opium poppy appeared and progressed at maximum and minimum temperatures ranged between 22.4 to 29.7 °C and 4.0 to 10.6 °C, respectively, coupled with relative humidity from 71 to 91%, irrespective of date of sowing of poppy. Maximum yield of latex, seed and husk was recorded in sowing done on 30 October, which was followed by 6 November sowing.
- Soil application of *Trichoderma viride* during *kharif* season as well as before sowing and seed dressing with the bio-agent significantly reduced the primary infection of downy mildew at Faizabad and Mandsaur. This resulted in significant increase in latex, seed and husk yield in the treated plots. At Mandsaur, treatment caused 13.31% increase in latex yield while 40.67% increase was recorded at Faizabad. The mycelia of *Trichoderma* were found to infect, penetrate and destroy the oospores of downy mildew.
- At Faizabad, it was noticed that within 72 hour of incubation remnants of dormant mycelia

of downy mildew organism – *Peronospora arborescence* started germinating by producing tufts of hyphae. These were responsible for initiating the primary infection. Secondary infection started with sporangia formed on the blighted spots over test seedlings. Oospore germination was also accomplished *in vitro*. About 20% of the oospores developed a sack like structures in which zoospores were formed. A small pore developed on the oospore cell wall through which the zoospores were released. In rest of the oospores protoplasts were detached from the cell wall and plasmolysed.

- Artificial inoculation technique to reproduce downy mildew symptoms was standardised at Faizabad. For primary infection heavily blighted leaf was used as inoculums in the pot and micro plot experiment. Pot inoculation was found to be better with higher success rate (62.57%). In case of secondary infection, seedlings were inoculated with conidia and disease symptoms were satisfactorily reproduced under net house conditions.
- Comparison of morphological features and biochemical constituents of both susceptible (Jawahar Aphim-16) and resistant (Ghazipur Local) cultivars of opium poppy at Faizabad showed that upon inoculation carbohydrate content decreased faster in susceptible compared to resistant cultivar. Similarly, in susceptible cultivar there was continuous decline in phenol content in the first 18 hour after inoculation, whereas there was significant increase in phenol content after 6 hour of inoculation in resistant cultivar and it was maintained up to next 6 hour Later the phenol content reduced and reached the normal level 24 hour after inoculation. Further, in susceptible cultivar polyphenol oxidase (PPO) activity showed steady decline after inoculation while in resistant cultivar, PPO activity showed increasing trend 12 hour after inoculation and reached to maximum level at 18 h after inoculation.

Palmarosa (*Cymbopogon martini* var. *motia*)

Palmarosa is a tall perennial tufted hedge, native of most parts of subtropical India and belongs to the family Poaceae. It is an important aromatic plant cultivated for its essential oils extracted from floral shoots and aerial parts. The palmarosa oil is rich in geraniol, which is used in perfumery, soap and cosmetics. It is native of most parts of subtropical India and grows well in a wide variety of soils which are devoid of water logging.

- Thirty-one clonal selections were made in Hisar. Wide range of variability was observed in selected clones for various morphological and yield parameters. Highest oil yield was in genotype RH 03-35 followed by RH 03-13 and on the basis of two cuts (harvesting), genotype RH-03-35 recorded the highest oil yield (291.11 litres/ha) followed by RH-03-30 (282.13 litre/ha) and RH-03-62-1 (280.8 litre/ha).
- In Palmarosa the improvement programme at Faizabad the germplasm IW – 31245 has been identified as good oil yielder with good geraniol content .

Patchouli *(Pogostemon cablin)*

Patchouli is bushy *herb* of the mint family, with erect stems, reaching two or three feet (about 0.75 m) in height and bearing small pale pink-white flowers. The plant is native to tropical regions of Asia and is now extensively cultivated in Caribbean countries, China, India, Indonesia, Malaysia, Mauritius, Philippines, West Africa and Vietnam. The scent of patchouli is heavy and strong. It has been used for centuries in perfumes and continues to be so today.

Pogostemon cablin, P. commosum, P. hortensis, P. heyneasus and *P. plectranthoides* are all cultivated for their oils and all are known as 'patchouli' oil, but *P. cablin* is considered superior. Leaves are harvested several times a year, and dried herb is exported for distillation of the oil.

- Investigations in under this project at Trichur have opened new vistas for successful cultivation of patchouli accessions in coconut garden as intercrop.

Podophyllum *(Podophyllum hexandrum)*

This species is belongs to the family Berberidaceae, also known as Bakrachamaka in Sanskrit, is traditionally used as purgative. It came to the limelight recently due to podophyllotoxin present in the resin of the whizome and root having tumour nectotising property. It is deistributed in the inner ranges of Himalayas from Kashmirs to Sikkim in elevation range of 3,000–4,000 m. It grows well in humus rich soil under shade.

- Extraction and estimation of podophyllotoxin (active principle of *Podophyllum*) has been done. It was separated from the mixture of lignans extracted by soxhelet appratus the rootstock (roots and rhizomes) in methanol followed by acidification and TLC.

Rosemary *(Rosmarinus officinalis)*

The plant is a small herb cultivated for its essential oil. It belongs to the family Labiatae. It is grown in comparatively cooler climate. Oil is extracted from the flowering top. Rosemary possesses stimulant and carminative antimicrobial properties due to its volatile oil. The oil is also extensively used in perfumery

- Standardization of propagation both by vegetative and sexual means was tried.
- Rosemary leaf essential oil is used in perfumery, cosmetics, flavouring meat dishes, etc. It is a perennial shrub and can be propagated both by seeds and shoot cuttings. The rooted plants are field transplanted during rainy season at a spacing of 90cm × 45cm. Fresh leaves from green branches are harvested giving an approx. yield of 150q/ha with oil content ranging from 1.5–2.0%.

Safed Musli *(Chorophytum borivillianum)*

Safed musli is an herbaceous pant widely distributed in the forests of Maharashtra, Gujarat, Rajasthan and Madhya Pradesh and belongs to the family *Liliaceae*. The dried root powder which contains saponins as active principles is used as tonic and aphrodisiac drug. Extensive collection of the material from the forests has dwindled the natural population and the growing demand of this crop leads to initiation of its commercial cultivation in *kharif* in places having warm and humid climatic conditions and adequate soil moisture throughout crop growth. Conservation and characterization of the genetic diversity, development of proper cultivation technology, easy quality estimation methods, post-harvest technologies, etc. are the major research areas in this crop.

- Germplasm of this species are being maintained at Anand, Hisar, Mandsaur, Udaipur, Indore and NRCMAP, Anand (Table 8.2).
- Variety 'Jawahar Safed musli 405' has been released from Mandsaur in 2004.
- Thirty three accessions were collected, maintained and evaluated at Indore. Each seven improved lines at Indore and Udaipur and 18 lines at Mandsaur were evaluated for various yield characters. At Indore, CBI-3 and MCB-412 were found promising. At

Mandsaur, MCB-405 (for fresh root yield and steroid content) and MCB-409 (dry root per plant) were found promising. MCB-405 also found superior at Udaipur in terms of fresh root yield.

- In safed musli, a wide spectrum of variability was observed in terms of qualitative and quantitative characters including fasciculated root whose multiplication ratio was 1:14. At Mandsaur, superior genotypes for fasciculated root yield and sapogenin content identified were: MCB-402, MCB-403, MCB-407, MCB-408, MCB-409, MCB-412 MCB-414 and MCB-418. At Udaipur, genotype RC-77, exhibited higher fasciculated root yield over the best check RC-64. AMVI-II had the highest yield and the multiplication rate at Anand. Whereas, nine genotypes, viz. RC-77, PC-2, RC- 64, PC-3, PC-25, PC-21, PC- 15, PC-31 and PC-30 at Udaipur and four genotypes, viz. CBI-7, HCB-6, HCB-4 and MCB-414 had significantly higher fasciculated root yield over the best check variety.
- Safed musli (in *kharif*) can very well fit in the cropping system with Asalio (in *Rabi*) with the highest net profit (Rs 5 lakh) with a cost: benefit ratio of 1:2.49 at Anand.
- In safed musli, it was established that harvesting should be done according to the usage of the fleshy root. And for drug purposes, early harvesting should be practiced to avoid loss in active principle. Whereas, fleshy roots meant for propagation should be harvested later to avoid loss due to drying.
- Morphological and biochemical characterization have been done in two species of safed musli, viz. *Chlorophytum borivilianum* and *Chlorophytum tuberosum. Chlorophytum borivilianum* contain higher amount of saponin (4%) and sapogenin (0.18%) as compared to *Chlorophytum tuberosum* (1.6 and 0.05%, respectively).
- In *Chlorophytum borivilianum,* in a study it has been established that elimination of inflorescence increased the number of fleshy root and as a result 41% increase in yield has been recorded compared to non-topped plants.
- In *Chlorophytum borivilianum,* chemical finger printing has been done to characterize qualitative comparison of the saponin profiles by HPTLC. It was found that two saponin fractions were prominent in two species of *Chlorophytum.*
- Seed germination in safed musli is too poor. Studies conducted at Anand revealed that for (increased) flowering, bunches with 4-5 fasciculate roots and for multiplication of fasciculate roots bunches with 2-3 fasciculate roots should be planted. It must be ensured that at the least a piece of crown remains attached to bunches. A single tuber with crown is good for maintaining the species/accessions. Removal of inflorescence (topping) improved the bulking of fasciculate roots which ultimately increased the yield.
- At Akola, the fresh and dry roots yield were significantly higher with dense plant population of 3.33 lakh (spacing 30cm × 10cm) per hectare. Application of organic manures at FYM 20 tonnes or 5 tonnes of vermicompost/ha significantly increased the fresh and dry roots yield and also the steroid content.
- Chlorosis was found to reduce number of leaves and fresh root yield per plant in safed musli at Akola region. The nutrient status in the leaves of chlorotic plants was also poor particularly in respect of iron, potassium and nitrogen contents as compared to green

plants. Different chemical applications significantly influenced the chlorophyll content of the leaves. Maximum chlorophyll content was obtained from humic acid complex+ ferrous sulphate applied plants. The foliar application of humic acid + urea, humic acid + ferrous sulphate and humic acid complex significantly increased the number of root per plant. Similar results were also obtained in terms of number of roots and fresh root yield.

- At Akola, the studies on nutrient uptake behaviour showed that the dry matter accumulation rate was rapid to very rapid from 30 days growth period up to 60 DAS, which was on account of increased number of leaves and length and size of the leaves. The content of macro as well as micronutrients was found to decrease with the advancement in the plant growth period and the lowest content of all the nutrients was noticed at physiological maturity stage (90 DAS). The major uptake (73-85%) occurred up to 45 days growth period.
- The highest fresh root yield was recorded with the harvesting at 180 DAP, however, the dry root yield was significantly highest with the harvesting at 240 DAP at Akola. The saponin content decreased successively with the increase in harvesting time and significantly lowest content was recorded due to harvesting at 240 DAP. However, significantly highest saponin yield was produced by the harvesting at 180 DAP.
- Post-harvest deterioration is another problem in safed musli. Reduction of saponin content in safed musli root powder was noticed after 6 months and it was significantly lowest at 12 months storage period and storage method of sun drying + open container at Akola.
- At Akola, it was observed that the saponin content was significantly lowest in the peeled roots, indicating the loss of saponin due to peeling of roots and subsequent washing with water. The highest magnitude of loss in Saponin content was 19.52%.
- At Udaipur, separation of roots for planting should be done around April 20. Optimum weight of each root pair^{-1} used for planting is 3g. Planting on top of 7" high ridges at plant density of 3.33 lakh/ha by adopting 30cm × 10cm spacings with 30 tonnes FYM/ ha gives higher yield. Soil application of elemental sulphur 150 kg/ha before 15 days of planting is beneficial to control chlorosis, when grown on calcareous soils. Dipping of unsprouted fasciculate roots for half an hour before planting in GA_3 (50 ppm) or kinetin (100 ppm) solution increases sprouting and ultimately yield by 16%. At Anand, 150 ppm cycocel application increased root length significantly over the control.
- A safe storage technique of planting material (from April — June) is treating the separated fasciculate roots with 0.1% Bavistin + 0.05% Streptocyclin solution for 20 minutes followed by drying for 4-5 days on blotting papers in shade and then storing in wooden boxes containing 4" layer of musli overlapped by 4" layer of soil. Root rot can be minimized by treating the roots using 10g Bavistin/kg roots before sowing. Treating the rootlets with *Trichoderma* mixture at 200g/kg rootlets also proved beneficial. The *Trichoderma* should be a resident isolate.
- At Akola, significantly highest root yield was recorded due to application of 5 t ha^{-1} vermicompost in 30cm × 10cm spacing. At Udaipur, increasing plant density from 1.66 to 3.33 lakh/ha significantly increased root yield. Response of FYM was recorded up to

30 tonnes/ha.

- Topping significantly increased the average length of leaves, number of tubers, weight of tubers and yield kg/ha when compared to non-topping. Topping increased the fleshy root yield by 41% over the non-topping.
- At Udaipur, significant loss in saponin content was noted during post harvest storage in safed musli roots that were either dried in sun or oven dried and stored in open or airtight bags during one year period. However, moisture content of the powder stored in the open bags increased in August due to moist weather due to rains and decreased there after.
- At Akola, trial on organic manure and spacing revealed that to get higher yield and monetary returns, safed musli could be planted with plant spacing of 30cm × 10 cm along with the application of FYM at 20 tonnes/ha or 5 tonnes/ha of vermicompost.
- At Anand, application of 5 tonnes/ha of poultry manure gave highest yield (4,532 kg/ha) followed by 20 tonnes/ha of FYM (4,455 kg/ha) and no fertilizer (3,928 kg/ha). Effect of plant spacing (plant population) was found to be maximum with 30cm × 10 cm (3,33,333 population) spacing on the yield (449 kg/ha).
- At Mandsaur, application of FYM at 20 tonnes/ha recorded better growth in terms of root growth and fasciculated root yield. Narrow spacing of 30cm × 10cm gave higher fasciculated root yield and better root growth than wider spacing.
- At Akola, safed musli when harvested at 240 DAP recorded highest fleshy root yield (675 kg/ha, dry root yield) with maximum saponin content at 180 DAP (46.67 kg/ha). 20 tonnes/ha of FYM and 75 kg/ha of N could increase the fleshy root yield of safed musli along with increase in GMR, NMR and input-output ratio.
- At Anand, Yield was recorded to be 438 kg/ha when inflorescence was topped as compared to non-topping which was 338 kg/ha. Spacing of 10cm × 10cm gave highest yield of 719 kg/ha followed by 20cm × 10 cm and 30cm × 10cm spacing.
- At Mandsaur, FYM and poultry manure alone or in combination with biofertilizers improved yield in Safed musli and its variety MCB-405 proved better than MCB-412 with respect to fasciculated root yield.
- At Udaipur, sprout suppressant MH when applied at 2000 ppm had an effect on germination (35.2%) at 30 days after harvesting as compared to control. However, germination ranged from 5.3–55.4 % in all the treatments in the order of 1500, 2000 and 2500 ppm at 20, 30 and 45 DAH.

Sage (*Salvia officinalis*)

It is a perennial species belonging to Lamiaceae is valued for its highly aromatic leaves, which are widely employed for seasoning of poultry stuffings, meats, soups, sausages and canned foods. The plants are raised through cuttings.

- Rooted cuttings should be transplanted at a spacing of 60 cm apart on 20 cm high ridges, for avoiding water logging. Harvesting of leaves can be done after 4-6 months of transplanting. Only one picking of leaves should be done during first year and after which two to three pickings may be made in a season. The yield of dry leaves is about

300 kg/ha (first year) and 1,500–2,000 kg/ha (second year). The crop remains productive for 4-5 years. Essential oil yield varies from 40-50 kg/ha/year.

Clary Sage (*Salvia sclarea*)

A member of Lamiaceae, this herb is grown in temperate regions. The flowering spikes yield essential oil, which is used in perfumery, for flavouring alcoholic and non-alcoholic beverages, liqueurs, candy ice-cream etc. It is propagated through seeds.

- Nearly 600-800g of seeds is sufficient for raising one-hectare area. Seed sowing can be done during November month and transplanting during February month at a spacing of 30cm × 45cm. About 45–50 q/ha fresh flowering spikes and 13–15 litres/ha essential oil is obtained.

Senna (*Cassia angustifolia*)

Senna is an important export-oriented medicinal crop belongs to *Leguminacae.* Leaves containing sennosides are used as laxative and find place both in Indian system of medicines and modern pharmacopoeia. It is grown as rainfed crop and is highly sensitive to high rainfall and waterlogging. Hence, it is not recommended for high rainfall zone. This is suitable for the drier regions and traditionally cultivated in Gujarat, Rajasthan and Tamil Nadu. Development of high-yielding varieties and good agricultural practices are researchable areas in this crop.

- Senna is the second largest export earner among the herbal medicines and dried leaves are largely exported.
- At Anand centre, floral biological aspects of senna were studied. Experiments were conducted to harvest high leaf yield with high sennoside content. Monsoon picking of leaves on 90:110:130 DAS and during Summer 70:50:110 DAS gave the desired return. As sennosides content goes down drastically after flower opening a variety, ALFT-2 was developed which came to this stage in about 70 days as against 35 days in the check variety. By this variety not only the quality was enhanced but also biomass production was increased.

Chirata (*Swertia chirayita*)

A critically endangered species and highly prized in the Indian system of medicine, is valued as a bitter tonic, blood purifier, hepatic stimulant etc.

- It can be propagated by seeds. Seedling growth is slow hence 4-6 leaved seedlings are pricked into polybags for further growth. After about 8-10 months in polybags, the seedlings are field transplanted at a spacing of 30cm × 45cm. In second or third year, plants develop aerial shoots which bear flowers. After seed set the entire plant dies and the whole plant is harvested. Periodically some radicle leaves should also be harvested.

Mushkbala, Tagar (*Valeriana jatamansi*)

The plant belongs to the family Valerianaceae. It is a perennial herb. It is native to Himalayan regions of India. Rootstock is thick and aromatic and used in medicines for the cure of diseases related to eyes, liver, blood and spleen. The root is used in preparation of cosmetics and hair oil. The chemical constituents responsible for its uses are valeopotriate and the essential oil present in rootstock. Distributed in Western Himalayas between 1,200–4,000m MSL, it is valued in traditional medicine for treating hysterical fits, nervous unrest, emotional troubles, epilepsy etc.

- A variety 'Dalhousi clone' has been released from Solan centre in 1994 for commercial cultivation. It is a high valepotriates containing chemotype (under trial at present) gives dry rootstock yield = 8-9 q/ha and valepotriates content = ~ 4.0 % and highly suitable for sub-temperate situation. The seed/planting material is available in Solan centre.
- Twenty-five accessions of Valeriana were catalogued at Solan. Maximum rootstock yield per plant was in VJ-2K2-UHF-19 (17.80g) and minimum was in VJ-2K2-UHF-28 (2.60 g). Separation of essential oil and valepotrates simultaneously by simple method was also developed.
- The phenomenon of gynodioecism, i.e., presence of female and bisexual flowers on different plants in the species has been identified. Three leaf shape types named as leaf type-I, leaf type-II and leaf type-III were observed. Seeds collected from females in the first year segregated into females and males in the ratio 1: 1.4. In the second year the ratio was 1: 1.31. However, the seeds collected from the bisexual flowered plants segregated in the ratio female and male 1:2.1 in the first year and 1:2.06 in the second year. Valepotriate content in female plants was 4.64% and that of bisexual flowered plant was 3.66%.
- About 189 plants of *Valeriana jatamansi* were collected and evaluated at Bharsar and the germplasm were classified on the basis of leaf shape into two groups.
- At Solan it was found that non-significant difference in seed germination in seed produced by female, bisexual (open pollinated) and bisexual (self-pollinated) flowers. Roots contain more essential oil (~1.87%) in comparison to rhizome (~ 0.15%).Seeds produced by female flowered plants can be used for raising populations with more number of female plants, which can give higher valepotriates yield per unit area. Under open conditions, the root stock biomass is significantly higher in comparison to shade conditions.
- It was also found that seed produced by bisexual flowered plants give progeny in the approximate ratio of 1:2 (female: bisexual). Seed produced by female plants give progeny in the approximate ratio of 1:1 (female: bisexual). Female plants contain more valepotriates as compared to bisexual plants. However, no such difference was found in essential oil.
- The available germplasm (about 30) at Solan centre have been evaluated and catalogued. It was found that all the populations were characterized by the phenomenon of gynodioecism, i.e., female and bisexual plants occurring separately. These female and bisexual plants could be differentiated only at the flowering stage and no clear difference could be noticed at the vegetative stage. Also amongst the populations, no significant inter population morphological difference could be observed. Isolation experiments conducted indicated bisexual plants set seeds while, as expected the female plants when segregated from bisexual flowered plant did not set seeds. These plants, i.e., female plants when allowed to grow sympatrically with bisexual plants set abundant seed. The bisexual plants act as pollen donor for the female plants. The progeny studies of these plants resulted that the female plants segregated into female and bisexual plant in the

ratio of 1: 1.31 and the progenies raised from bisexual plants segregated in to female and male in the ration of 1: 2.06. Valepotriate content was also varied significantly in the female and bisexual plants.

- Seed propagation is preferred due to ease of germination. Seeds are sown in May-June, and three to four month old seedlings are ready for transplantation during rainy season at spacing of 30cm × 30cm at Solan. Commercial harvesting of rootstock is possible after two years and yield of 30–70 q/ha of fresh rootstock and nearly 8-15 q/ha of dry rootstock can be obtained.
- At Solan, maximum germination of seeds (98%) and survival of seedlings in the field (97.96%) were observed in medium with FYM + site soil. The combination of site soil + FYM is the best growing media for raising the seedlings of *Valeriana jatamansi.* Sowing of seeds of *Valeriana jatamansi* during June was found optimum for getting maximum germination and survival which ultimately gave best plant growth.
- At Solan, maximum fresh aerial biomass (g/plant) of 5.38, 9.68, 24.44, 65.42 and 98.64 was recorded when planted at ridges as compared to furrows, at level beds and at sloppy terrace at periodical intervals of 6, 9, 12, 15 and 18 months, respectively. Similarly, transplanting on 16 August recorded maximum fresh aerial biomass (g/plant) of 6.63, 12.86, 31.61, 65.73 and 108.91 as compared to transplanting on 31st August and 15th September. Maximum rhizome yield of 23.18 g/plant was harvested after 18 months of transplanting from ridge planting and a harvest of 26.07 g/plant could also be obtained from transplanting on 16 August.
- At Solan, maximum aerial biomass, fresh underground biomass, fresh root yield and fresh rhizome yield could be obtained from a combination of Azotobacter + PSB + VAM followed by a combination of other biofertilizers. Valepotriate content (2.39%) was recorded to be maximum when applied with Azotobacter at 10 kg/ha followed by Azotobacter + PSB (2.35) and PSB at 10 kg/ha (2.34). The organic carbon (0.13 %), N (361.96 kg/ha), P (38.20 kg/ha) and K (187.20 kg/ha) were also maximum in combination of Azotobacter + PSB + VAM. Similarly, nutrient content and nutrient uptake were also found maximum in this combination.
- Use of different growing media had positive effect and use of vermicompost in site soil (1:1) gave maximum plant height, fresh aerial biomass and fresh underground biomass after 18 months of transplanting. Valepotriate content was maximum when grown in soil + saw dust followed by soil + FYM and soil + sand + FYM.

Vasaka (*Adhatoda*)

Vasaka belongs to Acanthaceae family and is evergreen perennial plant and commonly distributed throughout India. Leaves are medically useful. Vasicine present in the leaves are responsible for the therapeutic action. It is used as an expectorant for treating colds, sough, chronic bronchitis and asthma. Two species, i.e., *A. zeylanica* and *A. neddomei* are normally used as source for raw drug.

- At Trichur, sun drying was found to retain maximum vasicine content in chopped plant material (1.16%). However, vasicine content of root was higher in both *Adhatoda zeylanica* and A. *beddomei* when dried under shade. In terms of retaining the vasicine content, storing the dry materials in polybags, powdered materials in polybags (A. *zeylanica)* or in steel containers (A. *beddomei*) and hot water extracts in glass bottles were found best.

Vetiver (*Vetiveria zizaniodes*)

Vetiver is a widely distributed native perennial aromatic grass of Indian sub-continent. The oil extracted from the roots of the plant is used in perfumery as well as in many ayurvedic preparations.

- Evaluation of advanced germplasm was carried out at Trichur. Results revealed that IC82469 showed highest root yield whereas highest oil yield was recorded in IC78651 (4.8 litre/ha).

MILESTONES OF PROJECT

- Location specific varieties for 34 medicinal and aromatic plants with their package and practices have been developed.
- Package and practices and four varieties of isabgol for different zones have been developed.
- Forecasting system for downy mildew of isabgol has been developed.
- In safed musli, it was established that harvesting should be done according to the usage of the fleshy root. And for drug purposes, early harvesting should be practised to avoid loss in active principle. Whereas, fleshy roots meant for propagation should be harvested later to avoid loss due to drying.
- In *Chlorophytum borivilianum,* in a study it has been established that elimination of inflorescence increased the number of fleshy root and as a result 41% increase in yield has been recorded compared to non-topped plants.
- Cultivation of Kalmegh has also been recommended during *kharif* as profitable crop at Anand condition and as rainfed crop in medium black soil of Indore region.
- Vegetative propagation method has been standardized through semi hard wood laterals characterized by 50% browning of stem in *Gymnema sylvestre* (Madhunashini).
- At Solan centre agro-techniques have been developed for medicinal and aromatic plants like *Dioscorea deltoidea, Gentiana kurroo, Valeriana jatamansi, Picrorhiza kurroa, Gloriosa superba, Hypericum perforatum, Mucuna pruriens, Salvia officinalis, Salvia sclarea, Lavandula stoechas, Matricaria chamomilla, Rosmarinus officinalis, Swertia chirayita, Melissa officinalis, Mentha spicata* and *Tagetes minuta.*
- In *Gentiana kurroo* (Indian Gentian), commonly known as *karu* or *kutiki,* Solan centre of AINRPMAP has successfully domesticated and standardized propagation techniques for large-scale cultivation.
- Reproductive biology of Indian Gentian (Gentiana kurroo) in detail has been worked out. Higher fruit set has been observed in open-pollination (80.57%) compared to the controlled selfing (7.25%).

- *In-vitro* multiplication of Sarpagandha (*Rauvolfia serpentina*) has been developed by Hisar centre for the micropropagation and conservation of this valuable threatened plant species.
- Cultivation of Muskdana (*Abelmoschus moschatus*) proved to be a better option as rainfed crop in the shallow black soils of Malwa plateau of Madhya Pradesh.
- In sacred lotus, (*Nelumbo nucifera*), 42 accessions from different districts of Kerala and Tamil Nadu have been collected showing wide variability in flower colour, petal number, presence of transitional whorls, carpels in the torus.
- In Arjuna (*Terminilia arjuna*), an important medicinal tree, a successful air layering technique has been developed by application of IBA (6000 ppm) which has increased number of primary and secondary roots per layer as well as mean length of longest root.
- Cataloguing of characters for available germplasm of isabgol, senna and guggal accessions have been completed at Anand centre.
- Hybrid vigour in opium poppy has been identified for making available hybrid varieties for cultivation
- In liquorice cultural practices has been standardized for quality and high-yielding.
- Extraction and estimation methods for active principle of many of the medicinal and aromatic plants have been standardized.
- Reproductive biology and cytology of most of the MAP have been documented.
- Package of practices have been developed for cultivation of kalmegh at Anand and the variety AFLT – 2 has been released for commercial cultivation.
- Kalmegh as single crop and in double cropping, kalmegh followed by cress, gave highest total income when compared to other crops individually and in combinations.
- At Anand centre, the agro techniques for the following species have been standardized for *Plantago ovata* (isabgol), *Cassia agnustifolia* (Senna), *Commiphora wightii* (Guggal), *Andrographis paniculata* (Kalmegh), *Lepidium sativum* (Asalio), *Chlorophytum borivilianum* (safed musli), *Convolvulus microphyllus* (Shankhpushpi) and *Withania somnifera* (Ashwagandha)

TECHNOLOGY TRANSFER AND EXTENSION SERVICES

This project has succeeded in creating awareness of the importance and the varieties play a key role in production. Over 90 improved varieties were released for cultivation in different spices. The participating AICRP centres were periodically conducting farmers' day and disseminate the new technologies to them by lectures, exhibits, group discussion, etc. Besides these centres were also published a good number of technical and extension bulletins for benefit of famers. Further, the scientists from these centres have written a good number of popular articles in the local new papers and magazines. Radio talks by the expert scientists were also been broad coasted from time to time. A list of extension activities performed by different centres is presented in Table 8.5. All these activities leaded to an overwhelming response from the Indian farmers, private parties and NGOs for cultivation of some of medicinal plants like aloe, ashwaganhda, isabgol, kalmegh, safed musli, senna, etc. on commercial scale. However, the progress is slow because of the the lack of information on the demand of any medicinal plant. For example at one point of time many farmers, private firms and NGO accelerated the cultivation of Safed musli and there was huge production. As a result the prices slashed down created a panic situation to safed mulsi growing farmers.

Table 8.5. Details of extension activities performed by different AICRP-MAP centres

	Akola	*Anand*	*Bharsar*	*Faizabad*	*Hisar*	*Kalimpong*	*Mandsaur*	*Solan*	*Trichur*	*Udaipur*
Radio talk		2		10	5		60	5	1	40
TV talk		3		5	3		10	2		30
Polular Artcile	26	5		15	10			25	2	20
Media report		2		5	3			10	1	60
Extention bulletin	2			2	1				1	
Technical Bulletin	5	4		1	2				-	
Farmers day	40	10		3	30		5	20	5	15
Trainings for farmers	20	31		12	7		3	18	10	20
Field demonstrations	10	15		10	20		20		5	> 50
Distribution of planting material				10					>5	>30
Posters				7	2				1	20
Exhibitions	>10			>9	10		10	>12	3	8
Lectures				1					1	>25
Folders	5	5		2	2		8		1	8
Pamphlets	5	4		2	3		20		1	20
Field visit on campus				20	12		20		6	
Field visit off campus				15	10		30		5	

Table 8.6. Seed/planting material production (2005-06) at Solan centre

Scientific Names	*Common Name*	*Seed (kg)*	*Rooted Plants/ Cuttings (No.)*	*Seedlings (No.)*
Achillea millefolium	Gandana, millfoil	0.050		
Aegle mormelos	Bael		*125*	
Artemesia annua	Malaria buti	0.050		
Cymbopogon wintrianus	Lemon grass		*150 (slips)*	
Chrysanthemum cinerariifolium	Pyrethrum	0.050		
Digitalis lanata	Tilpushpi	0.300		
Echinacea spp.	Echinacea	0.250		
Eclipta alba	Bhringraj	0.350		
Gentiana kurroo	Karu		*500*	
Glaucium flavum	Yellow horned poppy	0.050		
Gloriosa superba	Kalihari		*5,000 (tubers)*	

(Contd)

Scientific Names	Common Name	Seed (kg)	Rooted Plants/ Cuttings (No.)	Seedlings (No.)
Glycerrhiza globra	Mulathi		20	
Lavandula stoechas	Lavender	0.050	50	
Matricaria chamomilla	Babuna	0.050		
Mucuna pruriens	Kaunch	0.250		
Mucuna species (white)	Kaunch	0.250		
Ocimum basilicum	Bhabri	0.350		
Ocimum killimandscharum	Kapur tulsi	0.050		
Ocimum sanctum	Pooja tulsi	0.925		
Pelargonium graveolens	Scented geranium		50	
Rosmarinus officinalis	Rosemary	0.050	60	
Salvia officinalis	Sage		25	
Salvia sclarea	Clarysage	0.150		
Solanum khasianum	Kantkari	0.050		
Solanum laciniatum	Kangaroo apple	0.050		
Spilanthes acmella	Akarkara	0.300		
Stevia rebaudiana	Madhupatri		1238	
Tagetes minuta	Jangli gainda	0.100		
Tylophora indica	Damabuti		5	
Valeriana jatamansi	Tagar			1,800
Withania somnifera	Aswagandha	1.450	76	3,800

MITIGATION TECHNIQUES TO ADDRESS CHALLENGES

- Isabgol, which is a traditional crop of western Rajasthan (having semi-arid to arid climate and loamy sand to sandy soils) is made popular and cultivable in South-Eastern Rajasthan (having sub-humid to humid climate and clay loam to heavy clay soils) by manipulation in agronomic practices like sowing time and method, seed rate, fertilizer and irrigation management etc. suggested by Udaipur. Downy mildew is a very severe disease of opium poppy in this region; a prophylactic measure is suggested on the basis of research carried out at this center by manipulation in various agronomic practices such as sowing time, plant population, irrigation, fertilizer etc.
- Quality of ashwagandha roots of Rajasthan state was considered inferior by the traders because of high crude fibre and low total alkaloid content. Manipulation in agronomic practices were recommended by this center such as optimum sowing and harvesting times, minimal irrigation, spray of plant growth retardants, fertilizer management, etc., which improved its quality by reducing crude fibre and increasing total alkaloid content.
- A large area of South-Eastern Rajasthan is having calcareous soils, where cultivation of Safed musli is made possible by recommending soil incorporation of elemental sulphur at (150 kg/ha) before 15 days of planting by Udaipur centre.
- A safe storage technique of safed musli is invented by Udaipur centre to keep planting material safe during storage (April-June), which otherwise is damaged by drying and diseases infection. For this purpose, treating the separated fasciculated roots with 0.1% Bavistin + 0.05% Streptocyclin solution for 20 minutes followed by drying for 4-5 days on blotting papers in shade and then storage in wooden boxes containing 4" layer of musli overlapped by 4" layer of soil is recommended.

- Breeding for quality improvement is the most important challenge in terms of increasing yield and active ingredients of different medicinal and aromatic crops. Research on these aspects is in progress in many of the centres.
- Quality seed production/multiplication of planting material of important medicinal and aromatic crops is a need of moment. Most of the centres are working on this aspect also (Table 8.6).
- Research recommendations on various crops have been generated but its dissimilation at farmers' fields is very poor in few centres due to lack of manpower. The efforts are taken to work on this aspect also.
- ***Successful domestication of Swertia chirayita:*** *Swertia chirayita* commonly known as chirata, is a high value endangered temperate medicinal plant which finds use in traditional therapies. The species is valued for its bitter principles as hepatic stimulant, blood purifier, tonic etc. The species is on the verge of extinction due to massive unscientific exploitation from its wild natural areas. The situation is so grave that it has been classified as critically endangered and has been put on negative export list of Ministry of Commerce, GOI. Most of the domestic demand is met from imports from Nepal and other areas. Due to scarce availability, most of the drug chirata that is used by domestic AYUSH industries is adulterated by other *Swertia* species like *S. paniculata, S. angustifolia* etc. Considering the demand and importance of genuine chirata, work on exploration and domestication of *Swertia chirata* was initiated here under AINRP on Medicinal Plants. The dedicated efforts of the scientists have resulted in successful domestication and development of basic agro-practices of the species. We are now in a position to supply genuine planting material of this species.
- ***Increased seed set in Gloriosa superba:*** Seed of *Gloriosa superba* is the source of colchicine used in treating gout, rheumatism etc. However seed set in nature is quite low. In order to increase seed set for making annual harvests economical, a method based on controlled pollination has been developed. By this method the seed set can be increased almost three times the normal yield. Approximately 300-350kg of dry seed/ha can be produced annually. The seed has approximately 0.75% colchicine content.
- ***High-yielding valepotriates strain in Valeriana jatamansi:*** *Valeriana jatamansi*, suitable for mid-hill zones, is high in demand as a source of valepotriates. The species is being excessively extracted from wild habitats presently. Chemotype with approximately 4 per cent valepotriates content has been isolated and is at an advanced stage of trials.
- ***Valepotriates and essential oil recovery from rootstocks of Valeriana jatamansi:*** Valepotriates and essential oil, the two important products of the species, are present in both the roots and rhizomes of this species although with varying concentrations. Essential oil is mostly concentrated in roots only (~ 1.8%) with some traces in rhizomes also. However, valepotriates are present both in roots and rhizomes in almost similar quantities (~3%). In order to plan proper recovery of the two valuable products from this species, it is suggested that, if the objective is to extract both essential oil and Valepotriates, then rhizomes should be processed for valepotriates extraction and roots for essential oil. If, however, only Valepotriates are to be extracted then the entire rootstock, i.e., roots as

well as rhizomes should be processed.

- ***Extraction of solasodine from Solanum laciniatum:*** Solasodine is found in high concentration in dark green coloured berries (about 3.5%) than in ripen berries (<0.70%). This suggests that only dark green coloured berries should be harvested for getting maximum yield of solasodine. Solasodine is also found to the level of about one per cent in the leaves. Solasodine can be extracted upto the level of about 90% purity by using some manipulations in the extraction procedure.
- Farmers of Mandsaur region having scarcity of water. The medicinal crop like isabgol and chandrasur has been introduced in the cropping sequence to face this problem. Ashwagandha is multiple use medicinal crop. Very often the increase foliage growth affected the root quality. The use of malic hydrozide combate this problem and increased more returns to the farmers, as well as pharmaceutical industries.
- The raised-bed technique in safed musli increases the root quality and quantity.
- Opium poppy is an important case crop of Malwa plateau which affect the economy of opium crop every year. This crop heavily infected by downy mildew disease, which reduce the latex yield. Mandsaur centre has developed resistant variety namely JOP-540 and JOP-539 which save the crop from this disease to great extent.

BENEFIT REAPED BY FARMERS

- The area under betelvine has increased significantly from 33,000 ha to above 50,000 ha over a period of 40 years of AICRP.
- Profitability has gone up due to effective disease management and crop management using INM and IPM.
- New varieties developed have also become popular among farmers and these varieties have replaced old varieties in significant area under cultivation.

HORTICULTURE

IN DIFFERENT AGRO-CLIMATIC CONDITIONS

HORTICULTURE

IN DIFFERENT AGRO-CLIMATIC CONDITIONS

— FOUR DECADES OF COORDINATED RESEARCH

(Volume• II: Vegetables, Potato, Tuber Crops, Mushroom and Floriculture)

H.P. Singh
M.S. Palaniswami

WESTVILLE

WESTVILLE PUBLISHING HOUSE
NEW DELHI

H.P.Singh and M.S. Palaniswami (2010) Horticulture in Different Agro-Climatic Conditions
— Four Decades of Coordinated Research (Volume II) pp 352, Westville Publishing House, New Delhi

First Edition 2010

ISBN 978-81-85873-51-0 (Vol. II)

[redacted]

Published by Mrinal Goel, Westville Publishing House
47, B-5, Paschim Vihar, New Delhi – 110 063
Tel: 011-25284742, Fax: 011-25267469
Email: westville_2002@yahoo.co.in
Printed at Salasar Imaging Systems, New Delhi

Foreword

All India Coordinated Research Project (AICRP) under National Agricultural Research System is an unique mechanism for the development of location and situation specific technologies, considering that the needs of plants are determined by agro-climatic conditions. India being endowed with varied agro-climatic conditions, ranging from humid western Himalayan region to arid western plain to sub humid Sutlej Ganga alluvial plain to humid/semi arid Western Ghats and Karnataka plateau, needs region specific hybrids/varieties and production system management for enhancing profitability. With this view, AICRP programmes started in seventies for various crop commodities have played a pivotal role in providing region specific cultivars and technologies which has boosted the production of crops and horticultural crops. Indian Council of Agricultural Research is implementing 77 AICRPs programmes having centres in all the agro-climatic zones, across the country. Through the programmes of AICRP, there has been speedier development of technologies and many of the crops have been adopted even in the regions where it was not grown before.

Horticulture, which includes fruits, vegetables, spices, tubers, mushrooms, plantation crops, medicinal and aromatic plants and flowers, has 13 AICRPs, for various crops with 251 centres located across the country in different agroclimatic zones. These coordinating centres have played a significant role in the exchange of knowledge, germplasm and development of new cultivars and technologies for production and health management. Efforts made in the past have resulted in substantial increase in production and productivity of horticultural crops. Thus, horticulture has emerged as the best option for diversification. The achievements of all these AICRPs in the past four decades have been scattered, which has been compiled and edited in the form of book entitled "Horticulture in Different Agro-climatic Conditions". I compliment Dr H.P. Singh and Dr M.S. Palaniswami, for their concerted efforts in bringing out this publication having achievements of these AICRPs in last 4 decades in two volumes. I am sure, this book will be of immense value, to all those who are concerned with Horticulture Research and Development.

S. Ayyappan
Secretary, Department of Agricultural Research and Education and
Director General, ICAR, New Delhi

Preface

Horticulture, which includes fruits, vegetables, tubers, spices, medicinal and aromatic plants, plantation crops, mushroom and flowers has made rapid stride during the last two decades owing to growing demands and technological innovations. Currently, horticulture contributes 30.4% to GDP of agriculture from nine per cent crop area and has become most important option for diversification. Silent revolution in eighties, due to innovative technologies has become foundation for Golden Revolution (development of horticulture), which is now conspicuously experienced across the country. Production and productivity of horticultural crops have increased manifold. India has emerged as the second largest producer of fruits and vegetables, largest producer and consumer of spices and cashewnut besides highest productivity of banana, grapes, papaya and cassava. Productivity of many of horticultural commodities continue to be higher than world average. These changes in the scenario of horticulture have been possible due to the technological innovation and support of the Government for the development through plan investment in technology led development referred as 'Technology Mission'.

All India Coordinated Research Project (AICRP) for various crops commodity has played a significant role in Indian agriculture, since India is endowed with varying climatic conditions and needs location and situation specific technologies. AICRP, a mechanism to develop location and situation specific technology was adopted in late sixties and after its success in delivering results, it was extended to many other crops including horticultural crops during IV and V Plan periods which were strengthened subsequently in different Plan period. There are 13 AICRPs under horticulture, covering tropical fruits, subtropical fruits, arid fruits, cashew, palms, spices, medicinal and aromatic plants, vegetables, potato, tuber crops, mushroom, betelvine and floriculture. These AICRPs along with the main institutes are involved in generating region specific/value added varieties, production, technologies, mitigation strategies for pests and diseases and transfer of technologies in different horticultural crops. Experimentation and validation under AICRPs have helped in introduction of new horticultural crops where it was not grown earlier. AICRPs have developed 298 varieties of vegetables, 132 fruits, 112 varieties of tubers and 129 spices and plantation crops and large collections are conserved at various centres. Nutrient needs, water management, pest and disease management have been worked out which have impacted horticulture development. Since many of the information are scattered it was felt essential to bring the 4 decades work together. Thus efforts have been made to bring out this publication.

In this book, technological innovations, and salient achievements emancipated from the 13 AICRPs, since their inception, are embodied under different chapters covering its history, objectives, functioning, budget, centres, accomplishments including genetic resources, biodiversity, recommended varieties/hybrids, productivity enhancing technologies, planting material production and distribution, on farm demonstration , and technology transfer. Therefore, the book entitled "Horticulture in Different Agro-climatic Conditions – Four Decades of Coordinated Research" brought out in two volumes encompassing various crops and discipline is considered to be a most valuable publication and shall be useful to all those concerned with horticulture.

We are thankful to all the Project Coordinators for their support and cooperation in our endeavour to document the information. Our sincere thanks are due to Dr S Ayyappan, Director General, ICAR for his support and encouragement. This book is intended to serve as handy compendium and reference to scientists, policy makers, progressive farmers and those who are interested in horticulture in different agro-climatic conditions.

H.P. Singh
M.S. Palaniswami

About the Editors

Dr H.P. Singh has served horticulture in various capacities namely, Project Coordinator (Fruits); Director, NRC for Banana; Horticulture Commissioner, Govt. of India; Chairman, Coconut Development Board; Dean (Agriculture), GBPUA&T, Pantnagar; Vice Chancellor, Rajendra Agricultural University, Pusa and now Deputy Director General (Horticulture), ICAR.

He has also served international organisations like the Chairman of APCC, Jakarta, National Director for internationally-aided projects and Chairperson of Steering Committee of BAPNET of Bioversity International and implemented various international programmes. Besides, excellent academic record, Dr Singh has contributed to research and agricultural education which has earned him three international, twenty one national awards and five fellowships including the Fellow of NAAS. He has developed many cultivars and technologies which has shaped the horticulture. Recognising his immense contribution, OUAT, Bhubaneshwar, has conferred D.Sc. (*Honoris causa*) last year.

Dr Singh is the architect of Mission mode approach for horticultural development, starting from planning to execution which has been due to his insight. He has provided leadership to horticulture and is credited to have authored and edited 39 books and 30 bulletins, besides 175 research papers. He has widely travelled in India and abroad. His efforts have provided a strong foundation for horticultural development, leading to ***Golden Revolution*** in the country.

Dr M.S. Palaniswami commenced his academic carrier in Tamil Nadu Agricultural University in 1975 soon after his Post-graduation in Agricultural Entomology and in 1977 entered into ARS (ICAR). Last 33 years he has been in ICAR, working in research and development of Tropical Tuber and Root Crops. He is the Project Coordinator (Tuber Crops) since September, 2003. He has more than 150 publications in national and international journals and conferences. He has authored five books and 25 book chapters.

The Ph.D student under his guidance was given Jawaharlal Nehru Award for the outstanding research thesis of 2006. More than 22 varieties could be released under his coordination. The AICRP Tuber Crops under his leadership received the Chaudhary Devi Lal Outstanding AICRP Award 2007 from the ICAR. As the Head of division Crop Protection in CTCRI, he managed it effectively during 1997–2004. He has handled the projects funded by USDA/USIF, NATP and AP Cess Fund. The USDA, Washington has awarded special appreciation certificate for his contribution in the USIF/USDA project. His contribution to insect pheromone and kairomone of sweet potato weevil and biotypes of *Bemisia tabaci* whitefly is noteworthy.

He was the Editor/Chief Editor of Journal of Root Crops during 1989-95. He is the life member of Indian society for Root crops, Indian society for vegetable science, Association of Advancement for Pest Management in Horticultural Ecosystem and Association for Advancement of Entomology (Vice President). Visited Colombia, France, Malaysia, Singapore and USA for attending official and academic meetings.

Contents

VOLUME – II

(Vegetables, Potato, Tuber Crops, Mushroom and Floriculture)

Contributors

D. Pandey
CISH, Remankhera
P.O. Kakori
Lucknow – 227 107
Uttar Pradesh

D. Prasath
All India Coordinated Research Project on Spices
Indian Institute of Spices Research
Calicut
Kerala

H.P. Singh
KAB -II
ICAR
New Delhi – 110 012

Johny A. Kallupurakkal
All India Coordinated Research Project on Spices
Indian Institute of Spices Research
Calicut
Kerala

Khayal C. Thakur
AICRP, Central Potato Research Institute
Shimla – 171 001
Himachal Pradesh

Manjit Singh
AICMIP
DMR, Solan
Himachal Pradesh

Mathura Rai
AICRPVC, IIVR
Varanasi – 221 005
Uttar Pradesh

M.S. Palaniswami
AICRP on Tuber Crops
CTCRI
Thiruvananthapuram – 695 017
Keralka

M. Anandaraj
All India Coordinated Research Project on Spices
Indian Institute of Spices Research
Calicut
Kerala

M.G. Bhat
AICRPC
National Research Centre For Cashew
Puttur – 574 202, D.K.
Karnataka

Om Prakash
CISH, Remankhera
P.O. Kakori
Lucknow – 227 107
Uttar Pradesh

Prakash S. Naik
AICRP, Central Potato Research Institute
Shimla – 171 001
Himachal Pradesh

P. Naveen Kumar
Indian Agricultural Research Institute
New Delhi – 110 012

P. Patil
AICRPTF, IIHR
Hessaraghatta Lake Post
Bangalore – 560 089
Karnataka

P. Manivel
AICRMAP
National Resarch Center for Medicinal and Aromatic Plants
Boriavi, Anand
Gujarat

R.L. Mishra
Indian Agricultural Research Institute
New Delhi – 110 012

R.P. Tiwari
All India Coordinated Mushroom Improvement Project
ICAR
Solan
Himachal Pradesh

R.T. Reena
AICRPTF, IIHR
Hessaraghatta Lake Post
Bangalore – 560 089
Karnataka

Saravanan Raju
AINP on Betelvine
NRCMAP
Boriavi, Anand
Gujarat

Satyabrata Maiti
AINP
National Resarch Center for Medicinal and Aromatic Plants
Boriavi, Anand
Gujarat

Suchitra Pushkar
Indian Agricultural Research Institute
New Delhi – 110 012

S.D. Shikamany
AICRPTF, IIHR
Hessaraghatta Lake Post
Bangalore – 560 089
Karnataka

S. Arulraj
All India Coordinated Research Project On Palms
CPCRI
Kasaragod – 671 124
Kerala

T. A. More
AICRP AZF, Central Institute for Arid Horticulture
Bikaner – 334 006
Rajasthan

V. Ramesh
AICRP on Tuber Crops
CTCRI
Thiruvananthapuram – 695 017
Kerala

Acronyms

AAU	:	Assam Agricultural University
AC&RI	:	Agricultural College & Research Institute
AICFIP	:	All India Coordinated Fruit Improvement Project
AICRP Palms	:	All India Coordinated Research Project on Palms
AICRP Potato	:	All India Coordinated Research Project on Potato
AICRP	:	All India Coordinated Research Project
AICRPAF	:	All India Coordinated Research Project on Arid Fruits
AICRPF	:	All India Coordinated Research Project on Floriculture
AICRPM	:	All India Coordinated Research Project on Mushrooms
AICRPMAP	:	All India Coordinated Research Project on Medicinal and Aromatic Plants
AICRPS	:	All India Coordinated Research Project on Spices
AICRPSTF	:	All India Coordinated Research Project on Sub-tropical Fruits
AICRPTC	:	All India Coordinated Research Project on Tuber Crops
AICRPTF	:	All India Coordinated Research Project on Tropical Fruits
AICRPVC	:	All India Coordinated Research Project on Vegetable Crops
ANGRAU	:	Acharya NG Ranga Agricultural University
APHU	:	Andhra Pradesh Horticultural University
BAC	:	Bihar Agricultural College
BAU	:	Birsa Agricultural University
BCKV	:	Bidhan Chandra Krishi Viswavidyalaya
CARI	:	Central Agricultural Research Institute
CAU	:	Central Agricultural University
CAZRI	:	Central Arid Zone Research Institute
CIAH	:	Central Institute of Arid Horticulture
CISH	:	Central Institute for Sub-tropical Horticulture
CITH	:	Central Institute for Temperate Horticulture
CPCRI	:	Central Plantation Crops Research Institute
CTCRI	:	Central Tuber Crops Research Institute
DAT	:	Days After Treatment
Dr. BSKKV	:	Dr. Balasaheb Sawant Konkan Krishi Vidyapeeth
EFY	:	Elephant Foot Yam
g	:	Gram
GAU	:	Gujarat Agricultural University
GBPUAT	:	Govind Ballabh Pant University of Agriculture & Technology

HaNPV	:	*Helicoverpa armigera Nuclear Polyhedrosis virus.*
HAU	:	Haryana Agricultural University
HC&RI	:	Horticultural College and Research Institute
HPKVV	:	Himachal Pradesh Krishi Viswa Vidyalaya
IARI	:	Indian Agricultural Research Institute
ICAR RC NEH	:	ICAR Research Complex for NEH Region
ICAR	:	Indian Council of Agricultural Research
IGKV	:	Indira Gandhi Krishi Viswavidyalaya
IIHR	:	Indian Institute of Horticultural Research
IISR	:	Indian Institute of Spices Research
JNKVV	:	Jawaharlal Nehru Krishi Viswa Vidyalaya
K_2O	:	Potassium
KAU	:	Kerala Agricultural University
kg	:	Kilogram
mg	:	Milligram
MPKV	:	Mahatma Phule Krishi Vidya Peeth
MPUAT	:	Maharana Pratap University of Agriculture and Technology
MSL	:	Mean Sea Level
N	:	Nitrogen
NAU	:	Navsari Agricultural University
NDUAT	:	Narendra Dev University of Agriculture & Technology
NPK	:	Nitrogen, Phosphorus and Potassium
NRC Litchi	:	National Research Centre on Litchi
NRC	:	Non-recurring Contingency
NRCC	:	National Research Centre on Cashew
NRCG	:	National Research Centre on Grapes
NRCM	:	National Research Centre on Mushroom
NRCO	:	National Research Centre on Orchids
NRCOG	:	National Research Centre on Onion and Garlic
NRCOP	:	National Research Centre on Oil Palm
NRCP	:	National Research Centre on Pomegranate
NRCSS	:	National Research Centre on Seed Spices
P_2O_5	:	Phosphorous
PAU	:	Punjab Agricultural University
ppm	:	Parts Per Million
q	:	Quintal
RAU	:	Rajendra Agricultural University
RC	:	Recurring Contingency
SKUAST	:	Sher-e-Kashmir University of Agricultural Sciences & Technology, Shalimar
t	:	Tonnes
TFP	:	Total Factor Productivity
TNAU	:	Tamil Nadu Agricultural University
TA	:	Travelling Allowance
TSS	:	Total Soluble Solids
UASB	:	University of Agricultural Sciences, Bangalore
UASD	:	University of Agricultural Sciences, Dharwad
UHS	:	University of Horticultural Sciences

1

Vegetables

Mathura Rai

All India Coordinated Research Project for the Improvement of Vegetable Crops was started during the Fourth Five Year Plan in 1970-71, to provide a national grid for multi-location testing of vegetable technologies developed by various research institutes and state agricultural universities. The headquarters of the project was at the Division of Vegetable Crops, Indian Agricultural Research Institute (IARI), New Delhi and Dr Vishnu Swarup joined the project in July 1971 as its first Project Coordinator. During 1986, level of the project was elevated to the Project Directorate. Further, in a significant development during 1992, headquarters of Project Directorate of Vegetable Research was shifted from IARI, New Delhi, to the present day Indian Institute of Vegetable Research (IIVR) at Varanasi. Initially during 1971, the project was started with following seven main centres and 10 sub-centres making a total of 17 centres.

CENTRES AND SUB-CENTRES

Main Centres

- Indian Agricultural Research Institute (IARI), New Delhi
- IARI Regional Vegetable Research Station, Katrain, Kullu, Himachal Pradesh
- Indian Institute of Horticultural Research (IIHR), Bangalore, Karnataka
- Punjab Agricultural University (PAU), Ludhiana, Punjab
- Rajendra Agricultural University (RAU), Sabour, Bihar
- Mahatma Phule Krishi Vidyapeeth (MPKV), Rahuri, Maharashtra
- Tamil Nadu Agricultural University (TNAU), Coimbatore, Tamil Nadu

Sub-Centres

- Govind Ballabh Pant University of Agriculture and Technology (GBPUA & T), Pantnagar, Uttar Pradesh
- Chandra Shekhar Azad University of Agriculture and Technology (CSAUA &T), Kanpur, Uttar Pradesh
- Assam Agricultural University (AAU), Jorhat, Assam
- Tamil Nadu Agricultural University (TNAU), Kovilpatti, Tamil Nadu

- Orissa University of Agriculture & Technology (OUAT), Bhubaneshwar, Orissa
- Jawaharlal Nehru Krishi Vishwa Vidyalaya (JNKVV), Jabalpur, Madhya Pradesh
- Department of Horticulture, Jammu and Kashmir, Srinagar
- Himachal Pradesh Agricultural University (HPAU), Palampur, Himachal Pradesh
- Andhra Pradesh Agricultural University (APAU), Lam, Guntur, Andhra Pradesh
- Rajasthan Agricultural University, Durgapura, Jaipur, Rajasthan

In the subsequent phase of development, several new centres were included under the project (Table 1.1). The basic aim behind these new inclusions was to cover broadly classified eight agro-climatic zones of India (Table 1.2) under the AICRP trials.

Table 1.1. Addition of participatory centres in subsequent phase of AICRP development

Centre	*Addition Period (five-year plan)*
Bidhan Chandra Krishi Vishwavidyalaya (BCKV), Kalyani	Vth plan
Marathwarda Krishi Vishwavidyalaya (MPKV), Ambajogai	Vth plan
Haryana Agricultural University (HAU), Hisar	Vth plan
Narendra Deo University of Agric. and Tech. (NDUA&T), Faizabad	VIth plan
Kerala Agricultural University (KAU), Vellanikkara	VIth plan
Andhra Pradesh Agricultural University (APAU), Hyderabad	VIIth plan
Gujarat Agricultural University (GAU), Junagarh	VIIth plan
University of Agricultural Science (UAS), Dharwad	VIIIth plan
Indira Gandhi Krishi Vishwavidyalaya (IGKV), Raipur	VIIIth plan

Table 1.2. Eight agro-climatic zones and states under each zone

Agroclimatic zone	*State*
Humid Western Himalayan Region	Jammu and Kashmir (J& K), Himachal Pradesh and Uttar Pradesh Hills
Humid Bengal - Assam Basin	West Bengal and Assam
Humid Eastern Himalayan and Bay Islands	Sikkim, Meghalaya, Manipur, Nagaland, Mizoram, Tripura, Arunachal Pradesh and Andaman & Nicobar Islands
Sub-Humid Sutlej Ganga Alluvial Plain	Punjab, Tarai region of Uttar Pradesh and Bihar
Sub-Humid to Humid Eastern and South Eastern Uplands	Eastern Madhya Pradesh, Orissa and Andhra Pradesh
Arid Western Plain	Rajasthan, Gujarat, Haryana and Delhi
Semi Arid Lava Plateaux and Central High Lands	Madhya Pradesh (excluding eastern area) and Maharashtra
Humid to Semi-Arid Western Ghats and	Karnataka Plateaux, Karnataka, Tamil Nadu and Kerala

However, Kovilpatti centre of Tamil Nadu Agriculture University was deleted during the Seventh Five Year Plan. Therefore, the total number of centres under the project became 25, located in different agro-climatic zones of the country. However, with the shifting of Project Directorate from IARI, New Delhi to Varanasi in 1992, Varanasi, also became one of the main centres of the project. Thus, presently there are 26 centres under the project which include 8 main and 18 sub-centres as given below.

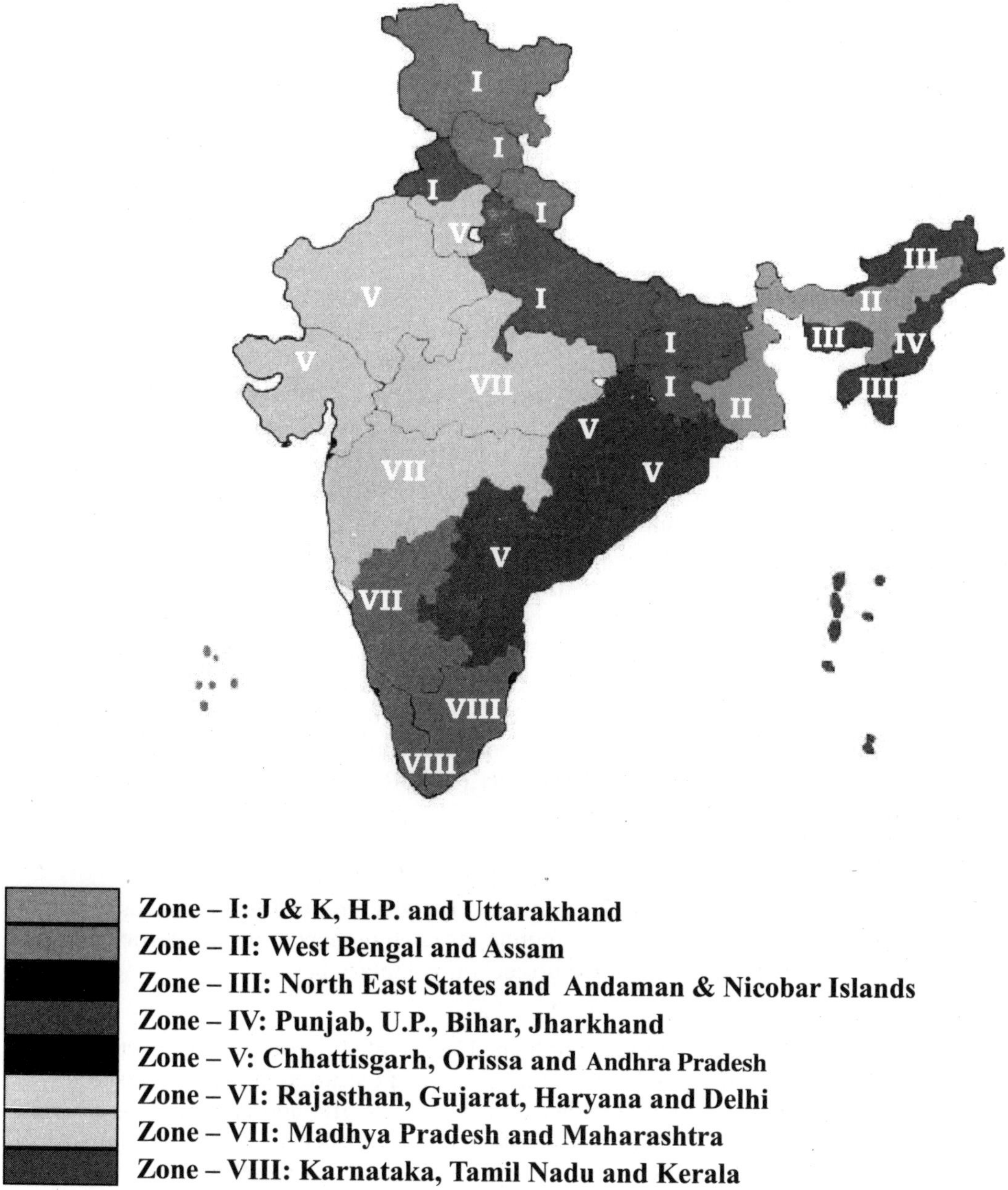

Fig. 1.1. Agro-climatic zones in India under AICRP on vegetable crops

Main Centres

- Indian Institute of Vegetable Research (IIVR), Varanasi
- Indian Agricultural Research Institute (IARI), New Delhi
- IARI Regional Vegetable Research Station, Katrain, Kullu, Himachal Pradesh
- Indian Institute of Horticultural Research (IIHR), Hassaraghatta, Bangalore, Karnataka
- Punjab Agricultural University (PAU), Ludhiana, Punjab
- Rajendra Agricultural University (RAU), Sabour, Bihar
- Mahatma Phule Krishi Vidyapeeth (MPKV), Rahuri, Maharashtra
- Tamil Nadu Agricultural University (TNAU), Coimbatore, Tamil Nadu

Sub-Centres

- Govind Balabh Pant Univ. of Agric. and Tech. (GBPUA & T), Pantnagar, Uttrakhand
- Chandra Shekhar Azad Univ. of Agric. and Tech. (CSAUA &T), Kanpur, Uttar Pradesh
- Assam Agricultural University (AAU), Jorhat, Assam
- Orissa University of Agriculture and Technology (OUAT), Bhubaneshwar, Orissa
- Y.S. Parmar University of Horticulture and Forestry, Solan, Himachal Pradesh
- Jawaharlal Nehru Krishi Vishwa Vidyalaya (JNKVV), Jabalpur, Madhya Pradesh
- Sher-E-Kashmir Univ. of Agri. Sci. and Tech. (SKUAS&T), Srinagar, J&K
- Acharya N.G. Ranga Agricultural University (ANGRAU), Hyderabad, Andhra Pradesh
- Rajasthan Agricultural University, Durgapura, Jaipur, Rajasthan
- Bidhan Chandra Krishi Vishwavidyalaya (BCKV), Kalyani, West Bengal
- Marathwada Agril. University (MAU), Ambajogai (Parbhani), Maharashtra
- Haryana Agricultural University (HAU), Hisar, Haryana
- Narendra Deo University of Agric. and Tech. (NDUA&T), Faizabad, Uttar Pradesh
- Kerala Agricultural University (KAU), Vellanikkara, Kerala
- Gujarat Agricultural University (GAU), Junagarh, Gujarat
- Acharya N G Ranga Agricultural University (ANGRAU), RRS, Lam, Guntur, Andhra Pradesh
- University of Agricultural Science (UAS), Dharwad, Karnataka
- Indira Gandhi Krishi Vishwavidyalaya (IGKV), Raipur, Chhattisgarh

Besides the above centres, about 22 centres involving central institutes, state agricultural and traditional universities and a few other public research organizations are participating in the cooperative research programme of the project on voluntary basis.

Objectives

The first major task of the All India Coordinated Vegetable Improvement Project (AICRP) was multi-location testing of vegetable varieties developed by different research institutes, at various centres in order to identify varieties suitable for specific agroclimatic zones. Later, research was undertaken on agro-techniques, like plant population density, fertilizer requirements and chemical weed control. Subsequently, research programmes were also undertaken on chemical control of diseases and insect pests. Initially, only 14 vegetables were covered under the research programme, however, with the subsequent addition of more crops, number of crops covered under the project is presently 24. Still, major emphasis is laid on 15 important vegetables, including tomato, brinjal, chilli, cauliflower, cabbage, pea, onion, muskmelon, watermelon, okra, French bean, pumpkin, carrot, radish and garlic.

Organization of workshop every year has become an integral activity of this project to discuss the results of experiments and to formulate future experimental plans at various participatory centres. Till date, total 25 workshops/group meetings have been organized (Table 1.3) in which active workers associated with different research projects participated and very vital recommendations have been made on various aspects.

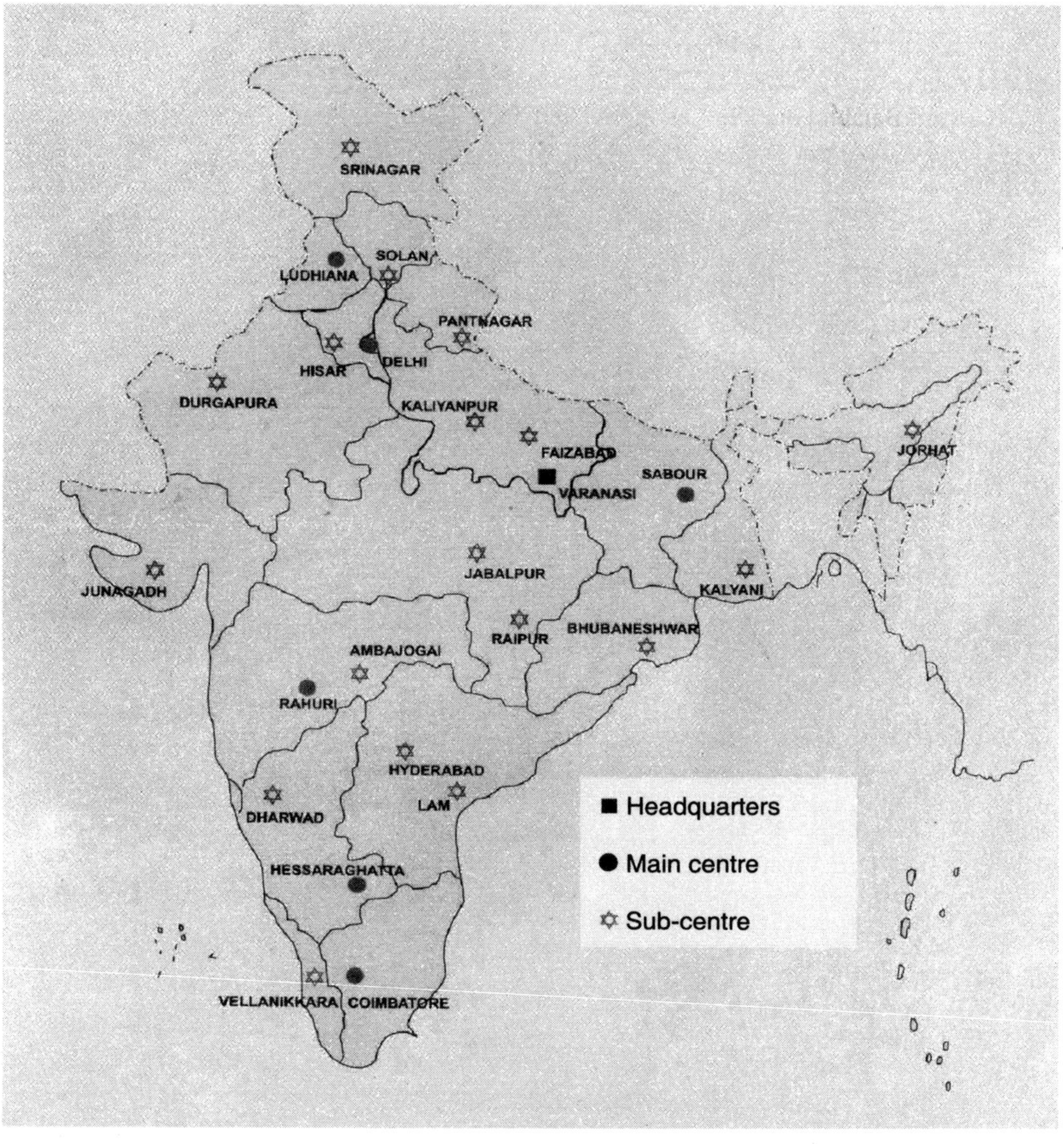

AICRP VC Centres

Genetic Resources/Biodiversity

Germplasm collection reported by different centres in the respective workshop/group meeting of the project

Crops	*Centres*	*Workshop/ Group Meeting (Year)*										
		I Feb. 1971	*II May 1972*	*III April 1975*	*IV May 1977*	*V March 1979*	*VI June 1981*	*VII May 1983*	*VIII June 1985*	*IX Jan. 1987*	*X Aug. 1988*	*XI June 1990*
Amaranthus	TNAU,Coimbatore	–	–	94	95	100	151	165	121	244	36	45
	KAU,Vellanikkara	–	–	0	–	–	–	42	–	–	–	–
	AAU, Jorhat	–	–	–	–	–	–	–	–	–	–	–
	IIHR, Hessaraghatta	–	–	–	–	–	–	–	–	–	–	–
	HARP, Ranchi	–	–	–	–	–	–	–	–	–	–	–
Bittergourd	CSUAT,Kalyanpur	–	–	–	–	–	–	–	–	–	–	–
	KAU,Vellanikkara	–	–	–	–	–	–	–	–	–	–	–
	JNKVV, Jabalpur	–	–	–	–	–	–	–	–	–	–	–
	HARP, Ranchi	–	–	–	–	–	–	–	–	–	–	–
Bottlegourd	NDUAT, Faizabad	–	–	–	–	–	–	–	–	–	–	–
	MPKV, Rahuri	–	–	–	–	–	–	–	–	–	–	–
	AAU, Jorhat	–	–	4	0	4	0	–	–	–	–	–
	JNKVV, Jabalpur	–	–	77	53	53	–	–	–	–	–	–
Brinjal	IIHR, Hessaraghatta	–	–	252	300	300	250	250	0	280	150	0
	JNKVV, Jabalpur	–	–	–	–	–	–	–	–	–	–	–
	OUAT,Bhubaneswar	–	–	32	39	45	52	55	0	103	126	160
	CHES, Ranchi	–	–	–	–	–	–	–	–	–	–	–
	KAU,Vellanikkara	–	–	–	–	–	–	–	–	–	–	–
	BAC (RAU), Sabour	–	–	92	0	115	115	140	125	100	100	95
	BCKV, Kalyani	–	–	0	–	–	39	0	0	71	0	56
	IGKVV, Raipur	–	–	–	–	–	–	–	–	–	–	–
	HARP, Ranchi	–	–	–	–	–	–	–	–	–	–	–
	CSUAT, Kalyanpur	–	–	0	0	160	0	–	–	–	–	–
	NBPGR, New Delhi	–	–	–	–	–	–	–	–	–	–	–
	IARI, New Delhi	–	–	350	240	240	0	–	–	–	–	–
	Tripura	–	–	–	–	–	–	–	–	–	–	–
	Aurangabad	–	–	0	–	46	–	–	–	–	–	–
	AAU, Jorhat	–	–	0	4	4	–	–	–	–	–	–
Cabbage	IARI–RRS, Katrain	–	–	99	18	328	80	80	0	0	0	239
	YSPUH&F, Solan	–	–	–	–	–	–	–	–	–	–	–
Capsicum	SKUT, Srinagar	–	–	6	8	8	–	–	–	–	–	–
	YSPUHF, Solan	–	–	–	–	–	–	–	–	0	0	0
	IIHR, Bangalore	–	–	–	–	–	–	–	–	–	–	–
	HPKV, Palampur	–	–	–	–	–	–	–	–	–	–	–
	IARI–RRS, Katrain	–	–	–	–	–	–	–	–	–	–	–
	TNAU, Coimbatore	–	–	111	53	53	63	12	–	–	–	–
Carrot	HAU, Hissar (tropical)	–	–	3	0	0	0	0	0	30	31	25
(Temperate)	IARI–RRS, Katrain	–	–	–	–	–	–	–	–	–	–	–
	SKU&T, Srinagar	–	–	0	7	7	0	0	0	9	9	11
	IIVR, Varanasi	–	–	–	–	–	–	–	–	–	–	–

Workshop/ Group Meeting (Year)													
XII Aug. 1992	XIII Dec. 1993	XIV Feb. 1995	XV March 1996	XVI May 1997	XVII June 1998	XVIII Oct. 1999	XIX Jan. 2001	XX April 2002	XXI May 2003	XXII May 2004	XXIII April 2005	XXIV April 2006	XXV May 2007
43	41	41	46	12	44	14	4	5	6	4	73	79	79
–	–	–	23	52	15	90	15	105	20	21	15	24	–
–	–	–	–	–	–	10	–	–	10	10	–	5	–
–	–	–	–	–	–	–	–	–	–	–	20	28	11
–	–	–	–	–	–	–	–	–	–	–	–	–	23
–	–	27	0	–	–	–	–	–	–	–	–	–	–
–	–	0	64	95	25	63	4	20	40	21	5	20	2
–	–	–	–	–	–	–	–	–	40	–	4	25	4
–	–	–	–	–	–	–	–	–	–	–	–	16	18
–	22	22	7	42	46	–	14	15	55	–	–	30	–
–	–	–	17	–	17	17	17	17	33	–	33	–	–
–	–	–	–	–	–	–	–	–	–	–	–	–	
–	–	–	–	–	–	–	–	–	–	–	–	–	–
151	22	0	–	–	–	–	–	–	–	–	4	56	51
72	–	–	–	–	–	–	–	–	–	–	–	–	–
170	181	180	150	0	54	76	84	17	27	–	16	10	60
–	–	–	–	–	–	–	28	54	–	–	–	–	–
–	–	–	–	–	–	–	–	–	–	–	10	–	15
105	105	105	–	–	–	–	–	–	–	–	–	–	–
83	90	0	–	–	–	–	–	–	–	–	–	–	28
–	–	–	–	–	–	–	–	–	–	–	–	64	76
–	–	–	–	–	–	–	–	–	18	30	29	40	8
–	–	–	–	–	–	–	–	–	–	–	–	–	–
–	–	–	–	–	–	–	–	–	–	–	–	–	60
–	–	–	–	55	30	–	–	54	50	25	27	25	–
–	–	–	–	–	–	–	–	–	–	–	–	7	
–	–	–	–	–	–	–	–	–	–	–	–	–	–
–	–	–	–	–	–	–	–	–	–	–	–	–	–
421	0	0	0	–	–	–	–	–	–	–	49	–	10
–	–	–	6	–	–	–	–	–	–	–	5	5	5
–	–	–	20	–	–	–	–	–	–		8	15	15
0	25	0	–	12	6	6	6	10	–	5	5	6	7
–	–	–	–	–	–	–	–	–	–	–	–	10	–
–	–	–	–	–	–	–	–	–	–	–	–	2	3
–	–	–	15	16	–	–	14	22	–	–	106	–	6
–	–	–	–	–	–	–	–	–	–		–		–
25	25	25	60	58	57	76	27	23	–	–	28	53	31
–	–	–	–	–	–	–	15	21	–	–	32	10	20
21	0	28	–	–	–	–	–	–	–	–	–	–	15
–	–	–	–	–	–	–	–	–	–	–	–	42	54

(Contd.)

Crops	*Centres*	*Workshop/ Group Meeting (Year)*										
		I Feb. 1971	*II May 1972*	*III April 1975*	*IV May 1977*	*V March 1979*	*VI June 1981*	*VII May 1983*	*VIII June 1985*	*IX Jan. 1987*	*X Aug. 1988*	*XI June 1990*
Cauliflower (Early)	BAC (RAU), Sabour	–	–	85	60	60	22	22	40	50	0	28
	IIVR, Varanasi	–	–	–	–	–	–	–	–	–	–	–
	CHES, Ranchi	–	–	–	–	–	–	–	–	–	–	–
	NDUA&T, Faizabad	–	–	–	–	–	–	–	–	0	6	0
Cauliflower (Mid)	BAC (RAU), Sabour	–	–	–	–	–	20	20	24	0	0	15
	IARI, New Delhi	–	–	–	–	–	–	–	–	0	41	302
	CHES, Ranchi	–	–	–	–	–	–	–	–	–	–	–
	IIVR, Varanasi	–	–	–	–	–	–	–	–	–	–	–
	GBPUAT,Pantnagar	–	–	0	–	–	–	30	0	95	95	105
	CSUAT, Kalyanpur	–	–	3	0	3	0	–	–	–	–	–
Cauliflower (Late)	IARI–RRS, Katrain	–	–	50	0	314	35	30	0	0	0	180
	BCKV, Kalyani	–	–	–	–	–	–	–	–	–	–	–
	IIVR, Varanasi	–	–	–	–	–	–	–	–	–	–	–
	YSPUHF, Solan	–	–	–	–	–	–	–	–	–	0	107
Cho–Cho	ICAR Complex– Barapani	–	–	–	–	–	–	–	–	–	–	–
Chillies	TNAU, Coimbatore	–	–	–	–	278	58	208	153	706	43	0
	APAU–RRS, Lam	–	–	101	0	324	200	223	386	223	263	0
	PAU,Ludhiana	–	–	–	150	150	–	–	–	–	–	–
	AAU, Jorhat	–	–					–	–	–	–	–
	KAU, Vellanikkkara	–	–					–	–	–	–	–
	TNAU–RRS,Kovilpatti	–	–	145	164	126	211	265	304	–	–	–
	BCKV, Kalyani	–	–	–	–	–	–	–	–	–	–	–
	UAS, Dharwad	–	–	–	–	–	–	–	–	–	–	–
	SKU&T, Srinagar	–	–	–	–	–	–	–	–	–	–	–
	IIVR, Varanasi	–	–	–	–	–	–	–	–	–	–	–
	HARP, Ranchi	–	–	–	–	–	–	–	–	–	–	–
	IIHR, Bangalore	–	–	–	–	–	–	–	–	–	–	–
	NBPGR, New Delhi	–	–	–	–	–	–	–	–	–	–	–
	CSAUAT, Kalyanpur	–	–	0	–	207	–	–	–	–	–	–
Cucumber	MPKV, Rahuri	–	–	15	21	21	0	0	50	0	0	24
	SKU&T, Srinagar	–	–	8	13	17	0	–	0	–	–	–
	IIVR, Varanasi	–	–	–	–	–	–	–	–	–	–	–
	CHES, Ranchi	–	–	–	–	–	–	–	–	–	–	–
	HARP, Ranchi	–	–	–	–	–	–	–	–	–	–	–
	IIHR, Hessaraghatta	–	–	–	–	–	–	–	–	–	–	–
	HPKV, Palampur	–	–	–	–	–	–	–	–	–	–	–
	YSPUHF, Solan	–	–	–	–	–	–	–	–	–	–	–
	BCKV, Kalyani	–	–	0	14	18	0	0	0	12	0	15
French bean	IIHR, Hessaraghatta	–	–	120	0	1300	460	460	460	500	0	0
	YSPUH&F, Solan	–	–	–	–	–	–	–	–	0	0	20
	UAS, Dharwad	–	–	–	–	–	–	–	–	–	–	–
	AAU, Jorhat	–	–	–	–	–	–	–	–	–	–	–
	IIVR, Varanasi	–	–	–	–	–	–	–	–	–	–	–

Workshop/ Group Meeting (Year)													
XII Aug. 1992	*XIII Dec. 1993*	*XIV Feb. 1995*	*XV March 1996*	*XVI May 1997*	*XVII June 1998*	*XVIII Oct. 1999*	*XIX Jan. 2001*	*XX April 2002*	*XXI May 2003*	*XXII May 2004*	*XXIII April 2005*	*XXIV April 2006*	*XXV May 2007*
16	16	16	21	21	23	30	10	10	11	12	14	10	10
–	–	–	–	–	34	–	14	70	67	112	46	94	94
–	–	–	–	–	–	–	35	8	–	–	–	–	–
–	–	–	–	–	–	–	–	–	–	–	–	–	–
12	12	13	25	25	30	–	–	–	–	–	–	7	7
0	0	0	–	–	–	–	–	–	–	–	–	–	–
–	–	–	–	–	–	–	–	13	–	–	–	–	–
–	–	–	–	54	33	55	33	34	44	54	44	58	51
0	0	0	–	–	–	–	–	–	–	–	–	–	–
–	–	–	–	–	–	–	–	–	–	–	–	–	–
0	0	0	42	0	–	–	34	35	–	–	32	–	10
–	–	–	–	–	–	–	–	–	–	–	–	–	–
–	–	–	–	–	–	–	60	60	–	–	14	–	–
0	10	0	0	0	6	20	7	8	–	6	5	5	3
–	–	–	–	–	–	–	–	–	4	–	9	–	–
16	0	108	79	85	85	85	85	89	6	95	65	98	98
0	0	70	310	250	246	–	14	238	246	256	291	–	11
–	–	–	–	31	–	–	–	–	–	–	–	–	–
–	–	–	–	17	–	35	–	–	10	10	–	10	–
–	–	–	–	46	16	34	34	–	5	8	–	–	–
–	–	–	–	–	–	–	–	–	–	–	–	–	–
–	–	–	–	–	14	14	6	11	12	–	8	24	51
–	–	–	–	–	40	–	–	19	–	17	13	20	–
–	–	–	–	–	–	110	20	–	172	219	15	25	20
–	–	–	–	–	–	–	–	–	–	–	–	25	37
–	–	–	–	–	–	–	–	–	–	–	–	36	53
–	–	–	–	–	–	–	–	–	–	–	–	23	34
–	–	–	–	–	–	–	–	–	–	–	–	–	48
–	–	–	–	–	–	–	–	–	–	–	–	–	–
0	0	0	9	0	19	–	15	–	10	10	10	–	21
–	–	–	–	–	–	–	–	–	–	–	–	–	
–	–	–	–	–	–	94	8	–	–	30	–	–	97
–	–	–	–	–	–	–	82	20	–	–	–	–	–
–	–	–	–	–	–	–	33		23	33	22	–	7
–	–	–	–	–	–	–	–	–	–	–	5	–	–
–	–	–	–	–	–	–	–	–	–	–	–	3	–
–	–	–	20	35	30	30		–	–	7	7	5	6
8	10	0	–	–	–	–	–	–	–	–	–	–	–
200	0	35	45	–	–	–	–	–	–	–	–	–	–
0	120	0	125	–	–	–	–	–	–	–	–	–	–
–	–	–	–	–	–	–	–	–	–	–	60	90	–
–	–	–	–	–	–	–	–	–	–	–	–	10	–
–	–	–	–	–	–	–	–	–	–	–	–	–	60

(Contd.)

Crops	*Centres*	*Workshop/ Group Meeting (Year)*										
		I Feb. 1971	*II May 1972*	*III April 1975*	*IV May 1977*	*V March 1979*	*VI June 1981*	*VII May 1983*	*VIII June 1985*	*IX Jan. 1987*	*X Aug. 1988*	*XI June 1990*
	GBPUAT,Pantnagar	–	–	–	–	–	–	–	–	–	–	–
	MPKV, Rahuri	–	–	18	39	39	0	0	0	88	0	38
Garlic	NHRDF, Nasik	–	–	–	–	–	–	50	120	92	23	83
	NBPGR, New Delhi	–	–	–	–	–	–	–	–	0	0	877
	NHRDF, Karnal	–	–	–	–	–	–	–	–	0	92	96
	NRC (O&G), Pune	–	–	–	–	–	–	–	–	–	–	–
	YSPUAT, Solan	–	–	–	–	–	–	–	32	–	–	–
Muskmelon	IARI, New Delhi	–	–	–	–	–	–	–	80	–	–	–
	NDUA&T, Faizabad	–	–	–	–	–	–	–	–	75	65	43
	ARS, Durgapura	–	–	–	–	0	0	50	50	50	50	40
	PAU, Ludhiana	–	–	5	0	84	0	0	0	75	0	34
	IIHR, Hessaraghatta	–	–	–	0	0	180	–	–	–	–	–
Okra	IIHR, Hessaraghatta	–	–	375	400	400	365	365	350	390	130	0
	OUATT,Bhubneshwar	–	–	0	–	12	12	14	15	22	25	27
	NBPGR, New Delhi	–	–	–	–	–	–	–	–	0	0	1000
	IIVR, Varanasi	–	–	–	–	–	–	–	–	–	–	–
	GAU,Anand	–	–	–	–	–	–	–	–	–	–	–
	UAS, Dharwad	–	–	–	–	–	–	–	–	–	–	–
	MPKV, Rahuri	–	–	–	–	–	–	–	–	–	–	–
	PAU, Ludhiana	–	–	0	101	126	100	–	–	–	–	–
	CSAUT, Kalyanpur	–	–	0	–	66	–	–	–	–	–	–
	BAC (RAU), Sabour	–	–	0	–	33	18	–	22	–	–	–
Onion	MPKV, Rahuri	–	–	32	54	54	0	0	0	159	132	135
	NBPGR, New Delhi	–	–	–	–	–	–	–	–	0	0	1464
	NRC (O&G), Pune	–	–	–	–	–	–	–	–	–	–	–
	IIHR, Hessaraghatta	–	–	–	–	–	–	–	–	–	–	–
	NHRDF, Nasik	–	–	–	–	–	–	–	–	–	–	–
	TNAU, Coimbatore	–	–	–	–	–	–	–	750	–	–	–
Paprika	SKUT, Srinagar	–	–	–	–	–	–	–	–	–	–	–
	IIHR, Hessasraghatta	–	–	–	–	–	–	–	–	–	–	–
	UAS, Dharwad	–	–	–	–	–	–	–	–	–	–	–
	IARI–RRS, Katrain	–	–	–	–	–	–	–	–	–	–	–
Pumpkin	IIHR, Hessaraghatta	–	–	90	130	130	120	120	120	130	0	0
	NDUAT, Faizabad	–	–	–	–	–	–	–	–	10	18	12
	APAU, Hyderabad	–	–	–	–	–	–	–	–	–	–	–
	IIVR, Varanasi	–	–	–	–	–	–	–	–	–	–	–
	TNAU, Coimbatore	–	–	166	160	160	–	–	–	–	–	–
	IARI–RRS, Katrain	–	–	0	25	25	–	–	–	–	–	–
	IARI, New Delhi	–	–	0	–	–	–	–	114	–	–	–
Peas	IARI, New Delhi	–	–	500	0	190	0	0	0	0	0	540
	JNKVV, Jabalpur	–	–	0	247	240	266	298	205	205	205	95
	MPKV, Rahuri	–	–	0	96	96	–	–	–	–	–	–
	NDUAT, Faizabad	–	–	–	–	–	–	–	–	–	–	–

Workshop/ Group Meeting (Year)													
XII Aug. 1992	*XIII Dec. 1993*	*XIV Feb. 1995*	*XV March 1996*	*XVI May 1997*	*XVII June 1998*	*XVIII Oct. 1999*	*XIX Jan. 2001*	*XX April 2002*	*XXI May 2003*	*XXII May 2004*	*XXIII April 2005*	*XXIV April 2006*	*XXV May 2007*
–	–	–	–	20	100	–	12	21	204	–	–	–	–
48	35	45	–	24	–	–	–	–	35	45	86	48	48
90	0	0	85	0	–	–	–	–	–	–	–	–	–
0	–	–	0	–	–	–	–	–	–	–	40	–	–
101	–	–	22	125	–	131	125	17	12	–	12	–	–
–	–	–	–	–	–	–	–	117	135	42	–	40	130
–	–	–	–	–	–	–	–	–	–	–	–	–	–
–	–	–	–	–	–	–	–	–	–	–	–	–	–
0	40	40	0	–	–	–	–	–	–	–	–	–	–
40	0	54	6	60	176	–	180	–	2	198	–	22	–
30	15	6	16	2	9	20	4	19	96	1	–	8	–
–	–	–	–	–	–	–	–	–	–	–	21	–	20
13]0	0	0	13	–	–	–		–	–	–	10	100	18
0	30	30	0	0	24	28	28	15	6	–	6	5	6
0	0	0	0	0	–	–		–	–	–	–		97
–	–	–	–	–	–	–	103	48	147	66	155	126	226
–	–	–	–	–	–	–	–	–	–	–	100	–	
–	–	–	–	–	–	–	–	–	–	–	–	6	20
–	–	–	–	–	–	–	–	–	–	–	–	100	100
–	–	–	–	–	–	–	–	–	–	–	–	–	–
–	–	–	–	–	–	–	–	–	–	–	–	–	–
–	–	–	–	–	–	–	–	–	–	–	–	–	–
166	109	111	180	180	230	–	150	19	–	–	–	–	–
0	0	0	0	–	–	–	–	–	–	–	–	–	–
–	–	–	–	–	–	–	–	46	387	387	158	83	48
–	–	–	–	–	–	–	–	–	–	–	10	–	–
0	0	0	569	141	–	144	–	–	19	–	–	25	–
–	–	–	–	–	–	–	–	–	–	–	–	–	–
–	–	–	–	–	–	–	–	–	–	–	7	15	10
–	–	–	–	–	–	–	–	–	–	–	–	7	11
–	–	–	–	–	–	–	–	–	–	–	35	46	–
–	–	–	–	–	–	–	–	–	–	–	–	–	13
0	40	–	–	–	–	–	–	–	–	–	–	–	–
0	6	9	4	22	20	–	16	34	25	–	–	9	–
–	0	0	–	62	6	1	65	34	39	–	36	–	–
–	–	–	–	–	–	–	–	–	–	–	70	–	–
–	–	–	–	–	–	–	–	–	–	–	–	–	–
– –	–	–	–	–	–	–	–	–	–	–	–	–	
–	–	–	–	–	–	–	–	–	–	–	–	–	–
0	0	0	0	–	–	–	–	–	–	–	–	–	–
0	81	0	100	0	–	–	–	–	–	–	–	–	–
–	–	–	–	–	–	–	–	–	–	–	–	–	–
–	–	–	–	22	–	–	–	–	–	–	–	–	–

(Contd.)

Crops	*Centres*	*Workshop/ Group Meeting (Year)*										
		I *Feb.* *1971*	*II* *May* *1972*	*III* *April* *1975*	*IV* *May* *1977*	*V* *March* *1979*	*VI* *June* *1981*	*VII* *May* *1983*	*VIII* *June* *1985*	*IX* *Jan.* *1987*	*X* *Aug.* *1988*	*XI* *June* *1990*
	PAU, Ludhiana	–	–	0	0	146	0	0	0	65	0	7
	Palampur	–	–	–	–	–	–	–	–	–	–	–
	IIVR, Varanasi	–	–	–	–	–	–	–	–	–	–	–
	CSUAT, Kalyanpur	–	–	–	0	412	0	0	434	0	0	0
Radish	College of Agril. (RAU), Udaipur	–	–	–	20	41	29	–	–	–	–	–
	CSUAT, Kalyanpur	–	–	–	–	22	–	–	–	–	–	–
Squash	IIHR, Hessaraghatta	–	–	20	30	30	20	–	–	–	–	–
C. maxima		–	–	–	–	–	–	–	–	–	–	–
C. pepo	IARI–RRS, Katrain	–	–	27	–	66	15	–	–	–	–	–
Tinda (Round gourd)	PAU, Ludhiana	–	–	77	9	9	11	–	–	–	–	–
	College of Agril. (RAU), Udaipur	–	–	35	35	35	0	–	–	–	–	–
	HAU, Hissar	–	–	–	–	–	–	–	–	–	–	–
	YSPUHF, Solan	–	–	–	21	21	–	–	–	–	–	–
Tomato	NBPGR, New Delhi	–	–	110	–	–	–	0	0	0	0	668
	IARI, New Delhi	–	–	0	53	53	0	0	–	–	0	–
	PAU, Ludhiana	–	–	0	–	–	–	0	23	280	0	3
	Aurangabad	–	–	0	–	192	–	–	–	–	–	–
	IIVR, Varanasi	–	–	–	–	–	–	–	–	–	–	–
	YSPUHF, Solan	–	–	–	–	–	–	–	–	–	–	–
	TNAU, Coimbatore	–	–	–	–	–	–	–	–	–	–	–
	IIHR, Hessaraghatta	–	–	200	53	650	400	400	420	750	800	0
	CHES, Ranchi	–	–	–	–	–	–	–	–	–	–	–
	CSUA&T, Kalyanpur	–	–	–	0	154	0	0	0	0	113	0
Turnip	USPUH&F, Solan			18	18	0	–	–	–	–	–	
Watermelon	ARS (RAU), Durgapur	–	–	–	–	–	0	400	40	40	40	20
	NDUAT, Faizabad	–	–	–	–			0	0	10	0	0
	College of Agril. (RAU), Udaipur	–	–	203	205	205	–	–	–	–	–	–
	PAU, Ludhiana	–	–	–	–	–	–	–	–	–	–	–
	IIHR, Hessaraghatta	–	–	–	–	–	–	–	–	–	–	–
	MPKV, Rahuri	–	–	–	100	100	–	–	–	–	–	–
Parwal	BAC (RAU), Sabour	–	–	–	0	28	–	27	0	24	16	23
	NDUAT, Faizabad	–	–	–	–	–	–	–	–	36	55	30
	IIVR, Varanasi	–	–	–	–	–	–	–	–	–	–	–
	CHES, Ranchi	–	–	–	–	–	–	–	–	–	–	–
	HARP, Ranchi	–	–	–	–	–	–	–	–	–	–	–
	BCKV, Kalyani	–	–	–	12	12	–	0	0	12	0	28
Ivy gourd	IIVR, Varanasi	–	–	–	–	–	–	–	–	–	–	–
	IGKVV, Raipur	–	–	–	–	–	–	–	–	–	–	–
	KAU,Vellanikkkara	–	–	–	–	–	–	–	–	–	–	–
	HARP, Ranchi	–	–	–	–	–	–	–	–	–	–	–

Workshop/ Group Meeting (Year)													
XII Aug. 1992	XIII Dec. 1993	XIV Feb. 1995	XV March 1996	XVI May 1997	XVII June 1998	XVIII Oct. 1999	XIX Jan. 2001	XX April 2002	XXI May 2003	XXII May 2004	XXIII April 2005	XXIV April 2006	XXV May 2007
0	209	34	34	40	–	–		2	5	4	6	–	–
–	–	–	–	–	–	–	–	–	15	34	6	–	
–	–	–	–	–	–	–	–	–	–	–	–	–	3
412	0	0	0	0	–	–	–	–	–	–	–	–	–
–	–	–	–	–	–	–	–	–	–	–	–	–	–
–	–	–	–	–	–	–	–	–	–	–	–	–	–
–	–	–	–	–	–	–	–	–	–	–	–	–	–
–	–	–	–	–	–	–	–	–	–	–	–	–	–
–	–	–	–	–	–	–	–	–	–	–	–	–	–
–	–	27	–	–	38		–	–	–	–			–
–	–	–	–	–	–	–	–	–	–	–	–	–	–
0	0	0	0	0	–	–	–	–	–	–	–	–	120
–	–	–	–	–	–	–	–	–	–	–	–	–	–
0	–	42	42	24	12	–	–	–	17	3	–	–	–
–	65	–	–	–	–	–	–	–	–	–	–	–	–
–	–	–	–	459	448	–	436	475	500	487	500	–	57
–	–	–	–	50	25	–	6	69	–	89	7	4	120
–	–	–	–	124	124	126	128	128	6	134	73	75	75
1000	22	0	10	–	–	–	–	–	–	–	8	–	–
–	–	–	–	–	–	–	38	128	–	–	–	–	–
260	0	0	0	–	–	–	–	–	–	–	–	–	–
–	–	–	–	–	–	–	–	–	–	–	–	–	–
20	30	0	36	0	36	31	–	35	–	1	–	10	–
12	0	0	0	–	–	–	–	–	–	–	–	–	–
–	–	–	–	–	–	–	–	–	–	–	–	–	–
–	–	–	–	–	–	–	–	–	1	–	–	–	–
–	–	–	–	–	–	–	–	–	–	–	–	–	10
–	–	–	–	–	17	–	–	–	–	–	–	–	–
0	22	18	52	–	16	20		7	8	9	–	17	–
0	30	40	34	34	34	28	10	28	28	–	–	25	–
–	–	–	–	180	146	167	167	119	225	23	–	8	–
–	–	–	–	–	–	–	56	–	–	–	–	–	–
–	–	–	–	–	–	–	–	–	15	17	51	59	13
28	24	–	14	17	21	21	12	9		–	–	12	–
–	–	–	–	–	45	39	39	60	18	23	–	–	–
–	–	–	–	–	–	–	35	–	–	–	–	–	–
–	–	–	–	–	–	–	–	21	18	5	7	–	12
–	–	–	–	–	–	–	–	–	–	–	–	–	2

(Contd.)

Crops	*Centres*	*Workshop/ Group Meeting (Year)*										
		I Feb. 1971	*II May 1972*	*III April 1975*	*IV May 1977*	*V March 1979*	*VI June 1981*	*VII May 1983*	*VIII June 1985*	*IX Jan. 1987*	*X Aug. 1988*	*XI June 1990*
Dolichos bean	IIVR, Varanasi	–	–	–	–	–	–	–	–	–	–	–
	CHES, Ranchi	–	–	–	–	–	–	–	–	–	–	–
	TNAU, Coimbatore	–	–	–	–	–	–	–	–	–	–	–
	HARP, Ranchi	–	–	–	–	–	–	–	–	–	–	–
	Tripura	–	–	–	–	–	–	–	–	–	–	–
	JNKVV, Jabalpur	–	–	–	–	–	–	–	–	–	–	–
	IGKVV, Raipur	–	–	–	–	–	–	–	–	–	–	–
Cowpea	HARP, Ranchi	–	–	–	–	–	–	–	–	–	–	–
	IIHR, Hessaraghatta	–	–	–	–	–	–	–	–	–	–	–
	IIVR, Varanasi	–	–	–	–	–	–	–	–	–	–	–
	IGKVV, Raipur	–	–	–	–	–	–	–	–	–	–	–
	NBPGR, New Delhi	–	–	–	–	–	–	–	–	–	–	–
Spine gourd	OUAT, Bhubaneswar	–	–	–	–	–	–	–	–	–	–	–
Total		0	0	3523	3047	8112	3292	3726	4438	4936	2664	6690

Workshop/ Group Meeting (Year)													
XII Aug. 1992	*XIII Dec. 1993*	*XIV Feb. 1995*	*XV March 1996*	*XVI May 1997*	*XVII June 1998*	*XVIII Oct. 1999*	*XIX Jan. 2001*	*XX April 2002*	*XXI May 2003*	*XXII May 2004*	*XXIII April 2005*	*XXIV April 2006*	*XXV May 2007*
–	–	–	–	–	19	–	17	28	–	43	42	–	110
–	–	–	–	–	–	–	28	30	25	–	–	–	–
–	–	–	–	–	–	–	–	6	–	–	31	12	–
–	–	–	–	–	–	–	–	–	–	41	–	–	14
–	–	–	–	–	–	–	–	–	–	–	–	31	
–	–	–	–	–	–	–	–	–	–	–	–	–	20
–	–	–	–	–	–	–	–	–	–	–	–	–	25
–	–	–	–	–	–	–	–	–	–	–	–	50	50
–	–	–	–	–	–	–	–	–	–	–	–	34	34
–	–	–	–	–	–	–	–	–	–	–	–	50	20
–	–	–	–	–	–	–	–	–	–	–	–	–	19
–	–	–	–	–	–	–	–	–	–	–	–	–	49
–	–	–	–	–	–	–	–	–	–	–	–	–	5
3730	1367	1162	2248	2528	2411	1561	2408	2458	2867	2634	2544	2200	2895

SOURCES OF AVAILABLE GERMPLASM COLLECTIONS

Crop	India	Abroad
Tomato	IIVR, Varanasi, IARI Regional Station, Katrain IARI, New Delhi, PAU, Ludhiana, HAU, Hissar GBPUA&T, Pantnagar, CSAUA&T, Kanpur NDUA&T, Faizabad, BCKVV, Kalyani, TNAU Coimbatore, OUA&T, Bhubaneshwar, GAU Anand, IIHR, Hessarghatta, KAU, Vellanikkara YSPUA&F, Solan, MPKVV, Rahuri	Austarlia, Bulgaria, Canada, Cuba, France, Germany, Italy, Netherland, Taiwan, USA, Russia, Denmark
Brinjal	IIVR, Varanasi, AAU, Jorhat, APAU, Hyderabad, CSAUA&T, Kanpur, GAU, Anand, GBPUA&T, Pantnagar, HAU, Hissar IARI Reg. Station, Katrain, IARI, New Delhi NDUA&T, Faizabad, MPKVV, Rahuri, OUA&T Bhubaneswar, TNAU, Coimabtore, IIHR, Hessarghatta, CHES, Ranchi KAU, Vellanikara	Australia, Bangladesh, Brazil, Denmark, France, Hungary, Nigeria, Sri Lanka, Phillipines, USA, Russia
Capsicum spp.	IIVR, Varanasi, IARI, Reg. Staition, Katrain IARI, New Delhi, NBPGR, Amravati, Lam APAU, Hyderabad, PKV, Akola, MPKVV, Rahuri OUA&T, Bhubaneswar, PAU, Ludhiana YSPUA&F, Solan, GBPUA&T, Pantnagar TNAU, Coimbatore, IIHR, Hessarghatta JNKVV, Jabalpur, DARL, Almora	Australia, Bulgaria, Czechoslovakia, Hungary, Italy, Netherland, Nigeria, Brazil, Taiwan, USA, Russia, Yeman, France, Germany, Sri Lanka
Pea	IIVR, Varanasi, IARI, Reg. Station, Katrain IARI, New Delhi, PAU, Ludhiana HAU, Hissar, GBPUA&T, Pantnagar CSUA&T, Kanpur, NDUA&T, Faizabad ARS, Durgapura, MPKVV, Rahuri JNKVV, Jabalpur, YSPUA&F, Solan IIHR, Hessarghatta, VPKAS, Almora IIHR, Hessarhatta	Nordic Gene Bank, Lund, Sweden, Taiwan, Netherland, Italy, USA, Russia, Czechoslovakia, Germany
Cowpea	IIVR, Varanasi, BCKVV, Kalyani, IARI, New Delhi NDUA&T, Faizabad, PAU, Ludhiana IIHR, Hessarghatta	IITA, Nigeria, USA, Brazil, Tanzania, Burkinpo Faso, Ghora, Kenya, Senegal, Haithi, Phillippines,
Frenchbean	IIVR, Varanasi, IARI, Reg. Station, Katrain IARI, New Delhi, YSPUA&F, solan, IIHR, Hessarghatta, ICAR, Res. Complex for NEH region, Barapani, HPKVV, Palampur VPKAS, Almora	USA, Columbia, Canada, Netherland, Poland, Phillipines, Nigeria, Australia
Allium species	NBPGR, New Delhi, MPKV, Rahuri TNAU, Coimbatore, NHRDF, Naski NHRDF, Karnal, UAS, Hebbal, IIHR, Hessarhatta GBPUA&T, Pantnagar, IARI, New Delhi PAU, Ludhiana, VPKAS, Almora YSPUH&F, Solan	Australia, Brazil, Denmark, France, Germany, Hungary, Indonesia, Iran, Italy, Japan, Hethetland, Nigeria, Polanad, Syria, Taiwan, USA, USSR

(Contd.)

Crop	India	Abroad
Cucubitaceous crops	IIVR, Varanasi, IARI, Regional Station, Katrain IARI, New Delhi, PAU, Ludhiana, HAU, Hissar GBPUA&T, Pantnagar, CSUA&T, Kanpur NDUA&T, Faizabad, BCKVV, Kalyani TNAU, Coimbatore, OUA&T, Bhubaneswar GAU, Anand, IIHR, Hessarghatta, KAU, Vellanikkara, YSPUA&F, Solan, MPKVV, Rahuri	China, Japan, USA, Russia, Ghana, Namibia, South Africa, Brazil, Peru, Bulgaria, Mexico, India, Phillipines, Thailand, Indonesia, Hungary, Guatemala, Afganistan, Taiwan UK, Canada, France, Germany, Italy, Japan, Korea, Netherland, USA, Russia, Denmark, China
Cole crops	IIVR, Varanasi, IARI, Regional Station, Katrain, GBPUA&T, Pantnagar, MOKVV, Rahuri, YSPUA&F, Solan, IARI, New Delhi, PAU, Ludhiana	
Root crops	IARI, Reg. Station, Katrain, IARI, New Delhi PAU, Ludhiana, HAU, Hissar, CSAUA&T, Kanpur TNAU, Coimbatore, IIHR, Hessarghatta YSMUH&F, Solan	Korea, Japan, France, Hungary, Italy, Netherland, Poland, Taiwan, UK, USA, Russia, Australia, Brazil, China, Egypt
Okra	IIVR, Varanasi, NBPGR, New Delhi, Akola, Trichur, Jodhpur, PAU, Ludhiana, GAU, Anand, MAU, Parbhani, IIHR, Hessarghatta, HAU, Hissar, TNAU, Coimabtore, IARI, New Delhi	Sri Lanka, Bangladesh, Phillipines, Sudan, Senegal, Nigeria, Cote, d Ivoive, Brazil, China, Indonesia, Australia, Popua, New Guinea, USA

VARIETIES

Tomato

Open-pollinated varieties

Following varieties have been identified and recommended for cultivation in various zones:

Variety	Remark	Developing Centres	Recommended Zones	Year of Identification
S-12	Small fruited	PAU	–	1975
Pusa Ruby	Small fruited	IARI	–	1975
HS-101		HAU	–	1975
SL-120	Large fruited	IARI	–	1975
Sweet-72	Large fruited	Gwalior	–	1975
T-1	Large fruited	CSAUA&T	–	1975
Pusa Early Dwarf	Small fruited	IARI	–	1977
Sioux	Small fruited	-	–	1977
Sel-152	Small fruited	PAU	–	1977
Punjab Chhuhara	Small fruited	PAU	–	1977
KS-2	Determinate	CSAUA&T	Uttar Pradesh	1985
AC-238	Indeterminate	GBPUA&T	–	1985
CO-3	Determinate	TNAU	I,IV,V,VI,VII	1987
Pb. Kesari	Determinate	PAU	I,IV,V,VI	1987
La Bonita	Determinate	NBPGR	I,IV,V,VII	1987
Pant T-3	Indeterminate	GBPUA&T	I,II,IV,V,VI,VII,VII	1987

(Contd.)

Variety	Remark	Developing Centre	Recommended Zones	Year of Identification
Arka Vikas	Indeterminate	IIHR	I,IV,V,VI,VII,VIII	1987
Arka Saurabh	Indeterminate	IIHR	I,IV,V,VI,VII,VIII	1987
Sel-7	Determinate	HAU	I,IV,V,VII,VIII	1990
Sel-1-6-4	Determinate	PAU	I	1995
Sel- 32	Determinate	HAU	II,VI,VII	1996
DT-10	Indeterminate	IARI	IV,VI	1996
BT-12	Indeterminate	OUAT	I,IV	1996
KS-17	Indeterminate	CSAUA&T	IV	1998
BT-116-3-2	Determinate	OUAT	V, VI	2001
NDT-3	Determinate	NDUA&T	IV, VII	2001
KS-118	Determinate	CSAUA&T	IV	2001
DVRT-2	Determinate	IIVR	VI	2001
BT-20-2-1	Indeterminate	OUAT	IV, VII	2001
NDT-9	indeterminate	NDUA&T	IV	2001
NDTS-2001-3	Determinate	NDUA&T, Faizabad	IV	2004
Mani Laima	Determinate	ICAR Res. Com. for NEH, Barapani	III	2004
IIVR Sel-1	Determinate	IIVR	V, VII	2005
IIVR Sel-2	Indeterminate	IIVR	IV	2005
BT-136	Indeterminate	Bhubaneswar	II, IV	2005
VLT-32	Indeterminate	VPKAS, Almora	IV	2005

Hybrid variety

Following hybrids have been identified for cultivation:

Hybrid	Growth Habit	Developing Centre	Recommended Zones	Year of Identification
ARTH-4	Indeterminate	Ankur Seeds	IV,VIII	1992
MTH-6	Indeterminate	Mahyco	VII,VIII	1992
ARTH-3	Determinate	Ankur Seeds	II,VII,VIII	1992
Pusa Hybrid-2	Determinate	IARI	I,IV,VI,VII	1993
FMH-2 (A. Vardhan)	Indeterminate	IIHR	I, VII	1993
NA-501	Determinate	Nath Seeds	IV,VII	1995
DTH-4	Determinate	IARI	VII	1995
KT-4	Indeterminate	IARI (Katrain)	IV	1995
NA-601	Indeterminate	Nath seeds	VI,VII	1995
FMH-2	Indeterminate	IIHR	IV	1995
BSS-20	Indeterminate	Beejo Sheetal	IV,VI,VII	1996
Avinash-2	Determinate	Novartis	VI	1998
HOE-303	Determinate	Novartis	IV	1998
Sun-496	Indeterminate	Sungro	IV, II, VII	1999
BSS-20	Indeterminate	Beejo Sheetal	All	2001
DTH-8	Determinate	IARI	IV	2001
CHTH-1	Determinate	IARI	IV	2001
ARTH-128	Indeterminate	Ankur	VII	2001
KTH-2	Indeterminate	CSAUAT	IV, V	2003
JKTH-3055	Determinate	J.K. Seeds	I, IV	2004
KTH-1		CSAUAT	IV	2004
Nun-7730	Indeterminate	NunHems	I, IV	2004
TH01462	Determinate	Syngenta	I, II, IV, VI,VII	2005

Resistant variety

Following disease resistant varieties of tomato have been recommended

Variety	*Resistant Against*	*Developing Centre*	*Recommended Zones*	*Year of Identification*
BWR-5 (Arka Alok)	Bacterial wilt	IIHR	II	1992
FMH-1 (A. Vardhan)	Bacterial wilt	IIHR	V, VII	1993
FMH-2	Nematode	IIHR	IV	1995
BT-10	Bacterial wilt	OUAT	V,VI	1995
H-24	TLCV (moderate)	PDVR	V	1997
BRH-2	Bacterial wilt	IIHR	VII	1998
LE-415	Bacterial wilt	KAU,	I, V, VIII	2004
H86	TLCV	IIVR	I,IV,V,VIII	2005

Brinjal

Open-pollinated variety

In brinjal, following open pollinated varieties have been identified based on their performance under various agroclimatic zones:

Variety	*Fruiting Type*	*Developing Centres*	*Recommended Zones*	*Year of Identification*
Pusa Purple Long	Long	IARI	IV,VI,VII,VIII	1975
Pusa Purple Cluster	Long	IARI	IV,V,VI,VII	1975
S-16	Long	PAU	IV,VI	1975
Pusa Kranti	Long	IARI	IV	1977
PB -129-5	Long	GBPUA&T	IV	1981
Pant Samrat	Long	GBPUA&T	IV,V,VI	1981
Arka Sheel	Long	IIHR	VIII	1981
Azad Kranti	Long	CSAUA&T	I,VI	1983
PB -91-2	Round	GBPUA&T	–	1985
ARU-1	Long	DARL	I	1985
T-3	Round	CSAUA&T	–	1985
H-4	Long	HAU	–	1987
Kat-4	Long	IARI, Katrain	VIII	1987
ARU-2C	Long	DARL	I,IV,VI,VIII	1987
K-202-9	Round	GAU	VI	1987
Aruna	Small round	Akola	VII	1988
H-7	Long	HAU	IV,VI	1990
NDB-25	Long	NDUA&T	II, IV, VII	1990
H-8	Round	HAU	II, IV, V, VI	1990
BB-26	Long	OUAT	V	1993
Punjab Barsati	Long	PAU	IV	1993
Sel-4	Long	APAU	V	1995
DBSR-31	Long	IARI	VI	1995
KS-224	Round	CSAUA&T	I,II,IV	1995
DBR-8	Round	IARI	VI	1995
DBSR-44	Small round	IARI	VI	1995

(Contd.)

Variety	Fruiting Type	Developing Centres	Recommended Zones	Year of Identification
AB-1	Round	GAU	III,VI,VII	1996
PLR-1	Small round	TNAU	IV,VI,VII	1996
BB-26	Long	OUAT	V,VII	1996
BB-13	Long	OUAT	VIII	1997
JB-64-1-2	Small round	JNKVV	VII	1998
KS-331	Long	CSAUA&T	IV,V	1998
JB-15	Long	JNKVV	I	1998
CHBR-1	Round	CHES (Ranchi)	IV	1998
DBSR-91	Small round	IARI	VII	1998
JB-64-1-2	Small round	JNKVV	VII	1998
Green Long	Green	RAU (Sabour)	IV	1998
Punjab Sadabahar	Long	PAU	IV , VI	2001
NDB-28-2	Long	NDUA&T	IV	2001
DBL-21	Long	IARI, New Delhi	IV	2004
KS-235	Round	CSAUA&T,	IV, V & VII	2004
ABSR-2	Small long	GAU, Anand	VII	2004
HABR-4	Round	HARP, Ranchi	IV	2005
IVBR-1	Round	IIVR	IV	2005
HABL-1	Long	HARP, Ranchi	I	2006
PB-66	Long	GBPUA&T, Pantnagar	VII, IV	2007

Hybrid variety

In brinjal, following hybrid varieties have been identified, based on their performance under various agroclimatic zones

Hybrid	Fruiting Type	Developing Centres	Recommended Zones	Year of Identification
Arka Kusumkar	Long	IIHR	VIII	1981
Arka Navneet	Round	IIHR	-	1981
Kat-4	Long	IARI (Katrain)	VIII	1987
Pusa Hybrid-6	Round	IARI	IV	1990
Pusa Hybrid-5	Long	IARI	IV, VII, VIII	1992
ARBH-201	Long	Ankur Seeds	IV, V, VI, VII	1993
NDBH-1	Round	NDUA&T	IV, VI, VII	1993
ABH-1	Small round	GAU (Anand)	IV, VI, VII	1993
MHB-10	Small round	Mahyco	IV, VI, VII	1993
MHB-39	Small round	Mahyco	IV, VI, VII	1993
NDBH-6	Long	NDUA&T	IV	1995
ABH-2	Small round	GAU	IV,VI	1995
ABH-2	Small round	GAU	VII	1996
Phule Hybrid-2	Small round	MPKV	VII	1997
Pusa Hybrid-9	Round	IARI	VI	1997
Phule hybrid-2	Small round	MPKV	VII	1997
ARBH-541	Long	Ankur Seeds	All	2001
PBH-6	Long	Pandey Beej	All	2001
JBH-1	Round	Junagadh	All	2001

(Contd.)

Hybrid	*Fruiting Type*	*Developing Centres*	*Recommended Zones*	*Year of Identification*
BH-1	Round	Ludhiana	IV	2001
BH-2	Round	PAU, Ludhiana	IV, V	2002
VRBHR-1	Round	IIVR, Varanasi	IV, VI	2002
IVBHL-54	Long fruited	IIVR	IV	2004
ARBH-786	Long fruited	Ankur Seeds	IV	2004
VNR-51	Small round	VNR, Seed	IV, VI	2005
Navina	Long fruited	VNR Seeds	IV	2007
HABH-17	Round	Harp, Ranchi	IV	2007

Resistant variety

Following bacterial wilt resistant varieties of brinjal have been identified for cultivation under specific zones

Variety	*Developing Centres*	*Recommended Zones*	*Year of Identification*
BB-7	OUAT	II, V	1990
BWR-12	IIHR	VIII	1990
SM-6-7	OUAT	VIII	1992
SM-6-6	KAU	I,VII,VIII	1996
BB-44	OUAT	V,VII	1996
CHES-309	HARP	I, VII	2001
BB-64	OUAT	IV, V, VII, VIII	2004

Chilli and Capsicum

Open-pollinated variety

Following open-pollinated varieties of chilli have been identified and recommended for cultivation under specific conditions and zones:

Variety	*Developing Centres*	*Recommended Zones*	*Year of Identification*
G-4	Lam	–	1975
G-5	Lam	–	1975
K-2	TNAU (K)*	–	1985
J-218	JNKVV	I, IV, V, VI, VII	1987
X-235 (LCA-235)	Lam	IV, V, VI, VIII	1987
X-235	Lam	I, V	1987
Muslawadi	MPKV	V	1987
Sel-1	IIHR	V, VII, VIII	1990
LCA 206-B	Lam	V,VI,VII,VIII	1990
AKC-86-39	Akola	VII	2001
BC-14-2	OUA&T	V, VI	2001
RHRC-Cluster Erect	MPKV	VII	2001
PMR-57/88-K	IIHR	VII	2002
LCA334	LAM	III, IV, V, VII	2002
ASC-2000-02	GAU, Anand	VII	2004
KA-2	IIVR	IV	2005
LCA-353	RARS, Lam	V, VIII, IV	2007
BC-25	OUAT, Bhubaneshwar	VI, VII, V	2007

*TNAU - Kovilpatti

Hybrid variety

Under hybrid trials following chilli and Capsicum hybrids have been recommended:

Vegetable	*Hybrid*	*Developing Centres*	*Recommended Zones*	*Year of Identification*
Chilli	HOE-888	Sandoz	IV,VIII	1997
Chilli	ARCH-236	Ankur Seeds	IV	1997
Capsicum	KT-1	IARI (Katrain)	I	1990
Capsicum	SEL-II	Shalimar	I	1997
Capsicum	Lario	Syngenta	I	2001
Chilli	ARCH-228	Ankur Seeds	IV, V, VI	2002
Capsicum	DARL-202	DARL	I, IV	2002
Chilli	CCH-2	IIVR	II, IV,V,VI	2005
Sweet pepper	KTCPH-3	Katrain	I,VI,VII	2005

Garden Pea

Open-pollinated variety

Following pea varieties have been identified for vegetable purpose

Variety	*Maturity Group*	*Developing Centres*	*Recommended Zones*	*Year of Identification*
Bonneville	Mid Season	IARI	–	1975
GC-141	Mid Season	Gwalior	–	1975
GL-195	Mid Season	Gwalior	–	1975
Arkel	Early Season	IARI	–	1975
Early December	Early Season	Gwalior	–	1977
IP-3	Early Season	GBPUA&T	–	1985
P-88	Early Season	PAU	–	1985
PM-2	Early Season	GBPUA&T	I	1987
Lincon	Early Season	IARI (Katrain)	I	1987
VL-3	Early Season	VPKAS	I,IV,VI	1987
VL-7	Early Season	VPKAS	IV	1992
Ageta-6	Early Season	PAU	I,IV,VI	1993
VL-6	Mid Season	VPKAS	IV	1993
PH-1	Mid Season	HAU	VII	1993
PH-1	Mid Season	HAU	V	1995
NDVP-8	Mid Season	NDUA&T	IV	1997
NDVP-10	Mid Season	NDUA&T	IV	1998
VL-8	Mid Season	VPKAS	I	1998
VRP-2	Early Season	IIVR	VI	2001
NDVP-12	Early Season	NDUA&T	IV	2001
VRP-3	Mid Season	IIVR	I	2001
Organ Sugar Podded	Edible Podded	PAU	VI	2001
VRP-5	Early	IIVR	I, IV, VIII	2005
CHP-2	Mid season	HARP, Ranchi	IV,VI	2005
VP-101	Early	VPKAS, Almora	IV, I	2007
PC-531	Mid season	PAU, Ludhiana	VI, VII, I	2007

Resistant variety

Following powdery mildew resistant varieties have been identified:

Variety	Developing Centres	Recommended Zones	Year of Identification
PRS-4	CSAUA&T	IV,VI,VII	1990
JP-4	JNKVV	IV,VIII	1990
JP- 83	JNKVV	VII	1992
NDVP-4	NDUA&T	IV	1995
DPP-68	Palampur	–	2001
KS-245	CSAUA&T	–	2001
NDVP-250	Faizabad	V	2002
DPP-9411	HPKV	I	2003
KTP-8	Katrain	I, IV, V	2005

Other Leguminous Vegetables

Open-pollinated variety

Following varieties of legume vegetables have been identified for cultivation:

Vegetable	Variety	Developing Centres	Recommended Zones	Year of Identification
Cowpea	L-1552	–	–	1983
Cowpea	Sel-61-B	IHHR	VII,VIII	1992
Cowpea	Sel-263	PAU	IV	1992
Cowpea	Sel-2-1	NDUA&T	IV	1993
Cowpea	IIHR-6	IIHR	IV,VII	1998
Cowpea	NDCP-13	NDUAT	II, III, IV, VII	2002
Cowpea	IVRCP-1	IIVR	IV	2004
Cowpea	CHCP-2	HARP, Ranchi	VIII	2005
Cowpea	IIVR CP-4	IIVR	IV, V, VII	2007
Dolichos bean	Deepaliwal	Akola	V,VII	1990
Dolichos bean	CHDB-1	HARP,	IV	2004
French bean	VL-Boni-1	VPKAS	North Indian Hills	1985
French bean	Arka Komal	IIHR	I,VII,VIII	1987
French bean	UPF-191	GBPUA&T	IV,VIII	1987
French bean	IIHR-909	IIHR	I	1997
French bean	CH-812	CHES	VII, III	2001
French bean	CH-819	CHES	I	2001
French bean	IVFB-1	IIVR	I, VII	2005

Okra

Open-pollinated variety

In okra, all the recommended open pollinated varieties so far, have been identified under disease resistance trial.

Hybrid variety

Under hybrid trials following okra hybrids have been recommended:

Hybrid	Developing Centres	Recommended Zones	Year of Identification
DVR-1	IIVR	IV, VII	1998
DVR-2	IIVR	VI	1998
DVR-3	IIVR	All	2001
DVR-4	IIVR	IV, V, VII	2001
HBH-142	HAU	IV, V, VII, VII	2005
SOH-152	Syngenta	IV, VII, VII	2005
SOH-1016	Syngenta	IV,VII	2007
NBH-180	Nuzi Vedu Seeds	VII	2007

Resistant variety

YVMV resistant/tolerant okra varieties have been identified:

Variety	Developing Centres	Recommended Zones	Year of Identification
P-7	PAU	All regions	1990
PB-57	Parbhani	All regions	1990
Sel-10 (A. Anamika)	IIHR	All regions	1990
Sel-4 (A. Abhay)	IIHR	II	1992
HRB-55	HAU	VI	1995
HRB-9-2	HAU	IV,VI	1996
VRO-3	IIVR	IV, VI	2001
VRO-4	IIVR	IV, V	2001
VRO-5	IIVR	VI	2002
VRO-6	IIVR	IV; V	2002
NDO-10	Faizabad	IV	2005
HRB-107-4	Hisar	VI,VIII	2005
IIVR-11	IIVR	VI,VII	2005
JNDOL-03-1	Junagarh	VII and VIII	2007

Onion

Open-pollinated variety

Following varieties have been recommended for cultivation under specified zones:

Variety	Remark	Developing Centres	Recommended Zones	Year of Identification
Punjab Selection	Keeping quality	PAU	–	1975
Pusa Red	Keeping quality	IARI	–	1975
Pusa Ratnar	Red	IARI	–	1975
S-131	White	IARI	–	1977
N-257-9-1	White	MPKV	–	1985
N-2-4-1	Red	MPKV	–	1985
Line-102	Rabi Season	IARI	I, IV, VI, VII	1987
Arka Kalyan	Rabi Season	IIHR	IV,VI,VII,VIII	1987
Arka Niketan	Kharif season	IIHR	VII	1987
Agri Found Dark Red	Kharif season	NHRDF	IV	1987
VL-3	Rabi season	VPKAS	Uttar Pradesh Hills	1990

(Contd.)

Variety	Remark	Developing Centres	Recommended Zones	Year of Identification
Agri Found Light Red	Rabi season	NHRDF	VI	1993
Punjab Red Round	Rabi season	PAU	IV	1993
Agri Found Light Red	Rabi season	NHRDF	VIII	1995
PBR-5		PAU	VI	1997
L-28	Rabi season	NHRDF, Nashik	IV & VII	2006
HOS-1	Rabi season	HAU, Hisar	VI	2006
B-780-5-2-2	Rabi season	NRC (O&G)	VI	2007

Garlic

Following varieties have been identified for cultivation:

Variety	Developing Centres	Recommended Zones	Year of Identification
G-41	NHRDF	IV,VII	1988
G-1	NHRDF	IV,VI,VII	1990
G-50	NHRDF	IV	1993
G-282	NHRDF	IV,VI,VII	1998
VLG-7	Almora	I	2001
DARL-52	DARL, Pithoragarh	I	2002
G-323	NHRDF, Nasik	VI	2002

Cauliflower

Open pollinated variety

Following open pollinated varieties have been identified for cultivation:

Variety	Maturity Group	Developing Centres	Recommended Zones	Year of Identification
Early Kunwari	Early group	PAU	–	1975
327-14-8-3	Sep. maturity	IARI	–	1975
351-4-1	Oct. maturity	IARI	–	1975
Improved Japanese	Nov. maturity	IARI	–	1975
Synthetic-1	Dec. maturity	IARI	–	1975
EC 12012	Jan. maturity	IARI (Katrain)	–	1975
Pusa Snowball	Jan. maturity	IARI (Katrain)	–	1975
K-1	Jan. maturity	IARI (Katrain)	–	1979
114-S-1	–	GBPUA&T	–	1981
Line 6-1-2-1	Dec. maturity	IARI	–	1985
Early Synthetic	Early group	IARI	IV, VIII	1990
235-S	Nov. maturity	GBPUA&T	II, VIII	1990
KT-25	Snowball group	IARI (Katrain)	I	2001
IVREC-2	Early	IIVR	IV	2005

Hybrid variety

Following hybrids have been identified for cultivation:

Hybrid	Maturity Group	Developing Centres	Recommended Zones	Year of Identification
Pusa Hybrid-1	–	IIVR	II and IV	1992
DCH-541	Early group	IARI	II, IV	2003
SYCFH-202	Early group	Syngenta	IV, VII	2004
Summer King	Early group	Sungro	I, IV	2004
SYCFH-203	Early group	Syngenta	IV,V,VII	2005

Cabbage

Open-pollinated variety

Variety Sel-8 has been recommended in 1985 for cultivation.

Hybrid variety

Following hybrid varieties have been identified for cultivation:

Hybrid	Developing Centres	Recommended Zones	Year of Identification
Pusa Synthetic	IARI (Katrain)	IV, I, II	1992
Shri Ganesh Gol	Mahyco	V	1992
Nath-401	Nath Seeds	I, IV, V, VI, VII	1993
BSS-32	Beejo Sheetal	VII	1995
Nath-501	Nath Seeds	VII	1997
Quisto	Novartis	IV	1998
KGMR-1	Katrain	I, IV	2005
Green Emperor	Tokita Seeds	I	2007

Carrot

Open-pollinated variety

During 1975, variety Sel-5 was recommended for cultivation. Other varieties identified are:

Variety	Developing Centres	Recommended Zones	Year of Identification
Sel-5	IARI	VII	1975
Pusa Meghali	IARI	VII	1992
SKAUC-50	SKUA&T, Srinagar	I	2006

Hybrid variety

So far, only one hybrid variety (Hybrid 1) developed at Mahyco, has been identified for its cultivation in zones I and VII.

Muskmelon

Open-pollinated variety

Following varieties have been identified for cultivation in specific agro-climatic zones:

Variety	*Developing Centres*	*Recommended Zones*	*Year of Identification*
Pusa Sarbati	IARI	IV	1975
Hara Madhu	PAU	IV,VII	1975
SI-45 (Pusa Madhuras)	IARI	IV,VI,VIII	1975
Arka Rajhans	IIHR	VIII	1975
Arka Jeet	IIHR	VIII	1975
Durgpura Madhu	Durgapura	IV,VI,VII,VIII	1975
NDM 15	NDUAT,Faizabad	IV	2002
IVMM-3	IIVR, Varanasi	IV	2006

Hybrid variety

Following hybrid varieties have been identified for cultivation:

Variety	*Developing Centres*	*Source*	*Year of Identification*
Hybrid M-3	IARI	IV	1990
MHY-5	Durgapura	VII	2001

Resistant variety

Following disease resistant varieties of muskmelon have been identified:

Variety	*Developing Centres*	*Source*	*Year of Identification*
DMDR-1	IARI	CGMV	2001
DMDR-1	IARI	CGMV	2001
DMDR-2	IARI	Downy mildew + CGMV	2001

Watermelon

Open-pollinated variety

Following varieties have been identified for cultivation:

Variety	*Developing Centres*	*Recommended Zones*	*Year of Identification*
Durgapura Meetha	Durgapura	IV,VI,VII,VIII	1975
Sugar Baby	IARI	V,VII,VIII	1975
Arka Manik	IIHR	IV,VII,VIII	1987
MHW-6	Mahyco	–	1999

Hybrid variety

In 1981, one hybrid (Arka Jyoti) developed at IIHR has been recommended for cultivation.

Cucurbits

Open-pollinated variety

Following open-pollinated varieties have been identified:

Vegetable	Variety	Developing Centres	Recommended Zones	Year of Identification
Ash gourd	IVAG-90	IIVR,	IV &VIII	2006
Ash gourd	PAG-72	Pantnagar	VIII	2006
Ash gourd	Pusa Ujwal	IARI	VIII	2007
Bitter gourd	Priya	KAU	II,VII,VIII	1992
Bitter gourd	RHRBG-4-1	MPKV	IV,VII	1998
Bitter gourd	KBG-16	CSAUA&T	IV	1998
Bitter gourd	PBIG-1	GBPU&T	IV	2001
Bottle gourd	Pusa Naveen	IARI	VII	1992
Bottle gourd	OBOG-61	GBPU&T	IV,VI	2001
Bottle gourd	NDBG-104	Faizabad	IV	2002
Bottle gourd	NDBG-132	Faizabad	VI	2004
Cucumber	CHC-2	HARP	IV	2001
Cucumber	CH-20	HARP	IV	2001
Cucumber	PCUC-28	Pantnagar	I, VII, VIII	2001
Pumpkin	CM-14	KAU	IV,V,VII	1987
Pumpkin	Pusa Vishwas	IARI	I,VIII,IV,V	1987
Pumpkin	Arka Chandan	IIHR	VIII	1987
Pumpkin	Arka Suryamukhi	IIHR	VIII	1987
Pumpkin	CM-350	KAU	VII, VIII	2001
Pumpkin	NDPK-24	NDUA&T	IV,VI	2001
Ridge gourd	CHRG-1	HARP	IV	2001
Ridge gourd	PRG-7	GBPU&T	VII	2001
Ridge gourd	IIHR-7	IIHR	VIII	2001
Sponge gourd	Sel-99	IARI	IV,VI	1995
Sponge gourd	CHSG-1	HARP	IV	2005
Sponge gourd	JSGL	Junagarh	VII	2005
Sponge gourd	KSG-14	CSAUA&T	IV	2006
Sponge gourd	PSG-40	Pantnagar	VII, I	2007

Hybrid variety

Following hybrid varieties have been identified in different cucurbitaceous vegetables:

Vegetable	Variety	Developing Centres	Recommended Zones	Year of Identification
Bitter gourd	Pusa Hybrid-2	IARI	IV, V, VI	2002
Bitter gourd	NBGH-167	Nirmal Seeds	IV	2004
Bottle gourd	PBOG-2	GBPU&T	VII	2001
Bottle gourd	PBOG-1	GBPU&T	–	2001
Bottle gourd	NDBH-4	NDUA&T	All	2001
Cucumber	PCUCH-1	GBPU&T	All	2001
Cucumber	Hybrid No. – 1	Century Seeds	I, IV, VII	2004
Cucumber	PCUCH-3	GBPU&T	I, IV	2005

TECHNOLOGIES PRODUCTION

Tomato

Fertilizer and spacing

- Tomato varieties, HS-101 and Pusa Ruby, should be planted at a spacing of 60cm × 30cm with the application of NPK @ 150:60:60/ha under Sabour conditions to get maximum net return per hectare.
- For Pusa Ruby, HS-101 and Sioux varieties, 75 cm × 45 cm spacing and for KS-2 and HS-101 varieties, 50 cm × 45 cm spacing proved most suitable and recommended for agro-climatic conditions of Faizabad and Kanpur.
- Application of NPK @ 150:60:60 kg/ha has been recommended for Pusa Ruby, Sioux and KS-2, to get maximum return and highest cost:benefit ratio under Kanpur conditions.
- In summer crop, mulching with sugarcane truss has been recommended to economize irrigation and maximize yield and profit under agro-climatic conditions of Northern Indian plains. An additional income of Rs. 1000/ha could be by application of mulch obtained. The mulching should be done at about 35 DAT after one manual weeding.
- Under tarai conditions, irrigation at 200-mm open pan evaporation has been recommended to get the economic yield during summer season.
- Application of NPK @ 150:60:60 kg/ha has been recommended for obtaining the most economical yield from Pusa Ruby, Pusa Early Dwarf and Arka Vikas varieties under the agro-climatic conditions of Bhubaneshwar. Out of the three varieties, Arka Vikas proved to be most promising.
- Planting of Pusa Early Dwarf (determinate) at 50 cm × 50 cm and Sioux (semi-determinate) and Pusa Ruby (indeterminate) at 50 cm × 45 cm spacing has been recommended for Durgapura conditions of Rajasthan. Pusa Early Dwarf gave yield of 300 q/ha with cost : benefit ratio of 1:3.49 at the spacing of 50 cm × 30 cm. While Sioux gave yield of 279 q/ha with a cost : benefit ratio of 1:3.45, when transplanted at 50 cm × 45 cm distance.
- Application of NPK @ 75:60:60 kg/ha has been recommended for Pusa Ruby (indeterminate), Punjab Chhuhara (semi-determinate) and Pusa Early Dwarf (determinate) under Jorhat conditions. These varieties gave Cost : benefit ratio of 1:5.8, 1:3.97 and 1:3.72, respectively under the said fertilizer levels.
- For obtaining most economic yield, application of N_2 @ 40 kg/ha as basal dressing + 20 kg/ha as topdressing + 20 kg/ha as foliar spray has been recommended under Hyderabad and Solan conditions. This treatment resulted in 243.45 q/ha yield of Pusa Ruby at Hyderabad and gave the cost : benefit ratio of 1:2.57. Yield from this treatment at Solan was 693.25 q/ha with cost : benefit ratio of 1:4.86.
- Application of pendimethalin @ 1.0 kg a.i./ha as pre- emergence spray along with one hand weeding at 45 DAT, gave the highest yield (214.42 q/ha) with cost : benefit ratio of 1:2.36 from Pusa Ruby at Jaipur, hence this treatment has been recommended for Jaipur region.
- Application of N_2 @ 180 kg/ha and P_20_5 @ 120 kg/ha showed the maximum Cost :

benefit ratio (1:2.99) in determinate hybrid ARTH-3 and therefore, recommended for Sabour conditions.

- Application of N_2 as 40 kg/ha as basal dressing + 40 kg/ha as topdressing gave highest yield with Cost : benefit ratio of 1:3.09. Hence, it has been recommended for Kanpur conditions.
- Direct sowing in field on raised beds during first week of October has been recommended for highest net income (Rs. 48,221/ha) and C:B ratio (1:2.45) under *tarai* region of Pantnagar.
- Application of N_2 @ 40 kg/ha as basal dressing + 20 kg/ha through topdressing and 20 kg/ha as foliar spray was found conducive and gave highest yield (413q/ha) and C:B ratio (1:4.96) in Sel.-7. Hence, it has been recommended for Varanasi conditions of eastern Uttar Pradesh.
- For Shakati, application of 40 kg N_2/ha as basal application along with 20 kg N_2/ha as foliar spray gave a Cost : benefit ratio of 1:2.8, therefore, has been recommended under Kerala conditions.
- Under Pantnagar conditions, application of PSB + recommended dose of NPK (150:90:60 kg/ha) gave the highest average yield of tomato, i.e. 653.4 q/ha along with the maximum C:B ratio 1:3.19. Hence, it is recommended for *tarai* region of Pantnagar conditions.
- At Kalyanpur, application of Azospirillum with 75% N and 100% PK gave the highest average yield 267.44 q/ha of tomato TYPE-1 with the maximum C:B ratio 1:2.20. Hence, it is recommended for Kanpur conditions.
- Root dipping of tomato seedlings with Azotobactor along with 75% N + 100% PK recorded highest fruit yield (635.38 q/ha) and C:B ratio (1:3:26). Hence, it is recommended for Srinagar conditions.
- Under Coimbatore conditions, seedling root dip with the Azospirillum in addition to 75% N + 100% PK recommended resulted in highest yield 427.2 q/ha along with the maximum Cost: benefit ratio of 1:4:11 in tomato. Hence, it is recommended for Coimbatore conditions.
- At Bhubaneswar, integrated application of FYM @ 40 tonnes/ha + half the dose of NPK through fertilizers resulted in maximum yield in tomato Utkal Deepti with C:B ratio of 1: 1.62. Hence, it is recommended for Bhubaneswar conditions.
- At Jabalpur, maximum yield (310.68 q/ha) and C:B ratio (1:2.58) of Jawahar Tomato-99 was obtained with the application of 20 tonnes/ha FYM and full dose of N:P:K (180:120:80 kg/ha) through fertilizers. Hence, it is recommended for Satpura hills and Kaymore plateau region of M.P.
- Application of 240 kg N + 60 kg P_2O_5 and 80 kg K_2O/ha in hybrid tomato Meenakshi at Faizabad gave maximum yield (748.98 q/ha) and C:B ratio 1:3.60. Hence it is recommended for Faizabad conditions. While at Junagadh, the maximum yield (285.26 q/ha) and C:B ratio (1:2.66) were recorded in ARTH–3 with the same treatment. Hence, it is recommended for Junagadh condition.
- Based on the highest yield (643 q/ha) and C:B ratio (1:10.33), application of nitrogen

@240kg/ha and P_2O_5 120 kg/ha is recommended for tomato hybrid ARTH–3 under Kanpur conditions. For Hisar conditions, application of nitrogen @180kg/ha and P_2O_5 @60kg/ha is recommended for obtaining highest yield (372q/ha) and C:B ratio (1:4.9) in the same hybrid.

- Application 40 kg N/ha as basal + 20 kg/ha as topdressing + 20 kg/ha as foliar spray resulted in maximum yield (277 q/ha) with a cost: benefit ratio 1:246 in tomato JT–99, hence it is recommended for Kymore Plateau and Satpura hill area.
- The application of 180 kg/ha nitrogen and 120 kg/ha P_2O_5 gave maximum yield (503q/ha) and C:B ratio (1:2.07) in tomato hybrid ARTH–3, hence it is recommended for adoption under Varanasi conditions.
- At IIVR, application of recommended doses of NPK (120:60:60)+ FYM @ 10 tonnes/ ha + Sulphur at the rate of 25 kg/ha + Azotobacter + mixture of all micronutritients (Zn, B, Mo, Fe, Cu and Mn) resulted in best mean yield 413.83 q/ha along with highest C:B ratio 1:3.65 of tomato cultivar H-86. Hence, it is recommended for Varanasi conditions.
- At IIHR Bangalore, cucumber hybrid Tripti recorded the highest yield of 269.2 q/ha with the C:B ratio 1:2.17 by application of FYM @ 10 t/ha + half NPK (30:25:40 kg/ha) + Biofertilizers. Hence, it is recommended for Bangalore conditions.
- At Hisar, maximum mean yield (432.3 q/ha) and C: B ratio (1:2.66) were obtained in tomato on application of recommended NPK + FYM @ 10 t/ha + S@25 kg/ha + mixture of all micronutrients (Zn, B, Mo, Fe, Cu, Mn) + *Azotobactor*. Hence, it is recommended for Hisar conditions of Haryana. The same treatment also recorded the highest yield (336.20 q/ha) along with the maximum C: B ratio of 1:2.63. Hence, it is recommended for Kalyanpur conditions.
- At Faizabad, highest mean yield (368.35 q/ha) along with highest, C.B. ratio 1:3.02 was recorded in tomato Narendra Tomato-6 with the application of NPK @ 120:60:60 kg / ha + FYM 10 tonnes/ha + Sulphur 25 kg/ha + *Azotobactor* + MM (mixuture of all micronutrients, Zn, B, Mo,Fe, Cu, and Mn). Hence, it is recommended for Faizabad conditions.

Liquid Fertilizers

- At Durgapura, the maximum yield of tomato Pusa hybrid–2 (486.9 q/ha) and Cost: benefit ratio (1:1.5) were recorded with 5 foliar sprays of water–soluble fertilizer NPK (15:15:30 @ 0.5%) at 10 days interval over and above to the recommended dose of NPK (120:80:60 kg/ha). Hence, it is recommended for Durgapura condition. However, at Kalyanpur 5 foliar sprays of water–soluble fertilizer (NPK in the ratio of 17:10:27) over and above to the recommended dose of NPK (120:60:80 kg/ha) gave the highest yield (247.7 q/ha) and Cost: benefit ratio (1:3.08) with tomato Type–1. Hence it is recommended for Kalyanpur condition.
- At Hyderabad, 5 foliar sprays (0.5%) of water soluble fertilizer NPK (15:15:30) gave maximum yield (393.8 q/ha) and C:B ratio (1:3.84) in tomato cv. Marutham. Hence, it is recommended for Hyderabad conditions.
- The maximum yield of tomato hybrid Tolstoi (745.12 q/ha) and Cost: benefit ratio (1:4.12)

was recorded at IIVR, Varanasi with the 5 foliar sprayings of water–soluble fertilizer NPK (19:09:19 @ 0.5% at 10 days interval after 30 DAT over and above to the recommended dose of NPK (150:80:100 kg/ha). Hence, it is recommended for Varanasi conditions.

- At Kalyanpur, 5 foliar applications of water–soluble fertilizer NPK (17: 10:27) at 10 days interval resulted in maximum yield (241.5 q/ha) and C:B ratio (1: 2.97) in tomato Type–1. Hence, it is recommended for Kanpur conditions.
- At Jabalpur, maximum yield of tomato (269.28 q/ha) along with Cost: beneft ratio (1:2.08) were obtained with 5 sprays of Multi K 13:0:45 applied at 10 days interval after 40 days transplanting. Hence, it is recommended for Satpura hills and Kaymore plateau region.
- At Sabour, the highest yield (494. 11q/ha) and C:B ratio 1: 2.13 of tomato was recorded with the five foliar sprays of NPK (15:15:15) @ 5g / litre in addition to recommended dose of NPK. Hence, it is recommended for Sabour conditions of Bihar.
- At Coimbatore, highest mean fruit yield 712.50 q/ha and maximum C: B ratio 1: 4.78 were recorded in hybrid tomato with five foliar sprays of NPK (19:19:19). Hence, it is recommended for Coimbatore conditions of Tamil Nadu.

Micronutrients

- At IIVR, Varanasi, maximum yield (827.85 q/ha) and C:B ratio (1:4.81) were recorded with hybrid tomato Tolstoi by 3 foliar application of micronutrients (mixture of B, Zn, Cu, Fe, Mn each @ 100 ppm and Mo @ 50 ppm at 10 days interval 30 DAT over and above to the recommended dose of NPK –150:80:100 kg/ha). Hence, it is recommended for Varanasi region.
- Three foliar sprays of Borax @ 100 ppm over and above to the recommended dose of NPK at 10 days interval 30 DAT at Hisar gave maximum yield (372.8 q/ha) and C:B ratio (1:4.13) in Hisar Arun. Hence, it is recommended for Hisar conditions.
- At Pantnagar, maximum yield (314 q/ha) and C:B ratio (1:2.15) along with high TSS (5.7%) and shelf life (7.6 days) were recorded in Pusa Hybrid–1 tomato with three foliar sprays of micronutient mixture (B, Zn, Cu, Fe, Mn, each @ 100 ppm and Mo @ 50 ppm) at 10 days interval starting from 40 days after transplanting. Hence, it is recommended for Tarai conditions of Pantnagar.
- At Coimbatore, foliar application of $ZnSo_4$ at 100 ppm concentration thrice starting from 40 days after transplanting resulted in the maximum yield (645.6q/ha) along with C:B ratio (1:4.86) in tomato hybrid–1 (COTH–1). Hence, it is recommended for Coimbatore conditions.
- Under Durgapura conditions, three foliar applications of Ferrous sulphate @ 100 ppm at 40, 50 and 60 DAT gave the maximum mean yield of 454.11 q/ha along with highest C:B ratio 1:2.44 of tomato Pusa Hybrid–2. Hence, it is recommended for Durgapura conditions of Rajasthan.
- Spray of 100 ppm boron on tomato under Solan conditions, is recommended for higher yield of better quality seeds.
- 20 kg Borax and 20 kg calcium per ha is recommended for getting maximum seed yield

and maximum return in tomato based on the experiment conducted for three years at Jabalpur.

Irrigation

- Maximum yield (460.98 q/ha) and cost: beneft ratio (1:4.21) were obtained in Narendra Tomato–2 at Faizabad with the application of irrigation at 75 mm CPE level. Hence, it is recommended for Faizabad conditions.
- At Bhubaneswar, application of irrigation at 50 mm CPE resulted in yield of 251.4 q/ha with the highest Cost: beneft ratio of 1:1.81 in tomato. Hence, irrigation scheduling at 50 mm CPE in tomato is recommended under Bhubaneswar conditions.
- At Sabour, highest yield (483.40 q/ha) and Cost: beneft ratio (1:4.65) were obtained with 75 mm CPE in tomato hybrid Avinash–2. Hence, irrigation in tomato at 75 mm CPE is recommended under Sabour conditions.

Nursery

- Seedlings of hybrid tomato Apoorva raised in potting plugs provides better vigour and plant stand than the other methods. Further, when seedlings were transplanted in the field gave higher yield (391.96 q/ha) and Cost: beneft ratio (1:1.63). Hence, it is recommended for Kanpur conditions of Uttar Pradesh.
- The raising of seedlings of hybrid tomato ARTH–3 at Coimbatore gave better results when it is grown in poly bags. But, there was no significant difference in the field performance of the seedlings raised in different methods. Hence, raising of seedlings in poly bags is recommended for Coimbatore conditions.
- At IIVR, Varanasi, raising of seedlings of tomato hybrid Avinash–2 in polybags gave better performance and become ready to transplant within 20 days after seed sowing. But when the seedlings were transplanted under field condition, maximum average yield (1026.6 q/ha) and Cost: beneft ratio (1:5.72) was recorded under nursery raised by traditional method compared to others. Hence, it is recommended for Varanasi condition.
- Based on the higher germination percentage, seedling vigour and economics (20 paise per seedling extra profit over control), seedling raising of hybrid tomato in 10cm × 6cm size polyethylene bags on raised bed with polyethylene layer is recommended under Pantnagar conditions of *tarai* area.

Organic Farming

- Application of 20 tonnes FYM + ½ recommended dose of NPK (150:80:100 kg/ha) in tomato hybrid Avinash–2 gave maximum yield (773.0 q /ha) along with C:B ratio (1:4.00), hence it is recommended for cultivation under the Varanasi region of Uttar Pradesh.
- At Faizabad, maximum yield (332.05 q/ha) and C:B ratio (1:3.45) of tomato Narendra Tomato–2 was recorded with green manuring + recommended dose of NPK (60:30:30 kg/ha) when the seedlings were transplanted at 60 cm × 45 cm. Hence, it is recommended for Faizabad condition.
- The highest yield (346.87 q/ha) and C:B ratio (1:2.35) was obtained with the application of FYM @ 20 t/ha + full recommended dose of NPK (150-60-60 kg/ha) in tomato S–7 at

Sabour. Hence , it is recommended for Sabour condition of Bihar.

- Application of FYM @ 40 tonnes/ha + full recommended dose of NPK (180:120:80 kg/ha) in tomato hybrid ARTH-3 gave maximum yield (329.69 q/ha) and C:B ratio (1:1.49). Hence, it is recommended for Jaipur conditions of Rajasthan.

Weed control

- For chemical weed control, pre-emergence application of oxyfluorofen @ 0.25 kg a.i./ ha, combined with one hand-weeding at 45 DAT has been recommended for Tamil Nadu conditions.
- The pre-planting incorporation of Goal @ 0.25 kg a.i. and basalin @ 1.0 kg a.i./ha has been recommended for weed control and getting economic yield under Sabour conditions.
- For weed control, Goal @ 0.25 kg a.i./ha as pre-planting incorporation or sencor @ 0.75 kg/ha as pre-emergence application followed by post-emergence spray of sencor @ 0.5 kg/ha, were found to be most effective and economical. The next best herbicide was basalin @ 1 kg a.i./ha as pre-planting incorporation. These treatments have been recommended for Sabour and Pantnagar regions.
- Application of pendimethalin @ 1.0 kg a.i./ha (PE) + 1 hand-weeding at 45 DAT has been recommended for effective weed control under mid hill conditions of Almora. Application of this treatment gave 114.68 q/ha yield with C:B ratio of 1:1.68.
- Trials were conducted for three years on effect of mixtalol. The results showed that spray of 4 ppm mixtalol four weeks after planting resulted in additional yields of 4.86%.

Crop Rotation

- The cropping sequences of paddy-tomato has been recommended for upland and medium lowland conditions of Jorhat, which gave maximum C:B ratio of 1:4.68.
- The cropping sequence of cowpea-tomato-cucumber resulted in maximum net return, followed by tomato-pea-okra. Hence, these cropping sequences have been recommended for Kaymore plateau and Satpura hills of Madhya Pradesh.

Protected Cultivation

- At Pantnagar, maximum yield of tomato hybrid Avinash-2 (1227.0q/ha) and C:B ratio (1:4.26) was recorded when seedlings were transplanted at 50cm × 30cm spacing maintaining single shoot per plant with staking under low cost polyhouse. Hence, it is recommended for cultivation under *tarai* region of Pantnagar.
- At IIHR, Bangalore, maximum yield of 1329.0 q/ha was obtained under protected cultivation of hybrid tomato SH-7611 with a spacing of 50cm × 20cm by following double stem training, which is recommended for Bangalore conditions.
- In naturally ventilated polyhouse, maximum yield (1399.64 q/ha) along with highest C:B ratio 1:3.92 in hybrid tomato Tolstoy was obtained with double stem training at 50cm × 20cm spacing. Hence, it is recommended for Varanasi conditions.

Staking and Pinching

- Planting of tomato hybrid ARTH–4 at 80cm × 30cm spacing on raised bed subjecting to staking and without pinching of shoots is recommended for Kanpur conditions.

- Planting of tomato indeterminate hybrid ARTH–4 at 80cm × 60cm spacing on raised beds along with staking is recommended for higher yield (334 q/ha) and C:B ration (1:3:41) under Pantnagar conditions. Planting at 80cm × 45cm spacing along with staking is recommended for maximum yield (439 q/ha) and C:B ratio (1:4:52) under Coimbatore conditions. For Hisar and Durgapura conditions, planting of tomato hybrid ARTH–4 at 80cm × 60cm is recommended. Pinching of side shoots up to 30cm was not found economical at any of the above locations.
- Planting of tomato hybrid ARTH–4 at 80cm × 45cm along with staking and pinching of side branches at 30cm from ground gave highest yield (372.36 q/ha) and maximum C:B ratio (1:2.84). Hence, it is recommended for cultivation under Sabour conditions.
- Planting of tomato hybrid ARTH–4 at 80cm × 45cm along with staking and without pinching gave maximum yield (470.80 q/ha) and maximum C:B ratio (1:2.92), hence it is recommended for cultivation under Ambajogai conditions.

Others

- Based on two years' study at IIHR, Bangalore, on tomato hybrid Arka Abhijit, retention of 2 or 3 flowers/truss, emasculation and simultaneous pollination on the day or a day before the anther dehiscence and a 4:1 female and male flower ratio is recommended for good fruit set, whereas staggered sowing of male parent at one week before and one week after female parent, is recommended for better availability of pollen–grain.
- It is recommended that seeds should be extracted from first three pickings for higher recovery of good quality seeds in tomato based on trials conducted for three years at Solan, IIVR, Ludhiana and IIHR.
- It is recommended that the seeds should be extracted from first picking in tomato for maximum test weight, germination (%) and vigour under Bhubaneswar conditions.
- Priming with KNO_3 (150ppm) produced maximum germinability and vigour in Tomato (Sel–7) seeds at IIVR
- On the basis of two years' studies seed treatment with carbendazim (2%) in tomato under Solan conditions is recommended for higher seed yield and quality.

Vegetable Biochemistry

- At Pantnagar centre, three years experiments with 2,4–D, atonik and mixtalol treatments showed that highest average yield (315 q/ha) was obtained when 5 ppm 2,4–D was sprayed three times (first spray in nursery one day before transplanting, second spray at first flower cluster appearance and third spray at fourth flower cluster appearance).
- Thirty two varieties were analysed for TSS, vitamin C, dry weight, acidity and lycopene. The maximum TSS was recorded in NA–501 (5.2), followed by NA–701, XLE 006, DTH–6 and ARTH–15, whereas maximum dry matter content was found in Pusa Hybrid followed by HOE–606. The highest vitamin C content was recorded in Selection 30 followed by PH–2. The highest acidity was recorded in Phule Hybrid followed by AN–2. Maximum lycopene content was recorded in BSS–39.

Brinjal

Fertilizer and Spacing

- In Pusa Purple Long, planting spacing of 60 cm x 40cm with 150 kg/ha of nitrogen (N_2) and 100 kg/ha of phosphorous (P_2O_5) were found optimum at IIHR, hence, this treatment has been recommended for Banglore conditions.
- Planting at spacing of 75 cm x 40 cm along with optimum fertilizer dose of 100 kg N_2 + 60Kg P_2O_5/ha has been recommended for varieties H–4 and Pusa Purple Long under Hisar conditions.
- Application of 50 kg/ha N_2 as basal dose and spacing of 75 cm x 60 cm has been recommended for the variety Pusa Purple Long, while for PBR 129–5, 100 kg/ha N_2 as basal dose with same spacing, has been recommended for Pantnagar region.
- Maximum yield (245 q/ha) and C:B ratio (1:3.59) were recorded with the application of 120 kg/ha N_2 in variety BB–1 planted at the spacing of 75cm x 50cm. Hence, this treatment has been recommended for Bhubneshwar conditions.
- At Jorhat application of FYM @ 10 tonnes/ha + rest NPK through fertilizers resulted in higher yield of brinjal (244.5 q/ha) as well as okra (131.5 q/ha) along with maximum C:B ratio (1:3.35) for whole cropping system. Hence, it is recommended for Jorhat conditions.

Liquid Fertilizers

- At Hisar, 3 foliar sprays of water-soluble fertilizer NPK (19:19:19 @ 0.5 % at 10 days interval after 30 DAT over and above to the recommended dose of NPK) in brinjal Hisar Shyamal gave maximum yield (259.8 q/ha) and C:B ratio (1:2.89). Hence, it is recommended for Hisar conditions.
- The highest yield of brinjal Pusa hyrid–6 (329.0 q/ha) along with C:B ratio of 1:3.20 were recorded with 5 foliar sprays of water-soluble fertilizer NPK (17-10-27 @ 0.5 %) at 10 days interval after 30 DAT over and above to the recommended dose of NPK-150:90:90 kg/ha. Hence, it is recommended for Pantnagar conditions.
- At Hyderabad, maximum yield of brinjal (365.6 q/ha) with C:B ratio (1:3.8) were obtained with 5 sprays of water-soluble fertilizer having the combination of NPK 15 : 15 :30. Hence, it is recommended for Hyderabad conditions.
- At Coimbatore foliar application of water soluble fertilizer NPK (19 :19: 19) 5 times at 10 days interval starting from 40 days after transplanting resulted in maximum yield (556 q/ha) and C:B ratio (1:5.05) in Brinjal Hybrid–1 (COBH–1). Hence, it is recommended for Coimbatore conditions.
- At Faizabad, in addition to the recommended dose of NPK (150:80:80) five foliar applications of water-soluble NPK fertilizer (19:10:27) after 30 days of transplanting at 10 days interval resulted in maximum yield (395 q/ha) and highest C:B ratio (1:3.53) of brinjal hybrid Suchitra. Hence, it is recommended for Faizabad conditions.
- At Vellanikkara, 5 foliar applications of multi-K at 10 days interval beginning 30 DAP resulted maximum yield (496 q/ha) and C:B ratio (1:1.6) in brinjal hybrid Neelima. Hence, it is recommended for Vellanikkara conditions.

- Trials conducted at Jabalpur on brinjal Jawahar Brinjal 64, maximum mean of yield 243.72 q/ha with highest C:B ratio, i.e., 1:3.02 was obtained by five foliar sprays of NPK (19:19:19) @ 5 g/litre in addition to recommended dose of NPK. Hence, it is recommended for Kymore plateau and Satpura hills.
- Spray of 100ppm ferrous sulphate on brinjal under Kanpur conditions is recommended for higher yield of better quality seeds.
- Foliar application of a mixture of micronutrients (i.e., 100ppm each of zinc sulphate, manganese sulphate, copper sulphate, ferrous sulphate, boric acid and 50ppm ammonium molybdate) is recommended for higher seed yield of brinjal in Maharashtra.

Weed Control

- For weed control, application of fluchloralin @ 1.5 kg a.i./ha or pre-emergence spray of alachlor @ 2.0 kg a.i./ha + one hand-weeding 30 days after transplanting (DAT) have been recommended under Pantnagar conditions. This treatment combination gave a C:B ratio of 1:1.17.
- Pre-emergence application of pendimethalin @1.0 kg a.i./ha + one hand-weeding at 30 DAT has been recommended for effective weed control under Sabour conditions.
- Application of fluchloralin @ 1.5 kg a.i./ha + one hand weeding at 30 DAT has been recommended for Hyderabad areas. The maximum yield under this treatment was recorded 213.54 q/ha with C:B ratio of 1:2.43.
- Application of fluchloralin @ 1.5 kg a.i./ha (PPI) supplemented with one hand-weeding at 30 DAT proved to be most effective and gave highest yield (758 q/ha) and C : B ratio (1:2.50). Hence, this treatment has been recommended for weed control in *kharif crop* under Kanpur conditions.
- Results on three years experiment showed that spray of 4 ppm mixtalol at 4–6 week after transplanting gave an additional yield of 7.1%.
- Application of Pendimethalin @ 1.5 kg a.i./ha + one hand-weeding at 30 DAT gave maximum fruit yield of brinjal JBL-1 (341.66 q/ha) and C:B ratio 1:3.14 at Junagadh. Hence, it is recommended.

Cropping Sequence

- Application of neem cake @ 5 q/ha and rest of the recommended NPK through chemicals fertilizer gave the highest mean yield of 415.86 q/ha along with maximum C:B ratio of 2.34 in brinjal. The residual effect of the same treatment was found to be the best for okra crop producing the highest mean yield of 133.71 q/ha with the maximum C:B ratio of 1:1.82. Hence, it is recommended for Kalyanpur conditions.
- At Bhubaneswar, application of FYM @ 10 tonnes/ha and recommended doses of NPK (125 : 50 : 75 kg/ha) gave highest fruit yield of brinjal (229. 36 q/ha) along with maximum C:B ratio of 1: 4.05. However, residual effects of treatment did not show significant improvements in succeeding okra yield.
- Under brinjal and okra cropping sequence at Faizabad, maximum mean yield of brinjal, i.e. 269.55 q/ha and C:B ratio 1:1.57 was recorded with the application of neem cake @ 5 q/ha + recommended dose of NPK through chemical fertilizers. The residual effect of same treatment gave highest C:B ratio 1: 2.16 in okra. Hence, this is recommended for Faizabad conditions.

Harvesting and Seed Extraction

- Harvesting of brinjal fruits at fully ripen stage is recommended for production of good quality seeds in cultivars Pant Rituraj and H-8 at IIVR, Varanasi, in Arka Neelkant at Bangalore, in BR-112 and H-8 at Hisar, in Pusa Purple Cluster and Pusa Kranti at Solan.
- Brinjal fruits under Karnataka conditions should be harvested 54-57 days after anthesis to obtain good quality seeds.
- On the basis of trial conducted at Solan, it is recommended that in brinjal (Pusa Purple Cluster) seeds may be extracted by using acid fermentation @ 10 ml/kg of pulp for 45 minutes whereas, in Pusa Kranti, fermentation in water for 48 hours should be used.
- Extraction of seeds from brinjal by fermentation in water for 48 hours is recommended for maximum recovery and good quality seeds on the basis of trials conducted at Hisar, Varanasi, Bangalore, Bhubaneswar and Vellanikkara.
- Acid fermentation @ 10 ml/kg pulp for 30 minutes is recommended in brinjal cv. JB-64, under Kymore plateau and Satpura hills conditions of Madhya Pradesh.

Others

- It is recommended that to avoid natural cross-pollination (NCP) in brinjal, crop should be planted at a distance of 200 feet based on the experiment conducted for four years at PAU, Ludhiana.
- On the basis of two years' studies, seed treatment with 0.25% HCl dip for 30 minutes in brinjal under Solan conditions is recommended for higher seed yield and quality.
- Maximum germination and minimum hard seed per cent in brinjal (Pb. Sadabahar) was observed at IIVR when after extraction the seeds were pre-treated for 30 minutes with 50 °C hot water or 0.25% HCl.
- The brinjal seed treatment with Trichoderma (5g/kg seed) is recommended to have maximum seed yield, germination percentage and diseases free seed under Maharashtra and Madhya Pradesh. conditions.
- Brinjal seed treatment with hot water (50 °C for 30 minutes) and sodium hypochlorite (0.5% for 30 minutes) is recommended for significantly improved seed germination, based on trials at Ludhiana.

Chilli and Capsicum

Fertilizer and Spacing

- Foliar application 50 kg N_2 (1.5% urea) and 50 kg N_2 as basal dressing (30 DAT) has been recommended under Coimbatore conditions.
- Application of NPK @ 90:60:40 kg/ha has been recommended for getting economic red ripe yield in variety Pant C-1 under agroclimatic conditions of Faizabad.
- Application of 120 kg N_2, 30 kg P_2O_5 and 10 kg K_2O/ha gave the highest dry chilli yield of 11.16 q/ha (Jawahar-219) with the C:B ratio of 1:2.17. Hence, it has been recommended for Jabalpur conditions for transplanted crop.
- Application of 120 kg N_2 and 60 kg P_2O_5/ha has been recommended to obtained maximum dry chilli yield and economic return from variety Pusa Jwala under Kanpur conditions.

- Application of *Azospirillum* @ 2 kg/ha as basal application in combination with 75% recommended dose of N_2 (i.e., 56 kg/ha), has been recommended under Tamil Nadu conditions.
- Application of 40 kg N_2 as basal dose + 10 kg N_2 as topdressing + 10 kg N_2 as foliar application in transplanted chilli, resulted in highest yield 230 q/ha with a C:B ratio of 1:5.78 at Jorhat. Hence, this treatment has been recommended for Jorhat conditions.
- Application of NPK @ 120:60:60 kg/ha resulted in highest yield (82.64 q/ha) from variety Pant C-1 with a C:B ratio 1:2.56 at Pantnagar. Hence, it has been recommended for *tarai* region of Uttar Pradesh. For Durgapura conditions, application of NPK @ 90:60:60 kg/ha has been recommended for getting maximum yield from Pusa Jwala.
- The cropping sequence of paddy-chilli has been recommended for upland and medium lowland conditions of Jorhat.
- Application of 75% recommended dose of N (150 kg/ha) + Azospirillum as seed treatment, seedling dip and soil incorporation at IIVR, Varanasi, gave maximum yield of green chilli LCA-235 (117.52 q/ha) along with maximum C:B ratio (1:1.77) followed by application of 50% recommended dose of N + Azospirillum. Hence, it is recommended for cultivation under Varanasi condition of Uttar Pradesh.
- At Kalyanpur, maximum yield (88.38 q/ha) of red ripe chilli Azad Mirch-1 along with C:B ratio (1:2.91) was recorded with the application of 100% recommended dose of NPK (120:60:80 kg/ha) + Azospirillum treatment. Hence, it is recommended for Kalyanpur condition.
- Maximum yield (86.39 q/ha) and C:B ratio (1: 2.82) were recorded with the application of *Azospirillum* along with 100% recommended dose of nitrogen in chilli Azad Mirch-1. Hence, it is recommended for Kanpur conditions.
- Application of 240 kg N, 60 kg P_2O_5 and 60 kg K_2O/ha under Hyderabad condition gave maximum dry chilli yield (16.42 q/ha) of G-4 and C:B ratio (1:1.92). Hence, it is recommended for cultivation under Hyderabad conditions.
- Application of 120 kg nitrogen and 60 kg P_2O_5/ha is recommended for getting maximum yield (136 q/ha) and C:B ratio (1:1.50) under tarai conditions of Pantnagar. For Faizabad conditions, application of 240 kg/ha nitrogen and 180 kg/ha P_2O_5 is recommended for obtaining the maximum yield (195.6 q/ha) and C:B ratio (1:4.22), while for Coimbatore conditions, application of 180 kg/ha nitrogen and 120 kg/ha P_2O_5 is recommended for highest yield (63.7 q/ha) and C:B ratio (1:3.98) in the same hybrid.
- The capsicum hybrid Bharat gave maximum yield (328 q/ha) and C:B ratio (1:5.15) with the application of 180 kg/ha nitrogen and 120 kg/ha P_2O_5, hence, it is recommended for Varanasi conditions.
- The highest yield of Capsicum hybrid Bharat (167.45 q/ha) along with C:B ratio (1:1.52) were recorded with three foliar sprays of water soluble fertilizers (15:15:30 @0.5%) at 10 days interval starting from 30 days after transplanting. Hence, it is recommended for Pantnagar conditions.
- At Srinagar, maximum yield (229 q/ha) of Capsicum Nishat-1 with a C:B ratio of 1:2.31 was obtained with three sprays of water-soluble fertilizers having the combination of

NPK 17:10:27. Hence, it is recommended for Srinagar conditions. At IIVR, Varanasi maximum yield of capsicum Indra (82.10 q/ha) and C:B ratio 1:4.08 were obtained with 5 foliar sprays of water-soluble liquid fertilizer having a combination of NPK 19:09:19 at 10 days interval after 40 DAP. Hence, it is recommended for Varanasi conditions.

- Three foliar sprays of micronutrients (mixture of B, Zn, Cu, Fe @ 100 ppm and Mo @ 50 ppm over and above to the recommended dose of NPK at 10 days interval 30 DAT gave maximum yields (226.34 q/ha and 83.13 q/ha and C:B ratio (1:5.05 and 1:3.50) at Faizabad and Kalyanpur. Hence, it is recommended for hybrid Bharat and California Wonder at Faizabad and Kalyanpur conditions, respectively.
- At Kalyanpur, three foliar sprays of mixture of micronutrients (B, Zn, Cu, Fe, Mn each @ 100 ppm and Mo @ 50 ppm) at 10 days interval starting from 40 days after transplanting, resulted in maximum yield (83.89 q/ha) and C:B ratio (1:3.43) in capsicum California Wonder. Hence, it is recommended for Kanpur conditions.
- Foliar application of mixture of all micronutrients (i.e., 100ppm zinc sulphate, manganese sulphate, copper sulphate, ferrous sulphate, borax and 50ppm ammonium molybdate) at 10 days interval starting from 40 days after transplanting is recommended based on three years data to obtain highest seed yield of 258kg/ha in Capsicum Nishat-1 under Srinagar conditions only.
- Spray of 100ppm of commercial formulation of micronutrients on chilli under Varanasi and Lam conditions is recommended for higher yield of better quality seeds.
- Foliar application of a mixture of micronutrients (i.e. 100ppm each of zinc sulphate, manganese sulphate, copper sulphate, ferrous sulphate, boric acid and 50ppm ammonium molybdate) is recommended for higher seed yield of chilli in Karnataka.
- For enhanced seed production of bell pepper California Wonder, spray of borax @ 0.5% at vegetative, flowering and fruit setting stage is recommended under Solan conditions.
- Application of nitrogen @ 100kg N/ha with planting at 60 cm × 30 cm spacing is recommended for obtaining maximum seed yield in paprika under Solan conditions with a cost : benefit ratio of 1 : 3.08.
- Planting of paprika (KTPL-19) at a spacing of 60 cm × 30 cm and application of 200kgN/ha has been recommended to obtain highest seed yield of 366kg/ha under Srinagar conditions.

Protected Cultivation

- Under polyhouse production of Capsicum hybrid Indra, based on highest yield (9.64 kg/m^2) and cost: benefit ratio (1:6.56); spacing of 60 cm × 30cm without training is recommended under naturally ventilated polyhouse under *tarai* conditions of Pantnagar.
- In naturally ventilated polyhouse, maximum C:B ratio 1:2.27 along with higher yield 451.55 q/ha was recorded in Capsicum hybrid Indra with the treatment combination of 60 cm × 30cm plant spacing without pruning. Hence, it is recommended for Varanasi conditions.

Weed Control

- At Coimbatore, highest weed control efficiency (74.7%) with a yield (21.75 q/ha) of dry chilli cv CO-1 was recorded with maximum C:B ratio (1:4.58) with the application of

Basalin @ 1.0 kg ai/ha + one hand-weeding at 45 days after transplanting. Hence, it is recommended for Coimbatore conditions.

Seed Treatment and Storage

- On the basis of two years' studies seed treatment with Trichoderma (5g/kg seed) in bell pepper under Solan conditions is recommended for higher seed yield and quality.
- For enhancing storage life of bell pepper seed, treating of seed with $CaCl_2$ @ 10^{-4} M solution was found best with respect to seed quality characters at Solan.
- At Vellanikkara, treating chilli seeds with antioxidants like $CaCl_2$ @10-5M helped in enhancement of its storage life.
- At Hyderabad, the treatment of chilli cultivar X-235 with PBA @ 10-6M or NaCl 10-4M was found best for enhancing storage life of these seeds.
- Based on three years' studies, chilli seeds treatment with 10^{-5}M $CaCl_2$ or PHBA is recommended to retain maximum germination up to 16 months under Bhubaneswar conditions.
- On the basis of two year's study at Solan centre, it is recommended that storage of bell pepper cultivar California Wonder seeds is best when stored in polythene bag (200-gauge) in refrigerator after treating with 0.3% thiram, whereas storage of onion cultivar Nasik Red seed is best when stored in polypet jar in deep freezer -20°C after treating with 0.3% thiram.

Growth Regulator, Pinching & Spacing

- On the basis of three years data, it is recommended that maximum yield of good quality seed in chilli var. Phule Jyoti can be obtained with the application of Planofix spray @ 2.5 ml/gallon of water + DAP1% + MOP 0.50% under Rahuri conditions.
- The maximum seed yield of 147kg/ ha in bell pepper cv California Wonder was recorded at Srinagar with spray of 20ppm NAA+ pinching of apical shoot when planted at 60 cm × 60 cm spacing.
- Application of 20ppm NAA along with apical pinching at 60cm × 45cm spacing is recommended for obtaining maximum seed yield in bell pepper cv. California Wonder under Solan conditions with a cost :benefit ratio of 1:2.08.

Others

- In Capsicum hybrid Pusa Deepti, a female : male ratio of 20:1, retaining 6 fruits/plant and emasculation a day before anthesis followed by pollination in the forenoon next day is recommended for higher yield of good quality seed, based on trials at Katrain.
- At Lam centre, the pooled data of three years revealed that in chilli, 1st July sowing with recommended dose of fertilizer and second picking recorded highest seed yield, 1000-seed weight and seed vigour.
- Transplanting of chilli seedlings in the last week of July along with the spray of Planofix (2.5 ml/litre of water) + DAP 1% + MOP 0.5% is recommended for higher seed yield based on three years' results at Jabalpur.

- Priming with KNO_3 (150ppm) produced maximum germinability and vigour in Capsicum (California Wonder) seeds at Solan.

Garden Pea

- Either of the three chemicals, viz. Lasso @ 0.75 kg/ha as pre-emergence spray or tribunil @ 1.5 kg a.i./ha pre-emergence or basalin @ 2 kg a.i./ha pre-plant incorporation along with one hand-weeding at 45 DAS was found to be most effective for weed control under Pantnagar conditions.
- Application of tribunil @ 1.87 kg a.i./ha as pre-emergence spray has been recommended for controlling weeds under agroclimatic conditions of Punjab.
- Application of Stomp (pendimethalin) @ 0.5 kg a.i./ha or Lasso (alachlor) @ 1.5 kg a.i./ha as pre-emergence incorporation in the field followed by one hand-weeding at 25 DAS has been recommended under Sabour conditions.
- In case of variety Azad P-1, Stomp 1.0 kg a.i./ha as pre-emergence application followed by one hand-weeding at 45 DAS, has been recommended for effective weed control for Kanpur region.
- Pre-emergence application of Pendimethalin @ 0.5 kg a.i./ha along with one hand-weeding at 25 DAS has been recommended for effective weed control under Sabour conditions. This treatment gave an yield of 55.30 q/ha with a C:B of 1:2.20.
- For getting the optimum yield and C:B ratio, application of four irrigations at pre-bloom, bloom, pod setting and after first picking stages has been recommended for Bonneville under Bangalore conditions.
- Three irrigations at pre-bloom stage, bloom and pod setting stages (30 DAS) have been recommended for Bonneville under Sabour conditions of Bihar.
- Processing qualities of seven advance lines were studied. Sensory tests were conducted and chemical compositions were worked out. On the basis of frozen storage test, FC-2 and 32-1-1-9 were found best in chemical composition and sensory qualities. Both of these varieties differ in grain quality, shelling percentage, moisture percentage ash, chlorophyll contents, ascorbic acid and different forms of sucrose.
- At IIVR, Varanasi maximum yield of Garden Pea cv. Azad P-3 (80.7 q/ha) along with maximum C:B ratio 1:1.32 was recorded with application of FYM @ 10 tonnes/ha + half recommended NPK. Hence, it is recommended for Varanasi conditions.
- At IIHR, Bangalore, the garden pea variety Arka Kartik recorded highest yield 105.2 q/ha with the maximum C:B ratio of 1:2.78 with the application of 10 tonnes/ha FYM plus half dose of N:P:K (20:30:25 kg / ha). Hence, it is recommended for Bangalore conditions.
- At Faizabad, application of neem cake @ 2.5 q/ha plus half of the recommended dose of NPK (15:30:20 kg / ha) recorded the highest mean yield of pod in garden pea Azad Pea-3 (102.10 q/ha) with maximum C:B ratio of 1:2.98. Hence, it is recommended for Faizabad conditions.
- At Bhubaneswar, integrated application of poultry manure @ 2.5 tonnes/ha + half recommended NPK (25:37.5:25 kg/ha) resulted in 44.16/ha pod yield and C:B ratio of 1:1.49. Hence, it is recommended for Bhubaneswar conditions of Orissa.

- For maximum yield (118.79 q/ha) and C:B ratio (1:2.73), 3 irrigations, i.e., at pre-bloom (30 days after sowing), at bloom (50% flowering) and at pod setting stages are recommended for Azad Pea-1 under Faizabad conditions.
- A fertilizer dose of 25 kg N + 60 kg P_2O_5 + 40 kgs K_2O/ha with a seed rate of 150 kg /ha. was found optimum for seed production of pea Arkel in Himachal Pradesh with a C:B ratio of 0.98.
- For seed production of early pea Arkel in Himachal Pradesh, sowing of seeds on 30 October with 20 cm row spacing was found suitable (cost: benefit ratio - 0.79).
- Priming with PEG (-1.0 Mpa) in Pea (Arkel) seeds at Jabalpur produced maximum germinability and vigour during three years' trial.
- A fertilizer dose of 25 kg N + 60 kg P_2O_5 + 40 kgs K_20 per ha with a seed rate of 150 kg/ha. was found optimum for seed production of pea Arkel in Himachal Pradesh with a cost: benefit ratio of 0.98.
- Application of 45:90:60 kg NPK/ha along with a seed rate of 75kg/ha in pea (Arkel) is recommended for higher seed yield under Jabalpur conditions with a cost : benefit ratio of 1 : 3.01.
- Seed treatment with black pepper (5%) is recommended for effective control of *Callosobruchus chinensis* in garden pea based on trials at Solan.
- For seed production of early pea Arkel in Himachal Pradesh, sowing of seeds on 30 October with 20 cm row spacing was found suitable (cost: benefit ratio - 0.79).
- The seed sowing on 25 November with 20cm row spacing is recommended for higher seed yield of Pea (Arkel) under Jabalpur conditions with a cost : benefit ratio of 1 : 3.35.

Other Leguminous Vegetables

- Frenchbean variety Contender gave maximum pod yield (129 q/ha) with a cost: benefit ratio of 1:2.23 with the application of 120 kg N_2 and 60 kg P_2O_5 /ha, hence this dose has been recommended for Solan conditions.
- Application of nitrogen @160 kg/ha and P_2O_5 @90 kg/ha is recommended for maximum green pod yield (65 q/ha) and C:B ratio (1:2.95) in Frenchbean variety Contender under Durgapura conditions. At Faizabad, application of 160 kg/ha nitrogen and 60 kg/ha P_2O_5 is recommended for highest yield (98 q/ha) and C:B ration (1:2.44) in the Frenchbean variety Patn Bean-2.
- Frenchbean variety Arka Komal gave maximum yield of 96.5 q/ha and CB ratio 1:1.38 along with the application of 120 kg N and 60 kg P_2O_5/ha, hence it is recommended for Varanasi conditions.
- Application of 160 kg N and 90 kg P_2O_5, at Hyderabad gave the highest yield of Frenchbean Contender (31.8 q/ha) with maximum cost: benefit ratio (1:1.63). Hence, it is recommended for green pod cultivation under Hyderabad conditions.
- The maximum marketable yield of Frenchbean Arka Komal (138.93 q/ha) along with C:B ratio (1:3.85) was recorded by the application of 160 kg nitrogen and 60 kg P_2O_5/ha. Hence, it is recommended for cultivation for green pod under Jabalpur conditions.

- At Kalyanpur, maximum yield of green pods of Frenchbean Pant Anupama (77.18 q/ha) and C:B ratio (1:2.77) was recorded with the application of 80 kg N and 60 kg P_2O_5/ha. Hence, it is recommended for Kanpur conditions of Uttar Pradesh.
- Application of 80 kg N + 40 kg P/ha with cost: benefit ratio 1:1.51 is recommended for higher seed yields of Frenchbean Contender under Jabalpur conditions.
- Maximum seed yield (15.75 q/ha) can be obtained with 160kg/ha nitrogen in Frenchbean Arka Komal under Ranchi conditions.
- Seed treatment with black pepper (5%) is recommended for effective control of Callosobruchus maculatus in Frenchbean, based on trials at Solan.
- Harvesting of cowpea seeds is recommended at 23 days after anthesis to obtain maximum recovery, germination and vigour based on two years data at IIVR.
- The cropping sequences, cowpea (*kharif*), - tomato (*rabi*), - cucumber (*zaid*) and *kharif* onion-potato-cowpea have been recommended for Jabalpur conditions in Madhya Pradesh.
- The cropping sequence of cowpea-tomato-cucumber resulted in maximum net return followed by tomato-pea-okra, hence these cropping sequences have been recommended for Kaymore plateau and Satpura hills of Madhya Pradesh.

Okra

Fertilizer and Spacing

- A dose of 150 kg each of N_2, P_2O_5 and K_2O with a spacing of 60cm × 45cm during rainy season for variety Pusa Sawani has been recommended for Jabalpur conditions.
- At IIVR, Varanasi, application of Pressmud @ 5 tonnes/ha + rest NPK through chemical fertilizers gave maximum yield of brinjal cv. IVBL-9 (106.7 tonnes/ha) and C:B ratio (1:4.09) and also maximum yield of okra cv. VRO-6 (12.5 tonnes/ha) and C:B ratio (1:2.76) in brinjal-okra cropping pattern trial. Hence, it is recommended for Varanasi conditions.
- Foliar application of a mixture of micronutrients (i.e. 100ppm each of zinc sulphate, manganese sulphate, copper sulphate, ferrous sulphate, boric acid and 50ppm ammonium molybdate) is recommended for higher seed yield of okra in Maharashtra.

Sowing Time and Spacing

- Sowing of okra (Harbhajan) on 15 May at 60cm × 30cm spacing is recommended for obtaining maximum seed yield with a cost : benefit ratio of 1 : 3.08 under Solan conditions.
- Maximum seed yield of okra (Hisar Unnat) can be obtained by sowing on 15 June at a spacing of 60cm × 30cm under Hisar condition.
- Under Vellanikkara condition, maximum seed yield of okra (Arka Anamika) can be obtained by sowing on 15 May at 45cm × 45cm spacing.
- Seed production of okra in kharif season is recommended for higher yield of good quality seeds in Maharashtra, Uttar Pradesh and Punjab.
- Maximum seed yield in okra cv. Utkal Gaurav can be obtained with 15th May sowing at 45cm × 45cm distance under Bhubaneswar conditions.

- Under Jabalpur conditions, seed production of okra cv. Parbhani Kranti can be profitably taken up at 60cm × 45cm spacing with sowing date of second week in March.

Liquid Fertilizer

- In okra cv. Aruna, maximum yield (161.4 q/ha) and C:B ratio (1:1.3) were obtained with 3 foliar sprays of water-soluble fertilizer having the NPK formulation of 19:19:19 at Vellanikkara. Hence, it is recommended for Vellanikkara conditions.
- At Hyderabad, maximum yield (78.3 q/ha) and C:B ratio (1:1.68) were obtained in okra cv. Arka Anamika with 5 foliar sprays of water-soluble fertilizers having NPK formulation – 15:15:30. Hence, it is recommended for Hyderabad conditions.
- At Coimbatore, maximum yield (202 q/ha) and C:B ratio (1:3.82) were obtained with 5 foliar sprays of water-soluble fertilizers having NPK formulation – 19:19:19 in okra hybrid Mahyco No.10. Hence, it is recommended for Coimbatore conditions.

Weed Control

- Lasso @ 2 kg a.i./ha followed by one hand-weeding at 5-6 weeks after sowing has been recommended for weed control under Pantnagar, Hisar and Sabour conditions.
- For weed control in variety Pusa Sawani, the application of fluchloralin @ 1.5 kg/ha as pre-plant incorporation with one hand-weeding at 45 days after sowing (DAS) has been recommended under Pantnagar conditions. Similar recommendation has also been made at PAU and HAU for Punjab and Haryana conditions, respectively. However, at Jabalpur, the application of Lasso @ 1.5 kg a.i./ha as pre- emergence + one hand-weeding at 45 DAS, proved most effective for weed control.
- For effective weed control with highest C:B ratio (1:2.52) in *kharif* crop, application of metolachlor @ 1.0 kg a.i./ha + one hand-weeding at 45 DAS has been recommended for Sabour conditions.
- For *kharif* crop, application of herbicide metalochlor @ 1.00 kg a.i./ha (PE) proved most effective resulting in maximum yield and C : B ratio (1:2.86). Hence, it has been recommended for chemical control of weeds in *kharif* crop in Varanasi region.
- Out of six rotations, okra-tomato-bitter gourd has the maximum return with cost: benefit ratio of 1:2.06 under Kalyanpur conditions. Next best rotation was cowpea-cauliflower-onion. Hence, this rotation has been recommended for central Uttar Pradesh.
- Application of metolachlor @0.75 kg ai/ha as pre-emergence spray followed by one hand-weeding 45 days after sowing is recommended for effective weed management and obtain economic yield (63 q/ha and C:B ratio 1:3.67) under Coimbatore conditions.
- Application of Dual @ 1 kg ai/ha as pre-emergence spray is recommended for effective weed management and obtain economic yield (122.29 q/ha) and C:B ratio (1:3.88) by okra Parbhani *Kranti* followed by Basalin @ 1.5 kg ai/ha as PPI. Hence, it is recommended for Kanpur condition of Uttar Pradesh.

Others

- Retention of first eight fruits per plant in okra is recommended for higher seed yield and quality under Solan conditions, whereas retention of 12 fruits per plant is recommended under Bhubaneshwar condition.

- Retention of first 10–12 fruits per plant in okra is recommended for higher yield of good quality seed under Srinagar and Chhatisgarh conditions.
- Two sprays of 150 ppm NAA at 30 and 50 days after sowing of okra are recommended for higher yield of good quality seed in Himachal Pradesh.
- For best quality okra seed, its harvesting after 35 days of anthesis under Bhubaneswar condition is recommended based on three years' data.
- It is recommended that okra fruits under Haryana conditions should be harvested at 54–57 days after anthesis to obtain good quality seeds.
- The prescribed isolation distance of 200m for certified seed production of okra is valid and should be maintained for quality seed production of okra as no NCP took place at the isolation distance of 200m for five years at Ludhiana.
- On the basis of two year studies at Solan centre, it is recommended that maximum seed yield in Okra Pusa Sawani is obtained when the seed was treated with Carbendazim (2g/kg seed) or with 0.25% HCl dip for 30 minutes.
- Based on two years' studies, presoaking of seeds in 50 °C hot water for 30 minutes or dipping in 0.25% HCl for 30 minutes is recommended under Srinagar and Kanpur conditions for higher germination of okra seeds.
- On the basis of two years' studies seed treatment with trichoderma (5g/kg seed) in okra under Varanasi, Kanpur and Jabalpur conditions is recommended for higher seed yield and quality, whereas seed treatment with imidacloprid (2.5g/kg seed) is recommended for Karnataka conditions.
- Treatment of seeds with 0.5% citric acid for 45 minutes is recommended for improved seed germination and reduced percentage of hard seeds in okra based on trials at Solan.

Onion

Bulb Crop

- In Pusa Red variety, use of 150 kg N_2, 60 kg P_2O_5 and 60 kg K_2O/ha with a planting distance of 15cm × 10cm has been recommended under Jabalpur conditions.
- At Rahuri centre, application of Tok E-25 @ 2 kg a.i./ha followed by one hand-weeding at 45 DAS have been recommended.
- Pre-plant incorporation of basalin @ 2 kg a.i./ha along with one hand-weeding at 45 DAT has been recommended for weed control in variety Pusa Red, to get highest profit under tarai conditions of Pantnagar.
- In order to get the economic bulb yield from variety N-53, application of 75 kg N/ha at 15cm × 10cm plant spacing has been recommended under agroclimatic conditions of Ambajogai. However, application of phosphorous was not found economical.
- Basalin @ 2 kg a.i./ha as pre-plant incorporation followed by one hand-weeding at 45 DAT proved most suitable for weed control and economic returns under Bhagalpur and Ludhiana conditions. Whereas at Jabalpur centre, application of Ranstar @ 1 kg a.i./ha has been recommended for effective weed control.

- The planting of variety Pusa Red at a spacing of 15cm × 10cm and application of N @ 150 kg/ha and P_2O_5 @ 60 kg/ha has been recommended for Sabour region.
- In case of variety Pusa Red, application of 150 kg N, 60 kg P_2O_5 and 30 kg K_2O with spacing of 15cm × 10cm has been recommended under Sabour conditions.
- Application of 50 kg N/ha as basal dressing + 30 kg/ha as top dressing + 10 kg/ha as foliar spray has been recommended for variety Pusa Red under Sabour conditions. A cost benefit ratio of 1:4.61 was obtained from this treatment.
- Application of fluchloralin @ 1.0 kg (PPI) + 1 hand-weeding at 45 DAT proved to be most effective for weed control under Durgapura conditions, which resulted in an yield of 178.88 q/ha with C:B ratio of 1:2.93.
- Application of N @ 150 kg/ha along with P_2O_5 @ 60 kg/ha at the spacing of 10cm × 10cm was found most effective and resulted in maximum yield (729 q/ha) and C : B ratio (1:2.91) from variety Agri Found Dark Red. Hence, it has been recommended for Junagadh conditions.
- Application of *Azospirillum* as seed treatment, seedling dipping and soil application along with recommended dose of N, have been recommended for obtaining maximum bulb yield and C:B ratio (1:2.35) in *kharif* crop from variety N-53 under Hyderabad conditions.
- For getting maximum yield (118.4 q/ha) and C.B. ration 1:1.33 from variety N-53, application of 150 kg N/ha at 15cm × 10cm of spacing has been recommended for Hyderabad conditions during kharif season.
- Application of N @ 120 kg/ha and planting distance of 45cm × 30cm resulted in maximum seed yield (8.20 q/ha) and C:B ratio (1:3.72). It has been recommended for onion variety Kalyanpur Red Round under Kanpur conditions.
- The maximum yield of onion cv. RO-1 (344.9 q/ha) and C:B ratio 1:0.91 were recorded at Durgapura with by the application of 100% recommended NPK (100:50:100) + Azospirillum. Hence, it is recommended for Durgapura conditions.
- Based on maximum yield (10.62 q/ha) and C:B ratio (1:7.66), planting of bulbs at 45cm × 30cm spacing and application of nitrogen @ 120 kg/ha is recommended for onion seed production under Tarai conditions of Pantnagar.
- At NRC for onion and garlic, the application of NPK @ 50:50:80 kg/ha as basal + 100 kg N in seven split through drip irrigation expressed better yield (29.6 tonnes/ha) and highest C:B ratio, i.e., 1:1.63 in onion. Hence, it is recommended for Pune conditions of Maharashtra.
- For maximum bulb yield (552 q/ha) and C:B ratio (2.22), application of pendimethalin @ 1 kgai/ha + one hand-weeding at 45 days after planting is recommended in onion crop under south Saurashtra conditions.
- For weed management in onion, application of oxyfluorfen @ 0.15 kg a:/ha as PE resulted in highest yield (11.26 tonnes/ha) and C:B ratio (1:3.40). Hence, it is recommended for Coimbatore conditions.

Seed Crop

- In case of variety N-53, 80 kg N, 50 kg P_2O_5 and 250q of FYM/ha with spacing of 45cm × 30cm has been recommended under Rahuri conditions.
- Planting of bulbs (variety Pusa Red) at 45cm × 30cm spacing with 80 kg N/ha has been recommended for seed production in Faizabad region. This treatment combination gave 7.27 q/ha of seed yield and cost: benefit ratio of 1:4.13.
- For obtaining highest yield (9.55 q/ha) and C:B ratio (1:1.83), application of 120 kg N per hectare and planting distance of 45cm × 30cm has been recommended for variety Agri Found Dark Red under Durgapura conditions.
- An application of 80 kg N (in two split doses), 60 kg P_2O_5 and 60 kg K_2O/ha and planting bulbs at a spacing of 45cm × 30cm have been recommended for Pusa Red to obtain maximum seed yield under the agroclimatic conditions of Sabour.
- For seed crop, planting of bulbs of Pusa Red at 45cm × 30cm and application of 80 kg N/ha has been recommended for Nasik area to get economic seed yield.
- For seed production of Pusa Red at Jabalpur centre, application of 120 kg N_2, 60 kg P_2O_5 and 80 kg K_2O/ha with a spacing of 45cm × 30cm has been recommended.
- Planting small sized onion bulbs (45cm × 30cm) for seed production on 11 October with a close spacing of 45cm × 30cm was found suitable for Rajasthan State.
- Trials conducted on seed pelleting and methods of planting on onion showed that higher yield, i.e., 25.6 tonnes/ha with the maximum C:B ratio, i.e., 1:2.40 was recorded with the pelleting of Trichoderma viridae @ 4g/kg seed under flat bed system in *kharif* planting. The same treatment is recommended for rabi season also.
- At Jabalpur centre, pooled data of three years, revealed that maximum seed yield of 9.35q/ha in onion cv. AFLR was recorded under treatment paired row planting with fertilizer mixture application before planting in individual holes.
- On the basis of three years data it is recommended that maximum yield (13.96q/ha) of good quality seed in Onion can be obtained by paired row planting on ridges under Rahuri conditions.
- Planting of onion bulbs variety Arka Kalyan and Arka Niketan in the month of November is recommended for highest seed yield with a cost: benefit ratio of 3.29 and 3.72, respectively under Bangalore conditions.
- Planting of onion bulbs variety Punjab Red Round in the second fortnight of October is recommended for highest seed yield under Ludhiana conditions.
- On the basis of three years, results at Rahuri, it is recommended that maximum good quality seed yield in onion (B-780) can be obtained in Maharashtra state by planting bulbs in the second fortnight of October and early-November.
- Planting small-sized onion bulbs (45cm × 30cm) for seed production on 11 October with a close spacing of 45cm × 30cm was found suitable for Zone III A of the Rajasthan State.
- At Srinagar, treatment of onion cultivar Yellow Globe with PBA @ 10-4 M was found best for enhancing storage life of these seeds.

- On the basis of two years' study at Solan centre, it is recommended that storage of onion cultivar Nasik Red seed is best when stored in polypet jar in deep freezer -20 °C after treating with 0.3% thiram.

Garlic

- Application of N150 kg + K_2O 60 kg/ha and N 100 kg + K_2O 50 kg/ha without P_2O_5 have been recommended for variety G-1 under Karnal and Pantnagar conditions, respectively. Application of NPK @ 100:80:50 kg/ha has been recommended for the same variety under Hisar conditions. The cost: benefit ratio at Hisar and Pantnagar were found to be 1:2.03 and 1:3.44, respectively.
- Application of N 100 kg + P_2O_5 80 kg + K_2O 60 kg/ha has been recommended for getting maximum economic yield under Kanpur conditions.
- Application of NPK @ 100:80:60 kg/ha in variety G-1, gave the highest yield (99.35 q/ha) with the C:B ratio of 1:6.04 at Jorhat. Therefore, this dose has been recommended for Jorhat conditions.
- Yield (152.45 q/ha) and C : B ratio (1:1.92) were found highest with the application of 100 kg N_2 , 80 kg P_2O_5 and 60 kg K_2O/ha in variety Local White. Hence, it has been recommended for Faizabad conditions.
- Application of 50 kg N and 60 Kg/ha K_2O is recommended for highest yield (88.0 q/ha) and C:B ratio (1:2.63) in garlic variety GG-2 under Junagadh conditions.
- The application of NPK @ 50:50:50 kg/ha as basal + 50 kg N in seven split through drip irrigation gave yield of 7.91 tonnes/ha with maximum C:B ratio 1:2.93. Hence, it is recommended for Pune conditions of Maharashtra.

Cauliflower

- In late group (snowball), spacing of 60cm × 40cm has been recommended for Bhubaneshwar and Kanpur conditions.
- A dose of 40 kg N /ha as basal + 40 Kg N/ha as topdressing + 30 kg N /ha as foliar spray has been recommended for variety Snowball-16 under Sabour conditions. However, for Pantnagar conditions, 60 kg N /ha as basal + 30 kg as foliar spray in three splits has been recommended for the same variety. Further, at Kalyanpur, the best response was with 60 Kg N/ha as basal + 60kg N/ha as top dressing in two equal splits.
- Plant spacing of 80cm × 60cm along with the application of 150 kg N and 60 kg P_2O_5/ha has been recommended for mid-season variety Pant Shubhra at Pantnagar to get the high economic yield. But in case of snowball group, 60 kg N/ha as basal dose + 30 kg N/ha in three splits as foliar sprays was found to be the best and has been recommended for Pantnagar conditions.
- Transplanting of mid-season variety Pusa Synthetic at the spacing of 60cm × 40cm and application of 150 kg N + 120 kg P_2O_5/ha has been recommended for obtaining economic yield under Kalyanpur conditions.
- For variety Pant Shubhra, recommendation of 100 kg N, 60 kg P_2O_5 and 60 kg K_2O/ha at

at a spacing of 60cm × 40cm has been recommended for Faizabad conditions.

- Application of 120 kg N , 60 kg P_2O_5 and 60 kg K_2O/ha with plant a spacing of 60cm × 60cm has been recommended for obtaining maximum yield of mid-season variety Pant Subhra under Kalyanpur conditions.
- The maximum yield of cauliflower cv. Pusa Snowball K-1 (304.5 q/ha) and C:B ratio (1:3.88) were recorded with the application of PSB @ 500 g/ha as seedlings root dip along with recommended dose of NPK through fertilizers. Hence, it is recommended for Pantnagar conditions.
- At Hyderabad, maximum yield (226 q/ha) along with C:B ratio (1:2.96) were obtained in late cauliflower with the application of VAM @ 15kg/ha + recommended dose of NPK. Hence, this is recommended for Hyderabad conditions.
- At Hisar, highest yield (347.8q/ha) and C:B ratio (1:5.53) were obtained with the application of PSB + 75% P and recommended dose of nitrogen and potassium in late cauliflower. Hence, it is recommended for Hisar conditions.
- At Faizabad, maximum yield (257.75 q/ha) and C;B ratio (1:3.19) were obtained in cauliflower cv. Snowball-16 with the application of *Azospirillum* plus recommended dose of NPK. Hence, it is recommended for Faizabad conditions.
- At Kalyanpur, application of *Azospirillum* with 75% of N and full dose P and K gave the maximum yield (293.95 q/ha) and C:B ratio (1:1.60) of cauliflower cv. Snowball-16. Hence, it is recommended for Kanpur conditions.
- At IIVR, Varanasi, application of Azospirillum + 75% N and full P and K in cauliflower Snowball-16 gave maximum yield (260.7 q/ha) along with C:B ratio (1:2.54). Hence, it is recommended for Varanasi conditions.
- Application of 100 kg/ha nitrogen and planting at 60x60 cm spacing gave maximum C:B ratio (1:3.92) in cauliflower variety Snowball-16, hence it is recommended for Srinagar conditions.
- At IIVR, Varanasi, 5 foliar sprays of water-soluble fertilizers having a combinations of NPK-19:19:19 at 10 days interval after 40 DAP, resulted in maximum yield (342 q/ha) in cauliflower Snowball-16. Hence, it is recommended for Varanasi conditions.
- At Bhubaneswar, maximum curd yield (178.03 q/ha) and C:B ratio 1:2.11 of cauliflower was obtained with three foliar sprays of NPK (19:19:19) in addition to recommended dose of NPK. Hence, it is recommended for Bhubaneshwar conditions.
- At Faizabad, maximum C:B ratio 1:3.15 and better yield of cauliflower Pusa Snowball K-1 was obtained with the foliar application of boron @ 100 ppm + Molybdenum @ 50 ppm. Hence, it is recommended for Faizabad conditions.
- At Bhubaneswar, higher mean yield of cauliflower (166.6 q/ha) along with maximum C:B ratio 1:4.09 were obtained with the foliar sprays of boron @ 100 ppm in addition to recommended NPK. Hence, it is recommended for Bhubaneswar conditions of Orissa.
- At Kalyanpur, higher curd yield of cauliflower (317.09 q/ha) along with the maximum C:B ratio 1:3.17 were recorded with the soil application of borax @ 10 kg/ha + Ammonium

Molybdate @ 2.0 kg/ha. Hence, it is recommended for Kanpur conditions.

- Application of N @ 120 kg/ha, 60 kg each of P_2O_5 and K_2O at the planting distance of 60cm × 40cm gave maximum seed yield (4.21 q/ha) with cost: benefit ratio of 1:3.15 in Improved Japanese at Jaipur. Hence, this treatment has been recommended for seed crop under Jaipur conditions.
- In variety Snowball-1, planting at 60cm × 45cm spacing coupled with application of N @ 150 kg/ha, proved to be most remunerative for seed production under midhill conditions of Almora.
- Application of GA 10 ppm at curd maturity produced highest 100-seed weight and germination in cauliflower at IIVR.
- At IIVR pooled data of two years revealed that highest seed yield in early cauliflower (Pusa Deepali) was recorded by sowing on 16 June while the maximum test weight was recorded with 31 July sowing.

Cabbage

- Variety Late Drumhead performed best under 75cm × 30cm spacing at OUAT, hence this spacing has been recommended for Bhubaneshwar conditions.
- Plant spacing of 75cm × 30cm and application of 180 kg N + 50 kg P_2O_5 + 50 kg K_2O/ha has been recommended for variety Pride of India under agroclimatic conditions of Jabalpur. Similarly, for Kanpur region, 70cm × 30cm spacing and application of 60 kg N/ha with basal dose of 50 kg P_2O_5/ha has been recommended.
- Planting of variety Pride of India at 60cm × 30cm spacing with the application of N @ 150 kg/ha, gave 275 q/ha yield with a cost : benefit ratio of 1:4:81 at Sabour. Therefore, these treatments have been recommended for Sabour region of Bihar.
- For getting optimum yield (~526 q/ha) and C:B ratio (~1:2.09), seed treatment with *Azospirillum* (500 g/ha) + soil application (5 kg/ha) + seedling dipping (1.0 kg/ha) + application of 60 kg N/ha, have been recommended for variety Pride of India under Solan conditions.
- Application of 180 kg N_2/ha + 60 kg/ha each of P_2O_5 and K_2O at the plant spacing of 60cm × 30cm gave highest yield (650 q/ha) with C : B ratio (1:3.61), therefore, this treatment has been recommended for variety Pride of India under Kanpur conditions.
- Maximum yield (274 q/ha) from variety Pride of India was obtained with the application of *Azospirillum* (seedling dipping @ 1.0 kg/10 litres water + soil application @ 5 kg/ha) supplemented with 75 % of the recommended dose of N (i.e., 180 kg/ha). Hence, these treatments have been recommended for Kaymore plateau and Satpura hills of Madhya Pradesh.
- Application of Azotobactor along with 75% of the recommended dose of nitrogen (150 kg/ha) gave maximum yield (360.5 q/ha) and C:B ratio 1:3.82, hence it is recommended for cabbage variety Pride of India under Srinagar conditions.
- Application of Azotobactor as seed treatment seedling, dipping and soil application of 75% recommended dose of N (140 kg/ha) in cabbage variety Pride of India gave maximum

yield (379.0 q/ha) and cost: benefit ratio (1:3.41). Hence, it is recommended for application of the *tarai* conditions of Pantnagar.

- At Hyderabad, 75% recommended nitrogen along with Azotobacter realized an yield of 376.78 q/ha with C:B ratio of 1:3.41 which is on par with 100% recommended dose of nitrogen (180 kg/ha). Hence, application of Azotobacter + 75% recommended dose of nitrogen can save 25% of nitrogen per hectare. Hence, it is recommended for Hyderabad condition.
- Application of 120 kg N/ha and planting at 60cm × 30cm spacing gave maximum yield 313 q/ha and C:B ratio (1:2.57) hence it is recommended for cabbage variety Pride of India under Kymore Plateau and Satpura hills conditions.
- At Durgapura, 3 foliar sprays of commercial formulation of multiplex @ 100 ppm over and above to the recommended dose of NPK at 10 days interval after 30 DAT gave maximum yield (321.7 q/ha) and C:B ratio (1: 0.62) in cv. Bajrang. Hence, it is recommended for Durgapura conditions.
- At IIVR, Varanasi, 3 sprays of micronutrient mixture of B, Zn, Cu, Fe, Mn @ 100 ppm and Mo @ 50 ppm over and above to the recommended dose of NPK (120:60:80 kg/ha) at 10 days interval gave maximum yield (765.98 q/ha) and C:B ratio (1:1.44) in cv. Quisto. Hence, it is recommended for Varanasi condition.
- At Jabalpur, maximum yield (360.40 q/ha) and C:B ratio (1: 2.75) of hybrid cabbage Krishna were recorded with 3 foliar sprays of $ZnSO_4$ at 100 ppm concentration. Hence it is recommended for Satpura hills and Kaymore plateau region.
- At Bhubaneswar, maximum yield (431.6q/ha) of hybrid cabbage along with C:B ratio (1:1.83) were obtained with 3 sprays of 100 ppm $ZnSO_4$ at 10 days interval starting from 40 days after transplanting. Hence, it is recommended for Bhubaneswar conditions. However, at Jorhat, application of boron at same concentration resulted in maximum yield (485 q/ha) and C:B ratio of (1:2.76) which is recommended for Jorhat conditions.
- The maximum mean head yield of 557 q/ha and C:B ratio (1:5.93) were obtained with three foliar sprays of micronutrient mixture (B, Zn, Cu, Fe, Mn, each @ 100 ppm and Mo @ 50 ppm) at 10 days interval starting from 40 days after transplanting in cabbage cv. Golden Acre at Srinagar. Hence, it is recommended for Srinagar conditions.
- The highest mean yield (400.7 q/ha) along with the maximum C:B ratio (1:2.92) were obtained in cabbage hybrid cv. Meenakshi with the application of ferrous sulfate @ 100 ppm in addition to recommended dose of NPK (180:60:60 kg/ha). Hence, it is recommended for Hyderabad conditions.
- Highest seed yield of 11q/ha in cabbage Golden Acre can be obtained at Srinagar with five weekly foliar applications of poly feed (19:19:19) @ 5 g/litre starting from 30 days of transplanting.
- Foliar application of water-soluble fertilizer 19:19:19 at 0.5% concentration is recommended for higher seed yield of cabbage in Himachal Pradesh.

Broccoli

- At Srinagar, application of vermicompost @ 5 tonnes/ha recorded the maximum C:B ratio 1:4.7 with higher mean yield of 156.61 q/ha in broccoli. Hence, it is recommended for Srinagar conditions of J&K.
- At Jorhat, planting of broccoli cv.KTS-1 at 45cm × 30cm spacing on 30 October resulted in maximum yield (158 q/ha) and C:B ratio (1:1.77). Hence, it is recommended for Jorhat conditions.
- At Kalyanpur, maximum yield (315.52 q/ha) and C:B ratio (1:2.27) were obtained when broccoli hybrid Fiesta was planted on 15 October at 45cm × 30cm spacing. Hence, it is recommended for Kanpur conditions. However, at Durgapura, the maximum yield (310.4 q/ha) and C:B ratio (1:8.5) were obtained at 45cm × 45cm spacing on the same date of transplanting, which may be recommended for Durgapura conditions.
- At Pantnagar, maximum yield (170 q/ha) and C:B ratio (1:3.04) were obtained in broccoli hybrid Fiesta when planting was done on 15 October at 45cm × 30cm spacing. Hence, it is recommended for *tarai* conditions of Pantnagar.
- At IIVR, Varanasi, planting of broccoli cv. Fiesta at 45cm × 30cm spacing on 15 October resulted in maximum yield (297.0 q/ha) and C:B ratio (1:4.25). Hence, it is recommended for Varanasi conditions.
- At Srinagar, maximum yield (186.7 q/ha) and C:B ratio (1:2.57) were obtained in broccoli hybrid Green Head with the planting on 30th August at 45cm × 30cm spacing. Hence, it is recommended for Srinagar conditions.
- At Faizabad, maximum yield (22.45 tonnes/ha) and C:B ratio 1:4.89 of broccoli Fiesta was obtained by 15 October planting at spacing of 45cm × 30cm. Hence, this date of planting and spacing is recommended for Faizabad conditions.
- At Ranchi, transplanting of broccoli during 15–30 October at 45cm × 30cm spacing recorded the maximum mean yield and C:B ratio 1:4.89. Hence, this date of planting and spacing are recommended for Ranchi conditions.

Carrot

- Application of N @ 150 kg/ha and planting of steckling at 60cm × 30cm for seed production from variety Pusa Meghali resulted in an yield of 12.55 q/ha with cost: benefit ratio of 1:4.03 under Faizabad conditions.
- Yield (10.86 q/ha) and C:B ratio (1:5.36) of variety Nantes were highest with the application of N 100 kg/ha and spacing of 45cm × 45cm at Ranichauri. Hence, it has been recommended for Uttar Pradesh hills. Similar dose of nitrogen and spacing has also been recommended for variety Pusa Keshar under Sabour conditions.
- At Hyderabad, maximum yield (316 q/ha) with C:B ratio 1:3.70 of carrot cv. improved Kuroda was obtained with the application of vermicompost @ 2 tonnes/ha + biofertilizer + half recommended dose of NPK (25 : 20 : 25 kg/ha). Hence, it is recommended for Hyderabad conditions.
- At Durgapura, maximum yield of carrot, i.e., 557.09 q/ha and C:B ratio 1:2.59 was

obtained under half recommended dose of NPK (30:20:60 kg/ha) + vermicompost @ 2 tonnes/ha + biofertilizer. Hence, it is recommended for Jaipur conditions of Rajasthan.

- At Varanasi, maximum yield (246.80 q/ha) of carrot cv. Early Nantes was obtained with the application of half dose of recommended NPK (60:30:30 kg/ha) + vermicompost @ 2 tonnes/ha + biofertilizer. Hence, it is recommended for Varanasi conditions.
- At Faizabad, maximum yield (212.91 q/ha) and C:B ratio 1:2.98 of carrot cv. Nantes was obtained with application of half NPK (40:30:30 kg/ha) + legume green manure @ 2.5 tonnes/ha + Biofertilizers. Hence, it is recommended for Faizabad conditions.
- At IIHR, Bangalore, maximum yield (248.7 q/ha) and C:B ratio 1:2.0 of carrot cv. Early Nantes was recorded with the application of half NPK (40:30:25 kg/ha) + FYM 10 tonnes/ha + biofertilizer. Hence, it is recommended for Bangalore conditions.
- At Srinagar, maximum yield (210.95 q/ha) and C.B. ratio 1: 4.11 was obtained with application of FYM @ 10 tonnes/ha + Vermicompost @ 2 tonnes/ha for organically produced of carrot var. Chamman. Hence, it is recommended for Kashmir valley.
- On the basis of trials at Hisar, use of bee hives is recommended for higher seed yield and C:B ratio (1:3.25) in carrot.
- Based on two years' study at Solan on carrot using pollinators *Apis mellifera* and *Apis cerana*, it is recommended that to obtain maximum yield/ha, open pollination in isolation should be adopted.
- Based on three years' study at Katrain using pollinators *Apis mellifera* and *Apis cerana indica*, carrot pollination in open conditions is recommended for higher seed yield compared to pollination by honey bees in cages.

Radish

Seed Production

- The maximum seed yield (4.12 q/ha) and C : B ratio (1:2.33) from Pusa Reshmi was obtained with the application of N @ 120 kg/ha with a constant dose of P_2O_5 and K_2O @ 60 and 50 kg /ha, respectively at the spacing of 75cm × 30cm. Hence, it has been recommended for Durgapura conditions of Rajasthan.
- The planting of stecklings at 45cm × 30cm distance along with N dose of 80 kg/ha has been recommended for getting economic seed yield from Pusa Rashmi under agroclimatic conditions of Bhagalpur. Under Jabalpur conditions, planting distance of 60cm × 30cm and application of 120 kg N/ha has been recommended in Pusa Chetki.
- Planting of stecklings at the spacing of 60cm × 30cm and application of N @ 120 kg/ha has been recommended for variety Pusa Chetki under Pantnagar conditions. Similar spacing and dose of N_2 has also been recommended for variety Japanese White under Srinagar conditions.
- Recommendation of 120 kg N, 60 kg P_2O_5 and 60 kg K_2O/ha and spacing of 60cm × 30cm has been made for the variety Japanese White for Srinagar region.

Turnip

Seed Production

- Dipping the roots in 1% ziram solution for half an hour before transplantation can effectively control root rot disease and it has been recommended for Himachal Pradesh.
- Katrain centre has recommended dipping of roots for half on hour in a solution of tetracycline (200 ppm) and bavistin (1000 ppm) before transplanting followed by two sprays of the said combination at fortnight interval at the onset of 5–10% flowering for the control of phyllody disease in variety Pusa Swarnima.

Muskmelon

- The spacing of 350cm × 60cm and application of NPK @ 100:60:60 kg/ha has been recommended for Durgapura Madhu variety under agroclimatic conditions of Durgapura.
- Planting of variety Hara Madhu at a spacing of 280cm × 90cm and application of NPK @ 100:60:60 kg/ha has been recommended for river bed areas of Faizabad.
- Various biochemical parameters were determined in 10 varieties. Maximum dry matter was recorded in Hara Madhu followed by PSGR-2-6-1 and MHCC-2-4-28, while highest TSS was recorded in M-3 hybrid followed by PSG-2-6-1. Acidity was found to be maximum in Punjab Hybrid along with best flavour. Both the above varieties have shown excellent chemical quality performance.

Watermelon

- The fertilizer dose of 100 kg N + 60 kg P_2O_5 + 60 K_2O/ha at planting distance of 320cm × 120cm has been recommended for getting maximum net return from variety Sugar Baby under river bed conditions of Faizabad.
- At Sabour centre, maximum yield (176q/ha) and C : B ratio (1:3.85) were obtained from the variety Sugar Baby with the application of 100 kg/ha N at 90cm × 200cm of spacing along with fixed dose of 60 kg each of P_2O_5 and K_2O. This treatment has been recommended for Sabour conditions.

Other Cucurbits

- Recommendation has been made to control powdery mildew of bottle gourd through spray of benomyl and bavistin (0.1%) at 10 days interval under Bangalore, Udaipur and Ludhiana conditions.
- At Pantnagar, maximum yield (513.3 q/ha) and C:B ratio (1:2.68) was recorded the plants were trained on arbour system without pinching of shoots in the bottle gourd hybrid Pant Sankar Lauki-1. Hence, it recommended for the Tarai conditions of Pant Nagar.
- In bitter gourd, N:P:K @ 60:60:30 kg/ha has been recommended under Coimbatore conditions. Application of N:P:K @ 90:60:60 kg/ha has been recommended under Faizabad conditions for local variety Jaunpuri Karela.
- From bitter gourd variety Priya, maximum yield (154 q/ha) with C:B ratio (1:2.71) was

obtained with the application of N_2 60 kg, P_2O_5 60 kg and K_2O_5 60 kg /ha. Therefore, this dose of NPK has been recommended for Saurashtra region of Gujarat.

- Maximum yield of 57.68 q/ha and C:B ratio 1:2.61 were recorded with the application of 90 kg nitrogen, 60 kg phosphorus and 60 kg potassium/ha in bitter gourd variety Hirkani. Hence, it is recommended for Rahuri condition.
- At Kalyanpur, maximum mean yield(161.25 q/ha) of bitter gourd cv. Summer Green and highest C: B ratio 1:2.13 were recorded with the foliar application of mixture of micronutrients (Zn, B, Mo, Cu, Fe and Mn). Hence, it is recommended for Kalyanpur (Kanpur) conditions.
- On the basis of three years' result at Vellanikkara, application of 80 kg/ha Nitrogen and 40 kg / ha phosphorous is recommended for higher seed yield in bitter gourd cultivar Preethi.
- Based on three years data at Vellanikkara, it is recommended that bitter gourd seeds should be harvested at 21 days after anthesis to obtain maximum recovery, germination and vigour.
- Priming with KNO_3 (150ppm) in bitter gourd cv Preethi seeds at Vellanikkara produced maximum germinability and vigour during three years' trial.
- At Varanasi, maximum mean yield of cucumber 218.76 q/ha along with highest C:B ratio 1:2.13 has recorded with the application of half NPK (60:30:30 kg/ha) + FYM @ 10 tonnes/ha + biofertilizer. Hence, it is recommended for Varanasi region.
- At Kalyanpur, maximum mean yield (223.18 q/ha) along with the highest C: B ratio 1:3.68 has recorded in cucumber with the application of half recommended NPK + FYM @ 10 tonnes/ha + biofertilizer. Hence, it is recommended for Kanpur conditions.
- At Hyderabad, maximum yield of fruit (111 q/ha) and C:B ratio (1:2.10) were obtained in cucumber cv. Poinsette with the application of FYM @ 10 tonnes/ha + biofertilizers in addition to recommended dose of NPK (100:50:50 kg/ha). Hence, it is recommended for Hyderabad conditions.
- In three years of experimentation at Pantnagar cucumber cv. Poinsette recorded the maximum mean yield of 14.12 kg/m^2 with the highest C:B ratio 1:31.01. Hence, this variety is recommended for protected cultivation.
- At Hisar, maximum yield (570.5 q/ha) and C:B ratio 1:2.61 of cucumber Cv. Phule Prachi was recorded. Hence, this variety is recommended for protected cultivation.
- The cropping sequence of cowpea-tomato-cucumber resulted in maximum net return followed by tomato-pea-okra, hence these cropping sequences have been recommended for Kaymore plateau and Satpura hills of Madhya Pradesh.
- Application of paddy straw mulch in pointed gourd is recommended for effective weed management, higher yield (127 q/ha) and C:B ratio (1:1.99) under Faizabad conditions.

TECHNOLOGIES PROTECTION

Tomato

Disease Management

- Control of buck eye rot by difolatan (0.3%) at 10 days interval has been recommended under Bangalore conditions.
- Dithane M-45 (0.2%) at weekly interval has been recommended to control late blight under Katrain conditions.
- At Rahuri centre, spray of dithane M-45 (0.25%) or difolatan (0.2%) has been recommended for the effective control of early blight. Three sprays should be given at an interval of 15 days, starting from the appearance of disease.
- The early blight in variety LE-79 at Dapoli centre was effectively controlled by the spray of any of three fungicides, viz. bavistin (0.1%) or dithane Z-78 (0.2%) or dithane M-45 (0.2%). Three sprays at 15 days interval starting from the first appearance of the disease have been recommended.
- Difolaton (0.1%) and blitox - 50 (0.3%) has been recommended by Ludhiana centre to control the blight disease in variety Punjab Keshari. These fungicides should be given at the interval of 15 days, starting from first appearance of disease.
- Application of carbofuran @ 1.25 kg a.i./ha in the seed bed at the time of transplanting followed by three sprays of endosulfan @ 0.05% a.i. or two sprays of endosulfan and one spray of monocrotophos @ 0.05% a.i. at 15 days interval after transplanting have been recommended for reducing the incidence and intensity of spotted wilt virus in variety Vaishali. These treatments had given a yield of 311–317 q/ha, which was an increase of 40% over control under Rahuri conditions.
- Four sprays of dithane M-45 (0.25%) at an interval of 15 days, starting from the appearance of disease have been found to be most effective at Pantnagar centre for the control of late blight in variety Pusa Ruby. Dithane M-45 can be substituted by Foltaf (0.2%). Both Foltaf and dithane M-45 were observed to be economical.
- Four sprays of cuman L (0.3%) or Flotaf (0.2%) at an interval of 15 days starting from the appearance of disease has been recommended for the control of early blight under Pantnagar conditions in Pusa Ruby variety. This treatment had given 250 q/ha yield, which was an increase of 38–40% over control.
- To control late blight effectively at Kalyani or similar agroclimatic conditions, it has been recommended to spray mancozeb (0.25%). Six sprays of mancozeb @ 0.25% at ten days interval, starting from 15 DAT was best among all the treatments, which reduces 63.53% disease incidence. The cost: benefit ratio was found to be 1:2.6.
- For the control of damping off, it has been recommended to treat the seeds with copper oxychloride @ 0.3% followed by soil drenching from same fungicide @ 0.3% twice. First drenching should be given just after germination of seeds and second 20 days after germination. All these treatment have been recommended for Coimbatore conditions.

- At Coimbatore centre, soil application of carbofuran 3G @ 1 kg a.i./ha + covering the nursery bed with nylon net of 400 mesh followed by three sprays of neem oil (2%) at 15, 25 and 35 DAT reduced mosaic incidence and gave highest yield with a cost: benefit ratio of 1:3.33.
- To control late blight effectively at Junagadh or similar agroclimatic conditions, it has been recommended to spray mancozeb @ 0.25% six times at 10 days interval starting from 15 DAT.
- Epidemiological studies conducted at Hyderabad revealed that as many as nine diseases were prevalent during *kharif* and 5 diseases during *rabi* seasons. The diseases which were of common occurrence in both the seasons were *Alternaria* leaf spot, powdery mildew, leaf curl, mosaic and TSWV. The centre experienced highest percentage of diseases during *rabi* season. The incidence of diseases, viz. *Alternaria* leaf spot, leaf curl, powdery mildew, TSWV and mosaic were 28.33, 9.83, 25.00, 13.00, 2.00%, respectively.
- At Vellanikkara centre, seed treatment and soil drenching with 0.2% captan was found most effective to control damping off, therefore, recommended for Kerala.
- Tomato seeds were infected by *Aspergillus niger*, Rhizopus nigricons and *Verticillium alboatrum* in Sabour conditions.
- For increasing germination and reducing seed colonization, seed treatment with captan @ 0.25% was most effective in Sabour conditions while carbendazim @ 0.2% at Coimbatore conditions.
- For the management of tomato wilt in Junagadh region, application of green manuring (Sunhemp) and antagonists either with neem cake or without neem cake is recommended as the best treatment combination.
- Four sprays of mancozeb @ 0.3% at 15 days interval are recommended for control of buck eye rot of tomato which gave only 11.45% fruit rot intensity and maximum yield 343.1 q/ha with cost: benefit ratio 1:7.52.
- Green manuring along with *Trichoderma viride* @ 5 kg/ha alone or in combination with neem cake @ 10q/ha has been found effective in minimizing tomato diseases, viz. *Rhizoctonia solani*, *Sclerotium rolfsii*, *Pythium* and *Phytophthora* spp. up to 50% and is recommended for Varanasi region.
- Highest germination percentage was recorded in tomato (74.60%) and lowest seedling infection (18.65%) by seed treatment with captan (0.2%) at Lam conditions.
- In tomato, Alternaria blight (*A. solani*) and buckeye rot (*Phytophthora parasitica*) were graded serious having varied incidence of 28-39.4% and 21.6-29.2%. Fruit rot of capsicum (*Phytophthora capsici*) was predominantly serious over the years at Solan valley with 18.9–23.4% incidence. Suitable recommendations should be adopted in the region.
- *In vitro* studies revealed that seed treatment by carbendazim @ 0.2% was the best for increasing the germination in tomato (86.5%) at Junagadh. The same treatment was also found best at Kalyanpur and Vellanikkara. However, captan @ 0.2% was better at

Hyderabad. Invariably the dominant associated seed mycoflora were Altrernaria, Fusarium, Aspergillus and *A. solani* to tomato seed.

- It was found that the incidence of buckeye rot of tomato in Solan region was positively correlated with rainfall and humidity. The multiple regression analysis revealed the significant effect of temperature, rainfall and humidity.
- Multiple fungal diseases comprise of buckeye rot and leaf blight under Solan region was effectively managed by four sprays of mancozeb @ 0.3% at 15 days interval. It gave maximum cost: benefit ratio of 1:6.2. The same fungicide was also very effective for management of tomato leaf blight in Junagadh region with 1:14.4 C:B ratio. Early blight, leaf blight and powdery mildew diseases of tomato were managed by mancozeb FP @ 0.5% giving 1:6 C:B ratio under Rahuri condition.
- Seed treatment @ 15 g/kg seeds and soil application of *Azotobactor chroococcum* @ 10 g/kg FYM/m^2 was best for the management of damping off caused by *Pythium aphaimidermatum* in tomato at Ranchi conditions.
- Seed treatment with captan @ 0.25% and soil drenching by 0.25% solution @ 6 litres/m^2 of nursery-bed was best for tomato. Seed treatment with *T. harzianum* @ 4 g/kg followed by soil application of *T. harzianum* (10 g multiplied in 1 kg FYM /m^2) was statistically at par at Hessardghatta conditions.
- Seed treatment with captan @ 0.25% and soil drenching with 0.25% solution @ 6 liters/m^2 was the best on the basis of pooled data for maximum seedling stand of tomato at Coimbatore. The same treatment was effective for tomato at Hyderabad center.
- Seed treatment with *T. viride* @ 5 g/kg was found effective to control seed-borne fungi and increasing seed germination (91.3 %) in tomato as compared to control (54.3%) at Rahuri.
- Tomato seed mycoflora are *Aspergillus* sp., *Penicillium* spp., *Alternaria* spp., *Rhizopus* spp. and *Fusarium* spp. The seed treatment with carbendazim @ 0.2 % was found most effective. This was at par with *Trichoderma* @ 4 g/kg seed in the management of seed borne fungi at Rahuri conditions.

Insect Pest Management

- Application of fenvalerate @ 30 g a.i./ha, 2-3 times after flower initiation at fortnightly interval was found effective and economical against fruit-borer under Rahuri conditions.
- Three sprays of fenvelerate @ 50 g a.i./ha and endosulfan @ 700 g a.i./ha alternately at first sight of egg or larvae/5 plants, reduces fruit-borer damage up to 25.67% and 29.89% in variety Punjab Chhuhara and Punjab Keshari, respectively.
- Marigold as a trap crop with only two sprays of endosulfan was found most effective to control fruit-borer (*H. armigera*). However, caution should be taken to complete picking and destruction of attached flowers to prevent emergence of second generation of *H. armigera*.
- Against leafminer, spraying of cyromaizin (250 g/ha) at 15 days interval and Sukrina (a neem formulation) @ 15 ml/litre was found very effective to reduce the leafminer infestation.

- On the basis of three years data, it is concluded that on initiation of egg laying by *Helicoverpa armigera*, 5 weekly releases of egg parasitoid, *Trichogramma pretiosum* @ 50,000/ha coupled with 3 sprays of HaNPV @ 1.5×10^{12} POBs/ha at 10 days interval and 3 fortnightly sprays of endosulfan 35 EC @ 700 g a.i./ha gave best control of tomato fruit-borer and resulted in the highest economic returns at Ludhiana.
- Tomato fruit-borer, *Helicoverpa armigera* can be effectively controlled by alternative use of egg parasitoid, *Trichogramma braziliensis* @ 2,50,000/ha and use of NPV at 200 LE/ha at 15 days interval starting from the first appearance of eggs on tomato at Hyderabad.
- Trap crop with marigold and spraying of polytrin-c @ 660 a.i./ha or endosulfan 700g a.i./ha at 28 and 35 DAT was found effective in reducing fruit-borer (*Helicoverpa armigera*) damage to tomato at Hyderabad.

Brinjal

Disease Management

- For the control of Cercospora leaf spot, bavistin (0.1%) was found to be effective at IIHR, hence recommended for Bangalore conditions.
- Based on three years trial at Sabour centre, seed treatment with bavistin (1 g/kg), seedling dip in 0.05% solution of bavistin followed by two sprays of bavistin @ 0.05%, starting from one month after transplanting at an interval of 10–15 days were found most effective, hence these treatments have been recommended for Bihar.
- Seed treatment with bavistin @ 18 kg/ha of seed, seedling dip in 0.05% bavistin for 30 minutes before transplanting followed by two sprays of 0.05% bavistin or three sprays of fytolan (0.3%) at 10–15 days interval starting after one month of transplanting, resulted in significant reduction of phomopsis blight and fruit rot. These treatments have been recommended for Sabour conditions.
- For effective control of brinjal blight disease in variety Pant Samrat at Pantnagar and in Pusa Purple Long at Sabour, it has been recommended to treat the seeds with bavistin @ 1 g/kg seed, followed by seedling dip in solution of bavistin (0.05%) for 15 minutes and application of two sprays of bavistin @ 0.05% (30 and 50 DAT). This treatment reduced the disease intensity by 50% and increased the yield by 32% over control under Pantnagar conditions. From this treatment fruit infection was reduced by 61% as compared to control, with an additional yield of 37 to 42%.
- Little leaf disease in variety Manjari Gota was effectively controlled under Rahuri conditions after raising the seedlings in seed beds treated with phorate @ 1.25 kg a.i./ha and seedling dip in solution of monocrotophos @ 0.05% for 8 hours. followed by application of phorate @ 1.25 kg a.i./ha 21 days after transplanting or raising the seedlings in seed beds treated with phorate @ 1.25 kg a.i./ha and seedling dip in solution of tetracycline hydrochloride @ 500 ppm for 15 minutes at the time of transplanting followed by spot application of phorate @ 1.25 kg a.i./ha at 21 DAT. These two treatments helped to control little leaf disease by 37 and 38% with a yield of 265.83 q/ha and 248.05 q/ha,

which were 35% and 27% increased over control, respectively.

- *Alternaria* and *Cercospora* leaf spot is most important disease of brinjal with maximum incidence 22.0 and 17.9% respectively during first fortnight of November in Coimbatore regions.
- *Aspergillus flavus, Curvularia lunata, Fusarium semitectum, Phomopsis vexans* and *Alternaria alternata* associated with brinjal seeds in Sabour conditions
- Brinjal seeds were associated with *Aspergillus fumigatus, Alternaria alternata, Fusarium* spp., *Curvularia* spp. *Rhizopus* spp. and *Penicillium* spp. The seed treatment with captan @ 0.2% was found most effective. This was at par with *Trichoderma* @ 4 g/kg seed in the management of seed-borne fungi at Rahuri conditions.
- Continuous survey for five years revealed that the crop is infected by 10 fungal diseases, one each bacterial, viral and phytoplasma diseases. Among the fungal diseases phomopsis blight rank first in relation to incidence and economic damage. The maximum incidence of phomopsis blight was observed after 105–135 days of transplanting. Next important diseases is Alternaria blight as per incidence recorded although disease not cause much economic loss as it appeared late in the season. Other important fungal diseases recorded are Sclerotinia blight and Collar rot.
- For increasing germination and reducing seed colonization, seed treatment with captan @ 0.25% was most effective for Sabour conditions, while carbendazim @ 0.2% at Coimbatore conditions.
- Highest germination percentage was recorded in brinjal (77.72%) and lowest seedling infection (18.23) by seed treatment with captan (0.2%) at Lam conditions.
- *In vitro* studies revealed that seed treatment by carbendazim @ 0.2% was the best for increasing the germination percentage at Kalyanpur and Vellanikkara for brinjal. However, captan @ 0.2% was better at Hyderabad. Invariably the dominant associated seed mycoflora were Altrernaria, Fusarium and Aspergillus. Phomopsis was specific to brinjal.
- Incorporation of green manure (Sesbania) and application of neem cake @ 10 q/ha coupled with seed, soil and seedling treatment with *Trichodrma viride* recorded only 6% disease incidence and gave maximum fruit yield (199.78 q/ha) with higher cost: benefit ratio of 1:20 in brinjal. The check treatment registered highest disease incidence of 30.4% with lowest fruit yield of 121.47 q/ha in agro-climatic conditions of Bhubaneswar.
- On the basis of five years of experimentation, it was concluded that four sprayings of carbendazim @ 0.1% starting first spray after a month of planting followed by subsequent sprays at 10 days interval is the best fungicide for management of Phomopsis blight of eggplant. It significantly reduced twig blight and fruit rot phase of Phomopsis and increased fruit yield. The cost: benefit ratio was recorded 1:5.2 at Sabour and 1:12.3 at Junagadh regions.
- Seed treatment @15 g/kg seeds and soil application of *Azotobactor chroococcum* @10 g/kg FYM/m^2 was best for the management of damping off caused by *Pythium*

aphaimidermatum in brinjal at Ranchi conditions.

- Seed treatment with captan @ 0.25% and soil drenching with 0.25% solution @ 6 liter/m^2 was the best on the basis of pooled data for maximum seedling stand of brinjal at Durgapura centre. This treatment was also best for brinjal at Sabour, Ludhiana and. Hyderabad.

Insect Pest Management

- Application of granules of carbofuran @ 1500 g a.i./ha during transplanting, followed by three sprays of cypermethrin @ 30 g a.i./ha at 10 days interval, were the most economical treatments to control jassids, aphids and fruit-borer at Rahuri, therefore, this treatment has been recommended for Rahuri region.
- The pest complex (jassids, white fly and fruit-borer) can be effectively managed with 5 sprays of cypermethrin (30 g a.i./ha) or deltamethrin (10 g a.i./ha) or endosulfan (700 g a.i./ha) at 14 days interval, starting from 10 DAT or two sprays of dimethoate (300 g a.i./ha) and three sprays of cypermethrin (30 g a.i./ha) under Sabour and Rahuri conditions.
- Three sprays of cypermethrin @ 30 g a.i./ha at 40, 55 and 70 DAT, effectively controlled fruit-borer at Hyderabad and Rahuri conditions. Soil application of carbofuran + three sprays of cypermethrin @ 30 g a.i./ha at 40, 55 and 70 DAT gave maximum net profit at Sabour conditions.
- To control various insect pests, BT @ 500 g (commercial) combined with endosulfan (35 E.C.) @ 350 g a.i./ha should be sprayed five times at 14 days interval, starting from 10 DAT under Ludhiana conditions.
- Five sprays of cypermethrin @ 30 g a.i./ha or ethofenprox @ 75 g a.i./ha at 15 days interval starting from 30 DAT imparted most effective and economical control of leafhopper and fruit-borer. Four sprays of carbaryl @ 800 g a.i./ha at 15 days interval, starting from 30 DAT was equally effective and economical for the control of pests.
- Use of leafhopper tolerant variety Punjab Barsati with a spraying schedule of cypermethrin @ 30 g a.i./ha at 20, 35, 65 and 80 DAT gave highest yield.
- Under Sabour and Hyderabad conditions, on the basis of net additional yield and net income, it has been recommended to control fruit borers by spray of cypermethrin @ 30 g a.i./ha at 20, 35, 50, 65 and 80 DAT.
- At Ludhiana, application of FYM @ 10 tonnes/ha and azotobacter @ 1.25 kg/ha followed by need based NSKE (4%) spray at 10 days interval is recommended for the management of pest complex on brinjal with highest C:B ratio.
- At Rahuri, application of NPK @ 100:50:50 along with 4 sprays NSKE (4%) was best in managing brinjal shoot-and fruit-borer with maximum C:B ratio (1:4) followed by application of FYM 10 tonnes/ha + Azospirillum @ 1.25 kg/ha along with 4 sprays of NSKE (4%) are recommended for BSFB management with the C:B ratio of 1:1.8.
- Barrier crop of maize sown 10 days before brinjal transplanting + shoot clipping at weekly interval followed by 5 sprays of NSKE 4% was effective against brinjal pests at Varanasi and Hyderabad centres.

- Cypermethrin at 50g a.i./ha at 20, 35, 50, 65 and 80 DAT was effective and most economical at Varanasi on variety 'Punjab Barsati' against the brinjal shoot-and fruit-borer.
- Five sprays of profenophos 50 EC (0.1%) at 20, 35, 50, 65 and 80 days after transplanting gave good control of brinjal shoot and fruit borer with C:B ratio of 1:4.39 under Ludhiana conditions.
- For Integrated pest management in brinjal under Rahuri condition the package comprised: (a) dipping of seedlings in imidacloprid (1 ml/litre) for three hours before transplanting; (b) clipping of infested shoots and removing infested fruits and destroying them; (c) spraying with NSKE 4% starting from flowering at fortnightly interval is recommended.
- At Kalyanpur, for control of brinjal shoot and fruit borer, maximum net income and C:B ratio was obtained by spraying cypermethrin @ 50g ai/ha at 20, 35, 50 and 65 DAT.
- Resistant/tolerant variety of locality, root dripping in imdacloprid 0.018 % suspension (1ml/litre) for 3 hours and weekly shoot clipping of infested shoots followed by foliar spray of NSKE 4% / profenophos 0.1% / cypermethrin 0.005 % alternatively was found most suitable for shoot-and fruit-borer management of brinjal at Sabour and Hyderabad centres.
- Root dipping in pesticide solution (Imidacloprid 200 SL @ 1 ml/litre for 3 hours) gave an effective control of leaf hopper on brinjal.
- Seedling root dip in imidacloprid 0.018% suspension (1ml/litre) for 3 hours + cypermethrin @ 0.005% foliar spray at flower initiation for thrice at 15 days interval was found relatively effective against shoot-and fruit-borer of brinjal at Sabour and Hyderabad centres.
- Shoot clippings followed by destruction of larvae gave good control of brinjal fruit borer.
- Sex pheromone based IPM technology involving mass trapping through use of plastic funnel baited with sex pheromone @ 100 traps/ha, shoot clipping at weekly interval from 20 DAT and 4 sprays of NSKE (4%) at 15 days interval from flowering was found to be effective for the management of brinjal shoot-and fruit-borer (Leucinodes orbonalis) at Sabour, Ludhiana, Rahuri, Anand and Katrain centres.

Chilli and Capsicum

Disease Management

- At Lam centre, experiments with three years trial were conducted and based on the results, following recommendations have been made for Andhra Pradesh: (i) control of fruit rot: optimum dose of dithane M-45 (0.25%). (ii) control of bacterial leaf spot and fruit rot: preventive sprays of paushamycin 200 ppm + copper oxychloride 0.3% in the rainy season (three sprays during October and November) and dithane M-45 (0.25%, 4 sprays from last week of November to January) at 15 days interval, (iii) control of fruit rot, thrips, mites and pod borer: five sprays of dithane M-45 alternatively combined with monocrotophos 0.05% or quinalphos 0.1% at 15 days interval from November to January.

- At Sabour centre, for the control of die-back and fruit rot, experiments were conducted continuously for three years. It was found that die-back and fruit rot can be controlled by seed treatment with ceresan wet 0.1% + four sprays of bavistin 0.05% at 10-15 days interval, starting from 30 DAT. Therefore, these treatments have been recommended for Sabour conditions.
- At Sabour centre, fruit rot and die-back were effectively managed by treating the seeds with bavistin @ 0.05% combined with one spray of 0.05% bavistin and three sprays of cuman L @ 0.15% at 15 days interval. Hence, this treatment has been recommended for Sabour conditions.
- At Rahuri centre, experiments were conducted continuously for three years on the control of die-back and fruit rot. For the control of die-back, fruit rot, powdery mildew and pod borer, 5-6 sprays of dithane M-45 (0.25%) + karathane (0.1%) + monocrotophos (0.1%) at the interval of 15 days, starting from 40-45 DAT was found effective.
- At Lam centre, in Sindhur variety, four sprays of dithane M-45 (0.25%) + karathane (0.05%) at 10 days interval has been recommended to control the powdery mildew and fruit rot.
- Two sprays of either topsin M (0.05%) or bavistin (0.05%) at the interval of 15 days have been recommended to control powdery mildew in variety Pusa Jwala under Bangalore conditions.
- Seed treatment with bavistin @ 0.05% followed by 4 sprays of bavistin @ 0.05% with first spray at 15 DAT and later sprays at three week interval, were found to be the most effective for maintaining die-back disease intensities in variety Pusa Jwala. This treatment had given maximum yield of 19.98 q/ha resulting in an increase of 72% over control at Sabour conditions.
- To control powdery mildew disease of variety Pusa Jwala at Rahuri or similar agro-climatic conditions, it has been recommended to give three sprays of sulfex @ 0.3% or topsin M @ 0.05% or karathane @ 0.1%. Three sprays should be given at 15 days interval, starting from soon after the incidence of disease. These fungicidal treatments reduced disease intensity, which resulted in increased yield by 43.50-53.19% over control.
- For the control of leaf curl disease, soil application of thimet @ 1.25 kg a.i./ha was found very effective. The application of thimet reduced leaf curl incidence and increased yield (35.77 q/ha) over control (21.26 q/ha). The soil application of furadon @ 1.25 kg a.i./ha was found almost equally effective in reducing leaf curl incidence. The yield observed from that treatment was 32.17 q/ha as against 21.26 q/ha in control.
- Soil application of phorate @ 1.25 kg a.i./ha at the time of transplanting + monocrotophos (0.05%) spray at the interval of 10 days had given the best results at Junagadh for the control of leaf curl disease, which resulted increased yield with a cost: benefit ratio of 1: 5.98.
- *Alternaria* leaf spot, *Phytophthora* leaf blight, *Cercospora* leaf spot, die-back, fruit rot, bacterial wilt and mosaic are the main diseases in Vellanikkara. Of these, *Alternaria* leaf

spot showed maximum severity that was recorded maximum codex value of 56.0 in September when temprature ranges from 23.2–30.6°C, RH 82–94% and high rainfall (30 cm).

- *Aspergillus flavus, A. niger, Cercospora* sp., *Fusarium equsettii* and *Colletotrichum capsici* associated with chilli seeds in Sabour conditions
- Chilli seeds mycoflora are *Colletotrichum capsici, Fusarium* spp. *Alternaria* spp., and *Penicillium* spp. The seed treatment with carbendazim @ 0.2 % was found most effective. This was at par with *Trichoderma* @ 4 g/kg seed in the management of seed-borne fungi at Rahuri conditions.
- *Colletotrichum gloeosporiodes, Cladosporium* sp., *Rhizoctonia solani, Alternaria solani, Curvularia lunata, Fusarium solani, Rhizopus* sp. and *Aspergillus* sp. in chilli were associated seed mycoflora. Carbendazim and captan @ 0.2% as seed treatment were found effective at Hessaraghatta as well as Solan conditions. Seed treatment with Trichoderma formulations @ 4g/kg reduced pathogenic seed fungi and seen inferior to fungicide.
- For increasing germination and reducing seed colonization, seed treatment with captan @ 0.25% was most effective in Sabour conditions while carbendazim @ 0.2% at Coimbatore conditions.
- Green manure + neem cake + *Trichorderma viride* lowered the wilt incidence and increased the yield in chilli at Hyderabad.
- Green manuring + neem cake @ 10 q/ha + antagonists (*Trichoderma viride*) 6 kg/ha recorded the minimum Fusarium wilt incidence and increased yield in chilli. The highest C:B ratio 1.03 in chilli was obtained in the treatment. This treatment combination is recommended for effective control of *Fusarium* wilt in chilli in Rahuri conditions.
- Highest germination percentage was recorded in chilli (86.72%), and lowest seedling infection (12.36%) by seed treatment with captan (0.2%) at Lam conditions.
- *In vitro* studies revealed that seed treatment by carbendazim @ 0.2% was the best for increasing the germination in chilli (88.5%) at Junagadh. The same treatment was also found best at Kalyanpur and Vellanikkara. However, captan @ 0.2% was better at Hyderabad. Invariably the dominant associated seed mycoflora were Altrernaria, Fusarium, Aspergillus. Colletotrichum was specific to chilli.
- Nursery management by use of nylon net (40 mesh), soil application of carbofuran @ 1 kg a.i./ha and seedling dip by imidacloprid @ 0.2% for 3 hours followed three foliar sprays of any systemic insecticide combined with wettable sulphuer alternatively along with border crop of maize/bajra/sorghum is the best combination for the integrated management of insect born viruses of chilli/capsicum. This combination gives maximum yield with highest C:B ratio of 1:6.8 in Coimbatore region and 1:7.6 at Hessaraghatta region.
- Seed treatment with captan @ 0.25% and soil drenching with 0.25% solution @ 6 liter/m^2 was the best on the basis of pooled data for maximum seedling stand of chilli, at Durgapura centre. The same treatment was effective for chilli at Hyderabad centre.

- Seed treatment with thiram @ 0.2% along with soil application of carbofuran @ 1.25 kg ai/ha and one spray of monocrotophos @ 0.05% along with mancozeb @ 0.3% at 30 day-after sowing in nursery beds followed by four sprays of carbendazim @ 0.1% at 15 days interval is best treatment with maximum (1:3.1) C:B ratio under Coimbatore conditions.
- Soil nursery covering with nylon net 60-100 mesh, soil application of carbofuran @ 1 kg ai/ha, seedling dip in imidacloprid @ 0.2% for 3 hours followed by soil application of carbofuran @ 1.5 kg ai/ha one week of transplanting + three spray of imidacloprid @ 0.1% at 10 days interval after 30 days of transplanting is recommended with maximum C:B ratio (1:11.7) from Rahuri.
- Two diseases have been identified as major problem, namely anthracnose (di-back) and damping off. Di-back symptoms appeared during last week of September to middle of October. Fruit rot symptoms also post-harvest loss. Bacterial leaf spot rank third economically important disease. Which appeared only on at early stage of September and in late stage of January. Incidence of Cercospora and Alternaria was very high at later stage of crop growth, i.e., in January. Other miner diseases recorded are root rot, Collar rot, Sclerotinia blight, Myrothiceum leaf spot, bacterial wilt. Enation leaf curl on chilli observed since last three years but incidence was very low.
- Application of chlorothalonil (0.2%) at 15 days interval started from 30 days of transplanting is recommended for managing disease complex of capsicum. It reduced leaf spot and fruit rot incidence up to 16.89 and 6.33% respectively and increased yield 143.65 q/ha as against 74.86 q/ha in control. The cost: benefit ratio was found to be 1:4.3. This recommendation has been advocated for managing bell pepper disease in Himachal Pradesh and similar agroclimatic condition.

Insect Pest Management

- Against pod borer, spray of cypermethrin @ 0.1 kg a.i./ha has been recommended under Lam conditions.
- Spraying of vertimec @ 20 g a.i./ha at 30,45,60 and 70 DAP was found effective in reducing chilli thrips incidence at Hessaraghatta, Rahuri and Hyderabad centres.

Integrated Approach to Control Disease and Pests

- At Ambajogai centre, fruit rot, powdery mildew and thrips on variety Pusa Jwala were effectively controlled by three sprays of dithane M-45 (0.25%) + karathane (0.1%) + metasystox (0.1%) at 15 days interval. The first spray has been recommended immediately after the appearance of the disease and pest. The next best treatment combinations were dithane M-45 (0.25%) + karathane (0.1%) + ekalux (0.1%) and dithane M-45 (0.25%) + karathane (0.1%) + Rogor (0.1%).
- At Coimbatore centre, soil application of carbofuran 3G @ 1.00 kg a.i./ha + covering the nursery bed with nylon net of 400 mesh followed by three sprays of 2% neem oil resulted in lesser mosaic incidence and highest yield with the cost: benefit ratio of 1:1.33. However, at Sabour centre, use of nylon net and soil application of furadon @ 1 kg a.i./ha

in seed bed and one application of furadon @ 5 kg a.i./ha at 6 DAT + 3 sprays of nuvacron 1 ml/litre. was found most effective in managing the disease with a cost benefit ratio of 1:5.64.

Garden Pea

Disease Management

- Three sprays of 0.2% dithane M-45 at an interval of 15 days has been recommended against *Alternaria* blight in Bihar. The first spray should be given immediately after appearance of disease.
- For the control of powdery mildew, four sprays with karathane (0.5%) at fortnightly intervals were found to be effective at Rahuri, Udaipur and Sabour.
- Four sprays of topsin M @ 0.1 % or dithane M-45 @ 0.25 % at 10 days of interval starting from the onset of the disease were found most effective for the control of *Ascochyta* blight in variety Lincon at Katrain. Spray of topsin-M or dithane M-45 gave increased yield of 16.01 and 15.90 q/ha over control, respectively.
- Incorporation of green manure (Sesbania) and application of neem cake @ 10 q/ha coupled with seed, soil and seedling treatment with *Trichodrma viride* recorded lowest disease incidence of only 3.76% and gave maximum green pod yield of pea, i.e., 33.23 q/ha, with maximum cost: benefit ratio of 1:18. The check treatment recorded highest disease incidence of 31.73 % and gave lowest pod yield of 10.71 q/ha in agro-climatic conditions of Bhubaneswar.

Other Leguminous Vegetables

- Bacterial wilt of guar variety Pusa Nav Bahar was effectively controlled by seed treatment from paushamycin 200 ppm for two hours before sowing and spray of blitox-50 (0.3%) at the first appearance of disease at Durgapura.
- For the management of collar rot in cowpea in Vellanikkara region, application of green manuring (Sunhemp) and antagonists either with neem cake or without neem cake is recommended as the best treatment combination.
- Green-mannuring (*Sesbania aculeta)* alone was found to be the most cost effective (C:B 1:6) in limiting the Frenchbean root rot at Solan apart from increasing yield.

Okra

Disease Management

- At IIHR, powdery mildew disease was controlled by five sprays of or calixin (0.05%) or four sprays of sulfex (0.3%) or four sprays of bengard (0.1%) at 10 days interval. Therefore, this treatment has been recommended for Bangalore conditions.
- Four sprays of triforine @ 0.15% at 10 days interval is best with maximum cost: benefit ratio (1:11.0) from Rahuri centre.
- Green manuring + neem cake @ 10 q/ha + antagonists (*Trichoderma viride*) 6 kg/ha recorded the minimum Fusarium wilt incidence and increased yield in okra. The highest

C:B ratio 3.03 in okra was obtained. This treatment combination is recommended for effective control of *Fusarium* wilt in okra in Rahuri conditions.

- Green mannuring (*Sesbania aculeta)* alone was found to be the most cost effective (C:B 1:6) in limiting the okra root rot at Hessaraghatta apart from increasing yield.
- On the basis three years report, it has been concluded that three sprays of carbendazim (0.1%) from 50 days after sowing at 10 days interval recorded very low disease incidence (12.65%) of Cercospora leaf spot with the highest yield (143.12 q/ha) as compared to control (49.78% and 90.50 q/ha). The same treatment showed highest cost: benefit ratio (4.14) as compared to control (2.28).
- The increased germination (90.8%) and suppression of mycoflora over control was recorded in carbendazim (0.25%) at Parbhani.

Insect Pest Management

- Carbaryl (1 kg/ha) spray at weekly interval for the control of fruit-and shoot-borer has been recommended. For the control of jassids, seed treatment with carbofuran 5% or soil application of carbofuran 1 kg/ha and disulfoton 0.5 kg a.i./ha has been recommended.
- For the control of shoot-and fruit-borer, fortnightly applications of carbaryl 1 kg/ha has been recommended.
- Experimental results of IIHR, Rahuri and Sabour centres revealed that application of monocrotophos (0.5 kg a.i./ha) and endosulphan could effectively control the fruit-borer and jassids. Residue analysis indicated that it was safe to consume fruits harvested after 4 days following monocrotophos spray.
- At Rahuri centre, most economic treatment for jassid and borer control was application of endosulfan @ 500 g a.i./ha followed by three sprays of fenvalerate @ 50 g a.i./ha at fortnightly interval.
- For control of jassid and fruit-borer of seed crop at Ludhiana, economic treatment was two sprays of dimethoate @ 500 g a.i./ha, followed by two sprays of cypermethrin @ 50 g a.i./ha at fortnightly interval.
- At Rahuri centre, economic treatment for jassid and borer control of seed crop was application of phorate granules @ 1.5 kg a.i./ha followed by two sprays of fenvalerate @ 50 g a.i./ha at fortnightly interval starting from 40 days after application of granules.
- For the control of shoot-and fruit-borer, two alternate sprays of endosulfan (500 g a.i./ha) with two sprays of deltamethrin (10 g a.i./ha) or fenvalerate or cypermethrin or permethrin (all @ 50 g a.i./ha) proved most effective under Sabour conditions of Bihar.
- For the control of jassids and shoot-and fruit-borer, four sprays with endosulfan @ 500 g a.i./ha at 14 days interval starting from 15 days after germination has been recommended under Rahuri conditions.
- Most suitable sowing period was considered as 15–30 July, when the crop suffered minimum and escaped insect pest attack in Ludhiana conditions. Whereas under Rahuri conditions June-July and January-February months were found most suitable, when the insect damage can be minimised to a greater extent.

- The action threshold of leaf hopper has been found to be 2.5 nymphs/leaf, which required four alternate sprays of endosulfan (500g a.i./ha) and deltamethrin (10g a.i./ha).
- Out of various insecticides and combination of insecticide schedules tried against insect pests, four applications of cartap hydrochloride @ 500g a.i./ha or alternate spray of endosulfan @ 500 g a.i./ha and deltamethrin @ 10g a.i./ha at 15 days interval was found most economical.
- For effective control of leafhopper and fruit-borer of variety Arka Anamika, spray of cypermethrin (0.005%) at 20, 35, 50 and 65 DAS was found to be most economical treatment and has been recommended.
- On the basis of economics studies on net additional income four monocrotophos sprays @ 500g a.i./ha at 20, 30, 50 and 65 DAS has been recommended for control of pest complex under Varanasi, Kalyanpur and Sabour conditions. Further, it has been recommended that where not absolutely needed, synthetic pyrethroids should not be used for management of pests.
- Alternate sprays of lambda-cyhalothrin @ 15g a.i./ha and profenophos @ 800g a.i./ha at 35, 50, 65 and 80 DAS had highest net returns of Rs. 11,350 with cost: benefit ratio of 1:4.93 and is suggested for the management of okra pests under Sabour conditions.
- For the management of leafhopper, whitefly and fruit borer of okra, soil application of neem cake @ 250 kg/ha at 20 days after sowing followed by three sprays of NSKE (4%) at an interval of 15 days from 45 days after sowing are recommended under Rahuri condition.

Onion

Disease Management

Bulb Crop

- Three sprays of miltox along with sticker triton AE at 15 days interval was found to be effective for the control of leaf blotch (*Alternaria porri*) at Rahuri. The first spray should be given immediately after the appearance of disease.
- Four sprays of dithane M-45 (0.25%) alone or four sprays of dithane M-45 (0.25%) + 0.05% monocrotophos at an interval of 15 days effectively reduced the thrips and purple blotch in variety Pusa Red at IIHR centre. At Karnal centre, six sprays of dithane M-45 (0.25%) + monocrotophos (0.05%) at 15 days interval were found effective against purple blotch, *Stemphyllium* blight and thrips.
- For the control of purple blotch, *Stemphyllium* blight and thrips, four sprays of dithane M-45 @ 0.25% + monocrotophos @ 0.05% or dithane M-45 @ 0.05% + metasystox @ 0.05%, starting from 15 DAT at 15 days interval have been found to be most effective in variety Punjab Red Round in Ludhiana conditions. These two treatments had given onion bulb yield of 284.60 and 265.57 q/ha, resulting in an increase of 74.17 and 55.14 q/ha respectively, over the control.
- At Sabour centre, four sprays of dithane M-45 @ 0.25% mixed with methyl dimeton

@ 0.05%, starting from 15 DAT at 15 days interval were found to be the most effective for the control of thrips, purple blotch and *Stemphyllium* blight disease intensity. This treatment had given maximum bulb yield of 217.56 q/ha, which was an increase of 83.29 q/ha over the control, therefore, it has been recommended for Sabour conditions.

- At NHRDF, Karnal and Nasik, four sprays at 15 days interval with mancozeb @ 0.3% + monocrotophos @ 0.05% was found best to reduce leaf blight disease intensity (2.85%) as compared to control (19.08%). The yield was 340.43 q/ha with cost: benefit ratio 1:8.09. The yield in control plot was 230.52 q/ha. Hence, this treatment has been recommended for Karnal and Nasik conditions.
- Onion seed treatment with carbendazim @ 0.2% along with drenching of soil twice using the same fungicide @ 0.1% at the time of germination and after 20 days of germination was found to be highly effective in reducing the damping off up to the extent of 30.44% with the maximum cost: benefit ratio of 1:3.0. Hence, this treatment has been recommended for Nasik region.
- At Karnal centre, seed treatment with thiram @ 0.2% + two soil drenching of thiram @ 0.2% at the time of germination and after 20 days after germination, gave minimum disease incidence (20.89%) and maximum cost: benefit ratio of 1:3.56.
- Leaf blight of onion is effectively managed by four sprays of mancozeb (0.3%) + cypermethrin (0.01%) at 15 days interval starting from 25 days of planting. The treatment recorded minimum disease incidence (27.87%) and highest seed yield (997 kg/ha). The cost: benefit ratio was found to be 1:15.45.
- Seed treatment with captan @ 0.25% and soil drenching with 0.25% solution @ 6 litres/ m^2 was the best on the basis of pooled data for maximum seedling stand of onion at Durgapura centre.
- The spraying of mancozeb (0.3%) when sprayed along with cypermethrin (0.01%) alternatively with sticker sandovit (0.1%) starting from 15 days after transplanting at 15 days interval was found most effective in controlling leaf blight disease of onion. The treatment reduced thrips population (87%) and leaf blight (68%) and increased yield 229% over control. The cost: benefit ratio was 1:24.
- Three sprays of mancozeb (0.25%) mixed with 0.015% cypermethrin starting with appearance of disease and subsequent sprays at 15 days interval have been found effective to control leaf blight disease of onion under Sabour condition. The treatment reduced 10.3% disease incidence and increase yield 271.11 q/ha. The C:B ratio was observed 1:9.6.

Seed Crop

- Four sprays of dithane M-45 (0.25%) alone or four sprays of dithane M-45 (0.25%) + 0.05% monocrotophos at an interval of 15 days effectively reduced the thrips and purple blotch in variety Pusa Red at IIHR centre. At AADF, Karnal, six sprays of dithane M-45 (0.25%) + monocrotophos (0.05%) at 15 days interval was found effective against purple blotch, *Stemphyllium* blight and thrips.

- Purple blotch in bulb crop under North Indian conditions (AADF, Karnal) was effectively reduced by four sprays of 0.25% dithane M-45 (mixed with 0.1% sticker triton) at 15 days interval. For seed crop, six sprays have been recommended to control the disease in North India.
- At Ludhiana centre, six sprays of dithane M-45 @ 0.25% mixed with monocrotophos @ 0.05% at 15 days interval, starting soon after the disease incidence, was found to be the most effective for control of purple blotch disease in Red Round. This treatment had given maximum seed yield of 5.66 q/ha resulting in an increase of 2.48 q/ha over control.
- For seed crop, six sprays of dithane M-45 @ 0.25% mixed with monocrotophos @ 0.05% at 15 days interval starting soon after the disease incidence were found to be the most effective for the control of purple blotch disease in variety Agri Found Light Red. This treatment maintained the purple blotch disease intensity at lower level (i.e. 39.25% as against control 60.85%) and gave maximum seed yield of 3.50 q/ha, which was an increase of 0.93 q/ha over control with maximum cost: benefit ratio of 1:3.72.
- Based on four years trial, it has been recommended to control purple blotch disease by providing three sprays of ziram @ 0.15% or mancozeb @ 0.2% or carbendazim @ 0.05%. This recommendation has been made for Junagarh areas.

Insect Pest Management

- Against thrips, four sprays of malathion (0.1%) at 15 days interval were found to be economical at Karnal, hence recommended for Karnal region of Haryana.
- Four sprays at 15 days interval starting from 30 DAT of cypermethrin (50 g a.i./ha) or permethrin (50 g a.i./ha) were found most effective against thrips. These two insecticides increased the bulb yield by 65.93 q/ha and 59.15 q/ha, respectively under Karnal conditions, therefore, recommended for Karnal conditions.
- One application of phorate @ 1 kg a.i./ha at 10 DAT followed by one spray with malathion (500 g a.i./ha) and fenvalerate (50 g a.i./ha) at 60 DAT was found effective against thrips at Sabour conditions.
- Four sprays of monocrotophos @ 350 g a.i./ha + sandovit sticker @ 0.5 ml/litre at 14 days interval, starting from 30 DAT effectively controlled thrips. Hence, this treatment has been recommended for Sabour conditions.
- For management of thrips, deltamethrin @ 10 g a.i./ha or monocrotophos @ 350 g a.i./ha should be sprayed thrice at 15 days interval, starting from 10 DAT. At Karnal conditions, application of cypermethrin @ 50 g a.i./ha at 15 days interval was found quite effective and manifested maximum marketable yield.

Garlic

Disease Management

- Four sprays of dithane M-45 @ 0.25% or cuman L @ 0.3% at 15 days interval starting from the first appearance of the disease, has been found to be most effective for the control of purple blotch and *Stemphyllium* blight in variety G-1 under Karnal conditions.

Spraying of dithane M-45 and cuman L, maintained purple blotch and *Stemphyllium* disease intensity at lower level (27.73%, 2.96% and 55.98%, 3.67% as against control 76.09%, 5.93%, respectively) with cost: benefit of ratio of 1:1.35 and 1:1.53, respectively.

Cauliflower

Disease Management

- At IARI, New Delhi and Katrain, seed dressing with 0.2% of thiride resulted in better germination and healthy nursery.
- *Sclerotinia* rot in seed crop of varieties Pusa Snowball-1 and Pusa Snowball K-I was effectively controlled by 4-5 sprays of 0.1% bavistin at an interval of 15–21 days at bloom and pod stages at Katrain. It has been recommended to start spraying soon after curd formation, as this disease is not a problem before pre- curd stage.
- At Sabour centre, seed treatment with 100 ppm plantomycin for 2 hours followed by seedling dipping in 0.5% sodium chloride solution was found highly effective against curd rot disease.
- For the effective control of the *Alternaria* blight disease (seed crop) under Sabour conditions, it has been recommended to treat the seeds with bavistin @ 1 g/kg and apply three sprays (curd, bloom and pods stages) of dithane M-45 @ 0.2% at an interval of 15 days soon after the appearance of the disease. This treatment had reduced the disease intensity by 42% as compared to control and increased the seed yield by 63%.
- Results obtained from the experiment conducted on the control of curd rot complex, revealed that 5 sprays of ridomil-MZ combined with streptocycline 100 ppm was most effective in controlling the disease in seed crop at Katrain. Application of this treatment resulted in 83.33% disease control with seed yield of 12.47 q/ha, hence recommended for Katrain conditions.
- Based on three years' data, the results revealed that four sprays of Indofil M-45 + Streptocycline 0.25% + 0.01% recorded 33.68 % disease incidence and cost: benefit ratio of 1:14.37 which was maximum as compared to other treatments.
- Seed treatment with *Pseudomonas fluorescens* @ 10 g/kg seed followed by seedling dip in 0.2% *Pseudomonas fluorescens* suspension for 30 minutes and three sprays of *Pseudomonas fluorescens* @ 0.2% at 10 days interval is best with highest (1:33.2) C:B ratio under Sabour conditions.
- Seed treatment with streptocycline @ 100 ppm for 15 minutes followed by seedling dip @ 100 ppm for 15 minutes before transplanting followed by three sprays @ 200 ppm at 10 days interval starting from 5 days after transplanting was the best for management of black rot of cauliflower in Bhubaneshwar and Kalyani region.

Insect Pest Management

- Four sprays of chlorothalonil 2000 ppm + steptocycline 100 ppm mixture was found very effective against curd rot complex of seed crop of late cauliflower. In this treatment, the incidence was only 5.26% against 49.54% in control. The yield significantly increased up to 8.38 q/ha as compared to 4.48 q/ha in control at Katrain centre. Therefore, treatment combination have been recommended for seed crop under Katrain conditions.

Cabbage

Disease Management

- Four spray of plantomycin (500 ppm) + blitox (3000 ppm) mixture were found more effective against *Erwinia* rot in seed crop at Katrain centre. This treatment maintained disease intensity at lower level and increased yield (14.42 q/ha) over control (8.71 q/ha). Hence, this treatment has been recommended for seed crop under Katrain conditions.
- Seed treatment with captan @ 0.25% and soil drenching with 0.25% solution @ 6 litres/ m^2 was the best on the basis of pooled data for maximum seedling stand of cabbage at Durgapura centre.

Insect Pest Management

- Against diamondback moth (DBM) and leaf webber, three sprays of fenvalerate @ 50 g a.i./ha was found to gave maximum cost: benefit ratio, therefore, this treatment has been recommended for Banglore conditions.
- The use of mustard as trap crop proved quite effective in restricting DBM incidence. Hence, it has been recommended to control DBM throughout India. Diamondback moth can be successfully managed by sowing mustard as a trap crop between the rows of cabbage, 15 days before cabbage planting along with 2-3 sprays of dichlorvos @ 1 kg a.i./ha in mustard and 2 sprays of cartap hydrochloride @ 500 g a.i./ha or 5% neem seed extract in cabbage, starting from primordial stage of the crop at 15 days interval.
- On the basis of reduction in population of diamondback moth, *Spodoptera* and aphids, two sowings of mustard as trap crop, first 13 days prior and second 25 days after cabbage planting in 25 rows, subsequent spray of NSKE (4%) at the primordial stage followed by two more sprays at 15 days interval have been recommended for management of pest complex under Hyderabad conditions.
- Cabbage hybrid, KIRCH-5, was found suitable for cultivation in plains under Varanasi conditions with less infestation of diamond back moth.
- Five sprays of pulverized neem seed powder extract (4%) and pongamia soap (1%) at 10 days interval from 20 DAT was found to be effective against diamondback moth in cabbage with a C:B ratio of 14.44 and 8.33 respectively under Hessaraghatta conditions.
- Fortnightly sprays of cypermethrin (@30 ml a.i /ha), dichlorvos (@ 500ml a.i /ha) and alternate sprays of Bt (Dipel ES @500 ml/ha) and endosulfan (700 g a.i /ha) were effective against pests of summer cabbage at Katrain.
- Four foliar sprays of NSKE (4%) at 10 days interval starting from 20 days after planting have been found effective for the management of diamondback moth (*Plutella xylostella*) and aphids in cabbage at Solan.
- Work carried out at IIHR, Bangalore, for three years indicated high level of resistance to cypermethrin in DBM. But appreciable resistance has not been observed to Bt and NSKE

4% and the level of resistance has not significantly increased over years. Thus, Bt and NSKE 4% are still (2003) very efficient in the management of DBM.

Muskmelon

Disease Management

- Spraying of dithane M-45 (0.3%) at fortnightly interval has been recommended to control downy mildew under Ludhiana conditions.
- Three sprays of calixin @ 0.05% or sulfex @ 0.25% effectively checked the powdery mildew disease at Rahuri at an interval of 15 days starting from 30–40 days after sowing.

Watermelon

Disease Management

- Recommendations have been made to control anthracnose by spraying benomyl and bavistin (0.2%) at 15 days interval under Bangalore conditions.
- Three sprays of 0.2% miltox or 0.2% dithane M-45 at 15 days interval have been recommended for the control of *Alternaria* blight in Maharashtra. The first spray should be given immediately after the appearance of disease.
- Seed treatment with Imidacloprid @ 5 g/kg seed + one foliar spray of Imidacloprid @ 0.075% after 30 days of emergence recorded least watermelon bud necrosis incidence (19.4%) with maximum yield of 162.4 q/ha as compared to 55.4% mean disease incidence and 88.7 q/ha yield in control at Hyderabad region. Four sprays of Acephate @ 1.5g/litre at 10 days interval is recommended in Hessaraghatta conditions. However, soil application of carbofuran @ 1.25 kg ai/ha is recommended in Durgapura conditions.

Other Cucurbits

Disease Management

- To control downy mildew disease of cucumber, five sprays of dithane M-45 or dithane Z-78 or aliette (all 0.3%) at 10 days interval have been recommended under Bangalore conditions.
- *Phytophthora* blight caused by *Phytophthora cinnamomi* was most destructive disease with 40% incidence in September. The *Didymella bryoniae*, *Pseudoperonospora cubensis* and nematode *Meloidogyne incognita* were recorded for minor incidence in pointed gourd in Varanasi region.
- Raised bed cultivation of pointed gourd along with four sprays of copper fungicide @ 0.5% starting from second week of May at 15 days intervals showed most effective in reducing Phytophthora vine rot (4.96%) and fruit rot incidence (5.03%) with maximum yield of 172.9 q/ha and C:B ratio 1:3.7 at Sabour.
- It has been observed that downy mildew in bitter gourd at Vellinikkara was positively correlated with RH, rainfall and number of rainy days. The linear function so obtained is recommended for disease prediction.
- Soil solarisation with white polythene for 30 days during peak period of summer and

seed treatment with *T. viride* @ 4 g/kg of seed is the best treatment combination for integrated management of Fusarium wilt of bottle gourd in Sabour region. However, seed treatment by Monceron @ 2 g/kg of seed is best at Ambajogai and Junagadh.

Insect Pest Management

- For effective and economical management of leaf miner in cucumber, erection of yellow sticky trap @ 1/hill, clipping of lower 2-3 infested leaves followed by application of neem soap @ 109/litre or NSKE (4%) at cotyledonary leaf stage and two foliar sprays of deltamethrin (0.005%) along with jaggery (2%) at 50% flowering stage is recommended under Anand and Rahuri condition with maximum C:B ratio.
- The IPM tactics involving erection of yellow sticky traps @ 1/hill after germination, clipping of infested basal leaves, application of neem soap @ 10 g/litre at two-leaf stage and application of deltamethrin @ 0.5 ml + 20 ml of jaggery per liter at 50% flowering or initiation of female flowers stage is most effective in the management of leaf miner (*Liriomyza trifolii*) and fruitfly (*Bactrocera tau*) in cucumber with C: B raio of 1:1.54 at Solan.
- Three years studies at IIHR, Bangalore, indicated that spraying of NSKE (5%) is effective in reducing the damage caused by leaf miner in cucumber. Another neem formulation econeem @ 7.5 ml/litre is also performing well and 5 applications of any of these neem products may be recommended at 10 days interval from 20 DAS.

MILESTONES AND LAMP POSTS OF PROJECT

- 1970-71 (IVth five year plan) – AICRP on Vegetables initiated at IARI, New Delhi
- 1986 – Level of the AICRP was elevated to Project Directorate of Vegetable Research (PDVR)
- 1992 – Headquarters was shifted to Varanasi from IARI, New Delhi
- 1999 – PDVR was upgraded to Indian Institute of Vegetable Research (IIVR)
- At present, there are 8 main and 18 sub-centres. Besides, about 22 centres are participating in the cooperative research programme of project on voluntary basis.

Achievements

- Development of 356 varieties/hybrids in different vegetable crops.
- Vegetable production increased from 20 million tonnes in 1947 to 113 million tonnes in 2006.
- Productivity has gone up from 10.5 tonnes/ha in 1991-1992 to 15 tonnes/ha in 2006.
- Vegetable food basket was enlarged from 14 vegetables to more than 70 vegetables.
- At export front, a sum of Rs 848.49 crore was realized with export of fresh vegetables in 2004-05.

Award

Looking at the contributions of this AICRP, it was bestowed with Chaudhary Devi Lal Outstanding AICRP Award in 2002.

PROJECT COORDINATORS

Following persons served as project coordinators from 1971 till 1999. In year 1999, the status of Project Directorate was elevated to a National Institute as Indian Institute of Vegetable Research, Varanasi and Dr G. Kalloo became the Director of the Institute. Since then there is no separate Project Coordinator of the project and this responsibility is being shouldered by the Director, IIVR, Varanasi. The Project Coordinators and list of workshops/meeting were:

- Dr Vishnu Swarup
- Dr S.S. Chatterjee
- Dr H.S. Gill
- Dr V.S. Sheshadri
- Dr G. Kalloo
- Dr. Mathura Rai

LIST OF WORKSHOPS/GROUP MEETINGS HELD UNDER AICVIP (VC)

Workshop	*Host Institute/University*	*Years*
Ist	Indian Agricultural Research Institute, New Delhi	1971
IInd	Punjab Agricultural University, Ludhiana	1972
IIIrd	University of Agricultural Sciences, Hebbal, Bangalore	1975
IVth	Orissa University of Agriculture & Technology, Bhubaneshwar	1977
Vth	Tamil Nadu Agricultural University, Coimbatore	1979
VIth	Mahatma Phule Krishi Vidyapeeth, Pune	1981
VIIth	Haryana Agricultural University, Hisar	1983
VIIIth	Rajendra Agricultural University, Sabour	1985
IXth	Narendra Deo University of Agriculture and Technology, Faizabad	1987
Xth	Kerala Agricultural University, Vellanikkara	1988
XIth	Dr Y.S. Parmar University of Horticulture and Forestry, Solan	1990
XIIth	Andhra Pradesh Agricultural University, Hyderabad	1992
XIIIth	Jawaharlal Nehru Krishi Vishwavidyalaya, Jabalpur	1993
XIVth	Indira Gandhi Krishi Vishwavidyalaya, Raipur	1995
XVth	Banaras Hindu University, Varanasi	1996
XVIth	Tamil Nadu Agricultural University, Coimbatore	1997
XVIIth	Mahatma Phule Krishi Vidyapeeth, Pune	1998
XVIIIth	Punjab Agricultural University, Ludhiana	1999
XIXth	Indian Institute of Vegetable Research, Varanasi	2001
XXth	Kerala Agricultural University, Vellanikkara (Kerala)	2002
XXIst	Gujarat Agricultural University, Anand (Gujarat)	2003
XXIInd	ANGRAU, Hyderabad	2004
XXIIIrd	BCKV, Kalyani, West Bengal	2005
XXIVth	UAS, Dharwad	2006
XXVth	CCSHAU, Hisar	2007

IMPORTANT VARIETIES/HYBRIDS IDENTIFIED THROUGH AICRP(VC)

Crop/Variety	*Remarks*
TOMATO	
Pusa Ruby	An early-maturing variety developed at IARI, New Delhi; derived through selection from the cross Sioux × Improved Meeruthi. Plants indeterminate (80–85cm), spreading, fewer branches. Fruits flattish-round, small-medium, uniform red, slightly lobed (4-5 locules), slightly acidic, suitable for processing. Gives an average yield of 30 tonnes/ha. This is popular in West Bengal, Assam, Rajasthan, Gujarat, Haryana, Delhi, Madhya Pradesh, Maharastra, Uttar Pradesh and Chattisgarh.
Hisar Arun (Sel-7)	Hisar Arun is an early-maturing variety developed at CCSHAU, Hisar; derived through modified pedigree method from the cross Pusa Early Dwarf × K-1. Plants are determinate, dwarf, erect, with cut leave and synchronized clustered flowers, bear 15–20 fruits. Fruits are round, red, medium, average weight of 65–70g, 4–6 locules with deep red flesh. First picking starts on 60–65 days after transplanting; gives an average of 15.4 tonnes/ha and 28.7 tonnes/ha early and total yield, respectively in 80–85 days of crop duration.
Kashi Hemant (IIVR Sel-1)	Determinate high-yielding, recommended for cultivation in Chhattisgarh, Orissa, Andhra Pradesh and Madhya Pradesh and Maharashtra Identified through AICRP(VC) and released by Central Varietal Release Committee.
Kashi Sharad (IIVR Sel-2)	High-yielding, recommended for Jammu and Kashmir, Himachal Pradesh and Uttra Khand. Identified through AICRP (VC) and released by State Variety Release Committee
Pusa Hybrid-2	This hybrid has been developed at IARI, New Delhi. Plants are determinate. Fruits are round, medium with good keeping quality; field resistance to nematodes; gives an average yield of 55 tonnes/ha. This is popular among farmers of Delhi, Haryana and Uttar Pradesh.
Avinash-2	This hybrid was developed at Syngenta India Seeds., Pune. Plants are determinate with very vigorous growth, dense foliage and profuse fruit setting. Fruits are medium, thick walled, round, glossy, dark red with average weight of

(Contd.)

Crop/Variety	Remarks
	70–90 g. First picking starts 55 days after transplanting; gives an average yield of 75 tonnes/ha; suitable for cultivation in TYLCV prone areas. This is popular among farmers of Rajasthan, Gujarat, Haryana, Delhi, Punjab, Uttar Pradesh, Bihar, Jharkhand, Maharastra, NEH Region, West Bengal and Goa.
Kashi Vishesh (H-86)	This variety is resistant to TLCV and has been developed using *L. hirusutum* f glabratum B'6013' as donor parent. Plant determinate having dark green foliage, has concentrated fruit characters and produce medium to large-sized fruits. Average fruit yield 480 q/ha. Identified for Jammu & Kashmir, Himachal Pradesh, Uttarakhand, Punjab, Uttar Pradesh, Bihar, Jharkhand, Chhattisgarh, Orissa, Andhra Pradesh and Karnataka, Tamil Nadu and Kerala. Identified through AICRP(VC) and released by Central Varietal Release Committee.
BRINJAL	
Kashi Prakash	Identified through AICRP(VC) for Uttar Pradesh, (IVBR-1) Bihar and Jharkhand and released by Central Varietal Release Committee.
Pant Samrat	This variety was developed at GBPUA&T, Pantnagar; derived through selection from local collection. Fruits are attractive; resistant to phomopsis blight and bacterial wilt; less infestation of shoot-and fruit-borer and jassids; good for rainy season cultivation; gives an average yield of 38.6 tonnes/ha. This is popular among the farmers of Jammu & Kashmir, Himachal Pradesh and Uttarakhand.
Pusa Purple Long (PPL)	Pusa Purple Long was developed at IARI, New Delhi; derived through single plant selection from a mixed cultivar (Batia), cultivated in Punjab, Delhi and western Uttar Pradesh Fruits are long (20–25 cm), purple, glossy and tender; suitable for summer and autumn season cultivation; gives an average yield of 28 tonnes/ha. This is popular among the farmers of North India.
Azad Kranti (KS-701)	This variety was developed at Vegetable Research Station, CSAUA&T, Kanpur; derived from the cross Banarasi Giant Long × Pusa Purple Long. Plants are medium tall, profusely branched with semi-erect stem, and dark green wavy margin lamina and bear 12–15 fruits. Fruits are medium, thick-

(Contd.)

Crop/Variety	*Remarks*
	long, smooth, taper at distal end, dark purple with green calyx and white flesh, weight ranges from 100 to 110g; first picking starts 60–65 days after transplanting; gives an average yield of 30.0 tonnes/ha in 140–150 days fruiting duration; crop maturity duration is 175–180 days; suitable for both late and early sown conditions; tolerant to fruit rot, little leaf, drought and frost. This is popular among farmers of Uttar Pradesh and Bihar.
MHB-10 (Hyb.) (Kalpataru)	Its plants erect, bushy, compact and spiny; fruits round to slight oval, shining reddish purple with white streaks (bicolour), green spiny calyx; weight of 60–70 g; first picking starts 70–75 days after transplanting; gives yield of 40-50 tonnes/ha. This is popular among the farmers of Punjab, Uttar Pradesh, Bihar, Jharkhand, Chatishgarh, Orissa, Andhra Pradesh, Maharastra, Goa and Tamil Nadu.
Kashi Sandesh(Hyb.) (VRBHR-1)	Semi-upright plant (height 71.0 cm) with green stems, purplish green leaves and purple flowers which appear 45 days after transplanting. Fruits purple, medium sized, round weighing 225 g. Picking starts 76 days after transplanting. Average yield 780 q/ha. This has been released and notified during XII meeting of Central Sub Committee on Crop Standard Notification and Release of Varieties for Horticultural Crops for cultivation in all brinjals growing parts of the country.
Kashi Komal (Hyb.)	Semi-upright plant, height 90-100 cm, stem and leaves light green; first flowering takes 35–40 days after transplanting. Fruits light purple, long, soft texture, average length 13 cm and diameter 3 cm. Picking starts 65–70 days after transplanting and gives an average yield of 800 q/ha. This has been released and notified by XII meeting of Central Sub-Committee on Crop Standard Notification and Release of Varieties for Horticultural Crops for all brinjal-growing parts of the country.
CHILLI	
Bhagyalakshmi (G-4)	This variety has been developed at RARS, Lam, Guntur; derived through selection in Thohian chilli from Sri Lanka. Plants are with narrow and dark green leaves. Fruits are olive green, turn dark red on ripening, long (7-8 cm), 3-3.5 cm girth with 38–40% seed content; fairly tolerant to

(Contd.)

Crop/Variety	*Remarks*
	diseases and pests; gives an average green fruit yield of 10 tonnes/ha under irrigated and 6 t/ha under rainfed conditions. This is popular among the farmers of Chatishgarh, Orissa and Andhra Pradesh.
Arka Lohit (Sel-1)	Arka Lohit has been developed at IIHR, Bangalore; derived through selection from a local collection (IHR 324). Fruits are dark green, smooth, straight, turn dark red at maturity, highly pungent (capsaicin 0.70 %); suitable for cultivation in irrigated and rainfed conditions; gives an average green fruit yield of 15 tones/ha in 180 days of crop duration. This variety is popular among farmers of Karnataka and Tamil Nadu.
Kashi Anmol (KA-2)	Negative selection was exercised to purify seeds of KA-2, (a Sri Lankan introduction). After several years of evaluation, KA-2 variety has been identified as promising for green fruit production under chilli-wheat cropping system. Recommended for Punjab, Uttar Pradesh, Bihar and Jharkhand, Identified through Central Varietal Release Committee.
Kashi Vishwanath (CCH-2, Hyb)	Suitable for green as well as red fruit production, recommended for West Bengal, Assam, Punjab, Uttar Pradesh, Bihar, Jharkhand, Chhattisgarh, Orissa, Andhra Pradesh and Rajasthan, Gujarat, Haryana and Delhi, Identified through Central Varietal Release Committee.
Lario	This hybrid has been developed at Syngenta India Ltd, Pune. Plant are medium-tall, compact, vigorous, inter-nodal length is longer than Indira, dense foliage provides adequate shelter to fruits and requires stacking. Fruits are of medium size, dark green, thick walled, fleshy and blocky (3-4 lobes) with average fruit weight of 150g. First picking starts 60–65 days after transplanting; gives yield of 30–35 tonnes/ha.
PEA	
VRP-2	It is an early-maturing variety. Plant grows up to 50 ± 5cm height with vigorous dark green foliage. The pods are straight, medium-sized (7-8 cm), light green filled with bold seeds. First flower appears at 38 ± 3 days after sowing and plant bears 12–14 pods which harvested 55–58 days after sowing. Green pod yield100 q/ha. Recommended for zone IV through AICRP (VC).

(Contd.)

Crop/Variety	*Remarks*
VRP-3	It is a mid-season, maturing variety This is the first variety, which flowers between early and mid season, i.e., 45–50 days after sowing. Yields 80–90q/ha. Recommended for zone I through AICRP (VC).
Kashi Nandini (VRP-5)	A high-yielding, early variety developed through pedigree selection from the cross P-1542 × VT-2-1. Plants are erect, dark green with 7-8 pods/plant. Plant height goes up to 43cm. Pods are long, attractive well filled with high shelling percentage. It is tolerant to leaf miner and pod borer. It gives an average yield of 65 q/ha. Recommended for Jammu and Kashmir, Himachal Pradesh, Uttarakhand, Punjab, Uttar Pradesh, Bihar, Jharkhand and Karnataka, Tamil Nadu and Kerala, Identified through Central Variety Release Committee.
FRENCH BEAN	
Kashi Param (IVFB-1)	Determinate high yielding, recommended for Jammu and Kashmir, Himacha Pradesh and Madhya Pradesh and Maharashtra Identified through Central Variety Release Committee.
Swarna Lata	Recommended for Jharkhand, Jammu and Kashmir, Himachal Pradesh and Uttarakhand, Identified through Central Variety Release Committee.
VL Boni-1	This variety has been developed at VPKAS, Almora. Plants are dwarf (30–40 cm) having white flower with purple tinge. Pods are round, light green, attractive, stringless and fleshy; first picking starts at 45–60 days; moderately susceptible to foliar diseases; gives 10–11 tonnes/ha pod yield. This is popular among farmers.
Swarna Priya	Recommended for Jharkhand, Sikkim, Meghalaya, Manipur, Nagaland, Mizorum, Tripura, Arunachal Pradesh, Western Madhya Pradesh, Maharashtra and Andaman and Nicobar Islands Identified through Central Variety Release Committee.
COWPEA	
Kashi Kanchan (VRCP-4)	Resistant to Golden Mosaic Virus and *Pseudocercospora cruenta*, recommended for UP, Identified by State Variety Release Committee.
Kashi Shyamal (IVCP-1)	Developed through selection from local collection. Plants are erect and bushy with green long pods having reddish

(Contd.)

Crop/Variety	*Remarks*
	colour at pod tip. It takes about 32 days to flower in *kharif* and 40 days during summer. It is tolerant to Golden Yellow Mosaic Virus under field conditions. It gives an average yield of 64 q/ha. Recommended for Uttar Pradesh, Punjab, Bihar and Jharkhand. Identified through Central Variety Release Committee.
Sel-263	Sel-263 is an early-maturing variety, developed through selection at PAU, Ludhiana. Plants are dwarf and can be grown in both spring and rainy seasons. Pods are green, thick, fleshy, tender and medium (20cm long); resistant to mosaic and golden mosaic viruses; gives an average pod yield of 22 tonnes/ha. This is popular among farmers of Punjab, Haryana and Rajasthan.
DOLICHOS BEAN	
Swarna Utkrisht	Tolerant to anthracnose, mosaic viruses and pod borer, recommended for Uttar Pradesh, Jharkhand, Bihar and Punjab. Identified through Central Varietal Release Committee.
OKRA	
VRO-5	It is bushy in nature, height up to 65 ± 8 cm in rainy season and 40 ± 5 cm in summer with 2-3 effective branching resistant to YVMV and ELCV, short internodal length (3.0 ± 0.5 cm). It bears 20 ± 3.5 ridges on fruits, 8-10 cm length at marketable stage and 13–15cm potential length with dark green colour having good marketable attributes. Flowering start on fourth to fifth nodes 39 ± 2 days after sowing. It ranked first at 12 centres, second at 2 and third at one centre out of 22 centres tested and green fruit yielded 171.97 q/ha at IIVR, Varanasi, 163.78 q/ha at MAHYCO, 156.74 q/ha at Vellanikkara, 142.2 q/ha at Ludhiana and 129.2 q/ha at IARI obtained during 1998-99 under multilocational trials.
VRO-6	Plants are tall (175 ± 15cm), during rainy season and 105 ± 10 cm in summer season, resistant to YVMV and ELCV with 1-2 effective branches in half curd shape. First flower appears 37 ± 2 days sowing on 4th nods during rainy season and third node during summer season. It bears 25±5 number of fruits of 8–10 cm fruit length at marketable stage and 16-18 cm potential length with 1.3–1.4 cm diameter. It ranked either first or second at Bhubaneshwar, Faizabad,

(Contd.)

Crop/Variety	*Remarks*
	Kalyanpur, Hisar, Kalyani, Sabour, Jabalpur, Ranchi (CHES), Mahyco, Ludhiana, IARI, Hyderabad, Akola and IIVR. It can yield up to 180 q/ha during rainy season and 135 q/ha during summer.
Varsha Uphar (HRB-9-2)	This is an YVMV resistant variety, developed at CCSHAU, Hisar; derived from the cross Lam Selection 1 × Parbhani Kranti. Plants are medium-tall (90–120cm), erect, short internode producing 2-3 branches with five-lobed leaves. Fruits are dark green, 18-20 cm long, smooth with five ridges; first picking starts about 45 days after sowing; gives an average fruit and seed yield of 9.8 t/ha and 1.7 tonnes/ha, respectively. This is popular among the farmers of Punjab, Uttar Pradesh, Haryana and Delhi.
Punjab-7 (P-7)	This is an YVMV resistant variety developed at PAU, Ludhiana; derived through bacross method from an inter specific cross *A. esculentus* cv. Pusa Sawani × *A. manihot* sub. *manihot* cv. Ghana. Plants are medium-tall (85–105cm) with short internodes, stem with pigmentation and deeply lobed leaves with less serrated margins, basal portion of the petiole is deeply pigmented. Fruits are medium-long (15–20cm), dark green, tender and 5-ridged, 15–21 g of weight, remain tender for 3-4 days; top of the fruit is blunt and slightly furrowed; resistant to jasside and cotton ball worm. First picking starts 54 days after sowing; gives an average yield of 9.5 tonnes/ha during rainy and 5 t/ha during spring season; average seed yield of 0.4–0.8 tonnes/ha. This is popular among the farmers of Jammu and Kashmir, Himachal Pradesh and Punjab.
Kashi Lila (IIVR-11)	Resistant to YVMV, recommended for Uttar Pradesh, Bihar and Jharkhand Identified through Central Varietal Release Committee.
Shitla Uphar (DVR-1)	Plants are medium tall resistant to yellow vein mosaic virus and enation leaf curl virus. Potential yield obtained 266 q/ha at Zuwari Agro, Hyderabad. Identified by AICRP(VC) for zones IV and VII.
Shitla Jyoti (DVR-2)	Plants are tall, resistant to YVMV and ELCV. Its potential yield was obtained 243.1 q/ha at Zuari Agro, Hyderabad, Identified through AICRP (VC) for zone VI and V. Central

(Contd.)

Crop/variety	*Remarks*
	Varietal Release committee has notified this hybrid for cultivation in zones VI and V.
ONION	
Agrifound Light Red	This variety has been developed by NHRDF, Nasik; derived through mass selection from a local stock of rabi season cultivar, grown in Dindori area of Nasik. Bulbs are light red, globular, tight skin, 4–6 cm in size with 13% of TSS; good keeping quality; gives bulb yield of 30–32.5 tonnes/ha in 160–165 days of crop duration. This is a popular variety in Punjab, Uttar Pradesh Bihar, Jharkhand, Chatishgarh, Orissa, Andhra Pradesh, Madhya Pradesh and Maharastra.
GARLIC	
Yamuna Safed (G-1)	G-1 has been developed at NHRDF, Nasik; derived through mass selection from local collection from Delhi (Azadpur market). Bulbs are compact, silvery white skin, creamy flesh, TSS 38%, dry matter 39.5%, 4.0–4.5 cm diameter, 12–15cm^2 size index with 25–30 cloves; cloves are sickle shaped and 0.8–1.00cm diameter; good storage quality; tolerant to insect pests and diseases like purple blotch, *Stemphylium* blight and thrips; gives an average yield of 15.0–17.3 tonnes/ha. This is a popular variety in Punjab, Uttar Pradesh, Bihar, Jharkhand, Chatishgarh, Orissa and Andhra Pradesh. Seeds could be obtained from NHRDF, Nasik or its regional branches.
Yamuna Safed-2 (G-50)	G-50 was developed at NHRDF, Nasik; derived through mass selection from a local collection from Karnal. Bulbs are compact, white creamy flesh, 10 bulb weight 160–240g; TSS 38–40%, dry matter 40–41 %, diameter 3.5-4.0 cm, number of cloves 35–40, 0.75-1cm diameter, 1.75–2.5 cm^2 clove size index; gives an average yield of 15–20 t/ha. This is a popular variety in West Bengal, Assam and Maharastra.
CAULIFLOWER	
Pusa Snowball K-1	This is a late-maturing variety developed at IARI Regional Station, Katrain; derived from an exotic collection through selection. Leaves are slightly puckered and wavy. Curds are selfed blanched, snow-white and compact; moderately resistance to black rots under field conditions; matures 90–95 days after transplanting; gives an average yield of 22.5 tonnes/ha. This is popular among the farmers of Jammu

(Contd.)

Crop/variety	Remarks
	Kashmir, Himachal Pradesh, Uttaranchal, West Bengal and Assam.
Kashi Kunwari (IVREC-2)	Early maturing, recommended for Punjab, Uttar Pradesh, Bihar and Jharkhand. Identified through Central Variety Release Committee.
Summer King	This hybrid has been developed by Sungro Seeds Ltd, New Delhi. Curds are creamy white, dome shaped, compact with average weight of 400–600 g and self-blanching; matures 55 days after transplanting; suitable for summer season cultivation. This is popular among farmers of Uttar Pradesh, Bihar, Haryana, Punjab and Rajasthan.
CABBAGE	
Quisto	This hybrid has been developed by Syngenta India Seeds, Pune. Heads are compact, dark green, heavy (1–3 kg), round and tender core with fine internal structure; ready for harvesting on 75–80 after transplanting; tolerant to Fusarium wilt; gives yield of 60–100 tonnes/ha. This is popular among the farmers of West Bengal, Assam, Punjab, Uttar Pradesh, Bihar, Jharkhand, Maharastra and Andhra Pradesh.
MUSKMELON	
Kashi Madhu (IVMM-3)	Tolerant to powdery and downy mildew, recommended for Uttar Pradesh, Punjab, Bihar and Jharkhand. Identified through Central Variety Release Committee.
CUCUMBER	
Swarna Ageti	Tolerant to powdery mildew, recommended for Bihar, Jharkhand, Uttar Pradesh, Andhra Pradesh, Delhi and Himachal Pradesh Identified through State Variety Release Committee.
Swarna Sheetal	Tolerant to powdery mildew, recommended for Jharkhand, Bihar, Uttar Pradesh, Maharashtra, Delhi, Andhra Pradesh, Himachal Pradesh and Kerala. Identified through State Variety Release Committee
Ridge gourd Swarna Manjari	Tolerant to powdery mildew, recommended for Jharkhand, Bihar, Orissa, Tamil Nadu, Maharashtra and Andhra Pradesh, Identified through State Variety Release Committee.

Varieties and Hybrids Indentified for Release in the Group Meeting of AICRP VC Held at Bangalore from 16-19 January, 2010

Varieties/Crops	*Entry*	*Source*	*Recommended Zone*
Brinjal	PB-70 (Green long)	GBPUAT, Pantnagar	IV, VI, VII, VIII
	DBL-02 (Purple long)	IARI, New Delhi	I, IV, VI
Tomato (Determinate)	Pant T-10	GBPUAT, Pantnagar	I, IV
Chill	PC-56	GBPUAT, Pantnagar	I, IV, V, VII, VIII
Dolichos Bean (Pole type)	IIVR Sem-8	IIVR, Varanasi	III, IV, V, VII
Bottle gourd	PBOG-89	GBPUAT, Pantnagar	I, IV, VII
Sponge gourd	VR-I	IIVR, Varanasi	IV
	Hybrids		
Crop	*F_1 hybrid*	*Source*	*Recommended Zone*
Tomato (Det.)	BCTH-4	BCKV, Kalyani	I, IV
Chilli	NCH-587	Nirmal seeds	IV, VII
	VNR-332	VNR Seeds	IV, VIII
Okra	JOH-05-9	Junagarh	V, VI, VII

Pusa Ruby

Hisar Arun

Kashi Hemant

Kashi Sharad

Pusa Hybrid

Kashi Vishesh

Avinash - 2

Varieties of tomato released for cultivation

Kashi Parkash

Pant Samrat

Pusa Purple Long

Azad Kranti

MHB-10 (Hyb.)

Kashi Sandesh

Kashi Komal (Hyb.)

Varieties of brinjal released for cultivation

Bhagyalakshmi

Arka Lohit

Kashi Anmol

Kashi Vishwanath

Lario

Kashi Kanchan

Kashi Shyamal

Sel 263

Varieties of chillies and cowpea released for cultivation

VRP - 2

VRP - 3

Kashi Nandini

Swarna Utkrisht

Varieties of pea

Kashi Param

Swarna Lata

VL Boni-1

Swarna Priya

Varieties of french bean

VRO - 5

VRO - 6

Varsha Uphar

Punjab - 7

Shitla Uphar

Kashi Lila

Shitla Jyoti

Agrifound Light Red Onion

Varieties of okra and onion

Yamuna Safed Garlic

Yamuna Safed - 2 Garlic

Pusa Snowball K-1 Cauliflower

Kashi Kunwari Cauliflower

Summer King Cauliflower

Quisto Cabbage

Kashi Madhu Musle Melon

Swarna Ageti Cucumber

Swarna Sheetal Cucumber

Swarna Manjari Ridge gourd

Varieties of garlic, couliflower, cabbage, musle melon, cucumber and ridge gourd

2

Potato

Prakash S. Naik and Khayal C. Thakur

Potato (*Solanum tuberosum* L.), a globally important crop, exceeded only by maize, rice and wheat, is consumed by over billions of people across the globe, of which half are in the developing countries. Since potato produces more edible energy and protein per unit time and area compared to many other food crops (Fig. 2.1), fits well into multiple cropping systems prevalent in tropical and subtropical agroclimatic conditions, provides profits and generates employment, is expanding rapidly in developing countries (Fig. 2.2).

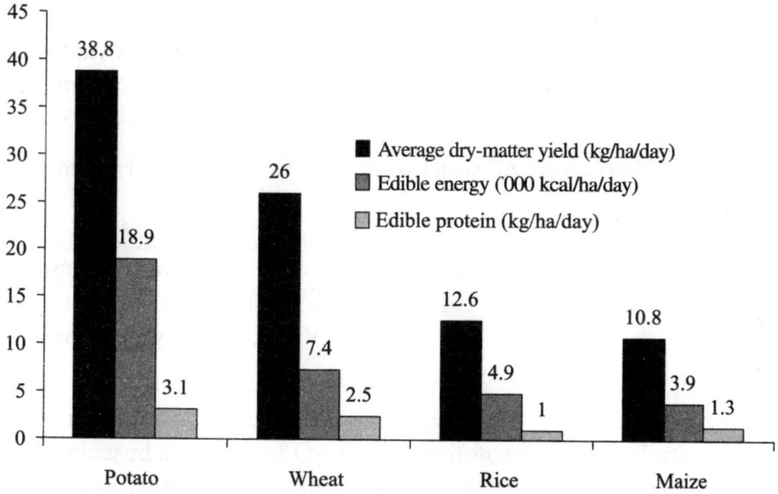

Fig. 2.1. Production of food and nutrients per hectare per day by major crops in India

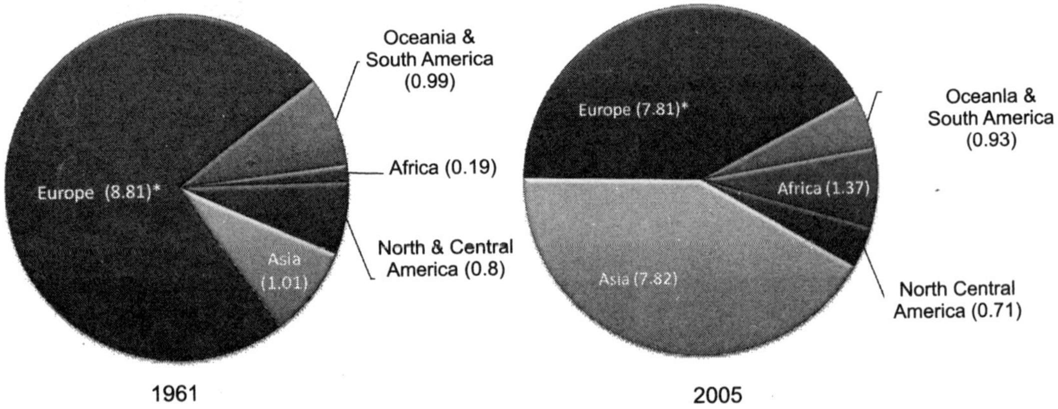

Fig. 2.2. Continent-wise distribution of potato area in 1961 and 2005 (*Source*: FAOSTAT)

BACKGROUND

Potato is not native to India and it was introduced in India from Europe in the beginning of early 17th century. The earliest reference of potatoes in India is found in an old account of the voyage of Edward Terry, who was chaplain to Sir Thomas Roe, British Ambassador to the court of Mughal Emperor Jahangir from 1615–1619. Potato is a temperate crop. Hence, its cultivation in India was initiated in selected hilly areas possessing congenial agroclimate for its cultivation. In those days, tuber seed of potato varieties was regularly imported from abroad by spending a large amount of foreign exchange. The early introductions were closely related to *Solanum tuberosum* ssp. *andigena*. During 1824–1939, systematic attempts were made to introduce new potato varieties in the country, mainly from Europe. These, however, could create very little impression as most of these varieties either failed to yield well under Indian conditions or degenerated and were lost. The failure of introduced varieties in India was mainly because these varieties were primarily bred to suit temperate long days of summer in Europe whereas potato in India is grown during short days of subtropical winters. Contrasting potato cultivation features in Europe and India are given in Table 2.1.

Table 2.1. Agro-climatic peculiarities of growing potatoes in subtropics

Agroclimatic parameters	*Subtropical countries*	*Temperate countries*
Growing season	Winter	Summer
Temperatures during planting and harvesting	High (Planting: 25–30 °C; Harvesting:10–20 °C)	Low (15–25 °C at both)
Crop duration	90–100 days	150–180 days
Photoperiod during growth	10.3 hours/day	14 hours/day
Day/night temperatures	25-30 °C/04-15 °C	25 °C/15 °C
Frosting	Common	Absent
Temperatures immediately after harvesting	Very high (Harvest is followed by summer)	Very low (Harvest is followed by winter)

Potato is grown in India in almost all the states and under very diverse conditions. Nearly 85% of potatoes are grown in vast Indo-Gangetic plains of north India during short winter days from October — March. About 7% area under potato cultivation lies in the hills where the crop is grown during long-summer days extending from April — October. Plateau regions of south-eastern, central and peninsular India, constitute about 8% area where potatoes are grown as a rainfed *kharif* crop during rainy season (July — October) or as irrigated *rabi* crop during winter (October — March). In a small area of about 4,000 ha covering mainly the Nilgiri and Palani hills of Tamil Nadu, the crop is grown round the year both as irrigated and rainfed crop. On the basis of soil, climate, varietal requirement, and other agronomic features, India has been divided into two major potato-growing zones and three special problem areas. These are given in Table 2.2.

Table 2.2. Potato-growing regions and their varietal requirements in India

Region	*States*	*Potato Seasons*	*Varietal Requirement*
ZONES			
North Indian Hills North Western Hills	Southern Jammu and Kashmir and Himachal Pradesh	Summer (April — September)	Long day adapted, resistant to late blight.
North Central Hills	Hills of Uttarakhand	Summer (April — August/September)	Long day adapted, resistant to late blight and bacterial wilt.
North Eastern Hills	Hills of Arunachal Pradesh, Manipur, Meghalaya, Mizoram, Nagaland and Tripura	Spring (January/February — May/June) and Autumn (August — November /December)	Same as North Central Hills.
North Indian Plains			
North Western Plains	Haryana, Punjab and Rajsthan	Autumn (October — January/February) and Spring (January — April /May)	Short day adapted, early bulking, moderately resistant to late blight and frost, slow rate of degeneration.
North Central Plains	Northwestern districts of Madhya Pradesh and western Uttar Pradesh	Autumn (October — February)	Short day adapted, medium maturing, moderately resistant to late blight, slow rate of degeneration.
North Eastern Plains	Plains of Assam, Bihar, North eastern Madhya Pradesh, Orissa, Eastern Uttar Pradesh and West Bengal	Autumn (November — March)	Short day adapted, early bulking, moderately resistant to late blight, slow rate of degeneration, red skinned medium to small sized tubers (Bihar, Madhya Pradesh, Orissa and Uttar Pradesj) or white medium tubers (Assam and West Bengal)
SPECIAL PROBLEM AREAS			
North Bengal Hills and Sikkim	North Bengal hills and Sikkim	Autumn (September — December); Spring (January — June)	Medium maturing, resistant to late blight, immune to wart, red skinned tubers
Plateau Region and peninsular India	Southern parts of Bihar, Gujarat, Karnataka, Andhra Pradesh, Madhya Pradesh Maharashtra and Orissa	*Kharif* (July — September) and *rabi* (November — February)	Early bulking, ability to tuberise under high temperatures, resistant to bacterial wilt, tuber moth and mites, slow rate of degeneration
South Indian Hills	Hills of Tamil Nadu	Summer (April — August); Autumn (September — December) and Spring (January — May)	Early bulking, resistant to late blight and cyst nematodes

A need was, therefore, felt that potato cultivation in India cannot depend on exotic varieties and technologies and the country must have its own research and developmental programmes for potato. A scheme for establishment of Central Potato Research Institute (CPRI) was, therefore, drawn up in 1945 under the guidance of Sir Herbort Stewart, the then Agricultural Advisor to the Government of India and CPRI was established in 1949 at Patna (Bihar). Hills being the ideal location for producing and maintaining healthy seed and using wide potato genetic base through hybridization for breeding improved varieties, on the recommendations of an expert committee, the headquarters of CPRI was shifted in hills at Shimla in 1956.

The idea of initiating All India Coordinated Research Project on Potato (AICRP-Potato) emerged during First Potato Workshop held in 1967 at Potato Experimental and Trial Center Jalandhar (now Central Potato Research Station, Jalandhar). The project was subsequently sanctioned by the Indian Council of Agricultural Research (ICAR) in July 1970 with the objective of studying the regional problems of potato production in a coordinated manner. The Headquarters of AICRP (Potato) is located at CPRI, Shimla. The project represents a national grid of integrated, cooperative and multidisciplinary research on potato involving CPRI, State Agriculture Universities (SAUs) and State Departments of Agriculture/ Horticulture.

Activities of the project were formally initiated during IV Five Year Plan in 1970 at five CPRI Research Stations (Jalandhar, Ootacamund, Patna, Rajgurunagar and Shillong) and 8 State Agricultural Universities/State Agricultural Departments based centers [Kanpur (Uttar Pradesh), Shopian (Jammu and Kashmir); Hassan (Karnataka); Kanke-Ranchi (Bihar); Chhindwara (Madhya Pradesh); Kalyani (West Bengal); Bhubaneshwar (Orissa) and Nadiad (Gujarat)]. In subsequent years, depending on the region specific research needs, some new centres were initiated and some old were closed.

Objectives

The mandate of AICRP (Potato) is to coordinate and monitor multilocational trials with improved potato hybrids, agronomic practices related to crop production vis-à-vis identification of remunerative potato-based cropping systems, plant-protection measures and post-harvest technology, all aimed at increasing production, productivity and utilization of potato in the country. The programme also acts as linkage between different stakeholders connected with potato production and utilization. Major activities under the programme are:

- Development/identification of improved potato varieties.
- Identification and evaluation of appropriate agronomic practices and profitable potato-based cropping systems for different regions.
- Studies on integrated management of nutrients and water.
- Monitoring of diseases/pests and other problems of potato in varied agroclimates and develop appropriate control measures.
- Identification of areas suitable for production of disease-free potato seed.
- Development of technologies for minimizing post-harvest losses.
- Extension and training programmes.
- Developing linkages with SAUs, State Departments of Agriculture/Horticulture, and other developmental/extension agencies.

Distribution of AICRP (Potato) Centres

Presently, 22 centres operate under AICRP (Potato) in different agro-climatic regions of the country, out of which 14 centres are located in SAUs, 7 centres are based at CPRI regional stations and there is one voluntary center at Ranichauri (Uttarakhand) (Fig. 2.3). Details of centres opened/closed since the inception of AICRP (Potato) are given in Table 2.3. Contact details for AICRP (Potato) centres are given in Annexure-I.

Table 2.3. Details of centres under the project since its inception in 1970

Location/Institute/University	*Latitude/Longitude*	*Established in*	*Current Status*	
HEADQUARTERS				
Shimla	PC Unit, CPRI, Shimla	31.06 N 77.13 E	1971 (IV Plan)	Continued
CENTRES				
Aurangabad	MAU, Parabhani	19.53 N 75.23 E	1987 (VII Plan)	Closed 1990
Bangalore	UAS, Bangalore	12.58 N 77.38 E	1975 (V Plan)	Closed 1990
Bhubaneshwar	OUAT, Bhubaneshwar	20.15 N 85.52 E	1971 (IV Plan)	Continued
Chhindwara	JNKVV, Jabalpur	22.03 N 78.59 E	1971 (IV Plan)	Continued
Deesa	SDAU, SK Nagar	24.14 N 72.13 E	1971 (IV Plan)	Continued
Dharwad	UAS, Dharwad	15.27 N 75.05 E	1993 (VIII Plan)	Continued
Dholi	RAU, Pusa	25.55 N 85.50 E	1994 (VIII Plan)	Continued
Faizabad	NDUAT, Faizabad	26.74 N 82.12 E	1987 (VII Plan)	Continued
Gwalior	CPRI Regional Station (SPU)*		2006 (X Plan)	Continued
Hassan	UAS, Bangalore	13.01 N 76.10 E	1972 (IV Plan)	Continued
Jalandhar	CPRI Regional Station	31.19 N 75.18 E	1993 (VIII Plan)	Continued
Jorhat	AAU, Jorhat	26.46 N 94.16 E	1971 (IV Plan)	Continued
Hisar	CCSHAU, Hisar	29.1 N 75.46 E	1975 (V Plan)	Continued
Kalyani	BCKV, Kalyani	22.34 N 88.24 E	1975 (V Plan)	Continued
Kanpur	CSAUAT, Kanpur	26.28 N 80.24 E	1971 (IV Plan)	Closed 1993
Kota	MPUA&T, Udaipur	25.10 N 75.52 E	1971 (IV Plan)	Continued
Kufri	CPRI Regional Station (SPU)*	31.06 N 77.13 E	1976 (V Plan)	Continued
Modipuram	CPRI Campus	29.01 N 77.45 E	1976 (V Plan)	Closed 2006
Ootacamund	CPRI Regional Station	11.24 N 76.44 E	1987 ((VII Plan)	Continued
Palampur	CSKHPKVV, Palampur	31.43 N 76.58 E	1971 (IV Plan)	Closed 1998
Pantnagar	GBPUAT, Pantnagar	28.48 N 79.09 E	1975 (V Plan)	Continued
Patna	CPRI Regional Station	25.37 N 85.13 E	1971 (IV Plan)	Continued
Port Blair	CARI, Port Blair	11.41 N 92.43 E	1987 (VII Plan)	Closed 1989
Raipur	IGKV, Raipur	21.15 N 81.41 E	1999 (IX Plan)	Continued
Rajgurunagar	Now NRC-OG	18 N 73 E	1971 (IV Plan)	Closed 2001
Ranchi	BAU, Ranchi	23.23 N 85.23 E	1971 (IV Plan)	Closed 1990
Shillong	CPRI Regional Station	25 N 91 E	1971 (IV Plan)	Continued
Srinagar	SKUAST (K), Srinagar	34.15 N 74.25 E	1972 (IV Plan)	Continued
VOLUNTARY CENTRES				
Babugarh	HRS, Govt. of UP	28.43 N 77.50 E	1983 (VI Plan)	Closed
Dharwad	UAS, Dharwad	15.27 N 75.03 E	1989 (VII Plan)	Closed 1993
Ludhiana	PAU, Ludhiana	30.55 N 75.54 E	1971 (IV Plan)	Closed 1989
Parbhani	MAU, Parbhani	19.53 N 75.23 E	1973 (IV Plan)	Closed 1987
Ranichauri	GBPUAT, Pantnagar	29.01 N 77.45 E	1984 (VI Plan)	Continued

* SPU = Seed Preparatory Unit

MAN POWER AND BUDGET

The AICRP (Potato) is a plan activity with the provision of 75% of the total outlay to be borne by the ICAR and remaining 25% by the respective SAUs with respect to SAU-based centres. The financial requirements of AICRP headquarters and ICAR Institute based centers are entirely met by the ICAR. There had been more than 30 fold increase in financial allocations for SAU-based centres from IV to XI five year plans. The sanctioned financial outlays for SAU-based centres during various plans are given in Table 2.3.

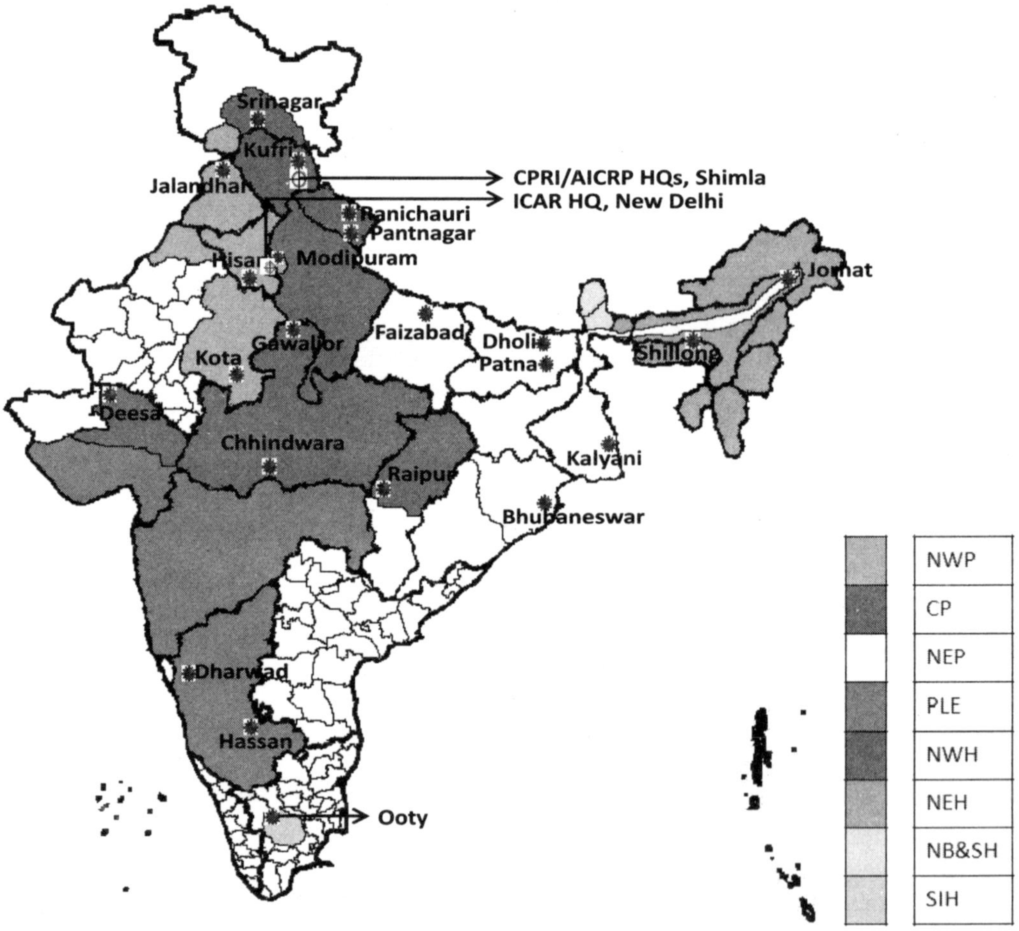

Fig. 2.3. Distribution of AICRP (Potato) centres in different potato production zones of the country. [North Western Plains (NWP); Central Plains (CP); North Eastern Plains (NEP); Plateau Region (PLE); North Western Hills (NWH); North Eastern Hills (NEH); North Bengal and Sikkim Hills (NB & SH); and South Indian Hills (SIH)].

Table 10.4. Plan-wise budget allocation (ICAR share) for SAU based centres

Head	*IV Plan*	*V Plan*	*VI Plan*	*VII Plan*	*Ad-hoc'90-92*	*VIII Plan*	*IX Plan*	*X Plan*	*XI Plan **
Pay and allowances	12.56	35.86	61.57	74.79	83.16	219.89	409.20	716.60	985.76
Contractual services	–	–	–	–	–	–	–	–	32.40
Traveling allowance	0.45	0.75	1.50	2.58	0.92	4.15	7.78	11.10	19.35
Recurring contingencies	2.60	3.50	10.50	21.72	11.52	37.60	71.05	105.59	126.00
Frontline demonstrations	–	–	–	–	–	–	–	–	6.15
Equipment	1.24	0.99	2.10	0.52	–	5.13	14.44	27.00	84.98
Works	–	–	0.90	1.50	–	6.56	4.20	6.37	95.36
Total	16.85	41.10	76.57	101.11	95.60	273.33	506.67	866.66	1350.00

* Proposed

Man power under AICRP (Potato) during various plan periods is depicted in Table 2.5. Currently there are 13 positions at AICRP (Potato) headquarters (Table 2.5a) and 88 positions at 14 SAU based centres (Table 2.5b). There are 38 scientists (7 Breeders, 8 Horticulturists, 10 Agronomists, 11 Pathologists and 2 Entomologists), 30 technicians, 9 administrative staff and 11 supporting staff at SAU based centres (Table 2.5).

Table 2.5. Man power under AICRP (Potato) in different plan periods

(a) Man power at AICRP (Potato) headquarters, CPRI, Shimla

Category	*Status of Posts in Different Plan Periods*							
	IV	*V*	*VI*	*VII*	*VIII*	*IX*	*X*	*XI*
Scientific	1	4	4	4	4	4	4	4
Technical	–	3	3	3	4	4	4	4
Administrative	1	4	3	3	3	3	3	3
Supporting	1	2	2	2	2	2	2	2
Auxiliary	–	1	1	1	*	*	*	*
Total	3	14	13	13	13	13	13	13

* The post of driver was re-classified in technical category during VIII Five Year Plan

(b)Category-wise man power at SAU based centres in different plan periods

Category	Status of Posts in Different Plan Periods							
	IV	*V*	*VI*	*VII*	*VIII*	*IX*	*X*	*XI*
Scientific	11	39	39	46	38	38	38	38*
Technical	32	47	47	41	30	30	30	30
Administrative	9	14	14	12	9	9	9	9
Supporting	8	13	13	12	11	11	11	11
Auxiliary	–	–	–	–	–	–	–	–
Total	60	113	113	111	88	88	88	88

* Plant breeders (9), Agronomists (9), Pathologists (11), Horticulturists (7) and Entomologists (2)

(c) Centre-wise sanctioned manpower in different categories at SAU based centers

Centre	*Sanctioned Man power*			
	Scientific	*Technical*	*Administrative*	*Supporting*
Bhubaneshwar	3	2	1	1
Chhindwara	3	2	b1	1
Deesa	4	3	1	1
Dharwad	3	2	–	–
Dholi	2	2	–	–
Faizabad	2	2	–	1
Hassan	2	2	–	–
Hisar	3	2	1	1
Jorhat	3	2	1	1
Kalyani	4	3	1	1
Kota	2	2	1	1
Pantnagar	3	2	1	1
Raipur	2	2	–	1
Srinagar	2	2	1	1
Total	38	30	9	11

MECHANISMS FOR EVALUATION OF PROJECT ACTIVITIES

Activities of the project are monitored through the visits of Project Coordinator, Principal Investigators (PIs) and scientists of the monitoring teams. Workshops/Group Meetings of AICRP scientists are held regularly to discuss results of the trials. During these meetings recommendations for release of varieties, agro-techniques and crop-protection measures are brought out. Based on previous results and keeping in view the emerging problems, new technical programmes are also formulated in these meetings. Till 1991, such meetings were used to be held annually, now these are held bi-annually. Since the inception of AICRP, 26 Workshops/ Group Meetings have been held in such a manner that venues are evenly distributed throughout the country (Table 2.6).

Table 2.6. Workshops/group meetings of AICRP (Potato)

Workshop/Group Meeting	*Date*	*Venue*
First	22–24 February, 1971	IARI, New Delhi
Second	4–7 September, 1972	CPRI, Shimla
Third	3–6 September, 1973	MPKVV, Rahuri, Maharashtra
Fourth	16–19 September, 1974	CPRI, Shimla
Fifth	22–24 January, 1976	CSAUA&T, Kanpur
Sixth	24–27 August, 1977	CPRI, Shimla
Seventh	25–26 November, 1978	CPRS, Jalandhar, Punjab
Eighth	23–25 November, 1979	CPRS, Modipuram
Ninth	26–28 August, 1981	CPRI, Shimla

(Contd.)

Workshop/Group Meeting	*Date*	*Venue*
Tenth	10–12 January, 1983	GBPUA&T, Pantnagar
Eleventh	9–12 January, 1984	GAU, Dantiwada, Gujarat
Tevellth	16–18 September, 1985	CPRS, Patna, Bihar
Thirtheenth	10–12 January, 1986	CPRI, Shimla
Fourteenth	7–9 August, 1987	BCKVV, Kalyani
Fifteenth	9–11 September, 1988	SKUAS&T, Srinagar
Sixteenth	17–19 August, 1989	HAU, Hisar, Haryana
Seventeenth	17–19 August, 1990	IARI, New Delhi
Eighteenth	20–22 August, 1991	NDUA&T, Faizabad
Ninteenth	26–28 August, 1993	GAU, SK Nagar, Gujarat
Twentiaeth	30 Aug.-1st Sept., 1995	UAS, Dharwad, Karnataka
Twenty-first	3–5 September, 1997	OUA&T, Bhubaneshwar, Orissa
Twenty-second	25–27 August, 1999	JNKVV, Jabalpur
Twenty-third	26–29 July, 2001	CPRI, Shimla
Twenty-fourth	5–7 September, 2003	UAS, Bangalore, Karnataka
Twenty-fifth	8–10 September, 2005	MPUA&T, Udaipur, Rajasthan
Twenty-sixth	7–9 September, 2007	RAU, Pusa (Samastipur), Bihar

Quinquennial Review Teams (QRTs) constituted by the ICAR also evaluate performance of AICRP regularly. Such evaluations have been done for the periods 1985–1994, 1995–2000 and 2001–05. Recommendations emanated from these QRTs have been highly useful in improving functioning and output of AICRP (Potato).

TECHNOLOGIES GENERATED

Genetic Resources

Central Potato Research Institute (CPRI), Shimla, is a national repository for potato germplasm. The present germplasm collections at CPRI, Shimla consists of more than 2,900 accessions (1,200 Tuberosum, 800 Andigena and 900 wild species) collected from 30 countries and are being maintained in *in vitro*, as well as in the field/glasshouse. A large number of potato genetic resources has been evaluated against biotic and abiotic stresses with the objective to identify desirable genotypes for use in different breeding programmes. The screening for resistance to late blight was mostly done under field conditions at Ootacamund, Shillong and Kufri; for common scab, powdery scab and purple top roll (PTR) at Shimla; for cyst nematodes at Ootacamund; for potato tuber moth at Shillong; for tolerance to soil salinity at Faizabad and Hisar; and for adaptability under rainfed conditions (*kharif*) at Dharwad.

The promising germplasm accessions having resistance/tolerance to different biotic and abiotic stresses are listed in Table 2.7. Germplasm lines with desired attributes for yield, tuber characters, better storability and resistance to biotic and abiotic stresses are used in hybridization programme at CPRI/SAU based centers.

Two genotypes, PS/F-220 and JN-189, possessing resistance to potato stem necrosis have been registered with the NBPGR, as novel parental lines vide registration numbers, INGR 04059 and INGR 07040, respectively.

Table 2.7. Promising genotypes having resistance/tolerance to biotic and abiotic stresses

Disease/Insect Pest/Stress and Location for Evaluation	*Promising Germplasm Accessions/Varieties/Hybrids*
Late blight (Ooty, Shillong, Shimla and Kufri)	CP* 1207, 1344, 1353, 1399, 1406, 1412, 1436, 1461, 1614, 1632, 1646, 1665, 1668, 1704, 1778, 1780, 1811, 1813, 1846, 1854, 1880, 1889, 1890, 1905, 1920, 1935, 1969, 1970, 1974, 1978, 1981, 1985, 1987, 1990, 1997 and 2014.
Common scab (Shimla)	CP 1418, 1582, 1683, 1815, 1843, 1878, 1921 and 2004.
Powdery scab (Shimla)	CP 1580, 1711, 1722 and 2042.
Purple top roll (Shimla)	CP 1981, 1985, 1989, 1991 and 1992.
Potato stem necrosis (Kota and Gwalior)	Resistant (R) (Disease incidence: <5% and Disease index: <0.25) JN-189**, HPS-25/13, PS/F-220, Kufri Sindhuri**, Kufri Anand, JX-214, JW-96, JN-1530, OP-1**, MP/97-637**. Moderately Resistant (MR) (Disease incidence: >5% to <10% and Disease index: >0.25 to < 0.5). JN-1536, JP-132, JX-118, MS/80-758, PH/F-1545, 85-P-621, TPSC-17, MS/90-512, MS/92-1090, MS/92-3146, JX-371, Kufri Chipsona-3, Kufri Chipsona-2.
Potato apical leaf curl virus disease(Hisar)	Resistant: Kufri Bahar up to 10% incidence: CP 1466Upto 20% incidence: CP 1375, 1416, 1421, 1564, 1578, 1602, 1664, 1668, 1679, 1689 and 1880.
Bacterial wilt(Hassan)	Tolerant: CP 1588, 1818, 2024, 2044, 2345, 2399, 3036, 3759, 3781 and 3784.
Potato tuber moth(Shillong)	Tuber resistance in CP 1701 and 1710.
Cyst nematodes (Ooty)	Resistance to *Globodera rostochiensis* JEX/A* 16, 260, 261, 269, 320, 376, 401, 424, 433 and 492 CP 1422, 1781, 1806, 1869, 1935, 1971 and 1981 SS* 255-5, 684-4, 1068-8, 1307-1, 1307-2, 1307-4, 1308-6, 1308-9, 1309-4, 1309-6, 1310-5, 1336, 1338 1427 and 1586-1. Resistance to *Globodera pallida* CP 1781 and 1806.
Soil salinity(Faizabad and Hisar)	Tolerance in CP 1430, 1473, 1477, 1536, 1906, 1949 and 2118.
Adaptability in *kharif* conditions(Dharwad)	Kufri Jawahar; CP 1665, 1686, 1691, 1740, 2129, 2242, 2289, 2335, 2336, 2348, 2352, 2372, 2389, 2401 and 2997.

* CP = *Solanum tuberosum* ssp. *tuberosum*; JEX/A = *Solanum tuberosum* ssp. *andigena* and SS = *Solanum* species.
** Dual resistance to stem necrosis and hopper burns.

Improved Varieties

Since the inception of AICRP (Potato) in 1970-71 (IV Plan Period) till date as many as 1,156 hybrids were evaluated under AICRP (Potato). Of these, 23 hybrids were recommended and eventually released as varieties for commercial cultivation in different parts of the country. Two more hybrids SM/93–237 and J/93–86, were recommended for release in 26th Group Meeting of AICRP (Potato) held at RAU, Pusa (Samastipur), Bihar, during 2007. Besides, a TPS population 92-PT-27 recommended in 2003 has also been released for commercial cultivation in areas where non-availability of quality seed is a major problem.

More than 85 % of potato in India is grown in north Indian plains; therefore, commensurate with this area, 15 cultivars were developed for this region. Cultivars have also been developed for north Indian hills and other special problem areas, viz. wart resistant cultivars for Sikkim and north Bengal hills and cyst nematode resistant cultivars for south Indian hills. The cultivars bred for the plains fall in 3 maturity groups, i.e. early (70–80 days); medium (90–100 days);

and late (100–110 days). The maturity groups for the hills, where the crop is raised under long photoperiods of summer, are early (110–120 days); medium (120–130 days); and late (130–140 days). All the cultivars developed for hills and newly released medium and late duration cultivars for the plains possess resistance to late blight. Cultivars with resistance/ tolerance to cyst nematodes, wart, viruses, frost and high thermo-periods have also been bred to address region specific problems. Of the 23 cultivars developed, 8 possess multiple resistances to different biotic and abiotic stresses. Besides, 3 cultivars, viz. Kufri Chipsona-1, Kufri Chipsona-2 and Kufri Chipsona-3 are suitable for processing. Potato varieties of early/medium maturity have enabled diversification of potato cultivation in the intensive agriculture in the country. Attributes of recommended hybrids and their salient features are given in Tables 2.8, 2.9 and 2.10 Morphological features of some important varieties are shown in Figure 2.4.

Table 2.8. Salient features of potato varieties recommended through AICRP (Potato)

Variety/Hybrid	*Year of Release*	*Tuber Characteristics and Yield Potential*	*Areas of Adaptability*
Kufri Anand (Hybrid MS/82-717)	1999	Tubers oblong, flattened, white with fleet eyes Sprouts red-purple. (35–40 tonnes/ha)	North Indian Plains
Kufri Arun(Hybrid MS/92-2105)	2004	Tubers medium size, oval, red skin, medium deep eyes and cream flesh. Sprouts red-purple (30-35 tonnes/ha)	North Eastern Plains
Kufri Ashoka(Hybrid PJ-376)	1996	Tubers-large size, oval-long, white, fleet eyes and white flesh. Sprouts red-purple. (25–30 tonnes/ha)	North Indian Plains
Kufri Badshah(Hybrid JF-4870)	1979	Tubers large, white, oval with fleet eyes, dull white flesh, tubers turn purple on exposure to light Sprouts red-purple. (30-35 tonnes/ha)	North Indian Plains & Plateau Region
Kufri Bahar(Hybrid E-3797)	1980	Tubers large, white, round-oval, medium deep eyes and white flesh. Sprouts light red-purple. (30–35 tonnes/ha)	North Indian Plains
Kufri Chipsona-1 (Hybrid MP/90-83)	1998	Tubers-medium to large, oval, white, high dry matter, fleet eyes. Sprouts white-green. (30–35 tonnes/ha)	North Indian Plains
Kufri Chipsona-2 (Hybrid MP/91-G)	1998	Tubers-medium size, round-oval, white, high dry matter, fleet eyes. Sprouts light red-purple. (30–35 tonnes/ha)	North Indian Plains
Kufri Chipsona-3 (Hybrid MP/97-583)	2005	Tubers white, oval, fleet eyes, cream flesh and high dry matter. Sprouts red-purple. (30–35 tonnes/ha)	North Indian Plains
Kufri Dewa(Hybrid C-3804)	1973	Tubers medium, round, white with purple splashes, medium deep purple picked eyes and white flesh. Sprouts red-purple. (20–25 tonnes/ha)	North Indian Plains
Kufri Giriraj(Hybrid SM/85-45)	1998	Tubers-medium to large, oval, white with fleet eyes. Sprouts red-purple. (20–25 tonnes/ha)	North Indian Hills
Kufri Himalini (Hybrid SM/91-1515)	2005	Medium size tubers, oval, fleet eyes with white cream flesh. Sprouts red-purple. (30–35 tonnes/ha)	North Indian Hills
Kufri Jawahar(Hybrid JH-222)	1996	Tubers-medium size, round-oval, white, fleet eyes and light yellow flesh. Sprouts red-purple (25–30 tonnes/ha)	North Indian Plains and Plateau Region

(Contd.)

Variety/Hybrid	Year of Release	Tuber Characteristics and Yield Potential	Areas of Adaptability
Kufri Kanchan (Hybrid SE/I-1307)	1999	Tubers-medium to large size, oval-long, red, light yellow flesh and shallow eyes. Sprouts purple (25–30 tonnes/ha)	North Bengal Hills and Sikkim
Kufri Lalima (Hybrid BS/C-1753)	1982	Medium size tubers, uniformly red coloured, round, medium deep eyes. Red-purple sprouts. (20–25 tonnes/ha)	North Indian Plains particularly Eastern Uttar Pradesh & Bihar
Kufri Lauvkar (Hybrid A-7416)	1972	Tubers large size, round, white, fleet eyes and white flesh. Red-purple sprouts. (20–25 tonnes/ha)	Plateau Region
Kufri Megha (Hybrid SSC-562)	1989	Tubers medium size, round-oval, white, fleet eyes. Red-purple sprouts. (25–30 tonnes/ha)	North Eastern Hills
Kufri Pukhraj (Hybrid JEX/C-166)	1998	Tubers-large, oval, slightly tapered, white, fleet eyes. Purple sprouts. (35–40 tonnes/ha)	North Indian Plains and Plateau Region
Kufri Pushkar (Hybrid JW-160)	2004	Tubers round to oval, white and fleet to medium deep eyes. Blue-purple sprouts. (30–35 tonnes/ha)	North Western Indo-Gangetic Plains
Kufri Shailja (Hybrid SM/87-185)	2004	Tubers medium size, white, high tuber dry matter, good keeping quality. Red-purple sprouts. (30–35 tonnes/ha)	North Indian Hills
Kufri Sherpa (Hybrid F-5242)	1983	Tubers medium size, flattened round, white, medium deep eyes and pale yellow flesh. Red-purple sprouts. (15–20 tonnes/ha)	North Bengal Hills and Sikkim
Kufri Surya (Hybrid HT/92-621)	2005	Tubers medium size, oval, white, shallow eyes, white flesh and high dry matter. Sprouts red-purple. (30–35 tonnes/ha)	North Indian Plains Plateau Region
Kufri Sutlej (Hybrid JI-5857)	1996	Tubers-large size, oval, white with fleet eyes. Sprouts white-green. (30–35 tonnes/ha)	North Indian Plains
Kufri Swarna (Hybrid SON-110)	1985	Tubers medium size and smooth, oval, white, fleet eyes, flesh white. Sprouts red-purple. (30–35 tonnes/ha).	South Indian Hills
TPS Population 92-PT-27	2007	Uniform and high yielding TPS population with resistance to late blight. Both parents flower under short days in the plains.	Eastern Region
Hybrid J/93-86*	2007	J/93-86 is an early maturing white tuber hybrid with moderate resistance to late blight. This hybrid yields higher than existing early cultivars both at 60 and 75 days after planting.	Entire Indian Plains
Hybrid SM/93-237*	2007	SM/93-237 is a medium maturing white tuber hybrid with high level of resistance to late blight. It yields higher than existing cultivars for hills.	All Indian Hills

* Hybrids recommended for release in 26th Group Meeting of AICRP (Potato) in 2007 (RAU, Pusa, Bihar)

Table 2.9. Duration and region-wise adaptability of the Indian potato varieties.

Variety	*Crop Duration*				*Regions of Adaptability**			
	Early	*Medium*	*Late*	*NIH*	*NIP*	*NB & SH*	*PLE*	*SIH*
Kufri Anand		✓			✓			
Kufri Arun		✓			✓			
Kufri Ashoka	✓				✓			
Kufri Badshah		✓			✓		✓	
Kufri Bahar		✓			✓			
Kufri Chipsona-1		✓			✓			
Kufri Chipsona-2		✓			✓			
Kufri Chipsona-3		✓			✓		✓	
Kufri Dewa			✓		✓			
Kufri Giriraj		✓		✓				
Kufri Himalini		✓		✓				
Kufri Jawahar	✓				✓		✓	
Kufri Kanchan		✓				✓		
Kufri Lalima		✓			✓			
Kufri Lauvakar	✓						✓	
Kufri Megha		✓		✓				
Kufri Pukhraj		✓			✓		✓	
Kufri Pushkar		✓			✓		✓	
Kufri Shailja		✓		✓				
Kufri Sherpa		✓				✓		
Kufri Surya	✓				✓		✓	
Kufri Sutlej		✓			✓			
Kufri Swarna		✓						✓
TPS Population 92-PT-27		✓		Eastern region				
Hybrid J/93-86**	✓				✓			
Hybrid SM/93-237**		✓		✓				✓

* NIH = North Indian hills; NIP = North Indian plains; NB & SH = North Bengal hills and Sikkim; PLE = Plateau region; SIH = South Indian hills; ** = Hybrids recommended for release in 26th Group Meeting of AICRP (Potato)

Table 2.10. Resistance to biotic and abiotic stresses associated with the Indian potato varieties

Variety	*LB**		*EB*		*X*	*Y*	*PLRV*		*W*	*CN*	*Frost*		*HT*
	*R***	*MR*	*R*	*MR*	*R*	*R*	*R*	*T*	*R*	*R*	*R*	*T*	*T*
Kufri Anand	✓												
Kufri Arun		✓											
Kufri Ashoka													
Kufri Badshah		✓		✓	✓								
Kufri Bahar													
Kufri Chipsona-1	✓												
Kufri Chipsona-2	✓											✓	
Kufri Chipsona-3		✓											
Kufri Dewa											✓		

(Contd.)

Variety	LB*		EB		X	Y	PLRV		W	CN	Frost		HT
	R**	MR	R	MR	R	R	R	T	R	R	R	T	T
Kufri Giriraj	✓												
Kufri Himalini	✓												
Kufri Jawahar		✓											
Kufri Kanchan	✓								✓				
Kufri Lalima				✓		✓							
Kufri Lauvakar													✓
Kufri Megha	✓												
Kufri Pukhraj		✓	✓										
Kufri Pushkar		✓											
Kufri Shailja	✓												
Kufri Sherpa		✓		✓					✓				
Kufri Surya		✓											✓
Kufri Sutlej		✓											
Kufri Swarna	✓									✓			
TPS Population 92-PT-27	✓												
Hybrid J/93-86#		✓											
Hybrid SM/93-237#	✓												

* LB = Late blight; EB = Early blight; X = Potato virus X; Y = Potato virus Y; PLRV = Potato leaf roll virus; W = Wart; CN = Cyst nematodes; and HT = High thermo-periods.

**R = Resistant; MR = Moderately resistant; and T = Tolerant.

= Hybrids recommended for release in 26th Group Meeting of AICRP (Potato) in 2007.

Region-specific breeding programmes have recently been initiated at 9 SAU based centers, viz. Bhubaneshwar, Deesa, Dharwad, Hisar, Jorhat, Kalyani, Pantnagar, Raipur and Srinagar. For these breeding programmes, hybridization will be conducted at CPRS Kufri (Shimla hills) under natural photoperiod in summer as well as at the respective centers during rabi season under artificially extended photoperiods.

Technologies for Enhancing Production and Productivity

True Potato Seed as an Alternate Planting Material

True potato seed (TPS) produced through sexual mode is used as an alternative planting material in some parts of the world. Since it is produced through fertilization, TPS is also called botanical seed and resembles seeds of other solanaceous plants like tomato, brinjal and chilly. A single fruit or berry contains on an average 150–200 seeds. The TPS has several advantages; it can be produced easily, is not bulky and thus easy to store and transport, only 100–150g TPS is sufficient to plant a hectare of potato crop and most of the viruses are filtered out during the process of sexual seed formation.

TPS is used for growing commercial potato crop by two methods: (i) transplanting of seedlings and (ii) use of seedling tubers. In the former method, potato seedlings are raised in nursery beds and transplanted in the field at 4-5 leaf stage. Potato crop raised from seedlings is harvested at maturity. This method requires about 100–125g TPS and 75 m^2 nursery bed area for growing seedlings sufficient for transplanting in one hectare area depending upon method adopted. It is important that irrigation is assured for growing the transplanted crop successfully.

In the second method, small-sized tubers are produced by allowing the seedlings to mature in nursery beds. Seedlings are dehaulmed at maturity and harvested in another 10–15 days when the skin becomes firm and well developed. These seedling tubers are stored and used as planting material for growing the commercial crop in the following crop season. About 50–60g TPS and 250–300m^2 nursery bed area is required for producing seedling tubers sufficient for planting one hectare. Stages of TPS production and utilization are shown in Fig.2.5.

Nearly 50 elite hybrid TPS populations were subjected to multilocation testing under the project and three of them, viz., HPS-I/13, TPS-C3 and 92-PT-27, were found promising and were recommended for release for commercial cultivation particularly in areas facing the problems of non-availability and high cost of quality tuber seed. Performances of these TPS populations when used as transplants as well as seedling tubers are given in Table 2.11.

Table 2.11. Performance of TPS populations 90 days after planting

TPS	*Families Centers and Yield (t/ha)*										
	BHN	*CHN*	*DES*	*DWD*	*FZB*	*JRH*	*MDP*	*PAT*	*RPR*	*SRN*	*Avrg.*
Performance of seedling tubers											
92-PT-27	10.2	11.2	34.0	3.3	24.3	11.9	16.8	28.3	22.5	22.7	18.5
HPS-I/13	9.7	8.7	29.8	4.8	25.1	16.5	13.2	23.6	18.6	19.1	16.9
TPS-C3	10.0	9.0	31.3	3.2	–	12.9	14.8	27.7	–	19.9	16.1
Performance of transplants											
92-PT-27	7.0	27.1	46.0	7.3	38.8	21.9	33.7	43.4	38.4	20.4	28.4
HPS-I/13	13.5	24.8	48.7	7.3	41.1	16.6	36.1	38.9	28.8	20.4	27.6
TPS-C3	7.5	12.8	–	7.6	36.1	16.1	–	39.8	–	18.9	19.8
Controls	20.8	15.1	56.2	14.2	29.3	17.1	30.2	35.1	30.9	22.2	27.1
	KBD	KBD	KBD	KCM	KJW	KPR	KAD	KSD	KBD	KJ	

Centers: BHN = Bhubaneshwar; CHN = Chhindwara; DES = Deesa; DWD = Dharwad; FZB = Faizabad; JRH = Jorhat; MDP = Modipuram; PAT = Patna; RPR = Raipur; SRN = Srinagar;
Varietal controls: KBD = Kufri Badshah; KCM = Kufri Chandramukhi; KJW = Kufri Jawahar; KPR = Kufri Pukhraj; KAD = Kufri Anand; KSD = Kufri Sindhuri; KJ = Kufri Jyoti

Agronomic Practices and Production Technologies

Studies were conducted on agronomic attributes like optimum planting dates, seed rate, seed size, spacing, nutrient requirement, water requirement, etc., in different agroclimatic regions in order to improve potato productivity and economize on the inputs so as to make potato farming economically, environmentally and technologically sustainable. Major achievements related to agronomic practices developed under AICRP (Potato) for potato in different regions are described as below.

Optimum Dates for Planting

Agronomic trials were carried out at different centers of AICRP (Potato) to identify optimum planting dates. Crop yields from staggered weekly plantings (2-3 weeks before and after the known planting periods) were assessed for 2–5 years and optimum dates of planting potato crop that give high yields in different regions of the country were identified. In Punjab, early potato crop could be planted between 10 and 20 September and in central Uttar Pradesh in first week of October. The crop could be harvested after 60–70 days to take advantage of high prices of fresh potatoes in market. Mulching of early potato crop with paddy straw, paddy husk, maize and *Dhaincha* stalks is beneficial in getting high yields. Mulching, however, had no beneficial effects at Hisar and Pantnagar. The early planting and early lifting of potato crop facilitates normal sowing of wheat on the same land giving additional benefit of multiple cropping. Nevertheless, optimum planting time of potato crop in *rabi* (winter season) ranges from first week of October to first week of December in plains of the country. In Dharwad area of Karnataka, 20–30 June is the optimum planting time for potato (Kufri Jawahar and Kufri Surya) for getting higher yields and economic returns. Optimum planting dates for different regions identified in these studies are given in Table 2.12.

Table 2.12. Optimum dates for planting potato crop in different regions.

State	*Center*	*Optimum Period for Planting*
Assam	Jorhat	1st to 2nd week of November
Bihar		
(Plains)	Patna	4th week October to 1st week Noveber
(Plateau-*Rabi*)	Ranchi	4th week October to 1st week November
(Plateau-*Kharif*)	Ranchi	4th week in July
Gujarat (Northern)	Deesa	1st to 2nd week of November
Haryana	Hisar	2nd to 3rd week of October
Karnataka	Bangalore	1st to 3rd week of November
(Plateau-*Rabi*)	Hassan	1st week November to 1st week December
(Plateau-*Kharif*)	Dharwad	4th week of June
Madhya Pradesh (Plateau)	Chhindwara	4th week Oct. to 2nd week Noveber
Maharashtra		
(*Rabi*)	Parbhani	1st week November
(*Rabi*)	Rajgurunagar	1st week November
(*Kharif*)	Rajgurunagar	1st to 2nd week of July
Orissa (Plains)	Bhubaneshwar	2nd week November
Punjab	Jalandhar	1st to 2nd week of October
Rajasthan (Western)	Kota	4th week October to 1st week November
Uttar Pradesh		
(Western)	Modipuram	1st to 2nd week of October
(Tarai)	Pantnagar	1st to 2nd week of October
(Central)	Kanpur	4th week October to 1st week November
West Bengal	Kalyani	1st to 2nd week of November

Optimum Seed Rate/Seed Size/Planting Distance and their Inter-relationship

Effect of seed size and planting spacing on yield: A series of experiments were conducted to find optimum plant densities for different seed sizes (by keeping the seed rate constant and adjusting inter-/intra-row spacing) to get high yields. The results showed that comparable yields could be obtained by planting medium-sized tubers (35–45 mm diameter) at 60cm × 20cm spacing and big-sized tubers (45–55 mm diameter) at 60cm × 25cm spacing keeping the plant populations at 83,000 and 67,000 plants/ha, respectively (Table 2.13). For small-sized seed (25–35mm diameter) a population of 1,11,000 plants/ha (60 cm × 15cm inter and intra-row spacing, respectively) was found to be the best.

Fig. 2.4. Morphological characters of some of the Indian varieties

1. Hybridization; 2. potato berries with TPS inside; 3. TPS extraction; 4. true potato seed (TPS); 5. seedlings in nursery for transplanting; 6. seedlings in nursery for seedling tuber production; 7. transplanting of TPS seedlings for ware potato production; 8. transplanted seedlings for seedling tuber production; 9. transplanted crop at 65-75 DAT; 10. harvesting of seedling tuber planted crop (F1C1); 11. potatoes produced from TPS derived planting materials.

Fig. 25. Stages of TPS production and utilization

Table 2.13. Effect of seed size and spacing on potato yield

Seed Size (mm)	Spacing (cm)	Plant Population	Average Yield (tonnes/ha)*
25-35	60 x 15	111,000	19.2
35-45	60 x 20	83,000	22.3
45-55	60 x 25	67,000	24.1

* Average of 15 centers (Bangalore, Bhubaneshwar, Chhindwara, Deesa, Hisar, Jalandhar, Kalyani, Kanpur, Kufri, Pantnagar, Patna, Parbhani, Rajgurunagar, Shillong and Srinagar)

Relationship between seed rate and seed size: The results of multilocation yield trials over the years revealed increase in potato yields with increasing seed rate. Although all the three tuber sizes, viz., 25g, 50g and 100g could be used for raising the crop, comparable yields could be obtained from different sizes by adjusting the seed rates. Use of very big tubers (100g) has had no added advantage on yield over smaller tubers. Comparatively higher seed rate was needed to produce seed size tubers than for ware tubers. The optimum seed rates for getting high yields from different sizes of tubers differed from location to location. The optimum seed rates for different seed tuber size at different centers are given below.

Seed Size (g)	Seed Rate (q/ha)	Optimum Yield
25	30	Deesa, Hisar, Kanpur, Ooty, Ranichauri and Srinagar
	37.5	Kalyani and Shillong
50	37.5	Deesa, Hisar, Kanpur, Ooty and Srinagar
	45	Kalyani, Ranichauri and Shillong
100	37.5	Kanpur and Ranichauri
	45	Deesa and Ooty
	52.5	Hisar, Kalyani, Shillong and Srinagar

Seed being the costliest input in potato production (nearly 50% of the total cost of cultivation), smaller size tubers (15g, 30g, and 45g) were also tried as seed at some centers and optimum seed rates supporting higher potato yields were determined. Optimum seed rates for different sizes of seed tubers are given below for some of the AICRP centers.

Seed Size (g)	Seed Rate (q/ha)	Optimum Yield
15	15	Jorhat
	20	Chhindwara, Jalandhar and Kalyani
	25	Hisar
30	20	Jalandhar
	25	Chhindwara, Jorhat and Kalyani
	30	Hisar
45	30	Chhindwara, Jalandhar and Jorhat
	35	Hisar and Kalyani

Nutritional Requirements of Potato Crop

Fertilizer needs of potato crop: The experiments on nutritional requirement of potato crop were conducted under AICRP (Potato) for several years in the different parts of country. The results showed that optimum requirement of nitrogen, phosphorus and potash varied between 80–240 kg N/ha, 50–180 kg P_2O_5/ha and 60–150 kg K_2O/ha, respectively (Table 2.14). The studies also revealed nitrogen is the major limiting nutrient in potato crop. However, the response of potato crop was poor to P_2O_5 at Bhubaneshwar (Orissa), Hisar (Haryana) and plateau areas of Ranchi (earlier in Bihar and now in Jharkhand) and to K_2O at Bhubaneshwar (Orissa), Hisar (Haryana), Rajgurunagar (Maharashtra), Ranchi (Chhota-nagpur), hills at Ooty (Tamil Nadu), Palampur (Himachal Pradesh) and Shillong (Meghalaya). The optimum requirement of major nutrients (NPK) at various locations varied in different experiments. The nearest values of N, P_2O_5 and K_2O supporting optimum yield in different parts of the country was derived from various experiments and are given in Table 2.14.

Table 2.14. Optimum fertilizer doses for potato in different regions

Region	*Fertilizer Dose (kg/ha)*			*Region*	*Fertilizer Dose (kg/ha)*		
	N	P_2O_5	K_2O		N	P_2O_5	K_2O
Assam	180	100	120	Madhya Pradesh	120	100	100
Bihar	240	100	120	Maharashtra	180	100	60
Central Uttar Pradesh	180	100	100	Meghalaya	160	120	60
Eastern Uttar Pradesh	180	100	150	Orissa	120	50	60
Gujarat	240	100	150	Punjab	240	100	120
Haryana	240	50	60	Rajasthan	120	100	100
Himachal Pradesh	120	100	60	Tamil Nadu (Hills)	80	180	60
Jammu & Kashmir	120	100	100	Uttarakhand (Hills)	120	100	100
Jharkhand	240	50	60	Uttarakhand (Plains)	180	100	120
Karnataka	180	150	150	West Bengal	180	100	120
				Western Uttar Pradesh	180	100	60

However, recently it has been observed that these doses need to be revised upwardly by about 50% in many regions to obtain economic yield of potato.

Nitrogen need of potato crops planted on different dates: The experiment involving combination of planting dates and levels of nitrogen showed that in the eastern part of Uttar Pradesh (Faizabad) nitrogen requirement of potato crop remained nearly the same in early (10 days before optimum date), normal and late plantings (10 days after the optimum date). At Deesa (North Gujarat) and Chhindwara (Satpura plateau) also the nitrogen requirement remained the same in early and normal planting but it was reduced by 20–25% in the late plantings at these centers (Table 2.15). Application of full dose of nitrogen was not found to give any economic advantage in the late planted crop at these locations.

Table 2.15. Nitrogen requirements of potato at different dates of planting

Centre	*Variety*	*Date of Planting*	*Optimum N Dose (kg/ha)*	*Yield (q/ha)*
Faizabad	Kufri Badshah	1st November	200	500
		10th November	200	463
		20th November	200	429
		30th November	200	376
Deesa	Kufri Badshah	1st November	300	396
		10th November	300	390
		20th November	300	365
		30th November	240	338
Chhindwara	Kufri Jawahar	15th October	160	201
		25th October	160	212
		4th November	120	202
		14th November	120	183
Kanpur	Kufri Badshah	15th October	200	250
		22nd October	200	278
		29th October	200	302
		5th November	200	296

Many farmers prefer urea (the cheapest source of nitrogen) over other nitrogenous fertilizers. Application of urea with neem cake was found to increase efficiency and efficacy of the fertilizer. Calcium ammonium nitrate (CAN) and ammonium sulphate are better sources of nitrogen. Soaking of seed tubers in a solution containing 1% each of urea and sodium bicarbonate followed by treatment with Azotobacter + Phosphobacteria can economize 25% N and P.

Suitable sources of P_2O_5 and economy of their use: The efficiencies of various sources of P_2O_5 were compared. Water-soluble phosphoric fertilizers, i.e., superphosphate and diammonium phosphates produced better results than water insoluble P fertilizers, viz. rockphosphate and basic slag. Soaking of seed tubers in 3.0% super phosphate + 0.5% urea + 0.2% mancozeb (fungicide) solution for 4 hours can meet partly or fully the P requirement of the crop in phosphorus deficient soils (Table 2.16).

Table 2.16. Effect of soaking seed tubers in phosphate solution on yield

Seed Tuber Treatment	*Center and Yield (q/ha)*				
	*HIS**	*RAN*	*PNT*	*PLM*	*Average Yield*
No soaking	233	252	241	59	196
Soaking in water	332	248	256	77	228
P soaking alone	367	275	266	77	246
P soaking + 50 kg P_2O_5 /ha	387	269	295	117	267
50 kg P_2O_5 /ha	366	272	265	92	249
100 kg P_2O_5/ha	369	321	262	97	262

* HIS = Hisar, RAN = Ranichauri, PNT = Pantnagar and PLM = Palampur

Soil and foliar application of nitrogen and potassium: The effect of soil and foliar application of nitrogen and potassium and their combination was studied at several centres. The results showed that application of nitrogen in split doses (i.e., half the dose at planting and the remaining half as topdressing at earthingup) was better than application of full dose of N at planting. Full dose of P_2O_5 and K_2O applied at the time of planting was the best. The spray application of urea (2 sprays of 1% solution) and potassium chloride (1% solution) can only supplement the soil application of nitrogen and potassium.

Micronutrient requirement of potato crop: The experiment was conducted at 14 locations. Both the soil application ($ZnSO_4$ and $MnSO_4$ @ 20 kg/ha, and $FeSO_4$ and $CuSO_4$ @ 10 kg/ha) and tuber soaking in 0.05% solution for 3 hours. were tried. Zinc, copper, manganese and iron were not found to be limiting factors for the potato crop at most of the locations. Only marginal increase in yield with the application of Zn (both by soil and soaking) was observed at Kalyani, Kanpur and Palampur, and with Fe at Kanpur.

Effect of sulphur application on potato yield was also studied in different parts of the country. Sulphur through gypsum and pyrites was applied @ 40 kg/ha. The results showed that sulphur application was not found to be effective in increasing potato yields in most parts of the country except at Hisar, Faizabad and Deesa (Table 2.17). At Hisar and Faizabad, the increase in potato yields with the application of gypsum in soil may be attributed to soil salinity amendment by gypsum.

Table 2.17. Effect of sulphur application on potato yield at different centers

Treatment	*Potato Yield (q/ha) at Different Centers*							
	*CHN**	*DES*	*FZB*	*HIS*	*KAL*	*PNT*	*RAN*	*Mean*
Urea + DAP + KCl	561	151	353	267	337	315	355	334
Amm. Sulphate + DAP + KCl	588	170	372	273	382	368	380	362
Urea + SSP + KCl	597	161	336	285	354	310	333	339
Urea + DAP + K_2SO_4	576	157	345	290	335	330	351	341
Urea + DAP + KCl + Gypsum	604	186	414	313	327	315	342	357
Urea + DAP + KCl + Pyrites	562	162	386	286	349	305	344	342

* CHN = Chhindwara; DES = Deesa; FZB = Faizabad; HIS = Hisar; KAL = Kalyani; PNT = Pantnagar; RAN = Ranichauri

Effect of farmyard manure potato: FYM mainly supplements nitrogen and some micronutrients, improves physical texture of the soil and has mulching effect that together enhances the yield of potato crop. In the crop rotation experiments potato responded more efficiently to FYM than wheat, maize, green gram, sunflower, French bean, jute, radish and muskmelon. The direct effect of FYM application on potato yields was better than its residual effect. FYM @ 10–15 t/ha also had a supplementary effect on P and K fertilizer application. Higher levels (30t/ha) of FYM application, however, could meet up to 100 kg/ha requirement of P_2O_5 and K_2O.

Fertilizer economy through biofertilizers: This study was aimed at ascertaining the effect of different biofertilizers and plant growth promoting bacteria on potato yield and their efficacy on economizing the use of chemical fertilizers, which is another costly input after seed in potato cultivation. The results obtained at different centers are given below.

Location	*Fertilizer Economy Through Biofertilizers*
Bhubaneshwar	Treatment of seed tubers with urea, sodium bicarbonate and biofertilizers was as effective as with *Bacillus* spp. However, biofertilizers treatment was more effective than plant growth promoting bacteria at 75% and 100% recommended fertilizers dose.
Chhindwara	The biofertilizer treatment tended to be more effective than PGB but the differences were non-significant. Response to treatment of seed tubers with urea, sodium bicarbonate and bio-fertilizers was highest.
Deesa	Biofertilizers tended to be more effective than PGB but the effects were non-significant. These treatments also yielded higher economic gains.
Dharwad	Though recommended dose of NPK gave higher yield but with the use of biofertilizers such as vermicompost, mycorrhiza and azotobacter the dose of NPK could be reduced to half the recommended dose.
Faizabad	Tuber yield was significantly improved by the biofertilizers and treatment of seed tubers with 1% urea and 1% sodium bicarbonate.
Hisar	*B. cereus* treatment and seed soaking in 1% urea + 1% sodium bicarbonate were equally effective in tuber yield whereas *B. subtilis* had negative effect on tuber yield.
Jalandhar	Use of biofertilizers and soaking of tubers in 1% of urea + 1% sodium bicarbonate had no significant effect on tuber yield. However application of biofertilizers and plant growth promoting bacteria had positive effect on tuber yield at 75% fertility level.
Jorhat	Biofertilizers were less effective than plant growth promoting bacteria (PGB). *B. cereus* was more effective than *B. subtilis*. PGB treatment influenced the plant height, number of stems and leaves per plant better than bio-fertilizer treatment.
Kalyani	Soaking treatment of seed tubers in 1% urea and 1% sodium bicarbonate + biofertilizer had positive effect and supported highest tuber yield. Treatment with *Bacillus* was equally effective but the differences were non-significant.
Kota	Recommended dose of NPK was best treatment in terms of tuber yield. Biofertilizers such as Mycorrhiza, phosphobacteria, *Azotobacter* had positive effect on potato yield. Tuber treatment with urea, sodium bicarbonate and biofertilizers found to increase tuber yield. Tuber treatment with *B. subtilis* produced significantly higher tuber yield over B. cereus and tuber soaking in 1% urea + 1% sodium bicarbonate.
Pantnagar	Biofertilizer showed their affectivity only at 75% of fertility level. *B. cereus* was more effective than *B. subtilis.*
Patna	Highest yield was obtained when seed tubers before planting were treated with urea, sodium bicarbonate and biofertilizers (Azotobacter + Phosphobacteria) together. Treatment of tubers with *Bacillus* spp. also had positive effect on yield. The biofertilizers treatment was more effective than plant growth-promoting bacteria in increasing tuber yield. Treatment of tubers with *B. cereus* together with 75% NP application increased total tuber yield.
Shillong	At 100% NP only *B. subtilis* produced significantly higher yield over soaking treatments where as at 75% NP the Azotobacter + phosphobacteria together with soaking of seed tubers in 1% urea + 1% sodium bicarbonate solution and *B. subtilis* both produced significantly higher tuber yield and supported higher net economic gains.
Srinagar	Azotobacter + phosphobacteria along with soaking of tubers in 1% urea + 1% sodium bicarbonate solution increased tuber yield significantly over soaking of seed tubers in 1% urea + 1% sodium bicarbonate only. *B. subtilis* had no significant effect on the yield.

At Chhindwara, Hassan, Hisar, Kalyani, Ooty and Patna, the dose of P could be economized by 25% with the use of phosphobacteria.

Water Requirement

Compared to other food crops, potato requires more water, especially during the active crop growth and tuber development to obtain higher yields. About 10% of area under potato cultivation in India falls under rainfed conditions. Potato crop in these areas generally experiences water stress due to uneven distribution of rainfall during initial stages of crop growth. Water stress during tuberization phase leads to severe yield losses. Therefore, potato productivity in rainfed areas is only 5–10 tonnes/ha. About 90% of potato in the country is grown under irrigated conditions. The experiments showed that optimum scheduling of irrigation in potato varied from 15–25mm Cumulative Pan Evaporation (CPE) depending on soil type and climatic conditions. Thus, during entire crop season 8-9 irrigations @ 50 mm are needed to obtain optimum yields. This calculates to about 4250 m^3 or 42,50,000 liter water per hectare for growing a successful potato crop.

Irrigation at wider intervals especially during tuberization stage has an adverse effect on potato yield. Mulching of potato crop resulted in higher yields in warmer regions of the country. At Ranchi, irrigation was required to be given at 6 days intervals. At Pantnagar, a minimum of five irrigations and at Rajgurunagar as many as 11 irrigations were needed to obtain optimum potato yields. At Shillong (Meghalaya) and Shopian (Jammu and Kashmir) irrigations at 50% and 75% available soil moistures, respectively did not improve potato yield much over the unirrigated crop, thereby, indicating that there was sufficient precipitation at both these places that met the water need of the crop. Irrigation at 30 mm CPE and 200 kg/ha nitrogen at Hisar and 20 mm CPE and 300 kg/ha nitrogen at Deesa were the best combinations for getting higher potato yields.

Low Cost Technologies

Potato is a high input crop. Nevertheless, in many parts of the country it is grown under sub-optimal input conditions. Seed and fertilizers are major and costly inputs in potato cultivation. The results of experiments conducted under the project with varying levels of these inputs at different locations for 2-3 years revealed that the highest yields at all the locations were obtained with the recommended seed rate and fertilizer doses. However, these inputs could be economized and higher net returns obtained in some regions as given below.

Centre	*Seed Rate and Fertilizer Doses Giving Highest Net Returns*
Bhubaneshwar	50% of recommended seed rate and fertilizer doses
Chhindwara	75% of the recommended dose of fertilizers (112.5 kg N, 60 kg P_2O_5 and 80 kg K_2O/ha) and 75% of the recommended seed rate (22.5 q/ha). This was closely followed by the treatment involving 50% of the recommended fertilizer dose (75 kg N, 40 kg P_2O_5 and 50 kg K_2O/ha) and 50% seed rate (15 q/ha)
Faizabad	50% recommended fertilizer dose (75 kg N, 40 kg P_2O_5 and 50 kg K_2O/ha) and 50% recommended seed rate (15 q/ha)
Ranchi	75% of the recommended seed rate (30q/ha) and the normal dose of fertilizers (150 kg N, 80 kg P_2O_5 and 100 kg K_2O/ha)

Potato Based Crop Rotations

High yield of potato in unit area and time is a valuable trait especially in the areas where climate permits more than one crop to be grown in the same field each year. In the Indian sub-tropics, potato is a short duration crop that can be grown in wide range of climatic conditions and soil types with wide flexibility in planting and harvesting time thus making it a most suitable crop for inclusion in intensive cropping systems in different regions of the country. Thus there is a great opportunity for vertical expansion in area and production of potato without affecting other crops. Some of the potato based remunerative crop rotations identified for different regions are given in Table 2.18.

Table 2.18. Remunerative crop rotations in various regions

Region	*Recommended Rotation*	*Region*	*Recommended Rotation*
Assam	Potato-moong-paddy (transplanted)potato- paddy (S*)- paddy K*) Black-gram -potato-okra	Jharkhand	Potato-moong-groundnut
Bihar	Potato-sunflower-rice	Madhya Pradesh	Potato-okra-soybean potato Moong-maize
Gujarat (North)	Potato-bajra-groundnut potato-groundnut-bajra Potato-groundnut-til	Orissa	Rice-potato-okra
Haryana	Onion-okra Potato-tomato-okra Potato-bottlegourd-dhaincha (GM)	Punjab	Potato-onion (for bulbs)-Groundnut potato-wheat-Cowpea
Himachal Pradesh (Kangra district)	Potato (spring)-maize-radish	Uttar Pradesh (Central)	Potato-bajra-groundnut
		West Bengal	Potato-jute-paddy Potato-paddy-okra

* S = summer; K = *kharif*

Economics of crop rotations: The net returns from various cropping systems, however, depend on the crops involved. The economics of predominant potato based crop rotations at different centres was worked out for 2-3 years and are given in Table 10.19.

Table 10.19. Remunerative crop rotations recommended for different locations.

Centre	*Recommended Rotation*	*Net Returns (Rs /ha)*			
		1st Year	*2nd Year*	*3rd Year*	*Average*
Deesa	Potato-bajra-groundnut	41,875	34,665	36,148	37,563
	Potato-groundnut-bajra	29,250	34,500	34,926	32,892
Ranchi	Potato-green gram-groundnut	19,058	23,385	19,590	20,678
Chhindwara	Potato-okra-soyabean	14,981	28,620	30,125	24,575
	Potato-moong-maize	19,559	20,539	29,597	23,232
Hisar	Onion-okra	52,405	60,841	49,761	54,336
	Potato-tomato-okra	46,940	38,038	38,988	41,322
Jorhat	Potato-moong-paddy (transplanted)	22,760	54,422	-	38,591
	Potato-summer paddy (transplanted)-Paddy (transplanted)	20,633	52,900	-	36,767

At Bhubaneshwar (Orissa), rice-potato-okra was the most profitable crop sequence, followed by rice-potato-jute, while in Dharwad area of Karnataka, potato-beans was found to be the most profitable crop sequence followed by potato-sorghum.

Planting/harvesting dates**:** It is imperative to identify suitable variety of a crop and find the appropriate time of planting and harvesting for proper adjustment of different crops in a sequence to harness better economic returns. Results obtained in studies undertaken at different locations using different combinations of varieties and dates of planting/harvesting are summarized in Table 2.20.

Table 2.20: Crop varieties, planting/harvesting dates and yields of best potato-based cropping sequences.

Centre	*Rotation*	*Variety*	*Date of*		*Yield (q/ha)*	
			Planting	*Harvesting*	*MP*	*BP*
Deesa	Potato	KCM	3rd wk Nov.	1-2nd wk Mar	335	–
	Bajra	GHB-30	2nd wk March	1-2nd wk Jun	51	95
	ground nut	GG-2	1st wk July	3-4th wk Oct	12	33
Jorhat	Potato	KBH	3rd wk Nov	1-2nd wk Mar	20	–
	greengram	T-44	2-3rd wk Mar	3-4th wk May	16	20
	paddy (transplanted)	Pankaj	3rd wk Jun	4th wk Oct	23	60
Ranchi	potato,	K. Lalima	4th wk Oct	1-2nd wk Feb	22	–
	moong,	Sunaina	3rd wk Mar	4th wk May	76	–
	groundnut	A/C 12–24	4th wk Jun	1st wk Oct	21	–
Hisar	potato	KBD	2nd wk Oct	2nd wk Feb	388	–
	tomato	Selection-7	2nd wk Feb	2nd wk Jun	276	–
	okra	Pusa Sawani	2-3rd wk Jul	4th wk Sept	68	–
Chhindwara	potato,	KBD	1st wk Nov	4th wk Feb	236	–
	okra,	Pusa Sawani	3rd wk Mar	2nd wk May	39	–
	soyabean	JS-72-44	1st wk Jul	–	5	–

MP = Main product; BP = By-product; KBH = Kufri Bahar; KBD = Kufri Badshah; KCM = Kufri Chandramukhi; T = transplanted

Fertilizer requirement: The sequential crops in the best potato based crop rotations identified in different regions were evaluated for their response to different fertilizer doses. Potato in each sequence received the recommended dose of N, P and K, whereas, in the other crops in rotation the fertilizer doses were altered to find the most economic dose giving yields at par with the recommended fertilizer doses in the respective crops grown before and after potato. Fertilizer requirements of the crops in best crop rotations at different centres are summarized in Table 2.21.

Table 2.21. Fertilizer requirements of rotational crops.

Centre	*Best Crop Rotation*	*Fertilizer Requirement of 2nd & 3rd Crops in a Sequence*
Chhindwara	Potato-okra-soybean	Normal yield in okra after potato is obtained even with half the recommended dose. Soybean (3rd crop) requires full dose when preceding okra crop received no fertilizers but is reduced to 50% of recommended dose giving normal yield if okra received either 50% or 100% of recommended dose.
Deesa	Potato-bajra (summer)-ground nut (*kharif*)	For higher net returns from the rotation, *bajra* after potato should be applied full dose of recommended NPK and the succeeding *ground nut* crop with half the recommended dose of NPK.
Hisar	Potato-bottle gourd-green manure (*dhaincha*)	Potato should be optimally fertilized and the succeeding crop (bottle gourd) may be applied either optimum NPK or only N (recommended dose) for normal yield and better gain from the rotation.
Jalandhar	Potato-onion-groundnut	Onion after potato needs full dose of NPK. The succeeding legume crops either groundnut or cowpea after onion needs no fertigation if the preceding crops (potato/onion) are fertilized fully. Residue incorporation of each crop could economize on 50% NPK to the succeeding crops.

(Contd.)

Centre	*Best Crop Rotation*	*Fertilizer Requirement of 2nd & 3rd Crops in a Sequence*
Jorhat	Potato-summer rice-*kharif* rice	All the crops in this rotation need to be fertilized with normal N, P & K but application of 125% NPK to *kharif* rice (transplanted) gives higher yield than optimum dose.
Patna	Potato-sunflower-rice	Sunflower yield after potato was not affected by the optimal fertilizer schedule but incorporation of preceding crop residue (potato) has positive effect on sunflower yield. Residue incorporation of each crop could economize on 50% NPK to the succeeding crops.

When residue incorporation of all the crops + biofertilizers along with 100% recommended doses of NPK in the form of inorganic fertilizers is applied to potato, only 75% recommended doses of nutrients through inorganic sources are needed for bajra in potato-bajra crop sequence in Gujarat. At Faizabad in Uttar Pradesh in rice-potato-wheat crop sequence, application of 100% recommended dose of NPK along with 20 tonnes FYM /ha to potato and only 75% recommended dose of nutrients to wheat grown after potato is required, while subsequent rice crop needed 100% recommended dose of nutrients in the sequence.

In Haryana, in potato-bottle gourd crop sequence, application of 20 tonnes FYM/ha along with incorporation of crop residues and use of biofertilizers to potato is needed, while subsequent bottle gourd requires only 75% recommended dose of nutrients.

At Modipuram in western Uttar Pradesh in potato-greengram sequence, application of 100% recommended dose of N + 20 tonnes FYM/ha gave higher yield of potato, while subsequent green gram in the sequence requires only 75% recommended doses of nutrients after this treatment.

Inter-cropping with Potato

Several crops suitable for inter-cropping with potato were identified at different centers (Table 2.22 and Fig. 2.6).

Table 2.22. Promising potato-based inter-cropping in different regions

Region	*Intercropping System*	*Region*	*Intercropping System*
Chhindwara	Potato + onion (for bulb) (1:1)*	Pantnagar	Potato + wheat (1:2)
	Potato + maize (1:1)		Potato + radish (1:1)
Dharwad	Potato + garlic (1:2)	Patna	Potato + wheat (1:2)
			Potato + radish (1:1)
Faizabad	Potato + mustard (3:1)	Rajgurunagar	Potato + sugarcane (1:1)
Hisar	Potato + fennel (1:2)	Ranchi	Potato + wheat (1:2)
	Potato + onion (for seed) (1:1)		
Kanpur	Potato + wheat (1:2)	Ranichauri	Potato + garlic (1:1)
	Potato + linseed (1:1)		
	Potato + onion (for seed) (1:1)	Srinagar	Potato + garlic (1:1)

* Proportion of number of rows of potato and intercrop (recommended inter-row and intra-row spacing for each crop). The rows are to be adjusted as per crop canopy and convenience. For example onion can be grown as single rows in furrows between two rows of potato; two rows of fennel are grown between two rows of potato; and three rows of linseed can be planted in place of every fourth row of potato.

Potato + onion (for bulb/seed): Onion in single rows was grown in furrows between the rows of potato. Fertilizer was given to both the crops as per recommendations. Though the yields of both the crops were reduced in intercropping, combination of potato and onion was found to be more remunerative than pure crops.

Potato + fennel: Fennel was successfully grown as an intercrop (even without fertilizer applications) in Haryana without adversely affecting the potato yields. Two rows of fennel (30cm x 20cm inter- and intra-row spacing) were grown between two rows of potatoes (60 cm x 20 cm inter- and intra-row spacing).

Potato + sugarcane: The yield of sugarcane increased when potato was planted as an intercrop. The highest yields of potato were obtained in double row plantings (twin rows per ridge with 30cm x 25cm inter- and intra-row spacing) between two sugarcane rows planted 90cm apart with recommended dose of fertilizers given to both the crops.

Potato + linseed: Three rows of linseed (at 25cm x 10cm inter and intra-row spacings) were planted in place of every fourth row of potato (planted at 50cm x 20cm inter- and intra-row spacing). Though the yield of potato was higher in the pure crop, inter-cropping was found to be more remunerative.

Potato + mustard: One row of mustard was planted after every three rows of potato (planted at 50 x 20cm inter- and intra-row spacing). The yield of potato was higher in the pure crop but the intercropping of potato and mustard was found to be more remunerative.

Potato + wheat: Two rows of wheat were sown between two potato ridges (60cm x 20cm inter- and intra- row spacing). Only potato crop was given the recommended fertilizer dose. A normal crop of potato was obtained. Besides, additional returns were obtained from wheat and the straw.

Economics of some of the important inter-cropping practices were worked out at different AICRP centers and are given in Table 2.23.

Table 2.23. Remunerative inter-crops identified for different regions

Centre	*Companion Crops*	*Yield (q/ha)*			*Net gain (Rs/ha)*		
		Potato Alone	*Inter Crop Alone*	*Potato + Inter Crop*	*Potato Alone*	*Intercrop Alone*	*Potato + Intercrop*
Hisar	Potato + fennel	287	31.0	280 + 17.0	54,467	46,130	68,613
	Potato + onion (for seed)	287	14.0	244+ 11.0	54,467	104,341	136,868
Faizabad	Potato + mustard	276	28.7	225+8.5	25,400	27,000	30,123
Chhindwara	Potato + onion (for bulb)	266	295.5	201+ 152.7	34,190	42,853	44,843
Kanpur	Potato + linseed	384	29.0	334 + 10.0	40,497	15,559	47,174
Ranchi	Potato + wheat	262	45.0	275 + 22.0	16,050	10,720	23,895
Rajgurunagar	Potato + sugarcane	149	1240.6	98+1404.6	9,630	28,352	33,752

Diversification of Potato Cultivation in Rice-wheat System

In the wake of shrinking arable land it may not be possible to put additional area exclusively for potato cultivation unless some other crops are sacrificed. Rice-wheat system in India occupies nearly 10 million hectares and larger area of which lies in the Indo-Gangetic plains where

Potato + maize

Potato + wheat

Potato + mustard

Potato + garlic

Potato + sugarcane

Potato + barley

Fig. 2.6: Some remunerative potato-based inter-cropping systems

Fig. 2.5. Important potato pests: 1. Late blight; 2. early blight; 3. phoma; 4. black scurf; 5. wart; 6. dry rot; 7. *Sclerotial* wilt; 8. brown rot infected tuber; 9. potato apical leaf curl virus disease; 10. severe mosaic; 11. stem necrosis; 12. common scab; 13. aphids; 14. mites; 15. whiteflies; 16. cut worm; 17. white grub; 18. potato tuber moth and 19. cyst nematodes

better irrigation facilities are available. Cultivation of potato as sandwich crop would be economically viable proposition for utilizing the time and space between rice and wheat crops. Suitable varieties of potato, rice and wheat for rice-potato-wheat system have been identified and are given below. With time, new improved varieties may replace these varieties. Therefore, farmers are advised to use recent improved varieties of potato, rice and wheat having similar maturity behaviour.

Center	*Potato*	*Wheat*	*Rice*
Jalandhar	Kufri Pukhraj	PBW-373	PR-106
Modipuram	Kufri Pukhraj	PBW-226	Saket-4
Patna	Kufri Ashoka	—	Proagro-6201
Faizabad	Kufri Ashoka	Malviya-234	Narendra-97
Hisar	Kufri Ashoka	PBW-373	Govind

Average yields of rotational crops in rice-potato-wheat system obtained at different centres in the Indo-Gangetic plains are given in Table 2.24.

Table 2.24. Yield of potato, wheat and rice varieties at different centres

Variety	*Crop*	*Average Yield (q/ha) of Different Crops*				
		Jalandhar	*Modipuram*	*Patna*	*Faizabad*	*Hisar*
Kufri Ashoka	Potato	302	221	205	254	216
Kufri Pukhraj	Potato	337	250	190	–	–
Kufri Jawahar	Potato	248	–	–	239	178
Kufri Bahar	Potato	–	199	–	–	–
Kufri Chandramukhi	Potato	–	–	152	239	–
Kufri Badshah	Potato	–	–	–	–	197
PR-106	Rice	53	–	–	–	–
Saket-4	Rice	–	55	–	–	–
Proagro-6201	Rice	–	–	62	–	–
Narendra-97	Rice	–	–	–	25	–
Govind	Rice	–	–	–	–	82
PBW-373	Wheat	59	–	–	–	38
Malviya-234	Wheat	–	–	–	60	–
PBW-226	Wheat	–	60	–	–	–

Pest Management

The potato crop is afflicted by a number of fungal, bacterial and viral diseases as well as by insect pests (Fig. 2.7).

Viruses and late blight are the most important diseases in India causing huge economic losses. In warm subtropical parts of the country the incidence and spread of viruses is rapid due to high populations of virus vectors (aphids, leafhoppers, thrips and whiteflies). Stem necrosis disease caused by tospovirus and spread by thrips is of recent origin in the states of central India. Potato apical leaf curl virus (PALCV) caused by begomovirus and spread by whiteflies is assuming alarming proportions in many potato growing states in north, central and eastern India. Both these viruses are serious problems mainly in early planted crops. Emergence of new late blight races and cultivation of late blight susceptible varieties is changing incidence pattern of late blight (frequency, severity, etc.) in the country. Early blight and leaf spots are more predominant in north eastern states and plateau region. These have also been reported from many other states where balanced fertilizers (particularly nitrogen) are not applied. Tuber borne diseases like black scurf, common scab, dry rot, soft rot and charcoal rot although are under control, they pose problems in areas where potato seed treatment and crop rotations are not followed. These are mainly management problems. Among bacterial diseases, bacterial wilt or brown rot is an important problem in some pockets of mid-hills and plateau region.

The scenario of diseases and insect pests keeps on changing due to changing weather conditions, development of pesticidal resistance and evolution of aggressive pathogens. Prevalence of cyst nematodes in Nilgiris, wart in Darjeeling hills, red ants in eastern plains and white grubs in hills indicate regional nature of insect pests.

Some of the major potato diseases and insect pests occurring in different states are summarized in Table 2.25.

Table 2.25. Important diseases and insect pests in different states of India

State	*Diseases*	*Insect Pests*
Andhra Pradesh, Karnataka and Maharashtra	Early blight and leaf spots, seed decay when used as cut pieces, sclerotial wilt, viruses, bacterial wilt (in pockets), charcoal rot (occasional), soft rot (in *kharif*) and late blight (occasional).	Leafhoppers, aphids, mites, whiteflies, potato tuber moth and cutworms (sporadic)
Arunachal Pradesh Meghalaya, Tripura Mizoram, Nagaland Manipur and Sikkim	Late blight (regular and severe) and bacterial wilt (in mid hills and at lower elevations).	White grub, cutworm (sporadic), red ants, leaf eating caterpillar and potato tuber moth
Assam	Early blight and leaf spots, viruses, late blight (frequent), dry rot (frequent) and bacterial wilt (only in some pockets).	Aphids, cutworms, red ants, white grub (sporadic), potato tuber moth and mole cricket.
Bihar and Jharkhand	Early blight and leaf spots, viruses, charcoal rot, dry rot, soft rot (in *kharif* in Jharkhand), late blight (frequent), black scurf and bacterial wilt (some pockets in Jharkhand).	Aphids, potato tuber moth and cutworms(sporadic)
Gujarat and Rajasthan	Seed decay when used as cut pieces, stem necrosis, early blight and leaf spots, viruses, black scurf, common scab and late blight (occasional)	Thrips, cutworms and *Heliothis* caterpil lars.
Haryana	Potato apical leaf curl virus (PALCV), early blight and leaf spots, late blight (frequent), black scurf and common scab.	Whiteflies, aphids and leafhoppers.
Himachal Pradesh	Late blight (regular and severe) and bacterial wilt (in mid-hills)	White grub, cutworms (sporadic) and potato tuber moth (in Kangra district)
Jammu and Kashmir	Late blight (regular and severe) and early blight and leaf spots.	Spring trail, white grub and cutworms (sporadic).
Madhya Pradesh and Chhattisgarh	Early blight and leaf spots, stem necrosis, PALCV, other viruses, black scurf and late blight (occasional but frequent in hills)	Thrips, whiteflies, aphids, leafhoppers and cutworms.

(Contd.)

State	*Diseases*	*Insect Pests*
Orissa	Early blight and leaf spots, PALCV, other viruses, soft rot, bacterial wilt, common scab and late blight (occasional)	Whiteflies, aphids, *Heliothis* caterpillars, potato tuber moth, epilachna beetle and cutworms (sporadic).
Punjab	Late blight (regular); common scab and charcoal rot (occasional).	Aphids, whiteflies, leafhoppers and cut worms (sporadic)
Tamil Nadu(Nilgiri and Kodai hills)	Late blight (regular and severe)	Cyst nematodes
Uttarakhand	Late blight (regular and severe in hills; regular in plains), early blight and leaf spots, black scurf, bacterial wilt (in mid hills) and common scab (in tarai region).	White grub (in hills), potato tuber moth and cutworms (sporadic).
Uttar Pradesh	Early blight and leaf spots, late blight (frequent), PALCV, black scurf, common scab, dry rot (eastern Uttar Pradesh) and charcoal rot (eastern Uttar Pradesh)	Whiteflies, potato tuber moth (eastern Uttar Pradesh), red ants (eastern Uttar Pradesh) and cutworms (sporadic).
West Bengal	Early blight and leaf spots, PALCV, other viruses, late blight (frequent in plains; regular and severe in Darjeeling hills), wart (in Darjeeling hills) and bacterial wilt (in midhills).	Whiteflies, aphids, leafhoppers, mole cricket, epilachna beetles, red ants and cutworms (sporadic).

Management of Late Blight

Late blight is an important disease of potato causing yield losses up to 80% in the susceptible cultivars under epiphytotic situation. Late blight is controlled through the use of resistant cultivars as well as chemical sprays. Occurrence of late blight has now become a recurring feature in the subtropical plains where more than 90% of potato is produced in the country. It is relatively easy to control late blight in resistant varieties. However, large areas are also planted with moderately resistant/susceptible varieties. Therefore, it became imperative to develop fungicide spray schedules for important potato-growing regions as well as for moderately resistant/susceptible varieties. Spray schedules developed for some of the varieties in different regions are given in Table 2.26. These spray schedules are to be followed immediately after appearance of late blight.

The forecasting models for late blight are important that enable to apply control measures including prophylactic sprays of fungicides at an appropriate time for disease management. Such models have been developed for north western hills and western Uttar Pradesh. Recently, AICRP (potato) has initiated programme for developing late blight forecasting models for other parts of the country as well.

Table 2.26. Fungicide spray schedules to be followed immediately after appearance of late blight in different varieties

Region	*Variety*	*Fungicide Schedule*
North western plains	Kufri Jyoti	One spray of mancozeb
	Kufri Jawahar	Two sprays of mancozeb
	Kufri Ashoka	One spray of metalaxyl
North eastern plains	Kufri Sutlej	No spray requiredl
	Kufri Lauvkar	One spray of metalaxy
	Kufri Jawahar	One spray of metalaxyl
	Kufri Lalima	One spray of metalaxyl/one spray each of metalaxyl + mancozeb
Central plains	Kufri Ashoka	One spray of mancozeb followed by one spray of metalaxyl

Management of Bacterial Wilt

The bacterial wilt of potato plants and the brown rot of tubers are caused by *Pseudomonas solanacearum* now known as *Ralstonia solanacearum*. In the eastern plains and plateau areas of Bihar and Madhya Pradesh the disease perpetuates mainly through infected seed tubers. In the mid-hills of the country and plains of Orissa the disease is both soil as well as seed borne.

Control of Bacterial Wilt through Crop Rotations

Good management of the disease was obtained through use of disease free seed; field application of stable bleaching powder @ 12 kg/ha with fertilizers at the time of planting; and ploughing and exposing the fields to high summer temperatures in the plains and winter temperatures in the hills before planting. Besides, some crop rotations were also found effective in reducing the bacterial wilt and brown rot incidence when used either alone or in combination with bleaching powder, ploughing, etc. Some of the results are described below.

Bhubaneshwar: Crop rotations wheat-green gram-potato and wheat-maize-mustard-paddy-potato reduced the bacterial wilt by 50% and 86% respectively. Above crop rotations together with the application of bleaching powder @12kg/ha at planting in the infected field further reduced wilt incidence to nil and increased yield of healthy potatoes by 43–54% in variety Kufri Chandramukhi (Table 2.27).

Table 2.27. Control of brown rot incidence at Bhubaneshwar

Treatment	*Wilt incidence (%) at 90 days*	*Tuber Yield (q/ha)*		*Total Yield*
		Healthy Tubers	*Diseased Tubers*	
Disease free seed (control)	33	123	23	146
Wheat-green gram-potato	16	161	10	171
Wheat-maize-mustard-paddy-potato	5	203	7	210
T2 + ploughing after harvesting	15	128	13	141
T2 + bleaching powder @ 12 kg/ha	0	217	2	219
T3 + ploughing after harvest	3	232	8	240
T3 + bleaching powder @ 12kg/ha	0	269	2	271

Hassan: Application of bleaching powder at the time of planting potato crop followed by summer ploughing (after harvesting potatoes) proved most effective in reducing the incidence of wilt as compared to other treatments (Table 2.28).

Table 2.28. Management of brown rot incidence at Hassan

Treatment	*Wilt Incidence (%) at 90 Days*	*Tuber Yield (q/ha)*		*Total Yield*
		Healthy Tubers	*Diseased Tubers*	
Potato-ragi - potato	27	160	15	175
Potato-ragi + summer ploughing – potato	24	173	19	192
Potato + bleaching powder-ragi – potato	13	222	14	236
Potato + bleaching powder+summer ploughing-potato	11	240	20	260

Shillong: Rotation of cabbage-maize-radish-maize-radish-potato along with winter ploughing was found better than other treatment schedules, which reduced bacterial wilt to nil and increased tuber yield in variety Kufri Jyoti (Table 2.29).

Table 2.29. Effect of different treatments on brown rot incidence at Shillong

Treatment	*Wilt Incidence at (%)*	*Tuber Yield (q/ha)*		*Total yield*
	90 Days	*Healthy Tubers*	*Diseased Tubers*	
Potato-potato-potato	12	124	24	148
Cabbage- maize-radish-potato	0.5	218	5	223
T2 + Winter ploughing after harvest	0	220	0	220
T2 + Bleaching powder @ 12 kg/ha	1	225	4	229

Biocontrol of bacterial wilt: Bioagents *Bacillus cereus* and *B. subtilis,* were evaluated for their efficacy in reducing the incidence of wilt in potato at Bhubaneshwar, Hassan and Jorhat centers and the results of the experiments are briefly reported below.

Hassan: Wilt incidence though high at Hassan was markedly reduced by tuber treatment with bio-agents *B. cereus*, *B subtilis* and avirulent *Ralstonia solanacearum* (Table 2.30).

Bhubaneshwar: *B. cereus* and *B. subtilis* proved effective and significantly reduced the disease and increased tuber yield in Kufri Chandramukhi.

Jorhat: Best control of wilt (0.9%) was provided by tuber treatment with *B. subtilis*. However, significant tuber yield was obtained with both *B. cereus* and *B. subtilis*.

Table 2.30. Efficacy of biocontrol agents in controlling bacterial wilt at Hassan

Treatment Tuber	*Variety*	*Wilt Incidence (%) at 60 Days*	*Yield (q/ha)*		*Total Tubers*	*Tuber rottage (%) at Harvesting*
			Healthy	*Diseased Tubers*		
Untreated healthy tubers	K. Jyoti	51	67	18	85	21
Tubers dipped in 10^6 CFU/ml suspension of avirulent *R. solanacearum*	K. Jyoti	22	110	9	119	8
Tubers dipped in 10^6 CFU/ml suspension of *B. cereus*	K. Jyoti	28	100	13	113	12
Tubers dipped in 10^6 CFU/ml suspension of *B. subtilis*	K. Jyoti	28	98	12	110	11
Tubers dipped in 10^6 CFU/ml suspension of *Enterobacter celocae*	K. Jyoti	39	92	14	106	17

Control of Tuber Borne Diseases

Black scurf (*Rhizoctonia solanii*) and common scab (*Streptomyces scabies*) are predominant tuber borne diseases in many potato-growing regions of the country. These diseases are both soil as well as tuber borne. A large number of treatments including tuber treatments with fungicides, soil amendments and their combinations, etc., were identified for proper management of tuber-borne diseases. The results showed that tuber treatment with organo-mercurial compounds (OMCs) was best in controlling both black scurf and common scab in potato (Table 2.31). However, since OMCs are carcinogenic, hazardous and environmentally unsafe hence, many treatments other than OMCs were tried, which involved tuber treatment with boric acid (3%) for 30 minutes; Aureofungin + streptocycline (40 ppm each) + $CuSO_4$ (1000 ppm) for 10 minutes; Benlate (0.1%) for 2 minutes; and Bavistin and Rovarol (0.2% each) for 30 minutes. Boric acid is a safer compound and easier to handle as compared to the hazardous OMCs. Seed tuber treatment with boric acid was at par with OMCs in effectiveness for controlling black scurf and common scab diseases as well as soft rot of tubers in storage. Among different methods tried, dip treatment of seed tubers in 3% solution for 30 minutes was found efficient and cost-effective. Irrespective of concentration used, all the boric acid treatments were similar in their effectiveness (Table 2.32).

Table 2.31. Effect of organo mercurial compounds and boric acid treatment in controlling tuber borne diseases

Treatment	*Disease Incidence (%)*					*Disease Index*				
	RAN	*JAL*	*PAT*	*KAL*	*MDP*	*PNT*	*DES*	*JAL*	*KAL*	*MDP*
Untreated diseased tubers	18	38	64	26	18	58	3.7	0.5	4.0	1.4
Emisan-6 (0.5%) 30 minites	6	13	4	13	5	26	1.6	0.2	2.9	0.4
Boric acid (3%) 30 minutes	10	13	24	11	8	46	1.5	0.2	2.5	0.8
Untreated healthy tubers	13	10	-	10	4	39	2.3	0.1	2.0	0.3

RAN = Ranichauri; JA = Jalandhar; PAT = Patana; KAL = Kalyani; MDP = Modipuram; PNT = Pantnagar; DES = Deesa

Table 2.32. Efficacy of boric acid for control of tuber-borne diseases

Treat-ment	*Average Disease Incidence*								*Average Disease Index*							
	JAL	*HIS*	*KTT*	*PNT*	*RNC*	*DES*	*KAL*	*BHN*	*JAL*	*HIS*	*KTT*	*PNT*	*RNC*	*DES*	*KAL*	*BHN*
T_1*	62	38	97	48	22	41	21	18	1.8	45	1.3	24	2.6	6.8	73	1
T_2	7	18	93	57	6	16	12	13	0.1	37	1.1	20	1.1	3.0	1	1
T_3	5	18	87	46	13	27	14	12	0.1	32	1.1	22	1.6	3.3	3	1
T_4	3	22	90	45	12	26	14	12	0.0	36	1.1	23	1.8	3.4	4	1
T_5	2	20	83	42	15	26	13	12	0.0	41	1.1	21	1.5	3.7	3	1
T_6	31	20	71	33	20	28	11	12	0.9	37	0.7	18	2.0	2.4	2	1

JAL = alandhar; HIS = Hisar; KTT = Kota; PNT = Pantnagar; RNC = Ranichauri; DES = Deesa; KAL = Kalyani and BHN = Bhubaneshwar.

*T_1 = Untreated (diseased); T_2 = Boric acid 3% dip for 30 minutes; T_3 = boric acid 3% spray; T_4 = boric acid 4% spray T_5 = boric acid 5% spray and T_6 = untreated healthy tubers

Studies on Degeneration of Seed Stocks Due to Viruses

All potato-growing regions in the country are not equally suitable for growing disease-free potato crop. The healthy seed stock of a variety, upon continuous use and exposures to aphids (vectors of viruses) becomes degenerated due to infiltration of viruses. The rate of degeneration in different commercial varieties vary from region to region and is governed by the susceptibility of cultivars to virus(es) and the activity of vectors that is determined mainly by the climate of a place. In a study under the project, it was observed that the productivity of seed stocks could be maintained at a desired level for couple of years by applying systemic insecticides at the time of planting, roguing of diseased plants and cutting haulms before aphid population cross critical level (20 aphids per 100 compound leaves).

The desired productivity could be maintained in all varieties for over five years in North western plains (Haryana and Punjab) and Western Uttar Pradesh; three to four years in North Gujarat; and four years in eastern plains of West Bengal. Among the earlier varieties Kufri Jyoti showed faster degeneration than Kufri Badshah in West Bengal. In Satpura plateau of Madhya Pradesh, the incidence of virus diseases was high and even the fresh breeders' seed in the area recorded high disease incidence in the very first exposure. In this region, loss of productivity in seed of Kufri Chandramukhi was faster than in Kufri Sindhuri which could maintain high productivity when the seed crop was given systemic insecticide at the time of planting and haulms were cut at proper time. The degeneration due to viruses among the recently released five potato varieties, viz. Kufri Ashoka, Kufri Anand, Kufri Jawahar, Kufri Pukhraj and Kufri Sutlej, was evaluated at Deesa, Faizabad, Jalandhar, Kalyani and Patna for three years. Among the varieties highest incidence of viruses was observed in Kufri Pukhraj (16.22%) and least in Kufri Jawahar (4.42%). Among locations Kalyani (WB) had highest incidence of viruses (28.07%) and least at Patna. This indicates that climate in the eastern part of the country is highly conducive for virus infestation in potato varieties due to higher vector activity during the crop season and varieties like Kufri Pukhraj and Kufri Anand or Kufri Sutlej would need frequent replacement of their seed stocks compared to Kufri Jawahar.

Surveillance of Aphids and Areas

Surveys were conducted in different parts of the country to monitor aphid population and to find aphid-free/low aphid period and places suitable for seed production. Periodic observations recorded on the appearance, build-up and peak population periods in insecticide sprayed and unsprayed plots have revealed that Jalandhar (Punjab), Hisar (Haryana), Kota (western Rajasthan), Kanpur (central Uttar Pradesh) and Faizabad (eastern Uttar Pradesh) are good areas for producing healthy seed of potato as the crop in these regions gets enough time to tuberize before the aphid population reaches critical level of 20 aphids per 100 compound leaves. Healthy tuber seed can also be produced at Kalyani (West Bengal) and Deesa (northern Gujarat) if the crop is given one spray of systemic insecticide in the 4th week of December. Pantnagar (*tarai* region of Uttar Pradesh); Chhindwara (Satpura plateau of Madhya Pradesh); Ranchi (Bihar plateau); Bangalore, Hassan, Aurangabad and Rajgurunagar (Deccan plateau); and Bhubaneshwar (Orissa) were not suitable for seed production as the aphid population crosses the critical level early in the crop season (Table 2.33).

New Emerging Diseases and Pests of Potato

Potato apical leaf curl disease: Potato apical leaf curl disease (PALCD) caused by apical leaf curl begomo virus (ALCV) is an emerging major disease, limiting potato production in the country. Sporadic incidence of this disease was noticed first time in December 1996 at Hisar in Haryana. It is now reported from Punjab and western UP as well. The virus is spread by whiteflies (*Bemisia tabaci*) and the progress of disease has positive correlation with whitefly population.

Table 2.33. Appearance and build-up of aphid population in different regions of the country

State	*Centre*	*Planting Time*	*Date of*		*Date of*		*Time of*
			M. persicae Appearance	*Reaching CL**	*A. gossypii Appearance*	Reaching CL	*Harvesting*
Assam	Jorhat	2 wk Nov.	1 wk Nov	< CL	–	–	3 wk Mar
Bihar	Patna	4 wk. Oct.	2 wk. Dec	2 wk. Jan	1 wk Dec.	4 wk Dec	1 wk Jan
Jharkhand	Ranchi	1 wk Nov.	1 wk Dec.	2 wk Jan.	2 wk Jan.	3 wk Jan	3 wk Feb
Gujarat	Deesa	1 wk Nov.	4 wk Dec	3 wk Jan.	3 wk Jan.	2 wk Feb	1 wk Mar
Haryana	Hisar	2 wk Oct.	1 wk Jan.	1 wk Feb.	Does not appear	2 wk Jan	
Karnataka	Hassan	1 wk Nov.	1 wk Dec.	3 wk Dec.	–	–	3 wk Feb
	Hassan (K)*	1 wk June	Always above critical level	1 wk Sept			
MP	Chhindwara	4 wk Oct.	2 wk Dec.	4 wk Dec.	1 wk Jan.	2 wk Jan	2 wk Jan
Maharashtra	Rajgurunagar [Rabi]	1 wk Nov.	1 wk Dec.	4 wk Dec.	2 wk Dec.	2 wk Dec	1 wk Mar
	Rajgurunagar (K)	1 wk July	1 wk Aug	2 wk Aug	3 wk Jul	4 wk Jul	1 wk Oct
Meghalaya	Shillong	2 wk Apr	1 wk May	2 wk Aug	2 wk May	4 wk May	1 wk Sept
Orissa	Bhubaneshwar	2 wk Nov.	1 wk Dec.	> CL	–	–	2 wk Feb
Punjab	Jalandhar	1 wk Oct.	1 wk Dec.	1 wk Jan.	1 wk Jan.	< CL	1 wk Jan
Rajasthan	Kota	1 wk Nov.	4 wk Jan.	2 wk Feb	4 wk Jan.	2 wk Feb	1 wk Feb
Tamil Nadu	Ooty	2 wk Apr.	3 wk Jun	< CL	–	–	1 wk Sept
Uttarakhand	Pantnagar	2 wk Oct.	1 wk Dec	3 wk Dec.	2 wk Dec.	4 wk Dec	3 wk Jan
Eastern UP	Faizabad	1 wk Nov.	2 wk Jan	4 wk Jan.	2 wk Jan.	3 wk Jan	2 wk Feb
West Bengal	Kalyani	1 wk Nov.	3 wk Dec	3 wk Jan.	3 wk Jan.	4 wk Jan	2 wk Mar

* K = *Kharif*; CL = Critical level.

In potatoes, apical leaf curl is a highly polycyclic disease because viruliferous whiteflies are very active fliers and feeders causing primary and secondary infections quickly within a single crop season.

Being a tuber-borne disease, its incidences have been reported from several parts of the country. The disease causes severe yield losses (≈ 30–40%) in early sown (first week of October) susceptible potato cultivars in north Indian plains. The disease incidence varies from year to year depending on weather conditions, vector population and cropping patterns of the area. Whiteflies are highly polyphagous and have been reported to feed on more than 500 plant species in India. High incidence of apical leaf curl prevails in Hisar, Sirsa, Ambala and Kurukshetra districts of Haryana (adjoining areas of cotton belt) as cotton crop harbours whiteflies that migrate to potato crop in October (after harvesting of cotton). In central, eastern and southern parts of the country, the whitefly population remains high throughout the potato season due to congenial higher temperatures. Recently, large population of whiteflies and apical leaf curl virus like symptoms in potato plants have also been observed at Chhindwara (Madhya Pradesh), Raipur (Chhattisgarh) and Bhubaneshwar (Orissa). Disease incidence remains low, where large areas are planted with variety Kufri Bahar, having a good degree of resistance/ tolerance against the disease.

Following management strategies have been developed to control PALCD menace in potato.

- Use of disease-free seed of resistant/tolerant varieties (a few Indian varieties have resistance/tolerance to ALCV. Kufri Bahar is the most resistant/tolerant variety).
- Delayed planting (last week of October/first week of November in north Indian plains) in susceptible varieties to skip whiteflies. However, delayed planting can lead to yield reduction, exposure to late blight in later part of the season and also disturb cropping sequence.
- Whiteflies can effectively be controlled by potato seed tuber treatment with Imidacloprid @ 0.004% a.i. conc. (10 minutes dip) followed by need based sprays of Imidacloprid @ 0.002% a.i. conc. at complete crop emergence and later on. Efficacy of Imidacloprid can further be improved by its application with summer oil (1%).
- Use of trap crops like oat and fennel sown and transplanted, respectively, 10–15 days earlier than potato can offer better protection against whiteflies.

Potato Stem Necrosis

Stem necrosis is a problem of relatively recent origin in potato-growing regions of Madhya Pradesh, Gujarat and Rajasthan. The disease is on the increase in its severity and spreading to new areas. The tomato spotted wilt virus (TSWV) has been found to be associated with stem necrosis symptoms. Studies were conducted at Chhindwara, Kota and Deesa to find the factors responsible for its appearance and methods of its management including identification of resistant/tolerant potato genotypes, delay in planting potato crop and management through insecticidal sprays.

Incidence of stem necrosis in different cultivars/genotypes: There was a significant difference among the varieties for susceptibility to stem necrosis disease (Table 2.34). Varieties Kufri Chandramukhi, Kufri Sindhuri, Kufri Anand and Kufri Jawahar and hybrids PS/F-220, JN-189 and HPS-25/13 were found to have little incidence of the disease as compared to most susceptible variety Kufri Bahar at 30 days after emergence at Kota (Table 2.34).

Table 2.34. Stem necrosis incidence, thrips population and yield in different genotypes at Kota

Genotype	*Disease Incidence (%) 30 Days After Emergence*	*Thrips Population 70 Days After Emergence*	*Tuber Yield (q/ha)*
K. Sindhuri	5	39	300
PS/F-220	4	33	281
JN-189	7	35	297
K. Anand	6	27	278
HPS-25/13	8	37	289
K. Bahar	64	40	225

Two genotypes PS/F-220 and JN-189 possessing resistance to potato stem necrosis have been registered with the NBPGR as novel parental lines vide registration numbers INGR 04059 and INGR 07040, respectively.

Control through delayed planting: High incidence of stem necrosis in potato plants was observed in the field at all the three locations when crop was planted before the optimum time but got reduced gradually with delay in plantings up to 25 October – 12 November, irrespective of the varieties used. The disease index followed the similar pattern (Table 2.35).

Chemical control: Foliar spray of insecticides (Monocrotophos and Methyl-demeton) applied at 15 days interval reduced the disease incidence significantly compared to unsprayed control/plots at Kota, Chhindwara and Deesa. Monocrotophos was found more effective than Methyl-demeton in reducing the population of thrips/whitefly, which are the main vectors of tospo virus causing stem necrosis in potato (Table 2.35).

Table 2.35. Effect of planting dates and insecticidal sprays on stem necrosis disease

Treatment	*Disease Incidence*			*Disease Index (%)*			*Tuber Yield (q/ha)*		
	Kota	*Deesa*	*CHN*	*Kota*	*Deesa*	*CHN*	*Kota*	*Deesa*	*CHN*
D_1 T_1*	25	0	31	0.65	0	0.9	282	105	243
D_1 T_2	29	0	34	0.78	0	1.02	267	97	227
D_1 T3	78	0	57	2.08	0	2.27	216	52	184
D_2 T1	6	3.3	27	0.15	1.3	0.82	308	344	276
D_2 T_2	7	2.8	29	0.23	1.3	0.85	287	336	263
D_2 T_3	59	7.3	48	1.7	1.8	1.83	260	322	193
D_3 T_1	–	0.3	18	–	0.3	0.18	–	547	304
D_3 T_2	2	1.3	23	0	0.5	0.22	253	476	286
D_3 T_3	12	2.3	36	0.53	1	0.72	237	469	211

*D_1 = 5th October, D2 = 25th October and D3 = 12th November;
CHN = Chhindwara
T_1 = Monocrotophos (Nuvacron) - 3 sprays @ 1ml/litre at 15 days interval
T_2 = Methyldemeton (Metasystox) - 3 sprays @ 1ml/litre at 15 days interval
T_3 = Control (No chemicals sprayed)

Integrated Pest Management

Diseases, insect pests and weeds are major constraints in achieving full yield potential in food crops. The losses due to these biotic stresses are around 40% and if we add about 20% post-harvest losses, the situation becomes more alarming. Indiscriminate use of chemical pesticides has resulted in emergence of more aggressive pests due to resistance development; residual problems in food and drinking water; and ecological imbalance due to elimination of beneficial microorganisms and insects. Therefore, for sustainability of potato crop, these biotic stresses need to be managed through eco-friendly measures supported by need based and judicious use of chemicals to achieve high economic returns without disturbing environmental balance and serenity. Integrated pest management (IPM) is one of the economically viable and environmentally safe key technologies to increase crop productivity in the country. Some of the important components of IPM practices in potato are given below:

- Use of resistant varieties.
- Use of healthy seed obtained from a reliable source.
- Hot weather and cold weather cultivation in plains and hills, respectively.
- Crop rotations with cereals, green manure and other non-solanaceous crops.
- Seed tuber treatment with 3% boric acid or 1% acetic acid + 0.05% zinc sulphate.
- Optimum dates for planting and harvesting.
- Use of botanical pesticides like neem products, biofumigants and biocontrol agents.
- Need based application of safer chemical pesticides.
- Curing of potatoes for 8–10 days before harvesting and avoiding injuries to the tubers during harvesting and transportation.
- Proper storage of potatoes.

In fact, a typical eco-friendly approach for sustainable management of potato pests involves biological control including deployment of host resistance, best cultural practices and need based use of chemical pesticides (Fig. 2.8). Such plant protection measures are also in harmony with international food safety and environmental protection protocols.

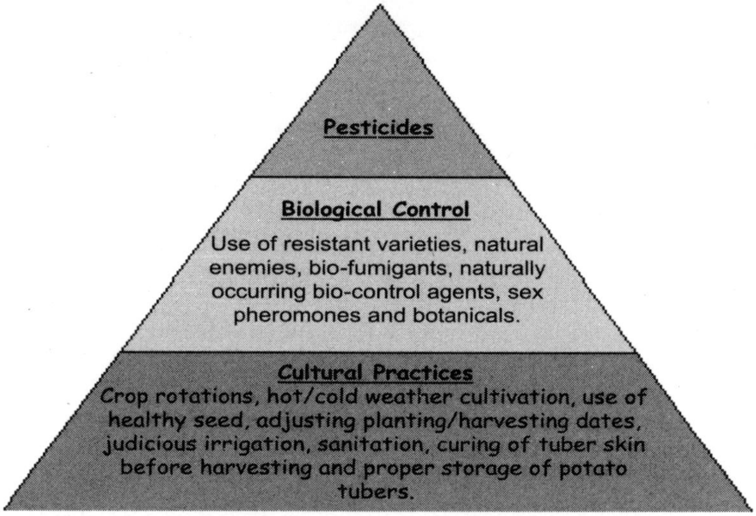

Fig. 2.8. Eco-friendly approaches for pest management in potato

Some of the IPMs available for management of important potato diseases and pests are given in Table 2.36. Improved varieties are important constituents of IPMs. Resistances of Indian potato varieties to different biotic and abiotic stresses are given in Table 2.36. In addition to the resistance to biotic stresses, maturity behaviour of the improved varieties also helps the farmers to escape biotic stresses. Maturity behaviours of Indian potato varieties are given in Table 2.9.

Table 2.36. Integrated pest management schedules for some potato pests

Pest	*Host Resistance*	*Agricultural Practices*	*Biocontrol Agents*	*Chemicals*
Late blight	Resistant varieties, short duration early bulking varieties to escape disease	Judicious irrigation, high ridging, adjustment of dates ofplanting	*Penicillium aurantiogris eum, Trichoderma harzianum and Aspergillus terrus*	Contact/systemic fungicides
Viruses	Tolerant varieties	Seed Plot Technique*	Meristem culture technique coupled with thermo- and chemotherapy	
Potato stem necrosis virus	Tolerant varieties	Late planting (to avoid thrips) and sanitation.	–	Contact/systemic insecticides for man agement of thrips.
Potato apical leaf curl virus	Tolerant varieties	Late planting (to avoid whiteflies), sanitation and use of trap crops.	–	Systemic insecticides for management of whiteflies.
Early blight and leaf spots	Resistant varieties	Balanced fertilizers, sanitation	–	Contact fungicides and urea
Soil and tuber borne diseases	Short duration varieties	Healthy seed, crop rotations, optimum dates of planting/ harvesting, hot/cold weather cultivation	*Trichoderma harzianum, T. viride, Bacillus subtilis,* non-pathogenic *Rhizoctonia etc.*	Tuber treatment with boric acid/contact or systemic fungicide.
Bacterial wilt	Short duration varieties	Healthy seed, crop rotations, hot/cold weather cultivation	*Bacillus subtilis, B. polymyxa* and *Pseudomonas fluorescens*	Bleaching powder
Wart	Resistant/Immune varieties	Healthy seed, crop rotations, hot/cold weather cultivation	–	–
Cyst nematodes	Resistant varieties	Healthy seed, crop rotations, hot/cold weather cultivation	–	Carbofuran
Potato tuber moth	–	Deep planting, proper ridging and cold storage of tubers	*Granulosis* virus, sex pheromones, Bt protein and leaves of neem, *Lantana*, and *Eucalyptus*	Contact insecticides

* Important components of seed plot technique are: green manuring, crop rotations, hot/cold weather cultivation, tuber treatment with boric acid, use of healthy seed, growing crop under aphid-free period, systemic insecticides at planting/ earthing up and roguing off types and diseased plants.

TRANSFER OF TECHNOLOGY

Different centres under AICRP (Potato) had been actively involved in dissemination of improved potato production technologies through demonstration trials, farmers' days, training programmes, mass media (newspaper, Radio/TV talks) etc. Seventy four radio/TV talks were delivered by scientists of AICRP (Potato) during last 10 years. The details of transfer of technology activities undertaken during the last one decade are given below:

Centre	*Extension Activities*
Bhubaneshwar	• On-farm trials on river bed potato cultivation in Mahanadi river area during 1994-95, 1995-96, 1996-97 and 1997-98. • On-farm trials were conducted every year in potato growing areas of different districts of Orissa.
Chhindwara	• 40 demonstration trials were conducted in farmers' fields on improved potato production technologies.
Deesa	• Farmers' day cum Exhibitions were organized at PRS Deesa in 1998, 2000 and 2001 which were attended by 2500 farmers from different parts of the state. • 13 group meetings and seminars were organized in potato producing areas which were attended by more than 5000 farmers.
Dharwad	• 41 on-farm trials were conducted on improved technologies of potato production and storage. • On-campus consultancy was provided to 327 farmers on the improved potato production technologies. • 13 Krishi Melas cum potato exhibitions were attended by 33,000 farmers.
Hassan	• 15 farmers' days/group discussions were attended by 1800 farmers; 10 on-campus and 50 off-campus training programmes as well as four potato exhibitions organinsed; 12 TV programmes telecasted; 150 scientific advisory field visits made; 250 advisory services provided to the farmers who visited the centre.
Hisar	• State level Farm Darshan and Kisan Mela were organized during 1995-2006 in the months of September and March, respectively. During these events farmers from Haryana and neighbouring states were guided about the improved potato production technologies.
Jalandhar	• Lectures were delivered on potato varieties and production and protection technologies in the potato seminars organized by KRIBHCO/ KRISHCO in 1997, 1999 and 2001.
Jorhat	• Two field days organized in March 2004 and 2005, which were attended by about 230 farmers.· Demonstration trials on potato production through TPS were conducted to educate farmers and agricultural officers. • On-farm trials on improved potato hybrids/varieties (3-4 trials every year) were laid out in different locations of Upper Brahmaputra Valley Zone of Assam.
Kalyani	• Nine demonstration trials were conducted in Hoogly district of West Bengal to educate farmers about improved potato varieties, production technologies, control measures of important potato diseases and use of TPS technology for potato production.
Kota	• Improved potato production technologies were disseminated through TV programmes and deliberations during farmer's days and Kisan Melas organized by ARS, KVK, CFCL, DSCL, Hariyali and Joint DAG, Kota.
Raipur	• On-farm trials conducted every year in the farmers' fields.
Srinagar	• On-farm/minikit demonstration trials and deliberations in the Kisan Goshthis, Kisan Melas, World food days and monthly TV workshops were effectively used to popularize improved potato varieties, crop production/protection technologies. • Field training and deliberations on Radio/TV covering agronomic practices, and management of diseases/pests were organized for farmers and field functionaries of Department of Agriculture during the past couple of years.

In addition, 30 extension bulletins/pamphlets in local languages were published by different centers of AICRP Potato.

IMPACT OF AICRP POTATO

Since the establishment of AICRP (Potato) in 1970-71 (IV Plan), it became possible to undertake multilocation testing of production technologies of potato in a coordinated manner in distinct agro-climatic regions. Use of new high-yielding varieties and improved production technologies made available through coordinated research under AICRP (Potato) has helped the farmers in different agroclimates to manage biotic and abiotic stresses effectively and enabled them to reap good dividends. The AICRP (Potato) had been instrumental in:

- Recommending 25 high-yielding potato hybrids for commercial cultivation in different agroclimates of the country.
- Early maturity, resistances to biotic/abiotic stresses and processing quality attributes in some of these varieties enabled diversification of potato cultivation in the country.
- Development of region-specific production technologies comprising integrated crop, nutrient and water management; remunerative potato based intensive cropping systems; use of low inputs, biofertilizers and biodynamics; development of agro-techniques for non-traditional areas and seasons; late blight forecasting models; replacement of organo-mercurial compounds by boric acid; development of eco-friendly IPM schedules to address region-specific pest problems; and improvisation of on-farm storage structures are likely to further improve the total factor productivity (TFP) of potato and foster socio-economic upliftment of resource poor farmers throughout the country.
- Concerted efforts have also been made under the project to transfer these technologies to end users through elaborate extension activities and strong linkages with the clientele.
- Increase in potato area and production has also facilitated emergence of potato based industries generating on-farm and off-farm employment in the country.

The above developments led to self-sufficiency in potato production and increase in area, production and productivity of potato in the country. Averaged over the last five years (2001–05) and as per FAO statistics, India with production of 23.27 million tonnes of potatoes from 1.27 million ha globally ranked 2nd and 3rd (after breakdown of Russian Federation) for potato production and area, respectively. Benefits reaped by the farmers/consumers and in turn by the nation as a whole due to the coordinated research under AICRP (Potato) are briefly described below:

Self-Sufficiency in Potato Production

India was a potato importing country until 1951. Our neighbouring countries continue to import potatoes from Europe even today. India has become self-sufficient in this respect, mainly due to the indigenous development of potato varieties, production and protection technologies for the diverse agroclimates in the country. The AICRP (Potato) has played a very significant and pivotal role in the development of improved potato varieties, production and protection technologies under its multilocation testing programme through a network of AICRP centres located in distinct agro-ecoregions of the country. This has also strengthened the establishment of National Seed Production Programme in the country, which alone is saving at least US $484 million equivalent to Rs 1936 crore annually. The indigenously produced tuber seed of potato varieties is available to the Indian farmers @ Rs. 5000–7000, i.e., US $139–194 per tonne as against US $ 1120–1200 per tonne for imported seed in the neighbouring countries.

Growth in Area, Productivity and Production of Potato

More than 95% of potato related research in the country is conducted by the CPRI and AICRP (Potato). The impact of development of improved varieties and region specific production/protection technologies by above units had been phenomenal in all aspects of potato production in the country. During the last five-and-a-half decades, potato crop has witnessed 5.59-fold increase in area, 15.13-fold increase in production and 2.70-fold increase in productivity in India (Fig. 2.9).

Increase in area, production and productivity of potato in different states during 1970-71 to 2004-05 is given in Table 2.37. At national level the % increases in area, production and productivity were 174, 392 and 80, respectively. However, these increases were not uniform across the country. The area under potato increased from (-67%) in Delhi to 882% in Gujarat. In Delhi, reduction in area can mainly be attributed to urbanization, however, in Himachal Pradesh (-25%), Orissa (-62%) and Tamil Nadu (-57%) the reduction may be due to economic non-viability of potato or adoption of more profitable crops other than potato. Potato production increased from (-88%) in Delhi to 1828% in Andhra Pradesh. The reductions in Delhi (-88%) and Orissa (-65%) were due to reduction in area and reduction in productivity of this crop in these states. Reduced potato production in Bihar (-5%), Chhatisgarh (-8%) and Uttarakhand (-28%) were due to reduction in productivity in these states, while in Tamil Nadu (-18%) the reduction was due to reduction in area under potato. Increase in productivity across the country (-63% in Delhi to 286% in Andhra Pradesh) was less than increase in area and production. In

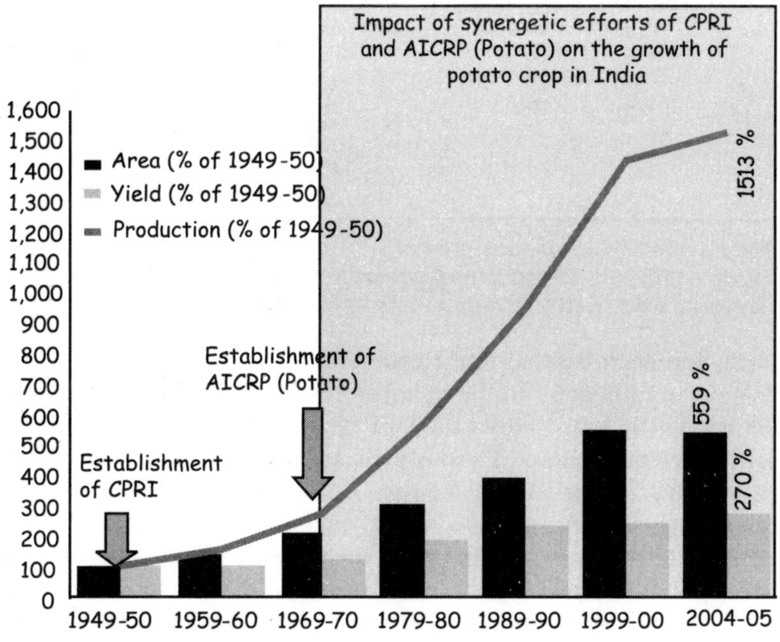

Fig. 2.9. Increase in area, production and productivity of potato in India

order to achieve uniform development of potato across the country, efforts are needed to improve its productivities in low-yielding states.

Table 2.37. Increase (%) in area, production and productivity of potato in different states during 1970-71 to 2004-05

State	*Increase in Area (%)*	*Increase in Production (%)*	*Increase in Productivity (%)*
Andhra Pradesh	400	1828	286
Arunachal Pradesh*	133	268	(-) 10
Assam	198	432	78
Bihar	37	(-) 5	(-) 31
Chhatisgarh*	23	(-) 8	(-) 26
Delhi	(-) 67	(-) 88	(-) 63
Gujarat	882	1035	16
Haryana	300	407	27
Himachal Pradesh	(-) 25	85	146
Jammu & Kashmir	8	141	122
Karnataka	526	894	59
Maharashtra	28	38	8
Manipur*	30	35	3
Meghalaya	25	115	73
Mizoram	150	178	11
Madhya Pradesh	192	285	32
Nagaland*	294	578	72
Orissa	(-) 62	(-) 65	(-) 6
Punjab	285	519	61
Rajasthan	138	971	349
Sikkim*	159	194	14
Tamil Nadu	(-) 57	(-) 18	90
Tripura	96	400	155
Uttar Pradesh	172	561	143
Uttarakhand*	4	(-) 28	(-) 31
West Bengal	410	671	51
All India	174	392	80

Source: Calculated from the data of DES, MoA, Govt. of India

* The % increase is between 1973-74 to 2004-05 for Nagaland and Manipur; 1980-81 to 2004-05 for Arunachal Pradesh; 1981-82 to 2004-05 for Sikkim; and 2000-01 to 2004-05 for Chhatisgarh and Uttarakhand.

National Economy, Industrialization and Employment Generation

Potato contributes significantly to the national economy. Its contribution can be judged from the fact that in 1995–96 it was grown on 0.6% of total cropped area and contributed 1.8% to the total value of agricultural output. Thus, with increase in area and production the economic contribution by potato has increased over the time.

Potato is a labour intensive crop and generates considerable employment during crop season as well as during post-harvest handling. In one of the experiments conducted by the All India Coordinated Agronomic Research Project, it was observed that person-days requirement per hectare in different crop sequences was highest when potato was included in crop rotation. It ranged from 313 person-days/ha in rice-potato-okra to 488 person-days/ha for maize-potato-green gram rotations.

In recent years demand for processed potato products in India and international market has risen at a fast pace due to increased urbanization, rise in per capita income, increase in number of working women and expanding tourism. With increase in potato production in the country and availability of suitable raw material, spurt in potato processing industries in India is visible during the last one decade. In addition to value addition, these industries also generate employment at various levels. Large quantities of potatoes are also processed at village level and provide a source of income for rural masses.

Increased potato production has also generated several other economic activities in the post harvest handling through cold storage, transportation and marketing.

Eco-Friendly Technologies

Since potato is high input crop, AICRP (Potato) had always been concerned for developing eco-friendly production and protection technologies. Evaluation of potato genetic resources is aimed to identify and use host resistances against major biotic and abiotic stresses. Several varieties having such resistances have been recommended for commercial potato production using only need based application of pesticides.

Region specific production technologies using integrated nutrient and water management, intercropping, crop rotations, biofertilizers, growth promoting bacteria and biodynamics have not only reduced use of chemical fertilizers but also made potato cultivation economically viable. The potato based intensive cropping systems have also enabled vertical expansion of area under potato cultivation.

Integrated pest management technologies developed under AICRP (Potato) have been extensively adopted by the farmers leading to minimal use of pesticides, reduction in yield losses and in some cases eradication of endemic pest problem like in case of wart in north Bengal hills. Possibility of using bio-control agents for control of major potato diseases and insect pests have opened up new avenues of employment generation among educated rural youth. Substitution of organo-mercurial compounds by boric acid for control of tuber borne diseases has been applauded world over.

Expansion of Potato Cultivation in Non-Traditional Areas

The identification of genotypes suitable for cultivation under rainfed condition in *kharif* season and release of heat tolerant variety Kufri Surya (Hybrid HT/92-621) has further opened new avenues for diversification of potato cultivation in non-traditional seasons and warmer areas.

Identification of New Areas for Production of Healthy Seed Potatoes

CPRI and AICRP-Potato in the joint surveys have identified new areas for production of quality potato seed. It was also observed that farmers can retain their own seed with little loss in productivity by following the seed plot technique for up to five years in the North-western plains as well as in the central plains and up to four years in Gujarat and West Bengal. In addition to these, the new areas identified for potato seed production are Hisar, Faizabad, Kanpur, Deesa and Kota.

In recognition of above achievements, the project was awarded Chaudhary Devi Lal Outstanding AICRP Award, 2005 by the Indian Council of Agricultural Research (Fig. 2.10).

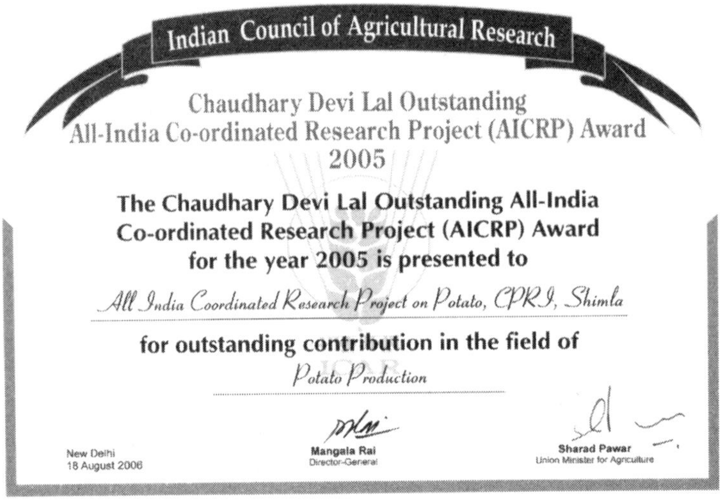
Indian Council of Agricultural Research

Chaudhary Devi Lal Outstanding
All-India Co-ordinated Research Project (AICRP) Award
2005

The Chaudhary Devi Lal Outstanding All-India
Co-ordinated Research Project (AICRP) Award
for the year 2005 is presented to

All India Coordinated Research Project on Potato, CPRI, Shimla

for outstanding contribution in the field of

Potato Production

New Delhi
18 August 2006

Mangala Rai
Director-General

Sharad Pawar
Union Minister for Agriculture

Fig. 2.10. Decoration of AICRP (Potato) with Chaudhary Devi Lal Outstanding AICRP Award 2005

MAJOR RECOMMENDATIONS

Major recommendations brought out under AICRP (Potato) since its inception in 1970-71 in different years are given below. Some of the herbicides and pesticides recommended earlier have been banned in India and old varieties of various crops have been replaced by new improved varieties, readers are therefore requested to check these while recommending or following the recommendations.

Potato Improvement

- Among the new cultures bred at CPRI, Shimla, A-7416 and C-3804 have shown sufficient promise for release. A-7416 is early in maturity and is suitable for both *rabi* and *kharif* seasons in the Deccan plateau. C-3804 combines frost tolerance with good keeping quality in ordinary stores at room temperatures and gives high yields in the tarai areas. These may be recommended to CVRC for release (1971)
- Results of varietal trials in hills have confirmed the superiority of newly released variety, Kufri Jyoti, as an early-maturing variety with consistently high yield and high degree of field resistance to late blight both in foliage and tubers. It has all the desirable characteristics. The potato areas in northern hills in Himachal Pradesh and Uttar Pradesh where late blight is an annual feature may be saturated with this cultivar as early as possible (1972).
- Kufri Chandramukhi has established its superiority as an early-maturing cultivar both in the PYTs conducted in seven locations and in the URTs in 15 locations. It is, therefore, recommended as an ideal variety for multiple cropping in plains of India. For raising of seed or a main season crop, it should be given prophylactic sprays with Mancozeb (2 kg/ha) against late blight (1973).
- On the basis of dry matter content and yield, culture JF-4870 is recommended for release. Data on all aspects of this culture may be collected by all the CPRI divisions. In future,

the Institute should have a Varietal Evaluation Committee which may put its recommendations for consideration at the workshop (1978).

- Hybrid, E-3797, recommended for release under the name Kufri Bahar, (1979).
- Three new cultivars recommended for release by CVRC. They are (i) Kufri Sherpa: resistant to wart and late blight gives good yields in Darjeeling hills and plains of West Bengal also suitable for NE hills and plains; (ii) Kufri Himalini: possesses high resistance to late blight and gives higher yield than Kufri Jyoti in the hill areas particularly in the North-eastern hills, and (iii) Kufri Lalima: suitable for Uttar Pradesh, Bihar, West Bengal, and Orissa where red tubers are preferred. Being a medium maturing cultivar, Kufri Lalima matures earlier than Kufri Sindhuri, hence fits better in the multiple cropping system (1981).
- SON-110: Suitable for Nilgiri and Kodaikanal hills, where cyst nematodes and late blight are major problems. It possesses resistance to both of them and gives about 30% higher yield than the commercial cultivar. Short tuber dormancy in the culture is considered advantageous for the area as three crops are grown in a year here. (1985)
- E-4486: A main crop culture, produces high yields and has wide adaptability in the Indo-Gangetic plains particularly in Haryana, Uttar Pradesh, Bihar and West Bengal also in Gujarat and MP; gives very good yields in areas where growing period is slightly longer; produces white, round-oval tubers of medium to large size. (1985)
- JH-222: Suitable for Punjab, Haryana, UP, Bihar, Madhya Pradesh and northern plateau regions; medium early in maturity, has a fast bulking rate and produces regular medium size tubers of white colour and pale yellow flesh, possesses a fair level of resistance to late blight, does not suffer from early blight and phoma diseases in the field; shows slow rate of degeneration; possesses good keeping quality; suitable for chipping. (1985, 1990)
- JI-5857: Recommended for release for Punjab, Haryana, Uttar Pradesh, Bihar and Madhya Pradesh. Appropriate simultaneous action for priority multiplication of disease-free stocks of this hybrid may be taken up by CPRI. (1991)
- It was recommended that the following 17 varieties may be denotified by CVRC. Kufri Alankar, Kufri Chamatkar, Kufri Dewa, Kufri Jeevan, Kufri Khasigaro, Kufri Kisan, Kufri Kuber, Kufri Kumar, Kufri Kundan, Kufri Muthu, Kufri Naveen, Kufri Neela, Kufri Neelamani, Kufri Red, Kufri Safed, Kufri Sheetman and Kufri Sherpa (1991).
- HPS-I/13 and TPS C3: It was recommended that two TPS families which have shown high yield in trials in Bihar, MP and Gujarat should be released so that a beginning could be made to introduce TPS technology of crop production in the identified priority areas (1991).
- The hybrids PJ-376 and JEX/C-166 were recommended for release. The former culture is suitable for early crop in the entire Indo-Gangetic plains and yields about 23–28 t/ha at 75 days, whereas, the latter is suitable as a main crop variety in the states of Punjab, Haryana, Uttar Pradesh, Rajasthan, Bihar, Madhya Pradesh and West Bengal and yields on an average 30 tonnes/ha at 90 days after harvesting. The proformae for their release will be submitted by the CPRI for consideration of the Central Variety Release Committee (1993).
- NW/E hills are in urgent need of a variety to replace Kufri Jyoti, released in 1964 which has now become susceptible to late blight. After carefully considering the data on yield, resistance to late blight and other ancillary characters, the group recommended hybrid SM/85–45 for an early release. The yield potential of this hybrid is about 26 tonnes/ha

as against 18 t/ha of Kufri Jyoti. The hybrid is suitable for entire hilly region (1997).

- In view of better yield performance of the hybrid D/79-56 than presently popular varieties, viz., Kufri Swarna and Kufri Jyoti in southern hills, and its resistance to cyst nematodes and late blight, this hybrid may be recommended for release. This will help in reducing damage due to cyst nematodes and late blight, besides further improving potato production in southern hills. The yield potential of D/79-56 is about 33 tonnes/ha in nematode infested soils as against 20 tonnes/ha of Kufri Swarna and 15 tonnes/ha of Kufri Jyoti. In nematode free soils, it yields about 39 tonnes/ha as against 24 tonnes/ha of Kufri Swarna and 23 tonnes/ha of Kufri Jyoti. The hybrid D/79-56 is recommended for southern hills (1997).
- Four hybrids suitable for processing were introduced for evaluation of tuber dry matter and chip colour. Of these, two hybrids, viz., MP/90-83 and MP/91-G have consistently performed well at Modipuram during 1993–97 and at Patna during 1996–97. Both these hybrids gave 2-4% higher tuber dry matter and excellent chip colour with yield at par with Kufri Jyoti, the popular variety used for chipping. In view of the immediate requirement of the processing industry and recommendations of QRT & RAC, these two hybrids were recommended for an early release. The hybrids MP/90-83 and MP/91-G have given 21.5% tuber dry matter on an average as against 18.7% of Kufri Jyoti. These hybrids are suitable for western Uttar Pradesh and Bihar states. The yield potential of these hybrids is about 32.5 tonnes/ha as against 31.6 tonnes/ha of Kufri Jyoti. Both of these hybrids possess a high degree of resistance to late blight (1997).
- TPS population 92-PT-27 allows TPS production in the plains without the use of additional light. The yield of this population is better than the standard variety in the eastern region. It produces more uniform tubers and better marketable yield than the existing TPS populations. This population may be recommended for release for commercial cultivation in eastern regions (2001).
- Hybrid PS/F-220 consistently showing resistance to potato stem necrosis (PSND) at Kota, a hot spot for the disease, is recommended for registration as elite germplasm line with NBPGR, New Delhi and for future use as a parental line in breeding programme (2003).
- Four hybrids given below were recommended for release as commercial varieties following in depth discussion in the group meeting (2003).

Hybrid	*Characteristics*
JW-160	A high-yielding medium maturing white tuber hybrid having field resistance to late blight and excellent keeping quality under ambient storage conditions is suitable for cultivation in plains and plateau regions of the country.
MS/92-2105	A red skinned high yielding hybrid with oval attractive tubers having field resistance to late blight is identified for cultivation in the Indo-gangetic plains and suitable for replacement of late blight susceptible varieties Kufri Sindhuri and Kufri Lalima.
SM/87 -185	A late blight resistant white tuber hybrid having higher tuber dry matter and better keeping quality over cv. Kufri Giriraj for cultivation in hilly regions of the country so as to avoid mono culture of Kufri Giriraj following breakdown of resistance for late blight in Kufri Jyoti.
HT/92-621	A white tuber heat tolerant hybrid suitable for cultivation under early planting conditions in the northern plains and warmer areas of the country having resistance to leafhopper and mites. In addition the hybrid is having high dry matter and is suitable for processing as French fries.

- Two hybrids given below have been recommended for release as commercial varieties after the detailed discussion in the group meeting (2005).

Hybrid	*Characteristics*
MP/97-583	Based on the performance over locations and processing quality (assessed by M/s. Frito-Lay), the hybrid MP/97-583 has been found to have high yield, high tuber dry matter and excellent chip color. It also produces acceptable quality chips after storage at 10-12°C up to 180 days and produces more processing grade tubers than the existing varieties Kufri Chipsona-1 and Kufri Chipsona-2. This hybrid is recommended for release in Indo-Gangetic plains, Kota and Indore region, under the name Kufri Chipsona-3.
SM/91-1515	Based on the over all performance over locations and better resistance to late blight than all available varieties, good tuber shape and no tuber cracking, the group meeting also recommended release of hybrid SM/91-1515 for cultivation in north western and north-eastern hills of the country.

- Two hybrids J/93-86 and SM/93–237 were recommended for release as varieties (2007).

Hybrid	*Details and Salient Features*
J/93-86	Hybrid bred at CPRS, Jalandhar by Drs Raj Kumar and GS Kang was introduced under AICRP in 2001. This hybrid was evaluated at 16 AICRP centers located in the plains at Bhubaneshwar, Chhindwara, Deesa, Dharwad, Dholi, Faizabad, Hassan, Hisar, Jalandhar, Jorhat, Kalyani, Kota, Modipuram, Pantnagar, Patna and Raipur for 5 years before recommending for release.J/93-86 is an early maturing white tuber hybrid with moderate resistance to late blight. This hybrid yields higher than existing early cultivars both at 60 and 75 days after planting. This hybrid is recommended for entire Indian plains.
SM/93-237	Hybrid bred at CPRI, Shimla by Drs TA Joseph, BP Singh, SK Kaushik, PH Singh and Vinay Bhardwaj was introduced under AICRP in 2002. This hybrid was evaluated at 5 AICRP centers located in the hills at Kufri, Ooty, Ranichauri, Shillong and Srinagar for 4 years before recommending for release.SM/93-237 is a medium maturing white tuber hybrid with high level of resistance to late blight. It yields higher than existing cultivars for hills. This hybrid is recommended for all Indian hills.

Potato Production

- The large-sized tubers from otherwise healthy crop can be cut into 50 g pieces with minimum two eyes in each piece and used as seed. The cut tubers may be treated with Mancozeb or Difolatan (500 g in 250 litre water for 10 minutes), kept in basket to drain out the excess solution and covered with wet gunny bag for 24 hrs. for suberization of cut surface. The suberized seed potatoes may then be planted after one or two days. The cost of the treatment may come to Rs. 5 for treating 100 kg seed tubers (1972, 1973).
- The beneficial effects of mulching on the early crop, when temperatures are high, have been established. The practice is also advantageous in that it would economise on water use, control the weeds, improve the response to phosphate, and help in germination in the cut seed potatoes after treatment with Mancozeb particularly when freshly harvested seed is used after breaking dormancy in early part of autumn (1972).
- The performance of hill seed used for planting by the middle of November is generally improved by the use of thiourea and GA for breaking tuber dormancy. This practice is

recommended in Karnataka, West Bengal, Orissa and Maharashtra, where potatoes are planted in November (1974).

- Mulching with paddy straw, paddy husk, maize and Dhaincha stalks consistently increase the yield of early crop. This practice economizes on the use of water and also reduces weed infestation (1974).
- Pre-emergence herbicides like Nitrofen, Alachlor and Simazine in areas where potato is not followed by wheat, and post-emergence herbicides like Paraquat and Propanil effectively control weeds in potato crop. This recommendation is of significance for raising seed crop as it can reduce spread of mechanically transmitted viruses (1974, 1976, 1977).
- Seed potatoes from cold storage from the northern states are as good or even superior to hill seed for raising winter crop in peninsular India, Orissa and West Bengal. Efficacy of hill seed which may be partially dormant at the time of planting in the above region could be improved through treatment with dormancy breaking chemicals. (1976).
- Seed potatoes need light irrigation (5 cm deep) at frequent intervals (15–20 mm cumulative pan evaporation) at Jalandhar and Kanpur. The intervals between irrigations could, however, be increased with the use of mulch especially during the pre-emergence period (1976).
- In case of main crop varieties and crops raised for table purposes, split application of N, half each at planting time and earthing up should be practised for getting maximum yields (1977).
- For obtaining maximum yields, medium plant population of about 67,000 plants/ha with a seed size of 4.5-5.5 cm diameter is recommended (1979).
- Use of FYM in the potato crop @ 300 q/ha can meet the nitrogen requirement of rotational crop grown after potato (1979).
- Use of Alachlor @ 3.0 litre/ha, Nitrofen @ 4.0 litre/ha and Patoran @ 2.5 kg/ha is effective in controlling weeds (1981).
- The optimum intra-row spacings for small (25 g), medium (50 g) and large size tubers (100 g) are 15, 20 and 30 cm respectively when inter row spacing is kept at 60cm (1983).
- Soaking of seed tubers in a solution of 3% single super phosphate + 0.5% urea + 0.2% Mancozeb for 4 hours in combination with soil application of 50 kg P_2O_5/ha can meet the full phosphate requirements of the potato crop in P deficient soils. The tubers should be dried before planting (1983).
- Whole tubers were found to be more productive than tubers cut to increase plant population per unit area. Hence, whole tubers may be used for planting the crop (1989).
- It was recommended that 25–50 g seed tubers may be used @ 30-37.5 q/ha for planting potato crop (1989).
- Potato can be successfully intercropped with wheat at Ranchi and onion for seed at Hisar (1989).
- It was recommended that herbicides Metribuzin (Sencor) as pre-emergence @ 1 kg/ha or Paraquat (Gramaxone) @ 2.0 lt./ha as post-emergence (when only 2–5 % crop emergence is recorded) may be applied to control weeds in potato crop (1989).
- At Hisar, second to third week of October and at Kota last week of October to first week of November were found to be the best time for planting potato. These periods were recommended for adoption in respective areas (1991).

- The optimum seed rates for JH-222 for 15, 30 and 45g seed tubers should be 20, 25 and 30 q/ha, respectively. The optimum seed rate for variety Kufri Badshah should be 22.5, 30.0 and 37.5 q/ha respectively (1991).
- The optimum fertilizer requirements for the potato crop at Kanpur (central Uttar Pradesh), Kota (western Rajasthan) and Deesa (north Gujarat) were found to be as under:

Location	*Variety*	*Fertilizer (kg/ha)*		
		N	P_2O_5	K_2O
Kanpur	Kufri Badshah	100	100	100
Kota	Kufri Bahar	128	100	100
Deesa	Kufri Badshah	240	200	150

It was recommended that above fertilizer doses should be used in respective regions for obtaining high crop yields (1991).

- When urea is to be used as a source of nitrogen in potato crop, the application of half the recommended nitrogen dose should be applied by broadcasting on moist soil followed by tillering and planking of the field 2-3 days before planting. The remaining half dose of urea should be applied as top dressing at earthing up time followed by light irrigation next day (1991).
- At Faizabad, use of 50% of recommended seed rate (i.e., 15 q/ha and 50% of recommended fertilizer dose (i.e., 75 kg N, 40 kg P_2O_5 and 50 kg K_2O) gave maximum net returns from the potato crop, while at Ranchi, 75% of the recommended seed rate (i.e., 30 q/ha) and full fertilizer dose (150 kg N, 80 kg P_2O_5 and 100 kg K_2O) were found economical. These were recommended for potato production under low input conditions (1991).
- At Kanpur (central Uttar Pradesh) intercropping of potato with linseed (oil seed) was found to be compatible and remunerative. Intercropping of this crop is recommended for the region to get high returns per unit area (1991).
- Following potato based crop sequences were found to be most remunerative and were recommended for adoption in respective regions (1991).

Centre/Region	*Rotation*
Chhindwara (Satpura plateau)	Potato – okra – soybean
Deesa (northern Gujarat)	Potato – bajra – groundnut
Hisar (Haryana)	Potato – tomato – okra
Jorhat (Assam plains)	Potato – moong – paddy (Transplanted)
Kanpur (central Uttar Pradesh)	Potato – bajra – groundnut
Ranchi (Chhota Nagpur)	Potato – moong – paddy

- Fluchloralin (Basalin) @ 1 kg a.i./ha as pre-planting and Oxyflurofen (Goal) @ 0.15 kg a.i./ha as pre-emergence were found to be effective herbicides and were recommended for weed control in potatoes (1991).

- For getting high yields in Deesa (North Gujarat), the potato crop should be planted in the first fortnight of November using 300 kg N/ha. If the planting is delayed by another 10 days the N should be applied @ 240 kg/ha (1993).
- Inter-cropping of mustard with potato (one row of mustard after every third row of potato) may be practiced in eastern Uttar Pradesh to get better returns than from the pure crops of mustard and potato. This will fetch about Rs. 3000 per ha more than from the pure mustard crop and about Rs. 4,600 per ha more than the pure potato crop (1993).
- Inter cropping of onion (for bulb) and potato, when onion is planted in furrows between two potato rows, can be practiced in Satpura plateau to get better returns than from the pure crops of onion and potato. The inter cropping will fetch about Rs.10,500/ha more than from the pure potato crop and about Rs. 2,000/ha more than pure onion crop (1993).
- It was recommended that in Punjab, potato crop should be planted during first fortnight of October for getting highest yield as well as net returns. It holds good for JH-222 which should be fertilized with 240 kg N/ha (1995).
- At Kota, the potato crop should be planted in the 1st week of November and fertilized (Kufri Badshah) with 150 kg N/ha, in the first week of November in Orissa and fertilized (Kufri Chandramukhi) with 120 kg N/ha and in the first fortnight of January in Palampur area of HP and fertilized (Kufri Jyoti) with 200 kg N/ha was found optimum (1995).
- At Jorhat 30 g seed size planted at the rate of 20 q/ha was found more economical than 40g seed size (1995).
- At Palampur (Himachal Pradesh), potato (spring)-maize-radish system was found to be highly economical than any other systems tried and was recommended for adoption in the region (1995).
- In the potato-okra-soybean system in Satpura plateau it was recommended that the potato should be fertilized with recommended dose of NPK and the crops following potato, i.e., okra and soybean should be fertilized with 50% dose of recommended NPK (1995).
- It was recommended that in Satpura plateau region the TPS crop from transplanted seedlings should be fertilized with recommended NPK dose (120:100:100 kg/ha and the seedlings should be transplanted on flat bed, rather than on the ridges. However, both in Satpura plateau and the tarai region of Uttar Pradesh the crop raised from seedling tubers should be fertilized with 150% of recommended dose using 10g size of seedling tubers for getting economical returns (1995).
- Recommended that in the event of shortage of potassium fertilizer at the time of planting potato crop, it can be applied half at planting and remaining half at earthing up. However, in general, the full dose of potash should be applied at the time of planting (1997).
- Recommended adoption of potato-jute-paddy sequence in Kalyani area of West Bengal in view of its giving higher net returns over wheat-jute-paddy system. In this system, variety Kufri Badshah was used for potato, JRO-524 for jute and IET/47-56 of paddy (1997).
- Recommended in the Satpura region of Madhya Pradesh inter- cropping of maize (after every row of potato with full N and half of PK at potato planting) in *rabi* along with potato giving optimum doses of NPK. This system will give additional returns to the farmers over sole crop of maize (1997).
- Recommended that in Jalandhar area of Punjab the net gains from potato production can

be enhanced with increase in fertilizer dose from 100 to 150% of recommended dose and reducing the seed size from 35–40g to 25–30g in variety Kufri Badshah by keeping the seed rate from 15 to 30 q/ha (1997).

- Recommended that at Deesa, Faizabad and Hisar, for obtaining good yields from the TPS crop raised from seedling tubers it may be fertilized with higher fertilizer dose (1.5 times higher than that recommended for the tuber crop). However, at Pantnagar, Chhindwara and Kalyani the TPS crop raised from seedling tubers may be fertilized with the same NPK dose as recommended for tuber crop (1997).
- Green manuring with Dhaincha to potato in green manure-potato-bottlegourd at Hisar was found beneficial not only in increasing the potato production but also gave more remunerative returns from the system. It is, therefore, recommended that the above potato based system may be popularized with Dhaincha as green manure with optimal recommended doses of fertilizers to each crop in the system as followed in Hisar area (1999).
- The potato based cropping system of potato-rice (summer)-rice (*kharif*) in Assam plains was not only found more remunerative than potato-moong-rice system, but it also saved on application of fertilizer. In this system, summer rice (second crop) grown after potato required 50% N only while *kharif* rice (third crop) required optimum NPK application compared to application of optimal dose of NPK to all the three crops in potato-moong-rice system. The above potato based cropping system is, therefore, recommended for adoption in the Assam plains (1999).
- For both the transplanted and seedling crops raised from true potato seed (TPS), application of 200 kg N, 100–150 kg P_2O_5 and 120–180 kg K_2O/ha is recommended for Bhubaneshwar, Jorhat, Chhindwara, Deesa, Faizabad and Dharwad areas. However, yield potential of transplanted seedling crop will be lower than the seedling tuber crop (1999).
- Following potato, wheat and rice varieties were identified and recommended for cultivation in rice-potato-wheat system at Jalandhar (Punjab), Hisar (Haryana), Modipuram (Western Uttar Pradesh), Faizabad (Eastern Uttar Pradesh) and Patna (Bihar) (2001)

Location/Region	*Potato*	*Wheat*	*Rice*
Jalandhar (Punjab)	K. Pukhraj	PB-373	PR-106
Hisar (Haryana)	K. Ashoka	PB-373	Gobind
Modipuram (Western UP)	K. Pukhraj	PBW-226	Saket
Faizabad (Eastern UP)	K. Ashoka	Malviya-234	Narendra-97
Patna (Bihar)	K. Ashoka	PBW-226	Proagro-6107

- At Deesa, potato-groundnut-til was better than potato-bajra-groundnut in terms of productivity. Potato and bajra should be fertilized with optimum NPK, while groundnut needs only half NPK (2001).
- In potato-onion-groundnut and potato-wheat-cowpea systems at Jalandhar, it was found that crop residue incorporation of each crop to succeeding crop could economize on NPK by 50% in both the succeeding crops after potato. In potato-onion-groundnut system, both potato and onion should be fertilized with optimum NPK while groundnut with

half NPK. In potato-wheat-cowpea system, potato should be fertilized with optimum NPK while succeeding crops of wheat with full N and cowpea with half NPK (2001).

- For maximizing yield of potato, doses of NPK need to be revised upwardly by 50% of the existing recommended doses at Jalandhar, Modipuram, Patna. Faizabad. Deesa. Chhindwara, Hisar, Kota and Dholi (2001).
- For maximizing the number of seed sized tubers, crop should be planted at 60cm × 10/15cm spacing and be raised with 150 kg N, 150 kg P_2O_5 and 80 kg K_2O/ha and haulm be cut at 80 days. About 80% of the produce was upto 40 g size at Hisar. Pantnagar, Faizabad, Deesa, Kota. Bhubaneshwar, Chhindwara, Jalandhar and Patna. The number of tubers produced by following this agro-technique was 40% higher than those produced by seed plot technique (2001).
- Intercropping of potato + garlic in the ratio of 1:1 at Ranichauri in Uttarakhand and Srinagar in Jammu and Kashmir and in the ratio of 1:2 at Dharwad in Karnataka was found profitable. It is recommended that the above practice may be adopted for getting better economic returns to the farmers in comparison to growing sole crop of potato and garlic (2003).
- Northern Gujarat is an important potato growing area in the country. Maximum tuber yield as well as net returns were obtained with fertilization of the crop with 275 kg N, 137.5kg P_2O_5 and 275kg K_2O/ha along with 25 tonne FYM/ha. It is, therefore, recommended that the above dose of inorganic and organic fertilizers may be applied for getting better returns from potato crop in the state (2003).
- Seed is the most important and expensive component in potato cultivation and accounts for 40–60% of the cost of cultivation. Studies conducted at many centres in the plains concluded that the seed crop should be planted at 60 cm × 10/15 cm spacing and haulm cutting should be done at 70/80 days after planting for maximizing the number of seed size tubers (up to 40g size) (2003).
- It is recommended that potato based cropping system i.e. paddy-potato-okra, which gave higher productivity and net returns at Kalyani (West Bengal) may be adopted in West Bengal (2003).
- Maximum tuber yield as well as net return were obtained from fertigation of potato crop with 187.5 kg N; 125 kg P_2O_5 and 125 kg K_2O/ha + 25 tonne FYM/ha at Kota (Rajasthan). It is, therefore, recommended that the above dose of inorganic and organic fertilizers may be applied for better returns from potato crop in the southern Rajasthan (2003).
- For economizing on N and P input it is recommended that potato tubers should be soaked in a solution containing 1% each of urea and sodium bicarbonate along with seed treatment with Azotobacter + Phosphobacteria and application of 75% recommended dose of N and P (2005).
- In rice-potato-wheat crop sequence, it is recommended that harvesting of potato (cv. Kufri Pukhraj) should be done at 90 days after planting and 100% N recommended dose should be applied to wheat with full nutrients to potato and rice in Western Uttar Pradesh. A growing period of 90 days for potato is recommended for achieving optimum yields in Indo-Gangetic plains (2005).
- The crop sequences viz., rice-potato-okra at Bhubaneshwar in Orissa and blackgram-potato-okra at Jorhat in Assam recommended for higher productivity and net returns (2005).

- In Orissa at Bhubaneshwar, rice-potato-okra is the most profitable crop sequence followed by rice-potato-jute while in Dharwad areas of Karnataka, potato-beans is the most profitable crop sequence followed by potato-sorghum (2007).
- In Gujarat (Deesa area) only 75% recommended doses of nutrients through inorganic sources are needed for bajra in potato-bajra crop sequence, when residue incorporation of all the crops + biofertilizers along with 100% recommended doses of NPK in the form of inorganic fertilizers is applied to potato (2007).
- At Faizabad in Uttar Pradesh in rice-potato-wheat crop sequence, application of 100% recommended doses of NPK through inorganic fertilizers along with 20 tonnes FYM / ha to potato and only 75% recommended dose of nutrients to wheat grown after potato is required, while subsequent rice crop needed 100% recommended dose of nutrients in the sequence (2007).
- In Haryana at Hisar in potato-bottle gourd crop sequence, application of 20 tonnes FYM/ ha along with incorporation of crop residues and use of biofertilizers to potato is needed while subsequent bottle gourd requires only 75% recommended dose of nutrients (2007).
- At Modipuram in western Uttar Pradesh in potato-green gram sequence, application of 100% recommended dose of N + 20 tonnes FYM/ha gave higher yield of potato, while subsequent green gram in the sequence requires only 75% recommended doses of nutrients after this treatment (2007).
- In Dharwad area of Karnataka 20–30 June is the optimum planting time for potato (varieties Kufri Jawahar and Kufri Surya) for getting higher yields and economic returns. (2007).
- At Chhindwara, Hassan, Hisar, Kalyani, Ooty and Patna, the dose of P could be reduced to 75% with the use of phosphobacteria (2007).

Potato Protection

- The crop raised with susceptible cultivars in North Indian hills may be sprayed with Bordeaux mixture (5:5:50) for controlling late blight. The first spraying may be given 10–15 days before the normal time of appearance of late blight or earlier, if weather becomes wet. The subsequent sprays may be given at 10-day intervals giving in all 5 to 7 sprays (1972).
- For controlling black scurf, tuber treatment with OMC (0.5% solution for 10 min.) and soil application of saw dust (25 q/ha) or PCNB alone (30 kg/ha) is recommended. Saw dust is to be applied about three weeks before planting. Normal dose of fertilizers to be applied to the crop. This treatment will also take care of common scab (1974).
- For preventing decay of cut seed tubers, the seed pieces may be treated with Mancozeb (0.2% sol.) or Captafol 80 WP (0.25% sol.) for 10 minutes or Banoyl (500 ppm) for 30 min. before planting (1974, 1976).
- For late blight in the hills, Triphenyl-tin-acetate or Captafol 80 WP (2.5 kg/ha) may be sprayed. A sticker (Triton AE 0.1%) should be used with the sprays. In the plains, Mancozeb (2.0 kg/ha), Triphenyl-tin-acetate (1.0 kg/ha) or Copper oxychloride (2.5 kg/ ha) sprays are effective against late blight (1976).
- For controlling early blight in the hills, Bordeaux mixture or Triphenyl-tin-acetate is suitable, whereas, in the plains Mancozeb, Triphenyl-tin-acetate and Captafol are effective (1976).
- Cutworms are effectively controlled by application of Aldrin 30 EC @ 6.0 l/ha (1977).

- Application of Carbaryl 50 WP @ 2 kg/ha is effective against cutworms (1977).
- Kelthane 18 EC applied @ 2.0 l/ha is effective in controlling mites (1977).
- Application of Carbaryl 50 WP @ 2.5 kg/ha or Quinalphos 25 EC @ 1.5 l/ha checks the incidence of PTM. Carbaryl also controls *Epilachna* and other defoliating insects (1977).
- Three to four sprays of Mancozeb in the plains and 5–6 sprays in the hills @ 2.0 kg/ha in 1000 lit. water are effective in controlling the foliar diseases, specially late blight and early blight (1981).
- Dipping potato seed tubers in 1.0% solution of acetic acid + 0.05% solution of zinc sulphate for 15 min. is effective in controlling black scurf (1981).
- Application of Aldicarb 10 G or Carbofuran 30 @ 2-3 kg a.i./ha half each at planting time and earthing up is effective against potato cyst nematode and root knot nematode (1981, 1985).
- For controlling early and late blights of potato, sprays of copper fungicides in the hills and Mancozeb in the plains are recommended at regular intervals (1983).
- The incidence of common scab can be reduced by treating seed tubers with OMC solution @ 0.25% for 30 minutes before storage (1983).
- Application of Phorate 10 G @ 1.0 kg (a.i.) and systemic insecticides like Methyl demeton, Dimethoate and Monocrotophos recommended in seed crop against aphids (1983).
- Spraying of 0.1% Methyl-demeton is recommended to check shoot borer damage (1983).
- For the control of cutworm Endosulfan 5% dust @ 25 kg/ha or Chlorpyriphos 20 EC @ 0.5 kg a.i./ha are recommended (1983).
- Where potato tuber moth is a problem as in Maharashtra, dusting of seed potatoes during storage with Quinalphos @ 250 g/q of potato seed is recommended (1983).
- For the control of tuber borne inoculum of common scab and black scurf, seed tubers should be dipped in 3% boric acid (pharmaceutical grade) solution for 30 minutes. The same solution may be used for 20 dips (1987).
- For the control of brown rot and bacterial wilt through cultural practices, seed tubers should be obtained from a bacterial wilt and brown rot free crop, preferably from area free from disease. Stable bleaching powder @ 12 kg/ha should be mixed with fertilizer and applied in furrows at planting time. After harvest, the field should be ploughed and exposed to summer temperatures during May and June in the plains and winter temperatures in December and January in this hill. Early planting latest by third week of February and early harvesting latest by third week of June should be adopted in the hills (1989).
- For the control of late blight a spray of Metalaxyl + Mancozeb (0.25%) should be given to the crop on the appearance of disease, followed by a second spray at 15-day interval in the hills. Thereafter, Mancozeb sprays starting after 15 days should be given at 7–10 day intervals depending upon the prevailing weather conditions (1991).
- The spraying of tubers with boric acid (3%) is as effect as boric acid or OMC dip treatment of tubers. The cost of treating one quintal tubers with boric acid spray is Rs. 18 as against Rs. 48 for boric acid dip treatment and Rs. 20 with OMC dip treatment. Boric acid spray treatment of tubers after harvest and before cold storage is, therefore, recommended for the control of tuber borne diseases (1993).
- It was found that if the seed of cv. Kufri Jyoti and Kufri Badshah maintained following

the seed plot technique viz. (a) application of thimet @ 10 kg/ha at planting and (b) haulm cutting when the aphid population reaches critical level of 20 aphids per 100 compound leaves, it did not show significant degeneration for 3 years at Kalyani. It is, therefore, recommended to the farmers of West Bengal that they can keep their seed healthy through seed plot technique for 3 years (1993).

- It was found that if the seed maintained following the seed plot technique. viz. (a) application of thimet @ 10 kg/ha at planting and (b) haulm cutting when the aphid population reaches critical level of 20 aphids per 100 compound leaves, the variety Kufri Chandramukhi, showed significant degeneration after 3 years whereas, that of cvs. Kufri Bahar and Kufri Badshah did not show any yield reduction even in 4 years. It is, therefore, recommended to the farmers in Gujarat that they can keep their seed of cv. Kufri Chandramukhi, for at least 3 years and that of cvs. Kufri Bahar and Kufri Badshah for 4 years (1993).
- For reducing the incidence of stem necrosis disease planting of the main potato crop during optimum planting period, i.e., during first fortnight of November in Rajasthan and Madhya Pradesh is recommended. However, in early planted crop (October) three sprays of monocrotophos 40 EC @ 1ml/litre. at interval of 15 days are recommended for effectively reducing incidence of the disease (1995).
- Application of two sprays of metalaxyl + mancozeb combination @ 0.25% at 15 days interval followed by weekly sprays of mancozeb @ 0.2% is recommended for the control of late blight in the hills. For Punjab region, application of one spray of metalaxyl + mancozeb combination @ 0.25% followed by mancozeb spray at 10 day interval is recommended. For other regions use of mancozeb alone @ 0.2% is recommended (1995).
- For the management of late blight in Faizabad region varieties K. Lalima and K. Jawahar, should be sprayed with metalaxyl based fungicide (8% metalaxyl, @ 2.5 kg/ha immediately after the appearance of the disease. K. Sutlej does not require any fungicide application (1999).
- Bio agents namely *B. cereus* and *B. subtilis* have given consistently significant control of bacterial wilt and significant increase in tuber yield at Hassan. They may be recommended for field use against bacterial wilt control in Karnataka (2001).
- New and safer pesticides i.e. imidacloprid and thiomethophos are recommended to replace methyl-demeton/monocrotophos and phorate for control of potato pests such as whiteflies, aphids, cutworm and white grubs (2003).
- In view of promising results in the control of black scurf disease in potato it is recommended that *Trichoderma viride* a known bio-agent now available commercially may be used extensively for its management (2003).
- Regular pests of potato like *Agrotis* sp., jassids etc. can be managed by the trap/barrier crops. Marigold and castor are some of the important barrier crops for *Heliothis* sp. It is recommended that these trap/barrier crops may be used for the management of above pests (2003).
- Following IDM schedules were recommended for the management of major potato diseases in the respective regions (2005).

Area	Treatment	Disease
Haryana (Hisar area)	Seed treatment with boric acid (3%), 1 foliar spray of mancozeb 0.2% and 2 foliar sprays of Methyl-demeton (metasystox 25 EC) 0.03% conc.	Black scurf, viruses, sooty mould
Tarai region of Uttarakhand, Eastern UP and Bihar	Seed treatment with boric acid (3%) and 2 foliar sprays each of mancozeb 0.2% and Methyl-demeton (metasystox 25 EC) 0.03% conc.	Black scurf, late blight, viruses and common scab.
West Bengal and Gujarat	Seed treatment with boric acid (3%) two sprays each of mancozeb (0.2%) and Methyl-demeton (metasystox 25 EC) 0.03% conc. and optimum date of planting in Gujarat.	Late blight, early blight, common scab and viruses.
Orissa and Assam	Disease free seed, seed treatment with boric acid, *B. subtilis*, crop rotation, 2 sprays of Methyl-demeton (metasystox 25 EC) 0.03% conc. and 1 spray of mancozeb (0.2%).	Black scurf, viruses and bacterial wilt
Rajasthan	Planting at optimum time, tuber treatment with 3% boric acid, 2 sprays of monocrotophos (0.05% a.i.) / imidacloprid (0.004% a.i.).	Stem necrosis and black scurf.
Madhya Pradesh	Optimum planting time, tuber treatment with 3% boric acid, 2 sprays each of mancozeb (0.2%) and monocrotophos (0.05% a.i.)/ imidacloprid (0.004% a.i), crop rotation and seed treatment with *Bacillus subtilis.*	Early blight, brown rot, and viruses.

- Seed treatment with Imidacloprid @ 0.004% a.i. (10 min. dip) + 2 sprays of imidacloprid @ 0.002% a.i. starting with germination and at 25 days after planting is recommended for the management of whiteflies and apical leaf curl in Haryana and Punjab region (2005).
- Following recommendation is made for IPM of potato pest complexes in different areas (2005).

Centre/Area	Pest Complex	Recommendations
Kalyani (West Bengal)	Cutworm, mole cricket and epilachna beetle	Spraying with chlorpyriphos (Dursban 20 EC @ 0.05% conc. at earthing up followed by one spray each with Nimbicidine @ 0.4% conc. and Imidacloprid @ 0.002% conc.
Jalandhar (Punjab)	Defoliating caterpillars and cutworm	Spraying with chlorpyriphos (Dursban 20 EC @ 0.05% conc. at earthing up followed by one spray each with Nimbicidine @ 0.4% conc.
Patna (Bihar) and Deesa (Gujrat) Hisar (Haryana)	Leafhoppers and mites	Crop meant for table purposes may be protected with higher dose of Nitrogen i.e. 200 Kg/ha followed by one spray with Nimbicidine @ 0.4% conc.

- Treatment of cut seed pieces with 0.2% mancozeb + 0.2% benzimidazole for 10 minutes is recommended for the effective control of seed piece decay at Dharwad. This treatment also resulted in significant increase in tuber yield (2007).

- Delayed planting of seed crop by 10–15 days, seed treatment (dipping of chitted tubers for 10 minutes) with 0.04% (imidacloprid 200 SL 4 ml/10 litres) of imidacloprid plus spray of imidacloprid @ 60 g a.i./ha at emergence and another spray of thiamethoxam 25WG @ 100 g a.i./ha after 15 days of emergence is recommended for the effective control of whiteflies and potato apical leaf curl disease (PALCD) in potato seed growing areas (2007).

Seed Production

- The time of appearance of *Myzus persicae* and its build up has been determined at 15 locations. In Bihar, the critical population builds up about three weeks later than in western Indo-Gangetic plains. In such area, the actual date of removal of haulms could be delayed up to 3rd week of January. The available data may be used to demarcate the areas suitable for seed production (1974).
- The farmers in the plains should be encouraged to grow their own seed during aphid free period instead of importing seed from hills because their own seed would be in a better germinable stage in comparison with the hill seed and would also be cheaper. This will reduce the chances of spread of soil borne diseases from the hills to the plains (1977).
- Haulm cutting in seed crop can be delayed by 10 days with the use of systemic insecticides like Metasystox (1983).
- Based on aphid population data of last several years, Hisar, Faizabad, Kanpur and Deesa centres have been identified as suitable places for seed production (1991).

Potato Storage

- The potato heap for storage under ambient conditions should be made immediately after the harvest and heap should be covered with at least one feet thick straw layer and properly protected from the rains. Potatoes of the heap should be disposed off as soon as the temperature in side the heap reaches 28–30 °C and monsoon rains set in (2003).

MILESTONES AND LAMP POSTS OF PROJECT

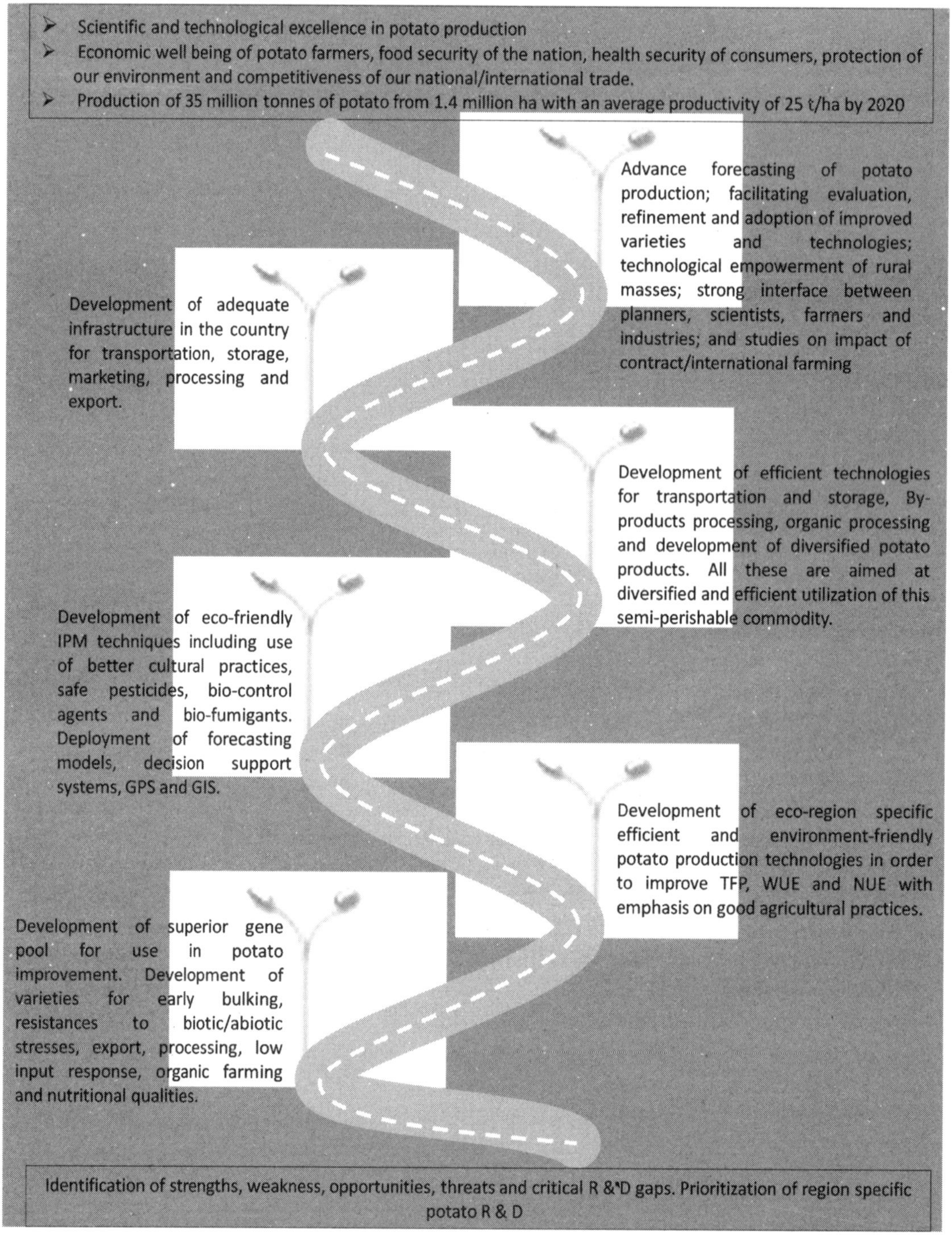

ADDRESSES OF ACIRP (POTATO) CENTERS ALL INDIA COORDINATED RESEARCH PROJECT ON POTATO

Shimla	Project Coordinator, Central Potato Research Institute Shimla – 171 001 (Himachal Pradesh) Telefax: 0177-2624398; E-mail: aicrpp_shimla@yahoo.com
CPRI Based AICRP Centers	
Gwalior	Head, Central Potato Research Station Gwalior – 474 006 (Madhya Pradesh) Tel.: 0751-2470302; Fax: 0751-2471967
Jalandhar	Head, Central Potato Research Station P.O. Model Town, P.B. 1, Jalandhar – 144 003 (Punjab) Tel: 0181-2791474; Fax: 0181-2790863
Kufri	Head, Central Potato Research Station Kufri – 171 012, Distt. Shimla (Himachal Pradesh) Telefax: 0177-2648234
Modipuram	Joint Director, Central Potato Research Institute Campus Modipuram, District Meerut – 250 110 (Uttar Pradesh) Tel.: 0121-2950297
Ootacamund	Head, Central Potato Research Station P.O. Muthorai – 643 004 (Tamil Nadu) Tel.: 0423-2550044; Fax: 0423-2550669
Patna	Head, Central Potato Research Station P.O. Sahaynagar, Patna – 801 506 (Bihar) Tel.: 0612-2224218; Fax: 0612-2223706
Shillong	Head, Central Potato Research Station Shillong – 793 009 (Meghalaya) Tel.: 0364-2560885; Fax: 0364-2560097
SAU Based Regular AICRP Centers	
Bhubaneshwar	Scientist in-charge, AICRP-Potato Orissa University of Agriculture & Technology Bhubaneshwar – 751 003 (Orissa) Tel.: 0674-2391068; Fax: 0674-2391068
Chhindwara	Scientist in-charge, AICRP-Potato Jawaharlal Nehru Krishi Vishwavidyalya Regional Agricultural Research Station Chhindwara – 480 001 (Madhya Pradesh) Tel.: 07162-225560; Fax: 07162-225493
Deesa	Scientist in-charge, AICRP-Potato, Potato Research Station Sardar Krushinagar Dantiwada Agriculture University Deesa – 385 535, Distt. Banaskantha (Gujarat) Tel.: 02744-220865; Fax: 02744-223552
Dharwad	Scientist in-charge, AICRP-Potato M.R.S. Department of HorticultureUniversity of Agricultural Sciences Dharwad – 580 005 (Karnataka) Tel.: 0836-2747366, 2448349 Fax: 0836- 2748377
Dholi	Scientist in-charge, AICRP-Potato Department of Plant Breeding, Tirhut College of Agriculture Dholi – 843 121 Muzaffarpur (Bihar) Tel.: 0621-2293227; Fax: 0621-2293227

Faizabad	Scientist in-charge, AICRP-Potato Division of Vegetable Crops ND University of Agriculture & Technology Narendra Nagar, P.O. Kumarganj, Faizabad – 224 229 (Uttar Pradesh) Tel.: 05270-262076; Fax: 05270-262123
Hassan	Scientist in-charge, AICRP-Potato Agricultural Research StationUniversity of Agricultural Sciences Bangalore Madenur, Distt. Hassan – 573 220 (Karnataka) Tel.: 08172-258825; Fax: 08172-258825
Hisar	Scientist in-charge, AICRP-Potato Department of Vegetable Sciences CCS Haryana Agricultural University, Hisar – 125 004 (Haryana) Tel.: 01662-289207; Fax: 01662-234952
Jorhat	Scientist in-charge, AICRP-Potato Department of Agronomy Assam Agriculture University, Jorhat – 785 013 (Assam) Tel.: 0376-2340030, Fax: 0376- 2310831, 2340001
Kalyani	Scientist in-charge, AICRP-Potato Research Complex BuildingBidhan Chandra Krishi Vishwavidyalaya Kalyani 741 235, Nadia (West Bengal) Tel.: 033-25824807; Fax: 033-25828460, 25870523
Kota	Scientist in-charge, AICRP-Potato Agricultural Research Station Umedganj FarmMaharana Pratap University of Agriculture & Technology PB No. 7, GPO Nayapura, Kaithoon Road, Kota – 324 001 (Rajasthan) Tel.: 0744-2844369; Fax: 0744-2844306
Pantnagar	Scientist in-charge, AICRP-Potato Department of Vegetable Sciences GB Pant University of Agriculture & Technology Pantnagar – 263 145 (Uttarakhand) Tel.: 05944-234396, 233076; Fax: 05944-233473
Raipur	Scientist in-charge, AICRP-Potato Department of Plant Breeding & Genetics Indira Gandhi Krishi Vishwa Vidyalaya Raipur – 492 006 (Chhatisgarh) Tel.: 0771-2442352, 2444324; Fax: 0771-2442131
Srinagar	Scientist in-charge, AICRP-Potato Department of Olericulture Sher-e-Kashmir University of Agricultural Sciences & Technology of Kashmir Shalimar Campus, Srinagar – 191 121 (J&K) Tel.:0194-2462124; Fax: 0194-2462124
SAU Based Voluntary Center	
Ranichauri	Scientist in-charge, AICRP-Potato GB Pant University of Agriculture & Technology (Hill Campus) Ranichauri – 249 199 (Uttarakhand) Tel.: 01376-252228; Fax: 01376-252128

3

Tuber Crops

M.S. Palaniswami and V. Ramesh

Tropical root and tuber crops are known as the energy banks of nature, serving either as primary or secondary staple to meet the calorie needs of about one-fifth of world's population. These crops have myriad and complex roles to play in food security and reduction of hunger and poverty. Among other things, farm households see the value of roots and tuber in their ability to produce higher edible energy production per hectare per day than other commodities. Root and tuber have the capacity to out produce cereals in terms of quantities of edible energy harvested per unit area per unit time. What is more important is that they produce under conditions where other crops may fail.

Tropical tuber crops are the third important group of food crops after cereals and pulses. These are rich in starch and calories and have great potential in meeting the food and energy needs. Burgeoning population, food shortage, ever increasing need to explore new frontiers in order to alleviate hunger and poverty, alternate sources for alcohol etc. have contributed a great deal to the importance of root and tuber crops as a source of food and energy. Tuber crops have a higher biological efficiency as food producers with higher dry matter production per unit area per unit time. These crops have gained importance not only as food crops but also for its scope for feed and agro-based industries.

Nevertheless, future growth rates calculated for cassava, potato and yam exceed those estimated for rice and wheat (Scott *et al.*, 2000). The versatility of the root and tuber crops will remain an enduring attraction for producers and consumers. More technically and economically efficient producers for producing raw material and finishing products and strengthened grower-to-processor linkages will be fundamental to achieving the full growth of these crops. In India with our vast biodiversity we have value-added germplasm repositories of cassava, aroids, sweet potato and yam as well as collections of several other root and tuber crops which will provide the genes to the year 2020 and beyond that, which can be used to stabilize/increase the yield and improve the quality of these crops.

Given the projected importance of roots and tubers in developing countries to 2020 and ahead, it is fundamental to retain these crops as an integral part of a global strategy to increase food production and utilization in Asia, Africa, and Latin America in the decades to come.

In India, cassava is being cultivated on an area of 0.24 million (m) ha, with a production of 6.7 million tonnes. Cassava has been an important crop in South India as a subsidiary food as well as industrial raw material especially in Kerala and Tamil Nadu.

More than 80% of the area under cassava cultivation is in Kerala and Tamil Nadu. Now it is slowly spreading to Maharashtra, Meghalaya, Assam, Sikkim, Mizoram, Karnataka, Andhra Pradesh and Rajasthan. The area under cassava cultivation in India has come down from 0.368 m ha in 1973-74 to 0.207 m ha in 2002-03. Though there was decline of 0.161 m ha, the total production had more or less maintained *status quo* with the result, the productivity had increased from 17 to 26 tonnes ha^{-1}. During the last 25 years (1980-81 to 2004-05) though the total area of cultivation in India was reduced by 33% under cassava the total production increased by 15% and productivity increased by 52%. Similarly, in Kerala also the cassava cultivation had come down from 0.13 m ha to 0.104 m ha, during the last 10 years. However, the production ranging only 23–26 lakh tonnes. In Tamil Nadu, 67,951 ha cultivated are mostly concentrated in Salem, Dharmapuri, Perambalur and Kanyakumari districts. The mean productivity of cassava in Tamil Nadu is 31.59 tonnes ha^{-1} which is the world's highest. In Andhra Pradesh, the cassava cultivation is largely restricted to East Godavari district (20,000 ha) where it is cultivated on a commercial scale as rainfed crop.

Sweet potato cultivation in India is largely in Orissa, West Bengal, Uttar Pradesh, Assam, Bihar, Jharkhand, Madhya Pradesh, Meghalaya, Maharashtra and Karnataka. 80% of the total sweet potato growing area is occupied by Orissa, West Bengal and Uttar Pradesh (Table 3.1). During the last 10 years, the area of cultivation of sweet potato has declined from 1,40,700 ha during 1995-96 to 1,00,000 ha during 2004-05. but at the same time the productivity has increased from 8 to 9 tonnes ha^{-1}. In Orissa , Bolangir, Koraput, Baragarh, Ganjam, Dhenkanal, Koenjhar and Sundergadh districts cultivate sweet potato in larger areas as rainfed crop. In Uttar Pradesh, Etah, Badaur, Farookabad and Sultanpur are largely involved in sweet potato cultivation. Sweet potato is cultivated in 3821 ha with a productivity of 8 tonnes ha^{-1} in Madhya Pradesh. In Maharashtra, sweet potato is grown in an area of 5100 ha. Its large scale cultivation has been reported to be restricted to Kolhapur division where the crop is taken in an area of 2800 ha. In Konkan region of Maharashtra, 200 ha are under sweet potato cultivation with an average yield of 13.5 tonnes ha^{-1}. In Karnataka sweet potato is largely grown in Belgaum, Dharwad, Bijapur, Uttar Kannada and parts of Raichur. There was a sharp decline in the area under cultivation in the country to half and 40% reduction in total production during 1980-81– 2004-05. However the productivity increased by 28.6%.

Among the important taro and yam growing countries in the world, the highest productivity in taro was observed in Egypt while in yams, the highest productivity was recorded in Solomon Islands. Cultivation of aroids and yams is spread all over India. In Tamil Nadu, about 900 ha each of taro and elephant foot yam is being cultivated and the productivity is 10 and 30 tonnes ha^{-1} respectively.

Taro and elephant foot yam are largely grown in Uttar Pradesh, West Bengal, Bihar, Chattisgarh, Jharkhand, Maharashtra, Gujarat, Andhra Pradesh and to a certain extent in Tamil Nadu and Kerala. In Konkan region of Maharashtra, 110 ha of area was under *Xanthosoma* and *colocasia* cultivation with an average productivity of 5.5 tonnes ha^{-1} during 2000. Yams are also cultivated in this part of Maharashtra in an area of 61 ha.

In Madhya Pradhesh, of the 11 agroclimatic zones, tropical root and tuber crops taro, elephant foot yam and yams are grown in three agroclimatic zones of the state namely Bundelkhand region, Kymore plateau, Satpura Hills and Northern Hills. Tikangarh, Datia, Chhatarpur and Gunda districts are known for tuber crops cultivation. In Orissa Ganjam, Dhenkanal, Koenjhar, Rajasthan, Banswara, Bhilwara, Bundi, Chittor, Dundarpur, Kota, Udaipur Mayurbhanj and Balasore are major tuber crops (aroids and other tubers) growing areas. These are grown on 2,381 ha. Taro cormel preparations are given as offering in the temple Puri.

Table 3.1. Area, production and productivity of sweet potato and cassava in India (2001-02)

State	*Area ('000 ha)*	*Production ('000 tonnes)*	*Productivity (tonnes/ha^{-1})*
		Sweet Potato	
Orissa	37.7	272.1	7.22
West Bengal	24.0	172.0	7.15
Uttar Pradesh	22.6	259.8	11.50
Assam	8.8	30.4	3.45
Bihar	5.0	75.6	15.12
Jharkhand	7.6	103.0	13.55
Madhya Pradesh	6.9	42.2	6.12
Meghalaya	5.2	17.6*	3.40
Maharashtra	4.7	66.7	14.20
Karnataka	3.5	26.4	7.63
Andhra Pradesh	1.3	13.4	10.40
Rajasthan	1.2	2.7	2.30
Tripura	1.1*	9.7*	9.08
Tamil Nadu	0.8	11.1	13.66
Kerala	0.8	9.0	10.98
India	**131.9**	**1130.3**	**8.57**
		Cassava	
Kerala	109.3	2471.8	22.61
Tamil Nadu	68.0	2146.5	31.59**
Andhra Pradesh	17.7 (13.1)	353.9 (79.5)	20.00 (6.05)**
Meghalaya	4.0	20.6	5.15
Assam	2.9	13.7	4.72
Nagaland	1.3	16.0	12.31
Karnataka	0.9	7.2	8.20
Sikkim	0.5	1.2	2.40
Mizoram	0.3	1.5	5.00
Rajasthan	0.1	0.2	2.00
India	**207.0****	**5426.2****	**26.21****

*2000 data **2002-03 data

Source: Anonymous 2005: Agriculture, Centre for Monitoring Indian Economy, March 2005, pp-283-91.

The All India Coordinated Research Project on Tuber Crops (AICRPTC) is the largest net work in the country on the tropical tuber crop research. The project was initiated during 1968 with two centres to begin with during fourth five year plan. At present, the AICRPTC has 10 centres located in different State Agricultural Universities and 10 + 2 voluntary centres in the ICAR Institutes. In addition the CTCRI at its headquarters, Thiruvananthapuram and its Regional centre, Bhubaneswar also serve as the voluntary centres of the project. The major tropical tuber crops under taken in the project are cassava, sweet potato, taro, tannia, elephant foot yam, giant taro, giant swamp taro, yams (different types), country potato, yam bean, winged bean, arrow root, canna, etc.

- Collection of germplasm of tuber crops from different regions of the country, particularly tribal/hilly areas and maintaining them as National Active Germplasm Site.
- Evaluation of germplasm for economically important traits including high yield, starch, carotene, short duration, tolerance/resistance to biotic and abiotic stress and sharing of promising entries among the centres.
- Carrying out regional trials to isolate improved high yielding varieties suitable to different agroclimatic conditions.
- To determine suitable agro-techniques for improved varieties of different crops in different regions.
- Evolve suitable and effective management tactics for major pests and diseases of tuber crops.
- To popularize and create awareness on the nutritional aspects of major tuber crops.
- To organize production and supply of healthy planting materials in major tuber crops growing areas in liaison with State Agriculture, Horticulture Departments and voluntary agencies like KVKs or NGOs.

Organizational Set Up

Various coordinating centres under the project

Coordinating Unit	*Year of Starting*	*Crops*
Rajendra Agricultural University, Dholi Muzaffarpur – 843 121, Bihar	1968	Sweet potato, *Colocasia*, Yams, Elephant foot yam and Yam bean
Tamil Nadu Agricultural University Coimbatore – 641 003, Tamil Nadu	1968	Cassava at Yethapur (Salem), Sweet potato, *Colocasia*, Elephant-foot yam and Yams at Coimbatore
Andhra Pradesh Horticultural University Tadepalligudem – 534 101, West Godavari Dist. Andhra Pradesh	1969	Cassava at PeddapuramSweet potato, *Colocasia* and Yams at Rajendranagar *Colocasia*, Yams and Elephant foot yam at Kovvur
Assam Agricultural University, Jorhat – 785 013 Assam	1971	Cassava, Sweet potato, *Colocasia*, Elephant foot yam and Yams
Dr Balasaheb Sawant Konkan Krishi Vidyapeeth Dapoli, Ratnagiri (Dt) – 415 712, Maharashtra	1975	Cassava, Sweet potato, *Colocasia*, Elephant foot yam and yams
Bidhan Chandra Krishi Viswavidyalaya Nadia, Kalyani – 741 235, West Bengal	1976	Cassava, sweet potato, *colocasia*, yam bean, elephant-foot yam and yams
Bisra Agricultural University, Kanke Ranchi – 834 006, Bihar	1987	Sweet potato, *Colocasia*, elephant foot yam, yam bean and Cassava

(Contd.)

Coordinating Unit	*Year of Starting*	*Crops*
Indira Gandhi Agricultural University, Kumharwand Jagdalpur (Baster) – 494 005, Chhattisgarh	1987	Cassava and Yams at Jadgalpur sweet potato, *Colocasia* and elephant-foot yam at Raipur
Narendra Dev University of Agriculture and Technology Faizabad – 224 229, Uttar Pradesh	1987	Sweet potato, *Colocasia,* elephant-foot yam and yam bean
Navasari Agricultural University, Navasari – 396 450, Gujarat	1994	Cassava, sweet potato, *Colocasia,* elephant-foot yam and yams
ICAR Research Complex for NEH Region Barapani – 793 103, Meghalaya.*	1975	Cassava, sweet potato and Aroids
Central Agricultural Research Institute Port Blair – 744 101, Andaman and Nicobar Islands.*	2000	Cassava, sweet potato and yams.
Central Agricultural University Iroisemba, Imphal – 795 004 Manipur	2006	Sweet potato, aroids and yams
Maharana Pratap University at Agriculture & Technology Udaipur - 313 001, Rajasthan	2006	Aroids and sweet potato
Universty of Agricultural Sciences Dharward, Karnataka	2007	Sweet potato and aroids

*Voluntary Centres

CTCRI, Thiruvanamthapuram and RC CTCRI, Bhubaneshwar are also involved in the programme.

Background Information

Centre/University	*Region**	*Soil Type/pH*	*Mean Annual Rainfall (mm)*
TNAU, Coimbatore	Eastern Ghats and Deccan Plateau	Red loam 7.0–8.0	620
ANGRAU, Hyderabad	Deccan Plateau	Sandy loam 7.8–8.0	750
Dr BSKKV, Dapoli	Deccan Plateau	Laterite 5.6–6.5	3086
NAU, Navasari	Central highlands and Kathiawar peninsula	–	2254
IGAU, Jagdalpur	Eastern Plateau and Eastern Ghats	Silty clay 5.5–6.2	1236
NDUAT, Faizabad	Northern Plain	Sandy loam 8.0–8.2	831
RAU, Dholi	Eastern Plain	Sandy loam 8.12–8.40	1320
BAU, Ranchi	Eastern Plateau	Red Yellow light grey catenary soils 6.5–7.0	1272
BCKV, Kalyani	Assam and Bengal Plains	Alluvial soil 5.5–6.0	1960
AAU, Jorhat	Assam and Bengal Plains	Sandy loam 4.8–5.0	2486
ICAR Research Complex, Shillong	North Eastern Hills	Apatani plateau sandy Loam-clay loam 5.10–5.64	2529

*Based on NBSS & LUP Pub. 24

Project Coordinator

At the beginning of the programme the Director of Central Tuber Crops Research Institute, Thiruvananthapuram, was coordinating the project on a part-time basis from 1968 and this continued till February, 1979 when a separate post of Project Coordinator was created.

Dr S.P. Varma, joined in March, 1979 as the Project Coordinator and was functioning from Rajendra Agricultural University, Dholi. Later on the Office of the Project Coordinator was shifted to Regional Centre, CTCRI at Bhubaneswar and again to CTCRI, Thiruvananthapuram, since 1993. Dr Varma continued till 15-8-1994. Dr P.G. Rajendran was the Project Coordinator during 16-8-1994 to 31-8-2003 with ably supported by two Scientists. Dr M.S. Palaniswami took over as Acting Project Coordinator from September 2003 and is the Regular Project Coordinator, w.e.f 4 August 2005.

Man Power and Budget

The approved man power for different AICRPTC centres are given below.

Centre	*Scientific*	*Technical*	*Administrative*	*Supporting*	*Total*
RAU, Dholi	4	4	1	6	15
ANGRAU, Rajendranagar	4	3	1	1	9
TNAU, Coimbatore	2	2	1	1	6
BCKV, Kalyani	3	2	–	1	6
AAU, Jorhat	2	2	1	1	6
Dr BSKKV, Dapoli	2	2	–	–	4
BAU, Ranchi	1	2	–	–	3
IGAU, Jagdalpur	2	2	–	–	4
NDUAT, Faizabad	1	1	–	–	2
NAU, Navasari	1	1	–	–	2
Total	22	21	4	10	57

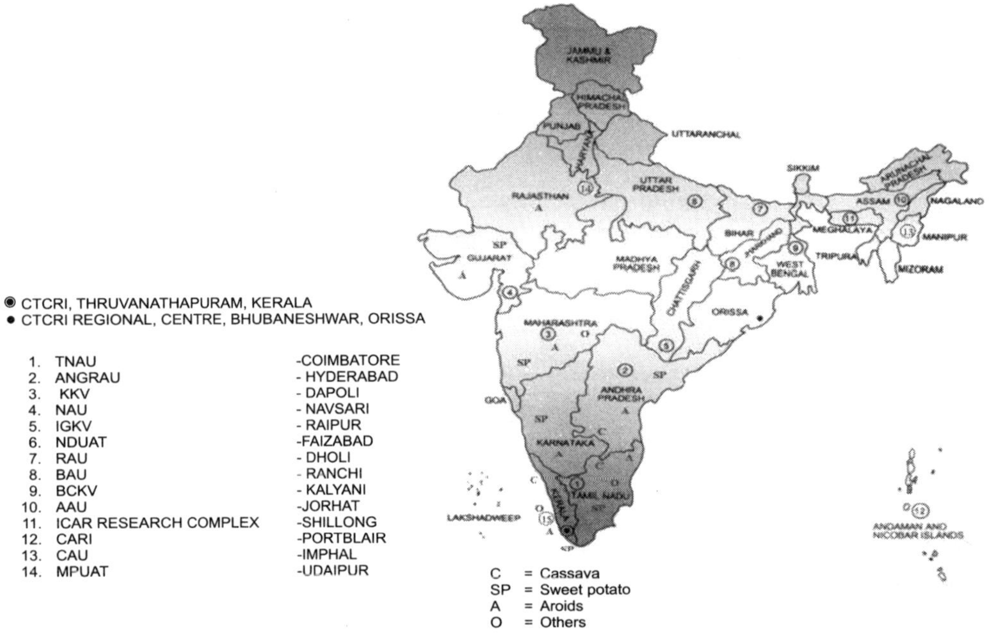

Fig. 3.1. Tuber crop-growing areas in India

Budget

Plan	(ICAR shares 75%, Rs lakh)	Sanctioned Man Power
IV plan	1.10	9
V plan	8.31	13
VI plan	32.65	47
VII plan	54.57	47
VIII plan	61.64	57
IX plan	328.85	57
X plan	434.00	57
XI plan	1263.53	57

Tropical Root Tuber Crops Covered Under Project

Crop	Scientific Name	Family	Edible Parts
Cassava	*Manihot esculenta* Crantz	Euphorbiaceae	Root, leaves
Sweet potato	*Ipomoea batatas* (L.) Lam	Convolvulaceae	Tuber
Greater yam	*Dioscorea alata* L.	Dioscoreaceae	Tuber
White yam	*Dioscorea rotundata* Poir	Dioscoreaceae	Tuber
Lessser yam	*Dioscorea esculenta* (Lour.) Burls	Dioscoreaceae	Tuber
Aerial yam	*Dioscorea bulbifera L.*	Dioscoreaceae	Tuber bulbils
Intoxicating yam	*Dioscorea hispida*	Dioscoreaceae	Tuber
Yam	*Dioscorea floribunda*	Dioscoreaceae	Tuber
Yellow yam	*Dioscorea cayenensis*	Dioscoreaceae	Tuber
Yam	*Dioscorea pentaphylla*	Dioscoreaceae	Tuber, bulbils
Cush-cushyam	*Dioscorea trifida*	Dioscoreaceae	Tuber
Arvi	*Colocasia esculenta var.* antiquorum (L.) Schott	Araceae	Corm, cormels, leaves, petiole
Bunda	*Colocasia esculenta var.* esculenta (L.) Schott	Araceae	Corm, leaves
Swamp taro	*Colocasia esculenta var.* stoloniferum (L.) Schott	Araceae	Stolon
Tannia	*Xanthosoma sagittifolium* (L.) Schott	Araceae	Corm, leaves, petiole
Elephant-foot yam	*Amorphophallus paeoniifolius* (Dennst.) Nicolson	Araceae	Corm, pseudostem
Giant taro	*Alocasia macrorrhizos* (L.) G. Donf	Araceae	Corm
Giant swamp taro	*Cyrtosperma chmissionis* (Schott) Merr.	Araceae	Corm
Country potato	*Solenostemon rotundifolius* (Poir) J. K. Morton	Labiatae	Tuber
Arrowroot	*Maranta arundinacea* L	Marantacea	Rhizome
Canna	*Canna edulis* Ker-Gawler	Cannaceae	Rhizome
Winged bean	*Psophocarpus tetragonolobus* (L.) DC.	Leguminoceae	Tuber, young pods
Yam bean	*Pachyrhizus erosus* (L.) Urban *P. ahipa, P. tuberosus*	Leguminoceae	Root
Vigna	*Vigna capensis* Walp	Leguminoceae	Root
Typhonium	*Typhonium* spp.	Araceae	Corm
Costus	*Costus speciousus* (Koenig) Sm.	Costaceae	Tuber
Shoti	*Curcuma zedoria* (Berg.) Roscoe.	Zingiberaceae	Rhizome
Jerusalem artichoke	*Helianthus tuberosus*	Compositeae	Tuber

Review of Progress and Monitoring

The project has been monitored and reviewed periodically. The Project Coordinator visits the centres for overseeing the deployment of sanctioned posts, utilization of funds, implementation of ICAR recommendations and technical programmes assigned to them. In addition QRT appointed by ICAR and DDG (H) with assistance of ADG (H) also review project at different centres. The progress and achievements of various experiments were analysed and compiled into annual reports. Workshops or Group Meetings were conducted at periodical intervals to assess and validate the results and findings recommended on merit for adoption. New technical programmes were discussed and developed for different centres. Progress reports and meeting proceedings brought out for implementation. So far 10 Workshops and ten Group Meetings were held under AICRP Tuber Crops. The details of Workshop/Group Meeting conducted are detailed below.

Workshop/Group Meeting	*Place*	*Year*
1st Workshop	IARI, New Delhi	1969
2st Workshop	CTCRI, Trivandrum	1971
3st Workshop	CTCRI, Trivandrum	1972
4st Workshop	CTCRI, Trivandrum	1975
5st Workshop	KAU, Trichur	1979
6st Workshop	RAU, Dholi, Bihar	1982
7st Workshop	CTCRI, Trivandrum	1983
8st Workshop	BCKV, Kalyani	1986
9st Workshop	AAU, Jorhat	1987
10st Workshop	APAU, Rajendrnagar	1989
1st Group Meeting	KKV, Dapoli	21-23, October 1991
2nd Group Meeting	CTCRI, Trivandrum	04-05, November 1993
3rd Group Meeting	TNAU, Coimbatore	04-06, October 1995
4th Group Meeting	CTCRI, Trivandrum	30 Sept - 01 October 1997
5th Group Meeting	BCKV, Kalyani	12-14 November1999
6th Group Meeting	IGAU, Jagdalpur	03-05 October 2001
7th Group Meeting	CTCRI, Trivandrum	24-25 May 2004
8th Group Meeting	NDUAT, Faizabad	04-06 May 2006
9th Group Meeting	NAU, Navsari	21-23 December 2007
10th Group Meeting	OUAT, Bhubaneswar	09-11 October 2009

GENETIC RESOURCES

The AIRCPTC centres were identified as National Active Germplasm sites of tuber crops for collection and maintenance of different accessions in and around each region. More than 24 of species tuber crops were taken care under the project. Currently, 4350 accessions of different species of tuber crops were being maintained in the ten State Agricultural Universities and two ICAR Institutes (ICAR Research Complex, Shillong and CARI, Port Blair) serving as National Active Germplasm Sites spread over 11 states and one union territory (Table 3.2).

Largest collection was in sweet potato (2020) and Dholi centre maintained 1166 sweet potato accessions. Cassava collection was mostly concentrated in Coimbatore at Yethapur (363 accessions out of 596). In *Colocasia,* there were 831 accessions (including arvi and bunda types). There were 70 collections of swamp taro mainly in Jorhat and Kalyani. In elephant foot yam, there were 195 accessions, mostly in Faizabad (51 nos.), Jagdalpur (34 nos.) and Rajendranagar (32 nos.). Among the yams, greater yam constituted larger collection (186) spread over the different centres. Yam bean and winged bean collections were mainly in Dholi centre (Fig. 3.2 and 3.3).

Cassava

The AICRP centres have a total of 596 germplasm accessions. Largest collection of 363 accessions is maintained by Coimbatore centre at Yethapur. Other major contributors are Jagdalpur (60), Rajendranagar (46) and Jorhat (30) (Fig.3.3). The 363 accessions at Yethapur (Coimbatore) evaluated for starch content revealed 67 accessions as having more than 25% starch content. The screening of cassava mosaic disease (CMD) incidence revealed that none of the accessions were resistant to CMD. Yethapur (TNAU) centre evaluated 316 accessions of cassava for yield and found that ME 833 and ME 767 were good yielders (about 5 kg/plant) with good starch content (30%).

Of the 46 accessions under cassava germplasm maintained in ANGRAU at Peddapuram station, the entry KPSLO 5/92 gave the highest fresh tuber yield of 23.17 tonnes/ha closely followed by ARK-D with 22.80 tonnes/ha and KPSL-2/92 (22.63 tonnes/ha) which were suitable as table varieties. The entry TCH-2 recorded highest tuber yield of 32.15 tonnes/ha followed by H-740/92 (31.27 tonnes/ha) and H-5/78 (29.6 tonnes/ha), which were suitable for industrial purpose.

Sweet Potato

A total of 2020 sweet potato accessions are being maintained at different centres (Table 3.2). Dholi centre maintained the largest collection of 1,166 accessions followed by Rajendranagar (156), Kalyani (143) and Jagdalpur (123) (Fig. 3.5). 15 new accessions were made from Dapoli (2), Kalyani (5), Faizabad (2), Ranchi (3) and Jadgalpur (3) during 2004–06.

Dholi centre evaluated 25 collections of sweet potato for yield and weevil susceptibility. The entry RS-92 recorded higher tuber yield with moderate weevil damage. At Kalyani, 10 genotypes of sweet potato (H-85/16, X-108-1, S-30/21, CO-3, OP-217, OP-219, S-30/25, H-268, S-30 and C-71) were evaluated for its proximate nutritional quality and mineral constituents of tuber on fresh weight basis (Table 3.3). 29 genotypes were indexed for seven viral diseases infecting sweet potato by symptomatology and serological methods (Table 3.4).

QRT in elephantfoot yam field at RAU, Dholi

IX Biennial group meeting at Navsari

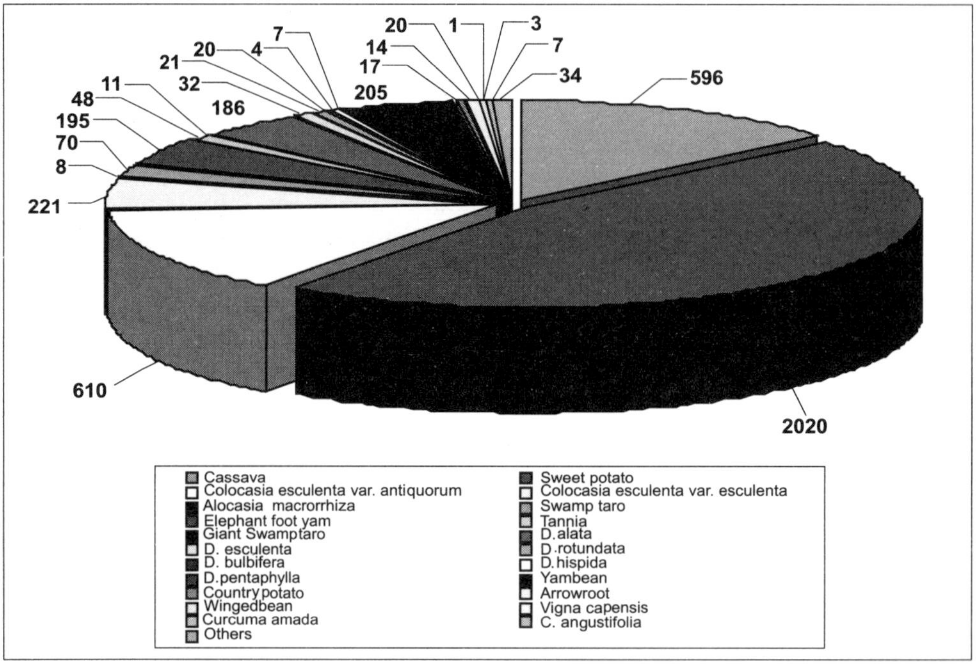

Fig. 3.2. Germplasm collection of tropical tuber crops maintained in AICRPTC centres

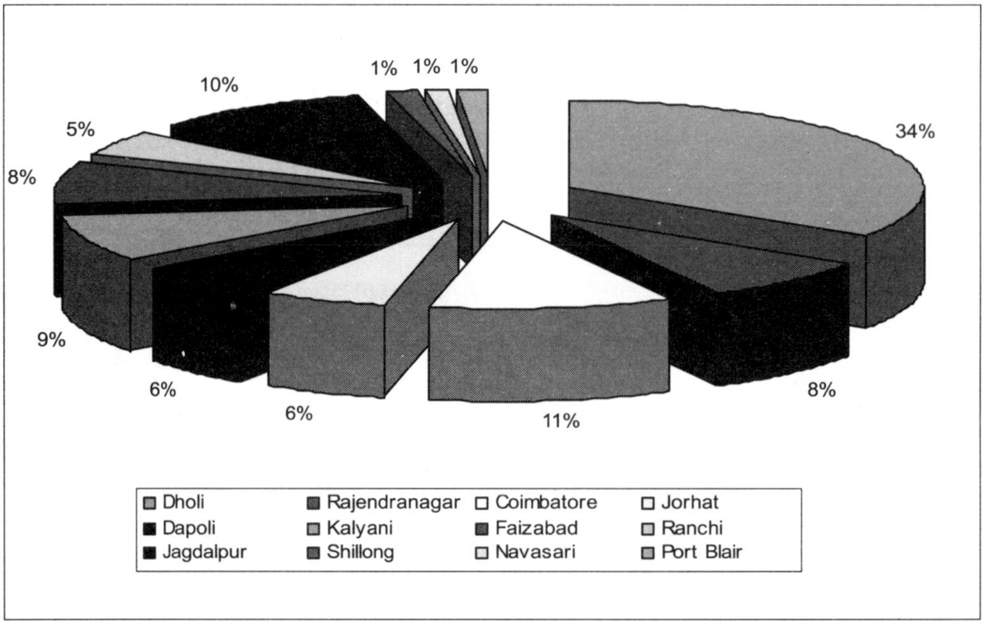

Fig. 3.3. Share of germplasm at different centres under AICRPTC

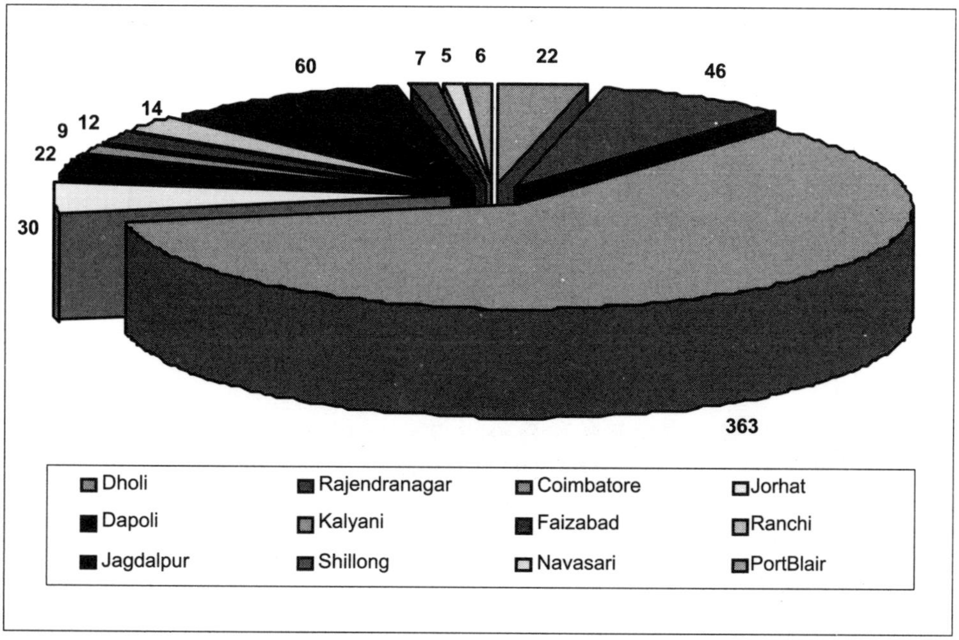

Fig. 3.4. Cassava germplasm (number of accessions) maintained at different centres

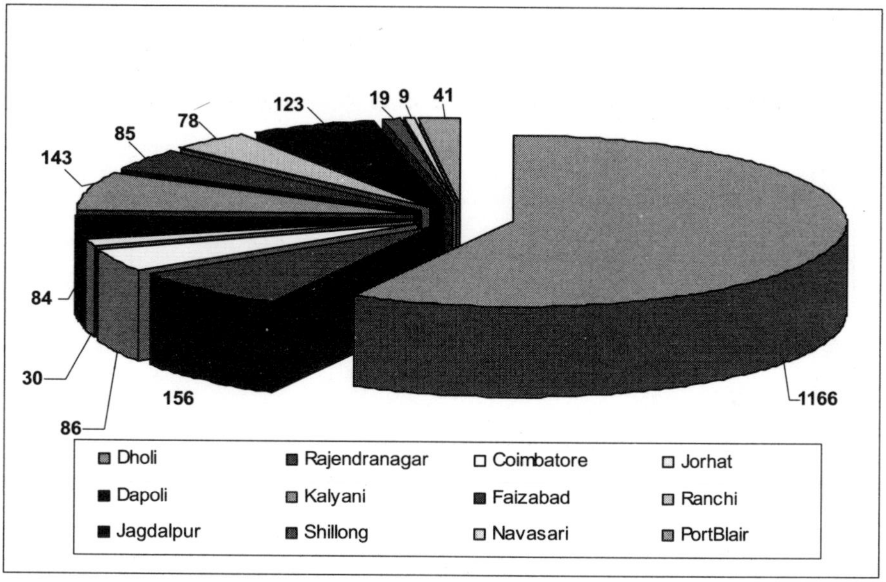

Fig. 3.5. Sweet potato germplasm at different centres

Table 3.2. Germplasm of different tropical tuber crops under AICRPTC Centres

Crops/Centre	*Dholi*	*Rajendranagar*	*Coimbatore*	*Jorhat*	*Dapoli*	*Kalyani*	*Faizabad*	*Ranchi*	*Jagdalpur*	*Shillong*	*Navasari*	*Port Blair*	*Total*
Cassava	22	46	363	30	22	9	12	14	60	7	5	6	596
Sweet potato	1166	156	86	30	84	143	85	78	123	19	9	41	2020
Taro	25	99	0	82	54	88	85	36	90	35	16	0	610
Bunda	75	0	8	0	0	15	55	41	26	0	0	1	221
Giant taro	0	0	0	7	1	0	0	0	0	0	0	0	8
Swamp taro	0	0	0	24	0	45	0	0	1	0	0	0	70
Elephant foot yam	18	32	5	13	8	22	51	6	34	1	3	2	195
Tannia	0	0	2	14	11	11	0	3	7	0	0	0	48
Giant swamp taro	0	0	0	0	0	11	0	0	0	0	0	0	11
D. alata	12	13	6	36	27	20	12	17	33	3	4	3	186
D. esculenta	10	0	5	5	5	2	0	0	0	0	0	5	32
D. rotundata	3	0	2	2	3	4	0	0	2	0	0	5	21
D. bulbifera	0	0	0	0	7	0	0	0	13	0	0	0	20
D. hispida	0	0	0	0	0	0	0	0	4	0	0	0	4
D. pentaphylla	0	0	0	0	0	0	0	0	7	0	0	0	7
Yambean	137	6	0	1	12	2	33	0	14	0	0	0	205
Country potato	1	0	1	1	2	0	2	3	6	0	0	1	17
Arrowroot	0	0	1	2	3	0	3	3	2	0	0	0	14
Winged bean	20	0	0	0	0	0	0	0	0	0	0	0	20
Vigna capensis	0	0	0	0	1	0	0	0	0	0	0	0	1
Curcuma amada	0	0	0	0	3	0	0	0	0	0	0	0	3
C. angustifolia	0	0	0	0	0	0	0	0	7	0	0	0	7
Others	0	0	1	0	7	0	0	8	16	0	2	0	34

Table 3.3. Proximate nutritional (%) and mineral mg 100 g^{-1} composition of tubers in certain sweet potato genotypes (at BCKV, Kalyani)

Genotype	*Nutritional Composition (%)*					*Mineral Composition (mg 100 g^{-1})*							
	Moisture	*Dry matter*	*Starch*	*Sugar*	*Crude protein*	*Potassium*	*Calcium*	*Phosphorus*	*Sodium*	*Iron*	*Zinc*	*Copper*	*Manganese*
H-85/16	74.60	25.40	16.01	6.24	0.93	276.20	1 10.00	80.20	12.27	1.61	0.37	0.17	0.14
X-108-1	78.20	21.80	17.43	3.30	0.79	291.40	90.20	63.00	16.73	1.68	0.34	0.16	0.23
S-30/21	71.80	28.20	20.23	5.40	0.78	269.00	88.90	82.00	5.92	1.39	0.39	0.17	0.16
CO-3	66.40	33.60	16.43	5.58	0.72	309.00	41.40	62.00	18.62	2.07	0.36	0.16	0.25
OP-217	74.40	25.60	12.70	5.30	0.55	261.00	100.10	80.00	7.38	1.79	0.41	0.15	0.21
OP-219	67.40	32.60	20.85	5.18	0.51	270.26	99.00	89.00	22.76	2.83	0.42	0.15	0.25
S-30/25	70.00	30.00	19.45	5.30	1.06	279.00	102.00	87.70	9.00	4.06	0.57	0.17	0.21
H-268	68.00	32.00	17.50	3.06	0.72	290.00	92.60	74.00	27.00	2.91	0.30	0.16	0.25
S-30	77.20	22.80	13.07	4.48	0.51	296.30	109.00	60.00	7.30	2.13	0.49	0.15	0.22
C-71	72.00	28.00	16.11	4.38	0.70	270.10	97.50	39.00	14.70	3.83	0.37	0.15	0.20
Mean	72.00	28.00	16.98	4.72	0.73	281.23	93.07	71.69	14.67	2.43	0.40	0.16	0.21

Table 3.4. Detection of sweet potato viruses in some cultivars by NCM-ELISA (at BCKV, Kalyani)

Cultivar	*Sweet Potato Viruses*						
	SPFMV	*SPMMV*	*SwPLV*	*SPCFV*	*C-6*	*SPMSV*	*SPCaLV*
Sree Bhadra	-	++	-	++	+++	+++	+++
S-73	-	-	-	-	+	-	-
X-38	-	+	-	-	-	+++	-
X-108	-	+	-	-	-	+	-
S-72	-	-	-	-	-	+	-
BCSP-9	-	-	-		++	-	-
BCSP-12	-	+		+	+		+
C-71	+	-	-	++	+	+	-
IB-400	-	-	-	+	++	-	-
IB 90-15-9	-	-	-	+	+	+	+++
H-85-179	-	+			+	-	-
Kalmegh	-	++	+	++	+++	++	++
OP-23	-	-	-	+	-	-	-
OP-34	-	-	-	-	+++	-	-
OP-59	-	-	-	-	+	-	-
S-25	+	-	-	+		-	-
S-30	-	++	+	++	++	+	++
S-43	-	+	-	-	-	-	-
S-162	-	-	-	+	+	-	-
Sree Vardini	-	++	-	++	++	+++	+++
X-9	-	+	-	+	+	-	+
X-25	-	-	-	-	+++	-	-
H-85/16	-	-	-	-	-	-	-
X-108-1	-	-	-	-	-	-	-
S-30/21	-	-	-	-	-	-	-
CO-3	++	-	-	+	+++	++	+
OP-217	-	-	-	++	-	-	-
S-30/25	++	-	-	-	-	-	-
H-268	-	-	-	-	-	-	-

Intensity of colour development in NCM-ELISA: +low; ++ Moderate; +++ high

Yams

A total of 270 accessions of various yams species, *viz. Dioscorea alata* (186 accessions), *D. esculenta* (32), *D. rotundata* (21), *D. bulbifera* (20), *D. hispida* (4), *D. pentaphylla* (7) are being maintained at different coordinating centres. Twenty two new collections were made during 2004–06 which included seven in *D. alata*, four in *D. rotundata*, four in *D. hispida* and seven in *D. pentaphylla.* Jagdalpur centre has maintained the largest collection of yams (59) which include collections of *D. hispida* and *D. pentaphylla* (Fig.3.6). The yam accessions Kovvur, Tallapalem and Yettipalli gave maximum tuber yield (50–56 tonnes/ha) at Rajendranagar centre. Jorhat catalogued 12 greater yam accessions with relevant characters described for each. At Rajendranagar, *Dioscorea* germplasm consisted of 15 types and during 2005-06, Tallapalem, a local collection has recorded maximum yield of 24.67 tonnes/ha followed by Kovvur and Aurangabad collection (22.67 and 20.22 tonnes/ha respectively).

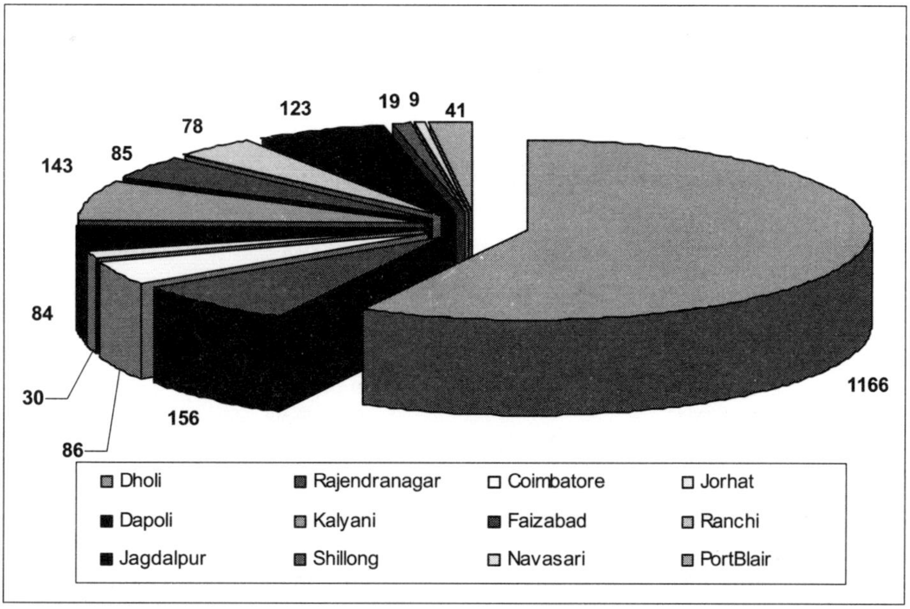

Fig. 3.6. Germplasm of yams maintained at different centres

Colocasia

A total of 901 germplasm accessions of *Colocasia* which included 610 accessions of *C. esculenta* var. *antiquorum* (taro or arvi), 221 accessions of *C. esculenta* var. *esculenta* (bunda) and 70 accessions of *C. esculenta* var. *stoloniferum* (swamp taro) was being maintained at different centres (Fig. 3.7). Nineteen collections of *Colocasia* (arvi) and four collections of *Colocasia* (bunda) were new additions during the year 2005-06, with major contribution (10 accessions) from Jorhat centre. Dholi, Kalyani, Faizabad and Jagdalpur also maintained good collection of both types of *Colocasia*. The swamp taro collection was mainly from Kalyani and Jorhat centres. At Kovvur, the germplasm is being classified into three groups *ie.,* short duration, medium duration and long duration types and evaluated for yield and morphological characters. In *Colocasia* germplasm, among the short duration types (less than 6 months), KCS-3 has recorded the highest cormel yield of 25.54 tonnes/ha followed by CA-44 (22.95 tonnes/ha) and CA-52 (22.73 tonnes/ha). In medium duration types (6-7 months), CA-42, No.6, CA-57, IGCol-4 and RNCA-1 proved superior in terms of cormel yield (27.6 tonnes/ha, 27.50 tonnes/ha, 26.7 tonnes/ha, 23.33 tonnes/ha and 22.08 tonnes/ha respectively). In long duration types (7 months and above), KCS-2 has recorded maximum cormel yield of 30.75 tonnes/ha followed by CA-135 and CA-1 (24.5 and 21.0 tonnes ha^{-1} respectively).

Elephantfoot Yam

A total of 195 accessions of elephant foot yam are being maintained at the different AICRPTC centres, of which 11 are new additions made during 2005-06. Kalyani centre has added 4 accessions to its collection, two each are added to the collections of Jorhat, Ranchi and Faizabad and one to Jagdalpur. Faizabad has the largest collection of elephant foot yam

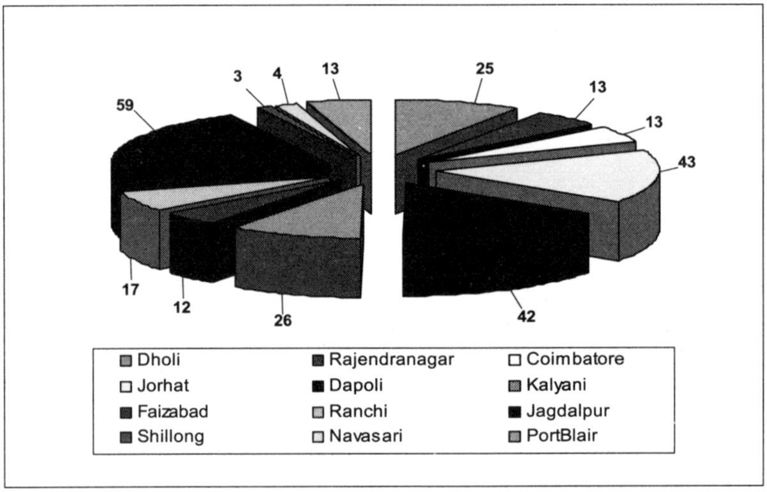

Fig. 3.7. Germplasm of *Colocasia* (accessions of both bunda and taro types) at different centres

genetic stock (51) (Fig. 3.8). At Kovvur sub-centre, there are 20 non-irritant and 13 irritant types. Among the non-irritant types, AC-13 has recorded maximum yield of 67.21 tonnes/ha followed by accession AC-6 (63.9 tonnes/ha) and AC-4 (58.0 tonnes/ha). Among the irritant types, Kovvur local has recorded maximum yield of 45.5 tonnes/ha followed by AC 24/1 (37.7 tonnes/ha).

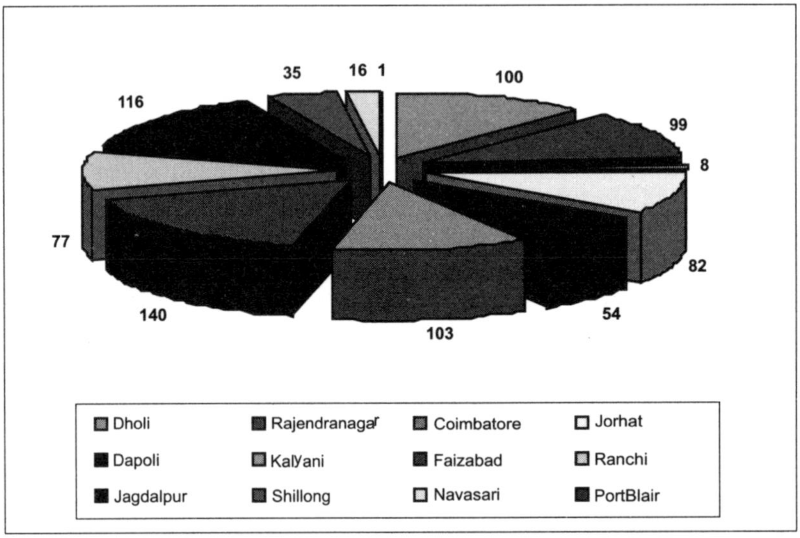

Fig. 3.8. Germplasm accessions of elephant foot yam at different centres

Xanthosoma

A total of 48 accessions are maintained at different centres of AICRPTC. Jorhat (14), Dapoli (11) and Kalyani (11) are major contributors to *Xanthosoma* germplasm.

GENERATION OF TECHNOLOGIES

Cassava

Varieties

The following are the varieties of cassava recommended for different states.

Variety	Recommended Region/ State	Yield (tonnes ha^{-1})	Duration (days)	Salient Features
H-97	Tamil Nadu, Kerala, Karnataka, Andhra Pradesh	25	260-300	Erect branching, dark grey stem, broad leaves, green petiole, cylindrical tubers, light brown tuber skin, white flesh, starch content 27-31%. Tolerant to cassava mosaic disease, performs well under drought conditions and field resistant to spider mite and scale insect.
H-165	Tamil Nadu, Kerala, Karnataka Andhra Pradesh, Maharashtra	33	220-240	Medium tall, erect branching, light grey stem, light brown emerging leaves, broad mature drooping lower leaves light green petiole closely arranged fusiform tubers, golden brown tuber skin, white flesh, starch content 23-25%. Yields good quality starch, easy harvestability, early bulking and suitable for raising as a rotation crop after paddy. Field resistant to cassava mosaic disease.
H-226	Tamil Nadu, Kerala Karnataka, Andhra Pradesh	30	260-300	Medium tall, erect branching, grey stem, green emerging leaves, broad leaves, green petiole, cylindrical tubers, tuber skin creamy with purplish patches, white flesh, starch content 28-30%. Preferred by industries, good cooking quality.
CO.1	Tamil Nadu	30	260-275	Plants medium tall (1.5-2.5 m), late branching, young stems green, emerging shoot pink, tuber skin whitish brown, creamy rind and white flesh, starch content 35%.
Sree Sahya	Assam, Andhra Pradesh Karnataka, Kerala, Gujarat, Maharashtra, North Eastern Region, Tamil Nadu	35	300-320	Tall, erect branching, dark, grey stem, stipules spiny and prominent, broad eaves, purple petiole, cylindrical, necked medium long tubers, light brown tuber skin, white flesh, starch content 29-31%. Tolerant to Cercospora, mites and scale insects and drought.
Sree Visakham	Assam, Andhra Pradesh Karnataka, Kerala, Maharashtra North Eastern Region, Tamil Nadu	35	270-300	Tall, erect branching, dark green stem, light brown emerging leaves, broad leaves, dark green petiole, fusiform compact tubers, brown tuber skin, light yellow flesh, starch content 25-27%. High *carotene* content, field resistant to Cercospora, less susceptible to spider mite and scale insect.

(Contd.)

Variety	Recommended Region/ State	Yield (tonnes ha^{-1})	Duration (days)	Salient Features
CO.2	Tamil Nadu	35	260-275	Highly branching plant type, tuber skin is brown with creamy white flesh. Good cooking quality, low HCN content, suitable for consumption and starch industry.
Sree Prakash	Kerala, Andhra Pradesh	30	210-240	Erect, non-branching, high leaf retention, reddish brown stem, dark green leaves, cylindrical tubers, shallow bulking, brown tuber skin and white flesh. Tolerant to *Cercospora* leaf spot, suitable for low land cultivation as rotation crop in paddy based cropping system.
CO.3	Tamil Nadu	36	240-270	Branching type, tuber skin dark brown with white flesh, good cooking quality, starch content 35.6%. High-yielding, field tolerant to CMD.
H-119	Tamil Nadu, Karnataka	30	210-220	Non-branching, High tuber starch (35.2%). High-yielding short duration crop and good cooking quality.
CO(TP)-4	Tamil Nadu	50	300	Tubers are long cylindrical with brown skin and white flesh. Starch content 40%. Suitable for both rainfed and irrigated conditions.
Sree Rekha	Kerala	45-48	240-300	Cream fleshed tubers with 28-30% starch content and excellent cooking quality. Suitable for cultivation under upland and lowland conditions.
Sree Prabha	Kerala	40-45	240-300	Light yellow fleshed tubers with 26-29% starch and excellent cooking quality. Suitable for cultivation under upland and lowland conditions.
Sree Jaya	Andhra Pradesh, Kerala	23	180-210	White fleshed tuber with 24-27% starch, 40-50 ppm cyanogen but excellent cooking quality. Early maturing variety.
Sree Vijaya	Jharkhand, Kerala, Assam	1823-35	180-210	Erect branching variety with 27-30 % starch content,40-60ppm HCN and excellent cooking quality and the tuber flesh colour is light yellow. Early maturing variety.

In Tamil Nadu, H-226, H-165, MVD-1 and CO.3 are popular whereas in Andhra Pradesh, H-165, Sree Prakash, Sree Jaya and H-226 are popular. Sree Vijaya is performing well in Assam and Jharkhand. A new entry H-740/92 is found to give higher tuber yield in Andhra Pradesh, Assam, Bihar and Jharkhand.

Production Technologies for Cassava

Time of Planting

Cassava is a rainfed crop and hence its successful planting and establishment depends much on rainfall. Time of planting for the different states is as shown as follows:

States/region	*Time of Planting*	
	Rainfed	*Irrigated*
Kerala, Maharashtra	April-May, September-October	December-January
Andhra Pradesh, Assam and Chattisgarh	June	March
Tamil Nadu	June	September

Land Preparation

Before planting the field is ploughed or dug to a depth of 25–30cm to establish a deep porous condition to improve rooting and tuberization.

Planting Material

Cassava is a vegetatively propagated crop and is raised by planting stem cuttings. Healthy stems of 6–10 months maturity, having a thickness of 2-3cm and free from diseases and pests should be selected for planting. About 10 cm from the lower mature portion and about 25–30cm from the upper immature top to be discarded. Immature cuttings or cuttings from virus infected plants should be avoided.

Stakes are to be prepared by cutting the stems with sharp implements like hack-saw or hand-saw to obtain a smooth circular cut so as to induce uniform rooting all around. Setts prepared within 15 days from stems stored with leaves gives better sprouting. Stem cuttings, 20cm long and planting at a depth of 10cm are optimum. The stem cuttings has to be planted vertically.

In non-traditional areas with rainfall limited to 4-5 months during south-west monsoon season, cassava setts can be raised in nursery with single/double budded cuttings under irrigated condition and after 20–25 days, transplanted to the main field.

Method of Planting

Mound method may be adopted in soils having clay content and restricted drainage, whereas ridge method may be followed on slopes to prevent soil erosion. In the mound method, stakes are planted on mounds of height 25–30 cm taken on land already prepared. In ridge method, ridges of 25–30cm height are prepared across the slope on which cassava stakes are planted. Ridge and furrow method of planting is suitable for irrigated cassava cultivation under Coimbatore conditions. The flat method of cultivation may be followed in level lands having good drainage. Pit method followed by mound method has been found superior to all other methods in level lands. In this method, pits (30cm × 30cm × 25cm) are opened first and soil is mixed with cattle manure and reformed into mounds. Cuttings should not be planted till the monsoon becomes active in the region. Under high rainfall zone of Kanyakumari district of Tamil Nadu, mound method of planting of cassava gives better yield.

Spacing and Plant Population

Cassava varieties are of three types; branching, semi-branching and non-branching. Branching and semi-branching types require more spacing and hence should be planted 90cm × 90cm apart, whereas non-branching types requires 75cm × 75cm spacing. The recommendation for different states is as of follows:

H 226

Sree Jaya

Sree Rekha

Sree Prabha

CO.2

CO.3

Popular varieties of cassava

States/Region	*Spacing*	
	Branching	*Semi-branching/non-branching*
Kerala, Andhra Pradesh	90 cm x 90 cm	75 cm x 75 cm
Maharashtra, Gujarat, Chattisgarh	90 cm x 75 cm	-
Tamil Nadu	90 cm x 75 cm	75 cm x 75 cm

Normally one stem cutting is planted per hill. As 6-7 cuttings of about 20 cm length can be obtained from each stem of a healthy plant, approximately 2000 stems are required for a hectare. Under 90cm × 90cm spacing, the total population will be 12,350 plants ha^{-1} and under 75cm × 75cm spacing, it will be 17,780 plants/ha.

The stem cuttings planted may sometimes fail to establish due to adverse weather conditions and poor quality of planting material. These gaps may be filled up within 15 days after planting preferably with longer setts of 30–40 cm length.

Removal of excess sprouts by nipping at the initial stage of establishment (20–25 days after sprouting) helps to prevent mutual shading and competition between plants. Retention of two healthy shoots on opposite sides is better than retaining only one shoot.

Manure and Fertilizers/Biofertilizers

Cattle manure or compost @ 10–15 tonnes/ha during the preparation of land or while filling up the pits to provide about 1 kg organic manure per pit is recommended.

The recommended dose of N, P_2O_5 and K_2O for high-yielding varieties is 100, 50 and 100 kg ha^{-1} respectively. For M4 and local varieties half the above dose is sufficient. For rainfed upland crop of Andhra Pradesh, 60 kg/ha each of N, P_2O_5 and K_2O is profitable as per the conclusion drawn based on three years experimentation (1979-81) at Peddapuram. At Jagdalpur, a fertilizer dose of 75:50:75 kg NPK/ha produced better yield. For short duration varieties like Sree Prakash and H-119, application of NPK @75:25:75 kg/ha is recommended under a close spacing of 75cm × 75cm for Tamil Nadu. A fertilizer dose of 100:50:100 kg NPK/ha and a spacing of 90cm × 90cm are recommended for Gujarat.

Being a long duration crop it is preferable to apply full dose of phosphorus, half dose each of nitrogen and potassium as basal and the remaining half dose of nitrogen and potassium after 45-60 days of planting along with the first interculture and earthing up. However, P application can be skipped for four years when the available P status of the soil is high. Thereafter a maintenance dose of 25 kg/ha per year only needs to be applied.

Green manuring *in situ* with fast growing vegetable cowpea varieties (B-61 or C-152) can substitute FYM application to this cropping sequence. The recommendation for different states is given below:

State	*Condition*	*Fertilizer dose NPK kg/ha*
Andhra Pradesh	Rainfed	60:60:60
	Irrigated	150:60:150
Chattisgarh		75:50:75
Tamil Nadu	Rainfed	75:25:75
	Irrigated	90:90:240
Kerala		100:50:100
Gujarat, Maharashtra		100:50:100

For acid laterite soils, lime application @ 2000 kg CaO/ha at the time of land preparation is beneficial to increase the yield and improve the quality of tubers. Sulphur @50 kg/ha and micro nutrients like zinc, boron and molybdenum @12.5, 10.0 and 1.0 kg/ha respectively as zinc sulphate, borax and ammonium molybdate are found beneficial for tuber yield enhancement.

Foliar spray of 1% $FeSO_4$ + 0.5% Zn SO_4 at 60 and 90 DAP is recommended for Chlorosis in Salem areas of Tamil Nadu. Application of AM fungi or Phosphobacteria (10kg/ha) along with half the recommended dose of P_2O_5 and full dose N and K and FYM (12.5 tonnes/ha) recorded a tuber yield of 28.03 tonnes/ha and it was comparable to that of recommended dose of NPK and FYM. This is recommended for Tamil Nadu and Andhra Pradesh.

Irrigation

Irrigation is normally not undertaken in Kerala as the crop is rainfed, whereas in most parts of Tamil Nadu and Andhra Pradesh the crop is irrigated. Supplementary irrigation to cassava during drought months can result in higher yields. Irrigation at approximately 25% available moisture depletion level throughout the growing season could double the tuber yield. It also improves the quality of tuber through improvement in starch content and reduction in cyanoglucoside content.

Cropping System

Cassava is grown both as a sole crop and also under multiple cropping systems. In Kerala, cassava is intercropped in 20% of coconut stands. In newly planted coconut gardens, cassava performs well up to 5 years with good yields. Based on tuber yield and quality, cassava varieties like Sree Visakham and Sree Sahya can be grown satisfactorily. Inter-cropping of cassava with bunch variety of groundnut and vegetable cowpea and sequential cropping of cassava and paddy are found to be profitable and suggested for generating more employment opportunities.

In Tamil Nadu, its crop is grown mostly in Salem-Dharmapuri -Erode-Coimbatore region and Kanyakumari district. In the former region the crop is cultivated as irrigated crop and tubers are mostly used as raw material for starch and sago industrial units. In the latter tract the cultivation practices are more or less similar to those in Kerala. The produce is used mostly for human consumption. In North-Eastern Hill region, cassava is grown mostly in mixed stands.

Inter-cropping

Cassava is a long duration crop that normally takes 8–10 months for harvest. It is planted at a spacing of 90cm × 90cm and it takes about 3-3½ months to develop enough canopy, and the available sunlight, water and nutrients between rows can be profitably utilized for raising short-duration intercrops. Legumes are the most suited intercrops in cassava. Inter-crops like groundnut (bunch types TMV-2 and TMV-7) and French bean (cv "contender") were found to be promising and economical. These crops can give an additional net income of Rs 3000–5000/ha within 3-3½ months. Intercropping of blackgram with cassava variety, CO.1 has been recommended for Tamil Nadu. Bellary onion (*Allium cepa* L.) is also found to be a profitable inter-crop in Tamil Nadu. In Andhra Pradesh, green gram or black gram is recommended which can increase tuber yield of cassava by 14% and can give 40% more return compared to cassava alone. Soybean and Frenchbean are suitable inter-crops in Assam. While intercropping, slight modifications in the agronomic practices become necessary, depending on the inter-crop.

Cultivation of Intercrops

Inter-crop/cultivar	*Duration (days)*	*Spacing (cm)*	*No. of rows*	*Seed rate (kg/ha)*	*NPK (kg/ha)*	*Yield(kg/ha)*
Groundnut						
TMV-2	100	30 x 20	2	40-45 (Kernel)	10:20:40	1200 dry pod
TMV-7						
Pollachi-2						
French bean	70	30 x 20	2	40	20:30:40	2000
Contender						
Top cross-4						
Cowpea (grain)	90	30 x 15	2	20	10:15:10	800
S-488						
Cowpea (vegetable)						
B-61	65	90 x 20	1	8	10:15:10	3000

Crop Protection

Mosaic, a viral disease is a serious problem in cassava. Most of the cultivars are susceptible to CMD leading economical yield loss. The disease spreads mostly due to the use of diseased planting material. Disease-free planting material should be identified well in advance during October-November. The plants with disease symptoms may be rogued out periodically.

In cassava, MNga-1 showed field resistance to mosaic at Salem, Tamil Nadu and Peddapuram, Andhra Pradesh where the disease prevalence was 21-56% in other varieties.

Harvest

Harvesting depends on duration of cultivar. Normally it becomes ready for harvest 9-10 months after planting. Duration for different varieties are as follows:

Sree Prakash, H-119,CI-590 Sree Jaya, Sree Vijaya	:	6-7 months
H-97, H-165	:	7-8 months
H-226	:	8-9 months
Sree Sahya , Sree Visakham Sree Rekha, Sree Prabha	:	9-10 months

Cassava mosaic infected crop

Harvesting is done by removing the soil from the base of the plants and pulling out the tubers. Average yield of cassava ranges from 10–15 tonnes/ha for local traditional varieties and 35–40 tonnes/ha for improved hybrids grown under the recommended package of practices.

The stem of cassava plants of one season forms the planting material for the subsequent season. The stems are stored after harvest in a vertical position under shade. As a prophylactic measure dimethoate (0.05%) spray is advised.

True Cassava Seed Programme

True cassava seed programme is useful in popularizing cassava in non-traditional areas where the crop is grown mainly for starch purpose and as food crop in tribal or poverty stricken areas of the country like Northern Bihar, Laterite zone of West Bengal, North Eastern states, Orissa, Madhya Pradesh and Maharashtra. The male sterile line 'Ambakkadan' which is a popular cultivar grown in Kottayam and Idukki districts of Kerala had recorded highest mean number of fruits per plant (168). The weight of 100 seeds range from 9–12.9g. Seeds can be stored under ambient conditions up to 6 months without loss of viability. Male sterile lines recorded very good germination (80–90%). It is better to raise the seedlings in the primary nursery before transplanting to field. The field establishment of seedlings range from 75–91%.

While transplanting the seedlings to the main field, the tap roots of the seedlings should be removed so that the tuber development is not affected. The seedling progenies were free from CMD compared to the clones. Progenies of CE-647, S-1 and Ambakkadan were good yielders at closer spacing 70cm × 70cm and the yields were at par with that of the clones of promising varieties. Ambakkadan seedlings recorded higher tuber yield of 2.2 kg/plant and the cooking quality was very good in both seasons. In the first clonal evaluation of seedling progeny lines, the progeny lines of CE-639, CE-630, CI-453, S2 and Ambakkadan had recorded distinctly higher yield than the controls. Studies conducted to determine optimum spacing and fertilizer requirement of seed crop of cassava indicated that closer spacing of 60 × 60cm was the most effective. Levels of fertilizer did not have significant effect on tuber yield.

Sweet Potato

Varieties

Variety	*Recommended Region*	*Yield (tonnes/ha^{-1})*	*Duration (days)*	*Salient Features*
H-41	Tamil Nadu, Karnataka, Kerala	20	120–130	Semi-spreading, leaf unifoliate with entire margin, profusely flowering, tuber skin reddish purple with white rind and flesh. Excellent cooking quality, cooks easily, tastes sweet and low in fibre content.
H-42	Tamil Nadu, Karnataka, Kerala	22	120–130	Semi-spreading plant type, leaf unifoliate with toothed margin, tubers fusiform, tuber skin purplish red and creamy flesh. Cooks easily, tastes sweet and almost free from fibre.

(Contd.)

Variety	*Recommended Region*	*Yield (tonnes* ha^{-1}*)*	*Duration (days)*	*Salient Features*
VL	Hilly areas of Uttar Pradesh Sakarkand-6	20	135–140	Late maturing with spreading vine having light green foliage, tubers elongated, bent on edges and purple in colour with light yellowish flesh. Tuber rich in carotene.
CO.1	Tamil Nadu	28	135	Moderately vigorous vine and is less spreading, leaves medium lobed and dark green, tuber skin light pink with white flesh. Starch 24%, soft texture on cooking with higher sugar content.
Rajendra Sakarkand-5	Plains of Bihar	20	105–120	Extremely spreading, leaves green and six lobed, number of tubers per plant 4-6, tubers cylindrical with long neck, white tuber skin and greenish white flesh. Can tolerate water logging for a week and tolerant to Cercospora and suitable for two tier cropping, deep rooted, suitable for fodder.
CO.2	Tamil Nadu	32	110–120	Compact plant type, shorter vine and less spreading, leaves 10 lobed and emerging leaves pink, tuber skin light pink with white flesh. Starch 29.5%.
CO.3	Tamil Nadu	32	110–115	A medium duration crop with bushy habit, emerging leaves pink and developed ones dark green, tuber skin light red with dark orange flesh. Rich in carotene, starch 30%, better consumer preference.
Varsha	Konkan region of Maharashtra	17	120	Semi-spreading with dark green vines, tubers fusiform with reddish purple skin and light yellow flesh. drought tolerant.
Sree Nandini	Kerala	20	100–105	Spreading with light green leaves, number of tubers per plant 3-4, tubers fusiform with creamy skin and white flesh. Drought tolerant, suitable for paddy fallows as a catch crop.
Sree Vardhini	Kerala	20	100–105	Semi-spreading having light brown emerging leaves, fusiform tuber with purple skin and yellow flesh. Early maturing dual purpose variety, tubers and foliage good for human consumption and as animal feed.
Samrat	Andhra Pradesh	20	110–120	Vines medium spreading, petioles violet, tubers globular with light pink skin and white flesh. Medium duration.
Rajendra Sakarkand-43	Assam, Karnataka, Andhra Pradesh, Maharashtra, Bihar	20	110–120	Medium spreading plant type, tubers 5-6 per plant, tuber skin brownish with white flesh. Tolerant to Cercospora leaf spot. Suitable for late harvesting which will fetch better price.
Rajendra Sakarkand-35	Assam, Bihar, Karnataka West Bengal	25	105–110	Medium spreading having cylindrical tubers, skin dull white with white flesh. Moderately tolerant to Cercospora disease, suitable for February planting in plains of Bihar, less mortality of vines even when planted in spring season.
Rajendra Sakarkand-47	Bihar, West Bengal, Uttar Pradesh	30	120–130	Spreading type with green leaves and petiole, emerging leaves purplish white, tubers cylindrical, deep bulking, purple skin with white flesh.

(Contd.)

Variety	*Recommended Region*	*Yield (tonnes/ha)*	*Duration (days)*	*Salient Features*
Rajendra Sakarkand-92	Bihar	25	75	Spreading, green leaves 6-7 cylindrical tuber. Red skin and white flesh, suitable for flood prone area of North Bihar.
Kiran	Andhra Pradesh	22	110–130	Medium spreading, palmate leaves with 4-5 shallow lobes, leaves light green, tubers necked fusiform and tuber skin red with orange flesh. Early bulking.
Sree Bhadra	Kerala, Maharashtra, Bihar Madhya Pradesh	20	90	Semi-spreading, greenish brown vine and petiole, dark brown emerging leaf, broad cordate leaves, tuber shape spherical, light pink tuber skin and creamy flesh. Early maturing with good cooking quality.
Sree Rethna	Kerala	20	90–105	Spreading, light greenish brown vine and petiole, brown emerging leaves, light green shouldered leaves, spherical round tubers, purple tuber skin and orange flesh, starch content 22-23%.
Gouri	Orissa	19	100–120	Semi erect, simple leaves with 3-5 lobes, profuse flowering, red skinned, deep orange flesh tubers, shape of tuber is obovate. High carotene content, can tolerate mid season stress, suitable for both kharif and rabi season.
Bidhan Jagannath	West Bengal	21–24	110–120	High in carotenoid content.
90/704	Jharkhand	10–14	110–120	High in carotenoid content.
Indira Madhur	Chattisgarh	22	120	
Narendra Sakarkand-9	Uttar Pradesh	26	120	Mid season variety, white skinned and white fleshed tubers with pleasant taste.
Narendra Sakarkand-10	Uttar Pradesh	22	120	Mid season variety,red skinned, white fleshed tubers. Can be grown as irrigated as well as rainfed.
RNSP-3	Telangana region of Andhra Pradesh	15	120	Suitable for kharif and rabi season.
Konkan Ashwini	Konkan region of Maharashtra	19	105	Short duration plant having purple colour elliptical long tubers. Viny plants having light green leaves with purple tinge at nodes on stem and alternate simple leaves. High yielding, short duration, elliptical long tubers with dark purple colour.
CO-CIP.1	Tamil Nadu	31	95–100	Shallow lobed leaves, Purple pigmentation at nodes, tubers round elliptic or long elliptic with attractive pink skin and pale yellow flesh.
Indira Nandini	Maharashtra, Chattisgarh		90–100	Short duration
Indira Naveen	Chattisgarh	24-31	110–120	

Orange fleshed genotypes S-1156, CIP SWA-2 and ST-14 are good yielders with high carotene (9541 ìg 100g) in Assam. Bidhan Jagannath, S 1221 and Kamalasundari perform excellently well in West Bengal and are rich in retinol equivalents and minerals. Short duration varieties ideal for intensive cropping system.

IGSP-C-15

Sree Kanaka

Kamalasundari

Indira Naveen

Indira Madhur

Lesser yam – Konkan Kanchan

Popular varieties of sweet potato and lesser yam

Production Technologies

Time of planting

It is mostly grown as a rainfed crop during kharif or rabi season and as an irrigated crop during other periods. The recommended time of planting in various states is given below:

State	Time of planting
Bihar, West Bengal, Assam, North Eastern States and Uttar Pradesh	Second fortnight of September to first fortnight of October
Bihar, Jharkhand	Middle of January to middle of February (with irrigation)
Chattisgarh, Madhya Pradesh and Chotanagpur Plateau of Bihar	August-September
Diara land in Bihar and Eastern Uttar Pradesh	November after recession of flood
Andhra Pradesh	August and November
Tamil Nadu	May-June (rainfed)September-October with irrigation

Planting material

Sweet potato is propagated through vine cuttings. To obtain vine cuttings, nurseries are raised from selected tubers. The nursery is prepared about 3 months prior to planting in the main field. For planting one hectare of land about 100 m^2 primary nursery areas required. The vines obtained from freshly harvested crop are also planted in the nursery to obtain sufficient planting material.

Vines obtained from nursery are found to be healthy and vigorous than those from freshly harvested crop and has recorded good establishment and tuberisation. The cuttings obtained from the apical and middle portion of vines, 20–30cm in length with at least 3-4 nodes and kept under shade for two days prior to planting in the main field, are advocated for better root initiation and establishment.

Land preparation and planting

Cultivation of sweet potato on mounds and ridges give good yield, the latter being followed widely by farmers. The land is well ploughed or dug to a depth of 15–20cm and brought to a fine tilth. Vines are planted at a spacing of 20 cm on ridges of height 20–25cm made 60cm apart. About 83,000 cuttings are required to plant one hectare of land. In case of mounds, 5 cuttings are planted in each mound at a spacing of 80cm × 75cm.

For Bihar, flat method of planting and for Andhra Pradesh, Uttar Pradesh and North - Eastern India the ridge method of planting are recommended. A spacing of 30cm × 30cm is recommended for the Indo-Gangetic Plain whereas 45cm × 45cm is ideal for diara land of Bihar. Under 2-tier cropping with genotype differing in growth and tuberization habit, i.e., Cross-4 or Kalmegh or Samrat (shallow-bulking type) and RS-5 (deep bulking type) in alternative rows, a closer spacing of 30cm × 15cm are also suggested.

Manuring and fertilizers/biofertilizers

In addition to FYM or compost @ 10 tonnes/ha, 40-60 kg/ha each of nitrogen and potash and 40 kg/ha of phosphorus is recommended for Bihar, West Bengal and Assam. While 60:60:60 kg NPK/ha is advocated for Andhra Pradesh, 60:60:90 kg NPK is recommended for Karnataka.

Full dose of P_2O_5 and half dose of N and K_2O as basal application and the remaining at 30–40 days of planting is recommended.

Application of recommended dose of N along with 2 kg *Azospirillum* ha^{-1} as vine dipping and 10 kg *Azospirillum* ha^{-1} as soil application is recommended to the farmers of Tamil Nadu, Assam, Bihar, West Bengal and Kerala. The recommended dose of N along with 2 kg/ha *Azospirillum* as vine dipping and 10 kg/ha as soil application is recommended for Andhra Pradesh. At Faizabad 30:40:60 NPK kg ha^{-1} with 30 kg N by vermicompost and 45:40:60 NPK kg/ha with 15 kg N by vermicompost gave maximum tuber yield.

Spraying Cycocel twice @ 500–1000 ppm on 30th and 45th days of planting recorded higher yields in Bihar, West Bengal, Assam, Maharashtra and Andhra Pradesh. The vegetative growth was restricted and there was substantial increase in tuber yield.

Irrigation

For proper establishment of the crop, sufficient soil moisture has to be ensured at the time of planting and seed beds remain moist for 4-5 days. During dry weather, irrigation at ten days intervals may be given. Irrigation would generally increase the yield and improve the grade and quality of marketable tubers.

In Tamil Nadu irrigation at 60% moisture depletion level increased the tuber yield by 24% over non irrigated crop. Whereas, for chalka soils of Telangana of Andhra Pradesh a moisture depletion of 33% was found ideal for irrigation. This works out normally at 15 days interval in rabi season. For dry zone of Telangana in Andhra Pradesh during rabi season 4-5 irrigations can give higher yields.

Crop rotation and cropping sequences

Raising sweet potato after kharif maize is recommended in eastern region. In Kerala, sweet potato can be grown in rotation with paddy in lowlands during January- April after harvest of second crop of paddy.

Ideal crop sequences can generate a net income of more than Rs. 25,000/ha.

Bihar–Maize (*kharif*)–Sweet potato (Autumn with early variety Kalmegh)–wheat (rabi) and Moong (Summer), Maize (pre-*kharif*)–Sweet potato (Autumn)–Onion (Rainy)
West Bengal–Moong (summer)–Taro (pre-*kharif*)–Sweet potato (Autumn)
Orissa–Midseason Paddy (rainy) followed by sweet potato.
Chattisgarh/Uttar Pradesh/Maharashtra–Vegetable cowpea–sweet potato

Pest Management

Sweet potato weevil is the most important pest causing very severe damage to the crop. It damages vines and tubers rendering them unfit for consumption. The following integrated pest management practices are to be adopted for effective control.

- Removal of sweet potato residue crop and tubers in the field and *Ipomoea* weeds in and around the field.
- Use pest free planting material and disinfecting the planting setts by dipping in dimethoate (0.05%).
- Installation of pheromone traps @1trap/100m^2 (10m × 10m) with BARC-CTCRI developed pheromone septa.

- Collection of weevils from the trap and changing water with detergent at periodic interval.
- Re-ridging the crop at 30 DAP and 60 DAP.
- Harvesting at 105–110 days.
- Destruction of crop residues and weevil affected plants.
- Continue pheromone trapping till a fortnight after harvest.

The IPM package against weevil revealed superiority over control and chemical treatments in Coimbatore, Rajendranagar, Faizabad, Dapoli, Dholi, Ranchi, Jorhat, Bhubaneswar, Thiruvananthapuram and Kalyani.

Eight sweet potato viruses have been indexed in different varieties using Nitro Cellulose Membrane NCM-ELISA at BCKV, Kalyani.

Harvesting

The crop attains maturity in 3– 4 months after planting depending on variety. In general, delay in harvest after 120 days increase sweet potato weevil infestation. Early variety like Kalmegh or Samrat can be harvested in 90 days to fit into the cropping system. In Assam, harvesting is done after 105 days. In Andhra Pradesh also it is advisable to harvest the varieties Samrat and RS 35 at 105 days.

Pheromone trap for sweet potato weevil management

Yams

Lesser Yam (Dioscorea esculenta)

Variety	Recommended Region	Yield (tonnes/ha)	Duration (days)	Salient Features
Sree Latha	Kerala, Bihar, Andhra Pradesh, Maharashtra, Assam, West Bengal	25	220-240	Vine dark brown, twinning to left, spiny and medium hairy petiole, oblong to fusiform tuber with grayish brown skin and creamy white flesh.
Konkan Kanchan	Konkan region of Maharashtra	18	190-210	High yielding, vine having light green coloured simple alternate leaves, spines at the base of petiole. Stem is thin, cylindrical and green in colour. Tubers are elongated, almost cylindrical with round ends. High yielding, equal size tubers, white and sweet in taste.
Sree Kala	Kerala	20	240-270	Vine twining to right, spiny medium hairy, purple coloured, tuber shape oval smooth with brown skin and white flesh. Good tuber shape and culinary quality.
Moa aloo (Local)	Assam	12	240	High yielder. Good tuber shape and excellent culi nary quality.

Sree Latha is a versatile variety suitable for Bihar, Assam, West Bengal, Kerala and Andhra Pradesh whereas Konkan Kanchan is a variety identified and recommended for Maharashtra. A fertilizer dose of 80:60:80 kg/ha of N, P and K is generally recommended. Full dose of P and half dose of N and K are applied 35–45 days after planting and the remaining dose of N and K is applied one month after the first application.

Greater Yam (D. alata)

Variety	Recommended Region/States	Yield ($t\ ha^{-1}$)	Duration (days)	Salient Features
Sree Roopa	Kerala	25	280-300	Stem green with purple ridges, green emerging leaves, digitate and smooth tuber, black tuber skin and white flesh. Good cooking quality.
Sree Keerthi	Kerala	25	280-300	Stem and petiole green, light brown emerging leaf, conical, rough tuber with brown skin and white flesh.
Konkan Ghorkand	Konkan region of Maharashtra	16	180	It is high-yielding and erect growing vine produces aerial tubers having purplish green colour stem and simple leaves.
Sree Karthika	Kerala	25-30	270-280	Good table variety
Orissa elite (Hinjilicut Local)	Orissa	20-24		Local selection with good culinary qualities.
CO.1	Tamil Nadu	32	255	Tubers are big white flesh, rich in carbohydrate (28%) and protein (2.5%).
Konwari aloo	Assam	18-25	280-300	High-yielding with better quality tuber.

African Yam (D. rotundata)

Variety	Recommended region	Yield (tonnes/ha)	Duration (days)	Salient features (days)
Sree Subhra	Kerala	35	300	Gresen petiole, vine twining right, spiny stem, green emerging leaves, cylindrical partly hairy tubers, brown tuber skin with white flesh, starch content 21-22%. Drought tolerant, good flavour.
Sree Priya	Kerala	35	300	Green petiole and vine twining to right and spiny, green emerging leaves, cylindrical smooth tubers with brown skin and white flesh. Drought tolerant, good flavour, suitable for intercropping in mature coconut gardens and with banana.

In India, greater yam is widely grown in all the states but major yam-producing states are Kerala, West Bengal, Bihar, Orissa, Assam, Andhra Pradesh, Gujarat and Maharashtra. They are raised as rainfed crops and the tubers are mainly consumed as vegetables. African yam or white yam is an introduction from Nigeria and two varieties identified, *viz.* Sree Priya and Sree Subhra are gaining popularity in Kerala. Seed tuber pieces of size 250 g are used for planting in *D. alata* and *D. rotundata* and single tubers of 100–125g in *D. esculenta.* Seed tubers are planted in mounds formed over pits at a spacing of 90cm × 90cm for *D. alata* and *D. rotundata* and at 75cm × 75cm for *D. esculenta*. Farm yard manure @ 10 tonnes/ha is incorporated in soil during land preparation and NPK @ 80:60:80 kg/ha is found to be optimum for the crop. Tuber yield of 25–30, 30–35 and 35–40 tonnes/ha is obtained for lesser yam, greater yam and white yam respectively with a duration of 250–280 days. In Andhra Pradesh *D. alata* is cultivated as non tribal crop while it is grown in tribal condition in Kerala and Maharashtra.

Leaf spot or blight in yams caused by *Cercospora* spp. and *Colletotrichum* spp. can be controlled by spraying cercobin or MBC 0.2%.

Colocasia-Taro and Bunda

Varieties

Variety	Recommended region (tonnes/ha)	Yield	Duration (days)	Salient features
Colocasia (Arvi or Taro)				
Satamukhi (Kovvur)	Andhra Pradesh, Maharashtra, Kerala, Karnataka, West Bengal, Bihar	15	180	Plant height 50-55 cm, medium spreading, 6-7 tillers, dark green sagitate leaf, light purple petiole. Field tolerant to blight.
Sree Pallavi	Kerala	15	205-210	Erect tall droopy leaves with glaucous green upper surface, light green lower surface, big conical corm with club shaped cormels. Field tolerance to leaf blight and dasheen mosaic and a tall variety with large number of small sized tubers.

(Contd.)

Variety	*Recommended Region*	*Yield (tonnes/ha)*	*Duration (days)*	*Salient Features*
Sree Rashmi	Kerala	15	205-210	Erect tall with droopy leaves having entire purple margin, green at upper surface and light green at lower surface, corm is cylindrical and medium sized conical cormels. This variety has acrid free leaves, corms and cormels, gives economy yield under low input level, field tolerant to dasheen mosaic.
Bidhan Chaitanya	West Bengal	15	180-240	Plant height 75-80 cm, with entire inclined waxy leaf with blund tip peltate shape and yellowish greeny colour, branched round to oval cormels. Tubers non-irritant with excellent cooking quality and very mealy and non aromatic.
Bidhan Joydeb	West Bengal	12	150-120	Medium spreading with a plant height of 60–70 cm with entire inclined cup shaped lamina having peltate shape and green in colour. Branched eliptical cormels. Tubers are non-irritant with excellent cooking quality very mealy and non aromatic.
Kadma arvi	Jharkhand	17	150	Erect, with emerging leaf colour light green, moderate flowering type. Resistant to leaf blight, viral infestation and pests, 18 cormels per plant. Tubers almost non-irritant with excellent cooking quality.
Ahina Katchu	Assam	17	175	High yielder better quality tuber.
Sree Kiran	Kerala	20	180	Hybrid
Sree Kripa	Kerala	20	180	Hybrid
PKS-1	Uttar Pradesh	20-31	150	Good for chips making.
RNCA-1	Telangana Region	14-18	180	Medium duration
Sonajuli	Orissa	11-12	180	Tolerant to blight, drought and salinity stress
NDC-1	Uttar Pradesh	12-15	170-180	Mid season variety with less acrid corms, which can be grown in both seasons summer and rainy.
NDC-2	Uttar Pradesh	12-15	180-190	Mid season variety with less acrid corm can be grown in both seasons
Jhankri	Orissa		11-12	Tolerant to blight, drought and salinity stress
White Gauria	Bihar	18	180	Early maturing variety, erect tall with droopy leaves. Moderately succeptible to leaf blight.
Indira Arvi-1	Chhattisgarh, Assam	22	210-240	Erect plant type with emerging leaves with green pigmented margin. Higher number of main marketable cormels (9-10). Cormels are non acrid with very good cooking quality. Tolerant to water logging condition. Corm shape round and cormel shape club with blunt end. Suitable for cultivation in kharif season. Field tolerant to pests and blight.
CO.1	Tamil Nadu	20	240	Erect plant type with less acrid tuber and good keeping quality. Tuber starch content 22.5% and protein 2.4%.
Muktakeshi	Orissa, Northeastern states	9-14	200	Resistant to blight
Bhavpuri	Andhra Pradesh	13 under Rainfed 28 under Irrigated	180-200	Rich in culinary qualities.

(Contd.)

Variety	Recommended Region	Yield (tonnes/ha)	Duration (days)	Salient Features
***Colocasia* (Bunda)**				
Narendra Bunda-1	Uttar Pradesh, Assam	35-40	180-210	Plant height 75-125 cm with erect growth habit with clock wise leaf arrangment and erect lamina orientation. Vine colour green with purple streaks on upper and lower portion. Cylindrical corm and eliptical cormels. Resistant to lodging. Corm and cormels are suitable for table purpose as it has very less acridity soft and easy to cook.
NDB-3	Uttar Pradesh, Assam	35-40	180-210	Corms cook fast.
NDB-21	Uttar Pradesh, Assam	45-50	150-180	Early maturing. Corms cook easily

Production Technologies

Arvi/bunda can be planted successfully during February and June. In Bihar, West Bengal, and eastern Uttar Pradesh, where irrigation is available, February planting is ideal. In Kerala, April — June is the ideal time for planting. February is the planting time in Andhra Pradesh, April in Assam and June in other parts of India.

A fertilizer dose of 80:60:80 kg/ha NPK has been found optimum for the crop and is being adopted in Bihar. In Tamil Nadu, 40:60:120 kg NPK/ha is recommended. A fertilizer dose of 80:60:80 for Uttar Pradesh and 120:60:120 for Andhra Pradesh are suggested. It is advisable to apply half dose of N and K and full dose of P at the time of planting. The remaining half dose of N and K applied in two split doses, i.e., first, 7–10 days after sprouting and second, a month later along with weeding and earthing-up.

In Bihar and North-eastern region, about 8 irrigations are required for summer crop. In Andhra Pradesh optimum time for irrigation of *Colocasia* is at 25% depletion of soil moisture or at 10 days intervals. Normally, it is harvested after 200–210 days. Harvesting 165 days after planting has been found ideal for Andhra Pradesh, West Bengal and Uttar Pradesh when it is grown as a summer crop.

In Bihar, intercropping of three rows of onion in between two rows of *Colocasia* has recorded an additional net return of Rs 20,000/ha under irrigated condition.

Kadma local is less susceptible to leaf blight (3.41% PDI) at Ranchi. Muktakeshi showed high degree of resistance to the disease at Dholi, Rajendranagar and Kalyani where as RCM-4-11 showed resistance at Dapoli. RNCA-1 at Rajendranagar and BCC-1 at Kalyani were also found to have leaf blight tolerance with high yield.

Timely application of Dithane M-45 (25%) reduce the intensity of *Phytophthora* leaf blight which is a serious disease of *Colocasia*. Early maturing cultivars, Muktakeshi ,Nadia local and Kadma local escape severe attack of *Phytophthora* leaf blight. The IDM against taro leaf blight conducted at Dapoli, Rajendranagar, Dholi and Kalyani revealed that IDM package (early planting, use of tolerant varieties, spraying of neem leaf extract and soil incorporation of *Trichoderma*) is effective in containing leaf blight and enhancing cormel yield economically. Neem leaf extract spraying (1 kg in 10 litre water) helps to suppress taro leaf blight in West Bengal.

Incorporation of biofertilizers like AM fungi (10 kg/ha) with or without Phosphobacteria (10 kg/ha) alongwith half the recommended doze of P_2O_5 (30 kg) and full dose of N and K

Application of karanj cake @500g/plant was found equally effective as that of 80:40:80 kg NPK/ha in producing maximum corm yield at Ranchi. A fertilizer dose of 120:60:60 kg NPK/ha with 5 kg FYM per pit of elephant foot yam gave maximum corm yield (35.8 t ha^{-1}) at Ranchi. Application of 125:60:100 kg NPK/ha with 50 kg N by FYM or vermicompost gave maximum corm yield (50.87 and 45.92 tonnes/ha respectively) at Faizabad. Vermicompost application was found to enhance the corm yield at Faizabad as well as at Kalyani. Application of 150 kg/ha each of N and K in three split doses (basal, 30 days and 60 days after planting) is effective in West Bengal. With cow pea as green manure 120:60:120 kg/ha NPK is also recommended for maximum yield for West Bengal.

At Navasari 75cm × 50cm spacing and seed tuber size of one kg gave maximum yield (68–96 tonnes/ha) when planted during early March. At Faizabad 60cm × 45cm or 60cm × 30cm spacing and 175 or 200g seed tuber size were found to be optimum for elephan-foot yam cultivation.

Cropping System

Moong – elephant-foot yam – sweet potato cropping sequence is recommended for West Bengal. In eastern parts of India, a cropping sequence of elephant-foot yam – wheat/bengal gram/pea or elephant-foot yam–okra/*Colocasia* is found to be profitable.

Vegetable cowpea as intercrop in elephant foot-yam is profitable for Uttar Pradesh, Bihar, Assam, Tamil Nadu and West Bengal giving an additional income of Rs. 50,000/ha. The fertilizer dose for intercropped elephant-foot yam in coconut gardens was found to be optimum for maximum yield at 80:60:100 kg NPK/ha at Port Blair.

Crop Protection

In elephantfoot-yam cultivation straw mulching the crop @ 2 kg/m was found to be effective in the weed management with enhanced corm yield at Kalyani, Thiruvananthapuram and Rajendranagar. At Dholi, sesame leaf mulching or black polythene mulching was effective. Cowpea mulching after harvest of green pods as green manuring also helped to reduce weeds and increase yield.

The trial on management of diseases on elephantfoot-yam showed that hexaconazole @ 0.01% + mancozeb @ 0.3 % spray at 50 DAP contained the disease incidence and enhanced the corm yield both at Kalyani and Rajendranagar. *Trichoderma* and *Pseudomonas fluorescence* (100 g in 20 kg FYM) applied in the pits were effective in containing the disease and it can be compared to the synthetic fungicides. Seed corm treatment with cow dung slurry, *Trichoderma* and application of hexa conazole is also equally effective.

Snails are a menace on tuber crops at Dapoli, Maharashtra. Yam bean as a border crop to elephantfoot-yam reduced the snail damage on the crop. Spraying of 5% yam bean seed and soap nut seed extracts were effective in checking snail population. Waste leaves of cauliflower acted as a trap by luring them.

Yam Bean

Pachyrhizus erosus L. is a starchy root with good sugar and ascorbic acid content. It is also known as potato bean. Rajendra Mishrikand-1 (RM-1) is released from Rajendra Agriculture University, Dholi, Bihar. It is recommended for Bihar and West Bengal with a tuber yield of 35 tonnes/ha and a duration of 110–140 days. The tuber contains 82% water, 1.5% protein, 10% starch and 5% sugar. It is widely used as salads especially during Pooja festivals. Seeds are sown during June-July (with a spacing of 30cm x 30cm and fertilizer dose of 80:40:80 kg NPK/ha) and harvest during January-February. Removal of flower buds (75 days after sowing) is essential for optimum tuber yield. Spraying 50 ppm of 2,4-D takes care of dehiscence of flowers and results in better yield. Seed crop is separately raised for seed purpose where in flowering and pod formation are allowed and pods are harvested at 240 DAP for seeds. Intercropping pigeon pea or maize in yam bean is recommended in Bihar.

Country Potato

It is cultivated in Tamil Nadu, Kerala and Karnataka. The tubers are rich in minerals and used as vegetables. Stems are used as planting materials. Nursery can be raised from healthy tubers. Stem cuttings are planted on ridges raised flat beds. Ideal planting time is July-August and can be harvested with in 4–6 months. The 60:60:150 kg NPK/ha is recommended for Tamil Nadu. Yield ranges from 10–12 tonnes/ha.

Tuber Crops as Intercrops in Orchards/Plantation

In Kerala, elephantfoot-yam is grown as an intercrop in coconut gardens. Elephantfoot-yam is identified as an ideal intercrop in litchi orchards of Bihar and mango orchards of Chattisgarh. In mango orchards of Raipur, arrowroot performed better with higher tuber yield. Sweet potato was found to be a good intercrop in cashew plantations at Dapoli in Konkan region of Maharashtra. This technology gives higher monitory returns for farmers.

TRANSFER OF TECHNOLOGIES

Distribution of planting material of released varieties and potential cultivars to farmers and NGOs

Crop	*RAU Dholi*	*BAU Ranchi*	*IGAU Raipur*	*TNAU Coimbatore*	*ANGRAU Rajendranagar*	*BSKKV Dapoli*	*AAU Jorhat*	*NDUAT Faizabad*	*NAU Navasari*	*BCKV Kalyani*
Cassava(cuttings)	380	–	500	30000	9540	3400	2800	500	–	500
EFY(kg)	3721	1646	2610	–	10,000	4	3607	–	Seed corm var.Gajendra	2769
Colocasia(Arvi) (kg)	1585	50	1270	–	1998	7.5	527	–	–	845
Colocasia (bunda)	–	–	–	–	–	–	401	–	–	–
Sweet potato (tuber/vinekg/ cuttings No.	745	830	700sq.m secondary nursery+ 45000 cuttings	–	1980	4637 cuttings	144kg tuber 12500 cuttings	–	–	14400 cuttings
Yams(kg)	300	–	100	–	–	143.5	–	–	–	–
Yambean(seeds-kg)	527.5	–	–	–	–	6.2	–	–	–	–

About 2,500 farmers/institutions have been benefited from different centres.

Dissemination of Technologies

Dapoli

About 1,250 farmers visited this centre during the last 5 years. Live tuber samples were displayed and information on cultivation practices and varieties were disseminated to farmers. About 100 college and school students visited the centre to know about the tuber crops. In the exhibition organized by the University, the centre displayed tuber crops specimens along with the information. Honorable Union Minister for Agriculture Shri Sharad Pawar visited the centre and appreciated the programme.

Dholi

The live specimens of high-yielding varieties of sweet potato, yambean, *Colocasia*, elephant-foot yam, cassava and *Dioscorea* were displayed in *Kisan Mela* held at various places. The method of cultivation and plant protection technologies of different tuber crops were shown through charts and posters for educating the farmers of localities. Leaflets/folders (in Hindi) describing details about production and protection technologies were distributed to farmers. A large number of farmers from different districts of north Bihar visited the tuber stall and enquired about tuber crops. His Excellency Governor of Bihar Sri Buta Singh visited the stall and appreciated. The planting material of tuber crops were also supplied to farmers. The tuber crops scientists of different disciplines also participated in *Kisan Gosthi* and replied the questions of farmers related to tuber crops. Dr P.P. Singh delivered two separate sessions on cultivation of yam bean and improved package of practices of taro through Doordarshan Kendra, Patna.

Ranchi

A number of training programmes/seminars were conducted for farmers on tuber crops at BAU, Ranchi. Regular advisory services were imparted by tuber crops scientists of centre to farmers. About 75 students were trained on all theoretical as well as practical aspects of tuber crops. On-centre (Ranchi) and off-centre trainings and technical consultancy on different tuber crops were imparted by scientists to farmers of Pithoria, Kadma, Sukurhuttu, Silli, Chatra, Hussir, Angara, Gola etc. Planting material were also distributed among them. Tuber crops farmers of Namkum (Ranchi) village were given one day training on improved cultivation package of different tuber crops during November 2004. Farmers' fair and Horticultural Show are regularly organized by BAU, Ranchi during January-February of every year. The live demonstration of high-yielding varieties of tuber crops, *viz.* elephant foot yam, sweet potato, cassava, *Colocasia* and *Dioscorea* were displayed in the stall. Improved production technologies in tuber crops cultivation were demonstrated through different aids of extension. A large number of farmers used to visit the stall and enquire about the tuber crops particularly about their therapeutic and nutritional value and propagation.

During 2005-06, four batches of about 100 farmers hailed from different places of Bihar, viz. Patna, Biharsarif and Vaishali were trained on cultivation practices of different tuber crops along with the other horticultural crops. The centre has started to exhibit improved varieties/technologies of different tuber crops to tuber farmers as live demonstration at Technology Park, BAU Ranchi. During 2005-06, five consecutive batches of 30 farmers hailed from Chatra, Koderma and Hazaribagh received training on nutritional importance of tuber crops.

Radio/TV talks were made on different tuber crops production technologies by scientists of the centre to popularize tuber crops.

Kalyani

One day farmers' training on tuber crops was organized on 18 March 2005 and a seminar on tuber crops was arranged on the same day. A folder (in Bengali) on agro-techniques and crop protection of important tuber crops in West Bengal was released for farmers.

Two training programmes were organised in Howrah district on taro-elephant-foot yam cultivation and sweet potato vine production. Udainarayanpur is found very potential zone for taro cultivation. Agricultural fair was held in Howrah district from 15–17 December 2005 and Kalyani Centre actively participated in the programme and arranged interaction with farmers. All the materials were displayed and farmers expressed high interest in elephant-foot yam seed corm production. The cultivars also have been provided to 'Ashalaya' Don Bosco, Nadia. Kalyani Centre opened a display stall of the tuber crops produce in Eastern Zone Agricultural Fair organised during 30 January — 2 February 2006.

Jorhat

"Improved cultivation practices on *Colocasia* and greater yam" in Assamese was broadcasted by AIR, Dibrugarh (Assam), on 19 December 2004 and centre participated in phone-in-line Programme, AIR, on 18 August 2004 to popularize tuber crops.

Faizabad

The AICRPTC centre Faizabad, arranged a tuber crop exhibition stall in the Gauri Ganj Kisan Mela in 2005 which was visited by Sri Rahul Gandhi, Honourable Member of Parliament. A group of Journalist trainees from Lucknow also visited the AICRPTC centre. They were imparted knowledge and given awareness on food and nutritional aspects of tuber crops.

Coimbatore

Thirty demonstration trials on improved technologies developed for tuber crops, 30 farmers' meetings and three farmers' days were conducted by the centre. Scientists imparted trainings to KVK farmers on tuber crops cultivation and the information was disseminated through Kissan call centre.

Rajendranagar

Latest technologies of tuber crops were imparted to officers of Department of Agriculture and Horticulture through 50 training programmes to farmers through 25 training programmes. Ten press notes were released on package of practices of tuber crops in Telugu. Fourteen radio talks and 12 TV programmes were broadcast on package of practices of tuber crops. The scientists participated in 9 Kisan Melas. Three leaflets were published in Telugu. Farmers from Maharashtra, Orissa and different parts of Andhra Pradesh visited the Centre.

Navsari

A total of 19 trainings/seminars which included 8 on-campus and 11 off-campus trainings on improved cultivation of tuber crops were conducted for farmers. A total of 222 farmers were benefited. The 9 farmers' day, 18 exhibitions, 10 scientists' visit to farmers' fields and 12 telephonic helpline were conducted for creating awareness on tuber crops.

The scientists of the centre were successful in popularizing the technologies for tuber crops. In elephant-foot yam, Gajendra is widely accepted variety. The farmers adopted the recommended tuber crops technologies. Many farmers grow elephant-foot yam as a commercial crop.

PUBLICATIONS

Reports/Bulletins

- Annual Reports of AICRP Tuber Crops – 1995 onwards
- Biennial Progress Reports of AICRP Tuber Crops – 1991 onwards
- Proceedings of Group Meetings – 1991 onwards
- Research Highlights (1968–95) AICRP (Tuber Crops), pp. 27
- Research Highlights (1996-97) AICRP (Tuber Crops), pp.11
- Crop Varieties, pp. 59
- True Cassava Seeds, (TCS), pp.22
- Production Technology for Cassava and Sweet Potato, pp.24

Technical Bulletins

- Scientific Cultivation of Tuber Crops by NDUAT, Faizabad (English).
- Cultivation of OI (Elephant-Foot Yam) for More Profit
- Better Cultivation Practices for *Colocasia* (Arvi)
- Profitable Cultivation of Sweet Potato
- Better Prospects of Cassava Farming in Jharkhand
- Scientific Cultivation Package for Yam bean

Extension Bulletins/Brochures

- "Agro-techniques and Crop Protection of Important Tuber Crops in West Bengal" in Bengali (BCKV, Kalyani).
- Delicious Items from Elephant-Foot Yam; Sweet Potato Recipes (two folders in Hindi) (NDUAT, Faizabad).
- Sweet Potato Cultivation; Elephant-Foot Yam Cultivation (two folders in Hindi) (RAU, Dholi).
- Three Pamphlets in Telugu for Sweet Potato, Taro and Elephant Foot Yam (ANGRAU, Rajendranagar).
- Package of Practices for Six Tuber Crops (Gujarati) (NAU, Navasari)

MILESTONES AND IMPORTANT EVENTS OF AICRP TUBER CROPS

1968
- In Fourth Five Year Plan All India Coordinated Research Project on Tuber Crops was approved and implemented with TNAU, Coimbatore and RAU, Dholi as centers. Director, CTCRI was holding the charge of Project Coordinator.

1969
- First Workshop on AICRP TC was held at IARI, New Delhi
- The APAU, Hyderabad centre was included

1971
- II workshop at CTCRI, Trivandrum
- The AAU, Jorhat was included as new centre
- Two Sweet potato and three Cassava varieties were released for Tamil Nadu, Karnakata, Kerala and Andhra Pradesh

1972
- III Workshop at CTCRI, Trivandrum

1974 • One sweet potato variety was released for Uttar Pradesh

1975 • IV Workshop at CTCRI, Trivandrum
- KKV, Dapoli was included as new centre
- The ICAR Research Complex for NEH Region, Barapani was included as voluntary centre.
- KAU, Thrissur centre was included

1976 • BCKV, Kalyani centre was included
- Two sweet potato varieties were released for Tamil Nadu

1977 • Three cassava varieties were released for Tamil Nadu, Assam, Andhra Pradesh, Karnataka, Kerala, Maharashtra and NE region.

1979 • Dr S.P. Verma, a full-fledged Project Coordinator joined in March with the Head-quarters at Rajendra Agricultural University, Pusa, Bihar.
- V Workshop at KAU, Thrissur
- The UAS, Bangalore was included as new centre at Ullal

1980 • One sweet potato variety was released for Bihar

1982 • VI Workshop at RAU, Dholi
- One sweet potato variety released for Tamil Nadu

1983 • VII Workshop, at CTCRI, Trivandrum
- One sweet potato and one lesser yam varieties were released for Maharashtra, Kerala, Bihar, Andhra Pradesh, Assam and West Bengal

1984 • One cassava and one lesser yam varieties was released for Maharashtra, Kerala and Tamil Nadu

1985 • One *Colocasia* (taro) variety released for Andhra Pradesh, Maharashtra, Kerala, West Bengal and Bihar

1986 • VIII Workshop at BCKV, Kalyani Center at KAU, Thrissur and UAS, Ullal were dropped

1987 • IX Workshop at AAU, Jorhat
- Three new centers, namely BAU, Ranchi, IGAU, Jadgalpur, NDUAT, Faizabad were included
- Three sweet potato, one cassava, two greater yam, two white yam and two *Colocasia* varieties were released for Kerala, Andhra Pradesh, Tamil Nadu and Assam

1988 • One lesser yam and one greater yam were released for Assam

1989 • The Headquarters of Project Coordinator was shifted from RAU, Pusa to RC, CTCRI, Bhubaneswar.
- X Workshop at APAU, Rajendranagar
- Farmers mela was conducted at Rajendranagar

1990 • Two sweet potato varieties were released for Assam, Bihar, Karnataka and West Bengal

1991 • First Group Meeting at KKV, Dapoli
- One Greater yam, one *Colocasia*, one elephant foot yam and one Coleus varieties were released for Tamil Nadu, Andhra Pradesh, Bihar, West Bengal and Assam
- Farmers mela was conducted at Dapoli

1993 • Second Group Meeting at CTCRI, Trivandrum

BCKV centre in tuber crops exhibition

Exhibition at Dholi

Tuber crops exhibition at Navsari

Dr H.P. Singh, DDG (Horti.) interacting with Shri A.K. Upadhyay, Special Secretary, DARE and others at tuber crops exhibition in Muzaffarpur

AICRPTC exhibition stall at Dholi

Dr. Mangala Rai, former DG, ICAR, visiting AICRPTC centre at Kovvur

- The headquarters of Project Coordinator was shifted to CTCRI, Trivandrum
- One sweet potato, one lesser yam and one Coleus varieties were released for Bihar, West Bengal, Uttar Pradesh and Kerala

1994
- NAU, Navasari was included as new centre
- Dr P.G. Rajendran took over as Project Coordinator in August
- One sweet potato variety was released for Andhra Pradesh

1995
- 3rd Group Meeting at TNAU, Coimbatore
- Research Highlights (1968–95) of AICRPTC were published
- One cassava variety released for Tamil Nadu, Karnataka and Maharashtra
- Farmers mela was conducted at Coimbatore

1996
- Two sweet potato varieties were released for Kerala, Maharashtra, Bihar and Madhya Pradesh

1997
- Fourth Group Meeting at CTCRI, Trivandrum
- 'Production Technology for Cassava and Sweet Potato' and 'Research Highlights 1997' were published
- Three taro varieties were released for West Bengal and Bihar

1998
- One sweet potato and one Elephant foot yam were released for Orissa and Kerala respectively

1999
- Fifth Group Meeting at BCKV, Kalyani
- Crop varieties recommended under AICRPTC (other than potato) was published
- Two sweet potato varieties, one Elephant foot yam and one Yam bean varieties were released for Andhra Pradesh, Tamil Nadu, West Bengal and Bihar
- Farmers Mela was conducted

2000
- CARI, Port Blair was included as voluntary centre
- 'True cassava seeds (TCS) for rapid spread of the crop in non-traditional areas' was published.
- One sweet potato, two cassava, one Tannia varieties were released for Maharashtra and Kerala

2001
- Sixth Group Meeting at IGAU, Jagadalpur
- Two sweet potato varieties and one Bunda variety was released for Bihar, Chhattisgarh and Uttar Pradesh
- Farmers mela was conducted at Jagdalpur

2002
- One cassava variety and one *Colocasia* (taro) varieties were released for Tamil Nadu and Orissa

2003
- Dr M.S. Palaniswami took over as Acting Project Coordinator in September
- One *Colocasia* (taro) variety was released for Chhattisgarh and Assam

2004
- Seventh Group Meeting at CTCRI, Trivandrum
- Five sweet potato, two greater yam and two *Colocasia* varieties were released for West Bengal, Jharkhand, Uttar Pradesh, Bihar, Orissa, Assam and Kerala

2005
- Dr M.S. Palaniswami took over as a regular Project Coordinator in August

2006
- Eighth Group Meeting at NDUAT, Faizabad
- A technical bulletin on 'Region Specific Technologies for Root and Tuber Crops' was published

- Scientific cultivation on elephant-foot yam, arvi, sweet potato, yam bean, (all in Hindi) was brought out as pamphlets
- Three sweet potato, three cassava, five taro, two Bunda and one elephant-foot yam varieties were released for Maharashtra, Chhattisgarh, Andhra Pradesh, Assam, Bihar and Uttar Pradesh
- Farmers mela was conducted at Kumarganj
- The CAU, Imphal and MPUAT, Udaipur, joined as voluntary centres
- Release of Annual Report 2005-06 by Hon'ble Minister of State for Agriculture, Government of India
- Indira Madhur a high carotene sweet potato released for Chhattisgarh

2007
- Four decades of AICRPTC at BCKV, Kalyani – bulletin released
- IX Biennial Group Meeting at NAU, Navsari 21–23 December
- Release of the Annual Report 2006-07 by Deputy Director-General (Hort.), ICAR
- Three decades of AICRPTC at Dr BSKKV, Dapoli – bulletin released
- Root and tuber crops exhibition at Navsari
- Preparation of XI plan for AICRP Tuber Crops. The UAS, Dharwad included as a new centre
- Indira Naveen and Indira Nandini sweet potato varieties were released by Chhattisgarh
- IGT-1 cassava for Bihar and Chhattisgarh, Kamalasundari sweet potato for West Bengal and Orissa, taro entries JCC-25 for Assam, KCS-2 for Andhra Pradesh, elephant-foot yam Sree Padma for Assam and Gajendra for Bihar, bunda entry NDB-1 for Assam, Jharkhand and swamp taro BCST-5 for Assam were recommended for cultivation

2008
- ICAR selected AICRPTC as the best AICRP for 2007
- Organized National Seminar on Amarplwpha – HUS Innovation Technologies at RAU Patna, August 19–21, 2008

2009
- Tenth group meeting of AICRPTC at OUAT, Bhubaneshwar, Oct, 9-11-2009

The Chaudary Devi Lal Outstanding All India Co-ordinated Research Project (AICRP) Award for the year 2007 was presented by ICAR to:
All India Co-ordinated Research Project on Tuber Crops, Thiruvananthapuram, Kerala for its outstanding contribution in the field of tuber crops horticulture.

RESEARCH GAPS AND PROGRAMMES IDENTIFIED FOR STRENGTHENING RESEARCH IN XI FIVE YEAR PLAN

Genetic Resources of Tropical Root and Tuber Crops (TRTC)

- Identifying unexplored areas in different regions based on geographical map of the area and exploring biodiversity of different species of tropical root and tuber crops (TRTC).
- Characterizing and cataloguing of genetic stock after eliminating duplicates.
- Molecular characterizaiton of important genetic stocks under each species of TRTC for its higher yields and for its biotic and abiotic stress tolerance/resistance.
- Network researching on *in vitro* conservation of genetic stock of different species of TRTC.

- Investigation on hither to unexplored minor tuber crops suitable for each region.

Varietal Development and Improvement

- Sharing promising varieties/ accessions among different centres for research work.
- Identification/development of biofortified accessions/varieties especially on β-carotene, protein, minerals such as Zn, Fe, Cu, Mn etc.
- Developing short duraton varieties with ideal plant type suitable for each region with economic traits. [Suitable varieties of cassava for sago/starch industires; suitable sweet potato for noodles, suitable vareities of TRTC for chops making, cassava varieties with profuse foliage for ericulture, determinate (dwarf) yam etc.].

Organic Farming

- Organic farming with reduced input of synthetic fertilizers and pesticides.
- Biofertilisers—identification of region specific suitable microbes for enhancing nutritional efficiency and as growth promotors and disease suppresors.
- Effective utilization of indigenious micro organisms suitable for each region for better management of tuber crops.
- Use of liquid biofertilizers in TRTC.
- *In situ* production of organic manure and vermicompost.

Tuber Crops Based Cropping System

- Intercropping in tuber crops and tuber crops as inter crops in plantation/orchards.
- Developing modalities for micro irrigation and fertigation system for efficient use of water and nutrients for different TRTC.

Pest and Disease Management Strategies

- IPM and IDM for major pests and diseases including use of resistant lines, biocontrol agents and chemecology-based components (Allemones/pheromones).
- Survey and surveillance for new pests and diseases.

Utilization and Scope for Rural Employment

- Exploiting pharmaceutical and nutraceutical values of different TRTC for benefit of the consumers (like β-carotene is a very good source fo vitamin 'A' and β-carotene rich cassava/sweet potato can be encouraged among vitamin 'A' deficient population. Similarly, elephant foot yam is very good for piles and rejuvenation etc.). Popularizing orange fleshed varieties of cassava/sweet potato in backward areas of Bihar, West Bengal, Jharkhand, Uttar Pradesh, Madhya Pradesh, Andhra Pradesh, Maharashtra and North-Eastern states.
- Encouraging small-scale entrepreneurship through TRTC *vis-a-vis-* sago, starch, noodle like food products.
- Ericulture using cassava leaves (profusely leaf bearing cassava varieties can be identified for ericulture) for production of eri silk.
- Storage of tubers against post harvest deterioration, due to physiological and due to

pests and diseases and improving shelf life of tubers.

- Utility of leaves of TRTC for food and feed.
- Biopesticides from yam bean and other TRTC

Extension Education and Awareness Programmes

- Area expansion under TRTC in different states especially non-traditional and under/ unutilized areas.
- Popularization of different tuber crops through farmers meet, seminar, exhibition, etc., and educating the nutritive, nutraceutical and pharmaceutical values of different TRTC for nutrition, health and wellness.
- Dissemination of knowledge on utilization of different TRTC, small scale/cottage industries of starch, gums, noodles, ericulture, biopesticides, etc.
- Developing a network data base of different centres/states with special reference to TRTC.
- Front line demonstration for transfer of technologies developed for each region.
- Human resource development-Training in molecular mapping and molecular marker-based QTL mapping.
- Off-campus and on-campus training for tuber crops farmers.

Planting Material Production and Distribution

- Organizing supply of healthy planting material of different species of TRTC utilizing minisetts/micropropagation technologies through the AICRPTC centres in liason with Agriculture/Horticulture Departments/KVKs.

4

Mushroom

R.P. Tiwari and Manjit Singh

The All India Coordinated Research Project on Mushroom (AICRPM) came into existence during VIth Five Year Plan on 1.4.1983 with its Headquarters at National Research Centre for Mushroom, Solan (Himachal Pradesh). The Director of NRC for Mushroom, Solan also functions as the Project Co-ordinator of the project. Initially the AICMIP started with six Centres one each at Punjab Agricultural University, Ludhiana (Punjab), G.B.Pant University of Agriculture and Technology, Pantnagar (Uttarakhand), C.S. Azad University of Agriculture and Technology, Kanpur (Uttar Pradesh), Bidhan Chandra Krishi Vishwa Vidyalaya, Kalyani (West Bengal), Tamil Nadu Agricultural University, Coimbatore (Tamil Nadu) and Mahatma Phule Agricultural University, Pune (Maharashtra). At a later stage during VIIth Plan, one new Centre at Indira Gandhi Krishi Vishwa Vidyalaya, Raipur (Chattisgarh) was added and two existing centres at Kanpur (Uttar Pradesh) and Kalyani (West Bengal) were dropped. However, three new centres during VIIIth Five Year Plan and 3 Co-ordinating and one co-operating Centres during IXth Five Year Plan were added to the existing list of centres by dropping one at Goa. At present, 10 Co-ordinating and one co-operating Centres are working under AICRPM programme with its Headquarters at NRCM, Solan, which are given:

All the Co-ordinating Centres and Headquarters at NRCM, Solan, work in co-ordination on common objectives and mandate. The technical programme for conducting participatory research at all the Centres and the Headquarters is finalized in a biennial workshop organised at any of the centres or Headquarters. During the biennial workshop the progress reports for the previous years are presented and discussed to monitor the progress made in the project during the previous years and the technologies generated are assessed for release and their adoption at all India level.

OBJECTIVES

- To conduct survey of naturally occurring wild mushroom with a view to identify and catalogue the edible species as well as for exploring of possibilities of promising strains.
- To evaluate the promising and high-yielding strains received from NRC stock for regional adaptability.

- To carry multilocational trial for finalization of standard production techniques for increasing yield of different edible mushrooms.
- To explore possibility of selection of cheaper strains available locally for substrate, compost and casing.
- To supply good quality spawn to the mushroom growers.
- Popularization of mushroom cultivation on large scale through training.

Ongoing Programmes

Crop Improvement

Wild germplasm collection, identification and conservation

Testing of different strains of white button mushroom, *Agaricus bisporus*.

Strainal evaluation of milky mushroom, *Calocybe indica*

Strainal evaluation of paddy straw mushroom, *Volvariella volvacea*

Strainal evaluation of oyster mushroom (*Pleurotus sajor caju* and *Pleurotus florida*).

Evaluation of yield potential of *Pleurotus fossulatus*.

Crop Production

Button Mushroom

Isolation and identification of thermophilic microorganisms from white button mushroom compost from different locations for their exploitation in faster composting process.

To test different locally available material for there feasibility as casing materials

Oyster Mushroom

Effect of chemical sprays on yield of *Pleurotus* spp.

Evaluation of yield potential of blue oyster mushroom (*Hypsizygus ulmarius*) on different substrates.

Paddy straw mushroom

Cultivation of paddy straw mushroom, *Volvariella volvacea*

Milky Mushroom

Cultivation of milky mushroom, *Calocybe indica*.

Shiitake Mushroom

Cultivation of shiitake mushroom, *Lentinula edodes*.

Post-harvest Technology

Washing treatment for button mushroom, *A.bisporus*.

Packaging for button mushroom, *A.bisporus* (washed with 0.05% KMS).

Drying of oyster mushroom (*Pleurotus* spp.).

Drying of milky mushroom (*Calocybe indica*).

Crop Protection

Survey and surveillance of diseases and insect pests in mushroom farms in different regions

MAN POWER AND BUDGET

Plan-wise budget (lakh) allocation for different centres

Head	*VI plan*	*Ad-hoc plan 1990-91*	*VII plan*	*VIII plan*	*IX Plan*	*X plan*
Pay and allowances	3.58	9.86	25.90	89.75	200.00	318.93
Traveling allowance	0.15	0.26	1.50	2.54	4.55	6.32
Recurring contingencies	1.65	1.18	6.70	17.16	40.86	18.63
Equipments	5.87	0.80	4.95	6.65	21.19	6.12
Works	-	1.78	8.50	17.20	15.55	-
Total	11.25	13.88	47.55	133.30	282.15	350.00

Centre-wise manpower in different centres

Centre	*Sanctioned Positions in Different Categories*			
	Scientific	*Technical*	*Admn*	*Supporting*
Ludhiana	2	1	1	2
Pune	2	1	1	2
Coimbatore	2	2	1	1'
Faizabad	2	2	1	1
Pantnagar	2	2	1	1
Raipur	2	1	1	1
Udaipur	2	2	1	1
Thrissur	2	1	-	-

MECHANISM FOR EVALUATION ACTIVITIES OF PROJECT

Activities of project are monitored through the visits of the Project Coordinator. Workshops/ Group meetings of AICRPM scientists are held regularly to discuss progress made under the project. During these meetings recommendations for release of varieties, cultivation technologies, and crop protection measures are brought out. Based on previous results and keeping in view the emerging problems, new technical programmes are also worked out in these meetings. Since the inception of project, 10 workshops/group meetings have been held in such a manner that venues are evenly distributed throughout the country.

WORKSHOPS/GROUP MEETINGS OF AICMIP

Workshops/Group Meetings of AICMIP	*Date*	*Venue*
Ist Biennial Workshop of AICRP on Mushroom	23-25 September 1986	National Centre for Mushroom Research and Training, Chambaghat, Solan
IInd Biennial Workshop of AICRP on Mushroom	12-13 February 1998	College of Agriculture, Pune
IIIrd Biennial Workshop of AICRP on Mushroom	18-19 December 1989	National Centre for Mushroom Research and Training, Chambaghat, Solan

(Contd.)

Workshops/Group Meetings of AICMIP	*Date*	*Venue*
IVth Biennial Workshop of AICRP on Mushroom	27-28 December 1991	Tamil Nadu Agricultural University, Coimbatore
Vth Biennial Workshop of AICRP on Mushroom	11-12 April 1994	National Centre for Mushroom Research and Training, Chambaghat, Solan
VIth Biennial Workshop of AICRP on Mushroom	29-30 December 1997	Rajasthan College of Agriculture, Udaipur
VIIth Biennial Workshop of AICRP on Mushroom	May 2000	National Research Centre for Mushroom, Chambaghat, Solan
VIIIth Biennial Workshop of AICRP on Mushroom	8-9 March2002	Tamil Nadu Agricultural University, Coimbatore
IXth Biennial Workshop of AICRP on Mushroom	25-26 October 2004	National Research Centre for Mushroom, Chambaghat, Solan
Xth Biennial Workshop of AICRP on Mushroom	26-27 October 2006	Indira Gandhi Krishi Vishwa Vidyalaya, Raipur (Chhattisgarh)

ESTABLISHMENT/HISTORY/BACKGROUND

Centres of AICRP on Mushroom

Location/Institute/University	*Latitude/Longitude*	*Year of Establishment*	*Current Status*
Headquarters			
• Solan, PC, NRCM, Solan (Himachal Pradesh)	30° 55" 0' N, 77° 7" 0' E	1983	Continued
Centres			
• Punjab, Punjab Agricultural University, Ludhiana (Punjab).	30° 41' 15" N, 76° 1' 45°E	11.01.1984	Continued
• Tamilnadu, Tamil Nadu Agricultural University Coimbatore (Tamil Nadu).	10° 59' 47" N, 76° 57' 40" E	1.04.1983	Continued
• Uttarakhand, G.B. Pant University of Agriculture and Technology, Pantnagar	29° 3' 0" N, 79° 31' 0" E	4.10.1984	Continued
• Maharashtra, Mahatma Phule Agricultural University, Pune (Maharashtra).	18° 32' N, 73° 51'E	22.08.1983	Continued
• N.D.University of Agriculture and Technology, Faizabad (Uttar Pradesh)	26° 31' 0"N, 82° 35' 0" E	01.10.1993	Continued
• Indira Gandhi Krishi Vishwa Vidyalaya, Raipur Chhattisgarh.	17° 48N and 24°5N80° 15-84°20E	1.03.1988	Continued
• Maharana Pratap University of Agriculture and Technology, Udaipur (Rajasthan)	24° 36' 16" N, 73° 49' 6" E	17.09.1993	Continued
• Kerala Agricultural University, Thrissur (Kerala)	80° 18' and 120° 48' N, 740° 52' and 770° 22' E	4.09.2000	Continued
• Megalaya, ICAR Research Complex for NEH Region, Barapani (Meghalya)	25° 34' 16" N, 91° 53' 48" E	4.09.2000	Continued
• Jharkhand, Horticulture and Agroforestry Research Programme (ICAR Research Complex for Eastern Region), Ranchi (Jharkhand)	22° 58' 0" N, 85° 32' 0" E	4.09.2000	Continued
• Dr.Y.S.Parmar University of Horticulture & Forestry, Nauni, Solan, (Himachal Pradesh)	30° 55" 0' N, 77° 7" 0' E	4.09.2000	Continued

Ludhiana

The work on mushroom research and development in the Department of Microbiology was initiated in early 1970's under an ICAR funded ad-hoc project, which was further supported by a state funded "Research and Development project" in 1979. Department of Microbiology, Punjab Agricultural University, Ludhiana, was subsequently selected as one of the coordinating centres, w.e.f. 1.4.1983 (sanction no. ICAR NE19 (1) HSMC dated 29.10.1982) of the All India Coordinated Mushroom Improvement Project with financial implication in the ratio of 75:25 (ICAR: Punjab State Government). Presently, the work carried out under AICMIP is being complemented by additional funding in the form of many ad-hoc projects funded by ICAR, UGC, DSR and self-supporting. Mushroom production is the only state-of-the-art economical solid state fermentation process to utilize the crop residues to provide high protein food, employment and improvement in the economic standards.

Punjab	
• Area	50, 36, 000 ha
• Districts	19
• Temperature	>30°C; 15-30°C; <15°C
• Agricultural Crops	Rice, wheat
• Mushrooms	40,000 tonnes

Pune

The Project was sanctioned by ICAR as per letter No. F-19(+) MC; dated 23 October 1982 and actually started functioning on 22.8.1983.The project is continued in XIth plan period. Maharashtra is the second largest state of the country with wide diversity in climatic conditions and soil types favouring the cultivation of various crops. Research on different aspects of mushroom cultivation in state was initiated at this college as early as in 1972. In recognition to the pioneering research work on mushroom in this University, the ICAR, New Delhi, sanctioned 6th AICMIP Center to this College in 1983. The funds for independent building and infrastructural facilities were received from VIIth five years plan. At present, a few farms of button mushrooms have started in Pune, Nashik, Sangli, Thane and Akola districts of Maharashtra. Oyster, locally known as "dhingri mushroom" is now very popular among common people because of simple technique of cultivation and less investment.

The popularity of milky mushroom among growers is also increasing and demand for milky mushroom is increasing in Mumbai and Pune a few growers have started growing of milky mushroom in the state successfully.

RESEARCH ACHIEVEMENT

A total of 72 research experiments were planned in AICMIP, since 1983 till date. Of which, 29 experiments were on button mushroom; 24 on oyster mushroom; 9 on black ear, milky, paddy straw and shiitake mushroom; 3 on germplasam collection and survey; 5 on post-harvest management and 2 on survey of insect-pests and diseases of mushrooms.

Coimbatore

With the sustained research, teaching and extension activities of the Department of Plant

Pathology at TNAU, mushroom cultivation has become one of the promising agricultural enterprises in Tamil Nadu since 1983. A mushroom theme park was established during 2002 and recently a new building for Mushroom Research and Training Centre has been established during 2006 with a total cost of Rs 26 lakh.

Ranchi

The research and training in mushroom production at this centre started in 2001 with a grant from Central Sector Scheme on Integrated Development of Horticulture by Ministry of Agriculture, Government of India and a laboratory for spawn production unit was established later on at centre under All India Coordinated Mushroom Project was also set to carry out location-specific trials on mushroom cultivation in Jharkhand.

Pantnagar

Mushroom research at G.B. Pant University of Agriculture and Technology, Pantnagar, was initiated in mid-seventies in the Department of Plant Pathology with a state financed 'Mushroom Development Project'. Later, All India Coordinated Mushroom Improvement Project was sanctioned in 1982 and started functioning from October 1984, which laid foundation of systematic research on mushrooms at Pantnagar. Mushroom Research Laboratory building constructed and equipped from the financial assistance of ICAR was formally inaugurated on 26 January 1989. It was strengthened by the University and named as Mushroom Research and Production Centre in 1998 and further with the addition of training unit, composting unit, a new research lab, class room, museum etc. It was renamed and recognized as Mushroom Research and Training Centre in April 2003 by the Board of Management of the University. The Centre is designed to provide R & D support for mushroom development programmes of the State/Nation.

Infrastructure and Activities

Presently, this Centre has a museum for display of different technologies, committee room to organize seminars and meetings, classroom and demonstration rooms to impart trainings to students and mushroom growers, four experimental and one pilot scale technology testing crop rooms, research laboratories, well-equipped spawn and composting units for spawn and compost supply to growers.

Museum: The Centre has a museum where all the activities are illustrated with the help of photographs, charts, nearly 200 dried and wet specimens of wild and cultivated mushrooms. Besides, Centre provides on payment video film, photographs, coloured slides and literature on cultivation technologies of different mushrooms.

Culture and Spawn Unit: Pure cultures of cultivated mushrooms are made available to researchers and spawn producers on payment basis. The Centre's commercial spawn production unit sales master as well as commercial spawn of different strains of cultivated mushrooms on advance booking.

Composting Unit: Centre has a well-equipped compost pasteurization unit for production of pasteurized compost. As per requirement of researchers and mushroom growers, the supply of pasteurized compost is being ensured on prior booking. It comprises a straw storage, pre-wetting platform, outdoor composting shed, two pasteurization tunnels each of 10 tonnes ready compost capacity, spawning room and boiler, etc.

Cropping Rooms: Four experimental crop rooms, one pilot scale testing crop room and two low-cost mushroom production rooms (huts) are available for conducting trials under varying conditions.

Committee and Class Rooms: The Centre has adequate training facilities which include a committee room with modern audio-visual gadgets. It also has classroom of 40 seats capacity for the training courses.

Faizabad

Mushroom research and teaching was started at N.D. University of Agriculture and Technology, Kumarganj, Faizabad, during 1980-81 as per state government plan. The AICRP on Mushroom (AICMIP) was sanctioned and implemented during 1993-94 and spawn laboratory was established during 2003-04. Since then research, teaching and extension work has been carried out as per the mandate.

Raipur

The All India Coordinated Mushroom Improvement Project, Centre Raipur, was established in the Department of Plant Pathology, College of Agriculture, Indira Gandhi Agricultural University on 1 March 1988 as an 8th Coordinting Centre of AICMIP. The entire Chhattisgarh region is divided into three district agroclimatic zones: (i) Chhattisgarh plains; (ii) Northern hill region and (iii) Bastar plateau.

In Chhattisgarh edible mushrooms collected from forest areas, were brought to local villages, towns and cities during rainy season for sale. They mainly collect and consume the species of *Termitomyces, Boletus, Volvariella, Tuber, Cantharellus,* etc. Survey conducted from 1988–97 revealed that in every village *Volvariella diplasia* is growing naturally in paddy straw stacks kept in field. During favaourable period, 5 quintals of fresh *Termitomyces* spp. per day is being sold in Raipur market alone.

Chhattisgarh is mainly dominated by rice and 80% of the area (4 m.ha.) is under rainfed rice system and considered as "rice bowl". As a result of which, about 50 lakh tonnes of paddy straw is produced every year. Economic mushroom production can help for their subsistence and designing of low-cost mushroom huts with conducive environment, help them for earning for their subsistence during off season.

A systematic work on mushroom cultivation (mainly oyster mushroom) was taken up only after the establishment of a centre under All India Coordinated Mushroom Improvement project at Indira Gandhi Agricultural University, Raipur, during 1988.

During the period awakening about mushroom and its cultivation was demonstrated in villages with the help of Women and Child Welfare Department of C.G., Forests etc. About 1,000 rural women from Raipur district alone were trained on mushroom production technology. Recently, a commercial unit of oyster mushroom production as "Chhattisgarh Mushroom" with a capacity of 90 kg fresh mushroom per day has been established at Tendua 29 km from Raipur. Thus, mushroom production technology in this part of the country is now again gaining a lot of interest for generation of employment and subsidiary income in rural sector of the society. About 15 Mushroom Spawn Production Units have been established in different parts of the state and fulfilling the requirements of mushroom spawn.

Vellayani

The ICAR sanctioned the AICRP on Mushroom Centre in January, 2001. Since then regular trials charted out by the Co-ordinator and team are being conducted at the centre. Initially Dr B. Balakrishnan stated the work as the Principal Investigator but later on it was taken over by Dr. Lulu Das in January 2006. She along with one research associate and one skilled worker is carrying out all the experiments of this Centre.

Barapani Centre

The Centre at Barapani, Meghlaya came in existence during ninth plan considering the availability of rich biodiversity in the area and great scope of mushroom cultivation. Being an ICAR institute no separate posts were kept for running of the Centre, however, funds were given in the form of contingencies only to meet the day to day requirement.

Udaipur

The Centre came in existence during the 8th five year plan with a full-fledged manpower and budget. The Centre has developed good facilities with respect to training and experimentation on different mushrooms. The strategic location of the Centre has helped in developing cultivation technology of oyster and milky mushroom and a large quantity of germplasm has been collected and preserved over the time.

Nauni

The Centre at Nauni, Solan, came in existence during ninth plan as a cooperating Centre. Manpower and budget was restricted to only the contingency support considering the availability of inftrastructure and trained manpower. The mushroom research at the Centre started in early 70's and very high level of research outputs has came out from this Centre over the last 3-4 decades of research.

GENETIC RESOURCES MANAGEMENT

Ludhiana

The *Agaricus*(34), *Agrocybe*(3), *Arrhenia*(1), *Bolbitius*(4), *Calvatia*(2), *Clitocybe*(1), *Conocybe*(1), *Coprinus*(13), *Cyathus*(1), *Geaster*(1), *Lepiota*(1), *Leucocoprinus*(1), *Lycoperdin*(1), *Morchella*(2), *Mycena*(1), *Pleurotus*(4), *Psathyrella*(3), *Psilocybe*(1), *Sphaerobolus*(1), *Stropharia*(1), *Termitomyces*(3), *Tricholoma*(1), *Tulostroma*(9) and *Volvariella* (4) accessions have been maintained.

Germplasm Collection

Genus	*Species*	*Strain*	*Accession*
Agaricus	*bisporus*	33	MTCC
	bitorquis	1	MTCC
Pleurotus	*sajor caju*	14	ATCC/MTCC
	ostreatus	11	ATCC/MTCC
	florida	10	ATCC/MTCC
	flabellatus	7	ATCC/MTCC
	cystidiosus	3	MTCC
	citrinopileatus	3	MTCC
	sapidus	2	ATCC/MTCC

(Contd.)

Genus	Species	Strain	Accession
	eryngii	2	MTCC
	fossulatus	1	ATCC/MTCC
	pulmonarius	1	MTC
	platypus	1	-
	salmoneostramineus	1	-
Volariella	volvacea	7	ATCC/MTCC
	diplasia	1	ATCC
Calocybe	indica	10	
Auricularia	polytricha	1	
Morchella	esculenta	1	MTCC
Hypsizygus	ulmarius	1	
Lentinus	edodes	2	
Ganoderma	lucidum	4	
Total 9 genera	22 species	117 strains	

64 strains have already been accessioned

The cultures are maintained through regular subculturing and are being cryopreserved in glycerol/DMSO in liquid nitrogen.

Pune

- Mushroom species, *viz. Pholiota, Lepiota, Chlorolepiota* and *Termatomyces* were collected from Lonawala, Igatpuri, Mahabaleshwar, Nashik and Goa and pure culture isolated.
- About 140 samples of wild fleshy fungi were collected from different parts of western Maharashtra and identified up to generic levels.
- Cultures maintained.

The following cultures are being maintained at the Centre

- *Agaricus bisporus*: P-1, U-3, S-11, CM-1 to CM-15, S-130, N-210, N-211, N-560.
- *Agaricus bitorquis*: NCB-6, NCB-13.
- *Pleurotus* spp.: *P.sajor-caju, P. florida, P. eous, P. citrinopileatus, P.ostreatus, P. eryngii, P. sapidus, P. membranaceous,* Blue oyster *(Hypsizygus ulmarius)* - CO_2.
- Speciality and other mushrooms
  - Milky mushroom (*Calocybe indica*)- CI-1 to CI-9, APK-2.
  - Shiitake mushroom *(Lentinula edodes)* – LS-5, LE.
  - Paddy straw mushroom (*Volvariella volvacea*)-PS 2
  - Wild fleshy fungi cultures: 23 (FFP-1 to FFP – 23)
- Samples preserved: 30
- Cultures deposited at NRCM, Solan: 6
  - FFP-1-OE-299-*Termitomyces* spp.
  - FFP-2-OE-300- *Agaricus* spp.

- FFP-6-OE-301- *Tricholoma* sp.
- FFP-7-OE-302- *Agaricus* spp.
- FFP-15-OE-303-*Panaeolus* spp.
- FFP-23 OE-304 *Ganoderma* spp.

Coimbatore

A total number of 22 fleshy fungi were observed during the survey. Among them, one each of *Agaricus bisporus, Ganoderma lucidum, Schizophyllum* sp., *Lycoperdon* sp., *Auricularia polytricha, Tricholoma gigantea, Hypsizygus ulmarius, Pleurotus florida, P. ostreatus, P. eous, P. platypus, P. djamor, P. flabeleatus., Lentinula edodes, Polyporus* sp., three isolates of *Calocybe indica* and three isolates of *Volvariella volvacea* were pure cultured and maintained in the germplasm bank of Mushroom Research and Training Centre, TNAU, Coimbatore.

The pure cultured mushroom cultures were tested for their bioefficiency and nutritional status and by continuous research; varieties were released for the benefit of farmers from time to time. Among these cultures *Volvariella volvacea* strain PS1 and WC 19 strain of *Calocybe indica* is under multilocational testing for further release as variety for commercial production

Pantnagar

Uttarakhand is gifted with rich flora of mushrooms. Under AICRPon Mushroom, a survey team surveyed the forest areas to collect the natural mushroom flora. Some of the fleshy fungi collected were as follows:

Table 4.1. Fleshy fungi collected at Pantnagar Centre

Mushroom Species	*Mushroom Species*	*Mushroom Species*
Agaricus silvaticus	Daldinia concentrica	*Auricularia* sp. (from silver oak)
Amanita phalloides	Deedalea confragosa	Auricularia *sp. (from silver oak)*
Amanita sp.	Earth Ball	*Auricularia* sp. big size
Auricularia sp.	Ganoderma lucidum	*Auricularia* sp. Small size
Boletus deulis	*Ganoderma* sp	*Auricularia* sp.(from mango tree)
Boletus sp. (brown)	*Geastrum* sp.	*R. delica* white coloured
Boletus sp.(light cream)	Puff Ball (big size)	*Geastrum triplex* (earth star)
C. lagopus	*Puff Ball (small size)*	Grifola *"Mandarin colour"*
Calocera viscose	Grifola sulfurea	Pleurotus cystidiosus *(2 isolates)*
Cantharellus sp.	*Grifola* "White"	*Pleurotus* djamor (dark pink)
Clavulina sp.	*Grifola* sp. "Orange white"	*Pleurotus* djamor (light pink)
Clitocybe sp.	Hericium erinaceus	*Pleurotus* sp. (creamy-white)
Clitopileus sp.	Leptonia sericella	*Pleurotus* sp. "off white"
Collybia confluens	Lycoperdon pyriforme	Pleurotus *sp. Funnel shaped*
Coltricia perennis	Lycoperson sp *white*	Scleroderma sp
Coltricia sp.	*Melanoleuca* sp.	*Scleroderma* sp. black spored
Coprinus comatus	Morchella esculenta	Scleroderma *sp.* dark grey spored
Coral fungi	*Nyctalis parasittica*	*T. edista*
Cordyceps sinensis	P. smithii	*Termytomyces* sp.
Coriolus sp.	*Russula* sp.(brown+purple)	*Trametes confragosa*

(Contd.)

Mushroom Species	Mushroom Species	Mushroom Species
Coriolus versicolor	*Russula* sp.(green colour)	*Tricholoma* sp.
Daedalea sp.	*Russula* sp.(grey colour)	Tricholoma sulphureum
Hericium sp.	*Schizophyllum commune*	*Tuber*
Hydnum sp.	Ramaria cristata	Volvariella diplacea
Hyqrophorpsis sp.	Russula atropurpurea	Volvariella volvacea
Laccaria omethystea	*Russula foetens*	X. polymorpha *dark grey- hard*
Lactarius blennius	*Russula lutea*	*X. polymorpha* grey-white
Lactarius deliciosus	*Russula nigricans*	*Xylaria* sp. black colour
Lactarius rufus	*Russula* sp. (red colour)	*Xylaria* sp. grey coloured and thin
Lactarius sp.	*Russula* sp. (white)	*Polyporus* sp.
Lactarius tabidus	*Phellinus ignarius*	Polyporus squamosus
Lepiota margaoni	*Pleurotus cervinus*	*Lepiota rhachodes*

- Of the collected mushrooms, five isolates of *Auricularia,* two species of *Pleurotus* and 10 isolates of *Ganoderma lucidum* have been brought under cultivation150 mushroom species from different locations of Uttarakhand have been collected and preserved.

Udaipur

A total of 400 species of wild fleshy fungi were collected during 1996- 2007 from different areas of Sawai Madhopur, Kota, Bundi, Sirohi, Banswara, Chittorgarh, Barmer, Jodhpur, Rajsamand, Sirohi, Pali, Dungarpur and Udaipur districts in rainy season. The specimens of wild mushrooms were preserved in dry form; their cultures along with database were deposited to National Research Center for Mushroom (ICAR) Chambaghat, Solan. Some common are: *Agaricus* spp., Pleurotus spp., *Phellorinia inguinans, Podaxis pistillaris, Auricularia polytricha, Schizophyllum communae, Daldania concentrica, Ramaria* spp., *Coprinus pieaceus, Polypore sulphureus, Tricholoma* spp., *Hericium erinaceum, Auricularia auricula judae, Pleurotus salmenostraminusc, Lapiota procera, Ganoderma lucidum, Xyleria polymorpha, Calocybe* spp., *Lycoperdon* spp., *Xylaria polymorpha, Volvariella volvacea, Laitoporus spp., Inonotus hispidus, Hydnellum scrobiculatum, Polypore* spp., *Podaxis pistillaris, Phellinus igniarius, Piptoporus betulinus, Pluteus cervinus, Pleurotus ulmarius, Pleurotus salmenostraminus Agaricus trisulphureus, Agaricus compestris, Tricholoma equestre Sclaroderma citrinum, Leucocoprinus* spp., *Mycenia* spp., *Pisolithus* spp. and *Volvariella volvacea,* etc.

Faizabad

On this aspect work regarding maintenance of cultures of experimental use and collection, culturing and supply of mushroom biodiversity has been done.

Raipur

The climatic conditions prevailing in Chhattisgarh make the natural habitat for a large number of mushroom flora. The detailed information for the presence of natural agaric flora in different agro-climatic zones of Chhattisgarh was started with the inception of All India Coordinated Mushroom Improvement Project in 1988 at IGAU, Raipur. The details are as under:

Vellayani

The cultures of strains of *Calocybe indica* were deposited in the National Mushroom Culture

Bank at NRCM, Solan, for which accession numbers were obtained. The other isolates of *Pleurotus eous, P. squarrosulus, Ganoderma, Schizophyllum, Termitomyces, Auricularia, Tricholoma,* etc., are being preserved in the laboratory of the AICRP center, Vellayani.

Ranchi

The survey was done in monsoon season in Ranchi and Hazaribag. The following fleshy fungi were collected and wet preserved.

Table 4.2. Fleshy fungi collected at Ranchi Centre

Common Name	*Scientific Name*
Oyster mushroom	*Pleurotus* sp.
Bada Khukhri	*Macrolapiota* sp.
	Termitomycetes spp.
Rugra	*Lycoperdon* sp.
Button mushroom	*Agaricus* spp.
Jamum Khukhri	*Boletus* spp.
Milky mushroom	*Calocybe* spp.
Medicinal mushroom	*Ganoderma* sp.
Namak Khukhri	*Termitomycetes* sp.
Bansh Khukhri	*Lentinus* sp.
Pagla Khukhri	*Volvariella* sp.

REGION SPECIFIC VARIETIES/VALUE ADDED VARIETIES

Ludhiana

At present, four mushroom varieties namely, *Agaricus bisporus, Calocybe indica, Pleurotus* spp. and *Volvariella* spp. have been recommended for their production in the State throughout the year. In addition, work on other mushrooms like, *Ganoderma lucidum, Hypsizygous ulmarius* and *Lentinus edodes* is also being undertaken.

Pune

- *Pleurotus sajor- caju* followed by *P. florida* are most suitable species for cultivation in Maharashtra during monsoon and winter seasons.
- Amongst different strains of *Agricus bisporus*, P-1, NCS-12 and 310 were observed to be high-yielding in Maharashtra.
- *Agaricus bitorquis* strain NCB-6 could be grown successfully at natural temperature (22-26°C) during monsoon under Pune conditions.
- Amongst different species, *Pleurotus sajor- caju* has given the highest yield (846 g/kg of dry substrate) followed by *P. florida*.
- Studies on evaluation of different strains of *Agricus bisporus* revealed that strains, viz. P-1 and NCS-12 have highest yield potential, i.e., 1649 and 1577 g/10 kg compost, respectively which were found to be at par with each other.
- The temperature tolerant strain of *Agricus bitorquis,* viz. NCB-6 and NCB-13 recorded the yield of 928 and 848g/10 kg compost at ambient temperature (22-26 °C), respectively.

Table 4.3. Fleshy fungi collected at Raipur Centre

Mushroom Species	*Habit and Habitat*	*Season*	*Distribution*	*Edibility*
Agrocybe erebia	Sporophore in groups on ground	Rainy season	Bilaspur	Non edible
Agrocybe sp.	Sporophore in groups or scattered in soil	Rainy season	Bilaspur	Non edible
Amanitopsis	Growing singly on leaf litter	Rainy season (August)	Barnavapara	Edible
Amanita citrina	Under shrubs	Rainy season	Bastar forest	Poisonous
Amanita sp.	Sporophore occurs on the soil under shrubs	Rainy season	Bastar forest	Not known
Armillaria mellea	Sporophore growing on deciduous stumps of dead trunk of tree	Rainy season	Raipur and Bilaspur	Non edible
Astreatus hygrometricus (Boda)	Scattered on the soil, Hypogeal myhorhizal association with root of sal tree	Rainy season (June -October)	Forest area of Chhattisgarh	Excellent
Auricularia polytricha	Solitary or gregarious on dead wood and logs	Rainy season (June -October)	All region of Chhattisgarh	Edible
Auricularia sp	On dead branches of mango, Custard apple, subabool, gulmohar	Rainy season (June -October)	Raipur	Edible
Boletus sp.	Sporophore solitary on sandy soil under dense shrubs on leaf litter	After good monsoon rains	In the area of Uddanti and Barnawapara Wild blife Sanctuary of Raipur	Edible but not consumed by local people
Bovista appendicellata	Sporophore scattered on leaf litter under the shed of broad leaves trees	Rainy season (June -October)	Bilaspur	Not known
Calocybe sp.	Sporophore solitary in soil under broad leaves trees	Rainy season (June -October)	Ambikapur forest	Edible
Clavatia cyantiformis	Solitary or in groups on leaf litter of shrubs	Rainy season (June -October)	Forest area of Korea District	Edible
Cantherellus sp. (Bans phutu)	Sporophore in groups, mycorrhizal association with bamboo trees	Rainy season during August month	In all bamboo growing areas of Chhattisgarh	Excellent, sold in local market @ 60-80/kg
Clavaria fumosa	Sporophore grows in groups on the ground on fallen branches	Rainy season	Sitanadi Wild Life Sanctuary	Not Known
Clavaria sp.	On wood in scattered	Rainy season	Tirathgarh, Narayanpur	Non edible
Clitocybe gestrpoa	Sporophore grows in groups under the shrubs on leaf litter	Rainy season	Tirathgarh, Chitrakut and other places of Bastar	Poisonous
Coprinus comatus (Shaggy cap)	Gregarious in grassland and road side	Rainy season	In all districts of Chhattisgarh	Edible

(Contd.)

Mushroom Species	Habit and Habitat	Season	Distribution	Edibility
Coprinus sp. (Ink cap)	Sporophore grows solitary or in groups	Rainy	Cow dung, damp soil, rotten paddy straw in heaps	Non-edible
Coriolus versicolor	On dead wood	Rainy	Narayanpur	Medicinal or therapeutic value
Crepodorus sp.	Sporophore grows solitary or in groups on dead wood of broad leaves trees	Rainy during August	Teerathgarh	Not known
Cyathus limbatus	Sporophore grows solitary or in groups in open grass and same times in soil	Rainy	-	Not known
Daldinia concentrica	Dead wood logs or live plants	Rainy	Teerathgarh, Narayanpur forest area	Not known
Flamulina sp.	On dead wood, mainly stumps	Rainy	Teerathgarh, Barnawapara Khajuri Nala and Uddanti Sanctuary	Edible
Ganoderma applanatum	Sporophore grows solitary or in groups on dead trunk of broad leaves tree	Rainy	All forest area of Chhattisgarh	Non-edible
Ganoderma lucidum	Sporophore grows solitary or in groups on dead trunk of broad leaves tree	July-October	All forest area of Chhattisgarh	Non-edible
Ganoderma suiji	Sporophore grows solitary or in groups	Rainy	Ambikapur, Bastar and Bilaspur forest area	Non-edible
Geastrum fimbriatum (earth stars)	Sporophore grows solitary or in groups in soil	Rainy	Labhandi research farm	Non-edible
Hygrocybe marchee	Sporophore scattered on grass land	Rainy	Churra, Gariabandh and Manpur area of Raipur	Not known
Hypholoma udum	Sporophore grows in groups on the trunk of deciduous tree	Rainy	Kevnchi, Lormi and Achanakmar forest	Not known
Hypoxylon fragiformae	Normally gregarious on rooting beech, grow on wood , crust like stromata	Rainy	Manpur, Sihava, Gariabandh forest	Non-edible
Laccaria laccata	From soil of mixed forest, sporophore in groups, singly or scattered	Rainy	Chitrakut, Teeratgarh, Narayanpur	Edible
Lactarius aceverimus	Solitary or in groups under broad leaves trees	Rainy	Narayanpur	Not known
Lactarius vellereus	Sporophore grows singly in scattered form under shrubs of broad leaves	Rainy	Narayanpur	Not known
Lactarius sp.	Grows singly on leaf litter	Rainy	Narayanpur	Not known
Lentinus sp.	Sporophore in bunches on dead wood logs	Rainy season	Korea	Edible

(Contd.)

Mushroom Species	Habit and Habitat	Season	Distribution	Edibility
Lepista nuda	In humus rich ground under broad leaves shrubs	Rainy	Achanakmar (Bilaspur)	Good
Lycoperdon maximum	Sporophore solitary or scattered in grass land	Rainy	Sandy loam soil	Edible
Marasmius creades	Sporophore gregarious in grass land	Rainy	Bastar forest	Edible
Melanoleuca alboflavida	Sporophore grows in soil or in leaf litter	Rainy	Barnawa forest area	Non-edible
Microglossum viridae	On the ground in grass or woods	Rainy	Narayanpur	Non-edible
Mutinus sp.	Sporophore in group under shade of old bamboo trees	Rainy	Bamboo rhizomes near Bilaspur area	Inedible
Pana olima foenislcic	Gregeraeous in grassy areas	Rainy	Achanakmar, Korea, Bilaspur	Non-edible
Phallorina sp	Sporophore solitary or in groups on soil under the trees	Rainy	Guru Ghasidas University Bilaspur	Edible
Pleurotus florida (Pihri)	Sporophore grows solitary or in groups on mango logs	Rainy	Narayanpur area	Edible
Pleurotus sajor caju	Sporophore in groups on dead wood logs	Rainy	Branches of mango trees	Good
Pleurotus stramonius	Sporophore in groups on dead branches of Mango	Rainy	Raipur	Good
Pleurotus sp.	Sporophore in groups on dry wood logs of many trees	Rainy	Bilaspur	Good
Podbrella microcarpa	In soil in singly or groups	Rainy	Nandghat, Bilaspur	Not known
Podaxis sp.	Sporophore solitary and grows on sandy soil	Rainy	Tarra village	Edible
Podoscypha mutzonata	Sandy loam in singly	Rainy	Nilanjee	Non-edible
Polyporus varius	Sporophore in groups on dead wood	Rainy	All districts of Chhattisgarh	Not known
Polyporus elegans	On dead wood	Rainy	Bilaspur	Non-edible
Ramaria Formosa	On wood logs of many trees	Rainy	Narayanpur	Non-edible
Russula delica	Sporophore grows solitary or in groups under shades of forest shrubs	Rainy	All districts of Chhattisgarh	Good
Russula lepida	Under broad leaves trees in soil, scattered	Rainy	All places of Chhattisgarh	Edible
Russula aeruginla	In groups under sal forest	Rainy	All places of Chhattisgarh	Edible
Russula atropurpurea	Under broad leaves trees specialy in acidic soil	Rainy	Bastar, Bilaspur	Edible
Russula mairei	Under shrubs or mixed forest	Rainy	Achanakma, Barnawapara	Non-edible
Russula virginca	Under shrubs or mixed forest	Rainy	Achanakma, Barnawapara	Edible
Rozites caperata	Under broad leavs on leaf litter	Rainy	Korea, Manendragarh	Non edible
Scleroderma verucosum (Dhela phutu)	Sporophore grows solitary or in groups in letrite soil	Rainy	Labhandi farm	Good
Scleroderma sp.	Sporophore grows singly or gregarious in the ground	Rainy	Bastar	Good
Stereum sp.	Sporophore on dead wood of broad leaves trees	Rainy	Korea, Ambikapur, Pendra, Marvahi	Non-edible
Strophoria semiglobata	Sporophore grows on soil and either solitary or gregarious	Rainy	Bilaspur, Ratanpur, Pali, Kathghora forest	Not known

- In strainal evaluation trial of *Agricus bisporus*, strain U-3 recorded highest yield (1611 g/10 kg compost), followed by strain Pant-31 (1478 g/10 kg compost).
- Highest average button mushroom yield level of 1863 g/10 kg compost has been obtained from strain CM-9 and it was followed by strain CM-7 (1718 g/10 kg compost).
- Out of 9 only 5 strains formed fruits in strainal evaluation of milky mushrooms. Among, five fruiting strain, strain CI-1 recorded maximum yield of 255 g/kg dry straw. Next best strain was CI-3 followed by CI-4 and CI-5.
- The strain CM-13 recorded significantly highest button mushroom yield, i.e., 1580 g/10kg compost followed by CM-10 and CM-16, i.e., 1480g and 1460g/10kg compost, respectively. The strain CM-14 recorded lowest yield, i.e., 1330 g/10 kg compost.

Coimbatore

Table 4.4. Varieties released at Coimbatore Centre

Oyster mushrooms	*Pure white*	*Co1*	*1986*
P. eous	Pink	APK 1	1995
P.djamor	White	MDU 1	1996
P.flabellatus	White	MDU 2	2000
P. ostreatus	Ashy white	Ooty 1	1998
P.sajor caju	Ash	M 2	1980
P.florida	White	PF	1989
Hypsizugus ulmarius	Bluish white	CO2	2004
Milky mushroom (*Calocybe indica)*	Milky white	APK 2	1998
Button mushroom (*Agaricus bisporus)*	White	Ooty 1	2001

Pantnagar

In Uttarakhand, three mushrooms are cultivated by growers throughout the year:

- Button mushroom (*Agaricus bisporus*) in October—February.
- Oyster mushroom (*Pleurotus sajor-caju*) in February—April and August—October.
- Milky mushroom (*Calocybe indica*) during April—June and July—September.

Faizabad Centre

Species and strains performing better in eastern Uttar Pradesh conditions are mentioned below:

- *Agaricus bisporus* CM-3, CM-13 and S-11.
- *Pleurotus florida, P. sajor-caju, P. sajor-caju* hybrid strain -1, *Hypsizygus ulmarius.*
- *Calocybe indica* strains CI-7, CI-6, CI-1, APK-2.
- *Volvariella volvacea* VV-3, VV-2, VV12, VV8.

Raipur

Indira Gandhi Agricultural University Raipur, under All India Coordinated Mushroom Improvement Project released a variety of Oyster Mushroom (*Pleurotus florida*) NMS 7 strain as Indira Sweta (OM 1) in 2006 by the State Release Committee for commercial cultivation throughout in Chhattisgarh.

Vellayani

Ananthan and interstock hybrid of *Pleurotus petaloides* developed in the AICRP, Vellayani centre was released by Kerala State Government and is widely acclaimed by farmers of Kerala. Eight native sub-species of *Pleurotus* have been domesticated.

A tropical milky mushroom *Calocybe indica* has been identified to be well suited to Kerala and is rapidly gaining popularity. This was first introduced from the Kerala Agricultural University to farmers. A native strain of *Pleurotus florida* is giving excellent yield through out the year and very popular among the farmers. It was found that *Auricularia polytricha* can be successfully cultivated in Kerala using paddy straw and mango wood chips and has been recommended for wide popularization.

Udaipur

The subtropical and tropical varieties of mushrooms like oyster and milky mushroom are popular in the region. Button mushroom is also being grown but during very limited period of the year. There are some big farms on button mushroom that are growing button mushroom throughout the year with environment controlled fecilities. The region is rich in mushroom biodiversity and collection and eating of mushroom from the wild is also common in the area.

Ranchi

The area represented by Ranchi Centre is rich in biodiversity and tropics and subtropical mushroom species are being grown in the area. The important mushroom speciews are: *Pleurotus* sp., *Calocybe indica* and *Volvariella* spp. Button mushroom is not much popular in the region.

Barapani

The area is again rich from biodiversity point of view and oyster, button and shiitake mushroom are more popular. The area has great scope of mushroom cultivation considering varied type of agroclimatic conditions prevailing in the region.

Nauni

The important mushrooms of region are button mushroom and oyster mushroom. Button mushroom is grown for 6 months under seasonal growing conditions and round the year under environment controlled conditions.

TECHNOLOGIES FOR ENHANCING PRODUCTION

Ludhiana

In order to enhance mushroom production in the state, continuous research efforts have been made on management practices through feasible formulations, supplementations, sprays and use of growth promoters. The mushroom productivity has been increased through strainal selection and their development using genetic techniques to grow them under different stress conditions of natural indoor environment. As a result, four mushroom varieties are successfully cultivated throughout the year in Punjab to get 2 crops of *Agaricus bisporus* (September— March), 3 crops of *Pleurotus* spp. (September—mid April), 3 crops of *Calocybe indica* (March—September) and 4 crops of *Volvariella* spp. (mid April—August).

Button mushroom – Ooty 1

Paddy straw mushroom
PSI strain

Milky mushroom – APK 2

Pleurotus eous – Pink oyster (APK1)

Stall in the State level Farmer's Day Exhibition – 2003 which won first prize for the best exhibit

Mushroom production technologies

Pune

- Wheat straw supplementated with soybean flour was recommended for getting higher yield of oyster mushroom.
- The highest yield of button mushroom, i.e., 1481 g/10 kg compost was obtained from wheat straw based compost followed by soybean straw based compost, i.e. 1031 g/10 kg compost.
- The use of coir pith alone as casing material resulted in highest average button mushroom yield, i.e., 1362 g/10 kg compost, followed by use of coir pith + FYM (1:1) as casing material, i.e., 1287 g/10 kg compost.
- The leaching of casing material for 16 hour recorded significant increase in yield of 1335 g button mushroom / 10 kg compost as compared to leaching of casing material for 8 hours.
- The 0.2% calcium chloride spray on mushroom resulted in the highest button mushroom yield of 1200 g/10 kg compost.
- Application of Phosphotika biofertilizer @ 1.0% at the time of spawning gave maximum average button mushroom yield of 1251 g/10 kg compost. Strain U-3 interacted well with Phosphotika biofertilizer.
- The application of Veradix-2 spray @ 0.15% at pinning stage recorded highest button yield of 1361 g/10 kg compost.
- Acceptable white colour of button mushroom was retained in dipping treatment in EDTA @ 200 ppm and 75 ppm concentration and by storing under ambient and refrigerated conditions respectively.
- The hot water treatment of the substrate resulted into significantly highest yield of 720 g/kg dry substrate then the other two methods of pasteurization (chemical and solar pasteurization) of substrate.
- The supplementation treatment with deoiled soybean meal @ 2% of dry weight of substrate resulted into giving maximum oyster mushroom yield of 733 g/kg dry substrate as compared to other supplementation treatment in both the species, *viz. Pleurotus Sajor caju* and *P. florida* of oyster mushroom. In another trial supplementation of substrate with rice bran @ 4% of dry weight of straw recorded highest yield of 715 g/kg dry substrate.
- Supplementation of wheat straw with 2% wheat bran + 2% rice bran was found significant in increasing yield of black ear mushroom.
- In *Pleurotus Sajor caju* significantly maximum yield (489 g/kg dry straw) was recorded in treatment of 0.1 M K_2HPO_4 and was at par with water spray. In *P. florida* also the effects of K_2HPO_4 and water spray were comparable. In both the species spray of urea gave lower yields.
- At refrigerated conditions treatment of 100 ppm EDTA and 100 ppm EDTA + 0.02% KMS showed retention of good milky white colour of button upto 48 hours. However, at ambient temperature desired colour was retained in 100 ppm EDTA + 0.02% KMS treatment up to 48 hours.

- The acceptable white colour was retained upto 48 hours when packed in 100 gauge PP bags without holes at both refrigerated and ambient temperature. However, acceptable white colour was retained upto 48 hours only when packed in 100 gauge PP bags with hole and stored at refrigerated condition. However, in both sets performance of 100 gauge PP and 100 gauge PE were more or less comparable.

Coimbatore

Button Mushroom

- Strains NCS 1 of *Agaricus bisporus* continued to be best strain suitable to Coimbatore conditions (gave 13.85 kg of mushrooms/100 kg of compost)
- FYM + loam soil was found to be the best casing material in case of *A. bisporus* and *A. bitorquis.*
- Strains S11 of *Agaricus bisporus* recorded 10.17-10.84 kg of mushroom for 100 kg of paddy straw compost prepared by LMC during winter months. It took 10 days for spawn run and 12 days for pinning.
- *A. bitorquis* (NCB 6) gave 10.6 kg of mushrooms / 100 kg of compost in paddy straw compost prepared by LMC.
- Maximum average fruit body weight was recorded in Pant 52 (12.74) followed by NCS-145 (12.69), U-3 (11.84), Pant 31 (11.50) and MS-30 (11.27), whereas maximum average fruit body weight was recorded with MS 39 followed by Pant 52 and NCS 145.
- Compost prepared using only wheat straw as base material was found better as compared to sugarcane bagasse and wheat straw in the ratio of 1:2. Maximum number of fruiting bodes were also recorded in wheat straw compost. Whereas, average individual fruiting body weight was more in sugarcane baggasse + wheat straw (1:2).
- Maximum number of fruiting bodies was obtained in formalin + DDVP (15 ml + 0.5 g/10 kg compost) treatment followed by formalin + DDVP (15 ml + 0.5 g/10 kg compost) and bavistin alone (0.5 g/10 kg compost) over control.
- Maximum mushroom yield was obtained when casing was done with FYM + spent compost + soil + sand in the ratio 1:1:1:1, followed by FYM + burnt rice husk (1:10), vermicompost + loam soil (1:1), spent compost and farmyard manure.
- Dipping the fruiting bodies of button mushroom (*Agaricus bisporus* strain S11) in EDTA 125 ppm solution showed best results by increasing the shelf life and maintaining good colour and texture both under refrigerated and ambient storage up to 72 hours as well as gill opening up to (32 ± 2 °C) after 48 hours. It ranged from 16–40% with increase in storage time.
- Packing of button mushroom strain S11 either PP bag or PE bag 100–125 gauge thickness (with 10 pinholes) was found to be good. Not much difference was noticed between PP and PE bags with respect to colour change and weight loss. However, storage under refrigerated condition reduced veil opening percentage as compared to ambient storage.

Oyster Mushroom

- In case of *Pleurotus florida* spraying 0.1 M K_2HPO_4 at pinhead formation stage increased the yield of 40 g per bed compared to control.

- The yield potential of Blue oyster mushroom *Hypsizygus ulmarius* var. CO (OM) 2 was compared with that of *P. florida*. The CO (OM) 2 recorded significantly superior yield 620 g / bed (B.E= 124 %).
- All *Pleurotus* spp, viz. *Pleurotus citrinopleatus, P. sajocajur, P. florida, P. platypus and P. solmoneo stramineus* can be cultivated through out year under Tamil Nadu conditions. Among the different species tested, *P. solmoneo straminues* recorded maximum yield with maximum bioefficiency during the month of November.
- Paddy straw or paddy straw + paper clippings was the most favoured substrate for *P. citrinopileatus, P. sajorcaju and P. platypus* followed by either cotton waste or sorghum stalks. *P. platypus* recorded higher yield in all the substrates than the other two species.
- Supplementation of paddy straw (65% moisture) with powdered, steam sterilized neem cake at 5% level increased the yield of *P. citrinopileatus* (26.2%), *P. sajorcaju* (23.5%) and *P. platypus* (15.5%). In all the cases, DCSR was also reduced considerably (2-6 days). Five per cent supplementation was better than 2%.
- Chemical sterilization of paddy straw using carbendazim 75 ppm + formalin 500 ppm increased the yield of *P. citrinopileautus* by 38 per cent over control while no significant difference between chemical sterilization and hot water treatment was observed with *P. sajorcaju.*
- Blanching of oyster mushroom (*Pleurotus florida*) for 2 min with 0.2% salt + 0.1 % citric acid before drying and storage (for 3 months) slightly improved the whiteness. Direct sun drying or cabinet with out chemical treatment or blanching increased the brittleness after 3 months of storage.

Paddy Straw Mushroom

- Supplementation of paddy straw with waste cotton and neem leaves at 5% level were recorded maximum yield and suitable for supplementation.
- Paddy straw soaked in cold water for 18 hours recorded maximum yield when compared to other treatments, *viz.* 48 hours and 24 hours soaking.
- Circular compact bed method gave significantly increased yield of 1020 g/bed (22.5 per cent bio-efficiency).
- Bed size measuring 45 cm diameter and 30 cm height recorded the maximum yield of 982 g / bed (19.64 percent bioefficiency) with more number of harvestable buttons (55.4) compared to other bed dimensions.
- Among the different strains of *V. volvacea* evaluated, strain PS 1 recorded maximum yield of 1263 g / bed (bio-efficiency of 25.3%), followed by VV 7 and PS 2 (23.8% and 23.7% bio-efficiency respectively).
- Amendment with waste cotton followed by neem cake and chicken manure increased the yield of *Volvariella volvacea* by 46, 33 and 30% over control.
- Soaking of paddy straw beyond 24 hours was not favourable for cultivation of *V. volvacea.*

Milky Mushroom

- Among the 16 different wild strains of *C. indica* evaluated for their yield performance, strains WC 19 gave significantly superior yields compared to all other strains.
- Among the different strains obtained from NRCM, Solan (CI 1 to CI 9) the strain CI 1, CI 7 and CI 8 gave maximum yield of 520 g, 505 g and 490 g per bed respectively.
- In general all the milky mushroom strains tested recorded increased bio-efficiency of above 172 per cent indicating the best suitability of climate prevailing in Tamil Nadu and Paddy straw as substrate.

Medicinal Mushroom

- Silver oak saw dust + wheat bran (20%) was found to be the best substrate for cultivation of *Lentinula edodse*. An average yield of 231 g with an average of 10.7 fruiting bodies was obtained, each mushroom weighed approximately 21.7g.
- Silver oak saw dust + wheat bran (8%) and tapioca thippi + wheat bran (4%) gave the maximum yield of 785.0 g and 730.5 g per kg of substrate respectively. The average weight of the fruiting body in the substrates ranged from 3.8 to 4.2g.
- Silver oak saw dust + wheat bran 8% supplementation of wheat bran was found to be the best substrate for the cultivation of *Auricularia polytricha* with an average yield of 811.7 g/ kg of substrate was obtained in this treatment.
- Rice bran (4%) followed by rice bran (2%), wheat bran (2%) favoured the yield of Jew's ear mushroom.

Pantnagar

Edible Mushrooms

- Compost formulation with sugarcane baggase + wheat straw, 2:1 developed resulting in reduced cost of cultivation of *Agaricus bisporus*.
- Developed chemical treatment (Formalin 15 ml + Bavistin 0.5 g/10 kg compost) of long method compost to avoid the moulds in cultivation of *A. bisporus.*
- Recommended supplementation of substrate with 2% mixture of neem cake + wheat straw + rice bran + soybean meal for *Pleurotus* species cultivation.
- Standardized cultivation of *Lentinula edodes* with substrates popular saw dust and wheat straw.
- Siderophore producing *Pseudomonads* have been screened out resulting in early fruiting and enhanced yield of *A. bisporus.*
- Standardized cultivation *Auricularia polytricha* using sterilized wheat straw supplemented with wheat bran (5%).
- Standardized cultivation of *Calocybe indica* using wheat straw as a substrate with casing of FYM + spent compost + sand (2:1:1).
- *Pleurotus sajor-caju* and *P. florida* recommended for commercial cultivation using locally available soybean straw / paddy straw/ wheat straw/ mustard straw/ mandua straw/ jhnagora straw.

- Standardized cultivation technology of *Hypsizygus ulmarius* using wheat straw supplemented with wheat bran.

Medicinal Mushrooms

Reishi Mushroom (Ganoderma lucidum)

- Of the 10 isolates of *Ganoderma* collected, three were examined and characterized biochemically and molecularly.
- The medium MEA, pH 5-6 and temperature range of 26 ± 2 °C favour maximum growth of *G. lucidum*.
- Standardized cultivation technology of *Ganoderma lucidum*, using sterilized wheat straw supplemented with wheat bran (5%).
- Extracted and determined the presence of polysaccharides and Ganoderic acids A, C2 and H from the fruiting bodies of *Ganoderma lucidum*.
- Determined antioxidative properties among different isolates of *G. lucidum*.

Shiitake Mushroom (Lentinula edodes)

- The medium PDA, pH 5.0 and temperature 25 °C favour the maximum growth of *L. edodes*.
- Standardized cultivation of *Lentinula edodes* with substrate wheat straw and popular sawdust supplemented with wheat bran 20%.
- Two isolates of *L. edodes* were examined and characterized biochemically.

Black ear/Jew's ear Mushroom (Auricularia)

- Five local isolates of *Auricularia* were examined and characterized.
- The medium wheat decoction agar, pH 6.5 and temperature 25 ± 2 °C favour the maximum growth of *Auricularia*.
- Paddy straw followed by rapeseed straw found to be the most suitable for yield of Auricularias.

Keera Ghas/Yarsha Gumboo (Cordyceps sinensis)

- Cultural growth of *Cordyceps* found to be initially white to light cream changing orange or pink to light purple and reverse purplish red or dark tan or blood red colour on different media.
- The medium SDYA and pH 6.0 favour the maximum growth.
- Determined antioxidative properties from the fruiting bodies of *Cordyceps sinensis*.

Udaipur

- About 118 farmers belonging to different regions of Rajasthan were trained through 14 raining programmes on mushroom cultivation technology and spawn production technology. A total of 125 letters were replied concerning the queries on cultivation of different mushrooms.
- About 118 Mushroom cultivation-training programmes was conducted and trained 1,277 farmers, farm women entrepurners and students were trained for mushroom cultivation. Total of 55 letters was replied about the mushroom regarding cultivation of button, oyster,

milky, spawn production and marketing of mushrooms. In addition to this technical and spot guidance was given to 320 farmers on various aspects of mushroom cultivation. Twenty radio talks and four TV programmes on mushroom production technology were also delivered. About 12.5 quintals of mushroom seeds were supplied to growers.

Faizabad

Besides assigned technical programmes, research work has been done on the basis of farmers, feedback. Problems have been shorted out and after research technologies have been transferred to farmers. Some of them are:

- Casing technology of *Calocybe indica* in comparison and contrast with *A. bisporus.*
- Utilization of spent substrate in mushroom cultivation.
- Timing of removing polythene cover of oyster beds.

Raipur

Button Mushroom

- ***Evaluation of different strain:*** NCS100 and NCH102 strain of *A.bisporus* were found to be the superior with regard to yield in several years trial.
- ***Effect of dipping treatment in EDTA solution on quality and shelf life of A. bisporus:*** Fruit bodies could be well preserved up to 3 days under ambient temperature and 11–16 days under refrigerated conditions in 200 ppm solutions of EDTA and 500 ppm of KMS.

Oyster Mushroom

- ***Evaluation of locally available substrates:*** Several locally available substrates were tried to evaluate the performance but paddy straw, wheat straw, soybean and mustard straw performed excellently well in almost all the trials conducted.
- ***Evaluation of mushroom species and strains:*** Of several species of *Pleurotus* tried, the performance of *P. florida and P. flabellatus* was excellent under experimental conditions as well as under farmers growing conditions. Among strains of *P. florida*, local strain sp. 26 was better than *P. florida* sp. 29.
- ***Methods of treatment of substrates:*** Of three methods evaluated, hot water treatment followed by chemical treatment of straw was effective in suppressing the contaminants and increasing the yield of *Pleurotus* spp.
- ***Standardization of spawning rates:*** Spawning @ 5% was better than 3% spawning rate in several trials laid out.
- ***Standardization of spawning methods:*** Layer spawning methods was found to be more suitable than thorough mixing methods and spot method.
- ***Effect of supplements:*** No clear cut trend was noticed when the straw was supplemented with various supplements. However, rice flour, rice bran, wheat bran, soybean deoiled meal and neem cake performed better than other supplements
- ***Effect of urea:*** Spraying of urea and treatment of substrates with urea were tried several times but there was no increase (sometime increase) in yield probably due to higher concentration of urea.

- ***Effect of hollow beds:*** Hollow beds resulted in substantial increase in yield of different species of *Pleurotus* probably due to increase in aeration and wetting of inner part of the substrates.
- ***Effect of paddy straw:*** Chopping of straw in the length of 6–9" were extremely helpful in significant increase in yield of *Pleurotus* spp. than straw length of 3" may be due to increased air space, surface area, effective penetration of water.
- ***Effect of effective microorganisms:*** Dipping of oyster mushroom bed in the solutions of 0.4 and 0.6% of effective microorganisms preparations manufactured by Dehradun based firm was found to significantly increase the yield by 50% or even more in several experiments.

Paddy Straw Mushroom

- ***Effect of hollow beds on yield:*** Paddy straw bundles put in a spiral fashion in a wooden pillar or cemented pillar was found to enhance the yield to great extent in paddy straw mushroom at Raipur in several trials. Frame method of paddy straw mushroom cultivation did also well in increase of yield than standard bed method of cultivation.
- ***Effect of composting of substrate:*** Partially composted cotton waste was found to be the most suitable substrate for production of paddy straw mushroom followed by paddy straw substrate based on five years study.

Milky Mushroom

- ***Effect of substrates:*** Wheat straw alone and in combination with paddy straw in the ratio of 1:1 was very good in increasing the yield of milky mushroom.
- ***Evaluation of strains:*** Four strains were evaluated on paddy straw substrate, the yield varied from 53 to 136g in different strains with BE from 20–27 %.

Black Ear Mushroom

- ***Effect of substrates:*** Wheat bran @ 4 % on dry weight basis was found to be suitable for good growth and higher yield of *Auricularia polytricha*.

Shittake Mushroom

- ***Effect of substrates:*** spawn run and bump formation was noticed but no initiation of fruiting body was observed.

Reishi Mushroom

- ***Effect of substrates:*** Sawdust and wheat straw took 30–35 days for spawn run and fruiting body initiation was observed after 8–10 days.
- ***Evaluation of strains:*** *Ganoderma lucidum* local strains, Chinese strains and Udaipur strains were tried with good amount of success.

Vellayani

The tropical climate of Kerala, its rich natural resource of mushroom flora and availability of large variety of agro-wastes make the state ideal for the promotion of cultivation of tropical mushrooms. Locally available alternative substrates were identified and their processing technology standardised for economic cultivation of oyster mushroom. Red banana pseudostem, mango wood chips and *Eleocharis plantagenia,* a wetland weed, are promising ones. An improvised air cooling system was developed for speedy sporocarp induction in dormant beds of oyster mushrooms during the cropping phase. Technique is cheap and can be adopted for small-scale growers. Low-cost technology for spawn production using cheap substrates like match work waste, potato blended gravel, etc., was developed. Techniques were standardised for blending coconut endosperm milk in culture media for enhanced growth and biomass production of *Pleurotus* sp.

The coir pith can be conveniently converted into an organic manure using three sub-species of mushrooms namely *Pleurotus eous, Schizophyllus commune* for retted and non-retted coir pith and *Pleurotus sajor-caju* for non-retted coir pith alone. Conversion of this waste material into a useful organic manure is a boom to farmers of Kerala and spawn of these varieties exclusively meant for coir pith decomposition and being sold out regularly to farmers.

Ranchi

Paddy Straw Mushroom

Strains, Vv-7, Vv-9 and Vv-11 of paddy straw mushroom (*Volvariella volvacea*) performed better than other strains and mushroom yield of 13.88, 15.8 and 12.66% biological efficiency (BE), respectively.

Milky Mushroom

The maximum BE was found in strain CI 6 (107%) followed by CI 7 (88.9%), CI 1 (80.2%), CI3 (73.8%) and APK2 (72.8%) (Fig.4.2). However, strain CI 6 and CI 7 were at par but significantly different from other strains. Thus, CI 6, CI 7 and APK 2 were better suited strains in fruit quality and yield in Jharkhand.

Oyster Mushroom

Supplementation with wheat bran @2% enhanced the yield significantly in *P. florida* (48.3%), *P sajor-caju* (44.9%) and in *P.ostreatus* (48.1%) over control (Fig.4.3). Hence, wheat bran @2% (w/w) supplementation is recommended in paddy–straw substrate for higher yield and profit. K_2HPO_4 significantly increased the yield of *P.sajor-caju*

Blue Oyster Mushroom

Maximum yield per kg dry straw was found in *Hypiyzygus ulmarius* (2232g), *Pleurotus florida* (1814g) and *Pleurotus sajor caju* (1725g). Thus only two species, viz. *Hypsizygus ulmarius* and *Pleurotus florida* can successfully cultivated in winter season.

Black Ear Mushroom

Black ear Mushroom was successfully gron with yield of 1567 g/ kg dry wt of straw in two flushes.

Nauni

- Compost inoculated with mixed inoculum of *Gilmaniella humicola + Scytalidium thermophilum + Thermomyces lanuginosus* and at the time of spawning enhanced the button mushroom yield over other treatment and the control.
- Casing material prepared with mixture of farmyard manure + spent compost + vermicompost gave highest button mushroom yield of 25 kg/100 kg compost and was followed by casing material prepared with FYM + spent compost. (24.50 kg/100 kg compost).

- Supplementation of compost with cotton seed cake and cotton seed meal at the time of spawning stimulated highest fungal population in compost and *Penicillium* sp. was the most common fungal species.
- Supplementation of compost with cotton seed meal, sesame meal and cotton seed cake enhanced the button mushroom yield.
- Supplementation of compost with peptone, glucose and malt extract led to maximum contamination from *Penicillium chrysogenum*, *Penicillium coryophilum*, *Rhizopus nigricans*, *Sepedonium chrysospermum* and *Mucor racemosus*.

Barapani

Button Mushroom

Strain, CM-14 of button mushroom (*Agaricus bisporus*) performed better than other strains with a mushroom yield of 16.2 % was recorded.

Oyster Mushroom

Supplementation with wheat bran (5 %) and rice bran (5 %) gave higher mushroom yield over control.Spraying treatment with 0.1 % urea gave significantly higher yield in oyster mushroom *(Pleurotus florida)*.

Paddy Straw Mushroom

Strain, Vv-6, of paddy straw mushroom (*Volvariella volvacea*) performed better than Vv-1, Vv-2, Vv-4 with a mushroom yield of 10.8 % biological efficiency (BE) was obtained.

Blue Oyster Mushroom

Average yield of 67 % was recorded in blue oyster mushroom (*Hypiyzygus ulmarius*) at Barapani Centre.

CROPPING SYSTEM

Ludhiana

A round the year cultivation cycle recommended by Punjab Agricultural University, Ludhiana is being followed by growers for cultivation under natural indoor environmental conditions following the schedule given above.

Pune

- Monsoon and rainy season are the most suitable season for all the species of *Pleurotus sajor-caju* was found to be higher yielding than other species.
- Different agricultural wastes were evaluated for yield potential of oyster mushroom. Cotton stalks and leaves, paddy and wheat straw are recommended for use in order of perferance considering availability and cost.
- Intial level of 1.5 % N in compost was proved to be optimum for getting higher yield of *A. bissporus* strains S-11.
- Wheat straw compost was found better than paddy straw compost for growing *Agaricus bisporus*.
- Different agro-wastes like wheat, paddy and soybean straws were evaulated seperately and also the combination of soybean + wheat straw (1:1) as base materials for compost.

The combination of soybean + wheat straw recorded significantly highest yield of *Agaricus bisporus* (1543g /10 kg compost) than other base material.

- In cultivation of *Agaricus bisporus,* additional of 100 g grain spawn / m^2 in casing material at the time of the casing found to give significant increase in mushroom yield. It recorded 30% increase over control.
- Cotton stalks and leaves recorded significantly higher yield of oyster mushroom (824g /kg dry substrate) followed by paddy and wheat straw yielding (721 and 717 g /kg of substrate) respectively.
- The wheat straw + 2 % wheat bran + 2 % rice bran supplementation is beneficial for getting more yield of black year mushroom.
- The highest fresh mushroom yield of *Calocybe indica* was obtained in mix substrate of wheat + paddy straw (1:1), i.e., 616 g /kg of dry substrate.
- The treatment of sawdust + 20 % wheat bran mixture was found to be better because of poor to good quality spawn run and recorded on an average 5 fruits / bed with an average weight of 20g.
- In the evaluation of yield potential of blue oyster mushroom, significantly highest yield was obtained in *Pleurotus sajor-caju* (551 g/kg dry substrate) and it was at par with yield of *P.florida* (545 g/kg dry substrate). The blue oyster recorded a yield 502 g/kg dry substrate.

Coimbatore

Outdoor Cultivation of Paddy Straw Mushroom (Volvariella volvacea)

Crops: banana, maize, sugarcane

- Paddy straw mushroom could be successfully cultivated in inter-row spaces of maize (30 DAS), sugarcane (8 months after planting) and banana (5 months after planting).
- Among different substrates tried, paddy straw always gave increased yield with bioefficiency of 9-16 % from 25 kg of substrate with a total crop cycle of 28-30 days.
- The cost: benefit ratio in intercropping in maize was to 1:1.37, in sugarcane field it was 1: 1.94 and in banana field it was 1:1.4 4.
- Inter-cropping paddy straw mushroom revealed that addition of N level from 13.2–16.9 %, P level from 9.5 – 16.67 % and K level from 5.90–11.82 % were noticed after 60 days *in-situ* biodegradation of mushroom spent substrates
- Overall reduction in weed population, their group and dry weight (9g /m^2) were noticed in places where mushroom beds had been maintained

Pantnagar

A relay cropping schedule developed for tarai region of Uttarakhand: two crops of *Agaricus bisporus* from September-March, four crops of *Pleurotus* spp. from September-November and February—April and three crops of *Calocybe indica* from March—October.

Table 4.5. Mushrooms suitable for different zones of Uttarakhand state as per climatic conditions

Zone	*Region*	*District*	*Mushroom Species & No. of Crops*
Zone-'A1'(100-600 m)	Tarai, Bhawar	U.S. Nagar, Haridwar, Nainital, Pauri- Garhwal, Champawat, Dehradun	Button – 2 crops (October—March) Dhingri – 2+2 crops (Aug.– Oct. & February—April) Milky – 2 crops (April—September) Shiitake – 1 crop (Nov.—Feb.) Black Ear – 2 crops (July—October)
Zone-'A2'(600-1000 m)	Lower Hill	Champawat, Nainital, Pauri Garhwal, Dehradun, Almora	Button – 2 crops (October—April) Dhingri – 2 crops (April—November) Milky – 1 crop (May—July) Shiitake - 1 crop (Nov.—Feb.) Black Ear – 2 crops (April—October)
Zone-'B'(1000-1500 m)	Medium Hill	Champawat, Nainital, Almora, Dehradun, Tehri Garhwal, Bageshwar, Paurigarhwal, Chamoli, Pithoragarh	Button – 2 crops (October-March) Dhingri – 2 crops (August—March) Milky– 2 crops (April—September) Shiitake – 1 crop (November—Feb.) Black Ear– 2 crops (July-October)
Zone-'C'(1500-2400 m)	High hill	Pithoragarh, Chamoli, Bageshwer, Uttar Kashi, Dehradun, Tehri Garhwal, Rudra Prayag	Button – 3 crops (Feb.—Nov.r) Dhingri – 2 crops (May—August) Shiitake – 2 crops (March—October)
Zone-'D' (>2400 m)	Higher hill	Pithoragarh, Chamoli, Uttarkashi	Button – 3 crops (February—October) Dhingri – 2 crops (June–July) Shiitake – 2 crop (March—October)

Faizabad

As per the climatic conditions of eastern Uttar Pradesh and the requirements of different species of mushrooms, research work was carried out and a yearly calendar of cropping system was prepared.

Raipur

Seven mushroom species, *viz.* oyster mushroom, paddy straw mushroom, milky mushroom, white button mushroom, shiitake mushroom, black year mushroom, reishi mushrooms were evaluated according to their suitability under prevailing climatic conditions of Chhattisgarh. A crop calendar was developed to grow different mushrooms and their species after conducting experiments for several years. It was suggested to grow different species of oyster mushroom during different parts of the year. *Pleurotus florida* and *P. flabellatus* were found to grow well almost round the year. *Pleurotus florida, P. eous, P. membranaceus* were found to grow well during monsoon season in Chhattisgarh plains, Bastar plataeu and Northern Hilly regions, while *P. sajor caju, P. colummbinus,* and *P. ostreatus* were found to perform well under Bastar plataeu and Northern Hilly regions during winter season.

Pleurotus eous, P. salmonea stramonius were found to be good during spring season in Chhattisgarh plains. *Calocybe indica, Volvariella volvacea, V. diplasia* and *Auricularia polytricha* were found to grow well during monsoon season under natural growing conditions, while button mushroom can be very well cultivated during winter season. Two crops of button mushroom were grown under Northern Hilly regions and Bastar plateau conditions while only one crop of button mushroom can be grown in Chhattisgarh plains during winter season. In spring/premonsoon season, *Volvariella volvacea* and *Calocybe indica* were nicely grown. Thus, the above mushrooms can be suitably grown depending upon the prevailing climatic conditions one after the other with higher yield simply by changing the types of mushrooms and their species almost the round year.

Vellayani

The traditional system of growing *Vovariella* has been discarded and polybag method of cultivation followed. For oyster mushroom too polybag method followed. For *Calocybe* spent mushroom substrate is also used for cultivation apart from the trials charted out by AICRP on Mushrooms. The *Calocybe* beds are split into halves before casing and are grown exclusively in blue silpaulin sheds to get best results. Now-a-days, due to non-availability of polythene covers a method of growing in waste paper basket is being tried and hope to get best results.

PEST AND DISEASE MANAGEMENT

Ludhiana

Pests and Diseases Encountered in Punjab

- Fungi — *Aspergillus, Chaetomium, Coprinus, Dactylium dendroides, Diehliomyces microsporus, Mycogone perniciosa, Penicillium, Populaspora byssina, Trichoderma, Verticillium fungicola*
- Bacteria — *Pseudomonas tolaasii*
- Insects/pest — *Lycoriella, Megaselia,* Mites and Springtails
- Nematodes — *Aphelenchoides composticola, Ditylenchus myceliophagus, Caenorhabditis* spp.

Most of the pathogens and pests of white button mushroom can be managed by sound hygiene and sanitation measures only which at the moment is the best available defense with a grower. The use of chemicals should be made only under extremely unavoidable circumstances. A grower can get reasonably disease-free crop at the farm if the following tips are routinely followed during various stages of mushroom cultivation:

- Reduce the dust load at mushroom farms by using maximum concrete on the area as dust is the primary source of a few mushroom pathogens.
- The location of the farm should not be accessible to toxic fumes and gases.
- Bulk ingredients used for compost preparation should either be fresh or stored properly. These should not be already wet, muddy or too broken down.
- Compost ingredients should be properly formulated, correctly supplemented if required and processed. Incorrect C/N ratio attracts unwanted weed/indicator molds.
- Choice of spawn and supplements should be well thought of. Try to procure and use spawn from a genuine source and it should be free of any contaminant.
- Casing soil should be properly prepared, stored and disinfected/pasteurized correctly. Any mixture suspected of containing spores should be avoided. Avoid over pasteurization of casing soil for longer duration. The casing soil preparation and casing operation should be carried out under strict hygienic conditions.

- The surface of the beds should be dressed properly of weed or cosmetic molds before casing and preferably spot treatment with 2 % formaldehyde be given. Some indicator molds like ink caps need to be weeded out before they autodigest and open to liberate spores.
- On occasions water chlorinated to 150 ppm chlorine may be used to reduce population levels of the bacterium on the mushroom caps. Condensation of water droplets and water splashes should be avoided.
- Special attention should be paid to maintain optimum air velocity. CO_2 should be 5000–15000 ppm during spawn run and the first few days following casing and 1000–1500 ppm thereafter.
- Prophylactic spray of Indofil M 45 (0.25-0.50 %) at casing, at pin head appearance and between second and third break mushrooms help reduce the growth and spread of a few fungal pathogens. Use of carbendazim (0.02-0.05 %) under certain conditions is also recommended.
- Picking of mushrooms should preferably be done once during the day and before the opening of the caps.
- At the end of the crop cycle, the production room should be properly pasteurized (cook out), cleaned out and prepared for another filling. The "Cook out" is essentially required in the event of nematodes infestation. Emptying or dumping off of the spent substrate should be done at a far off site from the mushroom producing area.
- As an aid to good sanitation and hygiene, use the recommended fungicides and pesticides as mentioned below in the event of appearance of disease and insect pests.

Treatment Suggested/Recommended

• Fungi	Formaldehyde 2 %, Indophil M-45, 0.2-0.5 %, Carbendazim 0.02-0.05 %, manual removal of infected patches and weed fungi
• Bacteria	Chlorinated water 75 ppm
• Insect/Pest	Malathion 0.05 %, Dichlorovos, 0.5 %, Lindane (g BHC) in compost
• Nematodes	Furadonn 3G (50 g/Qtl) in compost/casing, Pasteurization/cooking out where facility available.

Pune

- Chemical sterilization of wheat straw using 75 ppm bavistin with 500 ppm formalin observed to be equally good like hot water treatment.
- The chemical pasteurization of compost prepared by long method with DDVP (3 ml) + Bavistin (0.5g) per 10 kg compost recorded maximum yield of 1064 g/10 kg compost of button mushroom.

Coimbatore

IPM programme is generally followed in mushroom growing farms

IPM Programme

- Acquiring knowledge on the biology and behavior of the pests and diseases
- Monitoring the farm environment
- Assessment of the damage
- Making appropriate management decisions
- Use of IPM tools and techniques

Components of IPM

- Farm hygiene is given utmost priority
- Providing doors and windows and ventilators with 35 mesh nylon net to prevent of the flies in the cropping room
- Use of TNAU light traps and yellow sticky trap for monitoring and killing mushroom flies.
- At the end of each cropping, cooking out at 70–80ÚC at least for 2 hours before removing the beds to kill all stages of pests and diseases.
- The spent compost and casing materials should be safely disposed in isolated manure pits to check the breeding of flies.

Pantnagar

Major pathogens: *Verticillium, Sepedonium, Trichoderma, Fusarium, Mycogone perniciosa and Hypomyces rosellus*

Management

- *Pseudomonas fluorescence* proved effective in controlling mushroom diseases caused by *Verticillium, Sepedonium, Trichoderma* and *Fusarium.*
- Botanicals *Aegle mormelos* (0.5 %) and *Cleome viscosa* (1.0 %) were found effective in control of *Mycogone perniciosa* and *Trichoderma viride*, respectively.
- Chemical control of pathogens *Mycogone perniciosa* and *Verticillium fungicola* by sporgone (0.075 %) and *Hypomyces rosellus*, *Fusarium moniliformae* and *Sepedonium chryosporium* by Bavistin (0.025 %) + Formalin (0.2 %) have been worked out.

Udaipur

In survey and surveillance studies, several oyster mushroom farms were surveyed and in few farms infection of green mould, inky caps and *Aspergillus* spp. was recorded. In insects, Dipterien flies and spring tails were common. However, in button mushroom infection from brown plaster mould, *Coprinus* spp. and green mould was recorded. Infection from nematodes and Dipterien flies was also recorded from few farms.

It was also observed that in tribal area no disease and insect pests were found. Further only at 2-3 farms in Dhimadi village infection of *Coprinus* spp. And green moulds were observed in traces only.

Faizabad

Treatment and drying of straw in jute bags, and use of light traps for insect management, re-casing of mushroom beds against plaster moulds, and strict hygiene were used.

Raipur

- Bavistin and Subeej at 50 ppm were most effective in inhibiting the growth of most of the weed fungi whereas antracol was less effective.
- Bavistin 75 ppm and Formaline 500 ppm was most effective in inhibiting the growth of most of the weed fungi compared to other treatments.
- Minimum fungal count was observed in the substrate treated with riolon + formalin

(75 + 500 ppm) followed by bavistin +formalin. But maximum biological efficiency (71.25) was obtained with Topsin –M followed by Sultaf+formalin.

- Under *in vivo* conditions, rilon + formalin (75 + 500 ppm) recorded significantly higher yield (80.2 % BE) followed by Subeej+formalin (75 + 500 ppm-78.8 % BE) in comparison to 29 % BE in control.
- Neemazole @ 0.1 % was found to enhance the yield of *P. florida* (57.40) in comparison to control (16 %).

Vellayani

- Blue and yellow light traps were installed in mushroom houses which reduced the attack by flies. Gingelly oil rubbed on X-ray films were hung near the beds to trap the flies.
- Garlic paste put around the mushroom house were found to reduce the insect attack.
- Boiling was the most effective method of substrate treatment which resulted in the very low incidence of diseases in mushroom beds.
- Keeping the surroundings of the growing houses clean helped to tackle the pest problems.

TECHNOLOGIES TRANSFERRED

Ludhiana

A complete package of technologies has been transferred for the growers of the state and adjoining areas through various recommendations that have been included in University 'Package of Practices'.

Recommendations

- Round the year cultivation technology for 4 varieties mentioned above
- Low-cost compost and casing soil formulations
- New composting schedule of 24 days for LMC
- High yielding cultivars – PAU- S11 (*Agaricus bisporus*), PAU-2 (*Pleurotus ostreatus*), PAU-3 (*P. sojar caju*) and PAU-5 (*P. florida*) and *V. diplasia* selection and Mm-3 (*Calocybe indica*)
- Comprehensive disease control package for control of mycopathogens.

Extension Services Rendered

- Short training courses on mushroom production technology at PAU, KVK and Regional Research Stations on regular basis as per the mushroom cultivation season
- Visits to/by the growers and high dignitaries
- Sample analysis for pH, moisture, carbon to nitrogen ratio, nematodes, etc.
- Lectures/demonstrations

Extension Materials Prepared

- Bulletins – Punjabi and English
- Bilingual CD's
- Compendiums

- Audio cassettes, video/CD
- Pamphlets/handouts

Services to Mitigate

The challenges/problems encountered by the mushroom growers are addressed on personal contact basis and on-site visit/melas/demonstrations and through:

- Spawn supply
- Use of electronic and print media
- Audio/visual demonstrations
- Demonstration of popular literature
- Rendering advisory services
- Preparation of project reports
- On-line services

Pune

Extension Activities

(a) Trainings:

Number of training conducted	- 705
Bumber of beneficiaries	- 15,883

(b) Spawn supplied: - 41 tons

(c) Mass communication:

Popular articles	- 15
T.V. programmes	- 20
Radio talks	- 20
Frontline demonstration	- 5
Melavas/exhibitions	- 50
Visitors	- 95000

(d) Self-employment training programme:

The self employment training programme has been launched by Government of Maharashtra. Two batches of 20 participants each were provided for four month duration through training in all aspects of mushroom production.

Academics

(a) Teaching	:	Also engaged in UG and PG teaching.
(b) Awards	:	Dr S.S.Wange, received best poster presentation award during 2002.
(c) PG research	:	Research has been conducted by Ph.D. and M.Sc. students on various aspects of mushroom technology in the Department of Plant Pathology and Agricultural Microbiology, Mahatma Phule Krishi Vidyapeeth, Rahuri and Department of Plant Pathology and Agricultural Microbiology, College of Agriculture, Pune.

Coimbatore

The technologies on clean cultivation of mushroom, pest and disease management were transferred through the conduct of regular one day training programme on 5th of every month by this centre. Every month an average number of 40–50 participants attend the programme from different places of the state and even from other states also. So far 295 such programmes have been conducted and more than 20,000 people have participated in these programmes for the past 20 years.

Recently 30 advanced training on mushroom cultivation for five days and ten days were organized and about 105 people were benefited out of the trainings. For the first time in India, distance learning on mushroom cultivation has been organized at TNAU, Coimbatore, from 2005 onwards.

Extension Services Rendered

Spawn supplied	:	98,095	One-day training	:	295
Days training	:	28	10-days training	:	2
ODL trainings	:	2	Visitors to the unit	:	2000
Publications	:	17	Books Published	:	6

Ranchi

Popularization of Mushroom

During past six years 2002-03 to 2007 (up to October), a total of 166.7 quintal of spawn for oyster mushroom was made available to interested mushroom growers of Jharkhand and other states.

Training

Various training programmes were conducted by various organizations to popularize mushroom cultivation in the state. After creation of Jharkhand in 2000 several other organizations have also trained the farmers. A number of farmers were trained listed below. A mushroom divas was organized by HARP on 24 October 2003 to popularize mushroom cultivation in Jharkhand in which various interested mushroom growers participants. Recently mushroom training-cum-mela was organized on 3-4 November 2007 in which 97 persons were trained for oyster and button mushroom cultivation.

Table 4.6. Training programmes organized at Ranchi Centre

Name of Sponsoring Organization	*Number of Person Trained*
Holy Cross Krishi Vigyan Kendra, Hazaribagh	2448
ATMA Palamau	10 groups
SRI Morabadi, Ranchi	152
HARP, Plandu, Ranchi	1957
RK Mission, Dyvayan, Ranchi	357

Pantnagar

For quick dissemination of the research findings to the growers the centre pilots a variety of extension programmes:

- Demonstration and on-farm trials in adopted villages.
- Training courses for entrepreneurs, small growers, BPLs and extension workers.
- Organization and participation in exhibitions and farmers fairs.
- Consultancy services to entrepreneurs and advisory help to farmers.
- Supply of quality spawn to growers and cultures of improved strains to industry/spawn producers.
- Supply of pasteurized compost to growers.

Impact

Mushroom production during 1999–2006 had started showing upward trend. The mushroom production in the state reached to 5,340 tonnes in 2005-06 from 2,203 tonnes in 1999-2000.

It is evident from the data presented as above that there was 3,100 tonnes production during the year 2001 when Uttarakhand state was established after division from Uttar Pradesh. It increased substantially from 3,100 tonnes in 2001 to 5,340 tonnes in 2005. The spawn/ compost supplied to growers and increase in number of growers in 2006 indicated the production of the state must have reached more than 6,000 tonnes.

Mushroom Production Units

Mushroom production units in the states reached 903 in the year 2005 from 485 units during the year 2002.

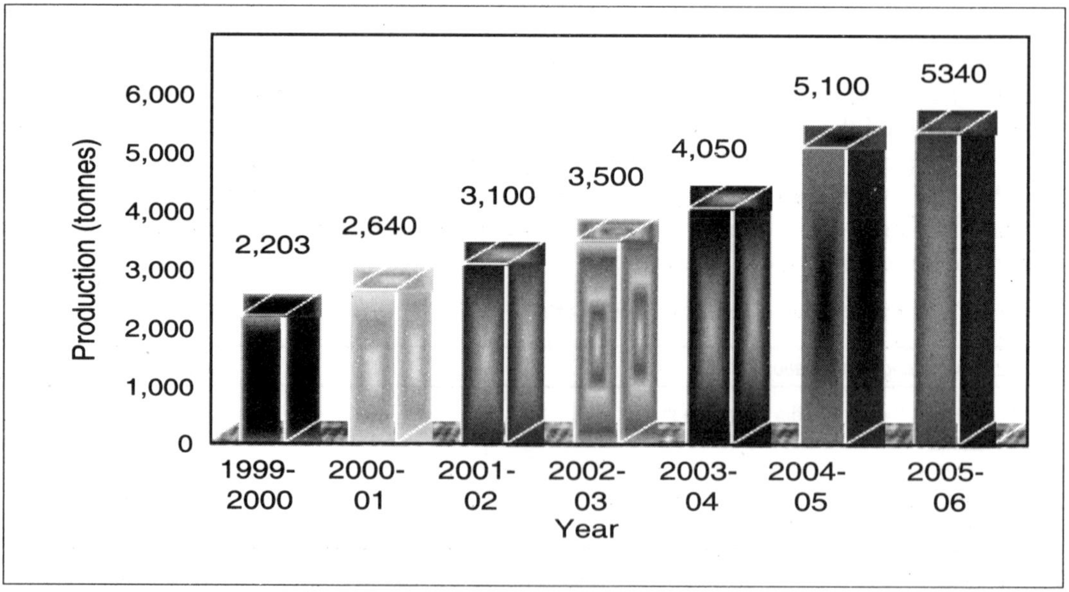

Fig 4.1. Mushroom production (tonnes) in Uttarakhand

Table 4.7. Mushroom production units in Uttarakhand

District	*Level*	*Units 2002*	*Units 2003*	*Unit 2004*	*Unit 2005*
Dehradun	Marginal Small Industrial	1010201	1310301	1510601	1670701
Haridwar	Marginal	03	09	11	31
Pauri Garhwal & UttarKashi	Marginal	04	06	09	15
Nainital	Marginal Small	13502	20003	21706	22406
Almora	Marginal	30	105	130	148
U.S. Nagar	Marginal Small	20304	28306	28806	29707
Total		485	741	825	903

Coordination and Linkages

Involved in testing of technology under "All India Coordinated Mushroom Improvement Project" within a network of 10 Centres located in different parts of the country. The Centre also maintains linkages with different departments like Microbiology, Plant Pathology, Food Science and Technology, Biotechnology, Post-Harvest Technology, Chemistry, Biochemistry etc. of the University and 30 Post-graduate and 14 Ph.D. degrees have been awarded on thesis research on different aspects of nutra and nutriceutical mushrooms.

Udaipur

About 118 farmers belonging to different regions of Rajasthan were trained through 14 raining programmes on mushroom cultivation technology and spawn production technology. A total of 125 letters were replied concerning the queries on cultivation of different mushrooms.

About 118 Mushroom cultivation-training programmes was conducted and trained 1277 farmers, farm women entrepurners and students were trained for mushroom cultivation. Total of 55 letters was replied about the mushroom regarding cultivation of button, oyster, milky, spawn production and marketing of mushrooms. In addition to this technical and spot guidance was given to 320 farmers on various aspects of mushroom cultivation. Twenty radio talks and four TV programmes on mushroom production technology were also delivered. About 12.5 quintals of mushroom seeds were supplied to growers.

Faizabad

Besides assigned technical programmes, research work has been done on the basis of farmer's feedback. Problems have been sorted out and after research technologies has been transferred to farmers. Some of them are:

- Casing technology of *Calocybe indica* in comparison and contrast with *A. bisporus.*
- Utilization of spent substrate in mushroom cultivation.
- Timing of removing polythene cover of oyster beds.

Raipur

Training Programmes

National Training Programmes

With a view to promote mushroom production and marketing, a self-employment generation scheme was launched on 1 January, 1997. Under this scheme, an open advertisement was given in the news papers regarding organization of training programme on charged basis so that only the interested entrepreneurs may come to join these training activity and involved in mushroom business. This training had very good impact and attracted the entrepreneurs from Madhya Pradesh, Orissa, West Bengal, Bihar, Rajasthan and Maharastra. These entrepreneurs after being exposed to 7-day training were able to establish either their own spawn lab or production units. Many of the entrepreneurs trained by us are now engaged in mushroom spawn production, crop production, marketing. The details of training are as under:

National training programmes organized at Raipur Centre

Training	*Period*	*Sponsoring Agency*	*No. of Participants*
Mushroom production technology to the entrepreneurs	01.1.99 - 07.1.99	Registration fee Rs. 1000/- per trainee	30 from Chhattisgarh, Rajasthan, Madhya Pradesh
Mushroom production technology to the entrepreneurs	10.11.99-16.11.99	Registration fee Rs. 1000/- per	28 from Chhattisgarh, Rajasthan, Maharastra
Mushroom production technology to the entrepreneurs	22.9.2000-28.9.2000	Registration fee of Rs. 1250/- per trainee	from Chhattisgarh, 32 Rajasthan, Madhya Pradesh
Mushroom production technology to the entrepreneurs	05.7.02 - 11.7.02	Registration Rs. 1250/-per person.	13 from different district of Chhattisgarh
Mushroom production technology to the entrepreneurs	07.1.04 - 13.1.04	Registration fee of Rs. 1250/- per head	39 from Chhattisgarh Madhya Pradesh, Bihar West Bengal, Orissa

State Level Training Pprogrammes

Several on campus and off campus training programmes on mushroom spawn production, cultivation of different mushrooms, mushroom preservation/processing were given to the farm women/men/rural youths under different programmes which benefited the farmers in variety of ways. The details of training programmes, duration, sponsoring agencies and numbers of beneficiaries are given below:

Adoption of Villages

Tarra and Dondekhurd villages were adopted under Institutional Village Linkages Programmes through Technology Assessment and Refinement and DBT Ad hoc project on "Transfer of mushroom production and processing technologies for the upliftment of rural farmers of Chhattisgarh" and are continuously visited by technical persons for imparting training on different aspect of mushroom, viz. Mushroom cultivation, spawn production, processing, marketing, etc. During training period, raw materials and other required equipments for above said purposes were also made them available for smooth running of their activities. The mushroom technology generated at IGAU Raipur was also evaluated in these two villages under farmers growing conditions and refinement of technology was done based on the active involvement of farmers. It is a continuous process adopted by us since the adoption of villages in 1997. The farmers (men/women) did frequently visit our laboratory and sought the necessary advice on implementation of mushroom technology in their villages.

Table 4.8. Training programmes organized at Raipur centre

Training	*Period*	*Sponsoring Agency*	*No. of Particioants*
Oyster mushroom production technology	12.6.99 – 13.6.99	Govt. Polytechnic College, Dhamtari	40
Cultivation of paddy straw mushroom (2), Preparation of mother spawn, cultivation of oyster mushroom	19.7.99 – 22.7. 99 4.8.99 – 7.8.99 6.3.2000 – 9.3.200013.3. 2000 – 16.3.2000	TAR through IVLP (ICAR)	15, 21, 28, 28
Mushroom production technology to entrepreneurs	5.11.99 – 11.11.99 10.11.99- 16.11.99	Registration fees of Rs. 1000/-	5, 28
Cultivation of paddy straw mushroom, preparation of mother spawn, Cultivation of oyster mushroom	22.7.2000 – 25.7.20002 .8.2000 – 5.8.2000 19.12.2000 – 22.12.2000 3.1.01 – 8.1.01	TAR through IVLP (ICAR), Tarra and Chatoud villages	15,15, 25, 25
Mushroom production technology	3.9.2000 – 9.9.200028.12.2000	Trainees themselves & Dept. of Horticulture, Raipur	6,28
Mushroom production technology	07.09.01	Mushroom project	38
Krishi Takniki Prashikshan Karyashala	22.08.01 – 23.08.01	Land & water Management	65
Paddy straw mushroom production technology, Spawn production technology, oyster mushroom production technology	23.07.01 – 26.07.01 26.11.01 – 29.11.01 01.12.01 – 06.12.01	TAR through IVLP at Tarra and Chatoud villages	30, 28, 28
Mushroom production technology	23.08.01 – 03.09.01	Trainees themselves	7
Mushroom production training	14.06.02 – 16.06.02	IGKV and Narmadanchal, Gramoudyog Mahila Kalyan Sansthan	27
Mushroom production technology	27.06.02 – 30.06.02	IGKV and Intensive Waste Land Development Programme, Mili Watershed, Saraipali	36
Mushroom production technology	14.07.02 – 19.07.02	IGKV and Tulsi lok Vikas Sansthan, Bhilai	107
Production of white milky mushroom, Mushroom spawn production, oyster mushroom production	9.08.02 – 11.08.02 11.02.03 – 14.02.03 17.02.03 – 22.02.03	TAR through IVLP at Tarra and Chatoud villages	30, 25, 24
Training programme on oyster mushroom	11.8.02 – 14.8.02 12.9.02	IGKV, Raipur and Mili WaterShed, Bagbehra	65, 57

(Contd.)

Training	*Period*	*Sponsoring Agency*	*No. of Participants*
Mushroom training programme	18.12.02	NSS Uunit of IGKV, Raipur	125
State level training programme on mushroom production technology to entrepreneurs	16.07.03 – 22.07.03, 8.10.03 – 14.10.03	Jointly organized by CCOST and IGAU, Raipur	35, 32
Cultivation of paddy straw mushroom	04.09.03 – 05.09.03	TAR throughIVLP	25
Mushroom production technology to women	9.2.04 – 12.2.04, 25.2.04 – 28.2.04, 16.3.04 – 19.3.04 28.5.04 – 31.5.04 21.7.04 – 24.7.04 1.12.04 – 4.12.04	Jointly organised by Swashakti Project, Raipur,C.G. and IGAU, Raipur,	33, 33, 32, 30, 31, 30
Mushroom processing technology, Mushroom spawn production technology, Mushroom production technology	14.6.05 – 16.6.05 20.6.05 – 22.6.05 18.7.05 – 21.7.05 19.9.05 – 21.9.05 23.11.05 – 25.11.05	Department of Biotechnology	20, 23, 13,37, 25
Mushroom production technology to the women	26.07.04 – 30.07.04	In joint collaboration with Govternment Polytechnic, Durg	32
Mushroom spawn production technology	12.8.04 – 18.8.04 24.11.04 – 30.11.04	Deptt. of Horticulture, Govt. of Chhattisgarh, Raipur	33, 24
Mushroom spawn production technology, Mushroom processing technology (3), Mushroom production technology	7.9.04 – 9.9.04 17.12.04 – 19.12.04 2.2.05 – 5.2.05 17.2.05 – 21.2.05 16.2.05 – 22.2.05	DBT Project, Govt. of India, New Delhi.	14, 18, 28, 32, 25
Mushroom production technology to the entrepreneurs by Department of Horticulture at Durg	1.1.05 – 4.1.05	Department of Horticulture, Govt. of Chhattisgarh, Durg	50
Mushroom production technology to entrepreneurs	25.1.05 – 31.1.05 9.03.05 - 15.3.05	Jointly organised by CCOST and IGAU, Raipur	23, 24
Mushroom production technology	14.2.05 – 15.2.2005	Trainees themselves	23

(Contd.)

Training	*Period*	*Sponsoring agency*	*No. of participants*
Mushroom processing technology	14.6.05 – 16.6.05 & 20.6.05- 22.6.05	DBT, New Delhi	20, 23
Mushroom spawn production technology	18.7.05 – 21.7.05	DBT	14
Mushroom production technology	1.8.05 – 11.8.05 19.9.05-21.9.05 23.11.05 – 25.11.05, 6.12.05 – 15.12.05	Self - Government Science College, Raipur DBT New Delhi	 7, 40, 20, 7
ICAR short course on "Emerging areas in mushroom diversity, production and post-harvest developments".	9.1.06 – 18.1.06	Indian Council of Agricultural Research, New Delhi	21 participants
Mushroom processing technology,Mushroom spawn production technology, Mushroom production technology	8.5.06 – 10.5.06 22.5.06 – 24.5.06 12.7.06 – 14.7.06 2.8.06 – 4.8.06 17.8.06 – 19.8.06	DBT Government of India, New Delhi.	20, 21, 13, 20, 20 women
Training on mushroom production technology	28.8.06 – 30.8.06	Self	2 students
Mushroom processing technology	6.8.06 – 8.8.06 5.10.06 - 7.10.06	Department of Horticulture Kabirdham	25 (11 W +14 M) 26 (25 W + 1 M)
Mushroom spawn production technology	19.9.06 – 25.9.06 5.12.06 – 11.12.06 2.1.07 – 8.1.07 24.1.07 – 30.1.07 12.2.07 – 18.2.07 21.2.07 – 27.2.07 27.2.07 – 3.3.07	Department of Horticulture,Raipur, Kabirdham, Dhamtari, Raipur	29 (24 W + 5 M) 29 (5 W + 24 M) 28 (4 W + 24 M) 25 (23 W + 2 M) 17 (M), 4 W + 25 M 4 W + 18 M
Mushroom production technology	27.12.06 – 30.12.06	CCOST, Raipur	60 (30 W from Matia and 30 M from Dondekala)
Mushroom production technology	27.07.07 – 30.07.07	CCOST, Raipur	60 women from Thatapur, Bhata and Mohbatta

Participation in Kisan Mela/Kisan Diwas

To create the awareness among common mass, mushroom stall with live demonstration of mushroom technology, technology demonstration through coloured posters, flexy boards, live mushroom recipes, processed mushroom products were depicted during Kisan mela/Krishak Diwas/ Kisan Gosthi/Rajyotsava/Krishi Pradarshini/NSS Camp/NSS Day, etc. The detail of activities and beneficiaries is as under:

Participation of Raipur Centre in Kisan Mela/Kisan Diwas

Particulars	*Date*	*No. of Participants*	*Activities*
Kisan Mela	30th October 99	750	Exhibits of mushrooms in the form of posters, charts, leaflets, live exhibits of different mushrooms and method were displayed. Mushroom recipes were prepared and displayed.
NSS Camp	9th January 2000	150	Lecture-cum-live demonstration on mushroom production technology was given
NSS camp	19th December 2000	200	Lecture cum demonstration on 'Cultivation Technology of Oyster Mushroom'
Krishak Mushroom Training	3th March 2001	200	Lecture cum demonstration on cultivation technology of oyster mushroom'
Training Programme & Crop seminar	8th March 2001	500	Lecture cum live demonstration on 'Mushroom preservation technology'
NSS Day	24th September 2001	450	Participated in exhibition
Kisan Mela	19th October 2001	700	Mushroom exhibits related to production, protection, preparation of recipes
Chhattisgarh Rajyotsava - 2001	1-7th November 2001	1,500	Mushroom stall was arranged with Department of Horticulture
NSS Camp	5th March 2002	150	NSS Unit, College of Agriculture, Raipur
Kisan Mela	21st December 2002	750	Organized by IGAU and Department of Agriculture at IGAU, Raipur
Chhattisgarh Rajyotsava	1-7th November 2002	20,000	Mushroom stall was arranged with Department of Horticulture
NSS Camp	18th March 2002	250	Lecture-cum-demonstration on 'Mushroom production technology'
Agricultural Exhibition	23-25th March 2003	3,000	Mushroom production and protection technology
Krishi Pradarshini and Kisan Mela	6-7th November 2003	1,800	Mushroom production technology
Krishi Pradarshini and Kisan Mela	22-23rd February 2004.	1,600	Production, protection and preparation of recipes
Krishi Pradarshini and Kisan Mela	12-13th March 2005	2,000	Mushroom production technology
Kisan Mela	22nd July 2004	550	Mushroom exhibition related to production, protection, preparation of recipes
Rajyotsava, 2005	1-7th November 2005	30,000	Mushroom production technology
Rajyotsava, 2006	1-7th November 2006	50,000	Mushroom production technology
Rajyotsava, 2007	1-7th November 2007	70,000	Mushroom production technology
Chhattisgarh Rajyotsava, 2007, District Kawardha	2-4th November 2007	20,000	Mushroom production technology
Exhibition during South Asian Conference	15-17th November 2007	250	Mushroom production technology

Organization of State Level Exhibition

A huge exhibition on "Production and Processing Technology of mushroom" was organized at Mushroom Research Lab of the Department of Plant Pathology, IGAU, Raipur from 20–22 January, 2002. Thirty six recipes of mushroom were displayed. Some of the important recipes were mushroom ice-cream, mushroom paneer, whey based mushroom chilli, mushroom kheer etc. which were very much liked by the peoples. The exhibition was attended by more than 1500 people. There were stalls from NHB, Dhanuka House, Chhattisgarh Mushroom, Entrepreneurs some of the recipes were available for sale by the Entrepreneurs and Chhattisgarh Mushroom. Most of the recipes were tasted by the VIP's and highly appreciated by them.

The exhibition was visited by the Hon'ble Agril. Minister Dr. Prem Sai Singh, Shri B.K.S. Ray, Principal Secretary and Agril. Productoin Commissioner, Govt. of Chhattisgarh, Raipur and Shri J.P. Negi, Managing Director, NHB, Gurgaon. The exhibition was inaugurated by Dr. V.K. Patil, Vice Chancellor, IGKV, Raipur. A bulletin "Mushroom ke Vyanjan" written by Ms. Ratna Naseene was released during inauguration of the exhibition. A lot of awareness was created regarding mushroom growing and its consumption during exhibition.

Extension Publication

Publication of bulletins/ Compendium/folders in Hindi on nutritional and medicinal values of mushroom, production technology of different mushrooms, mushroom recipes and processed products created lot of interest among the readers. Whenever and wherever these mushroom literature were displayed, it heavily attracted the viewers /audience. We have sold the literature costing Rs. 49,145/- from Mushroom Research Lab .Our literature is used by various NGO's for distribution to the trainees during training prgrammes on mushrooms organized by them. There was great demand of literature when displayed in World Trade Fare in New Delhi and Rajyotsav 2001, 2002, 2003, 2004, 2005 and 2006 organized by the Govt. of Chhattisgarh every year. Ten compendiums of lectures prepared by us during National level and State level training programme had great demand from the readers. Similarly, bulletin on Mushroom ke Swathista Vyanjan and Mushroom ke Vyanjan and Prasanskrit utpad are greatly demanded. So far Centre has published 10 Technical Bulletins, 13 compendium of Lectures, 13 Leaf folders, 30 popular articles and about 150 news items in various news papers.

Preparation of Video CD on Mushroom

A documentary film of 10 minute duration on cultivation of white button mushroom was prepared in overall technical guidance and direction of Dr M.P. Thakur, PI, DBT Mushroom Project in year, 2004-05. The video CD is very much useful to the farm women interested in mushroom growing. Similarly, a documentary film of 12 minutes duration on "Oyster Mushroom kee Vyavsayik Kheti" was prepared in overall technical guidance and direction of Dr M.P. Thakur, PI, DBT Mushroom Project in 2005-06. It has attracted a lot many peoples. These video films in form of CD's are available for sale.

Organization of National Seminars/Workshop

- A National Seminar on "Role of Antimicrobials for Sustainable Horticulture" was jointly organized by the Department of Plant Pathology and Indian Society of Mycology and Plant Pathology, Udaipur, on 16th Foundation Day of IGKV, i.e., 20th January, 2002.

- A 2 day National Workshop on "Awareness Creation on Biodiversity and Conservation of Mushrooms" was organized at the Department of Plant Pathology, Indira Gandhi Agricultural University, Raipur, in collaboration with National Biodiversity Authority of India, Chennai, during 1-2 December 2005.

Implementation of Ongoing Projects

Attempts were made to disseminate the knowledge of mushroom to growers through implementation of different projects as under:

Title of the Project	*Funding Agency*	*Duration*		*No. of Scientists/ Associates*	*Total Cost of Project (Rs in lakh)*
		From	*To*		
Exploitation of under utilized edible mushrooms of Chhattisgarh	NATP CGP 5	April 2002	-	2	12.00
Integrated Village Linkage Programme**	ICAR New Delhi.	April 2001	2004	6	44.00
All India coordinated Res. project on mushroom as incharge	ICAR New Delhi.	Since1988	till date	2	Amount varies from years to year
Transfer of mushroom production tech to the entrepreneurs as P.I.	CCOST Raipur	1st March 2003	31st March 2004	1	1.00
Transfer of mushroom production tech to the entrepreneurs as P.I.	CCOST Raipur	1st March 2004	31st March 2005	1	1.00
Transfer of mushroom production tech to the entrepreneurs as P.I.	CCOST Raipur	1st April 2006	31st March 2007	1	2.00
Transfer of mushroom production and processing technologies for the upliftment of rural farmers of Chhattisgarh	DBT, New Delhi	1st April 2004	31st March 2007	4	11.66

**As associated scientists along with extension team in transfer of mushroom technology at Tarra and Chatoud villages

Public Awareness Activity

Radio/TV talks and news coverage by different news channels like E TV (M. P.), Sahara TV, local channels, documentary film telecasted by ETV on different current topics under their "Annadata" programme had understanding of technical know how of different mushrooms. Direct telecast by these channels motivated and convinced many peoples to think, understand and know the importance of mushroom as a source of employment generation and additional source of income as well. The details of these activities are as under:

Awareness Creation

Mushroom Research Lab of the Department of Plant Pathology, IGAU, Raipur, is a model laboratory acting as a centre for source of culture tubes, mother spawn, commercial spawn, fresh mushroom, pickles, mushroom literature and consultancy since inception of the project in 1988. The details of the activities are presented below:

Mushroom products sold during different years at mushroom research lab,IGAU, Raipur

Year	*Fresh mushroom(kg)*	*Spawn(bags of 250 g)*	*Mother culture (bottles)*	*Culture tubes (No.)*	*Pickles (kg)*	*Amount (Rs)*
1996-97	181	933	–	–	–	9,539
1997-98	151	730	–	–	2.550	10,529
1998-99	97	998	21	02	2.850	13,390
1999-00	165	1,141	33	–	2.200	16,775
2000-01	195	2,612	36	16	1.450	33,539
2001-02	206	2,156	37	16	2.500	31,852
2002-03	163	3,860	52	32	–	49,624
2003-04	126	2,157	145	43	14,740 (Literature)	51,884
2004-05	210	1,697	187	86	–	41,632
2005-06	238	2,201	238	98	–	51,334
2006-07	150	2,063	97	51	965	53,685
Total						3,22,151

Spawn Makers

As a result of imparting training to entrepreneurs, they have established 'Mushroom Spawn Lab' in their areas and supplying commercial spawn to needy growers. They purchase pure cultures/mother spawn from our lab, then they multiply and sale to growers. A list of spawn makers and quantity of spawn sold are as under:

Mushroom spawn production in Chhattisgarh

Firm	*District*	*Quantity (quintals per annum)*
Mushroom spawn lab, IGAU,	Raipur	8-10
Chhattisgarh mushroom , Tendua, Abhanpur	Raipur	18-20
Vanved Network Pvt., Ltd. Sadar Bazar, Raipur	Raipur	20-25
Chhattisgarh Mushroom Swayatt Sahkarita Maryadit, Gudihari	Raipur	25-30
Shri Satguru Mushroom Centre	Raipur	8-10
Bose Mushroom Lab	Raipur	8-10
Roverent Mushroom	Durg	8-10
Chhattisgarh Resource Organization	Durg	20-25
Querishi Mushroom Farm	Durg	8-10
Deshmukh Spawn Lab	Durg	8-10
Om Mushroom Farm	Rajnandgaon	12-15
Biotech Lab, Govt. of Chhattisgarh	Ambikapur	5-7
Pradeep Shah Mushroom Lab	Ambikapur	10-12
Basant Gupta Mushroom Lab	Bilaspur	12-15
Raghav Mushroom Lab	Bilaspur	15-20
Preicious Mushroom Spawn Lab	Bilaspur	22-25
Rupak Mushroom Lab	Korba	40-50
Annapurna Mushroom Farm	Korba	8-10
Chakrabarty Mushroom Lab	Durg	8-10
Tamrakar Mushroom Lab	Durg	5-7
Total		300-325

Organizations Promoted

Mushroom Hitgrahi Sangh

It is a registered NGO working throughout Chhattisgarh absolutely for the cause of mushroom development. It has organized about 35 training programmes on 'Mushroom Crop Production Technology' and 'Mushroom Processing Technology' to rural women. The training was sponsored by Department of Horticulture, Department of Forest, Department of Women and Child Development, IFFCO, etc. About 1,050 rural women have been benefited by the activity of MHS.

Prerna Swawlamban Samiti

It is also a registered NGO working in Chhattisgarh especially for the upliftment of tribal people who are below poverty line. This NGO has very good linkage with the tribal Department. It is actively involved in transfer of mushroom technology only for the last one-and-a-half-years. It has trained about 1,150–1,200 tribal women/men in almost all districts of Chhatisgarh through organizing 25 training programmes. This NGO has used about 20 quintals of spawn for imparting training programmes and growing oyster mushroom.

Saheb Agrovet

It is a registered firm for sale of mushroom spawn. It purchases spawn from various sources and makes it available to growers, through its sale counter in main market, Mahalaxmi Market of Raipur. They have sold/utilized about 10,000 bags of mushroom spawn.

Sradddha Suman Samuh

It is a self-help group of 6 women of Tarra village (Dharsewa block of Raipur district) promoted by us on mushroom spawn production. This group started working since August, 2005 and sold about 3,500 bags (250 g/ bag) of spawn. They were also trained in mushroom processing and have sold about one quintal of mushroom papad which were consumed in mid- day meals and by other people in nearby villages. These women are also acting as a resource women for imparting mushroom training programmes organized by Department of Women and Child Development in nearby villages.

Sewa Shakti Samuh

It is a self-help group of 10 women of Dondekhurd (Dharsewa block of Raipur district) promoted by us and actively involved in mushroom growing and transfer of mushroom technology. The women of this group also act as resource persons for imparting mushroom training for village women.

Chhattisgarh Mushroom Swayatt Sahkari Samiti

It is a registered NGO and trained about 700–800 growers in Kanker and Jagdalpur area. It is involved in mushroom spawn preparation and production. They have also sold about 2000 bags of spawn to different states of India.

Consultancy on Establishment of Commercial Units

With our concerted efforts on transfer of mushroom technology, viz. training, live demonstrations, contributing articles, participation in kisan mela/ kisan diwas/ Rajyotsav, etc. a lot of awareness was created among common mass. Some entrepreneurs got convinced and established their commercial units of oyster mushroom production and some entrepreneurs have involved themselves in mushroom marketing. The details of their units are as under

Firm	*Place*	*Proprietors*	*Targets*
Chhattisgarh Mushroom	Tendua (Abhanpur) Raipur	Ms. Namrata Ghai	95 kg/day production
Chhattisgarh Mushroom processing unit	Tendua (Abhanpur) Raipur	Ms. Namrata Ghai	–
Om mushroom	Dongargarh, Rajnandgaon	Shri D. K. Shukla	90 kg/day production
Annapurna mushroom	Korba	Shri Ajay Vishwakarma	70 kg/day production
Mushroom Corner (Sale counter)	Raipur	Ms. Namrata Ghai	Sale of fresh, dry mushroom and processed product
Roverent mushroom (Sale counter)	Durg	Shri Rahul Gupta	Sale of fresh, dry mushroom and processed product

Future Prospects

- Creation of awareness about nutritional and medicinal values of mushrooms through mass media. Mushroom should be made essential in mid-day meals, in regular diet of school children, in daliya, in Dal Bhat Centres, in Aaganbadi programme, to pregnant ladies, old age people, in Government Canteens and Hospital Wards to alleviate undernourishment and malnourishment.
- A policy framework and programme is required to blend mushroom powder in Atta, dalia, or dal, papad, badi, murku to improve the nutritional qualities of food
- Mushroom processing units needs to be established to enhance shelf-life, prolong the availability of mushroom products and promote indirect marketing of mushrooms.
- Cheap and easy mushroom production technology needs to be evolved and transferred to people who are mainly dependent on forest produce including mushrooms.
- A strong need was felt to develop mushroom marketing channels to dispose of the fresh, dehydrated and processed mushrooms involving Self-Help Groups, NGO's and private organisations.
- Mushroom technology/knowledge should be effectively transferred through recent means of communication like Internet etc. to entrepreneurs, tribal women, school dropouts and consumers.
- Sincere efforts are required to promote mushroom consumption particularly in middle class family keeping in view of their nutritional and medicinal properties.
- The necessity of NGO's was strongly felt in mushroom production, processing, technology transfer, marketing, awareness creation and conservation of mushroom diversity.
- Efforts should be directed to increase mushroom production rather than training.
- Efforts should be made to trap solar energy for drying and processing of mushrooms.
- Efforts should be made to make mushroom sellable commodity like other vegetables.
- Good quality mushroom spawn laboratory, cold storage facility, and small scale processing units in a district place need to be established to facilitate mushroom production, processing and marketing.

- Targeted activity should be taken up in a identified areas looking to the availability of local resources.
- Work needs to be carried out as to how to minimize the cost and maximize the production of different mushroom?
- Designing and constructing mushroom huts/farms for different climates should be worked out by association with appropriate agencies taking care of local ethos and culture and also considering for maximum community involvement and community participation.

Vellayani

- A simple technique namely stapler method – a modified spawn production technique has been perfected in the centre and is being practiced by the farmers. A simple cultivation technique was evolved for the first time in Kerala for cultivation of *Calocybe indica* using spend compost of oystser mushrooms and technology transferred to farmers of the state.
- Regular trainings on cultivation of different varieties of mushrooms were conducted for unemployed youth, members of kudumbasree units, VHSE students, senior citizens, members of residents association.
- Advanced training on production of milky mushroom was given to more than three thousand growers from different parts of state. The TV programmes were recorded on oyster and other important mushroom production.
- Regular advisory services are being given from the centre to the mushroom growers of the state.
- Self-help programmes are being conducted with technical expertise of staff of the university. Mushroom units have been setup and spawn regularly supplied to these units.

BENEFIT FOR FARMERS/CONSUMERS

Ludhiana

The availability of production technology for growing different varieties of mushrooms round the year as per the PAU recommendations has enabled the growers in not only improving their dietary but also their economic standards. This has also helped to recycle the abundantly available agricultural residues, particularly paddy straw, into protein rich food, feed and fertilizer thereby solving environmental pollution problem resulting from their *in situ* burning. The increased production in the state which accounts for nearly 40% of the total national produce also reflects the increased level of domestic consumption.

A continuous effort has been made to improve upon the present mushroom cultivation practices and their popularization among the masses of both urban and rural areas. As a result, about 400 mushroom growers have adopted this venture for their livelihood. They can be classified into three categories: (i) small or domestic scale (ii) medium scale and (iii) commercial or large scale growers or industrial scale depending upon the straw processing capacity.

Grower	*Number of Growers Production*	*Straw Processed(q) (tonnes)*	*Mushroom*
Small			
Domestic	200	<50	5, 000
Medium Cottage	125	<100	
Commercial			
Seasonal	70	>100	
Industrial	4	>500	35, 000

A village in Amritsar, Dehriwal is worth mentioning here, which has more than 150 mushroom growers who are successfully cultivating white button mushroom. Large scale growers are concentrated in and around Amritsar, Patiala, Jalandhar, Ludhiana, Kapurthala, Hoshirapur and Gurdaspur districts. Some cold storage chambers have been converted to suit the mushroom cultivation in some of these districts.

Pune

Maharashtra is the second largest state of the country. The state has wide diversity in climatic conditions and soil type favouring the cultivation of various crops. Jowar, wheat, bajra and paddy are major foodgrain crops while sugarcane, cotton, banana, vegetables and fruits are major cash crops. Abundant waste plant materials are available and disposal of the same is sometime problem with farmers. If part of these crops waste is recycled and used for mushroom production, it will help to generate tremendous wealth in terms of high quality food and manure.

The mushroom flora of Maharashtra comprises nearly 232 species, which are distributed in 73 genera. Out of these 73 genera, 30 are reported to be edible while others are either toxic or the information regarding their edibility is not known at present.

Commercial production of button mushroom in Maharashtra was started in 1972. Initially it was done in potato cold storage with some local alterations in the structural designs. At present, 10 farms of button mushroom are started in Thane, Raigad, Pune, Jalgaon, Sangli and Akola districts of Maharashtra. The total mushroom production in Maharashtra was 300 tonnes during 1992-93 and it increased to 2,500 tonnes in 2005-06.

The oyster, locally known as Dhingri mushroom is popular amongst common people as compared to button mushroom, because of less investment required for its cultivation. Secondly, it can be grown on any substrate under natural conditions for 8-9 months in a year. More than 150 main growers are engaged in oyster mushrooms cultivation, marketing, spawn production and processing.

The popularity of milky mushrooms among growers is increasing and with increasing demand of milky mushroom in Mumbai, Pune and coastal Konkan region. A few growers have started growing of milky mushroom successfully.

This centre, viz. AICMIP, Pune centre, has been catering the needs of mushroom growers by providing opportunity for practical training in cultivation of oyster and button mushroom. Later on depending upon individual resources these people start commercial production of mushrooms. When these growers come across any specific problems in production, they again approached this centre for guidance in solving their problems. Spawn supply is also an important activity. Since, 1988 training-cum-demonstration is mushroom cultivation is being held on regular basis. Large number of postal querries received from different states, are also attended. For creating the awareness about importance of mushroom in our diet, various mass communication media like daily newspaper, magazine, AIR and T.V. has been successfully exploited. Demand for mushroom is increasing steadily in the state.

Summarized information on number and capacity of spawn, mushroom production as well as value-added mushroom product producing private units has been given below:

Spawn production		
No. of units	Type of mushroom	Production capacity (kg/year)
20	Button, oyster and milky	3,500
Mushroom production		
No. of farms	Type of mushroom	Production capacity (tonnes/year)
160	Button, oyster and milky	2,665

Value-added mushroom product production	
No. of Units	Mushroom value-added products produced
5	Mushroom soup powder mushroom biscuits, mushroom pickles, mushroom papad, mushroom powder capsule.

Coimbatore

Among different kinds of mushrooms, farmers of Tamil Nadu prefer to grow oyster and milky mushrooms because of their simple cultivation techniques. Total production of mushrooms is roughly about 18,000 tonnes/annum. In oyster mushroom, *Pleurotus florida, P. eous,* and *Hypsizygus ulmarius* are commercially grown in all parts of Tamil Nadu. About 400 spawn production units and more than 680 mushroom entrepreneurs are distributed through out the state.

The sustained efforts taken through All India Coordinated Mushroom Improvement Project for production and supply of quality spawn, mother culture and stock cultures of various kinds of mushrooms and timely supply has created a strong impact on mushroom production in this state

Many satellite mushroom growing units, self-help groups and spawn production centers have been established throughout the state due to the efforts taken up by TNAU centre.

Vellayani

Among different kinds of mushrooms, the farmers of Tamil Nadu prefer to grow oyster and milky mushrooms because of their simple cultivation techniques. Total production of mushrooms is roughly about 18,000 tonnes/annum. In oyster mushroom, *Pleurotus florida, P. eous,* and *Hypsizygus ulmarius* are commercially grown in all parts of Tamil Nadu. About 400 spawn production units and more than 680 mushroom entrepreneurs are distributed throughout the state.

The sustained efforts taken through All India Coordinated Mushroom Improvement Project for production and supply of quality spawn, mother culture and stock cultures of various kinds of mushrooms and timely supply has created a strong impact on mushroom production in this state.

Many satellite mushroom growing units, self-help groups and spawn production centers have been established through out the state due to the efforts taken up by TNAU centre.

MILESTONES AND LAMP-POSTS

Ludhiana

The project has several milestones and lamp-posts during last many years.

- Several recommendations already listed above have been generated over the years which are being followed.
- Infrastructure for short method of composting (pasteurization tunnel) created
- A mushroom compost turner has been locally designed and fabricated as prototype. The same has been used as a model and a 2nd one has fabricated at NRCM
- Spawn production unit has been developed and four spawn producing labs have been created in different districts (Patiala, Sangrur, Jalandhar and Hoshiarpur) by the State Department of Horticulture which depend on PAU laboratory for master spawn.
- Two test demonstration facility in the form of walk in growth chambers (climatically controlled) have been created.
- Specialized mushroom training courses (four nos.) are a regular feature in the University training schedule every year.

Pune

Maharashtra is the second largest state of the country with wide varity and diversity in climatic conditions and soil types favouring the cultivation of various crops. Abundant agro-industrial wastes are available with the farmers. The possibility of good quality mushroom in good quantity is possible due to favourable reasons such as abundant and cheap raw materials, cheap labours and various agro-climatic conditions.

Since its establishment, Pune Centre played a key role in the development and dessiminating mushroom production technologies and in imparting trainings to develop skilled man power. However, there are some key issues which needs to be tackled on priority basis.

- Little awarness about its nutritional and medicinal values and taboos and superstition about mushrooms.
- Illegal and unrecognized mushroom training centers.
- Low domestic consumption of mushrooms.
- Less number of recognised mushroom labs in the states.
- Low and inconsistent yield due to non availability of high yielder and good quality strains.
- High input cost due to non-availability of abundant indigenous technologies.
- Lack of marketing channels and proper prices.
- Lack of coordination and collaborations among growers and researchers.
- Keeping in view the mandates and objectives assigned, this centre propose to undertake research activities on following broad areas (future lamp posts) mainly.
- Survey, collection, isolation and conservation of wild fleshy fungi.

- Genetic improvement/hybridization studies in cultivated edible mushrooms using traditional methods.
- Improvement and refinement in production technologies of new edible and medicinal mushroom.
- Research on accelerated and improved composting by using microbes.
- Testing and utilization of newer waste materials for mushroom cultivation and as casing material.
- Recycling of spent compost and straw using useful microbes.
- Improvement and development of spawn production technology to increase shelf-life and suitable for bulk transport.
- Integrated pest and disease management for important pest diseases in mushroom production.
- Research on post-harvest technology of mushrooms, to improve shelf life, packaging, drying and value added products.
- Effective transfer of technology using publications and distribution of literature, through electronic media, holding of mushroom melas, demonstrations, exhibition and training programmes.
- Transfer of technology through village adoption, collaboration with NGO's and SHG's (self help groups).

In order to meet the above stated board research areas and objectives as well as proper and timely execution of programmes, this centre should be provided with the following inputs:

- Creation of state of art laboratory, modern polyhouses and growing rooms with ultramodern climatic control facilities.
- Up-gradation of existing project labs and installation of new sophisticated instruments.
- Sanction of additional scientific and technical staff to undertake interdisciplinary work.
- Training and exposer of scientists and technical staff to modern techniques and recent scientific advancement as well as other related areas in best laboratories to upgrade their skills.
- Provision of separate vehicle for survey and timely contact with growers in remote areas.

Finally for proper execution of various programmes proposed above this centre will require substantial increase in financial grants as well as reduction in bureaucratic hindrances.

Above proposed broad areas of research programme, if implemented properly and sincerely will:

- Boost mushroom production
- Generate mass employment
- Generate foreigen exchange
- Results in economic upliftment
- Results in poverty alleviation
- Results in removal of malnutrition
- Help in recycling of agro-waste thus help in reduction of environmental pollution.

Coimbatore

- Evolving high-yielding varieties of *Calocybe indica* and *Volvariella volvacea*
- Exploring the possibility for biopesticidal molecules and pharmaceutical values of macro basidiomycetes
- Creating new entrepreneurs for increasing the mushroom production.

Faizabad

In 1980, mushroom production was almost nil, which is now around 4,000 tons.

- Number of farmers growing mushroom regularly is over 5,000 in eastern Uttar Pradesh.
- Mushroom demand is more than supply.
- Over 100% biological efficiency of *Calocybe indica* at farmers' fields in a duration of 90–100 days harvesting.
- Over 35% harvest of button mushroom at farmer's fields in a duration of 90–95 days harvesting.
- Day-by-day increasing demand of spawn.

Raipur

- Release of Indira Sweta, a variety of oyster mushroom *Pleurotus florida* for commercial cultivation in Chhattisgarh.
- Development of spiral method of paddy straw mushroom cultivation in wooden /cemented poles which almost doubled the yield and enhanced the period of growing.

VELLAYANI

A mobile mushroom unit is to be set up with all amenities. Members of the faculty can visit and conduct on the spot training supply spawn to meet requirement and inspect the progress of work done at the grower's area. Each individual dwelling must be able to produce atleast one kilogram mushroom each day to meet nutritional requirement. Ready-made mushroom beds are to be supplied to the customers. People prefer ready-made beds since they lack time and patience to make their own mushroom beds. So supplying ready made mushroom beds warmly welcomed by all households and that way mushroom growing could be promoted.

Meeting of growers are to be conducted once in three months and their experiences shared and shortcomings of any or any new technology devised will be considering for the smooth conduct of the project. Processing zone are to be set up in the rural areas engaging women and value added products of mushroom can be prepared as part of cottage industry. Establishment of a depository herbarium for tropical species of mushrooms. Publication of a series of monographs in tropical species of mushrooms.

Leaflets and brochures have to be prepared regularly with up to date details to popularize mushroom cultivation in Kerala. Popular articles in magazines and media coverage are very essential to promote mushroom cultivation. Ready to cook series of mushroom recipes have to be marketized to cope with the high-speed life style.

Raipur

- Development of mushroom crop calendar for cultivation of different mushrooms and their species during different seasons of year and prevailing agroclimatic conditions of the Chhattisgarh State.

- Collection, identification and characterization of 102 wild mushroom flora from 16 districts of Chhattisgarh.
- Isolation, preservation and maintenance of about 28–30 species of mushroom flora from Chhattisgarh.
- Identification of *Pleurotus florida Pleurotus flabellatus Pleurotus sajor –caju* as most promising species for cultivation in the entire Chhattisgarh.
- Identification of paddy, wheat, paddy and wheat straw for oyster mushroom, paddy straw and cotton substrates for paddy straw mushroom, paddy, wheat, paddy and wheat straw for milky mushroom and paddy straw, wheat straw for white button mushroom cultivation
- Collection, identification and characterization of eight local species of *Ganoderma lucidum* (Reishi Mushroom) collected from forest areas of Chhattisgarh.
- In strainal evaluation study, CI 10 of *Calocybe indica*, Vv 11 of *Volvariella volvacea*, NCS 100 (CM 3) and NHS 102(CM 10) of *Agaricus bisporus* were found to be the most promising strains in several years testing under prevailing climatic condition of C.G.
- Guiding of Three Ph. D. and 15 M.Sc. (Ag) Theses and 35 project works to undergraduate students on different aspects of mushrooms.
- Organization of Three National Seminars/Workshops and One ICAR Short Course on Mushroom.
- Publication of 35 research papers and presentation of about 65 research papers in International /National Conferences.
- Establishment of 20 Mushroom Spawn Production Units in the state of Chhattisgarh.
- Establishment of six commercial units of oyster mushroom production in Chhattisgarh.
- Development of literature in simple Hindi language (11 bulletins, 10 compendiums, 13 leaf folders, 2 CD's) for the rural peoples, illiterate women/men, etc.
- Training of about 4,000 entrepreneurs from SEVEN states of the country, *viz.* Rajasthan, West Bengal, Bihar, Maharastra, Madhya Pradesh, Orissa, Chhattisgarh states of India.
- Organization of Five National Level Training Programmes on Mushrooms and 79 sponsored training programmes on different aspects of mushrooms.
- Contribution of 75 popular articles in different magazines of National Repute.
- Delivering of about 150 talks in different training programmes/Kisan Mela/Kisan Gosthi/ Kisan Sammelan organized by different agencies.
- Contribution of about 250 press releases on different aspects of disease and mushrooms to give it a wider publicity.
- Delivering of 35 talks on television telecasted by Raipur Doordarshan Kendra, Raipur, ETV, and local channels on different aspects of mushrooms and plant disease management.
- Delivering of 25 talks on All India Radio on different aspects of mushrooms and plant disease management.
- Attended about 100 VIPs in Mushroom Research Lab to get them acquainted with mushroom technology.
- Organisation of three day state level exhibition only on mushroom production technology, mushroom recipes and mushroom processing technologies.

5

Betelvine

Satyabrata Maiti and Saravanan Raju

BETELVINE is a native of South East Asia and believed to have Malaysia as centre of its origin. It is an evergreen perennial dioecious creeper cultivated in India and elsewhere for its leaves that are used as masticatory. The plant is grown exclusively as a commercial cash crop in Assam, Tripura, Meghalaya, West Bengal, Orissa, Bihar, Uttar Pradesh, Madhya Pradesh, Maharashtra, Andhra Pradesh, Karnataka, Tamil Nadu and Kerala. It is estimated that approximately 53,140 ha are under cultivation. Sixty per cent of cultivation is in eastern states, which include West Bengal, Assam, Orissa and North-Eastern States, viz. Meghalaya and Tripura. Nearly, 30% of area under cultivation is restricted to Andhra Pradesh, Tamil Nadu, Karnataka and Kerala.

The crop is highly labour intensive and particularly suited to small holdings. Once established, it becomes a perennial source of employment and cash flow for day-to-day requirements of farmer. Betelvine is cultivated primarily in two systems.

- *Under natural forest eco-system:* Fast growing plants such as *Sesbania grandiflora, Erythrina variegeta* or *Moringa oleifera* are used to provide support as well as shade to betelvine plants.
- *Under artificially created shade known* as *'Boroj, Bareja, etc.* Besides above mentioned two systems a third system known as mixed cropping with arecanut, coconut, jackfruit, mango etc. is followed in a limited scale.

Though the area under this crop is much restricted as compared to other commercial or cash crops, it is grown in certain specific areas in each state where it is considered to be of importance. In general, the cultivated area is restricted to coastal areas or in river banks as the water requirement of the crop is high. The crop is highly labour intensive, hence particularly suited to smallholdings of marginal and small farmers. Exceptions are in Maharashtra, where large-sized cultivations are in existence.

Peculiar feature of this crop is the existence of a large number of varieties recognized by growers and traders. It is likely that the same cultivar may be called under different name in different states/places. National Botanical Research Institute, Lucknow, identified 85 cultivars of betelvine from different parts of the country on the basis of morphological and chemical parameters and identified six distinguishable cultivars namely Bangla, Deswari, Kapoori, Meetha, Sanchi and Khashipatti. There are reports that farmers in Sangli get an yield as high as eight to ten million leaves per ha per year. The net income is around half a lakh rupees per year. Strategies have, therefore, to be evolved to augment productivity. Policy, decision would be to organize research for targeted yields and quality parameters. It seems to prove that betel leaves, apart from being a stimulant and refreshing agent, has medicinal properties as recognized in Ayurvedic texts. The leaf is used to get rid of cough, throat and chest congestion. It hastens saliva secretion and neutralises acidity when chewed with lime. Main constituents of betel leaf are vitamin B and C, carotene, chlorophyll, and minerals besides possessing intrinsic qualities. Betel leaf chewing plays a positive role in dental and beauty care. It has religious significance and essentially required on festivals and religious/marriage occasions.

For too long, our focus has been on commodity rather than on market. The new liberalized trade scenario of WTO will offer fresh challenge for the industry. Stringent quality stipulations are imposed in all consumables in respect of residual toxicity of insecticide and fungicide used in protecting the crop quality. The dawn of the new millennium has brought about new opportunities for betelvine growers and chewers/consumers. Extensive research over last two decades conducted by AICRP in eight centres in eight states in eight agroclimatical regions by a team of 22 scientist in 26 research project has yielded useful results.

HISTORY

The All India Coordinated Research Project on Betelvine was formulated and sanctioned by the ICAR considering the importance of the betelvine crop in 1981 on recommendation of the special team constituted by the ICAR with Dr L.K. Joshi, Jawaharlal Nehru Krishi Vishwa Vidyalaya, Late Dr R. K. Hegde, University of Agricultural Sciences and Dr B.B. More, Mahatma Phule Agricultural University. It was started initially with six centres, viz. University of Agricultural Sciences (Dharwad), Orissa University of Agriculture & Technology (Bhubaneswar), Acharya N. G. Ranga Agricultural University (Chin Thalapudi), Tamil Nadu Agricultural University (Vellore), Jawaharlal Nehru Krishi Vishwa Vidyalaya (Jabalpur) and Mahatma Phule Agricultural University (Rahuri). The project was initially funded from CESS fund of the ICAR with a total allocation of Rs. 26.25 lakh spread over three years. Subsequently, three more centres namely Bidhan Chandra Krishi Vishwavidyalaya (Kalyani), Assam Agricultural University (Jorhat) and Indian Institute of Horticultural Research (Hessaraghatta) were included in 1983 with an additional grant of Rs. 10.251akh spread over two years. In addition, National Botanical Research Institute (Lucknow) was identified as voluntary participating centre. In the subsequent years, Rahuri centre of MPAU was shifted to Sangli and Vellore centre of TNAU to Sirugamani. The Headquarters of the project was located at the Indian Institute of Horticultural Research (Bangalore) for coordination among the centres. The project continued in the VII Plan with a total allocation of Rs. 54.57 lakh for the entire plan period. During this plan period Rajendra Agricultural University (Pusa) was included in the project from April 1988 and National Botanical Research Institute (Lucknow) from April 1989. The UAS, Dharwad centre, was dropped because of its ill performance. In VIII Plan, initially the budget was Rs. 164.25 lakh which was further revised to Rs. 200.00 lakh. Annual plan budget for subsequent years 1996-97 and 1997-98 were Rs. 47.40 lakh and Rs. 52.64 lakh, respectively.

Objectives

The project was initially started as disease management programme since it was the immediate need of the farmers for the spread of the crop and attaining high productivity. In subsequent years the project was made holistic with thc inclusion of other activities such as crop improvement, crop production, other components of crop protection and post-harvest technology. The project took very specific activities namely, survey, epidemiology, control, collections of various germplasm for screening and maintenance of germplasm in non-traditional betelvine growing area. The base of the betelvine research was expanded in the VIII Plan by incorporating agro-technique component in such a manner that the project could develop an effective and economically viable Integrated Pest Management strategy for this crop. In VIII Plan, component of crop improvement was incorporated at Project Coordinator's Cell with a view to make crop improvement possible via traditional breeding route. The AICRP on Betelvine has been converted into All India Networking Project (AINP) during X Five Year Plan Period (2002–07) to fulfill the recommendations of Dr Jain Committee approved by the GB of ICAR with the following objectives.

- Collection, maintenance, evaluation and cataloguing of betelvine germplasms
- To reappraise the pest and disease situation of betelvine in West Bengal.
- To undertake detailed studies on all pest and diseases problems in betelvine with aim of developing methods of control
- To identify source of resistance to the pests and diseases by collecting and testing the available germplasms in different parts of the country.
- To develop agro-technique for the development of effective and economically viable integrated pest management strategy of betelvine.

Table 5.1. Estimated area under betelvine in major states in India

State	*Area*
Andhra Pradesh	2,723
Assam	7,850
Bihar	3,166
Karnataka	6,682
Kerala	1,280
Madhya Pradesh	600
Maharash tra	1,419
Orissa	4,007
Tamil Nadu	4,795
Uttar Pradesh	2,214
West Bengal	18,203
Others	600

Betelvine centres in operations are as follows

AICRP Centres	Institution/SAUs	State	Year of Starting
Bapatla	Andhra Pradesh Agricultural University	Andhra Pradesh	1981
Jorhat	Assam Agricultural University, Jorhat	Assam	1985
Pusa	Rajendra Agricultural University	Bihar	1985
Dharwad	University of Agricultural Sciences	Karnataka	1981
Jabalpur	Jawaharlal Nehru Krishi Vishwa Vidyalaya	Madhya Pradesh	1981
Sangli	Mahatma Phule Krishi Vidyapeeth	Maharashtra	1981
Bhubaneshwar	Orissa University for Agriculture and Technology	Orissa	1981
Sirugamani	Tamil Nadu Agricultural University	Tamil Nadu	1981
Kalyani	Bidan Chandra Krishi Vishwa Vidyalaya	West Bengal	1985
Islampur	Rajendra Agricultural University	Bihar	2004
Bangalore	Indian Institute of Horticultural Research	Karnataka	2004

The AINP on Betelvine will be phased out by end of 10th Plan and from 11th Plan, the project will be merged with AINP on MAP, with existing infrastructures and sanctioned staff strength.

GENETIC RESOURCES AND CROP IMPROVEMENT

Betelvine is believed to be originated in central and eastern Malaysia and has spread throughout tropical Asia. It was taken at a later date to Madagascar and East Africa. Betelvine (*Piper betle* Linn.) is a perennial, evergreen creeper grown in shady conditions having moderate temperature with high humidity. It is cultivated in India, Bangladesh, Srilanka, and to a limited extent in Pakistan, Malaysia, Thailand, Indonesia, Maldives, Vietnam and Papua New Guinea. In India, it is commercially cultivated over 50,000 ha as a cash crop. This crop has great market value both inside and outside India and leaves are exported to Pakistan, Bangladesh, Indonesia, Malaysia, Burma and Thailand.

Betelvine or betel leaf is one of the oldest plants respected as holy in India and embedded in tradition and religion. Leaves are consumed fresh in combination with betel nuts (*Areca catechu*). As medicine, it is used in Indian System of Medicine to cure indigestion, stomach ache, diarrhoea, flatulence and to heal wounds, scales and burns, swelling due to sprains, bruises, respiratory disorders, sore throat, constipation, boils and gum sores. Research in recent years has shown betel leaves to strengthen the immune system and to inhibit certain cancer cells. But its greatest role around the world is most likely as the leaf that holds the betel quid together, which is made from *Areca catechu* seeds, hydrated lime, and optional herbs and spices.

Germplasm plays the key role for the development of every living organism. Different combinations of genes create the genetic diversity that has been the basis of crop improvement. Importance of germplasm collection was soon realized with the creation of All India Coordinated Research Project on Betelvine in 1981 and collection of germplasm on a systematic scale was initiated in nine states of India.

Centres of the All India Coordinated Project took the responsibility of collecting genetic variability available in their states and thereafter exchanging among the participatory centres for conservation, evaluation and utilization. In addition, the then headquarters of the project located at Indian Institute of Horticultural Research, Bangalore, also collected some germplasm from Kamataka which was made available to the centres.

Betelvine is a dioecious plant, propagated vegetatively and grown commercially in India for its leaves used for chewing. Till the discovery of male and female flowers and fruiting in India, it was believed that only male plants were grown in India. Selection by farmers for quality, taste, flavour, texture, leaf size and shape was a continuous process. Therefore, accessions collected so far are nothing but cultivars grown in isolated pocket by farmers and these are assorted male and female types with farmers' identified economic traits.

The collections so far made since 1981–2000 are listed with their local names and collection source as provided by the participating centres. It is observed that duplication of cultivars with a different name is very common. Attempts are made to identify such duplication. The nomenclature of cultivar is most confusing due to growers' craze to name them after village or towns. Balasubrahmanyam *et al.* (1994) stated that even one particular variety was given different names at different places, e.g., Desawari as Bilhari, Sanchi as Kalipatti, Kapoori as Safedi, etc., while Desi refers to Bangla in Bengal, Kapoori in Maharashtra and Desawari in Madhya Pradesh. In Assam, Assamiya pan is also synonymously known as Gach pan, i.e., APb 1. Ganguly (1984) recognized 15 types under cv. Bangla with reference to taste, texture and morphology. Das and Chattopadhyay (1984) classified Betelvine of Midnapur district of West Bengal into 17 types comprising of cultivars of Bangla, Meetha and Sanchi on the basis of morphological, anatomical, micrometrical and phytochemical parameters. Narayan Reddy (1984) classified the cultivars into two group, viz. pungent and non-pungent types from a collection of 55 cultivars of different parts of India. Balasubrahmanyam *et al.*(1994b) stated out that 85 collections at National Botanical Research Institute, Mahoba, belonged to five distinguishable groups, viz. Bangla, Desawari, Kapoori, Meetha and Sanchi.

A total 321 betelvine germplasm are collected by various centers from all over the country and are maintained at various ANIP centres. The collected germplasm materials are evaluated and catalogued by the respective centres for growth, yield and quality characters along with the susceptibility to various pests and diseases.

Classification of Germplasm

It is now conspicuous that though a considerable number of germplasm have been assembled, the identification of cultivars through one or a set of stable marker phenotypic characters is necessary. As mentioned earlier *Piper betle* Linn. is a dioecious plant; sex determination by examination of floral parts is the conventional method. However, sexual dimorphism in leaf shape identified by Maiti and Biswas (1991) can be used successfully for sex determination in places where flowering is absent. Male plants have leaves which are narrowly ovate with 1.84 ± 0.21 length: breadth value and female plants have cordate, ovate to round leaves with 1.26 ± 0.13 length: breadth value. Cultivars, which are designated as Bangla types, are female. Balasubrahmanyam *et al.* (1994) recognized Bangla type to possess 7-9 secondary veins, while 4–6 secondary veins in Kapoori type/Desawari type; in reality all betelvine cultivars are found to bear minimum 7 veins (hence named in Sanskrit as 'Saptasira' which originate at a point, opposite and divergent halfway to the tip and leaf margin and not parallel as said. Assamiya pan (APb 1) conspicuously possesses 8-9 veins, which are more prominent in dorsal side. The point of origin of veins are also not opposite rather more prominently alternate halfway towards the tip. These could be marker character of Assamiya pan and its related ecotypes.

Another unique character of Assamiya pan is the overlapping of lower left lobe by right lobe in ventral view due to which the leaves get an angular spread which is different than that found in Bangla types. It was observed that female cultivars bear longer petioles than that of the Kapoori or male types. Assamiya pan is exceptional which bears a short internode than the other Bangla cultivars, those are invariably female. Surprisingly, the flower bearing capacity of betelvine cultivars is very sparse. The climatic conditions of Bangalore proved to be very much congenial for flowering. Many cultivars, through exchange programme from its adapted-native place to another environment have shown that the environmental effect considerably changes phenotypic characters visibly in respect of leaf size, leaf colour, petiole length, texture and taste.

The Kapoori cultivars were found to bear small (80 cm^2) leaves. Bangla types, which were grown in the bareja produced medium (81–150cm^2) to large (>151cm^2) leaves. While those naturally grown in open conditions as long duration perennials invariably produce large leaves. The environmental conditions were found to playa greater role in determining leaf size. Leaf was also found to bear a unique characteristic, i.e., the lobes at base, which were never equal. In most of the cases, it was observed that either of the right or left lobe was bigger in size. Assamiya pan and other related ecotypes possess the overlapped lobes with distinct leaf shape and petiole length individually. The leaf tip was also found to vary either shortly acute to acute, obtuse or acuminate. However, it was a common character. The stomata counts may serve as a good index for differentiating variety/cultivar as evidenced from the findings. However, the sampling of the leaves needs to be standardized. Rawat *et al.* (1989) made an attempt to classify germplasm chemo–taxonomically. However, no conclusive remark can be made at this stage without screening all the available germplasm.

APb 5 possesses the sweet flavoured taste which is uniquely identical within its related ecotypes. The cultivars of Assam and neighbouring hill states are grown in open system with live support crops. The classification of betelvine seems to vary widely from each other. Based on the data on evaluation by the different centres accessions can be classified on the basis of following characters and can be utilized as per the need.

Sex

- (a) Male
- (b) Female

Leaf size

- (a) Large
- (b) medium
- (c) small

Petiole length

- (a) Long petiole
  - (i) straight or (ii) angled
- (b) Short petiole
  - (i) straight or (ii) angled

Leaf lobe

- (a) Lobe overlapped and prominent

(b) Lobes apart from each other and prominent
(c) Less conspicuous lobe

Venation type
(a) Opposite and alternate at distal distribution end
(b) Opposite to each other except mid-vein

Number of vein
(a) 5-6
(b) 7 and
(c) more than 7

Branching habit
(a) With few branches
(b) With numerous branches

Stomata number
(a) High
(b) Low

Value
(a) High
(b) Low

Essential oil composition
(a) Specific rotation of essential oil in alcohol
(b) Saponification values after phenol removal
(c) Phenol percentage

Soil pH requirement for growth
(a) Acid soil loving
(b) Alkaline soil loving ego Mitha pan

Leaf pungency
(a) Highly pungent, e.g., Assamiya pan, Naga pan, Gan pan
(b) Moderately pungent eg: Kapoori and sanchi varieties
(c) Low pungent, e.g., Mitha pan
(d) Non pungent

Method of cultivation
(a) Open type, e.g., Kapoori varieties
(b) Closed type/Bareja eg: Bangla varieties

Similarly, accession SPb 10 was released in Tamil Nadu as Sirugamani 1 after testing at multilocations. The variety produced profuse branching and out yielded the local varieties. Hybridization programme was initiated at IIHR, Hesseraghatta, Bangalore, in 1996-97 and developed a large number of hybrids over the years, out of which four hybrids were selected for multilocation testing in the centres. The evaluation of hybrids is in progress.

(i) Flowering and Fruiting in Betelvine

In betelvine, flowering and fruiting was studied at IIHR, Bangalore. Sexual dimorphism for leaf character was noticed. The female leaves are cordate in shape and males being ovate. The length and breadth ratio of leaf in both sexes varies with values of 1.84 ± 0.21 for males and 1.26 ± 0.13 for females. Betelvine is a non-agamosperm with true seed formation only through pollination. This opened up the field for heterosis breeding. Profuse flowering in both female and male clones of betelvine was observed throughout the year. Crossing between female and male genotypes were carried out by dusting pollen on female catkins (10.00 to 10.30 am). A set of female parents, which flowered, was crossed with the male parents available in germplasm. Successful crosses developed fruits, which matured in 90-100 days after crossing. The seeds extracted immediately were tested for germination. Germination is satisfactory with good radical and plumule emergence. Seedlings were grown in plots, which include single cross, poly cross, and also open pollinated progeny. Variation in the progeny at seedling stage is being observed and crossing programme is being continued. Four selected hybrids are being evaluated in four centres.

(ii) Hybridisation Programme

Flowering was observed in 16 female clones and 13 male clones. Continuous flowering was observed in female clone SGM1 and male clone Swarna Kapoori. Kappoori Arvi and Kapoori Bhambla started flowering during March. A total of 72 different crossings were made.

(iii) Hybrid Evaluation Trial

The trial was conducted at TNAU (Sirugamani). It was found that only GN_1 hybrid (Godi Bangla × Kapoori Nasik) exhibited normal vigour. Other three hybrids did not show any hybrid vigour rather growth was inferior to ruling varieties. GN1 produced marketable leaves of 34.68 lakh leaves ha^{-1} weighing about 295.0g/100 leaves. The other three hybrids, viz. P_1, P_2 and GB (OP) expressed hybrid depression and produced unmarketable leaves of poor quality, thus not suited for cultivation. GN1 hybrid will be tested in the farmers' fields for its adaptability.

REGION SPECIFIC VARIETIES/VALUE ADDED VARIETIES

SGM 1

The variety SGM 1 is a pure line selection from Phalghat cultivar (accession spb 10) which was released in 1994 by TNAU centre after multilocations testing. It is a female clone. The vine is erect and short statured with multilateral growth habit (25–31 barnches per vine). The stem is thick and short internodes produces attractive yellow leaves. This variety is characterized by its profuse leaf production and the leaves are pungent. It is field tolerant to pest and diseases and ideal for homestead gardens. The leaf yield is 110 lakh per ha in 2 years. It grows well on live supports like *Sesbania grandiflora* in open conditions with double row system having irrigation channels on both sides of row or raised bed.

Utkal Sudam

The variety Utkal Sudam which is released for cultivation in the eastern states of India is the selection named Kpb 2 (Ghanaghatte). The accession showed high adaptability in the states like Assam and Orissa. Kpb 2 out yielded local Bangla types in the initial varietal evaluation trials in their respective states.

SGM-2 (Tamil Nadu)

SGM 2 is a pure line selection from Panchampatti, Dindugal female type (spb 12). It is released by TNAU centre for cultivation in 2004. The vines are erect type with multilateral

Hybridization in betelvine

Betelvine cultivation under live support system

Artificial shade for betelvine cultivation

vine production (17–20 vines branches per vine). It has thick stem and produce attractive dark green leaves with long petiole (4-5cm). The leaves are slightly pungent. The internodes are shortened and produce profuse leaf production and it can yield 100lakh leaves per ha in 2 years. SGM 2 is tolerant to various pests and bacterial leaf spot disease. It is suited for open conditions with *Sesbania grandiflora* as live support.

Bidhan Pan (West Bengal)

A clonal selection of IC 360989 by BCKV centre was released as Ghanaghatte (Bidhan Pan 1) by West Bengal Variety Release Committee in 2002. The variety is Bangla type having cordate leaves and very short internodes. This variety produces higher leaf yield with tolerance to pest and diseases.

Krishna Pan (Maharashtra)

This variety was relased for cultivation in the plains of Maharastra state in 2001 by MPKV centre. It is a selection from local type grown in Shirpur (District Dhule) known as Shirpurkata having more number of *Kali and phaphada* leaves. The leaves are oblong, thicker and light green in colour. It yields 71 lakh leaves per ha and moderately resistant to foot rot, however, susceptible to betelvine bug and nematodes.

TECHNOLOGIES FOR ENHANCING PRODUCTION

Leaf Yield and Quality

(a) Nitrogen Sources, Levels and Method of Application

The leaves are the economically important produce from betelvine, hence the response of betelvine for the applied nitrogen in terms of yield and quality was studied at different centres. Band placement of 200 kg N as CAN produced significantly higher yield at BCKV. Keeping quality, however, was deteriorated with the use of chemical fertilizer. At ANGRAU, the highest yield was obtained at 200 kg N/ha as urea, though the economic level was 100 kg N /ha onlv. Effect of different sources was not evident. Yield and keeping quality were not affected due to various sources and levels of nitrogen at JNKVV, however leaves per vine were perceptibly

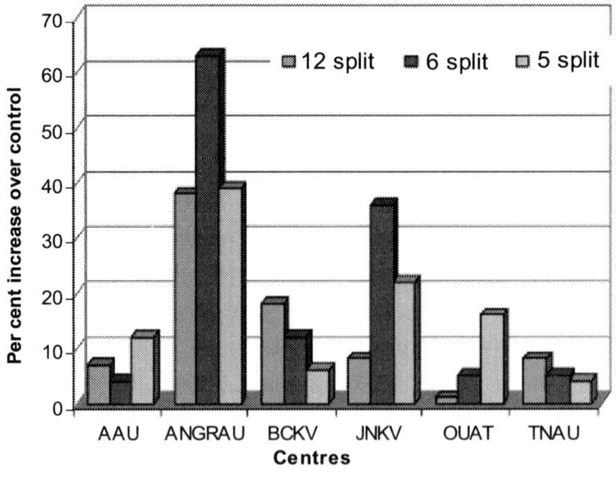

Fig. 5.1. Effect of split doses of nitrogen on leaf yield

higher with linseed oil cake, urea and their combination treatments. Nitrogen applied through organic sources recorded lower disease incidence, but did not improve the keeping quality. Dibbling of fertilizer was found to be superior to band placement. At MPAU, 200 kg N /ha was beneficial to improve yield of leaves whereas at QUAT maximum elongation of vines and production of leaves was achieved with 150 kg N/ha applied as urea + mustard cake in1: 1 ratio. Further lower doses of nitrogen through neem cake and mustard cake reduced disease incidence. In TNAU, different sources and levels did not influence the yield but 150 kg N /ha through neem cake + urea (1: 1) resulted in the highest yield with the lowest foot rot incidence. Four to twelve split applications of nitrogen were found to produce highest leaf yield with the lowest foot rot incidence.

(b) Organic and Inorganic Fertilization

Application of FYM at full dose was found to increase all the growth and yield parameters significantly in addition to reducing the disease incidence in MPAU, QUAT, RAU and TNAU. FYM application also increased the keeping quality of leaves in MPAU, QUAT and RAU. However, in ANGRAU application of neem cake + urea (1: 1) and in JNKVV and AAU, oil cake + urea (l: 1) gave maximum vine growth and leaf yield. Keeping quality of leaves was also improved by neem cake + urea in ANGRAU. Whereas, in AAU and JNKVV centres organic fertilizer (full dose) improved the keeping quality of leaves with low disease incidence (Fig. 5.2).

(c) Slow Releasing Organic Nitrogenous Fertilizers

Neem coated urea (NCU) was superior to tar coated urea and prilled urea in terms of favourable growth and yield, improved quality and lower pest incidence. However, at BCKV, NCU did not show any positive effect on yield but influenced keeping quality significantly and decreased *Phytophthora* incidence. At RAU, yield and quality were not influenced due to NCU but decreased incidence of bacterial leaf spot and Anthracnose. NCU increased leaf yield at ANGRAU.

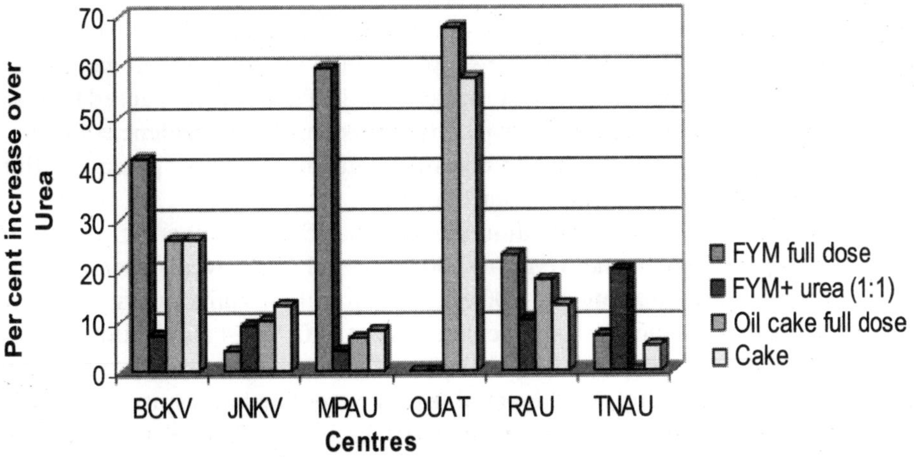

Fig. 5.2: Comparison of organic and inorganic fertilizers on yield

(d) Levels of Potash on Yield

The results from different centres indicated that potash had a positive effect on yield, keeping quality and disease incidence. The dose of 100kg K_2O ha at ANGRAU, JNKVV, MPAU and NBRI, 125 kg at BCKV and QUAT and 75 kg at TNAU were found superior in increasing the leaf yield.

(e) Levels of Phosphorus on Yield

Application of phosphorus improved the yield at ANGRAU (100 kg/ha), JNKVV (125 kg/ha), MPAU (100 kg/ha) and QUAT (100 kg/ha). However, no effect was recorded at AAU and BCKV.

(f) Micro-nutrients on Yield

Studies on micronutrients showed that application of manganese (0.5%) and zinc (0.25%) at ANGRAU. zinc (0.5%) at BCKV, molybdenum (0.1 %) at JNKVV and manganese (0.5%) at QUAT improved the leaf yield. No significant result was obtained in TNAU.

Integrated Crop Management

At AAU, the incidence of disease was significantly less in the treatments where *Trichoderma* and Bordeaux mixture as applied. Final soil status indicated the fixation of phosphorus as well potash insoil leading to soil build up of both the nutrients. The retention of K was higher in plant population with combination of *Trichoderma* and Bordeaux mixture. The farmers' practices resulted in lower return compared to INM and IPM approach because of adopting lower plant population as well disease, nutrient and irrigation management aspects.

The results showed that the treatment consisting of optimum plant population (1,00,000 plants/ha) + recommended fertilizers (neem cake + urea (1:1) at 200 kg N, 100kg P_2O_5 and 100kg K_2O/ha), irrigation (100% replenishment of CPE) and application of Bordeaux mixture (4 drenches and 8 sprays) recorded significantly higher leaf yield of 39.15 lakhs/ha at ANGRAU, Bapatla.

At BCKV, highest fresh mass of 100 leaves (407.14g) and leaf yield (44.44 lakh ha^{-1}) were recorded in treatment consisting of plant population of 1,50,000 vines/ha with application of 200kg N in splits of organic form + 100kg P_2O_5 + 100 kg K_2O + Irrigation at 100% replenishment of CPE + Bordeaux mixture application (4 D+ 8S) + recommended insecticides if required. In the same treatment, lowest incidence of leaf and foot rot caused by *Phytophthora* sp. and leaf spot caused by *Colletotrichum capsici* was observed. Incidence of leaf and foot rot and leaf spot were recorded highest in control treatment (farmers' practice). But the cost: benefit ratio being highest at farmer's practice.

At JNKV, maximum vine elongation per month (15.58) number of leaves (36.25/ha^{-1}), length of inter-node (7.54cm) and weight of 100 leaves (618.65g) were observed with treatment consisting of best plant population + 200kg nitrogen in four splits in organic form + 100kg P_2O_5 + 100kg K_2O + irrigation at 100 % replenishment of CPE + 4 applications of *Trichoderma viride* + Sanitation + recommended insecticides. The maximum disease incidence (15.30%) and minimum keeping quality (9.82days) of leaves were noted in farmers practice.

At OUAT, application of 150kg nitrogen + 100kg P_2O_5 +125kg K_2O + irrigation at 100% replenishment of CPE + Bordeaux mixture application (4D + 8S) + recommended insecticides gave higher vine elongation (28.2cm), number of laterals (2.0) and the leaf yield (45 lakhs/ha) with larger leaves (14.11cm), fresh weight of 100 leaves (292g) and less diseases incidence. The shelf life of leaves was better in farmers practice with an organo leptic rating of 2.6.

At RAU centre, the treatments having 50 lakh plants/ha + 200: 100:100kg/ha N:P_2O_5: K_2O (4-splits) as organic + irrigation + 4 application of *Trichodermaviride* + sanitation + organic insecticide and 1.50 lakh plants/ha + 200: 100:100kg/ha N:P_2O_5: K_2O (4 -splits) as organic + irrigation + 3 drenching of Bordeaux mixture(1%) and 6 sprays of BM (0.5% + sanitation + organic insecticide gave superior yield over farmer s practice. Higher disease incidence and shorter shelf life was noticed under farmers practice. Although other parameters such as vine growth, fresh weight and leaf area remained un-affected, integrated crop management fetched higher benefit as compared to farmers practice.

Nutrient Uptake Studies

Nutrient uptake studies carried out at different centres showed that application of nitrogen in the form of FYM or oilcake equal to recommended dose of N provided better crop performance with superior betel leaf quality. The performance of betelvine crop under different centres with the optimum nutrient application.

At BCKV centre, the results showed that yield and growth parameters were maximum in cases of nutrient dose NPK (kg/ha) = 150:50:50 (N = oilcake). Chemical analysis of plants showed that plants from treatment - NPK(kg/ha) = 200:100:100 (N = Oilcake + urea) contained highest N and P in per cent. But where NPK (kg/ha) = 200:100:100 (N = FYM) were applied, plants contained highest K. Soil analysis shows that lowest residual P and K were obtained in treatment NPK (kg/ha) = 150:50:50 (N = Oilcake + urea) were applied. Low residual N in soil was found in treatment - NPK ((kg/ha) = 200:100:100 (N = oilcake + urea).

At JNKVV, N, P and K uptake was highest under the treatment where NPK applied at the rate of 200 (oil cake) :100:100 kg/ha. The uptake of N, P_2O_5 and K_2O were 156.25, 68.56 and 82.55 kg/ha whilst they were lowest (120.25, 38.25 and 48.56 kg/ha) under NPK application @ 150 (FYM): 50: 50 kg/ha.

Application of NPK @ 200:100:100 kg/ha as FYM resulted in highest uptake of nutrients. N, P and K uptake were 272.30, 45.20 and 136.76 kg/ha respectively. However, soil nutrient status maximum under the application of NPK @ 200:100:100 kg/ha as oilcake : urea. The treatment receiving nutrient dose of 200:100:100 as N, P_2O_5 and K_2O significantly superior yield as compared to the treatments which received comparatively lower dose of nutrient whether applied in the form of organic or combination of organic: inorganic. It was also noted that crop with significantly better vine growth, marketable leaves, leaf area, fresh weight and return/ rupee investment was produced with higher supply of nutrient. A correlation between higher dose of nutrient, nutrient-uptake, nutrient balance and production was also recorded. Inclusion of inorganic form of nutrient resulted in poor keeping quality and less disease resistance. The shelf life of leaves was significantly higher in control plot (crop with no supply of nutrient) where disease severity was also reduced, but yield was significantly poor and showed negative balance of nutrient in the soil.

At TNAU centre, the treatment consisting of 150 as FYM, 50 P_2O_5 and 50 K_2O kg/ha application recorded the highest leaf yield of 40.19 lakh leaves per ha and recorded maximum organic carbon content (0.53%) during second year at TNAU centre. The available N was high in the treatment having NPK of 150 (oilcake + urea) : 50:50 kg/ha.

Efficacy of Biofertilizers

Various biofertilizers like azotobacter, phosphobacter and azosprillum were applied in combination with other organic manures like vermicompost and inorganic fertilizers to study the efficacy of biofertilizers on growth, yield, shelf-life and organoleptic quality of betelvine. Two levels of azotobacter (5 kg and 10 kg), azotobacter 5 kg + phosphobacter, vermicompost, MOC + urea (1:1) were applied in combination with NPK to betelvine for their effect on vine growth, yield and keeping quality of betel leaf.

Application of bio fertilizers along with organic and inorganic fertilizers produced higher yield in BCKV, OUAT and TNAU. Application of oilcake + urea resulted in higher yield at AAU and JNKV centres. At MPKV, application of Azotobacter 5 kg/ha along with Phosphobacter @ 5kg/ha resulted in increased vine growth, leaf yield and shelf-life. At Pusa (RAU), application of vermicompost registered the highest production of crop in term of total number of leaves, fresh weight of hundred leaves, and leaf area and all other treatments did not result in appreciable increase in leaf yield. It was also noted that increase in urea dose (200 kg/ha) resulted in maximum disease incidence and poor keeping quality of leaves.

Amongst different treatments, vermicompost (10 t/ha) had marked effect on number of branches, vine growth (12.15cm), leaves (25.15 lakh/ha) with more number of branches (22.15) over other treatments. Though phytophthora foot rot severity is lesser in all microbial biofertilizer in comparison to oil-cakes and urea.

Bio-regulators on Yield and Keeping quality of Leaves

Tricontanol (0.05%) sprayed thrice at 30 days interval had favourable influence on growth and yield parameters in all the centres (Fig. 5.3). However, an increase in keeping quality of leaves with the tricontanol spray was observed only in MPAU. QUAT and TNAU. A significant increase in cost benefit ratio was also observed in MPAU and RAU.

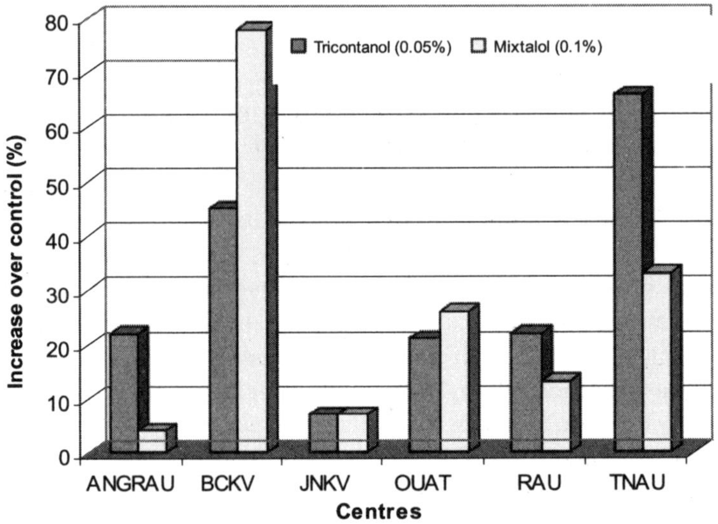

Fig.5.3: Effect of growth regulators on leaf yield

Assessment of Optimum Level of Organic Carbon Content

The balance of organic carbon in soil is of pivotal importance in maintaining soil properties and dynamics of soil microflora in betelvine cropping system. Assessment of organic carbon content in soil is carried out to find out the optimum level for sustainable cultivation of betelvine in various regions of our country where betelvine is cultivated commercially.

At AAU centre, optimum level of organic carbon was assessed in different cultivation practices followed in Assam, i.e., closed type, semi closed type and open type in three locations (Ambagan, Lengeri, Borhola) at 15 different sites. Per cent organic carbon varies from 0.56 to 0.78 in different betelvine gardens of Assam and it was medium in terms of available nitrogen. Soil type at Borhola village was predominantly silty clay loam with high organic carbon content (ranges from 0.67 to 0.78%). The available P_2O_5 (43.8-48.8 kg/ha) and K_2O (105.5-108.2 kg/ha) were highest among the locations.

A survey was conducted at Bhander Khola, (N) during December,2006. During the survey period, crop characteristics, soil characteristics, fertilizer application and disease and pest infestation were recorded from 10 farmer's field and soil samples were collected from those barejas for analysis as per schedule. The farmers' of Bhanderkhola, mainly used mustard oil cake. None of them are using super phosphate, murate of potash and micronutrients except one for supplying the nutrients to the crops. The organic carbon (%) content, total N (%), total P_2O_5 /kg/ha and total K_2O kg/ha of different borajas ranged from 0.49–0.78%, 0.014–0.042%, 4.4–196.46 kg/ha and 168.3–382.8 kg/ha respectively. The pH of above soil ranged from 6.78-8.13.

Ten soil samples were collected from three Betelvine growing areas to assess the optimum level of organic carbon content in betelvine cultivation by JNKV centre. It was found that betelvine one of the farmer had applied nitrogen in the from of urea and oil cake to the tune of 265 kg/ha which resulted in maximum organic carbon content (1.3%) among the locations tested and got maximum leaf yield (42.52 lakh/ha). The field which received 155 Kg nitrogen in the from of oil cake and urea along with 0.6% organic carbon content got minimum leaf yield (33.25 Lakh ha). Roving survey was carried out at different farmers fields of Anlo village in Niali block by OUAT centre and the organic carbon content was ranging from 0.59% to 0.80% among the locations.

Survey was carried out in 10 different adapted farmers' field at Poigaiputhur and Thanneerpalli by the TNAU centre. Observations were made out in betelvine variety SGM BV-2. Among the fields studied the betelvine recorded the highest leaf yield (40.45 lakh/ha) during second year and an increased shelf-life of 12.76 days during second year was the one having highest organic carbon content.

Irrigation Schedules and Methods on Yield of Leaves

Application of irrigation water at an IW /CPE ratio of 1.00 with irrigation depth of 5 and 3 cm at RAU and BCKV respectively were found to be better for growth and leaf yield. However. irrigation did not influence on keeping quality of leaves, incidence of bacterial leaf spot and uptake of major nutrients at RAU.

Furrow irrigation after each row in ANGRAU and after two rows in MPAU were reported to be superior methods for higher leaf yield. Irrigation methods did not have any effect on keeping quality of leaves and disease incidence.

Drip Irrigation on Growth and Leaf Yield

At JNKVV, 100% replenishment of pan evaporation rate through drip irrigation produced highest leaf yield (25.55 lakh/ha) at JNKVV. However, at MPKV center 125% evaporation replenishment produced highest yield (42.32 lakh/ha). In this treatment water saving was in the tune of 34.26 per cent.

Packing and Storage

Packing of leaves without petiole was found to be a better practice for improving shelf-life when compared to packing with petiole in all the centres. In general keeping quality during storage was better in winter than summer and rainy seasons.

Evaluation of Live Support Crop

Superiority of *Moringa* sp. as live support crop was recorded uniformly in two locations. *Sesbania grandiflora* was also good when growth parameters were compared. *S. sesban* was not found suitable since it started producing branches from the base of the plants and also had slow growth rate.

Effect of Plant Population

At RAU, Pusa, it was observed that plant population density had no effect on vine elongation and number of leaves. However, marketable leave increased significantly due to rise in plant density, but resulted in reduction in leaf size, shelf-life and benefit: cost ratio. The plant population @1.5 lakh plant/ha was quite economical and superior to other treatments.

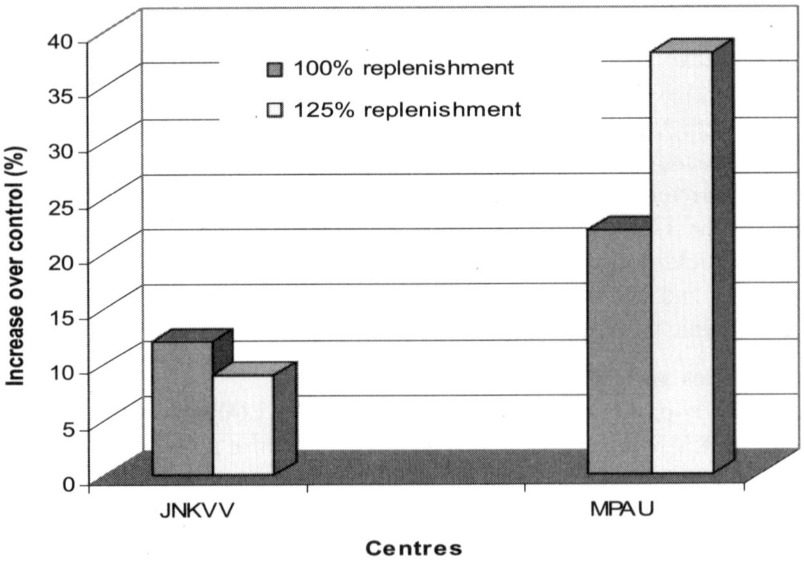

Fig 5.4. Effect of drip irrigation on leaf yield of betelvine

At Islampur (RAU), amongst four plants spacing/plant populations, maximum vine elongation, populations (9.35), number of branches (18.35) and number of leaves/ha (26.75 lakh/ha) recorded with 1.5 lakh /ha plant population. The *Phytophthora* foot rot incidence increased with increasing plant population.

Crop Regulation

To increase the harvest of betel leaves in the lean period, a technique named staggered lowering was standardized. The results showed that significant improvements in crop growth and yield was noticed by this technique in various centres. At AAU centre, there was no significant difference in vine elongation, number of branches per plant, fresh weight and keeping quality of the leaves. However, significant difference was observed on yield, disease incidence and net return during February and September lowering and July and November lowering compared to other lowering periods. February and September lowering gave highest net return (2.49 lakh/ha/year).

At BCKV, the results showed that leaf yield was found highest (33.30 lakh/ha) in treatment where July and November lowering was done and it was statistically at pare with February and June lowering. The lowest (27.81 lakh/ha) leaf yield was recorded in August and December lowering and it was statistically at par with September and January lowering. The lowest (6.66%) leaf spot disease was recorded in September and January lowering and it is at par with February and June lowering (8.02%).

STRATEGIES FOR DISEASE AND PEST MANAGEMENT

Disease Survey

Two types of survey namely: I) Fixed Plot Survey at a monthly interval and II) Roving survey at pre and post monsoon period was conducted. Studies revealed that *Phytophthora* leaf and foot rot, anthracnose and bacterial leaf spot and stem rot were important in most of the locations. In Uttar Pradesh and Assam, disease incidence was in general low and none of the disease were found alarming at farmers field. In Andhra Pradesh, Sclerotial rot and Fusarium wilt were serious. In Orissa, Sclerotium rot incidence was around 27.5% in open type of cultivation. A serious out break of bacterial stem rot was observed in Salem area in Tamil Nadu during 1989-90. A species of *Phytophthora* (*P. capsici*) was isolated and identified as pathogenic to betelvine. A new disease caused by *Chlamydomyces palorum* was reported from Assam causing up to 22.5% leaf and stem infection at pre-monsoon months in Tinusukhia district.

Epidemiological of *Phytophthora* Foot Rot Disease

(a) Phytophthora Stem and Leaf Rots

Phytophthora stem and leaf rots cause severe damages to betelvine gardens all throughout the country. This is the most and destructive fungal disease that produces both wet rot symptoms on leaves and wilting due to foot rot. At first circular dark brown spot appear which becomes wet and rot under continuous high humid conditions. Otherwise dark brown necrotic spots with alternate light brown zonations appear. The loss of luster of the leaves in the foot rot condition followed by complete wilting and drying of the vines occur in a quick sequence.

Results revealed that at ANGRAU center, disease incidence showed a significant and negative relationship with maximum temperature (r = - 0.4972). The other weather factors studied did not show any significant correlation with PDI. Experiments conducted at BCKV showed that x-coefficients for all the weather parameters like temperature (maximum and minimum), relative humidity (maximum and minimum) and rainfall had positive significant effect on per cent disease incidence of stem rot in variety Ghanagete. The results of correlation studies for Simurali Bhabna revealed that x-coefficients of T_{max} (1.45), RH_{max} (2.29), RH_{min} (0.31) and rainfall (0.008) had positive significant effect on the percent disease incidence of stem rot. In Simurali Deshi variety, except T_{min} (0.047) all the other weather parameters had positive significant effect on per cent disease incidence of stem. In Harishpur Bangla, coefficient of all the weather parameters like T_{max} (2.21), RH_{max} (3.19), RH_{min} (0.46) and rainfall (0.001) were positively correlated with per cent disease incidence of stem rot, while T_{min} (0.29) had negative effect.

Results of JNKVV revealed that the temperature, relative humidity and rainfall had direct influence over the disease incidence. At OUAT, results showed that a significantly positive linear relationship of minimum Temperature (0.620), afternoon RH (0.594), and number of rainy days in a week for causing the vine rot disease. Experiments carried out at RAU showed that important contributing factors are maximum relative humidity above 90 per cent and minimum temperature 25.8–27.5 °C which prevailed during the period accounted for 20.05 and 12.10% respectively of the total variation in per cent disease index.

(b) Anthracnose

The leaf spot symptom is irregular in shape and size, light to dark brown surrounded by diffuse chlorotic yellow halo. Marginal leaf tissue becomes balck, necrotic and gradually spread towards the leaf centre. Occasionally diffused yellow halo also develops. In the anthracnose satge, circular, black lesions that occur rapidly increases in size and gridle the stem culminating in the death of the vine.

Results of correlation studies at AAU center revealed that minimum temperature, bright sunshine hours and rainy days had positive correlation and the other factors showed negative correlation with the disease incidence. Results of RAU showed that maximum and minimum temperature, maximum and minimum relative humidity and amount of precipitation during observation had positive effect, whereas number of rainy days and number of cloudy days per week had negative effect on disease development in terms of per cent disease incidence of anthracnose of betelvine. Important contributing environmental factors for disease incidence were maximum relative humidity above 90 %, maximum and minimum temperatures ranging between 34.1 and 25.8 °C which have accounted for 23.84, 14.81 and 9.2 % respectively of the total variation in disease incidence.

(c) Bacterial Diseases

Bacterial leaf spot, leaf blight or stem canker disease are becoming a serious in many of the betelvine growing parts of the country. Brown leaf spots, surrounded by translucent water soaked area appear on the lower surface and yellow halo appear on corresponding upper surface. Elongated brown spots of variable length appear on vine that results in stem canker. In severe infection, stem cracks and large area of leaf lamina are covered causing bighted leaf.

The correlation studies carried out at JNKVV revealed that maximum relative humidity and rainfall had the positive correlation on the development of disease.

Varietal Resistance to Diseases

The betel vine varieties grown in different centres were screened under artificially created epiphytotic conditions for studying the resistance to important diseases. Halishar sanchi was found to posses multiple resistance against *Phytophthora* foot rot, *Anthracnose* and bacterial leaf/stem infection and basal rot of betelvine. The Pachaikodi and Kapoori exhibited high resistance to Phytophthora rots and bacterial leaf/stem infections at JNKVV. All Kapoori varieties showed resistance to powdery mildew at IIHR, Bangalore. Resistance sources identified at different centres are given in Annexure VIII.

***Phytophthora* Leaf and Stem Rot Control**

Studies on the efficacy of systematic fungicide formulation, viz. phosphoric acid, Fosetyl-A, and Chlorothalonil in control of the disease in comparison with Bordeaux mixture (1%) four sprays at monthly interval. *Phytophthora* leaf and stem rot can be controlled by spraying Bordeaux mixture (0.5%) followed by drenching Bordeaux mixture (1%).

Chemical Control of *Anthracnose*

Bordeaux mixture (0.5) was found superior in controlling *Anthracnose* disease in Assam and West Bengal. However, in Orissa mancozeb (0.1% ai) was found most efficient in controlling the disease. In Tamil Nadu, Bitertanol (0.025% ai) was superior that was followed by mancozeb (0.1% ai) and bordeaux mixture (0.5%).

Chemical Control of Bacterial Leaf Spot

Superiority of bacterinol (0.1% n ai) with bordeaux mixture (0.5%) treatment was observed next to streptocycline (250 ppm) treatment.

Biological Control of Foot Rot

Trichoderma harzianum inoculated neem cake in four applications, at quarterly interval was the next best to bordeaux mixture in reducing the disease incidence as well as improving leaf yield at AAU, BCKV, JNKVV and TNAU. Cost: benefit ratio was also next to bordeaux mixture in AAU and TNAU, while in OUAT, VAM and *Trichoderma* (quarterly) applications were second to bordeaux mixture in terms of disease incidence and yield.

Integrated Management of *Phytophthora* Foot Rot

Results indicated that diseases can be best managed by the treatment having Sanitation + 1 application of Bordeaux mixture (1%) at the monsoon + biological agent (*T. harzianum*) after 1 month + 1 additional Bordeaux mixture after 2 months of first B. mixture at AAU, ANGRAU, BCKV, MPKV, OUAT and RAU.

Studies on Spray Schedule for *Phytophthora* Leaf and Stem Rot

Studies on spray schedule for *Phytophthora* leaf and stem rot at three centres, viz. BCKV, OUAT and JNKVV reported the superiority of four drenching + eight spraying of bordeaux mixture uniformly starting from June to September at monthly and fortnightly intervals respectively. However, in Tamil Nadu, three drenchings with six sprayings starting from June to August was found to be the best.

Standardization of Inoculums for Mass Multiplication of *Trichoderma*

At AAU, 60 g of inoculum was found to be best in producing highest spore concentration

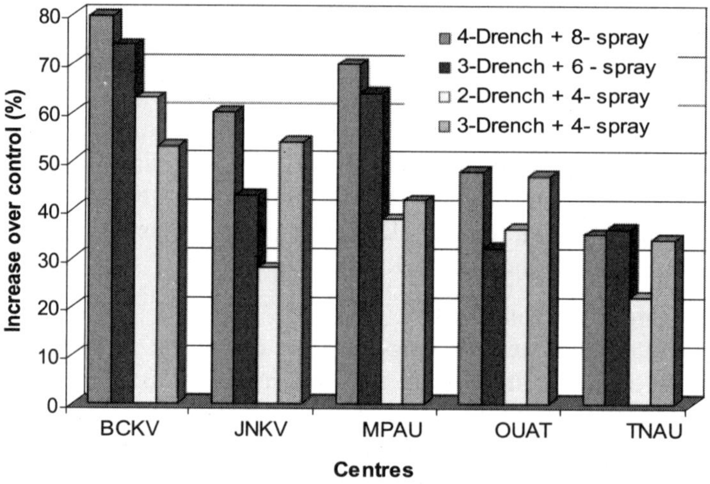

Fig. 5.5: Effect of spray schedule on leaf and stem rot

and colony forming units when 10kg mustard oil cake was inoculated and incubated for 60 days. Similar result was also obtained in RAU. However, rate of growth of both the spore concentration and the colony forming units were slowed down after 40 days of incubation. At BCKV, a detailed study was carried out to find out suitable substrate as well as their quantity. Results showed that oat or barley used as substrate for the production of mass culture of *T. harzianum*, were best and 40g of unit inoculum of both the low cost cereals can be used and which maybe distributed to the farmers for mass culturing in oil cake.

Three applications of *Trichoderma viride* inoculated oil cake at quarterly interval was found second best in reducing nematode population and leaf yield in AAU center after carbofuran treatment compared as control. However, both the treatments, carbofuran and three applications of *Trichoderma viride* were at par in producing leaf yield at JNKVV.

Rhizospere Competence and Survival period of *Trichoderma* sp. Garden

At ANGRAU, rhizosphere competence and survival of *Trichoderma* spp. in betelvine garden trial showed that there was significant increase in *Trichoderma* population of rhizosphere soil up to 40 days after addition of talc formulation of *Trichoderma viride*, after 60 days there was decreasing trend in population.

STRATEGIES FOR MANAGEMENT OF INSECT PESTS AND MITES

Survey of Insect Pests and Mites

In Assam incidence of whitefly, mealybug and mites were observed. Whitefly infestation was in peak during April — July. In Andhra Pradesh, severe damage by *Spodoptera litura* (up to 30%) and red spider mite (up to 8%) was observed during October. An incidence of 5–15% mirid bug (*Disphinctus politus*) was recorded around Tuni area during monsoon months which caused 40–50% damage during post-monsoon months in Visakapatnam districts. In West Bengal, blackfly was observed to be serious in Midnapur and Bankura districts. In Orissa, whitefly and red spider mite were prevalent during summer months, whereas betelvine bug, aphids and jassids were confined to winter months. In Tamil Nadu, incidence of scales (*Lepidosaphas cornatus*) red spider mite, mealybug and *Spodoptera litura* were observed.

Population Fluctuation of Insects of Betelvine

In APAU, tobacco caterpillar incidence started during the second fortnight of September and was maximum during November-December. Infestation was nil from January onwards. Build-up of white mite started during second fortnight of November and was maximum during December-January. Lower temperature was favourable for the development of the pest.

At BCKV the population of white and black fly was maximum during August-September and June-August respectively. The second peak in population of white and black fly was also observed during December-January and September—November respectively.

Highest scale insect population was seen during April-May in TNAU. Mite population was high during March. Both scale and mit population was positively correlated with maximum and minimum temperatures.

Crop Loss Assessment due to Insect Pests and Mites

Tobacco Caterpillar

To assess the cumulative loss in leaf yield by tobacco caterpillar, a trials were undertaken by ANGRAU centre. The results showed that plots protected against tobacco caterpillar with neem oil 0.5% spray significantly lower mean leaf damage of 4.37% where the unprotected plots recorded 18.73%. The cumulative number of leaves harvested per ha during the period of pest infestation was 5.24 and 4.30 lakh in protected and check plots respectively.

Mites

Similar trial by ANGRAU centre as indicated above was conducted against mites by spraying wettable sulphur 0.3% two times at fortnight interval during December. The plots protected against mite damage recorded a mean mite incidence. The protected plots had lower mite damage with a mean mite incidence of 4.49 as against 14.02 in untreated plots.

Scale Insect

At TNAU, the results revealed that the plots protected with Chlorpyriphos 0.04% spray recorded the least number of linear scale insect/leaf ranging from 0.01-0.04 with a mean of 0.03 as against 0.26–1.20 with a mean of 0.85 in untreated plots from April—August in 2006.

Black Fly and White Fly

Assessment of crop loss due to the attack of aleyrodid fly complex (white fly and black fly) on betelvine variety Simurali Deshi (Bangla) at BCKV revealed that there was a reduction in direct leaf yield to the tune of 10.52% for a mean population of 31.7flies/vine. Overall monetary loss was 36.84% for five month period.

Use of Plant Origin and Safer Pesticides

At APAU, the effect of pesticides of plant origin (tobacco decoction 2% and neem seed extract 5%) in controlling mites was on par with dicofol (0.05%) and dichlorovos (0.05%). However, at TNAU wettable sulphur (0.5%) and phosphomidon (0.05%) was significantly better than plant origin pesticides in controlling mites.

Screening of Betelvine Cultivars Against Whitefly

At BCKV, the results showed that there were great differences among the entries for whitefly infestation level. Twenty three betelvine cultivars were screened in field condition against whitefly. Kalipatti and Simurali Sanchi and Halisahar Sanchi were moderately resistance; Harishpur Bangla, Bilhari, Godi Bangla and Kadwa showed moderately susceptible. The remaining 16 cultivars were highly susceptible.

Population Fluctuation of Mite Species in Betelvine Garden

The mite species of nine groups have been recognized from recent observations on betelvine at BCKV. Among the species observed only destructive nature of *Brevipalpus phoenicis* has been observed. The characteristic symptoms of this mite on betelvine were blackish lesions at the ventral surface especially near the margins of the leaf. Gradually the blacken spots spread all over the leaf that eventually enhanced maturity and drying of the leaf. The matured leaves were almost infected than the younger leaves. The phytophagous mite, *Brevipalpus phoenicis* was observed to persist in the *baraj* condition throughout the year maintaining varying population level. Higher population was recorded in March—June and October. The highest population was observed during May in both the young and mature leaves. The small reddish, dorso-ventrally flat mite was found to feed at ventral surface of leaves producing characteristic patches on feeding site that reduces growth and vigour of the plant as well as reduces market price of the leaves.

Management of Mites in Betelvine Garden

At ANGRAU, the plots treated with Abamectin 0.3 ml/l recorded minimum damaged leaves (5.25%) followed by wettable sulphur 0.3% + neem oil 0.5% (6.16%), Dicofol 5 ml/litre (6.65%) and wettable sulphur 0.3% + NSKE 5% (7.06%) where as untreated plots recorded 21.33% damaged leaves. Higher leaf yield of 7762.5 bundles/ha was recorded when Abamectin 0.3 ml/litre was sprayed with an extent of 47.93% increased leaf yield over check.

Management of Red Spider Mite

The trial was conducted with SGM-1 variety at TNAU centre. The population of red spider mite was counted at peak incidence on three leaves viz., lower, middle and upper leaves in 10 randomly selected vines for each replication and the treatments were imposed. Wettable sulphur + NSKE 5% recorded maximum reduction of red spider mite (83.50%). The combination of wettable sulphur 0.15% + azadirachtin 0.03% was found to be best treatment for effective management of mites and thereby reducing number of damaged leaves (4.82 lakh/ha). Based on C:B ratio the conventional miticide, viz. wettable shlphur was found to be highly effective for the management of mites in betelvine.

Screening of Betelvine Cultivars

Tobacco Caterpillar

At ANGRAU centre, pungent varieties of betelvine recorded lower incidence of tobacco caterpillar with 3.71–31.47% damaged leaves as against 6.59–37.77% damaged leaves in non-pungent varieties. Under laboratory conditions, higher survival of larvae ranging from 26.6–33.3, 66.0–80.0 and 86.6–93.3% was recorded in non-pungent cultivars as against the lower survival of 0.0–13.3, 13.3–26.6 and 26.6–53.3% in pungent cultivars in third and fourth and fifth instar larvae respectively.Among pungent varieties, Kalipathi (Maharashtra) (31.47%), Nov Bangla (Orissa) (23.72%) and Black leaf (A. P.) (22.10%) recorded above 20% of damaged leaves. Among non-pungent varieties Kapoori (Vuyyuru) Andhra Pradesh (6.59%) and Patchaikodi Tamil Nadu (11.24%) recorded less percentage of damaged leaves.

Mites

Mite incidence ranged from 18.51–53.33% and 30.95–67.79% in pungent and non-pungent cultivars respectively. Among pungent germplasm, lower mite incidence was recorded in Ramtek Bangla (Maharashtra) (18.51%) and Awamipan West Bengal (19.17%). Similarly in non-pungent germplasm, Meethapan (12.90%) and Gangeri Andhra Pradesh (22.03%) recorded lower infestation due to mites.

Germplasm Screening

Non preference found to play key mechanism towards infestation by pests, viz. linear scale insect and red spider mite. Phenlic compound methyleugenol found to influence the dynamics of red spider mite in kapoori and SGM BV2.

Management of Insect Pests

Leaf Eating Caterpillars

At ANGRAU, mean incidence revealed the significant superiority of all treatments tested over check in reducing the incidence of leaf eating caterpillars. Minimum larval population and percent plant damage was recorded in the plots treated with chlorpyriphos 0.05% which was followed by Monocrotophos 0.05% spray. Among the safer insecticides NSKE 5% spray + NSKE 5% soil drenching recorded 7.66, followed by meem oil 0.5% spray + neem oil 2% soil drenching (7.91) and cartap hydrochloride 0.1% (8.16). All are on par with each other and found better in reducing the incidence of leaf eating caterpillars against 13.08 larvae in untreated check. Similarly, minimum percent of damaged plants were observed in plots treated with cartap hydrochloride @ 0.1% (35.36%) followed by NSKE 5% + NSKE 5% soil drenching (37.46%) and neem oil 0.5% spray + neem oil 2% soil drenching (39.53%).

Stem Borer

Against stem borer also chlorpyriphos 0.05% proved to be effective with lowest percent of infested plants (7.37%), followed by Monocrotophos (0.05%) (7.86%) at AGRAU centre. Among safer insecticides cartap hydrochloride @ 0.1% spray recorded significantly lower incidence of stem borer (11.35) which was followed by neem oil 0.5% spray + neem oil 2% soil drenching and NSKE 5% spray + NSKE 5% soil drenching. All the three treatments were on par with each other. Reduction in pest damage over check to an extent of 43.67% was recorded in the plots treated with cartap hydrochloride where as the plots treated with chlorpyriphos 0.05% and Monocrotophos (0.05%) recorded 63.42% and 60.99%.

At TNAU centre, neem seed kernel extract 5% spray thrice at monthly interval was found to suppress the steem weevil *Alcidodus bubo* followed by *Beaveria bassiana* 5g (1.8 × 10^6 spores per spray) twice at 15 days interval both in field trial and large scale testing at farmers' field.

Fixing ETL for Important Pests

There were significant differences in number of scales/vine with respect of various treatments. Maximum population/vine occurred to vines infested with 10 linear scales/Rvine followed 7, 5, 3 and 1 linear scales/vine. All treatments were significantly different. In the control plots where betelvine was protected by spraying Chlorpyriphos 0.04% recorded only 2 percent damage was observed and the leaf yield was highest (62.67 leaves/vine). The mean leaf yield (61.33 and 60.00 leaves /vine) in plants released with initial load of 1, 3 and 5 scales respectively per vine was on par with control indicating that betelvine could tolerate low levels of linear scale population without adversely affecting yield. The leaf yield recorded from vine released with 5 linear linear scales were drastically lower and on par with 1 and 3 released scales. The leaf yield recorded from that from vines released with 7 and 10 scales were significantly lower than 5 scales/vine. The results of the present study indicate that *L.cornutus* is an important pest of betelvine which cause significant yield loss when the population of scales is above 5 scales/vine (0.25 scales/leaf). At higher population, the marketable yield loss is substantial. It is evident from the study that the betelvine can tolerate infestation of *L.cornutus* of the attack is upto 5 scales/2m vine (0.25 scales/leaf) without affecting yield where the age of the crop is one and half year after lowering.

Management of Insect on Sesbania (Live support on betelvine)

At ANGRAU centre, mean data on leaf damage indicated a minimum leaf damage of 8.18% was recorded in plots treated with chlorpyriphos 0.05% followed by cartap hydrochloride with 9.78% other effective treatments were *Nomuraea rileyi* @ 5 g/litre., NSKE 5% and neem oil 0.5%. An increased leaf yield of 23.83% over check was recorded by controlling the pest with cartap hyclorochloride 0.1% during pest infestation.

STRATEGIES FOR MANAGEMENT OF NEMATODES

Survey for Nematodes

Most predominant plant parasitic nematodes associated with betelvine are *Meloidogyne incognita* and *Rotylenchulus* sp., *M. incognita* was most destructive throughout the betelvine growing regions.

Interaction of Root-Knot Nematode with Pathogenic Fungus

No significant additive effect was observed when *Fusarium solani* was inoculated simultaneously with root-knot nematode. However, highest disease incidence was recorded when *Colletotrichum capsici* and root-knot nematode were inoculated simultaneously.

Population Threshold of Nematode

Meloidogyne incognita population at 1,000 per pot was found to reduce leaf size and fresh weight of 100 leaves at APAU. Population of reniform nematode (*Totylenchulus* sp.) was found optimum at 10,000 per plot to produce maximum root damage. In case of *Helicotylenchus incises*, 1,000 per pot population caused significant reduction in plant growth parameters at TNAU Coimbatore.

Biological Control of Root-Knot Nematode

Field application of *Paecilomyces lilacinus* inoculated oil cake thrice at 500 kg/ha reduced root-knot index and produced maximum vine growth and leaf yield in AAU, whereas in OUAT and JNKVV it was next best to carbofuran (1.5 kg ai/ha) application in reducing the root knot index, however, both the treatments were equally good for leaf yield.

Host Resistance to Nematode

Awami pan showed resistance to root knot nematode in AAU. Halishar sanchi (WB) recorded lowest root knot index in OUAT. Harapatta exhibited moderate resistance in RAU. However, none of the varieties showed resistance in JNKVV.

IPM of Root-Knot Nematode

Application of oil cakes (500 kg/ha) with soil drench of carbofuran (0.1%) and three application of *P. lilacinus* inoculated oil cakes (500 kg/ha) was on par with carbofuran (1.5 kg/ha) in AAU, JNKVV, TNAU, RAU, OUAT centres with respect to root knot index and leaf yield.

Pathogenicity of Rotylenchulus Reniformis

Increase in the nematode population from 100/kg of soil onwards decreased the growth and yield of the vine JNKVV, OUAT and RAU. Maximum reduction in growth and yield was observed when the inoculum rate was 5,000 nematodes/kg of soil.

RECOMMENDATIONS OF AICRP BETELVINE

Genetic Resources

- The betelvine varieties (SGM-1 Ghanaghatte selection Sudam) released by AICRP Centres may be characterised through DNA finger printing and deposited at the respective SAUs wherever the facilities are available and also at NBPGR, New Delhi. A national number may be obtained for each variety registered with NBPGR and recorded at respective centres.
  The varieties, viz. *Maghi* (RAU), Hallishar (BCKV) and *Kattaki* (JNKVV) qualify release for which formalities may be proceeded with.
- Germplasm collection must be completed on time-targeted basis. A cess fund project may be provided involving AAU, Jorhat and IIHR Centre, Bangalore.
- The following Centres may be mandated to evaluate, characterise and cataloguing the germplasm of betelvine noted against each center, viz.
  AAU Jorhat – North-Eastern region
  NBPGR, Trichur – Southern region
  IIHR, Bangalore – Southern region
- The centres collecting the germplasms will divide the material into three samples, one for NBPGR for cryopreservation one for AAU to be identified for cataloguing and the third for retaining at the collecting centres.
- NBPGR should enhance the effort to collect exotic germplasms from the countries growing betelvine/centres of diversity. Samples of the germplasm should be made available to AAU (Jorhat) and IIHR (Bangalore) for maintenance.
- The Betelvine Centre at IIHR (Bangalore) will take up sole responsibility of crop improvement because of its geographical advantages in flower production.

- Every registered betelvine variety and elite collection, as well as the quality attribute profile of essential oils should be studied by BCKVV, Kalyani and JNKVV, Jabalpur centres and incorporated in the descriptor. The scientist's post at NBRI, Lucknow Centre should be redeployed to BCKVV Centre to carry out the work.

Varietal Improvement

- Hybrids or betelvine did not perform well in any of the Centres for which the causes need to be studied critically.
- Mass multiplication of improved varieties should be taken up for distribution among the growers.
- Hybridisation should be intensified at the IIHR centre for developing new varieties for high yield, better quality and disease resistance. Inter-specific hybridisation for disease resistance, profuse lateral production, drought/shade tolerance, and scope of chemical mutagen and irradiation should be explored.
- Efforts may also be made to locate marker genes for sex, intrinsic quality, productivity, resistance, lateral production, etc.

Crop Production

- Optimum plant densities should be arrived at on the basis of phytosanitation, tapping of solar energy and individual plant productivity.
- Ideal live standards free from arthropod pests and diseases to train betelvine should be identified.
- In *Kapoori* variety optimum height for maximum leaf production should be arrived at.
- Inter-cropping in betelvine gardens may be worked out to generate additional income and employment. Crops like ginger, turmeric and pachouli may be thought of.
- On-farm trials to popularise the packages of suitable agrotechniques in major growing areas may be taken up. Additional funds may be provided by the ICAR for the purpose.
- The most efficient oilcake in different centres/region should be found out by regional screening of available oil cakes.
- Optimum ratio of organic including vermicompost vs inorganic should be arrived at in view of high pest insect and disease incidence and poor storability of leaves. Integrated plant nutrient management (IPNM) should receive attention at all the centres for sustainable production.
- Water management through drip/micro/sprinkler including fertigation to economise water and nutrients should receive priority.
- Research on targeted yield in betelvine with integration of technologies developed should be the future goal.
- Lateral production potential in *kapoori* types should be exploited to utilise 'bush betelvine' for commercial cultivation both in pots and in fields.
- Micronutrient requirement in betelvine should be critically assessed for productivity and quality.

Crop Protection

- Of all the diseases that affect betelvine production, *Phytophthora* foot rot and leaf rot, *Anthracnose*, still continue to be most devastating. A denotification system to identify the fields with sick soil in high-inoculation population of noxious fungi and nematodes may be worked out.
- IPM approach in insect pest and disease management has of late received some attention in the AICRP on betelvine. An appropriate IPM technology may be tuned for tackling major arthropod pests and diseases for different states in location-specific forms.
- Research on use of botanicals and biopesticides for tackling arthropod pest and disease problem should receive priority.
- Nematodes appear to be predisposing factor for *Phytophothora* diseases for which interaction studies are called for.

Post-harvest Processing and Quality Upgradation

- Field fungi and storage decomposers cause enormous loss to Bangla types during storage and transit. Specific research program is needed to develop technology to reduce this post-harvest loss.
- Studies on the existing practices adopted in post-harvest processing in betelvine may be undertaken by every network centre except IIHR, Bangalore, to identify the factors that affect the quality and undertake field study to prevent unhygienic practices.
- While preparing package of practices post-harvest packaging should be included to bring the awareness among the farmers about the importance of the quality.
- The present practice of bleaching in Bangla types is hazardous for which new methods may be developed in the lines of IIT (Kharagpur) at BCKVV, Kalyani and OUAT Bhubaneshwar. Separate funds may be provided by the ICAR for the purpose.
- Research on product development like Pan preserve, powder or flakes may be attempted. Oil research deserves to be priority area and therefore warrents special attention. Special attention may be given for assessing residual toxicity of pesticides in betel leaf in pre- and post-harvest stages, as initiated as BCKV, which may be encouraged by additional fund from the ICAR.

RESEARCH GAPS AND FUTURE THRUSTS

Conservation, Evaluation and Cataloguing of Germplasm

Work on collection of local germplasm. their maintenance and evaluation is in progress in all the coordinating centres including IIHR. Until now this study has not been given due emphasis, consequently it has been patchy and pragmatic to a large extent. more or less confirmatory of the superiority of the existing varieties/clones in large scale cultivation in the area. There should be a definite programme of collection of indigenous germplasm. Each centre should maintain as any lines. clones or varieties and evaluate them under standard conditions as far as practicable. For this purpose certain guidelines will have to be evolved. Standard proforma for collection of data on a comprehensive basis including morphologic characters. cold tolerance. resistance to diseases and pests should be developed. Data such collected should be amenable to computer analysis. hence proforma should be so designed. After initial screening. varieties/clones/lines so selected should be tested at first zonally, as the physical requirements o.f the crop are different in different zones/locations.

Establishment of Gene Bank

As there is considerable confusion regarding varieties/clones. It is necessary to establish gene banks preferably two in appropriate places to maintain different varieties and clones for future studies.

Standardization of Agro-techniques

Work already in progress in different centres should be supplemented by studies on soil, both in open type/bareja system of cultivation. Recommendations on application of fertilizers and micro-nutrients should be based on soil test/crop response studies. Physical and chemical properties of soils should be determined in respect of commercial plantations in different major locations in each state. Studies should be undertaken to develop location specific integrated nutrient management. It has been observed that problems of irrigation and drainage are existing. Besides knowledge regarding optimum water regime in bareja and open type is lacking. As soil water level has an important role in the incidence and severity of diseases, and nematode infestation, this aspect has a special significance.

Cropping Sequence

In open types of shorter duration of 2-3 years. the desirable cropping sequence has to be worked to minimize the population of soil borne pathogens. In barejas, in many places mixed or inter cropping with vegetables, is practiced. Role of such crops in relation to diseases pests and nematodes in perenniation and spread of inoculum needs to be found out.

Integrated Insect Pest and Disease Management

Should be worked out on location specific basis. keeping in view of prevalent diseases and pests, calendar of operations for disease and pest management should be prepared accordingly, particularly with reference to key pests and diseases. In the development of this technology, emphasis should be laid on biological methods which have shown promise. In control of pests, pesticides of biological origin, or those with low level residual toxicity should be tried. 'Red' label pesticides should be avoided. This suggestion is being made not only to minimize the risk of consumers who take raw betel leaves and possible environmental hazards, but also to meet the possible requirements of importing countries where rigid standard regarding pesticide residues is prescribed.

Screening for Resistance to Biotic and Abiotic Factors

Should be undertaken under more rigid control conditions to have reliable data on the performance of the varieties in different locations and the results have to be compared and evaluated at all India level through Workshops/Symposiums. Until now results have not been consistent. For this purpose, ICAR may provide necessary funds for the construction of glass house and development of other facilities in each centre.

Epidemiology of Diseases and Insect Pests

Work carried so far indicates broadly weather conditions under which these diseases and pests have been found to be more prevalent and damaging. Such studies should be undertaken on epidemiology using weather parameters under microclimatic conditions. The data collected should be statistically analyzed so that important weather factors can be determined and forecast mechanism can be worked out at least in a few cases of pests and diseases.

Quality Parameters of Betelvine Leaves

Leaf samples for the major varieties, clones collected under germ plasm should be analyzed. Regarding major chemical parameters particularly essential oils like saffrol, eugenol, terphenol and anithol, which are contributory to pungency, sweetness and flavour which determine quality or acceptance of varieties.

6

Floriculture

R.L. Mishra, P. Naveen Kumar and Suchitra Pushkar

The systematic efforts to initiate planned research in floriculture was materialized with the starting of All Indian Coordinated Research Project on Floriculture during IV Five Year Plan in 1970-71 at 5 centres located at Indian Agricultural Research Institute (IARI), New Delhi; Indian Institute of Horticultural Research (IIHR), Hessaraghatta, Bangalore; Indian Agricultural Research Institute Regional Station, Katrain; Botanical Survey of India (BSI), Howrah (Kolkata) and BSI, Shillong. During V Five Year Plan, four more Centres, viz. Punjab Agricultural University (PAU), Ludhiana; Dr Y.S. Parmar University of Horticulture & Forestry (Dr YSPUHF), Solan, Kerala Agricultural University (KAU), Vellanikkara, and Mahatma Phule Krishi Vidyapeeth (MPKV), Pune, were added. The two centres of the project located at BSI Shillong and BSI Kolkata were transferred to ICAR Research Complex for NEH Region, Shillong and Bidhan Chandra Krishi Viswavidyalaya (BCKV), Kalyani, during April, and October 1977, respectively.

COORDINATING CENTRES

There were 35 voluntary centres, besides these 9 regular centres during 1973–77 and voluntary centres were Agricultural Institute, Allahabad; University of Agricultural Sciences, Bangalore; Lal Bagh Botanical Garden, Bangalore; Orissa University of Agriculture and Technology, Bhubaneswar; The Agri-Horticultural Society, Kolkata; Hill Fruit Research Station, Chaubattia; Tamil Nadu Agricultural University, Coimbatore; Sim's Park, Coonoor; Llyod Botanic Garden, Darjeeling; Directorate of Parks and Gardens, Gandhi Nagar; Haryana Agricultural University, Hissar; Andhra Pradesh Agricultural University, Hyderabad; Jawaharlal Nehru Krishi Vishwavidyalaya, Jabalpur; Parks and Gardens, Jamshedpur; Assam Agricultural University, Jorhat; IARI Regional Station, Kalimpong; C.S. Azad University of Agriculture and Technology, Kanpur; National Botanic Gardens, Lucknow; Government Botanical Garden, Ootacamund; G.B. Pant University of Agriculture and Technology, Pantnagar; College of Agriculture, Sabour; Fertilizer Corporation of India, Sindri; Bhabha Atomic Research Centre, Trombay; University of Udaipur, Udaipur; Banaras Hindu University, Varanasi and Botanical Survey of India, Yercaud.

During VI Five Year Plan, four new centres were started to strengthen the coordinated project at Rajasthan Agricultural University (RAU), Udaipur; Horticultural Experiment and Training Centre (HETC), Chaubattia (Ranikhet); Tamil Nadu Agricultural University (TNAU), Coimbatore and Orchid Sanctuary, Kalimpong (under BCKV, Kalyani). Under Tamil Nadu Agricultural University, Coimbatore, two sub-centres were operating at Horticultural Research Stations located at Kodaikanal and Yercaud. During VII Five Year Plan, the Coordinated Project was further strengthened with the addition of two more centres at Andhra Pradesh Agricultural University (APAU), Hyderabad and Sher-E-Kashmir University of Agricultural Sciences and Technology (SKUAST), Srinagar, to give better representation of the agro-climatic regions not covered by the project earlier. In order to cover the coastal region of eastern India, Regional Plant Resource Centre (RPRC), Bhubaneswar, was also included in the Coordinated Project during the VIII Plan period, which started functioning from 1.4.1996. During the VIII Plan, one existing centre, viz. Horticultural Experiment and Training Centre (HETC), Chaubattia, was dropped from the Coordinated Project. In 1997, one institute based centre has been established at National Research Centre for Orchids, Pakyong, Sikkim. One sub centre at Kodaikanal (under TNAU) was dropped in 1997.

During IX Plan, 3 new centres were approved to represent three more states, *viz.* G. B. Pant University of Agriculture and Technology, Pantnagar; Assam Agricultural University, Jorhat, (Assam) and Birsa Agricultural University, Ranchi (Jharkhand). The AICRP on Floriculture, has got 21 Centres at present which includes 15 budgetary, 4 Institutional and 2 Voluntary, the details of which are given in Table 6.1.

Table 6.1. List of coordinated centres of AICRP on Floriculture

Budgetary Centres	*Year of Starting*	*Crops Dealing*
Acharya N.G.Ranga Agricultural University, Hyderabad	1987	Gladiolus, chrysanthemum, tuberose
Assam Agrilcultural University, Jorhat	2000	Gladiolus, chrysanthemum, orchids, anthurium, tuberose, gerbera
Bidhan Chandra Krishi Viswavidyalaya, Mohanpur	1972	Carnation, orchid, tuberose, gerbera and anthurium
Birsa Agricultural University, Kanke, Ranchi	2000	Gladiolus, chrysanthemum, and gerbera
Dr Y.S.Parmar University of Horticulture & Forestry, Solan	1975	Gladiolus, carnation, chrysanthemum
G.B. Pant University of Agrilcultural & Technology, Pantnagar	2000	Rose, gladiolus, chrysanthemum, tuberose
Horticultural Research Station, Yercaud (TNAU)	1982	Gladiolus, orchid, anthurium, gerbera
Kerala Agricultural University, Vellanikkara	1975	Orchid, anthurium, gerbera
Mahatma Phule Krishi Vidyapeeth, Pune	1975	Rose, gladiolus, carnation, chrysanthemum, tuberose, gerbera
Orchid Sanctuary, Kalimpong (UBKV)	1985	Gladiolus, carnation, orchids, anthurium, alstroemeria
Punjab Agricultural University., Ludhiana	1975	Rose, gladiolus, chrysanthemum, tuberose, gerbera, lilium
Rajasthan College of Agriculture, Udaipur (MPUAT)	1980	Rose, gladiolus, chrysanthemum, tuberose
Regional Plant Resource Centre, Bhubaneswar	1994	Rose, chrysanthemum, orchid
Sher-e-Kashmir University of Agricultural Sciences & Technology, Srinagar	1987	Gladiolus, chrysanthemum, tulip, daffodils, lilium, alstroemeria
Tamilnadu Agricultural University, Coimbatore	1982	Chrysanthemum, orchid, anthurium

(Contd.)

Budgetary Centres	Year of Starting	Crops Dealing
Institute Based Centres		
Indian Agricultural Research Institute, New Delhi	1971	Rose. Gladiolus, carnation, tuberose
Indian Agricultural Research Institute, Regional Station, Katrain	1971	Gladiolus, carnation, gerbera, tulip, daffodils, lilium, alstromeria
Indian Institute of Horticulture Research, Hessaraghatta Bangalore	1971	Rose, gladiolus, carnation, chrysanthemum, orchids, anthurium, tuberose, gerbera
ICAR Research Complex for NEH Region, Umiam (Meghalaya)	1972	Orchid, anthurium, gerbera
Voluntary Centres		
National Botanical Research Institute, Lucknow	1971	Rose, gladiolus, chrysanthemum, tuberose
University of Agricultural Sciences, Bangalore	1977	Carnation, anthurium

The Coordinated Centres of AICRP on Floriculture are covering 17 states of India, *viz.* Andhra Pradesh (ANGRAU, Hyderabad), Assam (AAU, Kahikuchi), Delhi (IARI, New Delhi), Himachal Pradesh (Dr YSPUHF, Solan and IARI R.S., Katrain), Jammu & Kashmir (SKUAST.,Srinagar), Jharkhand (BAU, Ranchi), Karnataka (IIHR, Bangalore and UAS, Bangalore), Kerala (KAU,Vellanikkara), Maharashtra (MPKV,Pune), Meghalaya (ICAR Research Complex for NEH Region, Umiam), Orissa (RPRC,Bhubaneswar), Rajasthan (MPUAT,Udaipur) Tamil Nadu (TNAU, Coimbatore and HRS,Yercaud), Punjab (PAU,Ludhiana), Uttar Pradesh (NBRI,Lucknow), Uttaranchal (GBPUAT,Pantnagar) and West Bengal (BCKV, Kalyani & UBKV, Kalimpong) including 6 hill stations. This project covers 14 State Agricultural Universities of our country. These centres are presently working on 12 ornamental crops viz. rose, gladiolus, carnation, chrysanthemum, orchid, anthurium, tuberose, gerbera, lilium, alstroemeria, tulip and daffodils consisting of 64 different research projects.

Objectives

- Collection of germplasm of floricultural plants including the wild species from different parts of the country and abroad and its maintenance and assessment under different agroclimatic conditions.
- Development of new and improved varieties including F_1 hybrids.
- Standardisation of agro-techniques for the production of cut flowers, bulbs and plants for domestic and export markets.
- Standardisation of propagation techniques including embryo/tissue/ meristem culture for quicker multiplication.
- Development of suitable pre and post-harvest technology for cut flowers.
- Identification and control of major diseases and pests of the commercial flower crops.

Man Power

The strength of personnel working in the project has been increased from time to time depending on the work load assigned to coordinated centres. At present (X Five Year Plan), the sanctioned strength is 75, which includes 39 scientific, 11 technical, 1 administrative and 24 supporting staff (Table 6.2). There was an increase of 160% in the sanctioned staff position of X Plan over V Five Year Plan. The distribution of scientific personnel discipline-wise is presented in the Table 6.3.

Table 6.2. Staff position of AICRP on Floriculture

Plan	*Scientific*	*Technical*	*Others*	*Total*
IV	*	*	*	*
V	16	2	10	28
VI	29	8	21	58
VII	32	8	21	61
VIII	34	8	26	68
IX	39	11	25	75
X	39	11	25	75

* Staff was appointed only at ICAR based centres and BSI-Calcutta

Table 6.3. Distribution of scientific personnel of AICRP on Floriculture

Plan	*Breeding*	*Pathology*	*Physiology*	*Total*
IV	*	*	*	*
V	13	1	2	16
VI	26	1	2	29
VII	28	2	2	32
VIII	30	2	2	34
IX	34	3	2	39
X	34	3	2	39

* Staff was appointed only at ICAR based centres and BSI-Calcutta

Budget

The coordinated project was started with a very modest budget of Rs 7.83 lakh during IV Five Year Plan. This in the course of time was increased to 808.44 lakh during X Five Year Plan, which includes a state share of Rs 180.79 lakh. This is distributed under heads like Pay and Allowances (Rs. 608.89 lakh), Travelling Allowances (Rs. 34.02 lakh), Recurring Contingency (Rs. 120.15 lakh) and Non-Recurring Contingency (Rs. 45.38 lakh). The sanctioned budget during different Five Year Plans is presented in Table 6.4.

Table 6.4. Budget details of AICRP on Floriculture

Plan	*No. of Centres Including P.C. Cell*	*Amount (Rs lakh) ICAR Share*
IV	6	7.83
V	10	21.61
VI	14	68.04
VII	16	73.84
VIII	16	181.39
IX	23	420.00
X	22*	625.75
XI	21	1722.13

* NRC for Orchids, an Institutional Centre was dropped during X Five Year Plan

WORKSHOPS/GROUP MEETINGS

The All India Coordinated Research Project on Floriculture conducted a number of group meetings from time to time for the finalization of research programme for the next two years, keeping in view of the changing priorities and thrust areas of rapidly expanding floriculture industry. The first workshop was organized at IARI, New Delhi, in February, 1971 and the second at Agri-Horticultural Society of India, Calcutta in June, 1973. From 1973–77, there was no workshop/group meeting as there was no Project Coordinator in position. The list of workshops/group meetings held so far under the project is given in Table 6.5.

Table 6.5. The list of Workshop/Group Meetings of the AICRP on Floriculture

Workshop/Group Meeting	*Year*	*Date*	*Place*
I	1971	18–19 February	IARI, New Delhi
II	1973	29–31 January	AHSI, Calcutta
III	1978	16–18 February	NBRI, Lucknow
IV	1980	17–19 March	PAU, Ludhiana
V	1982	22–24 March	IARI, New Delhi
VI	1983	21–24 September	IIHR, Bangalore
VII	1985	19–22 March	RCA, Udaipur
VIII	1986	29 Sept.– 1st Oct.	MPKV, Pune
IX	1988	18–20 March	Dr. YSPUHF, Solan
X	1990	23–25, April	KAU, Vellanikkara
XI	1992	16–18, November	TNAU, Coimbatore
XII	1994	14–16, December	APAU, Hyderabad
XIII	1996	26–29 December	RPRC, Bhubaneswar
*BSS	1997	16 October	IARI, New Delhi
XIV	1999	6–9 December	MPKV, Pune
XV	2001	22–24 January	RPRC, Bhubaneswar
XVIGM-TP**	2003/2004	17–18 July, 29-30 April	UAS, BangaloreIARI, New Delhi
XVII	2005	12–14 July	Dr. YSPUHF, Solan
XVIII	2007	15–17 November	RCA, Udaipur

*Brain Stroming Session

**Group Meting for finalization of Technical Progrmme for 2003–05 Group Meeting on Crops

Orchid	1977	10-11, Oct	ICAR Research Complex for NEH Region, Umiam, Barapani (Meghalaya)
Carnation	1983	10-11 March	PAU, Ludhiana

RESEARCH PROGRAMME

The All India Coordinated Research Project on Floriculture was started operating with 60 projects on 25 crops during the IV Plan. The utility and efficiency of these programmes were reviewed critically during the end of each Five Year Plan and the number of crops was subsequently reduced to 15 and 5 during the VI and VII Plan periods, respectively. The number of research projects was reduced to 43 and 26, respectively, during the above plans for increasing the functional efficiency of the project. Since 1982, intensified research projects were taken up on 5 crops which were identified for their export potential, viz. rose, gladiolus, orchid, carnation and chrysanthemum. In view of the changed priorities, research work was also initiated during VII Five Year Plan on few seasonal flowers and bulbous plants. During the VIII Plan, two important commercial crops, *viz.* gerbera and anthurium were included in the Coordinated Project. During IX Plan, seasonal and bulbous plants were dropped and there were 8 crops namely rose, gladiolus, chrysanthemum, carnation, orchid, anthurium, tuberose and gerbera. In the present X^{th} Five Year Plan, work is planned on 4 more crops, viz. lilium, alstroemeria, tulip and daffodils which have great potential in northern hills. Thus, at present there are 64 research projects covering 12 crops.

Crop-wise Research Programme

The crop-wise research programme, which was allotted to different centres during the XVII Group Meeting held at Dr Y.S. Parmar University of Horticulture and Forestry, Solan, in 2005 is as follows.

Rose

- Collection, maintenance and assessment of germplasm under open field and protected conditions
- Breeding of new cultivars suitable as cut flower for export and for domestic market
- Integrated nutrient management under open and protected conditions (use of biostimulants, Panchagavya and Manchurian tea along with vermicompost)
- Micronutrient studies
- Pre-harvest technology through organic culture (use of biostimulants)
- Post-harvest technology (pulsing solutions, holding solutions and wet storage)
- Disease management (powdery mildew, black spot, *Botrytis* grey mould)
- Insect pest management (mites and thrips)

Gladiolus

- Collection, maintenance and assessment of germplasm
- Breeding and testing of new cultivars
- Integrated nutrient management (use of biostimulants, Panchagavya and Manchurian tea along with vermicompost)
- Pre-harvest technology through organic culture (use of biostimulants)
- Post-harvest technology (wet storage, pulsing and holding solutions)
- Disease management (*Fusarium* wilt, *Botrytis* blight).

Carnation

- Collection, assessment and maintenance of germplasm
- Integrated nutrient management (Polyhouse) (use of biostimulants, Panchagavya and Manchurian tea along with vermicompost)
- Pre-harvest technology (use of biostimulants)
- Post-harvest technology (pulsing solution, holding solution, wrapping material and wet storage)

- Disease and insect pest management (*Fusarium* wilt, *Helicoverpa* and mites).

Chrysanthemum

- Collection, assessment and maintenance of germplasm
- Breeding (for loose and cut flower, and pot culture) and testing (under open field and polyhouse) of new cultivars
- Integrated nutrient management (open and polyhouse) (use of biostimulants)
- Panchagavya and Manchurian tea along with vermicompost)
- Micronutrient studies
- Chrysanthemum blooming through photoperiod control
- Pre-harvest management (use of biostimulants)
- Post-harvest technology (wet storage, wrapping material, holding and bud opening solutions)
- Management of leaf spot/leaf blight disease.

Orchid

- Collection, evaluation and maintenance of orchid germplasm (Sympodial and monopodial)
- Integrated nutrient management (use of Panchagavya, Manchurian tea and vermiwash) (epiphytic & terrestrial)
- Standardization of biofertilizer doses
- Micronutrient studies
- Standardization of post-harvest technology (Pulsing solutions, holding solutions and lining material).

Anthurium

- Collection, evaluation and maintenance of germplasm
- Breeding and testing of new varieties
- Standardization of propagation techniques (field propagation and tissue culture)
- Integrated nutrient management (use of panchagavya, manchurian tea and vermiwash)
- Standardization of slow release fertilizers
- Standardization of micronutrient doses
- Standardization of biofertilizer doses
- Pre-harvest management through organic culture
- Post-harvest management (Pulsing and holding solutions)
- Control of bacterial blight and *Phytophthora* rot.

Tuberose

- Collection, evaluation and breeding of germplasm
- Standardization of agro-techniques (planting time and fertilizer schedule)
- Integrated nutrient management (use of bio stimulants panchagavya and manchurian tea

along with vermicompost)
- Pre-harvest management through organic culture
- Post-harvest technology (holding solutions and wrapping material)
- Integrated pest management (root knot and foliar)
- Management of diseases (*Sclerotium* wilt and leaf blight).

Gerbera
- Collection, evaluation and maintenance of germplasm
- Breeding and testing of new cultivars for cut flowers for domestic use and export
- Standardization of growing structures (effect of shading)
- Standardization of agro-techniques (spacing and nutrition)
- Integrated nutrient management (use of biostimulants, panchagavya and manchurain tea along with vermicompost)
- Pre-harvest technology through organic culture
- Post-harvest technology (holding and pulsing solutions)
- Integrated disease and pest management (leaf spot/leaf blight, foot rot/root rot, whitefly and leaf miner).

Tulip
- Collection, maintenance and assessment of germplasm
- Post-harvest studies.

Daffodils
- Collection, maintenance and assessment of germplasm
- Post-harvest management.

Lilum
- Collection, maintenance and assessment of germplasm
- Production technology for bulb formation.

Alstroemeria
- Collection, maintenance and assessment of germplasm.

Centrewise Research Programme

Acharya N.G. Ranga Agricultural University, Hyderabad

Gladiolus
- Collection, maintenance and assessment of germplasm
- Pre-harvest management through organic culture
- Post-harvest management (holding and pulsing solutions, and wet and dry storage).

Chrysanthemum
- Collection, assessment and maintenance of germplasm

- Testing of newly evolved varieties
- Integrated nutrient management
- Micronutrient deficiency studies
- Pre-harvest management through organic culture
- Post-harvest technology (wet storage, wrapping material, holding and bud opening solutions).

Tuberose

- Collection, evaluation and maintenance of germplasm
- Testing of newly evolved varieties.

Assam Agricultural University, Kahikuchi

Gladiolus

- Collection, maintenance and assessment of germplasm
- Pre-harvest management through organic culture
- Post-harvest management (wet storage, dry storage, pulsing solutions and holding solutions).

Orchids

- Collection, evaluation and maintenance of germplasm (tropical group)
- Standardization of post-harvest technology (pulsing and holding solutions, and lining material).

Anthurium

- Collection, evaluation and maintenance of germplasm
- Integrated nutrient management (organic cultivation, slow release fertilizers and micronutrients)
- Control of bacterial blight and *Phytophthora* rot of anthurium.

Tuberose

- Collection, evaluation and maintenance of germplasm
- Testing of newly evolved varieties
- Standardization of agro-techniques (planting time and fertilizer schedule)
- Pre-harvest management through organic culture
- Post-harvest technology (holding solutions and wrapping material)
- Management of diseases (*Sclerotium* wilt and leaf blight).

Gerbera

- Collection, evaluation and maintenance of germplasm
- Standardization of agro-techniques (spacing and nutrition)
- Pre-harvest technology through organic culture
- Post-harvest technology (holding and pulsing solutions)

- Disease and pest management (leaf spot / leaf blight, foot rot / root rot, whitefly and leaf miner).

Bidhan Chandra Krishi Viswavidyalaya, Kalyani

Carnation

- Disease management (*Fusarium* wilt).

Orchids

- Collection, evaluation and maintenance of orchid germplasm (sympodial- tropical group)
- Integrated nutrient management (use of Panchagavya, Manchurian tea and vermiwash) (epiphytic and terrestrial)
- Standardization of biofertilizer doses.

Anthurium

- Collection, evaluation and maintenance of germplasm
- Integrated nutrient management (use of panchagavya, manchurian tea and vermiwash)
- Standardization of slow release fertilizers
- Standardization of micronutrient doses.

Tuberose

- Collection, evaluation and maintenance of germplasm
- Standardization of agro-techniques (planting time and fertilizer schedule)
- Pre-harvest management through organic culture
- Post-harvest technology (holding solutions and wrapping material)
- Integrated pest management (root knot and foliar).

Gerbera

- Collection, evaluation and maintenance of germplasm
- Testing of newly evolved varieties
- Standardization of growing structures (effect of shading)
- Standardization of agro-techniques (spacing and nutrition).

Birsa Agricultural University, Kanke, Ranchi

Gladiolus

- Collection, maintenance and assessment of germplasm
- Breeding and testing of new cultivars

Chrysanthemum

- Collection, assessment and maintenance of germplasm.

Gerbera

- Collection, evaluation and maintenance of germplasm (only testing)
- Testing of new cultivars for cut flowers for domestic use and export.

Dr Y.S. Parmar University of Horticulture and Forestry, Solan

Gladiolus

- Collection, maintenance and assessment of germplasm
- Breeding and testing of new cultivars
- Pre-harvest technology through organic culture (use of biostimulants)
- Post-harvest technology (wet storage, pulsing and holding solutions).

Carnation

- Collection, assessment and maintenance of germplasm
- Integrated nutrient management (polyhouse) (use of biostimulants, panchagavya and manchurian tea along with vermicompost)
- Pre-harvest technology (use of biostimulants)
- Post-harvest technology (pulsing solution, holding solution, wrapping material and wet storage).

Chrysanthemum

- Collection, assessment and maintenance of germplasm
- Testing (under open field and polyhouse) of new cultivars
- Micro-nutrient studies
- Chrysanthemum blooming through photoperiod control.

Tulip

- Collection, maintenance and assessment of germplasm
- Post-harvest studies.

Daffodils

- Collection, maintenance and assessment of germplasm
- Post-harvest management.

Lilium

- Collection, maintenance and assessment of germplasm
- Production technology for bulb formation.

Alstroemeria

- Collection, maintenance and assessment of germplasm.

G.B.Pant University of Agriculture and Technology, Pantnagar

Rose

- Collection, maintenance and assessment of germplasm under open field and protected conditions.

Gladiolus

- Collection, maintenance and assessment of germplasm

- Breeding and testing of new cultivars
- Pre-harvest technology through organic culture (use of biostimulants)
- Post-harvest technology (wet storage, pulsing and holding solutions).

Chrysanthemum
- Collection, assessment and maintenance of germplasm.

Tuberose
- Collection, evaluation and breeding of germplasm (only testing)
- Pre-harvest management through organic culture
- Post-harvest technology (holding solutions and wrapping material).

Gerbera
- Testing of new cultivars for cut flowers for domestic use and export

Horticultural Research Station, Yercaud (under TNAU)

Gladiolus
- Collection, maintenance and assessment of germplasm.

Orchids
- Collection, evaluation and maintenance of orchid germplasm (sympodial and monopodial)
- Integrated nutrient management (use of panchagavya, manchurian tea and vermiwash) (epiphytic and terrestrial)
- Standardization of biofertilizer doses.

Anthurium
- Collection, evaluation and maintenance of germplasm
- Breeding and testing of new varieties
- Integrated nutrient management (use of panchagavya, manchurian tea and vermiwash)
- Standardization of slow release fertilizers
- Standardization of micronutrient doses
- Standardization of biofertilizer doses
- Pre-harvest management through organic culture
- Post-harvest management (pulsing and holding solutions).

Gerbera
- Collection, evaluation and maintenance of germplasm.

Kerala Agricultural University, Vellanikkara

Orchids
- Collection, evaluation and maintenance of orchid germplasm (sympodial and monopodial)
- Integrated nutrient management (use of panchagavya, manchurian tea and vermiwash) (epiphytic and terrestrial)

- Micronutrient studies
- Standardization of post-harvest technology (pulsing solutions, holding solutions and lining material).

Anthurium

- Collection, evaluation and maintenance of germplasm
- Breeding and testing of new varieties
- Standardization of propagation techniques (field propagation and tissue culture)
- Integrated nutrient management (use of panchagavya, manchurian tea and vermiwash)
- Standardization of slow release fertilizers
- Standardization of micronutrient doses
- Pre-harvest management through organic culture
- Post-harvest management (Pulsing and holding solutions)
- Control of bacterial blight and *Phytophthora* rot.

Gerbera

- Collection, evaluation and maintenance of germplasm
- Standardization of growing structures (effect of shading).

Mahatma Phule Krishi Vidyapeeth, Pune

Rose

- Assessment of germplasm under open field and protected conditions
- Integrated nutrient management under open and protected conditions (use of biostimulants, panchagavya and manchurian tea along with vermicompost)
- Pre-harvest technology through organic culture (use of biostimulants)
- Post-harvest technology (pulsing solutions, holding solutions and wet storage)
- Disease management (powdery mildew and black spot)
- Insect pest management (mites and thrips).

Gladiolus

- Collection, maintenance and assessment of germplasm
- Breeding and testing of new cultivars
- Integrated nutrient management (use of biostimulants, panchagavya and manchurian tea along with vermicompost)
- Pre-harvest technology through organic culture (use of biostimulants)
- Post-harvest technology (wet storage, pulsing and holding solutions)
- Disease management (*Fusarium* wilt).

Carnation

- Collection, assessment and maintenance of germplasm
- Pre-harvest technology (use of biostimulants)

- Post-harvest technology (Pulsing solution, holding solution, wrapping material and wet storage)
- Disease and insect pest management (*Helicoverpa* and mites only).

Chrysanthemum

- Management of leaf spot/leaf blight disease.

Tuberose

- Collection, evaluation and breeding of germplasm
- Standardization of agro-techniques (planting time and fertilizer schedule)
- Pre-harvest management through organic culture
- Post-harvest technology (holding solutions and wrapping material)
- Management of diseases (*Sclerotium* wilt and leaf blight).

Gerbera

- Collection, evaluation and maintenance of germplasm
- Breeding and testing of new cultivars for cut flowers for domestic use and export
- Standardization of agro-techniques (spacing and nutrition)
- Pre-harvest technology through organic culture
- Post-harvest technology (holding and pulsing solutions)
- Integrated disease and pest management (leaf spot/leaf blight, foot rot/root rot, whitefly and leaf miner).

Orchid Sanctuary, Kalimpong (under UBKV)

Gladiolus

- Collection, maintenance and assessment of germplasm.

Carnation

- Collection, assessment and maintenance of germplasm
- Integrated nutrient management (polyhouse) (use of biostimulants, panchagavya and manchurian tea along with vermicompost).

Orchids

- Collection, evaluation and maintenance of orchid germplasm (only sympodial)
- Micronutrient studies
- Standardization of post-harvest technology (pulsing solutions, holding solutions and lining material).

Anthurium

- Collection, evaluation and maintenance of germplasm
- Breeding and testing of new varieties
- Standardization of propagation techniques (field propagation)
- Integrated nutrient management (use of panchagavya, manchurian tea and vermiwash)

- Standardization of slow release fertilizers
- Standardization of micronutrient doses.

Alstroemeria
- Collection, maintenance and assessment of germplasm.

Punjab Agricultural University, Ludhiana

Rose
- Collection, maintenance and assessment of germplasm under open field and protected conditions
- Breeding of new cultivars suitable as cut flower for export and for domestic market
- Integrated nutrient management under open and protected conditions (use of biostimulants, panchagavya and manchurian tea along with vermicompost)
- Micronutrient studies (open)
- Pre-harvest technology through organic culture (use of biostimulants)
- Post-harvest technology (pulsing solutions, holding solutions and wet storage)
- Disease management (*Botrytis* grey mould).

Gladiolus
- Collection, maintenance and assessment of germplasm
- Breeding and testing of new cultivars
- Pre-harvest technology through organic culture (use of biostimulants)
- Post-harvest technology (wet storage, pulsing and holding solutions)
- Disease management (*Fusarium* wilt, *Botrytis* blight).

Chrysanthemum
- Collection, assessment and maintenance of germplasm
- Breeding (for loose and cut flower, and pot culture) and testing (under open field and polyhouse) of new cultivars
- Micronutrient studies
- Pre-harvest management (use of biostimulants)
- Post-harvest technology (wet storage, wrapping material, holding and bud opening solutions)
- Management of leaf spot/leaf blight disease.

Tuberose
- Collection, evaluation and breeding of germplasm
- Pre-harvest management through organic culture
- Post-harvest technology (holding solutions and wrapping material)
- Management of diseases (*Sclerotium* wilt and leaf blight).

Gerbera

- Collection, evaluation and maintenance of germplasm
- Testing of new cultivars for cut flowers for domestic use and export
- Integrated disease and pest management (leaf spot/blight only).

Lilium

- Collection, maintenance and assessment of germplasm
- Production technology for bulb formation.

Maharana Pratap University of Agriculture and Technology, Udaipur

Rose

- Collection, maintenance and assessment of germplasm under open field and protected conditions.

Gladiolus

- Collection, maintenance and assessment of germplasm
- Pre-harvest technology through organic culture (use of biostimulants)
- Post-harvest technology (wet storage, pulsing and holding solutions).

Chrysanthemum

- Collection, assessment and maintenance of germplasm
- Testing (under open field and polyhouse) of new cultivars.

Tuberose

- Collection, evaluation and breeding of germplasm (testing only).

Regional Plant Resource Centre, Bhubaneswar

Rose

- Collection, maintenance and assessment of germplasm under open field and protected conditions
- Integrated nutrient management under open and protected conditions (use of biostimulants, panchagavya and manchurian tea along with vermicompost)
- Pre-harvest technology through organic culture (use of biostimulants)
- Post-harvest technology (pulsing solutions, holding solutions and wet storage).

Chrysanthemum

- Collection, assessment and maintenance of germplasm
- Testing (under open field and polyhouse) of new cultivars
- Micro-nutrient studies
- Chrysanthemum blooming through photoperiod control
- Pre-harvest management (use of biostimulants)
- Post-harvest technology (wet storage, wrapping material, holding and bud opening solutions).

Orchids

- Collection, evaluation and maintenance of orchid germplasm (only sympodial)
- Integrated nutrient management (use of panchagavya, manchurian tea and vermiwash) (epiphytic and terrestrial)
- Standardization of post-harvest technology (pulsing solutions, holding solutions and lining material).

Sher-e-Kashmir University of Agricultural Sciences and Technology, Srinagar

Chrysanthemum

- Collection, assessment and maintenance of germplasm
- Testing (under open field and polyhouse) of new cultivars.

Gladiolus

- Breeding and testing of new cultivars
- Pre-harvest technology through organic culture (use of biostimulants)
- Post-harvest technology (wet storage, pulsing and holding solutions).

Tulip

- Collection, maintenance and assessment of germplasm
- Post harvest studies.

Daffodils

- Collection, maintenance and assessment of germplasm
- Post-harvest management.

Lilium

- Collection, maintenance and assessment of germplasm
- Production technology for bulb formation.

Alstroemeria

- Collection, maintenance and assessment of germplasm.
- Tamilnadu Agricultural University, Coimbatore.

Carnation

- Pre-harvest technology (use of biostimulants)
- Post-harvest technology (pulsing solution, holding solution, wrapping material and wet storage).

Chrysanthemum

- Collection, assessment and maintenance of germplasm
- Testing (under open field and polyhouse) of new cultivars
- Integrated nutrient management (open and polyhouse) (use of biostimulants panchagavya and manchurian tea along with vermicompost)
- Micronutrient studies.

Orchids

- Integrated nutrient management (use of panchagavya, manchurian tea and vermiwash) (epiphytic and terrestrial)
- Standardization of biofertilizer doses
- Micronutrient studies.

Anthurium

- Collection, evaluation and maintenance of germplasm
- Integrated nutrient management (use of panchagavya, manchurian tea and vermiwash)
- Standardization of slow release fertilizers
- Standardization of micronutrient doses
- Standardization of biofertilizer doses
- Pre-harvest management through organic culture
- Post-harvest management (pulsing and holding solutions).

Tuberose

- Collection, evaluation and maintenance of germplasm.

Indian Agricultural Research Institute, New Delhi

Rose

- Collection, maintenance and assessment of germplasm under open field and protected conditions
- Breeding of new cultivars suitable as cut flower for export and for domestic market
- Micronutrient studies (open).

Gladiolus

- Collection, maintenance and assessment of germplasm
- Breeding and testing of new cultivars.

Tuberose

- Collection, evaluation and maintenance of germplasm
- Standardization of agro-techniques (planting time)
- Integrated pest technology (root knot and foliar).

Indian Agricultural Research Institute, Regional Station, Katrain

Gladiolus

- Collection, maintenance and assessment of germplasm.

Carnation

- Collection, assessment and maintenance of germplasm
- Disease and insect pest management (*Fusarium* wilt, *Helicoverpa* and mites).

Gerbera

- Collection, evaluation and maintenance of germplasm.

Tulip

- Collection, maintenance and assessment of germplasm
- Post-harvest studies.

Daffodils

- Collection, maintenance and assessment of germplasm
- Post-harvest management.

Lilium

- Collection, maintenance and assessment of germplasm
- Production technology for bulb formation.

Alstroemeria

- Collection, maintenance and assessment of germplasm.

Indian Institute of Horticulture Research, Hessaraghatta, Bangalore

Rose

- Collection, maintenance and assessment of germplasm under open field and protected conditions
- Micronutrient studies (open)
- Disease management (powdery mildew, black spot)
- Insect pest management (mites and thrips).

Gladiolus

- Collection, maintenance and assessment of germplasm
- Breeding and testing of new cultivars
- Integrated nutrient management (use of biostimulants, panchagavya and manchurian tea along with vermicompost)
- Disease management (*Fusarium* wilt, *Botrytis* blight).

Carnation

- Collection, assessment and maintenance of germplasm
- Disease and insect pest management (*Fusarium* wilt, *Helicoverpa* and mites).

Chrysanthemum

- Collection, assessment and maintenance of germplasm
- Breeding (for loose and cut flower, and pot culture) and testing (under open field and polyhouse) of new cultivars
- Integrated nutrient management (open and polyhouse) (use of biostimulants panchagavya and manchurian tea along with vermicompost)

- Micro-nutrient studies
- Pre-harvest management (use of biostimulants)
- Post-harvest technology (wet storage, wrapping material, holding and bud opening solutions)
- Management of leaf spot/leaf blight disease.

Orchid

- Micronutrient studies.

Anthurium

- Collection, evaluation and maintenance of germplasm
- Breeding and testing of new varieties
- Standardization of propagation techniques (field propagation and tissue culture)
- Post-harvest management (pulsing and holding solutions).

Tuberose

- Collection, evaluation and breeding of germplasm
- Integrated pest management (root knot and foliar).

Gerbera

- Collection, evaluation and maintenance of germplasm
- Breeding and testing of new cultivars for cut flowers for domestic use and export
- Standardization of growing structures (effect of shading)
- Integrated nutrient management (use of biostimulants, panchagavya and manchurain tea along with vermicompost)
- Pre-harvest technology through organic culture
- Post-harvest technology (holding and pulsing solutions)
- Integrated disease and pest management (whitefly and leaf miner).

ICAR Research Complex for NEH Region, Umiam

Orchids

- Collection, evaluation and maintenance of orchid germplasm (Sympodial)
- Integrated nutrient management (use of panchagavya, manchurian tea and vermiwash) (epiphytic and terrestrial).

Anthurium

- Collection, evaluation and maintenance of germplasm
- Standardization of propagation techniques (field propagation and tissue culture).

Gerbera

- Collection, evaluation and maintenance of germplasm
- Testing of new cultivars for cut flowers for domestic use and export

- Standardization of growing structures (effect of shading)
- Pre-harvest technology through organic culture
- Post-harvest technology (holding and pulsing solutions).

National Botanical Research Institute, Lucknow

Rose

- Collection, maintenance and assessment of germplasm under open field and protected conditions
- Breeding of new cultivars suitable as cut flower for export and for domestic market.

Gladiolus

- Collection, maintenance and assessment of germplasm.

Chrysanthemum

- Collection, assessment and maintenance of germplasm
- Breeding (for loose and cut flower, and pot culture) of new cultivars
- Chrysanthemum blooming through photoperiod control.

Tuberose

- Collection, evaluation and breeding of germplasm
- Standardization of agro-techniques (planting time and fertilizer schedule).

University of Agricultural Sciences, Bangalore

Carnation

- Integrated nutrient management (polyhouse) (use of biostimulants, panchagavya and manchurian tea along with vermicompost).

Anthurium

- Standardization of biofertilizer doses.

SALIENT RESEARCH ACHIEVEMENTS

Development of Varieties

Rose

The IARI, New Delhi, developed 41 varieties of roses under HT group (Abhisarika, Anurag, Arjun, Bhim, Charugandha, Chitra, Chitralekha, Chitwan, Dr. B.P. Pal, Dr. Benjamin Pal, Dr. Bharat Ram, Ganga, Gulzar, Hans, Jawahar, Madhosh, Mother Teresa, Mridula, Mohini, Mrinalini, Nehru Centenary, Noorjehan, Preyasi, Priyadarshini, Pusa Bahadur, Pusa Christina, Pusa Garima, Pusa Gaurav, Pusa Manhar, Pusa Mansij, Pusa Priya, Pusa Sonora, Rajkumari, Raktagandha, Raktima, Shreyas, Soma, Sujata, Surabhi, Surekha and Vasant), 22 varieties under Floribunda group (Arunima, Banjaran, Chandrama, Deepika, Deepshikha, Delhi Princess, Kavita, Lahar, Manasi, Navneet, Nav Sadabahar, Neelambari, Prema, Pusa Veerangana, Pusa Barahmasi, Pusa Pitamber, Rupali, Shabnam, Shringar, Sindoor, Suchitra and Usha), and one climbing cv. Clg. Sadabahar.

The IIHR, Bangalore, developed Dr. G.S. Randhawa and Kiran under HT. Yercaud developed YCD-1, YCD-2 and YCD-3 Floribunda roses. NBRI Lucknow developed 10 HT roses, viz., Kronenberg, Light Pink Prize, Mrinalini Light Pink Mutant, Mrinalini Stripe, Pink Montezuma, Salmon Beauty, Summer Holiday Mutant, Winter Holiday Mutant, Sylvia White and Girija; 11 Floribunda rosés, viz., Angara, Curio, Pink Contempo, Pink Imperator, Salmon Beauty Mutant, Sharada, Sukumari, Tangerine Contempo, Twinkle, Yellow Contempo and Zorina Pink Mutant, one Miniature cv. Windy City Mutant; and one climbing rose cv. Clg. Cri.

Gladiolus

Centres of AICRP on Floriculture, released some 81 gladiolus varieties, viz. Aarti, Agnirekha, Anjali, Apsara, Archana, Arka Suvarna, Arka Kesar, KKL-1, Arun, Basant Bahar, Bindiya, Chandni, Chaubattia Ankur, Chaubattia Arunima, Chaubattia Shobhit, Chaubattia Tripti, Subhangini, Chirag, Darshan, Dhanvantari, Dhiraj, Gazal, Hans, Indrani, Jwala, Kalima, Kohra, Kum Kum, Manhar, Manisha, Manmohan, Mayur, Meera, Mohini, Mridula, Mukta, Nazrana, Neelam, Phule Ganesh, Phule Neelrekha, Phule Prerna, Phule Tejas, Pitambar, Poonam, Priyadarshini, Punjab Dawn, Pusa Archana, Pusa Suhagin, Pusa Gulaal, Pusa Gunjan, Pusa Jyotsana, Pusa Kamini, Pusa Lohit, Pusa Mohini, Pusa Rangmahal, Pusa Sukanya, Pusa Swapnil, Pusa Urmi, Pusa Urmil, Pusa Urvashi, Pusa Swarnima, Pusa Kiran, Pusa Shubham, Rashmi, Sadabahar, Sagar, Sanyukta, Sapna, Sarang, Shabnam, Shagun, Shakti, Shobha, Shringarika, Shweta, Sindhur, Smita, Suchitra, Sunayna, Suryakiran, Tambri, Triloki and Vandana.

Chrysanthemum

More than 150 chrysanthemum varieties have been developed by various AICRP centres (Lucknow, Ludhiana, Coimbatore, Hessaraghatta and Kalyani) and these are Agnishikha, Ajay, Alankar, Anamika, Appu, Apsara, Apurva singer, Aruna, Arka Ganga, Arka Ravi, Arka Suvarna, Arun Kumar, Arun Singar, Asha, Ashankit, Basant, Basanti, Basantika, Baggi, Bindiya, Birbal Sahni, Co.1, Co.2, Chandrakant, Chandrika, Colchi Bahar, Cosmonaut, Dhawal, Diana, Flirt, Gairik, Gauri, Gulal, Guldasta, Gul-e-Sahir, Haldighati, Hemant Singar, Hemanti, Himani, Himanshu, Indira, Jaya, Jayanti, Jhalar, Jubilee, Jugnu, Jwala, Jyoti, Jyotsana, Kalima, Kapish, Kanak, Kaumudi, Kansya, Khumani, Kirti, Kiran, Kum Kum, Kunchita, Kundan, KSL-16, Lal Kila, Lalima, Lalpari, Lilith, NBRI Little Darling, NBRI Mini Jessie, NBRI Mini Indiana, NBRI Kusum, Lohita, Maghi Brown, Maghi Pink, Maghi Yellow, Manbhavan, May Day, Mayur, Meghdoot, Mini Queen, Mohini, Mother Teresa, Navneet, Navneet Yellow, Neelima Nirbhaya, Nirbhik, Nirmal, Pancho, Pankaj, Peet Singar, PCO.1, PCO.2, Phuhar Pingal, Pitaka, Pitamber, Priya, Prof. Harris, Puja, Purnima, Ragini, Rakhee, Rangoli, Ravi Kiran, Red Gold, Rimjhim, Rohit, Santi, Shabnum, Sharad Har, Sharad Kranti, Sharad Kumar, Sharad Mala, Sharad Mukta, Sharad Sandhya, Sharad Shobha, Sharad Singar, Sharad Sheela, Shefali, Shveta, Shizuka, Shukla, Shyamal Red, Shymal White, Sonali, Subarna, Suhag Singar, Sujata, Swarna Singar, Swarnim, Sweta Singar, Tamra, Taruni, Tulika, Tushar, Usha, Vandana, Varsha, Vasantika, White Charm, White Prolific, Yellow Charam, Yellow Gold, Yellow Prolific, Yellow Star, etc.

Other Ornamentals

The IIHR, Bangalore, developed 2 promising hybrids, *viz.* IIHR-164 (purple) in *Vanda* and IIHR-38 (purple violet) in *Dendrobium* orchids. Coimbatore developed 3 varieties, viz.

Co.1, Co.2 and Parimullai under *Jasminum auriculatum,* and 2 varieties, viz. Co.1 Pitchi, and Co.2 Pitchi in *J. Grandiflorum.* The IIHR, Bangalore, developed one jasmine var. Arka Surabhi. In tuberose, IIHR, Bangalore, developed 4 varieties, *viz.* Prajwal, Shringar, Suvasini and Vaibhav; NBRI Lucknow 2 varieties, viz., Rajat Rekha and Swarna Rekha; and Pune centre one variety, Phule Rajni. In China aster, Pune centre developed Phule Ganesh White, Phule Ganesh Pink, Phule Ganesh Purple and Phule Ganesh Violet; and IIHR centre Kamini, Poornima, Shashank and Violet Cushion. In marigold, IARI centre released Pusa Narangi Gainda and Pusa Basanti Gainda. Apart from these, many hibiscus and bougainvillea varieties were developed by IIHR centre. NBRI centre also developed many bougainvilleas.

PRODUCTION TECHNOLOGY

Rose

Rosa indica var. *odorata* rootstock was found best for Coimbatore, Delhi, Lucknow, Ludhana and Pune conditions; *R. multiflora* and IIHR - Thornless for Bangalore, Hyderabad and Kalyani conditions; and *R. chinensis* (Titri) for Udaipur conditions.

Depending upon the varieties used, 30cm × 20cm spacing was found best for Super Star and Gladiator under Pune and Udaipur conditions; 30cm × 30cm spacing for Montezuma at Kalyani conditions; and 30cm × 40cm for Super Star under Delhi, and for Sonia Meilland under Udaipur conditions, as quality flower production was maximum there. The use of 650 kg N, 800 kg P_2O_5, 700 kg K_2O, 10 kg elemental sulphur, 50 kg each of $MgSO_4$/ha and $CaSO_4$/ha, and November and January spraying of 1% $FeSO_4$, 0.5% $ZnSO_4$ and 0.2% boron at Delhi condition for Super Star roses at 30cm x 30cm spacing, gave maximum yield. At Kalyani with cv. Montezuma, NPK 100:100:75 g/m^2; at Pune with cv. Gladiator, NPK 60:10:20 g/m^2; at Bangalore with cv. Happiness, NPK 50:100:75 g/m^2; and with the same variety at Yercaud, NPK 75:150:50 g/m^2 gave maximum flower yield.

For commercial cultivation of rose, NPK 400:200:200 ppm/plant/week at 'run off' stage is recommended. Coimbatore centre commended NPK [$(NH_4)_2$ NO_3, H_3PO_4 and KNO_3] 150:40:150 for cv. First Red under protected conditions. At Ludhiana, 60:40:40 g/m^2 NPK with cv. Super Star, is recommended.

Delhi centre found 10cm pruning from 23 September to 1 October in rose cvs. Queen Elizabeth and Happiness, and 45cm leaving 4 shoots, from 13 October to 17 November in cv. Super Star as the optimum.

Against weeds, Oxyflurofen at 0.5kg a.i./ha drenching at Kalyani, and 1kg a.i./ha at Ludhiana; Oxyflurofen (1kg a.i./ha) or Diuron (2 kg a.i./ha) at Delhi in cvs. Super Star and Raktagandha, and Diuron (2.5 kg a.i/ha) or Simazine (3 kg a.i./ha) or Round Up' (1 kg a.i./ha) at Yercaud and Bangalore conditions, had been found quite effective.

Gladiolus

Pune centre with cv. Sancerre at 25cm × 10cm spacing and Hessaraghatta, Kalyani and Ludhiana centres found 4.5-5.0cm gladiolus corms at 20cm × 20cm spacing and 8cm depth quite effective.

For weed control in glaldiolus field, application of Basalin or Stomp @ 1.87 l/ha, followed by hoeing 75 days after planting, proved quite effective.

NPK application of 45-50: 10-30: 10-20 g/m^2 along with 10kg FYM or 300:150:200 kg/ha has been recommended for different regions of the country.

Carnation

Under polyhouse conditions in Solan, when cvs Impala, Super Star, Veleta and Fantasia were supplied with 4 hour extended lighting, induced early flowering with maximum floral stalk. Also they recommended April planting resulting in earlier flowering (160 days) while October planting delayed it (213 days) but gave better stalk length and floral diameter. Also they recommended 'pinch and a half method' for earlier and regulated flowering. Planting at 15cm × 15cm spacing (4,44,444 plants/ha), twice pinching (40 and 60 days after transplanting) and application of NPK @ 30-40:20:10 g/m^2 gave highest yield with nitrogen in 3 splits (basal, one month after and two months after transplanting).

Chrysanthemum

Solan centre developed the technology for year round flower production through staggered planting, pinching, growth retardants, photoperiod regulation and cultural practices.

Orchids

A potting medium consisting of charcoal, brick pieces and coconut fibre in equal parts for epiphytic orchids like *Aerides multiflorum* and *Dendrobium moschatum*; loam soil, river sand, leaf mould, charcoal dust and old mortar (1:1:1:0.5:0.5) for terrestrial orchids like *Cymbidium aloifolium and Phaius tankervillae*; and brick pieces, coir dust and charcoal for *Vanda* spp. have been found quite suitable. In epiphytic orchids, *viz. Aerides multiflorum* and *Phalaenopsis amabilis,* hardwood charcoal chunks were found best as compared to tree fern fibres, over burnt brick pieces and coconut husk while *in Cattleya bowringiana and Dendrobium moschatum,* tree fern fibre was found best.

Isabgul as gelling agent and polyprophylene bags as culture vessels gave outstanding response for orchid micropropagation. In *Cattleya bowringiana and C. trianni*, long day treatment suppressed complete flowering while supplementary illumination (16 h day length) induced earlier flowering by 47–59 days in *Dendrobium, Phalaenopsis amabilis* and *P. Schilleriana.*

A fertilizer solution containing potassium nitrate (2.63 g), ammonium sulphate (0.44 g), magnesium sulphate (2.04g), mono-calcium phosphate (1.09g), calcium sulphate (4.86g), ferrous sulphate (0.50g) and manganese sulphate (2ml, of 1% solution) in 4.5 l water at 7 days interval improved growth and flowering in *Dendrobium moschatum, Aerides multiflorum, Vanda* spp. and *Dendrobium* cv. Sonia. For commercial cultivation of *Dendrobium* cv. Sonia-17,0.2% spraying twice weekly of NPK (20:30:30) is recommended. In *Rhyncostylis gigantea*, 500 ppm each of N. P, K, and in *Aerides multiflorum* and *Dendrobium moschatum,* 1000 ppm N and 500ppm each of P and K at fortnightly interval spraying resulted in better growth and flowering. At Yercaud conditions, in *Epidendrum radicans* and *Coelogyne* spp., VAM applied near root zone at the time of planting and foliar application of NPK (10:5:10) @ 0.2% gave best results.

Anthurium

Top cuttings, nodal cuttings and suckers can successfully be propagated on 100% cocopeat. Spadix explant could successfully be established after $HgCl_2$ 0.1% for 10 minutes surface sterilization and culturing on MS medium supplemented with 2, 4-D (2 mg/litre) and kinetin (0.3 mg/litre) for organogenesis at Vellanikkara.

Shading at 75% level responded best. NPK at 30:10:10 or 30:20:20 at 0.2% + GA_3 200ppm + *Azospirillum* + VAM is recommended for commercial cultivation of anthurium.

Tuberose

The varieties Prajwal and Vaibhav have been recorded best at most of the centres, however, the performance of single type tuberose is recommended for cultivation at different locations.

Planting of medium (2.6-3.0cm) and large bulbs (3.0–3.5cm) at 20cm × 20cm has been recorded best. However, for early flowering and more flower yield, 45 days rest period is suggested.

Gerbera

Pune centre recorded 2-row planting of gerbera cv. Sangria at 30cm x 20cm spacing under naturally ventilated or tunnel type polyhouses as best though under open conditions, 30cm × 30cm spacing with 10g/m² NPK is recommended. Srinagar centre recorded 15g N, 15g P and 10g K/m² as best at 30cm × 20cm spacing. Pune centre also recommended 100 ppm N, 40 ppm P_2O_5 and 150 ppm K_2O per day along with 10kg FYM/m²/year in gerbera cv. Golden Gate under a naturally ventilated polyhouse. They also recorded highest number of flowers in var. Ornella under polyhouse with 15g N, 20g P and 20g K/m².

Jasmine

Terminal nodal cuttings in *Jasminum grandiflorum,* and terminal and semi-hardwood in *J. sambac* and *J. auriculatum* are best suited for mist propagation in the sand medium. Most species rooted well when cuttings contained 4 leaves, treated with 4000 ppm IBA and planted in vermiculite. April—September was the best period for taking the cuttings.

Pinching of new shoots produced after pruning delayed flower bud formation by 14 days in *J. Auriculatum* and 17 days in *J. grandiflorum.* Pruning of *J. grandiflorum* in third week of December at 90 cm height retaining 10 shoots was found best at Hessaraghatta. However, instead of pruning, chemical defoliants, *viz.* PCP (pentachlorophenol) 3000 ppm or potassium iodide 4000 ppm were found best. In *J. grandiflorum*, cycocel at 1000 ppm twice (one in the first week of February and second in first week of March) increased flower production *vis-a-vis* essential oil content.

Spacings of 1.5m × 1.5m for *J. grandiflorum var.* White, 50cm × 50 cm for *J. sambac* and 75cm × 75cm for *J. auriculatum* were found optimum. However, 1.8m × 1.8m spacing (3,086 plants/ha) was found most economical in case of *J. grandiflorum.*

Other Crops

In China aster, 30cm x 30cm spacing and NPK 30:30:30 g/m² or 300:200:200 kg/ha has been reported best. In marigold, 40cm × 30cm (83,333 plants/ha) for African and 15cm × 15cm for French marigold with a NPK dose of 30:20:20 g/m² is recommended. Pinching at 40 and 60 days after transplanting is also recommended.

Post Harvest Management

Rose

Storage of rose cut flower at 4–8 °C for 24 hour followed by simulated transit has been recommended by Ludhiana centre. Delhi centre recommended pre-cooling at 4 °C for 24 hour for reduction of respiration rate and increasing of vase life of cut roses. Wet stored cut roses, irrespective of the cultivar and temperature, showed increased fresh and dry weight.

Ludhiana centre with Raktagandha cut roses, recommended 3% sucrose + 200 ppm 8-HQC + 200 ppm citric acid. Pune found increased vase life of 6.85 days in Gladiator rose when pulsed with 3% sucrose + 300 ppm aluminium sulphate. Further, Pune centre found increased vase life with pre-transit treatment of 3% sucrose + 300 ppm citric acid + 1mM silver thiosulphate. Pulsing of Raktagandha roses with 3% sucrose for 18 h at 20 °C, extended the vase life. Ludhiana, Lucknow and Pune found enhanced vase life in cut roses when pulsed with 300 ppm aluminium sulphate + 3% sucrose, followed by holding solution treatment of 300 ppm aluminium sulphate and 1.5% sucrose.

RESEARCH ACHIEVEMENTS (CENTRE-WISE)

Kerala Agricultural University, Vellanikkara

During X Plan, centre has collected and maintained 115 orchids consisting of wild species of 20 genera, 93 anthurium and 33 gerbera.

Orchids

Both sympodial and monopodial orchids were found to perform well under Vellanikkara conditions. Earliness and efficiency of spike production was highest in 25% double shade level and had maximum life in vase. But number and breadth of individual floret was maximum in 50% single shade level.

Potting media trials carried out using two dendrobium hybrids, Emma White and Sonia Bom Jo, indicated that in Emma White, media containing tile bits + coconut husk + charcoal recorded maximum weight and number of shoots, maximum spike production, length of spike and number of florets, whereas tile bits recorded maximum number of leaves and internodal length, maximum spike production and produced longest spike with maximum longevity on the plant in the Sonia Bom Jo. Largest flowers were produced in charcoal in Sonia Bom Jo and charcoal + tile bits in Emma White.

In the integrated nutrient trial involving Manchurian tea, panchagavya and vermiwash along with half the recommended dose of fertilizer, alone and in combination, plants sprayed with full dose of recommended fertilizers and those with half dose of recommended fertilizers along with manchurian tea, vermiwash and panchagavya recorded better vegetative growth as indicated by stem length, stem girth and number of leaves.

Biofertilizer trial indicated that NPK@ 20:10:10 (0.2% spray) along with Azospirillum recorded maximum values for all floral characters. Post-harvest longevity of flowers of Dendrobium Sonia 17 was extended to 21 days by the application of nutrient solution (0.2%) containing P and K (10:30:30 and 10:20:20), compared to 15 days in 20:20:20 and 17 days in 30:10:10.

Renappa had longest vase-life after pulsing. In this treatment, wilting of first floret started after 20 days (pulsing with triadimefon 10ppm). Last floret was retained up to 44 days by BA 50ppm. In Sonia 28, maximum extension of longevity was by 10% sucrose (14 days). BA 100ppm extended senescence up to 32.5 days, compared to 18.5 days in control. Triadimefon 20ppm (20 days as against 9.5 days) was more effective than other treatments. The vase-life could be extended from 15 days (control) to 32 days by holding in a solution containing 8

HQC 200ppm + sucrose 6 per cent. In treatments involving BA complete senescence of inflorescence was delayed considerably. Maximum vase life was obtained by covering with butter paper and keeping $KMnO_4$ in side cardboard boxes.

Anthurium

In grown-up plants, only etiolated plants produced two suckers and others did not respond. When top cutting was used as propagule, the treatments IBA (1000ppm) and charcoal were found to produce two suckers each. All the treatments given in the basal portion of top cut plants produced suckers, kinetin (200ppm), giving maximum number of suckers. In stem splits, all treatments produced one sucker, each.

In trial with different planting media, plants in sand + coconut husk came to flowering earlier in both Tropical and Gino Orange. Spike production was the maximum for this treatment, in Tropical and in coconut pith in Gino Orange, indicating that spike production is correlated with leaf production. The longest spike and maximum flower size were recorded for the same media in Gino Orange. The treatment coconut pith also gave maximum vase-life.

Trial using organic supplements, viz. panchagavya, manchurian tea and vermiwash indicated that height of the plant, spread of plant, length and width of leaf, as well as length and girth of petiole were highest in with common basal dose and common basal dose + 3% panchagavya. Leaf production was more with common basal dose + 3% manchurian tea + 3% vermiwash and T8 (common basal dose + 3% manchurian tea + 3% vermi wash + panchagavya). Sucker production was seen slightly improved by manchurian tea and panchagavya (T4) application.

Tuberose

Tuberose, viz. Local, Single, Double, Suvasini, Shringar, Hyderabad Single and Hyderabad Double evaluated had good flowering. 'Double', March-April as well as October-November plantings gave good vegetative growth in terms of plant height, number of leaves and leaf length. Floral characters like length of spike, rachis length, size of flowers, number of flowers, duration of flowering and vase life were also maximum in these months of planting.

Silver nitrate 25ppm + sucrose 2% gave maximum vase-life (8.7 days) which is significantly superior to all other treatments except citric acid 300ppm + sucrose 2%.

Gerbera

Flowering was better in gerbera Yanara, Essandre, Juvena, Tamara, Lindessa, Estonil, Dana Ellan, Goliath, Macmara Pink, Sazou, Snow White and Avant Garde.

China aster

China Aster, viz., Ostrich Blume Mixed, Princess Mixed, California Mixed, Crego Giant Pink, Crego Giant White, Princess Pink, Crego Giant Blue and Crego Giant Violet, tested at this centre, had no significant difference in flower production. However, most attractive flowers were those of Crego Giant Violet.

Marigold

Variety NBG was found to be best in weight of flowers, size and seed yield. The spacing in *Tagetes patula* (French marigold) revealed that maximum number of flowers was under closer spacing (10cm × 20cm). For maximum height, spread and number of flowers, planting during rainy season (July—September) was found to be the best.

Rose

The flower production was maximum for wild rose, followed by B.R. Kent. There was no significant difference among the varieties with respect to flower size and weight of flowers.

Tuberose

The nutritional trial revealed that different levels of NPK tried could not exert any influence on flower size. When bulb size increased, flower size also increased. However, flower yield was not directly related to bulb size.

Hibiscus

Performance of four varieties of hibiscus, viz. two types of *H. schizopetalus* and one type of *H. mutabilis* collected from different zones of Kerala, Tamil Nadu and Karnataka were evaluated. Considerable variation was found both in vegetative and floral characters.

Rooting in stem cutting was assessed under two treatment conditions; with leaves and without leaves. Retention of leaves had significant influence.

In quick dip method the best rooting was recorded by NAA 3000 mg/litre, followed by IBA 5000 mg/litre and IAA 10000 mg/litre. In prolonged dip method, cuttings treated with NAA 50 mg/litre, IBA 75 mg/litre and IAA 100 mg/litre were better. Of the two methods, quick dip method was significantly superior.

Distinct variation was observed in the morphology of flowers in the 36 types/ varieties or species studied. Pollen production per anther varied from 87 to 500 and pollen fertility from 4.6–97.4%. *Hibiscus mutabilis* set fruits naturally. Self-compatibility was noticed in this.

Bougainvillea

The germplasm collection included 49 cultivars of bougainvillea. Among these, Jayalakshmi, Thimma (variegated), Mary Palmer, Mahara Crimson, Cherry Blossom, Blondy and Louis Wathen flowered profusely and could be propagated easily by cuttings. Rose Usha and Dr. H.B. Singh were slow growing and could be recommended for pot culture.

Dipping the cuttings in IBA 500 mg/litre solution for a period of six hours gave maximum percentage of rooting, both in rainy and summer season.

The propagation of shy rooting types can be done successfully by micropropagation techniques. The rooting of shoot apices of Mahara substantially increased to above 85% under *in vitro* conditions, compared to under field conditions.

Jasmine

The germplasm collection included 21 genera/species/varieties. In terms of flower number, flower weight, ease of propagation, etc. *Jaminum auriculatum* CO-1 and *Jasminum sambac* Gundumalli were found to be best.

Wider spacing of 1.5 m × 1.5 m in *Jasminum grandiflorum* var. White gave higher yield of flowers in terms of number and weight. Flower production in *Jasminum auriuclatum* was maximum when plants were pruned at a height of 40cm from ground level, 45 days after cessation of flowering. Rooting of cuttings of *Jasminum auriculatum* was almost 10 times more under mist condition. Among the growth regulators tried, IBA 3000 mg/litre under mist was significantly superior to all other treatments.

Under Kerala conditions, the best season for simple layering was from May—September.

Gladiolus

Among four varieties of gladiolus tested at this centre, Nazrana, with its better number of spikes, weight and length of spike and number of florets per spike, was most promising.

The earliest flowering was observed in Beverly Ann (81.2 days) and duration of flowering was maximum (10 days) for Agnirekha. Vase-life was longest (9 days) for Peach Blossom.

All the hybrids from Kalyani Centre performed well under Vellanikkara conditions. Earlier flowering (64.33 days) was noticed in Hybrid 88/50 with 12 florets/spike and maximum vase life (8.4 days). The Hybrids 88/35 had maximum number of florets/spike (13.4) and minimum vase life (6.8 days).

In general, floral characters were superior in the varieties American Beauty and Friendship, whereas, vase life was the maximum in Agnirekha and Manoser Red. True Yellow produced heavy corms. Of the two seasons tried, November planting was found to be better than April planting, in order to get quality spikes as well as good corm and cormel yield.

Orchids

Crossing was continued in *Dendrobium* hybrids, and pod set was observed on flowers pollinated on third and fourth days after flower opening. Pod setting was maximum (91.00%) in cross Candy Stripe × Kanchana Green. When *Dendrobium moschatum* and *D. densifolium* were used as male parents, pod setting could be obtained in all the varieties (Pink Tips, Emma White, Hieng Beauty, Candy Stripe, Kanchana Green and New Pink), but the response of Hieng Beauty was poor (20% pod setting).

Vacin and Went and half MS were found to be good media for *in vitro* seed germination and further seedling development of *Dendrobium* hybrids.

When the protocorms developed from the selfed seeds of the embryos of *Dendrobium* hybrid 'New Pink' were subjected to gamma irradiation, considerable differences could be noticed in the nature of response among treatments. A very high rate of protocorm formation and plantlet differentiation was observed in the treatment with 25 Gy (2.5 KR) gamma rays. Techniques for clonal multiplication of *Phalaenopsis* through *in vitro* methods was standardized for basal portion of *in vitro* produced plantlets and inflorescence segments of field grown plants. Addition of coconut water at 15% level favoured (Protocorm like bodies) formation. When PLB's of *Phalaenopsis* were subjected to gamma irradiation, a very high rate of PLB formation and plantlet development could be noticed in 10Gy (1.0KR) treatment.

Anthurium

The effect of holding solutions containing $AgNO_3$, BA, CCC and sucrose on post-harvest life of cut flowers of anthurium var. Agnihotri was studied. BA at 25 mg/litre recorded maximum vase-life (22 days).

Mahatma Phule Krishi Vidyapeeth, Pune

Agro-techniques for cut flowers of rose, gladiolus, aster and tuberose under open field conditions and later on rose and gerbera under protection were standardized. Simultaneously, collection of germplasm in important flower crops was also undertaken.

Only one polyhouse measuring $400m^2$ formed the initial protected infrastructure for cultivating flower crops. Four polyhouses covering an area of 1500 m^2 were erected subsequently in November, 1997 for conducting research on agro-techniques on rose and gerbera.

Tuberose hybrid GK-T-D-2, with longer rachis length, appropriate distribution of florets, more floret weight and longer vase-life has been recently released Phule Rajani. This is a unique tuberose cultivar and highly appreciated by the floricultural growers of Maharashtra. A number of tuberose hybrids for specific purposes like loose flower production, cut flower, as pot plants, etc., have been identified and will be released for commercial cultivation shortly.

The AICFIP, Pune centre is one of the leading centres undertaking research in floricultural crops in the country. Apart from release of nine cultivars, a total of 44 need-based recommendations on various aspects of cultivation of floricultural crops have been given by the centre since inception.

About 22 varieties of rose, 54 of gladiolus, 8 of carnation, 51 of chrysanthemum, 10 of tuberose, 46 of gerbera and 4 of china aster were maintained at centre.

Five varieties including 4 gladiolus, i.e., Phule Ganesh, Phule Prerna, Phule Neelrekha and Phule Tejas and tuberose variety Phule Rajani were released during X Plan period.

Rose

Twelve days refrigerated storage of cut roses of First Red is recommended.

For rose cultivation under polyhouse conditions 10 kg FYM/m^2 /year as basal soil application and; 400mg nitrogen, 200mg phosphorus and 200 mg potash, per plant per week through drip is recommended for production of more number of flowers and higher monetary returns.

For cultivation of rose under polyhouse conditions red soil + compost + sand (2:1:1) as media; 40g N, 32g P_2O_5 and 60 g K_2O per m^2 per year through straight fertilizers as fertilizer dose, of which 20g N, 32g P_2O_5 and 30g K_2O per m^2 per year as basal dose and remaining 20g N and 30g K_2O per m^2 per year through fertigation (in 52 equal splits applied weekly through out the year) and nutrient sprays of water soluble fertilizer (19:19:19 % NPK) @ 2 g/ l of water once in a month from September to January is recommended.

For the control of black leaf spot disease of rose spraying the crop with 0.05 % Propiconazole or 0.2% Captan or 0.2% Copper-oxy-chloride + 0.1% Sticker, six times at an interval of 10 days during rainy season is recommended.

Under protected conditions it is recommended to give five sprays of 0.05% Difencanozole or 0.05% Tridemorph or 0.05% Dinocap at eight days interval, starting the first spray immediately at the disease appearance for effective control of powdery mildew, improving number of quality flower and increasing monetary returns in rose.

Pulsing treatment of Al_2 (SO4)3 (300ppm) + Sucrose (3%) is recommended for extended the vase life of cut flowers of rose cv. First Red (7.73 days).

Holding solution of Sucrose (1.5%) + Citric Acid (300ppm) having pH 5.6 is recommended for increasing vase life of rose flower cv. First Red (8.16 days).

Twelve days refrigerated storage of cut roses cv. First Red is recommended.

Chrysanthemum

Varieties Indira, Sonali Tara and Mutant No. 9 were found suitable for loose as well as cut flower production. It is recommended to give six sprays of 0.2% Chlorothalonil at 15 days interval, starting the first spray immediately at the disease appearance, for effective control of leaf spot/leaf blight disease, increasing flower yield and monetary returns.

Wrapping the flowers of cv. Sonali Tara in cellophane paper and holding for 24 hours at room temperature (20-25 °C) is recommended for long distance transport (6.3 days).

Gladiolus

The gladiolus cultivars Suchitra, Sapna, Tropic Seas and Pusa Suhagin were found superior in respect of floret production (16, 10, 12, and 13 florets/spike) and spike length (92.5, 89, 93.2, and 93.2 cm). They have high rate of corms and cormel production and are recommended for commercial cultivation.

For the control of *Fusarium* wilt disease of gladiolus, it is recommended to dip the corms and cormels twice, i.e., before storage and before planting, for 15 min in the fungicidal solution, followed by drenching twice at monthly interval with 0.3% Captan or 0.3% Thiram. To control the storage rot of gladiolus corms, it is recommended to dip the corms in 0.2 % Mancozeb or 0.3% Thiram or 0.2% Benomyl for 15 min after harvest, drying them in shade and storing in a cold storage. Pulsing treatment of 10% sucrose + 200 mg/litre 8-HQC for 24 hours is recommended for extending the vase life of gladiolus *cv.* Sancerre.

Wrapping the gladiolus spikes of cv. Sancerre in cellophane paper and holding for 24 hours at room temperature (20-25 °C) is recommended for long distance transport (7.42 days).

Tuberose

For production of more flower stalks with maximum florets of tuberose, planting of bulbs in the month of April is recommended. For higher flower production and monetary returns of tuberose, application of 30 t FYM, 200 kg nitrogen, 150 kg phosphorus and 200 kg potash per hectare is recommended. Half dose of nitrogen with full dose of phosphorus and potash should be applied at the time of planting and remaining half dose of nitrogen be applied 45 days after planting.

Wrapping the tuberose stem cv. Shringar in polythene sleeves (50 gauge) or brown paper for increasing the vase life (6.50 and 5.62 days) and holding for 24 hours at room temperature (20–25 °C) is recommended for long distance transport.

Gerbera

For cultivation of gerbera under polyhouse conditions, application of 10 kg FYM/m^2/year as basal dose and 20:20:15g NPK/m^2 month is recommended for production of more number of flowers cv. Sunway. The vase solution of Silver Nitrate 25 ppm or Citric Acid 300 ppm or Aluminum Sulphate 300 ppm is recommended for enhancement of vase life of gerbera flowers (8.50, 7.70 and 6.13 days respectively) cv. Sunway.

China aster

In China aster *cv.* Ostrich Plume Blue, the planting at 30cm × 20cm spacing in November is recommended to obtain more yield of flowers (33.13 flowers/plant) and seed (122.26kg/ha). For China aster, a fertilizer dose of 400 kg N + 100 kg P_2O_5 + 200kg K_2O/ha was found optimum for more yield of flowers.

Carnation

In carnation, *cv.* Scania recorded higher yield, better stem length and bigger flower size and has been recommended for cultivation.

In carnation, sand was found to be better medium for profuse rooting of the cuttings.

Dr Y.S. Parmar University of Horticulture and Forestry, Solan

The centre was created during 1975. Solan centre is situated at an altitude of 1276m amsl, latitude of 770 11' 30'' E and longitude of 300 52' 00'' N. The climate is sub temperate, semi humid characterized by cold winters. These climatic conditions are conducive for the growth and flowering of crops like gladiolus, carnation, chrysanthemum and bulbous crops like Lilium, Alstroemeria, Tulip and Daffodils. The centre is actively engaged in the improvement of these crops; genetically as well as in the cultural practices.

Rose

Various cultivars of rose species, climbers and ramblers were evaluated and cultivars like Cuddles, Rose Mantle, Black Jade, Angle Face and Paradise have been found promising under Solan area. The root-stock recommended are *Rosa indica* var. odorata and *Rosa multiflora.*

Gladiolus

Varieties like Nova Lux, Wind Song, White Prosperity, Apollo, Mayur, American Beauty, and Priscilla have been recommended for cultivation.

Carnation

Cultivars like Red Corso, Espana and Pamir were recommended for cultivation. From the trials after 1998 cultivars like Impala and Super Star, among the standards and Purple Choppin, Fantasia and Metro among the sprays were recommended.

Chrysanthemum

Ajina Purple, Ashoka, Baggi and Mountaineer, have been recommended for commercial cultivation. Ravi Kiran and Yellow Gold gave good results. Round the year flowering of Ajay has successfully been achieved with planting in April, June, August and October.

Total germplsm collection consists of 30 cultivars standard type and rest spray types. Out of this the promising Standard. Like Sonar Bangla, Snow Ball, Mountaineer, Tata Century, Thai Ching Queen, Purnima, Fiji, Fiji Yellow, are recommended for cultivation and among spray types the important recommended are Surf, Nanako, Birbal Sahni, Baggi, Ajay, White Bouquet and Kundan.

Alstroemeria

Three cultivars, Alladin, Serena and Pluto, were collected.

Narcissus

About 23 cultivars/species were collected and evaluated.

Lilium

About 4 Asiatic hybrids namely, Prato, Elite, Alaska and London were collected.

Punjab Agricultural University, Ludhiana (1975)

Germplasm collection of 183 rose varieties including 116 Hybrid Tea, 57 Floribuds, 5 Miniature, 3 Polyantha and 2 Shrub roses were maintained. About 62 varieties of gladiolus, 253 of chrysanthemum, 13 of tuberose and 7 of gerbera were collected.

Rose

One Hundred and ten varieties of rose were collected and evaluated. In general floribudas performed better than H.T. Out of varieties evaluated so far, most outstanding varieties were observed as Montezuma, Happiness, Queen Elizabeth, Raktagandha, Christian Dior, American Heritage, Kiss of Fire, Papa Meilland, Meilland, Iceberg, Pascali and McGredy Sunset.

Three following varieties were released for commercial loose flower production in Punjab.

- ***Baggi:*** A mid-season variety with decorative type snow-white flowers
- ***Birbal Sahni:*** An early variety with pompon type snow-white flowers
- ***Ratlam Selection:*** A mid-season variety with decorative type creamish- white flowers

Hybrids Developed

Punjab Gold

One attractive hybrid namely, Punjab Gold (Flirt × Gule Sahir) was selected. It is a dwarf type and suitable for pot culture. Flowers are semi-double, golden yellow in colour but red at the time of opening.

Punjab Anuradha

It is a double korean hybrid (Flirt × Gule Sahir) hybrid; flowers with a tinge of coppery red at bud stage changing to golden yellow at full bloom; suitable for pot culture.

Post-Harvest Technology

Cut stems of Birbal Sahni packed in corrugated paper and subjected to simulated transit for 24 hours showed marked increase in vase life when treated with preservative solution (sucrose 1.5 per cent + 8-HQC, 200 mg/litre) after transit.

Control of Diseases

Control of Alternaria leaf spot: Foltaf (0.2%) was found to be the best fungicide for control of alternaria leaf blight.

Control of Phoma leaf spot: Indofil M-45 (0.2%) was found to be better than Blitox (0.2%) and Bavisttin (0.1%) for control of Phoma chrysanthemum leaf spot of chrysanthemum. Ajay, A-4, A-106, Viking, Hazoor White Bouquet, Banglori yellow, Lemons, A-80 produced ratoon during from April to mid-May.

Breeding New Cultivars

The crosses involving parents White Bouquet, Gul-e-Sahir, Nanaku, Reegan White, Vyking, Ajay, Flirt, Regal Times, Reegan Emperor, Puncho and Mother Teresa. The following six hybrids were released for commercial cultivation in Punjab.

- ***Anmol:*** Anmol, having bright yellow anemone type of flowers is suitable for growing in pots.
- ***Royal Purple:*** with purplish pink anemone type of flowers is suitable for pot culture.
- ***Yellow Delight:*** an early-flowering variety with bright yellow pompon type of flowers is suitable for cut flowers.
- ***Garden Beauty:*** with maroon type of flowers is suitable for garden display.
- ***Winter Queen:*** with pink spoon type flowers is suitable for garden decoration.
- ***Autumn Joy:*** an early-flowering , highly floriferous variety with pink decorative flowers is suitable for cultivation for garden display.

Another hybrid of chrysanthemum, Punjab Bahar, has also been developed.

Punjab Bahar: This variety is an open-pollinated seedling of Panchoo and has red coloured buds, yellow coloured double flowers and is suitable for pot culture.

Tuberose

A total of 13 following varieties were collected and maintained: Shingar, Suhasini, Prajwal, Vaibhav, Hyderabad Single, Hyderabad Double, Calcutta Single, Calcutta Double, Swarna Rekha and Rajat Rekha, Sikkim Selection, Local Single and Local Double.

Basal doses of biofertilizers (Azospirillum + Phosphobacteria + VAM) as well as *Trichoderma viride* and sprays with Panchagavya and Manchurian mushroom tea did not show consistent effects on growth as well flowering parameters.

Gerbera

Collection, evaluation and maintenance of germplasm of gerbera Rosalin, Ornella, Sangria, Pink Elegance, Himalaya, Testarose, Tonneke and Golden Gate were evaluated. The flowering was early (62.33 days) in cv Rosalin and Tonneke. The maximum number of flowers/plant (27.66) and maximum vase life (13.00days) was observed in Himalaya.

Production Technology

The basal doses of biofertilizers (Azospirillum + Phosphobacteria + VAM) as well as *Trichoderma viride* and sprays with Panchagavya and Manchurian mushroom tea did not show significant effects on growth and flowering of plants.

Management of diseases: Seven fungicides were tested against leaf spot of gerbera cv. Glory during 2001-02 and 2002-03 under polyhouse conditions. None of these reduce disease or improve yield significantly over the control. The disease did not appear during 2003-04. Again during 2004-05 and 2005-06 the severity of the disease was not affected in a significant manner by any of the fungicidal treatments.

There was maximum vase-life was observed in solution containing sucrose (2%) + NaOCl (50 ppm) as compared to control.

Pulsing of stems with NaOCl and $CaOCl_2$ for one minute significantly increased vase life over control. In these two treatments there was no stem break. Hence, NaOCl and $CaOCl_2$ (1000 ppm) chlorine) have been suggested as pre-transit pulsing treatments for gerbera.

ICAR Research Compex for NEH Region, Umiam

Orchids

- Indigenous and exotic species were collected, evaluated and maintained
- Substrate media for pot culture of epiphytic and terrestrial orchids were standardized
- Nutritional requirement for *Cymbidium elegans, Dendrobium chrysanthum* and *Paphiopedilum villosum* were standardized.
- Suitable media for *in vitro* seeds germination and subsequent plantlet regeneration of *Cymbidium, Dendrobium and Phius* etc has been identified.

Gerbera

- Cut flowers cultivars were identified such as Autumn, Ambica, Cream Beauty
- Agro techniques were standardized (nutritional studies)

- High-yielding varieties Sweet Heart, Red Monarch, Salmon Spray were identified
- Five mutant were identified from Alesmera namely Mutant No. 13, 19, 21, 29, 30
- Substrate media for seed germination was identified (FYM+sand+soil @ 1:1:1 v/v)
- Vase solution was identified i.e Sucrose 3% + PBZ (5ppm) extended vase life from 7–11.5 days in different varieties.

Gladiolus

- Suitable varieties Blue Bird, Oscar, Morning Kiss, Lovely Melody, Snow Princess, Her Majesty, etc., were identified for cut flower and corm/cormel production.
- *In vitro* shoot multiplication and cormel production was standardized.

Rose

- Suitable rootstock such as *Rosa multiflora* was identified for budding.
- Banjaran, Sugandha and Rumba were found superior for growth and flowering.

Acharya NG Ranga Agricultural University, Hyderabad

Chrysanthemum

- In Chrysanthemum, a total of about 215 germplasm collections comprising various types and colour are maintained and evaluated.
- Among the germplasm lines remarkable variability was recorded for floral characters like days to first bud appearance, duration of flowering and number of flowers per plant. The days for first bud appearance from the collection ranged from 47 days to the maximum of 90 days with variety Basanthi and Paragon respectively.
- Yellow chrysanthemums are commercially preferred. On the basis of research findings for flowering duration the yellow cultivars, viz. Co-1, Basanthi, Punjab Gold, Raichur and Silper are recommended for cultivation for extended period of flower supply to market. However, Basanthi and Raichur are recommended.
- In white flower varieties Ratlam Selection was early to flower recording maximum number of flowers per plant.
- For improvement in keeping quality of cut flowers, holding solution containing Sucrose 2% + $Al_2(SO_4)_3$ 300 ppm was found effective in extending the post harvest shelf life of cut chrysanthemums.
- Packing cut flowers in polyethylene sleeves exerted superiority for packaging studies.

Gladiolus

- A total of 45 germplasm collections are maintained and evaluated for growth and floral characters. Flowering was early with Bindiya, whereas it was late in Anjali.
- In varietal studies, higher spike length was recorded in Hybrid 94-101, while more number of florets/spike was with Shubangini.
- Vase-life of cut spikes was significantly superior with holding solution comprising sucrose 4% + $Al_2 (SO_4)_3$ 300ppm + NaOCl 25 ppm.

Tuberose

- The germplasm pool of tuberose comprises varieties.
- Among the tuberose varieties the varietal selection made from the centre, viz. Hyderabad Single and Hyderabad Double proved to be superior for all the characters. These two selections were added in the AICRP technical programme for evaluation at national level.

Assam Agricultural University

About 10 varieties of tuberose, 30 of gladiolus, 10 of gerbera, 7 of anthurium and more than 30 of unidentified orchids have been collected.

- New varieties of tuberose namely, Shringar and Suhashini have been recommended for farmers of Assam.
- About 1,25,000 number of tuberose bulbs of Shringar and Suvasini have been made available for farmers through FLD and direct sale
- About 2500 number of suckers of gerbera variety Red Monarch have been produced and Proposed for FLD.

Bidhan Chandra Krishi Viswavidyalaya, Kalyani

Carnation

The cultivars Alcardio, Harmy Yellow, West Pretty, Peachy Mombo, Supermix White and Salamanca were found to be better performers. Single pinching in Spray carnations and pinch one and a half in Standards were better. Fertilizing with 500 ppm N once in fortnight combined with pinching one and a half in Standards or single pinching in Sprays improved quality and production. Standards receiving 500 ppm N+ single pinching and sprays treated with 500 ppm N + double or one and a half pinching performed better. Application of Benomyl (0.1%) + captaf (0.2%) was effective against *Fusarium* wilt of Carnation. Benomyl 0.2% at 15 days interval was very effective.

Orchids

Among 25 orchids, Sonia R-1 and Sonia R-2 were promising. *Dendrobium* var. Anjana, *Dendrobium* var. Minisonia, *D.* var. Sonia 17 and *D. moschatus* performed better. *Phalaenopsis* var. PSY and *Phalaenopsis* var. PCY were good performers regarding their floral characters.

Tuberose

Calcutta Double attained the highest plant height coupled with larger and heavier spikes. Rajat Rekha and Swarna Rekha were dwarf plants with smaller spikes. Calcutta Single produced maximum number of spikes per square meter. Hyderabad Single, Hyderabad Double, Prajwal and Vaibhav were promising genotypes from qualitative point of view. Hyderabad Single and Suvasini performed better. March planting of tuberose bulbs is proved profitable for production of flower spikes with more number of florets/spike.

Gerbera

Calcutta Orange, Calcutta Pink among local and Casava, Cabana, Hybrid Red, Osiris and

Thalasa among the hybrids were promising for flower production. About 25% shade improved stalk length whereas flower size is maximum under 75% shade level. Plants treated with N (15 or 20g/m^2), P (15 or 10g/m^2) and K (20g/m^2) had better growth and yield.

Birsa Agricultural University, Ranchi

The center was maintaing rose germplasm of 110 cultivars including 102 Hybrid Tea, 5 of Floribunda and 3 of miniature. About 37 of cultivars of gladiolus, 40 of chrysanthemum cultivars, Acce-ll0 and 18 of varieties of gerbera were collected and maintained.

Rose

A total of 110 cultivars including 102 H. T, 5 Floribundas and 3 Miniatures were collected, maintained and evaluated.

Gladiolus

The centre is maintaining 29 varieties of gladiolus. The performance of these varieties were tested under Ranchi agroclimatic condition. The cultivars Jyotsna, Jacksonville Gold Sancerre and Candyman were found to be promising with regards sprouting, plant height, length of spike, number of floret, size of flower, duration of flowering and vase life as compared others.

Chrysanthemum

Germplasm consisting of 32 existing varieties and 7 new varieties collected from Solan (Discovery and Puma sunny) and NBRI, Lucknow (Jayanti, Little Darling, Sunil, Maghi and Flirt) were maintained and assessed during the period under report.

Gerbera

Germplasm consisting of 12 existing accessions were being maintained. Ten new varieties of gerbera collected from BCKV Mohanpur were added to the existing collection.

Varietal recommendation

Based on last four years performance, gladiolus Candymon and Jacksonville Gold of double typed have been identified for this region.

Spray type cultivar, Alfred Wilson, Punjab Anuradha and Snowball type, Snowball White and Kikubiori are suitable chrysanthemum for this region.

G.B. Pant University of Agriculture and Technology, Pantnagar

The Pantnagar centre collected and maintained the germplasm of rose consisting 30 Hybrid Tea, 11 of Floribunda and 3 of Miniature. Besides that centre has collected 27 varieties of gladiolus, 9 of gerbera, 28 of chrysanthemum and 11 of tuberose.

Rose

Based on the overall performance of germplasm and post-harvest studies, varieties Happiness and Super Star in Hybrid Tea group and Yellow Contempo and Suryakiran were found to be promising for *tarai* region. In testing, it was found that maximum percentage of bud success was in Pusa Priya (Hybrid Tea group) and Prema (Floribunda group).

Gladiolus

Based on the overall performance of germplasm and post-harvest studies, varieties

Eurovision, Sobha, I.I.H.R, Hyb.87-22-1. Jester, Trader Horn and American beauty were found to be promising for tarai region. Amongst 18 newly introduced varieties, Hybrid-10 and Subhangini were found best almost in all desirable characters viz. earliness of flowering, spike length, number of florets per spike, duration of flowering, vase life and yield of spike per plant.

Tuberose

Maximum florets per spike was recorded in cv. Double and maximum spike length was observed with cv. Single while in case of duration of flowering was maximum with cv. Suvasini. The floral preservatives of Sucrose (5%) + Al_2SO_4 (300 ppm) and Sucrose (5%) + Cobalt Chloride (100 ppm) were found maximum vase life of tuberose cv. Suvasini. Spikes wrapped in newspaper stored for 24 hours at 4°C had maximum opened florets, percent opened florets and longest vase life. The maximum vase life was observed in cv Suvasini in comparison to Swarnrekha, Single, Double, Kalyani Single, Kalyani Double and Shringar) significantly highest in Suvasini.

Rajasthan College of Agriculture, Udaipur

Gladiolus

Western Song, Hybrid 94–101, Snow Princess, Aarthi, Friendship Pink, Sancerre, Gulal, Chandni, Shabnum were recommended for commercial production.

Chrysanthemum

Arun Kumar, Ajina Purple, Guesser and DollyWhite was found to be the best among the smaller flowered cultivar and among the larger flowered cultivar Shagun, Jaya, Harvest and Red Gold were found promising in this region. Ravi Kiran among newly evolved varieties was found best under Udaipur conditions.

Regional Plant Resource Centre, Bhubaneswar

Regional Plant Resource Centre is maintaining 285 cultivars of rose comprising 177 Hybrid Tea, 59 Floribunda, 42 miniatures and 7 climbers. A total of 136 cultivars of chrysanthemum comprising 63 spray types and 73 standard types are being maintained. Fifty-two orchid species and 22 varieties of hybrids belonging to *Dendrobium, Vanda, Aranda, Arenthera* are being maintained at the orchidarium of the centre.

Rose

Hybrid Tea rose cultivars like Montezuma and Mainu Parle are recommended for commercial production under Bhubaneswar conditions to meet the domestic demand.

Among the Hybrid Tea cultivars evaluated 'Pusa Gaurav' performed best under Bhubaneswar conditions. This variety can be grown successfully in open field for cut flower as well as loose flower production. Among the Floribundas 'Neelambari' is good for loose flower production where as 'Arunima' and 'Banjaran' are good for display in the garden.

Chrysanthemum

Spray type cultivars 'Flirt' and 'Yellow Gold' are recommended for commercial cultivation under Orissa conditions.

Standard type cultivars like Snow Ball, Eva Turner, Kikubiory, Lily Gallon and Alfred

Simpson' were found suitable as cut flower as well as for exhibition purpose. National testing of Chrysanthemum: The spray type cultivars 'Chandrika', 'Ravi Kiran' and 'Yellow Gold' were found suitable for cut flower as well as garden display. The cultivars like 'Mother Teresa', 'Punjab Gold' and 'Punjab Anuradha' have short spray length and are suitable for pot culture.

The chrysanthemum cultivars Chandrika and Yellow Gold were suitable for loose flower and cut flower and can be taken up for commercial cultivation.

Orchids

Vermiwash (3%) and panchagavya (3%) in addition to 50% of recommended fertilizer dose was most effective in influencing all the characters studied. The maximum number of flowers/plant was achieved with this treatment.

- Maximum vase life of 26 days was observed in pulsing solution containing HQ (500 ppm) and sucrose (5%) for 6hrs. The spike length (25.13 cm) and water uptake (16.2 ml) was maximum with the same treatment.
- A maximum vase life of 25 days was achieved when the stems were kept in holding solution containing 25 ppm silver nitrate, 400 ppm hydroxy-quinone and 5% sucrose.

Sher-E-Kashmir University of Agricultural Science and Technology, Srinagar

Srinagar centre maintained 64 of gladiolus varieties, 98 of chrysanthemum, 34 of varieties of carnation, 10 of daffodils, 5 of alstroemeria, 4 tulip varieties and 4 of Lilium.

In the concluded and on-going projects on commercially important flowers about two hundred and forty varieties of different crops, viz, rose, gladiolus, chrysanthemum, carnation, gerbera, daffodils, tulips, lris, alstroemeria and lilium has been collected, maintained, multiplied and assessed for different vegetative and floral characters for use as cut flowers, bedding, pot culture, garland, garden decorations and for conventional breeding

Breeding and Testing of New Cultivars

Since establishment of the division the improvement work carried out on commercially important flowers has resulted in development of 12 SKUAST-K Gladiolus hybrids. All the hybrids performed better in respect of floral advantage characters and will continue for future evaluation and multiplication. Regarding testing the new cultivars Swarnima, Shagun, Hybrids 87-22-1 and Hybrid 94–55 performed better in most of quality parameters than other cultivars under test against check variety 'White prosperity'.

Tamil Nadu Agricultural University, Coimbatore

During this period, germplasm consisted of 67 chrysanthemum incuding 60 varieties and 7 accessions. About 20 varieties of anthurium and 14 of tuberose were collected. At Yercaud, 120 gerbara including 37 varieties and 83 accessions were collected. Anthurium consisted of 114 accessions and 26 varieties, 32 varieties of gladiolus and 36 of orchids were collected during the period under report.

Coimbatore

About 72 accessions of chrysanthemum were evaluated. Acc.4, Acc.6, Shineme Getsu Acc.76, Acc.3 and Acc. 103 found promising. In testing, 20 varieties evaluated, it was observed that Indira, Red Gold and CO.2 found promising. Under tropical conditions, Acc.103 was found to be most promising with the highest flower yield. In testing of newly evolved varieties under different agro climatic regions, the cultivars Indira and Red Gold were found promising.

Anthurium: 20 collections were, maintained and evaluated. The cultivars Temptation and Glamour were found promising.

Tuberose: 14 accessions were evaluated, Acc. 10 produced the highest number of spikes/ plant (14.7) closely followed by Acc. 2 (14.0).

Crop Management

In Chrysanthemum var. CO. 1, effect of artificial short days and long days analysed. It was observed that 14 hrs light was best for number of branches, suckers, early lowering, flower yield. 50% of recommended dose fertilizers (500 g/m^2 – twice a year) + 3% Panchagavya produced highest yield (2288.9 g/plot) with good quality.

Growing media consisted of leaf mould + cocopeat – enhanced branches, suckers, flowers, longevity of flowers, early flowering in Anthurium and Temptation. Application of FYM 200 g + Decomposed coir pith + *Azospirillum* + Phosphobacteria + Trichoderma 20 g + spraying with 3% Panchagavya + 3% Manchurian tea resulted in production of the maximum number of flowers (8.1) of good quality in the variety Red Dragon. Basacote plus 6 m @ 6 g / pot (once in 6 months) induced the highest no. of spikes /plant (5.5).

Yercaud

Gerbera: About 83 accessions evaluated, a pre-release culture 'GJ 23' identified . It has recorded highest number of flowers per clump (43.0) and the largest flower diameter (10.7 cm).

Anthurium: 147 collections evaluated, maximum number of flowers per plant (8.16); spathe length (14.28cm); spathe width (12.32cm); vase life (17.37 days) were observed in Honduras while maximum flower stalk length (60.13cm) was eound in Verdon Red. In Sweet Heart, maximum spadix length (7.87cm).

Carnation: Tempo, Arthur Sim and Tasma were found promising

Orchids: The following orchids were found promising for the Shervaroy conditions: *Acanthephippium bicolour, Arachnanthe clarkia, Coelogyne flaccida, Cymbidiums, Oncidium, Paphilianthe tere, Paphiopedilum, Rhyncostylis retusa* and *Vanda coerula.* Year round flowering is observed in *Epidendrum radicans* and in *Spathoglottis plicata.*

Gladiolus: The varieties Dharshan, Western Song, Pusa Swarnima, Legend, Red Ginger, Thombolina were found promising

Crop Management

Anthurium: Vegetatively propagation done by treating nodal cuttings with IBA 1500 ppm. 80% shadenet produced high yield of quality flowers. NPK 20:20:20 @ 5% and BA 250 ppm improved flower yield and quality. Spraying NPK 20:40:40 @ 0.5% + Calcium @ 0.1% + Magnesium @ 0.1% + BA 250 ppm at weekly intervals improved flower yield and quality.

Orchids: Growing media of coconut fibre for *Cymbidium sp.*and Loam soil for *Spathoglottis plicata* were recommended. NPK @ 20: 10:10 at 0.2% and spraying of GA_3 @ 50 ppm wasobserved better for *Epidendrum radicans.* Shade requirement for *Liparis sp* (75% shadenet) and *Epidendrum radicans* (25% shadenet) was best.

Uttar Banga Krishi Viswavidyalaya, Kalimpong

Twenty varieties of gladiolus, 30 of carnation, 34 of orchid species including 180 genera, 24 anthurium and 4 of alstroemeria were collected.

- The evaluated vegetative characters have been proved apparently better with the treatment containing Loamy soil: coconut fibre: cocopeat (1: ½: ½) in Cymbidium soulhunt 200 during 2000-03. During 2003-04 a ratio of 1:1:1: ½ mixture of loamy soil leaf mould, river sand, charcoal dust and old mortar was found to be best suitable for terrestrial orchids.
- The M_5 (coconut fibre : charcoal : Cocopeat = 1: 1: 1) was the best growing media for epiphytic orchids (*Aerides odoratum)* in respect of numbers of spikes, number of flowers per spike per plant during 2003-04.
- Standardization of nutrient and growth regulators: The qualitative and quantitative parameters were found better with the combination of F_3 (NPK - 30:10:10 – 0.2%) and G_2 (BA 100 ppm) in *Cymbidium elegans* during 2001–03.
- The holding solution consisting $AgNO_3$ (25 ppm) + HQ (400 ppm) + Sucrose (5%) was found to be the best treatment for cymbidium orchid during 2003-04.
- Among the 18 varieties of *Anthurium andreanum*, the varieties namely: Sunset Orange, Salmon, Maringue, Kalimpong Red and Agnihotri performed better and recommended these varieties for commercial cultivation for this region.
- The recommended and better performed varieties for this zone are Chipper White, Summer Sunshine, Preskilla, Big Time Supreme, Juster Ice-Gold, American Beauty Swarnima and IIHR-Hybrid-87-22-1 during 2005-06.
- Only two varieties namely "Aladin" and "Pluto" have established. Among two varieties, "Pluto" performed better under the climatic condition of Darjeeling hills.

Indian Agricultural Research Institute, New Delhi

About 900 varieties of rose, 75 of gladiolus, 70 of chrysanthemum, 17 of tuberose are being maintained at this centre.

Rose

Seven new rose varieties were released. These are Pusa Mohit, Pusa Abhishek (striped ad spotted bud sport of Jantar Mantar), Pusa Manhar, (Jantar Mantar × Lahar), Pusa Muskan (Pink Parfait × Alinka), Pusa Urmil (Bud sport of Jantar Mantar) Pusa Ranjan (Pink Parfait × Iceberg) and Pusa Mohit (thornless). Pusa Mohit has been released at State level by the government of National Capital Territory of Delhi.

Five exotic varieties Grand Gala, First Red, Confetti, Nobeless and Tinake were grown under low-cost polyhouses for developing indigenous production technology. First Red was found to be the best variety with regard to its characters like stalk length, number of flowers produced per unit area and bud size. Among them Confetti was more susceptible to aphid infestation.

Gladiolus

Five varieties Pusa Urmi, Pusa Jyotsna, Pusa Gulal, Pusa Shabnum and Pusa Urvashi were rleased 2002.

Two hybrids namely Kataksh and Ojaswini (Tarzan) were identified.

Angila, Green Willow, Archana, IIHR-87-22-1 and all IARI varieties like Pusa Suhagin, Urmil, Shwetha, Chandni, Neelam and Shagun were promising under Delhi conditions.

Tuberose

About 17 varieties collected and being maintained at this centre

Prajwal, Hyderabad Double were found to be most consistent in the performance. Variety Prajwal recorded the highest plant height, flowering duration, and number of florets per spike and diameter and weight of individual daughter bulb.

Indian Agricultural Research Institute Regional Station, Katrain

More than 25 varieties of gladiolus, 9 of carnation, 8 variety of Lilium, 2 of alstroemeria were collected and maintained.

Gladiolus

Cultivar 'White Friendship' performed well in our local conditions. Cultivars Jester, American Beauty, No.22 and RS performed well in our local conditions. Cultivar Swarnima performed well. The maximum plant height (107.6 cm) and spike length (70.3 cm) were recorded in 'Top Brass'. Number of florets was recorded maximum (17.8) in 'Jester' while maximum floret size (10.1 cm) was in 'Shagun'. 'Tiger Flame' produced maximum (3.1) spikes/plant.

Carnation

Impalla performed well under open conditions. Cultivars Prandello, Farida, Red Corso, Dona, Sun Rise, Disagree – 9080, Salya, Bright Rendovous and Raggio-de Sole were found to be promosing at Katrain. Minimum disease incidence was recorded in Red Apharadite. In carnation cv. Impalla, minimum disease incidence of *Fusarium* wilt was recorded with Carbendazin (0.1%) and Mancozeb (0.25%) as per 0-4 scale.

Lilium

Cultivars 'Apledroon', 'Prato', 'London' and 'Pollyanna' in Asiatic hybrids and 'Expression', 'Woodriffis Memory' and 'Siberia' in Oriental hybrids performed well under our local conditions Sand was the best media for scale propagation of lilium. IBA at low concentration, i.e., 200 ppm was found the best for more production of scale bulblets. Soil + sand were the best media for getting more number of flower buds and to increase the bulblet size during bulblet propagation of lilium.

Alstoremeria

Three varieties of Alstroemeria, No.14, Pluto, and Aladdin, were collected and evaluated for their performance. Among these "Pluto' was performed well under our local conditions.

Daffodils

Keeping quality of cut daffodil cultivar 'King Alfred' spikes was found to be better with a pulsing solution consisiting of sucrose (4%) + citric acid (1000 ppm) and the same solution at lower concentration, i.e., sucrose (2%) + citric acid (300 ppm).

Indian Council of Agricultural Research Complex for NEH Region, Umiam

Germplasm collection of included 22 varieties of gladiolus, 17 of orchids, 24 of anthurium and 23 of varieties of gerbera.

- A cheap method of micropropagtion of orchids has been developed using isubgol as gelling agent and polypropylene bags as culture vessels.
- Agro-techniques for orchids, gladiolus and gerbera have been standardized.

Indian Institute of Horticultural Research, Hessaraghatta, Bangalore (Division of Ornamental Crops)

Tamil Nadu, Andhra Pradesh, Maharashtra, Uttar Pradesh and several other states. About 305 roses, 35 of carnation, 250 of chrysanthemum, 2 species and 110 of gladiolus, 11 tuberose, 28 of gerbera and 20 exotic varieties and 15 local strains of Anthurium were collected and maintained at the centre.

Tuberose

Two high yielding hybrids Prajwal and Vaibhav were released. Prajwal is a hybrid bearing single type flowers on tall stiff spikes. The flower buds are slightly pinkish. It yields more loose flowers than other varieties while Vaibhav is a hybrid with double flowers on medium spikes. Flower buds are greenish and spike yield is higher compared to other varieties.

Rajat Rekha, Swarna Rekha and Sikkim Selection exhibited very poor flowering and hence observed to be not suitable for commercial cultivation. Shringar, Suvasini, Prajwal and Vaibhav excelled in their performance compared to other varieties in yield.

Chrysanthemum

The following varieties were developed and released at the institute during the period under report.

Usha Kiran: This is an induced mutant from Kirti. It takes 121 days for flowering. It produces 170 flowers per plant. Flowers are semi double type with prominent disc, each being 5 cm diameter. Flowers are attractive yellow colour. This is suitable for bedding/potting and loose flower purpose. It has shelf-life of 7 days.

Arka Ganga: This is a cross between Flirt and Red Gold: Bears attractive double Korean white flowers with pink tinge. Flower yield is higher than 'Local white' variety. Good for loose and cut flower purposes.

Arka Ravi: It is a cross between Mundial and Hybrid 87–17–1. It produces attractive peach colour Korean type of flowers. It is suitable for cut flower purpose. It has a vase life of 12 days.

Anthurium

The variety Pistache and Simba shows highly vigorous with plant growth, long stalk, high spathe diameter and disease free. The Singapore Red is highly floriferous yielding 11 flower/plant/year but highly susceptible to bacterial blight. The variety Tropical showed the highest vase life.The species *Anthurium ornatum* showed highly vigorous plant growth compared to *A.andreanum* One F_1, a cross between *A. ornatum* (selection from IIHR) which is a fragrant variety and variety Marshal produced a flower with andreanum type with fragrance.

Gerbera

Varieties, Devil, Bony, Alberino, Cassiana and Arianna were the best among the 23 RAMCO varieties evaluated and Sangria was best of the 5 varieties of KF Biotech. Overall yield of all these varieties ranged from 25–37 flower/plant/year.

Four double selections were evaluated for 3 years which yielded an average 35 flowers/plant/year and were highly suitable for open conditions.Two semidouble varieties were evaluated for 2 years. The flowers are on par with commercial check variety Sangria and yield an average of 35 flower/plant/year.

National Botanical Research Institute, Lucknow

Rose

Three rose varieties namely 'Mrinalini Stripe', 'First Prize Lighter', 'Sylvia Lighter', were developed and released. Holding solution consisting of aluminium sulphate (300 ppm) + sucrose (3%) were found most effective in respect of vase life, bud opening and size of flower.

Gladiolus

Three gladiolus varieties were developed and released in the name of 'URVASHI', 'NEELIMA' and 'NBRI TAMBRI'.

The performance of spikes was observed better in newspaper sleeves after 24 hours of storage followed by corrugated packing material which resulted maximum floret opening and increased floret size. However, unwrapped (control) showed poor opening performance in six gladiolus varieties, viz. Candy Man, Lavender Puff, Violet Queen, NBRI Tambri, Snow Princes and Tiger Flame.

Chrysanthemum

The NBRI, Lucknow has developed 11 chrysanthemum varieties: Mother Teresa, Sadbhawana, Y-2K, Kargil' 99, Shanti, NBRI Kusum, NBRI Diana, NBRI Little Darling, NBRI Mini Jessie, Mini Queen and Rangoli through open pollinated seedling selections.

A combination of sucrose (2%) + silver nitrate (25 ppm) + citric acid (75 ppm) has been found most effective in opening of buds in cv. Dolly White, Kundan, Shyamal, Shanti, Jubilee, Flirt, Pink Gin, Sharad Mala which were harvested at colour show stage, followed by the combination of sucrose (2%) + 8HQC (500 ppm) was also respond better in cvs Ajai, Sharad Har, Kundan and Apsara over control.

Performance of sprays (cut stems) was observed better in newspaper lots after 24 hours of storage in Shanti, Jubileee, White prolific and Kundan as leaves remain green turgid. The cut stems of Lal Pari and Little Darling also respond better in corrugated sleeves due to turgidity of leaves after 24 hours of storage. Unwrapped (control) showed poor response in all the cultivars.

University of Agricultural Sciences, Bangalore

University of Agricultural Sciences, Bangalore, is a Voluntary Centre started during 1979 without any financial help from AICRP on Floriculture with two projects on Chrysanthemum and Rose. However, during 1999-2000 an amount of Rs. 40,000 per year was sanctioned towards recurring contingency for conducting experiments under the said projects.

About 50 variety of chrysanthemum, 9 of carnation and 6 of variety of anthurium were collected.

Chrysanthemum

In evaluation trial, Punjab Anuradha, Yellow Star, Arka Ravi, Nilima, Ravi Kiran were found to be better in respect of flowering performance.

Carnation

Damina, Madam Collette, Cindrella, Kalahari and Bonny were found to better carnations. Growth of carnaiton Desio was better when inorganic fertilizers were supplemented with the vermi compost, Panchagavya and Manchurian tea (50% RDF + vermicompost + 3% Manchurian tea + 3% Panchagavya) with early flower bud initiation (141.83 days), maximum diameter of flower bud (1.99 cm), largest flowers (6.54 cm) and number of flowers /m^2/year (132.33)

Anthurium

Application of biofertilizers (*Azosprillum*, PSB, VAM each at 2g/plant), GA_3 200 ppm along with application of NPK (30:10:10 at 0.2%spary) resulted in more number of flowers/ plant (7.33), maximum spathe length (11.13cm), spathe width (9.35 cm), spadix length (5.79 cm) and longevity of the spikes on plant (70.21 days).